Aziridines and Epoxides in Organic Synthesis

Edited by
Andrei K. Yudin

Related Titles

Gerald Dyker (ed.)

Handbook of C-H Transformations

Applications in Organic Synthesis, 2 Vol

ISBN 3-527-31074-6
2005

Dennis G. Hall (ed.)

Boronic Acids

Preparation and Applications in Organic Synthesis and Medicine

ISBN 3-527-30991-8
2005

Jens Christoffers, Angelika Baro, and Steven V. Ley (eds.)

Quaternary Stereocenters

Challenges and Solutions for Organic Synthesis

ISBN 3-527-31107-6
2005

Paul Knochel (ed.)

Handbook of Functionalized Organometallics

Applications in Synthesis

ISBN 3-527-31131-9
2005

Martin Hiersemann, Udo Nubbemeyer (eds.)

The Claisen Rearrangement

Methods and Applications

ISBN 3-527-30825-3
2005

Francois Diederich, Peter J. Stang, and Rik R. Tikwinski (eds.)

Acetylene Chemistry

Chemistry, Biology, and Material Science

ISBN 3-527-30781-8
2004

Aziridines and Epoxides in Organic Synthesis

Edited by
Andrei K. Yudin

WILEY-VCH Verlag GmbH & Co. KGaA

The Editor
Andrei K. Yudin
St. George Street 80
M5S 3H6 Toronto
KANADA

Cover
Grafik-Design Schulz, Fußgönheim

Library of Congress Card No.:
applied for

British Library Cataloguing-in-Publication Data
A catalogue record for this book is available from the British Library.

Bibliographic information published by Die Deutsche Bibliothek
Die Deutsche Bibliothek lists this publication in the Deutsche Nationalbibliografie; detailed bibliographic data is available in the Internet at <http://dnb.ddb.de>.

Typesetting: Typomedia GmbH, Ostfildern
Printing: Betz-Druck GmbH, Darmstadt
Binding: J. Schäffer GmbH, Grünstadt

Printed in the Federal Republic of Germany
Printed on acid-free paper

ISBN-13 978-3-527-31213-9
ISBN-10 3-527-31213-7

To Jovana

Foreword

Epoxides have fascinated me since my days as an undergraduate at the Massachusetts Institute of Technology. I vividly remember taking a course in organic chemistry, watching an inspiring (if unconventional) professor, Barry Sharpless, perform a demonstration in which a cage that contained a collection of gypsy moths was opened, allowing them to respond to the presence of a nearby sample of (+)-disparlure (an epoxide-containing sex pheromone for the gypsy moth). The result was memorable, and it was in fact this class that led to my decision to pursue a career in organic chemistry.

Of course, (+)-disparlure is only one of the many natural products that contain either an epoxide or an aziridine. Important and intriguing biologically active compounds such as the mitomycins, azinomycins, and epothilones also bear these functional groups.

Interest in epoxides and aziridines has been amplified because, not only are they significant synthetic endpoints, but they are also tremendously useful synthetic intermediates. Due to the strain associated with the three-membered ring, they are "spring-loaded" for reactions with nucleophiles, allowing a wide array of powerful functionalizations to be achieved. Thus, ring-openings of aziridines and epoxides have been applied industrially to produce a variety of bulk chemicals, including polyethylenimine, ethylene glycol, and epoxy resins. Furthermore, aziridines and epoxides serve as versatile intermediates in natural product and pharmaceutical synthesis. Reactions with a broad range of nucleophiles proceed cleanly with excellent regioselectivity and/or stereoselectivity, furnishing products that bear useful amino and hydroxyl groups.

Discovering effective new methods for the synthesis of aziridines and epoxides, as well as developing novel transformations of these heterocycles, has been an extremely active area of research in recent years. The publication of this book, Aziridines and Epoxides in Organic Synthesis, is therefore timely, since there have been no monographs on this topic in quite some time. Prof. Andre Yudin has brought together a set of insightful reviews by leading researchers that nicely illustrate a rich diversity of chemistry. The twelve chapters cover a broad spectrum, including methods for the synthesis of aziridines and epoxides, functionalization reactions, applications in natural product synthesis, and biosynthesis studies. I anticipate that this highly readable book will be the "go to" resource for those

Aziridines and Epoxides in Organic Synthesis. Andrei K. Yudin

ISBN: 3-527-31213-7

interested in learning about the state-of-the-art in this important field. Equally significantly, the monograph will no doubt inspire further exciting developments in this area.

Gregory C. Fu, Cambridge, MA
October 2005

Table of Contents

Aziridines and Epoxides in Organic Synthesis. Andrei K. Yudin

ISBN: 3-527-31213-7

Preface

Aziridines and epoxides are among the most versatile intermediates in organic synthesis. In addition, a number of biologically significant molecules contain these strained three-membered rings within their structures. The synthetic community has been fascinated with prospects of selective synthesis and transformations of aziridines and epoxides. Recent years have witnessed a number of important advances in this area and I felt that a book that summarizes these achievements would be a valuable addition to the chemistry literature. I was very glad to receive enthusiastic support from my colleagues from around the World. Roughly divided into equal number of chapters dedicated to epoxides and aziridines, this volume will serve as a useful resource. The synthesis part covers additions to aldehydes and imines, olefin transformations, cyclizations, and metal catalysis. The applications encompass chemistry of vinyl aziridines and epoxides, aziridinecarboxylates and phosphonates, metalated epoxides and aziridines, asymmetric ring opening chemistry, complex target-oriented synthesis, and click chemistry. Another important area discussed in this book is the biosynthesis of aziridines and epoxides.

This project has turned into a wonderful compilation of outstanding manuscripts and I am very grateful to the authors who contributed to it. Last, but not least, I want to express my gratitude to Dr. Evgenii Blyumin, Iain Watson, and Lily Yu for their valuable editorial comments at the revision stages.

Andrei K. Yudin
Toronto, November 2005

Aziridines and Epoxides in Organic Synthesis. Andrei K. Yudin

ISBN: 3-527-31213-7

List of Contributors

Editor

Andrei K. Yudin
Chemistry Department
University of Toronto
80 St. George Street
Toronto, ON M5S 3H6
Canada

Authors

Hans Adolfsson
Department of Organic Chemistry
Stockholm University
The Arrhenius Laboratory
106 91 Stockholm
Sweden

Varinder K. Aggarwal
Synthetic Chemistry
School of Chemistry
Cantock's Close
Bristol BS8 1TS
UK

D. Michael Badine
3 ch de la Dole
1279 Chavannes de Bogis
Switzerland

Daniela Balan
Department of Organic Chemistry
Stockholm University
The Arrhenius Laboratory
10691 Stockholm
Sweden

Christopher D. Bray
School of Chemistry
University of Nottingham
University Park
Nottingham NG7 2RD
UK

Bang-Chi Chen
Discovery Chemistry
Bristol-Myers Squibb Pharmaceutical
Research Institute
Princeton NJ 08543
USA

Paolo Crotti
Department of Bioorganic Chemistry
and Biopharmacy
University of Pisa
via Bonanno, 33
56126 Pisa
Italy

Aziridines and Epoxides in Organic Synthesis. Andrei K. Yudin

ISBN: 3-527-31213-7

Franklin A. Davis
Department of Chemistry
Temple University
Beury Hall (016-00)
Philadelphia PA 19122
USA

Valery V. Fokin
Department of Chemistry
The Scripps Research Institute
BCC-315
10550 N. Torrey Pines Rd.
La Jolla CA 92037
USA

Sabine Grüschow
LSI
University of Michigan
210 Washtenaw Ave.
Ann Arbor MI 48109–2216
USA

David M. Hodgson
Department of Chemistry
University of Oxford
Chemistry Research Laboratory
Mansfield Road
Oxford OX1 3TA
UK

Eric N. Jacobsen
Department of Chemistry
Harvard University
12 Oxford Street
Cambridge MA 02138
USA

Philip A.S. Lowden
School of Biological and Chemical Sciences
Birkbeck College
University of London
Malet Street, Bloomsbury
London WC1E 7HX
UK

Vijayalakshmi A. Moorthie
6 Colsterdale
Carlton Colville
Suffolk NR33 8TN
UK

Lars P.C. Nielsen
Department of Chemistry
Harvard University
12 Oxford Street #312
Cambridge, MA 02138
USA

Berit Olofsson
Organic Chemistry
Arrhenius Laboratory
Stockholm University
106 91 Stockholm
Sweden

Mauro Pineschi
Department of Bioorganic Chemistry and Biopharmacy
University of Pisa
via Bonnano, 33
56126 Pisa
Italy

Hiroaki Ohno
Graduate School of Pharmaceutical Sciences
Osaka University
1–6 Yamadaoka, Suita
Osaka 565–0871
Japan

David H. Sherman
LSI
University of Michigan
210 Washtenaw Ave.
Ann Arbor MI 48109–2216
USA

Peter Somfai
Organic Chemistry
KTH Chemistry
Royal Institute of Technology
10044 Stockholm
Sweden

Peng Wu
Department of Chemistry
University of California at Berkeley
Hildebrand Hall #1460
Berkeley, CA 94720
USA

J. B. Sweeney
School of Chemistry
University of Reading
Reading RG6 6AD
UK

Ping Zhou
Chemical Sciences
Wyeth-Ayerst Research
Princeton NJ 08543
USA

1
Asymmetric Synthesis of Epoxides and Aziridines from Aldehydes and Imines

Varinder K. Aggarwal, D. Michael Badine, and Vijayalakshmi A. Moorthie

1.1
Introduction

Epoxides and aziridines are strained three-membered heterocycles. Their synthetic utility lies in the fact that they can be ring-opened with a broad range of nucleophiles with high or often complete stereoselectivity and regioselectivity and that 1,2-difunctional ring-opened products represent common motifs in many organic molecules of interest. As a result of their importance in synthesis, the preparation of epoxides and aziridines has been of considerable interest and many methods have been developed to date. Most use alkenes as precursors, these subsequently being oxidized. An alternative and complementary approach utilizes aldehydes and imines. Advantages with this approach are: *i*) that potentially hazardous oxidizing agents are not required, and *ii*) that both C–X and C–C bonds are formed, rather than just C–X bonds (Scheme 1.1).

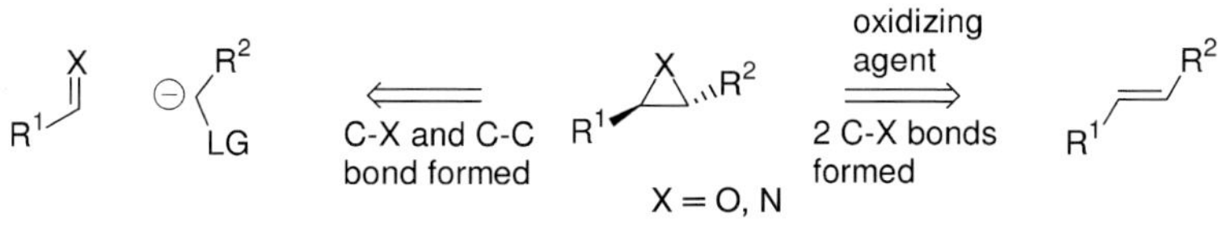

C-X and C-C bond formed

oxidizing agent

2 C-X bonds formed

X = O, N

Scheme 1.1

This review summarizes the best asymmetric methods for preparing epoxides and aziridines from aldehydes (or ketones) and imines.

1.2
Asymmetric Epoxidation of Carbonyl Compounds

There have been two general approaches to the direct asymmetric epoxidation of carbonyl-containing compounds (Scheme 1.2): ylide-mediated epoxidation for the construction of aryl and vinyl epoxides, and α-halo enolate epoxidation (Darzens reaction) for the construction of epoxy esters, acids, amides, and sulfones.

Aziridines and Epoxides in Organic Synthesis. Andrei K. Yudin

ISBN: 3-527-31213-7

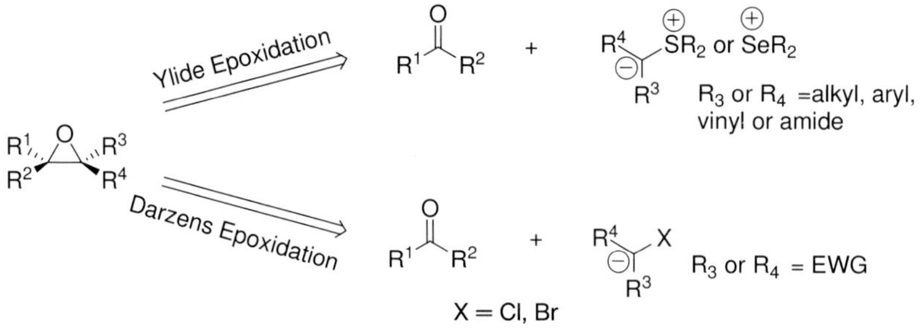

Ylide Epoxidation

SR_2 or SeR_2

R_3 or R_4 =alkyl, aryl, vinyl or amide

Darzens Epoxidation

X = Cl, Br

R_3 or R_4 = EWG

Scheme 1.2

1.2.1
Aryl, Vinyl, and Alkyl Epoxides

1.2.1.1 **Stoichiometric Ylide-mediated Epoxidation**

Solladié-Cavallo's group used Eliel's oxathiane **1** (derived from pulegone) in asymmetric epoxidation (Scheme 1.3) [1]. This sulfide was initially benzylated to form a single diastereomer of the sulfonium salt **2**. Epoxidation was then carried out at low temperature with the aid of sodium hydride to furnish diaryl epoxides **3** with high enantioselectivities, and with recovery of the chiral sulfide **1**.

Using a phosphazene (EtP_2) base, they also synthesized aryl-vinyl epoxides **6a-c** (Table 1.1) [2]. The use of this base resulted in rapid ylide formation and efficient epoxidation reactions, although it is an expensive reagent. There is potential for cyclopropanation of the alkene when sulfur ylides are treated with α,β-unsaturated aldehydes, but the major products were the epoxides, and high selectivities could be achieved (Entries 1–4). Additionally, heteroaromatic aryl-epoxides could be prepared with high selectivities by this procedure (Entries 5 and 6) [3]. Although high selectivities have been achieved, it should be noted that only one of the two enantiomers of **1** is readily available.

The Aggarwal group has used chiral sulfide **7**, derived from camphorsulfonyl chloride, in asymmetric epoxidation [4]. Firstly, they preformed the salt **8** from either the bromide or the alcohol, and then formed the ylide in the presence of a range of carbonyl compounds. This process proved effective for the synthesis of aryl-aryl, aryl-heteroaryl, aryl-alkyl, and aryl-vinyl epoxides (Table 1.2, Entries 1–5).

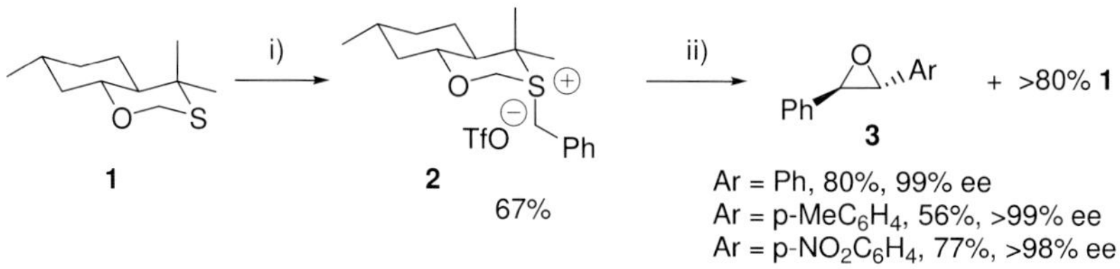

1 → i) → **2** (TfO⁻) 67% → ii) → **3** + >80% **1**

Ar = Ph, 80%, 99% ee
Ar = p-MeC_6H_4, 56%, >99% ee
Ar = p-$NO_2C_6H_4$, 77%, >98% ee

i) BnOH, Tf_2O, pyridine, CH_2Cl_2, -10 °C; ii) NaH, CH_2Cl_2, ArCHO, -40 °C, 24 - 48 h.

Scheme 1.3

Table 1.1 Synthesis of aryl-vinyl epoxides by use of chiral sulfide **1** a phosphazene base.

i) BnOH, Tf_2O, pyridine, CH_2Cl_2, -10 °C; ii) EtP_2, **5a-e** (R^2CHO), -78 °C, CH_2Cl_2.

Entry	R^1 (ylide)	R^2CHO	Epoxide: epoxycyclop.: cyclop.	Epoxide *trans: cis*	Epoxide *ee trans* (*cis*) (%)
1	Ph	**5a**	77:11:12	100:0	97
2	*p*-$MeOC_6H_4$	**5a**	100:0:0	77:23	95 (98)
3	Ph	**5b**	100:0:0	97:3	100
4	Ph	**5c**	100:0:0	97:3	100
5	Ph	**5d**	–	100:0	96.8
6	Ph	**5e**	–	100:0	99.8

Until this work, the reactions between the benzyl sulfonium ylide and ketones to give trisubstituted epoxides had not previously been used in asymmetric sulfur ylide-mediated epoxidation. It was found that good selectivities were obtained with cyclic ketones (Entry 6), but lower diastereo- and enantioselectivities resulted with acyclic ketones (Entries 7 and 8), which still remain challenging substrates for sulfur ylide-mediated epoxidation. In addition they showed that aryl-vinyl epoxides could also be synthesized with the aid of α,β-unsaturated sulfonium salts **10a-b** (Scheme 1.4).

1.2.1.2 Catalytic Ylide-mediated Epoxidation

The first attempt at a catalytic asymmetric sulfur ylide epoxidation was by Furukawa's group [5]. The catalytic cycle was formed by initial alkylation of a sulfide (**14**), followed by deprotonation of the sulfonium salt **15** to form an ylide **16** and

Table 1.2 Application of the chiral sulfide **7** in asymmetric epoxidations.

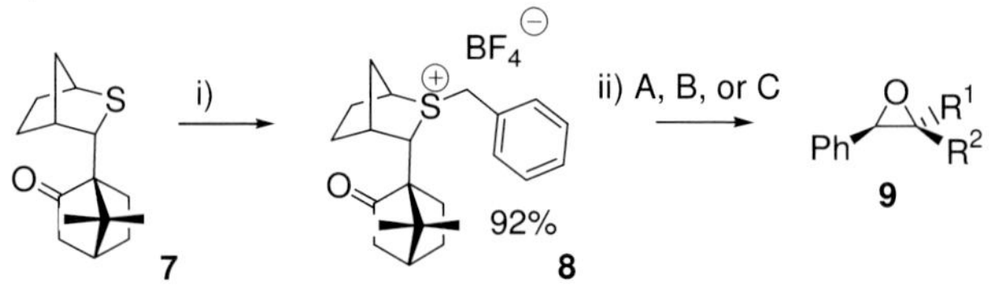

i) BnBr, $AgBF_4$, CH_2Cl_2; ii) **A**: KOH, R^1R^2CO, $MeCN:H_2O$ (9:1), rt; **B**: EtP_2 R_1R_2CO, CH_2Cl_2, -78 °C; **C**: KHMDS, THF, -78 °C.

Entry	R^1COR^2	Method	Yield (%)	d. r. *trans* : *cis*	*ee* trans (%)
1	PhCOH	A	75	98:2	98
2	2-PyrCOH	B	88	98:2	99
3	C_4H_9COH	C	87	90:10	>99
4	CH_2=C(Me)COH	B	52	>99:1	95
5	(*E*)-MeCH=CH_2COH	B	90	>99:1	95
6	cyclohexanone	B	85	–	92
7	$MeCOC_6H_4$-p-NO_2	B	73	>1:99	71
8	MeCOPh	B	77	33:67	93 (50)

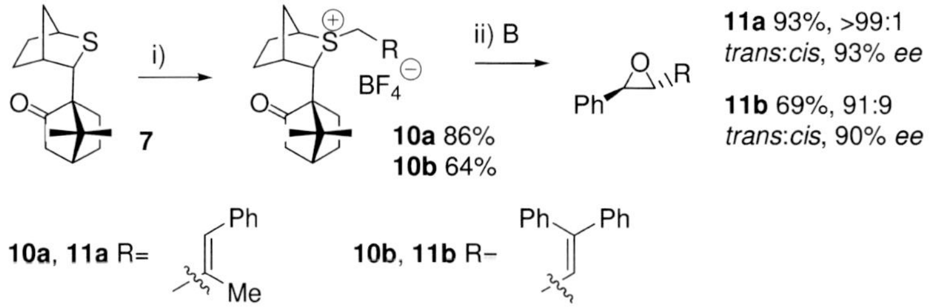

11a 93%, >99:1 *trans*:*cis*, 93% *ee*

11b 69%, 91:9 *trans*:*cis*, 90% *ee*

10a 86%
10b 64%

i) RCH_2OH, HBF_4, Et_2O; ii) **B**: EtP_2, PhCHO, CH_2Cl_2, -78 °C.

Scheme 1.4

subsequent reaction with an aldehyde to furnish the epoxide with return of the sulfide **12** (Scheme 1.5). However, only low yields and selectivities resulted when the camphor-derived sulfide **12** was employed. Metzner improved the selectivity of this process by using the C_2 symmetric sulfide **13** [6].

Although reactions required 2 days to reach completion in the presence of stoichiometric amounts of sulfide, they became impracticably long (28 days) when 10% sulfide was employed, due to the slow alkylation step. The alkylation step was

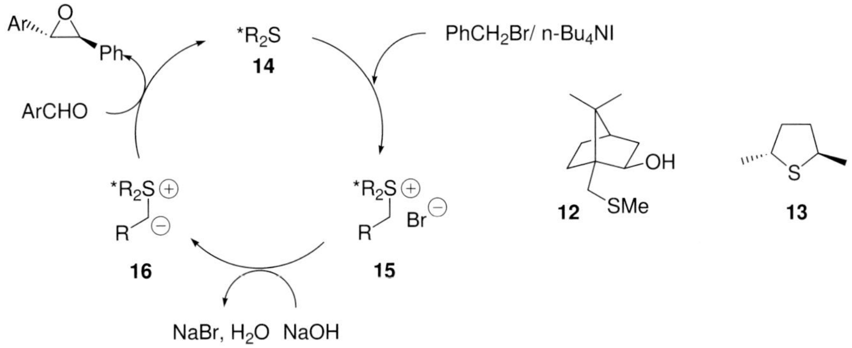

Scheme 1.5

accelerated upon addition of iodide salts, however, and the reaction times were reduced (Table 1.3). The yields and selectivities are lower than for the corresponding stoichiometric reactions (compare Entry 1 with 2, Entry 4 with 5, and Entry 6 with 7). The use of iodide salts proved to be incompatible with allylic halides, and so stoichiometric amounts of sulfide were required to achieve good yields with these substrates [7].

Metzner et al. also prepared the selenium analogue **17** of their C_2 symmetric chiral sulfide and tested it in epoxidation reactions (Scheme 1.6) [8]. Although good enantioselectivities were observed, and a catalytic reaction was possible without the use of iodide salts, the low diastereoselectivities obtained prevent it from being synthetically useful.

Table 1.3 Catalytic ylide-mediated epoxidations.

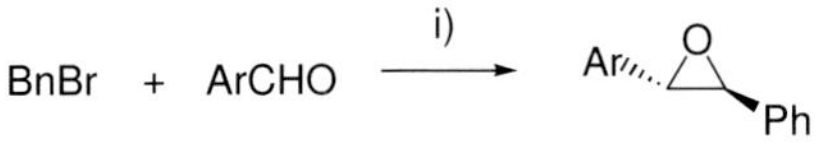

i) NaOH,n-Bu_4NI, **13**, t-BuOH-H_2O (9:1), rt.

Entry	Ar in ArCHO	Eq. 13	Time (days)	Yield (%)	d.r.	*ee* (%)
1	PhCHO	1[a]	1	92	93:7	88
2	PhCHO	0.1	4	82	93:7	85
3	*p*-ClC_6H_4	0.1	6	77	80	72
4	cinnamyl	1[a]	2	93	98:2	87
5	cinnamyl	0.1	6	60	89:11	69
6	2-thiophenyl	1[a]	4	90	91:9	89
7	2-thiophenyl	0.1	6	75	88:12	80

[a] Without *n*-Bu_4NI.

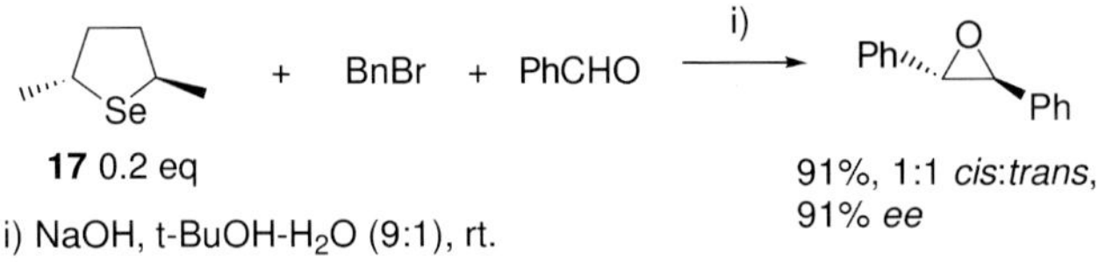

17 0.2 eq + BnBr + PhCHO →(i) epoxide (Ph, Ph)

91%, 1:1 *cis:trans*, 91% *ee*

i) NaOH, t-BuOH-H_2O (9:1), rt.

Scheme 1.6

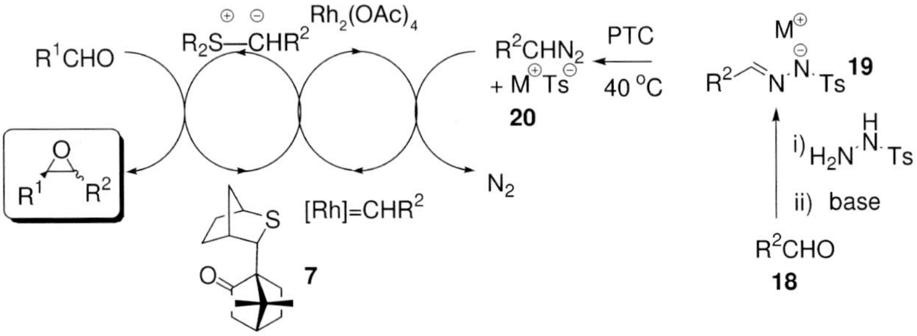

Scheme 1.7

Aggarwal and co-workers have developed a catalytic cycle for asymmetric epoxidation (Scheme 1.7) [9]. In this cycle, the sulfur ylide is generated through the reaction between chiral sulfide **7** and a metallocarbene. The metallocarbene is generated by the decomposition of a diazo compound **20**, which can in turn be generated *in situ* from the tosylhydrazone salt **19** by warming in the presence of phase-transfer catalyst (to aid passage of the insoluble salt **19** into the liquid phase). The tosylhydrazone salt can also be generated *in situ* from the corresponding aldehyde **18** and tosylhydrazine in the presence of base.

This process thus enables the coupling of two different aldehydes together to produce epoxides in high enantio- and diastereoselectivities. A range of aldehydes have been used in this process with phenyl tosylhydrazone salt **19** (Table 1.4) [10]. Good selectivities were observed with aromatic and heteroaromatic aldehydes (Entries 1 and 2). Pyridyl aldehydes proved to be incompatible with this process, presumably due to the presence of a nucleophilic nitrogen atom, which can compete with the sulfide for the metallocarbene to form a pyridinium ylide. Aliphatic aldehydes gave moderate yields and moderate to high diastereoselectivities (Entries 3 and 4). Hindered aliphatic aldehydes such as pivaldehyde were not successful substrates and did not yield any epoxide. Although some α,β-unsaturated aldehydes could be employed to give epoxides with high diastereo- and enantioselectivities, cinnamaldehyde was the only substrate also to give high yields (Entry 5). Sulfide loadings as low as 5 mol% could be used in many cases.

Benzaldehyde was also treated with a range of tosylhydrazone salts (Table 1.5). Good selectivities were generally observed with electron-rich aromatic salts (Entries 1–3), except in the furyl case (Entry 7). Low yields of epoxide occurred when a hindered substrate such as the mesityl tosylhydrazone salt was used.

Table 1.4 Tosylhydrazone salt **19** in catalytic asymmetric epoxidation.

RCHO + PhCH=N–N(–)Ts Na(+) → [1 mol% $Rh_2(OAc)_4$, 5-20 mol% sulfide **7**, 5-10 mol% $BnEt_3NCl$, CH_3CN, 40°C, 24 - 48 h] → trans-epoxide (Ph, R)

Entry	Aldehyde	Sulfide equiv.	*t* (h)	Yield (%)	*trans:cis*	*ee trans* (*cis*) (%)
1	benzaldehyde	0.05	48	82	>98:2	94
2	3-furaldehyde	0.05	48	77	>98:2	92
3	valeraldehyde	0.2	48	46	75:25	89
4	cyclohexanecarboxaldehyde	0.05	48	58	88:12	90 (74)
5	*trans*-cinnamaldehyde	0.05	48	70	>98:2	87
6	3-methyl-2-butenal	0.2	24	21	>98:2	87

Table 1.5 Use of a range of tosylhydrazone salts in catalytic asymmetric epoxidation of benzaldehyde.

PhCHO + ArCH=N–N(–)Ts Na(+) → [1 mol% $Rh_2(OAc)_4$, 5-20 mol% sulfide **7**, 0-20 mol% $BnEt_3NCl$] → trans-epoxide (Ar, Ph)

Entry	Ar	Solvent	t (°C)	(mol%) 7	Yield	*trans:cis*	*ee* (%) *trans* (*cis*)
1	*p*-MeC_6H_4	CH_3CN	40	5	74	95:5	93
2	*p*-$MeOC_6H_4$	CH_3CN	30	20	95	80:20	93
3	*o*-$MeOC_6H_4$	CH_3CN	30	5	70	>98:2	93
4	*p*-ClC_6H_4	CH_3CN	40	20	81	>98:2	93
5	*p*-$CO_2MeOC_6H_4$	CH_3CN	30	20	80	>98:2	73
6	*p*-$CO_2MeOC_6H_4$	CH_3CN/ H_2O (9:1)	30	20	>10	100:0	86
7	3-furyl	toluene	40	20	46	63:37	63 (31)

With electron-deficient aromatic substrates (Entries 4 and 5), high yields and selectivities were observed, but enantioselectivities were variable and solvent-dependent (compare Entry 6 with 7 and see Section 1.2.1.3 for further discussion). With α,β-unsaturated tosylhydrazone salts, selectivities and yields were lower. The scope of this process has been extensively mapped out, enabling the optimum disconnection for epoxidation to be chosen [10].

1.2.1.3 Discussion of Factors Affecting Diastereo- and Enantioselectivity

The high diastereoselectivities observed in aryl-stabilized sulfur ylide-mediated epoxidation can be understood by considering the intermediate betaines (Scheme 1.8). In reactions with benzaldehyde it was found that the *trans* epoxide was derived from the non-reversible formation of the *anti* betaine **23**, whilst the *cis* epoxide was generated by the reversible formation of the *syn* betaine **24** [11]. This productive non-reversible *anti* betaine formation and unproductive reversible *syn* betaine formation results in the overall high *trans* selectivities. Of course, the extent to which the intermediate betaines are reversible will depend upon the stability of the betaines, the stability of the starting aldehyde **22**, the stability of the starting ylide **21**, and the steric hindrance of the aldehyde/ylide [12, 13]. A less stabilized ylide will exhibit less reversible *syn* betaine formation and will result in a lower diastereoselectivity (compare Entry 1 with 2, Table 1.5; the less stabilized *p*-methoxybenzyl ylide gives a lower diastereoselectivity than the *p*-metylbenzyl ylide).

There are four main factors that affect the enantioselectivity of sulfur ylide-mediated reactions: *i*) the lone-pair selectivity of the sulfonium salt formation, *ii*) the conformation of the resulting ylide, *iii*) the face selectivity of the ylide, and *iv*) betaine reversibility.

To control the first factor, one of the two lone pairs of the sulfide must be blocked such that a single diastereomer is produced upon alkylation. For C_2 symmetric sulfides this is not an issue, as a single diastereomer is necessarily formed upon alkylation. To control the second factor, steric interactions can be used to favor one of the two possible conformations of the ylide (these are generally accepted to be the two conformers in which the electron lone pairs on sulfur and carbon are orthogonal) [14]. The third factor can be controlled by sterically hinder-

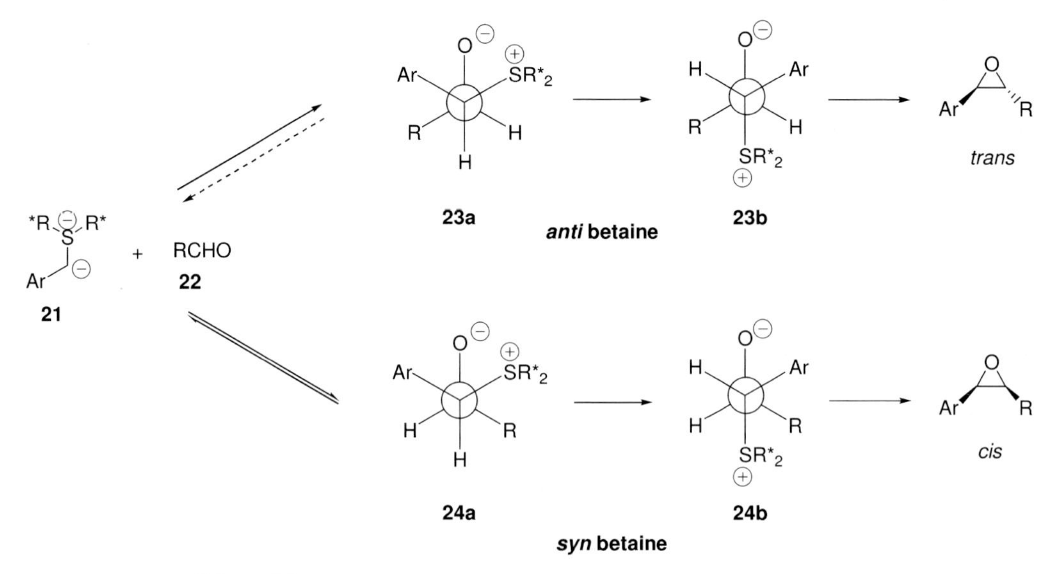

Scheme 1.8

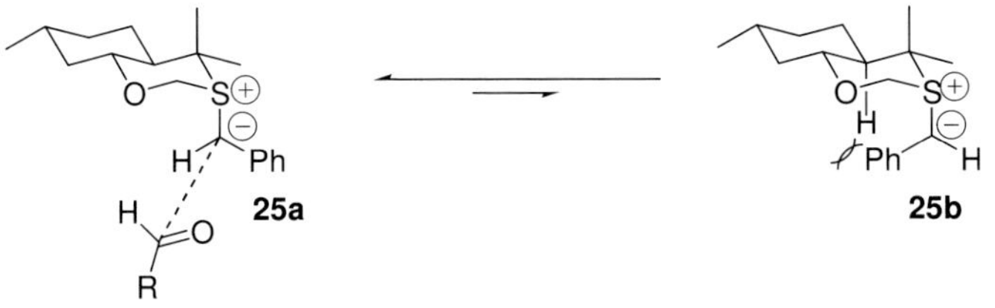

Scheme 1.9

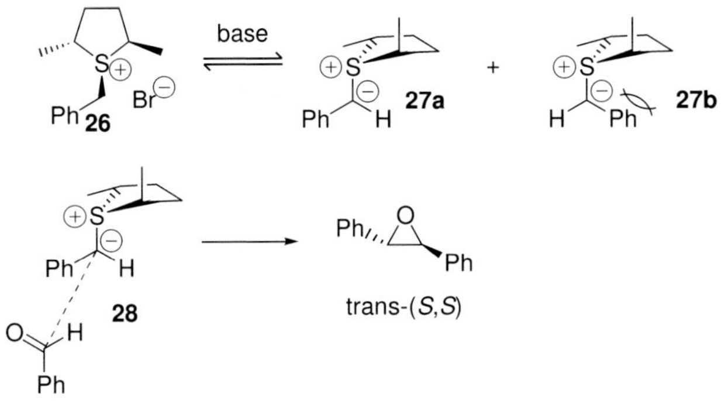

Scheme 1.10

ing one face of the ylide, thus restricting the approach of the aldehyde to it. By considering these first three factors, the high selectivities observed with the sulfides previously discussed can be broadly explained:

For oxathiane **1**, lone pair selectivity is controlled by steric interactions of the *gem*-dimethyl group and an anomeric effect, which renders the equatorial lone pair less nucleophilic than the axial lone pair. Of the resulting ylide conformations, **25a** will be strongly preferred and will react on the more open *Re* face, since the *Si* face is blocked by the *gem*-dimethyl group (Scheme 1.9) [3, 15].

The C_2 symmetry of sulfide **13** means that a single diastereomer is formed upon alkylation (Scheme 1.10). Attack from the *Si* face of the ylide is preferred as the *Re* face is shielded by the methyl group *cis* to the benzylidene group (**28**). Metzner postulates that this methyl group also controls the conformation of the ylide, as a steric clash in **27b** renders **27a** more favorable [16]. However, computational studies by Goodman revealed that **27a** was not particularly favored over **27b**, but it was substantially more reactive, thus providing the high enantioselectivity observed [17].

In the case of sulfide **7** the bulky camphoryl moiety blocks one of the lone pairs on the sulfide, resulting in a single diastereomer upon alkylation. One of the conformations (**29b**) is rendered less favorable by non-bonded interactions such that conformation **29a** is favored, resulting in the observed major isomer (Scheme 1.11). The face selectivity is also controlled by the camphoryl group, which blocks the *Re* face of the ylide.

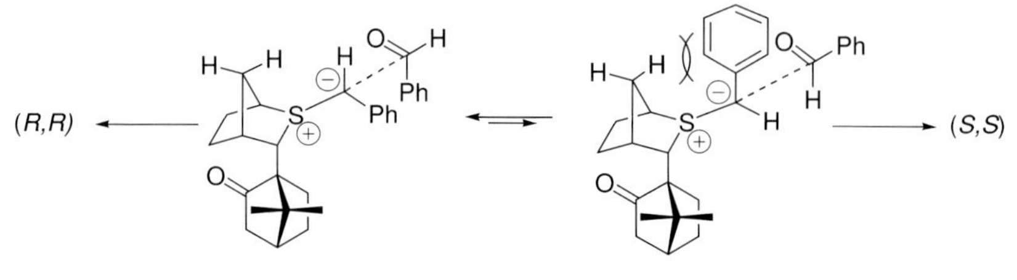

Scheme 1.11

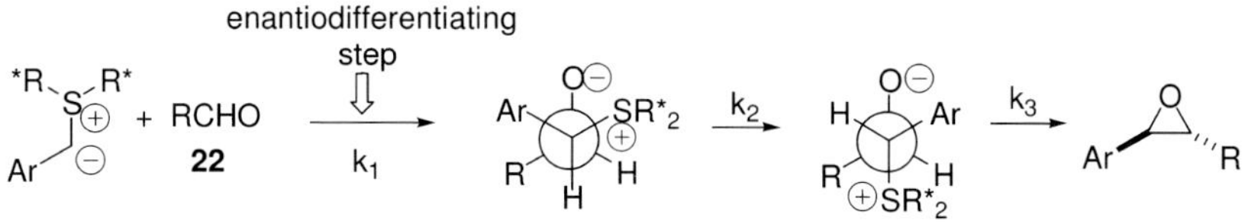

Anti betaine formation non-reversible

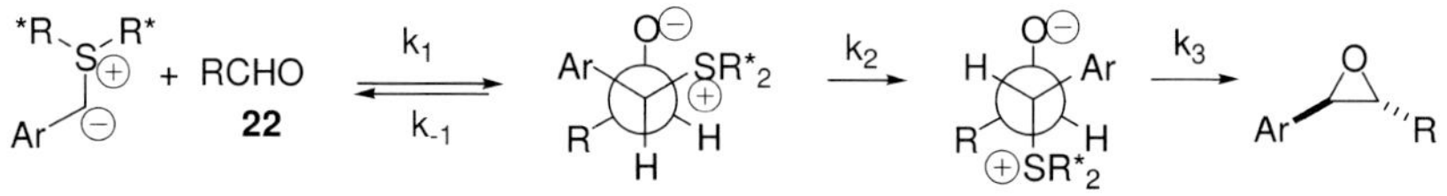

Anti betaine formation reversible

Scheme 1.12

The fourth factor becomes an issue when *anti* betaine formation is reversible or partially reversible. This can occur with more hindered or more stable ylides. In these cases the enantiodifferentiating step becomes either the bond rotation or the ring-closure step (Scheme 1.12), and as a result the observed enantioselectivities are generally lower (Entry 5, Table 1.5; the electron-deficient aromatic ylide gives lower enantioselectivity). However the use of protic solvents (Entry 6, Table 1.5) or lithium salts has been shown to reduce reversibility in betaine formation and can result in increased enantioselectivities in these cases [13]. Although protic solvents give low yields and so are not practically useful, lithium salts do not suffer this drawback.[18]

The diastereo- and enantioselectivity are clearly dependent on a number of factors, including the reaction conditions, sulfide structure, and nature of the ylide.

1.2.2 Terminal Epoxides

One class of particularly challenging targets for asymmetric epoxidation is that of terminal epoxides. Aggarwal and co-workers found that zinc carbenoids generated

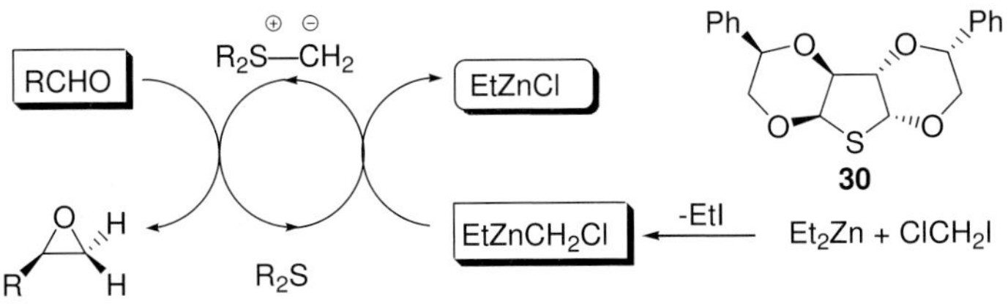
RCHO
R_2S—CH_2
EtZnCl
Ph
Ph
O
S
30
$EtZnCH_2Cl$
-EtI
Et_2Zn + $ClCH_2I$
R
H
R_2S

Scheme 1.13

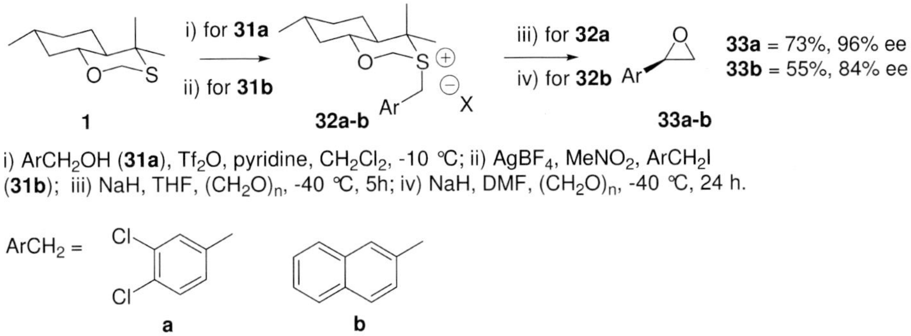
i) for **31a**
ii) for **31b**
iii) for **32a**
iv) for **32b**
33a = 73%, 96% ee
33b = 55%, 84% ee
1
32a-b
33a-b
Ar
X
i) $ArCH_2OH$ (**31a**), Tf_2O, pyridine, CH_2Cl_2, -10 °C; ii) $AgBF_4$, $MeNO_2$, $ArCH_2I$ (**31b**); iii) NaH, THF, $(CH_2O)_n$, -40 °C, 5h; iv) NaH, DMF, $(CH_2O)_n$, -40 °C, 24 h.
$ArCH_2$ =
Cl
Cl
a
b

Scheme 1.14

from Et_2Zn and $ClCH_2I$ could efficiently transfer a methylidene group to a sulfide, and, in the presence of aldehydes, produce epoxides in good yield (Scheme 1.13) [19, 20].

Unfortunately, the highest enantioselectivity so far obtained for the synthesis of styrene oxide by this route is only 57% *ee* with Goodman's sulfide **30** [21]. Thus methylidene transfer is not yet an effective strategy for the synthesis of terminal epoxides.

Another way to disconnect a terminal epoxide is to add a functionalized ylide to paraformaldehyde. This was the route explored by Solladié-Cavallo, who treated two aromatic ylides with paraformaldehyde at low temperatures and obtained good selectivities (Scheme 1.14) [22]. It would thus appear that this is the best ylide-mediated route to terminal aromatic epoxides to date.

1.2.3
Epoxy Esters, Amides, Acids, Ketones, and Sulfones

1.2.3.1 Sulfur Ylide-mediated Epoxidation

In general sulfur ylide-mediated epoxidation cannot be used to form an epoxide with an adjacent anion-stabilizing group such as an ester, as the requisite ylide is too stable and does not react with aldehydes [23]. With the less strongly electron-withdrawing amide group, however, the sulfur ylide possesses sufficient reactivity for epoxidation. The first example of an asymmetric version of this reaction was by

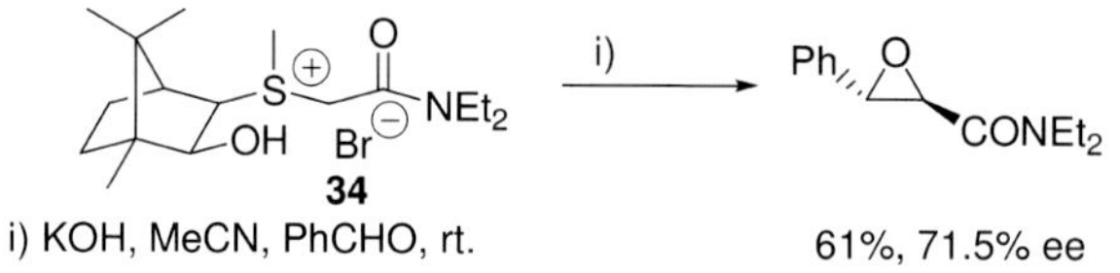

i)
Ph
O
CONEt2
OH
Br
NEt2
O
S
34
i) KOH, MeCN, PhCHO, rt.
61%, 71.5% ee

Scheme 1.15

Table 1.6 Use of the sulfonium salt **36** in low-temperature epoxidations.

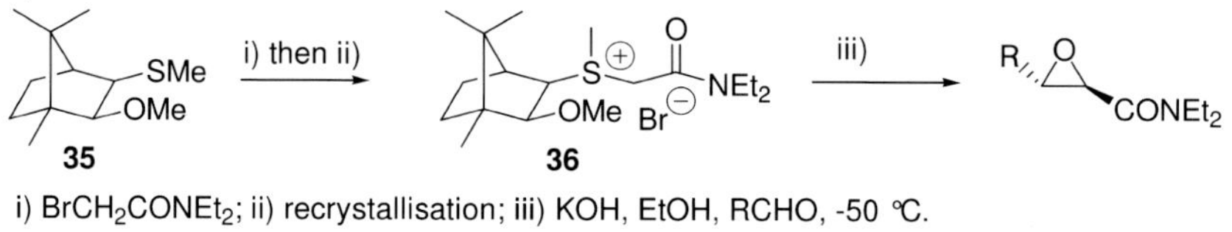

SMe
OMe
35
i) then ii)
S
O
NEt2
OMe
Br
36
iii)
R
O
CONEt2
i) BrCH2CONEt2; ii) recrystallisation; iii) KOH, EtOH, RCHO, -50 °C.

Entry	R	Yield (%)	*ee* (%)
1	Ph	93	97
2	*p*-ClC_6H_4	87	99
3	*p*-MeC_6H_4	88	98
4	3-pyridyl	87	95
5[a]	dodecyl	84	63
6[b]	*t*-butyl	87	93

[a] –30 °C. [b] –20 °C.

Dai and co-workers, who used sulfonium salt **34** in epoxidation reactions to give glycidic amides (Scheme 1.15) [23].

Improved selectivities were achieved by the Aggarwal group, who used sulfonium salt **36** (Table 1.6), with the same parent structure, in low-temperature epoxidation reactions [24]. In most cases complete diastereocontrol was accompanied by high enantioselectivities; aromatic and heteroaromatic aldehydes were excellent substrates (Entries 1–4). Aliphatic aldehydes gave variable results: mono- and trisubstituted aldehydes gave moderate to high enantioselectivities (Entries 5 and 6), whilst secondary aliphatic aldehydes gave very low enantioselectivities. Although tertiary amides are difficult to hydrolyze, they can be cleanly converted to ketones by treatment with organolithiums.

As the formation of betaines from amide-stabilized ylides is known to be reversible (in contrast with aryl- or semistabilized ylides, which can exhibit irreversible *anti* betaine formation; see Section 1.2.1.3), the enantiodifferentiating step cannot be the C–C bond-forming step. B3LYP calculations of the individual steps along the reaction pathway have shown that in this instance ring-closure has the highest barrier and is most likely to be the enantiodifferentiating step of the reaction (Scheme 1.16) [25].

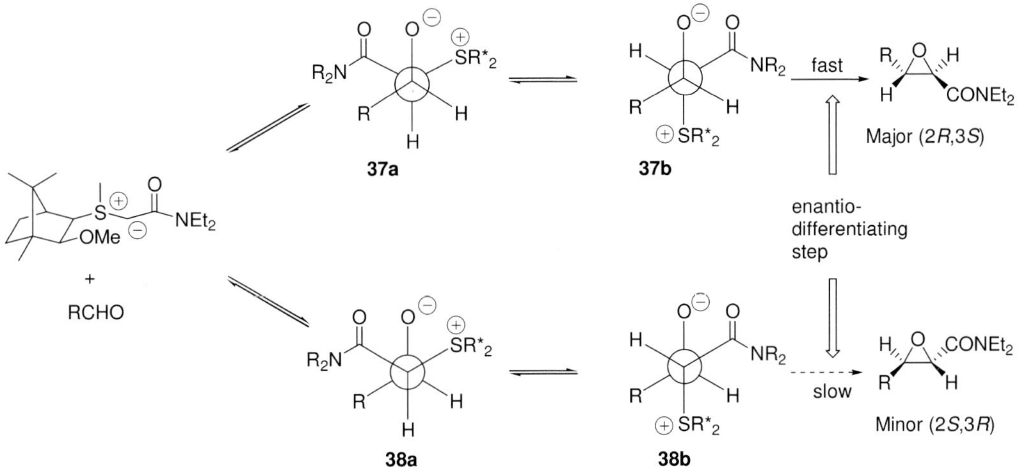

Scheme 1.16

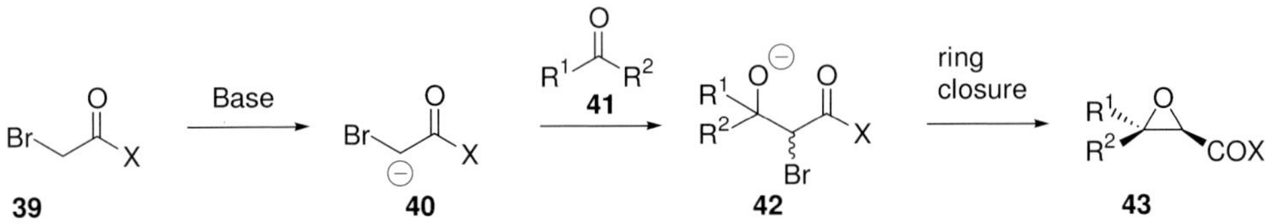

Scheme 1.17

1.2.3.2 **Darzens Reaction**

Epoxides bearing electron-withdrawing groups have been most commonly synthesized by the Darzens reaction. The Darzens reaction involves the initial addition of an α-halo enolate **40** to the carbonyl compound **41**, followed by ring-closure of the alkoxide **42** (Scheme 1.17). Several approaches for inducing asymmetry into this reaction – the use of chiral auxiliaries, reagents, or catalysts – have emerged.

1.2.3.3 **Darzens Reactions in the Presence of Chiral Auxiliaries**

Although chiral auxiliaries have been attached to aldehydes for asymmetric Darzens reactions [26, 27], the most commonly employed point of attachment for a chiral auxiliary is adjacent to the carbonyl to be enolized. Indeed, many groups have investigated this strategy, and a variety of chiral auxiliaries have been employed. As the initial step of the Darzens reaction is an α-halogen aldol condensation, it is perhaps unsurprising that existing asymmetric aldol chemistry should have been exploited and adapted to the Darzens reaction. Prigden's group investigated the use of 2-oxazolidinones developed by Evans (Table 1.7) [28, 29], treating a variety of metal enolates (tin(II), tin(IV), zinc, lithium, titanium, and boron) with both aliphatic and aromatic aldehydes. The best results by far were obtained with

Table 1.7 2-Oxazolidinones as chiral auxiliaries in Darzens reactions.

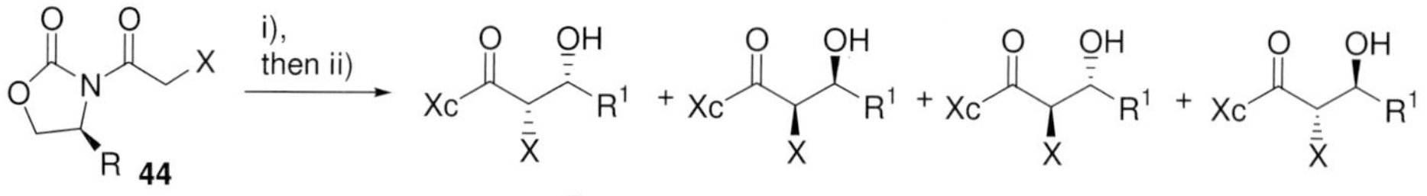

44 → i), then ii) → 45 (*syn B*) + 46 (*syn A*) + 47 (*anti A*) + 48 (*anti B*)

X_c = chiral auxiliary

i) enolate formation; ii) R^1CHO.

Entry	R^1	X	R (X_c)	M[a]	Yield (%)	d.r. *syn*:*anti*	Enantioselectivity *syn* B:A	*anti* A:B
1	*n*-C_5H_{11}	Cl	*i*-Pr	B	62	>50:1	>50:1	–
2	*i*-$PrCH_2$	Cl	*i*-Pr	B	55	>50:1	>50:1	–
3	*i*-Pr	Cl	*i*-Pr	B	52	>50:1	>50:1	–
4	Ph	Cl	*i*-Pr	B[b]	68	>99:0	>99:0	–
5	Ph	F	*i*-Pr	Sn^{IV}	50	<0:99	–	94:5
6	Ph	Br	*i*-Pr	Sn^{II}	67	85:15	79:6	6:9
7	*o*-$NO_2C_6H_4$	Br	Ph	Sn^{II}	44	0:100	–	0:100
8	*o*-$MeOC_6H_4$	Br	Ph	Sn^{II}	65	21:79	0:21	0:79
9	*o*-*t*-BuC_6H_4	Br	Ph	Sn^{II}	58	0:100	–	0:100

[a] B refers to $B(n\text{-}Bu)_2$, Sn^{IV} refers to $Sn(n\text{-}Bu)_3$, Sn^{II} refers to $Sn(OSO_2CF_3)$. [b] B refers to BEt_3.

the use of boron enolates, which furnished the *syn* adducts with very high diastereo- and enantioselectivities (Entries 1–4).

Through the use of a tin(IV) enolate with benzaldehyde it was possible to generate the *anti* A diastereomer **47** with high selectivity (Entry 5). With tin(II) enolates a highly substituent-dependent outcome was observed. Low selectivities resulted with *para*-substituted aromatic aldehydes, but good selectivities were observed for *ortho*-substituted aromatic aldehydes (Entries 7–9). Simultaneous re-

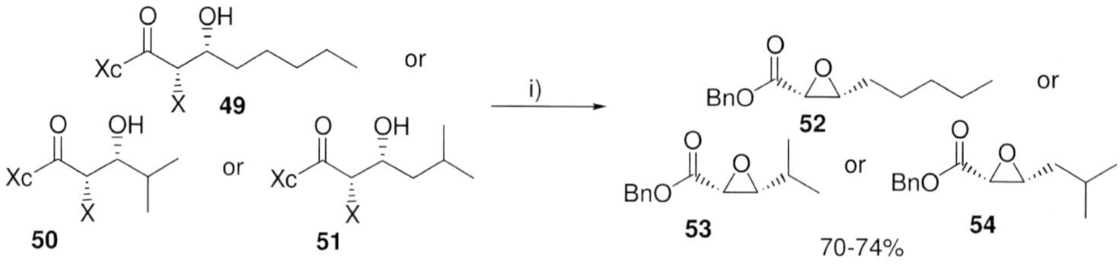

49 or **50** or **51** → i) → **52** or **53** or **54** (70-74%)

i) BnOLi, THF, -78 to -20 °C. X_c = chiral auxiliary

Scheme 1.18

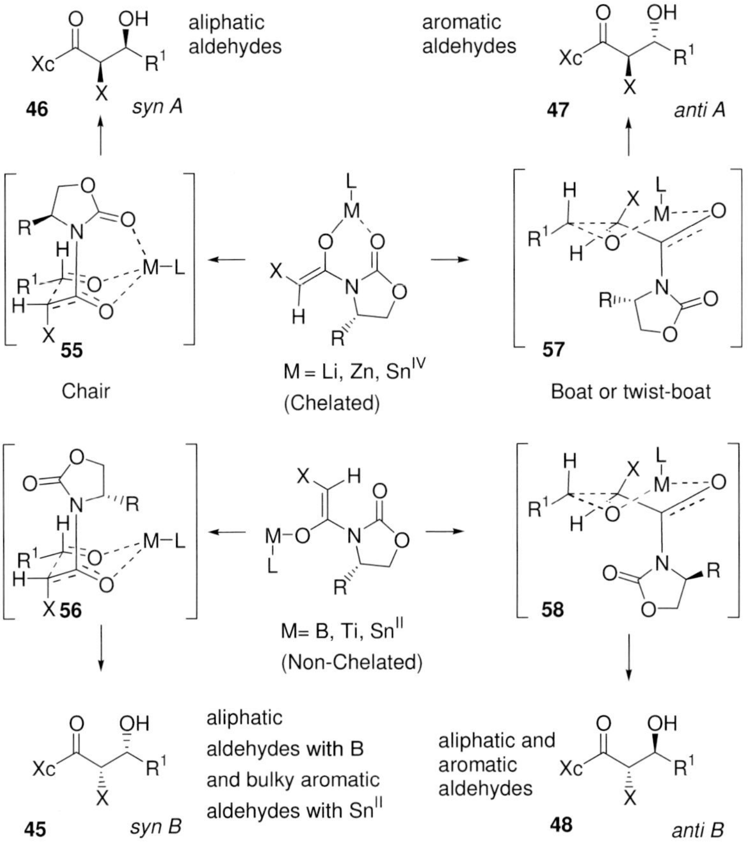

Scheme 1.19

moval of the auxiliaries and ring-closure cleanly furnished the corresponding epoxy esters without epimerization (Scheme 1.18).

The results were interpreted by considering enolates with tin (IV), zinc, or lithium counter-ions to react via three-point chair transition states **55** with aliphatic aldehydes to give predominantly the *syn* A adducts **46**, whilst tin (II), boron, and titanium enolates reacted via non-coordinated chair transition states **56** with aliphatic aldehydes to give the opposite *syn* B adducts **45** (Scheme 1.19). Aromatic aldehydes reacted with tin (IV), zinc, or lithium enolates through chelated twist-boat transition states **57** to give the *anti* A halohydrins **47**, whilst boron and titanium enolates still reacted via the nonchelated chair-like transition states to give the *syn* B **45**. The tin (II) enolate exhibited borderline selectivities. It reacted with aromatic aldehydes to give the *syn* B diastereomer **45** as with boron and titanium enolates, but with *ortho*-substituted aromatic aldehydes, an *anti* B (**48**) selectivity was observed, indicating that a twist-boat transition state **58** was being favored.

Thus, by varying the enolate counter-cation and the aldehyde, it was possible to

Table 1.8 Use of 8-phenylmenthyl esters to induce asymmetry in the Darzens reaction.

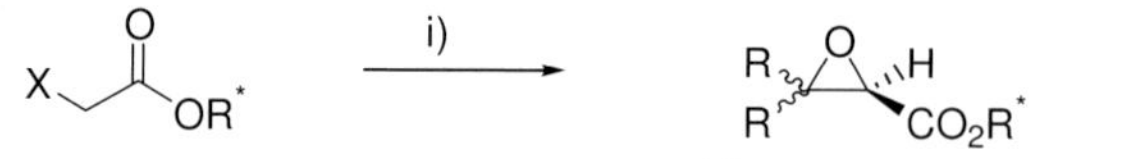

i) R_2CO, t-BuOK, CH_2Cl_2, -78-0 °C. R* = (-)-8-phenylmenthyl.

Entry	R_2CO	X	Yield (%)	*cis:trans*	de (*cis*) %	de (*trans*) %
1	acetophenone	Br	56	5.6:1	>95	21
2	propiophenone	Br	43	4.2:1	>95	>95
3	cyclohexanone	Cl	45	–	96	
4	acetone	Cl	64	–	87	
5	benzophenone	Cl	45	–	77	
6	benzaldehyde	Cl	90	2.8:1.0	38	33

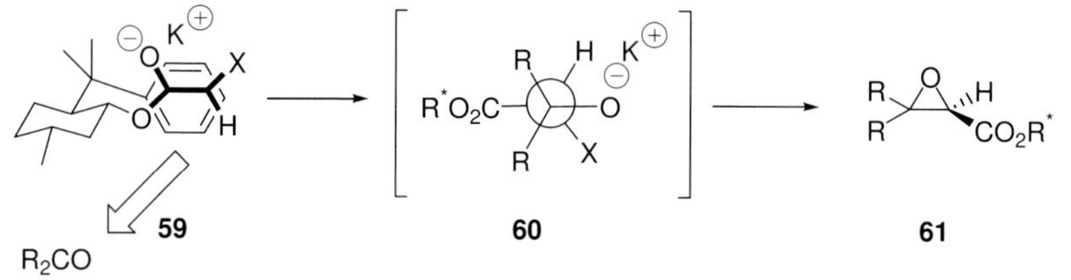

Scheme 1.20

access a range of halo-aldol adducts, which could also be cyclized to the required epoxy esters without epimerization (Scheme 1.18).

Ohkata [30, 31] and co-workers have employed an 8-phenylmenthyl ester to induce asymmetry in the Darzens reaction (Table 1.8). Moderate to high diaster-

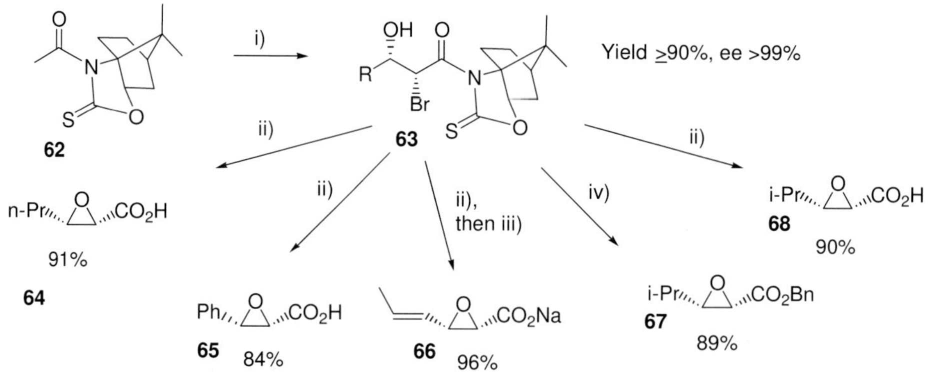

i) $TiCl_4$, DIPEA; then Br_2, DIPEA, RCHO, -78 °C; ii) CH_3CN/H_2O, NEt_3; then MeOH, K_2CO_3; iii) Na_2CO_3, CH_3CN/H_2O iv) DMAP, B_nOH ; then KF, LiF, n-Bu_4NHSO_4, CH_2Cl_2.

Scheme 1.21

Table 1.9 Scope of the indanyl-derived auxiliary **69**.

i) $ClCH_2COCl$, pyridine; CH_2Cl_2, 0 °C; ii) $TiCl_4$, iPr_2NEt, then RCHO,$TiCl_4$, MeCN or NMP, CH_2Cl_2, -78 °C, 2h; iii) $TiCl_4$, iPr_2NEt, then RCHO $TiCl_4$ CH_2Cl_2, -78 °C, 2h; iv) K_2CO_3, MeOH; v) K_2CO_3, (aq), DMF.

Entry	Aldehyde (RCHO)	Yield (%) (71 + 72)	*anti* 71: *syn* 72
1	*i*-BuCHO	90[a]	99:1
2	BuCHO	70[a]	96:4
3	PhCHO	47,[a] 62[b]	96:4
4	PhCHO	64	10:90
5	$BnOCH_2CHO$	86	1:99
6	$BnOCH_2CH_2CHO$	79	4:96
7	(*R*)-BnOCH-(Me)CHO	94	1:99
8	(*S*)-BnOCH-(Me)CHO	82	99:1
9	(±)-BnOCH-(Me)CHO	95[c]	5:95

[a] NMP (2.2 equiv.) used as additive. [b] MeCN (2.2 equiv.) used as additive. [c] 2 equiv. of aldehyde used, 30 min reaction time.

eoselectivities resulted from its reaction with ketones to furnish trisubstituted aliphatic and aromatic epoxy esters, but only low selectivities resulted in its reaction with benzaldehyde.

The high enantioselectivity observed was interpreted in terms of the face selectivity of the (*Z*)-enolate **59** (Scheme 1.20). The phenyl moiety is thought to stabilize the enolate through a π-π interaction and effectively shield its *Re* face such that the incoming ketone approaches preferentially from the *Si* face.

Yan's group has used the camphor-based chiral thioamide **62** in asymmetric Darzens reactions (Scheme 1.21) [32]. The addition of the titanium enolate of **62** to a range of aldehydes resulted in the formation of essentially single diastereomers of halo alcohols **63**. Treatment of these with aqueous potassium carbonate resulted in the formation of the corresponding aryl (**65**), alkyl (**64** and **68**), and vinyl (**66**) epoxy acids without epimerization. If the thioamide adduct was instead treated with DMAP and benzyl alcohol, followed by KF and LiF in the presence of *n*-$Bu_4N^+HSO_4^-$, the epoxy ester **67** was formed [33]. In all cases the *cis* epoxide predominated; the selectivity was thus complementary to sulfur ylide chemistry, which almost always favors the *trans* epoxide.

Ghosh and co-workers have recently used the indanyl-derived auxiliary **69** (Table 1.9) in titanium enolate condensations with a range of aldehydes [34]. Of the four possible diastereomers, only the *anti* **71** and *syn* **72** were produced (the alternative *anti* and *syn* diastereomers were not detected by ^1H or ^{13}C NMR). The use of monodentate aliphatic aldehydes resulted in the formation of *anti* diastereomers **71** with high selectivities with the aid of acetonitrile or *N*-methylpyrrolidinone (NMP) as an additive (Entries 1 and 2). The use of bidentate aldehydes resulted in high *syn* diastereoselectivities without requiring the use of an additive (Entries 5 and 6). Interestingly, benzaldehyde exhibited *anti* selectivity in the presence of an additive (Entry 3), but *syn* selectivity in its absence (Entry 4). Additionally, a double asymmetric induction using (2*R*)- and (2*S*)-benzyloxypropionaldehyde was attempted (Entries 7 and 8). In the matched case ((2*R*)-), only the *syn* diastereomer **72** was produced, but in the mismatched case ((2*S*)-) the *anti* diastereomer **71** was obtained instead. It was thus possible to perform a kinetic resolution on two equivalents of racemic aldehyde (Entry 9) and to obtain the *syn* diastereomer **72** (through reaction of matched (2*R*)-aldehyde) with high selectivity. The (2*S*)-aldehyde was isolated in 40% yield and in 98.7% *ee*. Treatment of the halo-aldol adducts with potassium carbonate in DMF resulted in the formation of the epoxides (**74**). Simultaneous epoxide formation and removal of the auxiliary could be effected by treating the adducts with potassium carbonate in methanol to give the epoxy acids (**73**).

1.2.3.4 **Darzens Reactions with Chiral Reagents**

Clearly it is advantageous to be able to use achiral starting materials and a chiral reagent to induce an asymmetric reaction, thus obviating the need to attach and remove a chiral auxiliary and permitting the recovery and reuse of the chiral reagent.

Corey used a chiral bromoborane **75** (1.1 equiv.) to promote the addition of *tert*-butyl bromoacetate (**76**) to aromatic, aliphatic, and α,β-unsaturated aldehydes to give the halo alcohols **77** with high enantio- and diastereoselectivities (Table 1.10) [35].

Additionally, the sulfonamide precursor to **75** could be recovered and recycled to regenerate the bromoborane **75** [36]. The resulting aldols could then be cyclized to the epoxy esters by treatment with potassium *tert*-butoxide (Scheme 1.22).

A valine-based chiral oxazaborolidinone **80** (generated *in situ* from Ts-L-Val and $BH_3 \cdot THF$) was used by Kiyooka and co-workers [37] to catalyse the reaction be-

Table 1.10 Chiral reagent **75** in asymmetric Darzens reactions.

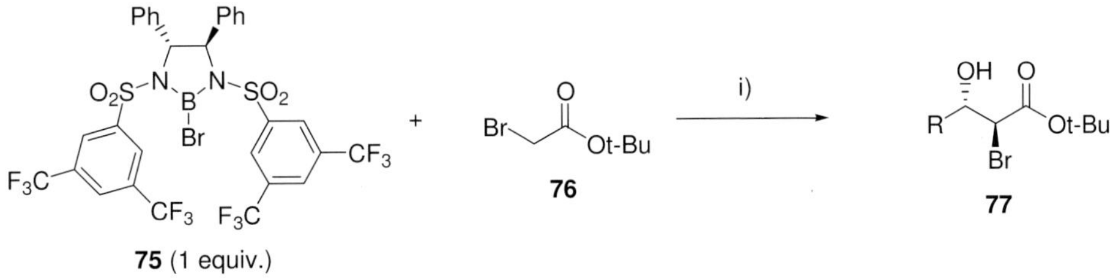

75 (1 equiv.) + **76** → i) → **77**

i) Et_3N, -78 °C, 1:2 toluene/hexane; then RCHO, -78 °C.

Entry	R of RCHO	Yield (%)	*anti*:*syn*	*ee* (%)
1	Ph	94	91:1	98
2	(*E*)-PhCH=CH	96	99:1	98
3	$PhCH_2CH_2$	72	95:5	91
4	cyclohexyl	65	98:2	91

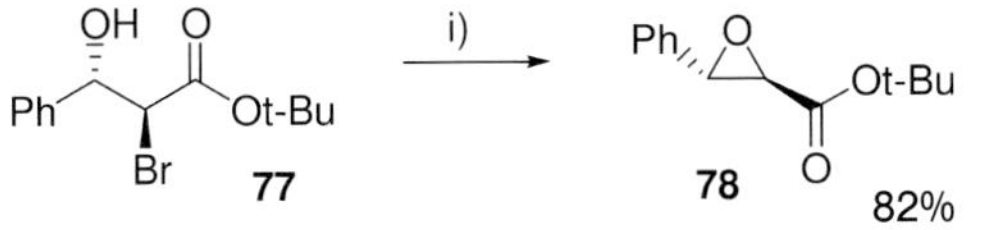

77 → i) → **78** 82%

i) KOt-Bu, t-BuOH.

Scheme 1.22

Table 1.11 Chiral induction through the use of the valine-based **80**.

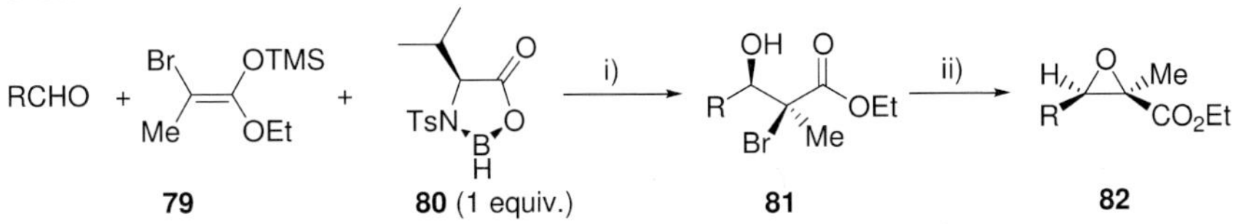

RCHO + **79** + **80** (1 equiv.) → i) → **81** → ii) → **82**

i) CH_2Cl_2, -78 °C, 15 h; ii) NaOEt, EtOH, rt, 2 h.

Entry	R of RCHO	Yield 81	*syn* % : *anti*	% *ee* (*syn*)	Yield 82 (%)
1	Ph	82	7:1	95	87
2	*i*-Pr	68	16:1	97	74
3	$PhCH_2CH_2$	80	9:1	96	81
4	*n*-Pr	85	10:1	98	78
5	$TBSOCH_2CH_2$	87	15:1	95	93

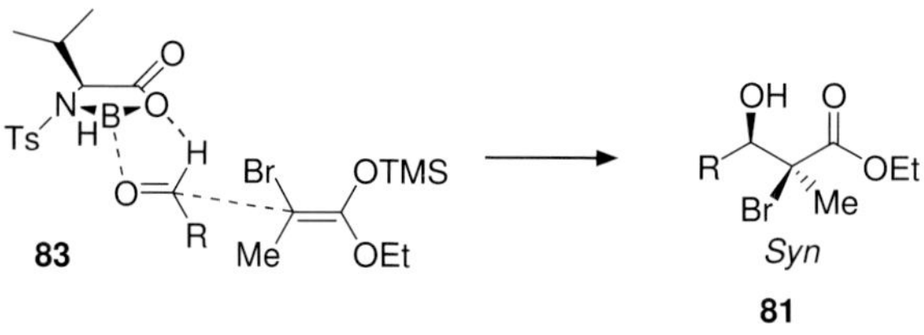

Scheme 1.23

tween β-bromo-β-methylketene silyl acetal **79** and a range of aldehydes (Table 1.11). Good diastereoselectivities and excellent enantioselectivities resulted in the formation of the halo alcohols **81**, which could be converted into the trisubstituted aryl or alkyl methyl epoxy esters **82** by treatment with sodium ethoxide.

A transition state assembly as depicted in Scheme 1.23 was proposed in order to interpret the observed selectivity. Electronic effects are thought to be operative, as the methyl and bromo substituents in transition state **83** are sterically similar.

1.2.3.5 Darzens Reactions with Chiral Catalysts

Of course, the most practical and synthetically elegant approach to the asymmetric Darzens reaction would be to use a sub-stoichiometric amount of a chiral catalyst. The most notable approach has been the use of chiral phase-transfer catalysts. By rendering the intermediate enolate **86** (Scheme 1.24) soluble in the reaction solvent, the phase-transfer catalyst can effectively provide the enolate with a chiral environment in which to react with carbonyl compounds.

Early work on the use of chiral phase-transfer catalysis in asymmetric Darzens reactions was conducted independently by the groups of Wynberg [38] and Colonna [39], but the observed asymmetric induction was low. More recently Toké's group has used catalytic chiral aza crown ethers in Darzens reactions [40–42], but again only low to moderate enantioselectivities resulted.

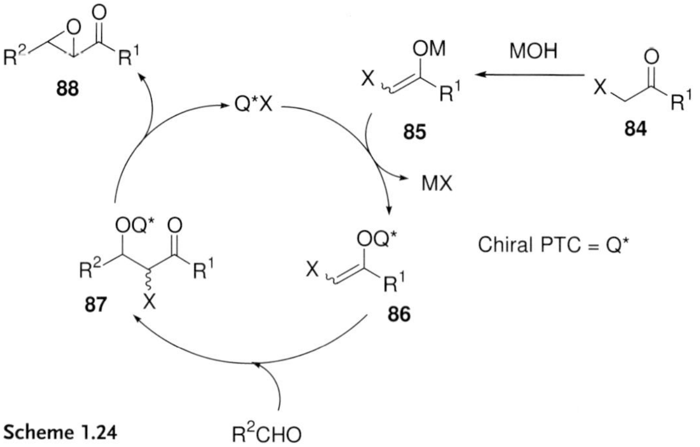

Scheme 1.24

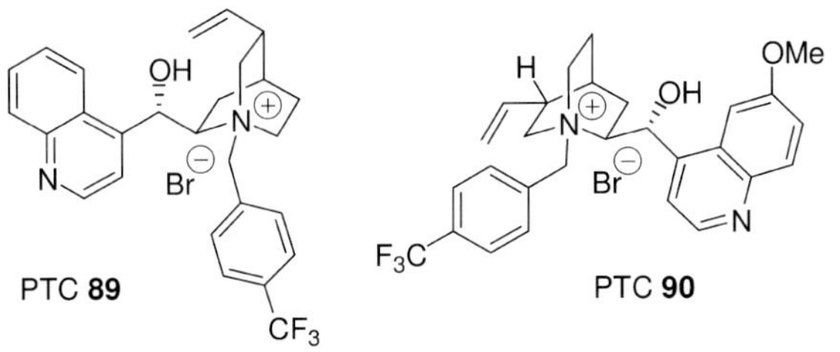

PTC **89** PTC **90**

Scheme 1.25

Arai and co-workers have used chiral ammonium salts **89** and **90** (Scheme 1.25) derived from cinchona alkaloids as phase-transfer catalysts for asymmetric Darzens reactions (Table 1.12). They obtained moderate enantioselectivities for the addition of cyclic **92** (Entries 4–6) [43] and acyclic **91** (Entries 1–3) chloroketones [44] to a range of alkyl and aromatic aldehydes [45] and also obtained moderate selectivities on treatment of chlorosulfone **93** with aromatic aldehydes (Entries 7–9) [46, 47]. Treatment of chlorosulfone **93** with ketones resulted in low enantioselectivities.

Table 1.12 Cinchona alkaloid-derived phase-transfer catalysts for asymmetric Darzens reactions.

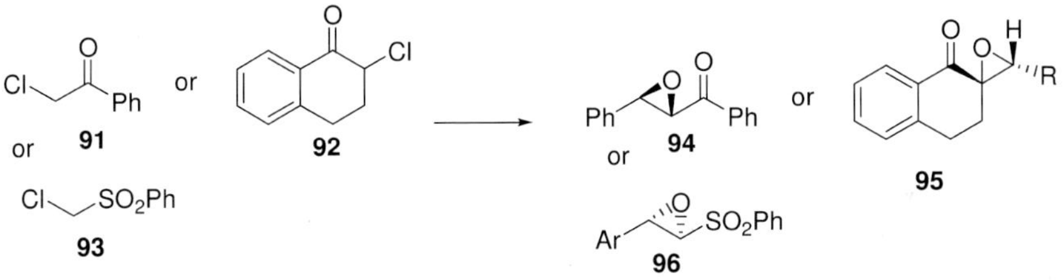

Entry	R^2CHO	Halide	Method	Yield	*ee* (%)
1	*i*-PrCHO	**91**	I	80	53
2	EtCHO	**91**	I	32	79
3	PhCHO	**91**	I	43	42
4	*i*-PrCHO	**92**	II	99	69
5	*t*-BuCH$_2$CHO	**92**	II	86	86
6	PhCHO	**92**	II	67	59
7	PhCHO	**93**	III	85	69
8	*p*-MeC$_6$H$_4$CHO	**93**	III	84	78
9	*p*-BrC$_6$H$_4$CHO	**93**	III	80	64

Method I: PTC **89** (10 mol%), LiOH · H$_2$O, *n*-Bu$_2$O, 4 °C, 60–117 h;
Method II: PTC **90** (10 mol%), LiOH · H$_2$O, *n*-Bu$_2$O, rt, 43–84 h;
Method III: PTC **90** (10 mol%), KOH, toluene, rt, 1 h.

Table 1.13 BINOL-derived phase-transfer catalysts for asymmetric Darzens reactions.

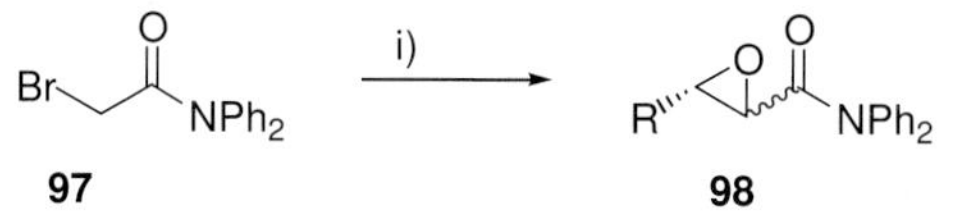

i) PTC **91** (2 mol%), $RbOH \cdot H_2O$, CH_2Cl_2, rt, 6.5-24 h.

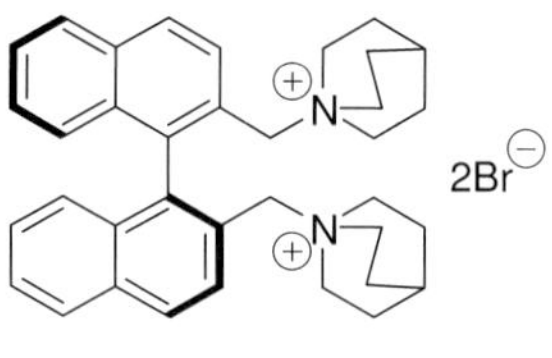

PTC **99**

Entry	RCHO	Yield (%)	*cis:trans*	*ee cis* (%)	*ee trans* (%)
1	PhCHO	81	2.3:1	58	63
2	*m*-BrC_6H_4CHO	93	2.4:1	51	60
3	*p*-$MeOC_6H_4CHO$	quant.	2.2:1	57	67

More recently, the same group has used a simpler and more easily prepared chiral ammonium phase-transfer catalyst **99** derived from BINOL in asymmetric Darzens reactions with α-halo amides **97** to generate glycidic tertiary amides **98** (Table 1.13). Unfortunately the selectivities were only moderate to low [48]. As mentioned in Section 1.2.3.1, tertiary amides can be converted to ketones.

1.3 Asymmetric Aziridination of Imines

Asymmetric transformation of imines into chiral aziridines remains less well developed than the analogous transformation of aldehydes into epoxides [49, 50, 51]. The reported methods can be divided into three conceptual categories involving

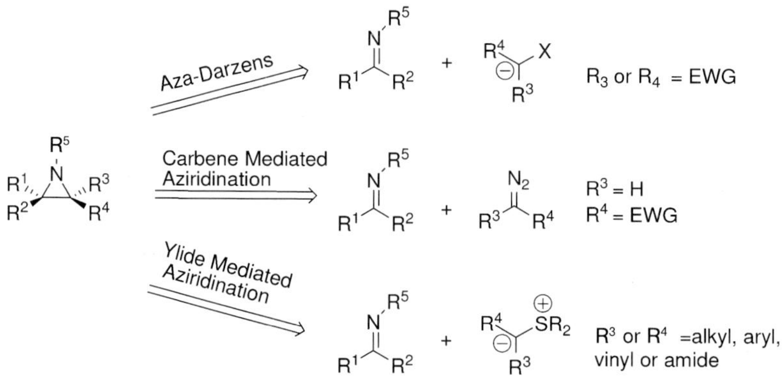

Scheme 1.26

reactions of imines with: *i*) α-halo enolates (aza-Darzens), *ii*) carbenes, or *iii*) ylides (Scheme 1.26). Categories *i*) and *ii*) are employed to prepare aziridines bearing electron-withdrawing groups such as esters or amides. Category *iii*), the ylide methodology, on the other hand, provides a route to aryl, alkyl, vinyl, and terminal aziridines, as well as ester- or amide-substituted aziridines. The most common method of asymmetric induction reported has been with the aid of chiral auxiliaries. There have been attempts at reagent-controlled induction, which has been most successful in the sulfur ylide methodology. However there exist only two examples of asymmetric catalysis: a sulfur ylide-mediated aziridination by Aggarwal and a Lewis acid-catalyzed diazoacetate decomposition by Wulff.

1.3.1
Aziridines Bearing Electron-withdrawing Groups: Esters and Amides

1.3.1.1 Aza-Darzens Route

The aza-Darzens reaction is analogous to the Darzens synthesis of epoxides (see Section 1.2.3.2) but employs imines in the place of aldehydes (Scheme 1.27).

Davis has employed the enantiopure sulfinimine *N*-(benzylidene)-*p*-toluenesulfinimine in reactions with α-halo ester enolates to obtain aziridine-2-carboxylates in good yields and with high diastereoselectivities (Scheme 1.28) [52]. The selectivities are consistent with a six-membered chair-like transition state **100**, containing a four-membered metallocycle. It is assumed that the enolate of the unsubstituted α-bromoacetate has the *E* geometry resulting in the *cis* aziridine, while the enolate of the substituted α-bronoacetate adopts the *Z*-geometry resulting in the *trans* aziridine.

Davis has also employed a similar procedure for the synthesis of aziridine-2-phosphonoates, involving the addition of *N*-(2,4,6-trimethylphenylsulfinyl)imine to anions of diethyl α-halomethyl phosphonates (Scheme 1.29) [53, 54]. Aziridines

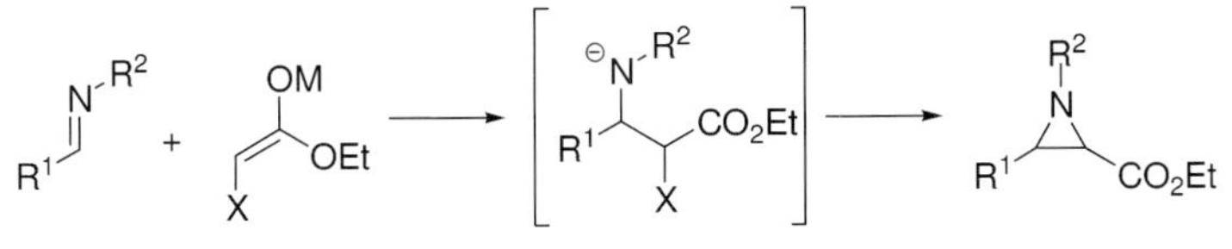

Scheme 1.27

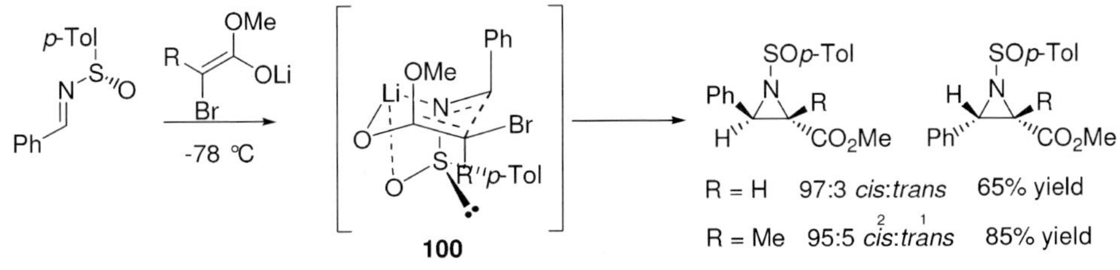

Scheme 1.28

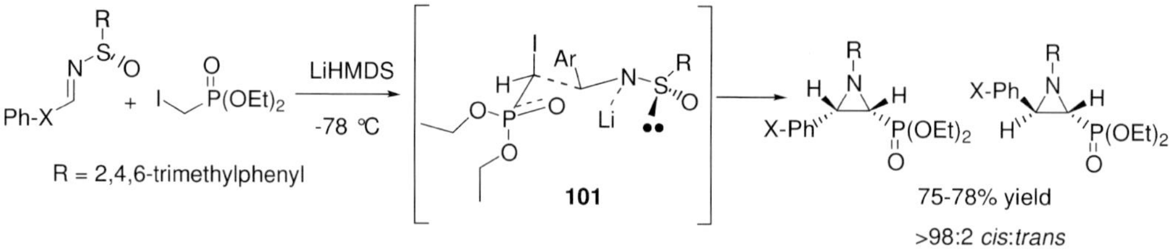

LiHMDS, -78 °C

R = 2,4,6-trimethylphenyl

101

75-78% yield

>98:2 *cis*:*trans*

Scheme 1.29

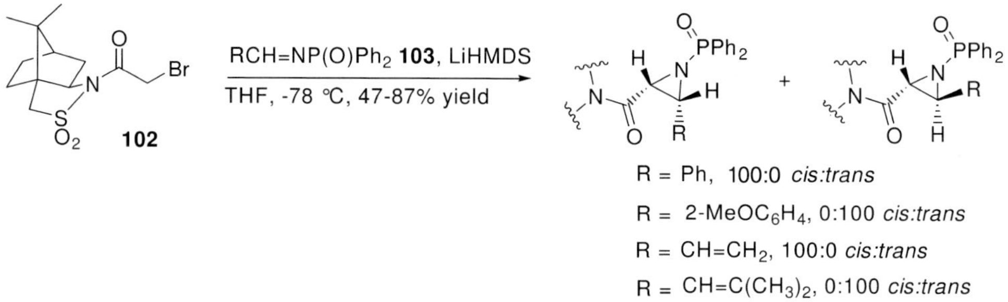

$RCH{=}NP(O)Ph_2$ **103**, LiHMDS

THF, -78 °C, 47-87% yield

102

R = Ph, 100:0 *cis*:*trans*

R = 2-$MeOC_6H_4$, 0:100 *cis*:*trans*

R = $CH{=}CH_2$, 100:0 *cis*:*trans*

R = $CH{=}C(CH_3)_2$, 0:100 *cis*:*trans*

Scheme 1.30

were obtained as single *cis* diastereomers (>98:2) in 75–78% isolated yields. The high selectivity is believed to arise from two types of steric interaction in the transition state. Attack by the anion at the C=N bond opposite the bulky sulfinyl group is highly favored and, secondly, the iodo group needs to occupy the axial position in the transition state **101**, as it would then have fewer steric interactions with the ethoxy phosphonate groups.

The substrate scope is limited, as electron-withdrawing groups (X = *p*-NO_2 or *p*-CF_3) on the aromatic substituent are not tolerated. However, this route does provide valuable intermediates to unnatural α-amino phosphonic acid analogues and the sulfimine can readily be oxidized to the corresponding sulfonamide, thereby providing an activated aziridine for further manipulation, or it can easily be removed by treatment with a Grignard reagent.

An alternative approach is to have the chiral auxiliary on the enolate. Sweeney has reported the addition of bromoacyl sultam **102** to phosphonyl imines **103**, which afforded the *cis*- or *trans*-aziridines with high levels of diastereoselectivity depending on the imine substituent (Scheme 1.30) [55].

1.3.1.2 Reactions between Imines and Carbenes

Synthesis of aziridines by treatment of carbenes with imines was reported by Jacobsen [56]. A metallocarbene **104** derived from ethyl diazoacetate and copper fluorophosphate was treated with N-arylaldimines to form aziridines with reasonable diastereoselectivities (>10:1 in favor of *cis*) but with low enantioselectivities (about 44% *ee*). This was shown to result from a competitive achiral reaction path-

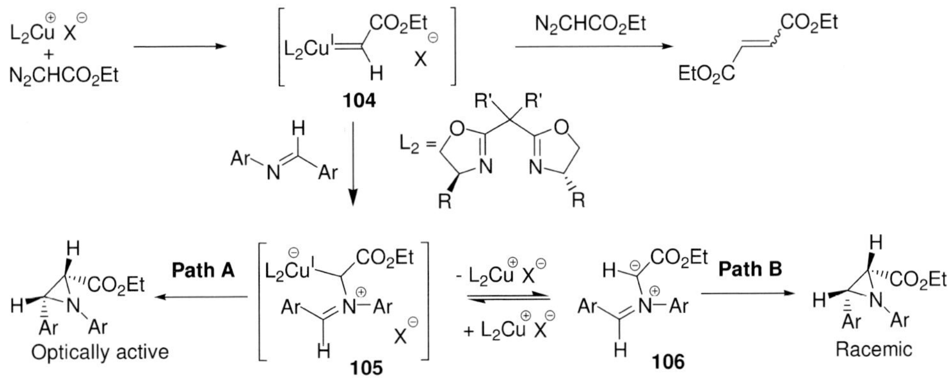

Scheme 1.31

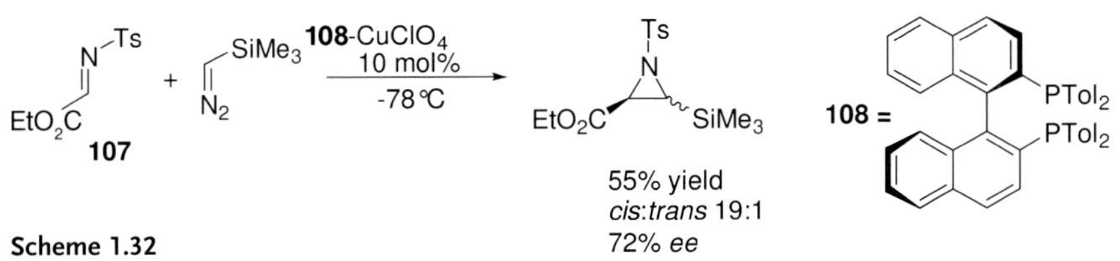

Scheme 1.32

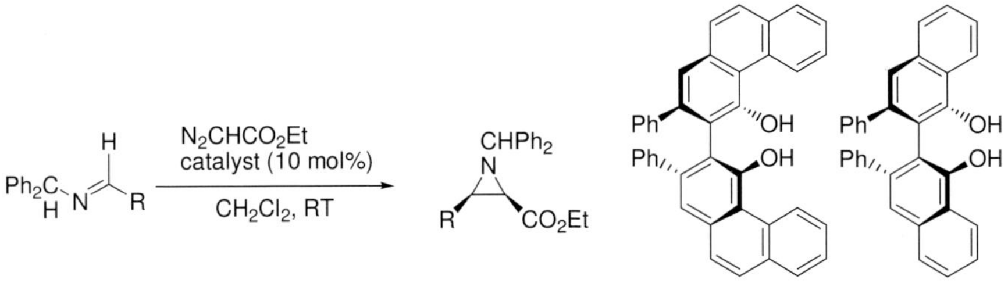

(S)-VAPOL **109** (S)-VANOL **110**

Scheme 1.33

way (Scheme 1.31). Path A goes through the chiral metal species **105**, yielding non-racemic aziridine, whereas path B goes through a planar azomethine ylide **106**, yielding the racemic aziridine. The reaction showed limited scope, as it was quite sensitive to the electronic properties of the imine.

Jørgensen has recently reported similar enantioselective reactions between *N*-tosylimines **107** and trimethylsilyldiazomethane (TMSD) catalyzed by chiral Lewis acid complexes (Scheme 1.32) [57, 53]. The *cis*-aziridine could be obtained in 72% *ee* with use of a BINAP-copper(I) catalyst, but when a bisoxazoline-copper(I) complex was used the corresponding *trans* isomer was formed in 69% *ee* but with very poor diastereoselectivity.

Table 1.14 Wulff's asymmetric aziridination synthesis.

Entry	R	Catalyst ligand	Yield of *cis*-aziridines (%)	*ee* of *cis*-aziridine (%)	*cis:trans* aziridine
1	Ph	**109**	77	95	>50:1
2	Ph	**110**	85	96	>50:1
3	*p*-BrC_6H_4	**109**	91	98	>50:1
4	*o*-MeC_6H_4	**109**	69	94	40:1
5	3,4-$(OAc)_2C_6H_3$	**110**	85	96	>50:1
6	1-naphthyl	**109**	87	92	>50:1
7	2-furyl	**110**	55	93	>50:1
8	*n*-Pr	**110**	60	90	>50:1
9	*t*-Bu	**110**	77	97	>50:1
10	*c*-C_6H_{11}	**109**	74	94	38:1

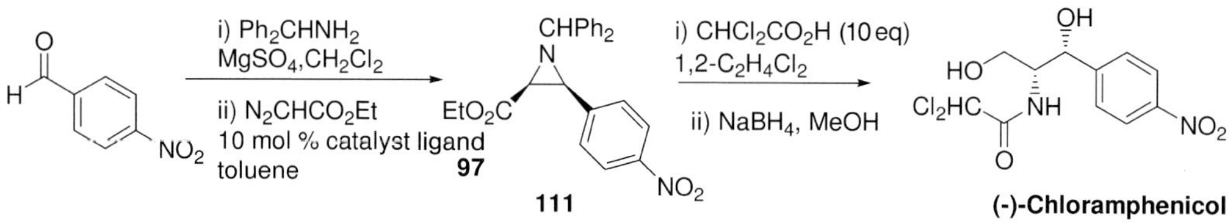

i) Ph_2CHNH_2, $MgSO_4$, CH_2Cl_2; ii) N_2CHCO_2Et, 10 mol % catalyst ligand **97**, toluene → **111**; i) $CHCl_2CO_2H$ (10 eq), 1,2-$C_2H_4Cl_2$; ii) $NaBH_4$, MeOH → **(-)-Chloramphenicol**

Scheme 1.34

The most successful approach in this reaction category has been the use of chiral boron Lewis acid catalysts, in the addition of ethyl diazoacetate to imines reported by Wulff (Scheme 1.33) [59–60].

Catalysts prepared either from VAPOL (**109**) or from VANOL (**110**) ligands and triphenylborate were found to catalyze the asymmetric aziridination efficiently. Good to high yields, excellent enantioselectivities, and *cis* diastereoselectivities were observed with all the reported substrates, which included aromatic, hetero-aromatic and aliphatic imines (Table 1.14).

This is by far the most versatile route to the synthesis of ester-substituted aziridines, especially as the benzhydryl group can easily be cleaved by hydrogenolysis. Wulff has applied this methodology to a short asymmetric synthesis of the antibiotic (−)-chloramphenicol in four steps from *p*-nitrobenzaldehyde (Scheme 1.34) [61]. In this case it was found that treatment of the aziridine **111** with excess dichloroacetic acid gave the hydroxy acetamide directly, so no separate deprotection step was required.

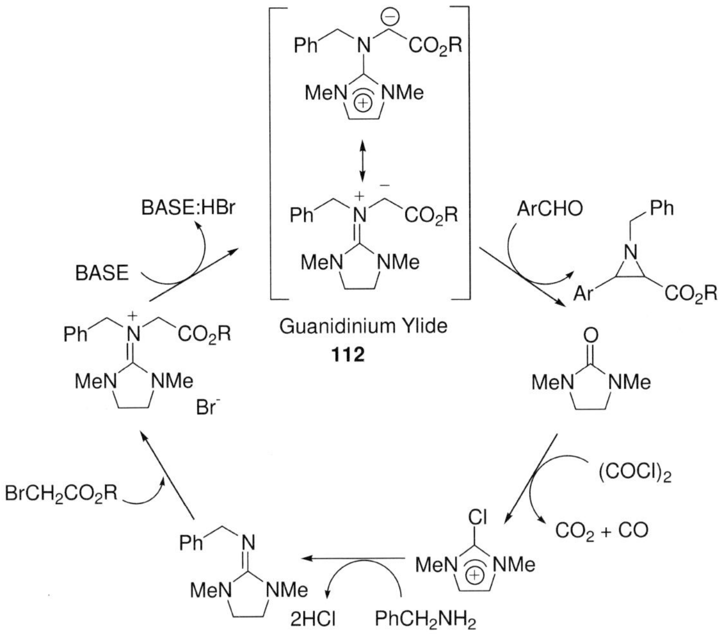

Scheme 1.35

1.3.1.3 **Aziridines by Guanidinium Ylide Chemistry**

A novel guanidinium ylide-mediated procedure has recently been reported by Ishikawa [62]. Though not an imine transformation, it does employ an imine precursor in the form of an aldehyde. Guanidinium ylides react with aldehydes to form aziridines (Scheme 1.35). The mechanism for the formation of the aziridine is believed to involve [3+2] cycloaddition between the guanidinium ylide **112** and the aldehyde, followed by stereospecific extrusion of the urea with concomitant aziridine formation.

Table 1.15 Chiral guanidylium ylides for asymmetric synthesis of aziridines.

Entry	Ar	Yield of *trans*-aziridines (%)	*ee* of *trans*-aziridine (%)	*trans*:*cis* aziridine
1	3-[(1-Boc)indolyl]	70	95	92:8
2	2-[(1-Boc)indolyl]	87	76	91:9
3	3,4-OCH_2OPh	82	97	93:7
4	C_6H_6	31	77	34:66
5	*p*-ClC_6H_4	35	59	41:59

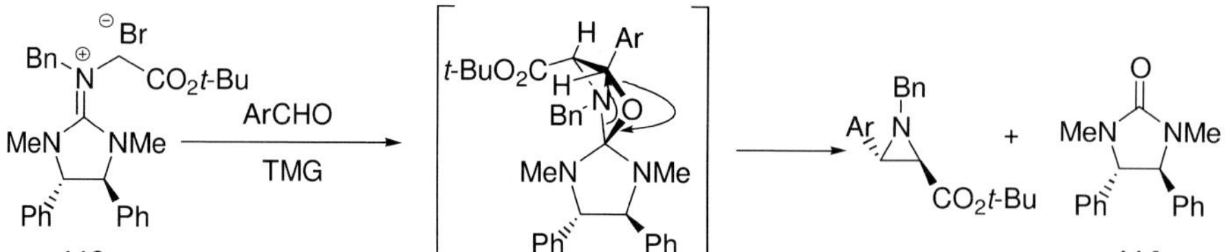

Scheme 1.36

This reaction was found to be applicable to aryl, heteroaryl, and α,β-unsaturated aldehydes, providing aziridine-2-carboxylates, sometimes with high *trans* diastereoselectivity. Excellent enantioselectivity was observed with use of a chiral guanidinium ylide (Scheme 1.36), but simple phenyl substituents on the aldehyde gave poor yields (Table 1.15). The enantioselectivity is controlled by the facial selectivity in the [3+2] cycloaddition (Scheme 1.36). The other product of the reaction was the chiral urea **113**, which could be recovered in high yield and reconverted into the guanidinium salt **114**. Guanidinium ylide chemistry provides a complementary methodology to sulfur ylide chemistry, which currently dominates non-metal-mediated asymmetric aziridination.

1.3.2 Aziridines Bearing Alkyl, Aryl, Propargyl, and Vinyl Groups

The usual route to aziridines bearing alkyl, aryl, propargyl, and vinyl groups, as well as to terminal aziridines, is through reactions between ylides and imines. The reaction between an ylide and an imine forms a betaine **115**, which ring-closes to form an aziridine through elimination of the heteroatom-containing leaving group originating from the ylide (Scheme 1.37). The heteroatom-containing group **116** derived from the ylide can thus be recovered and reused. The main class of ylides used in asymmetric aziridination reaction are sulfur ylides.

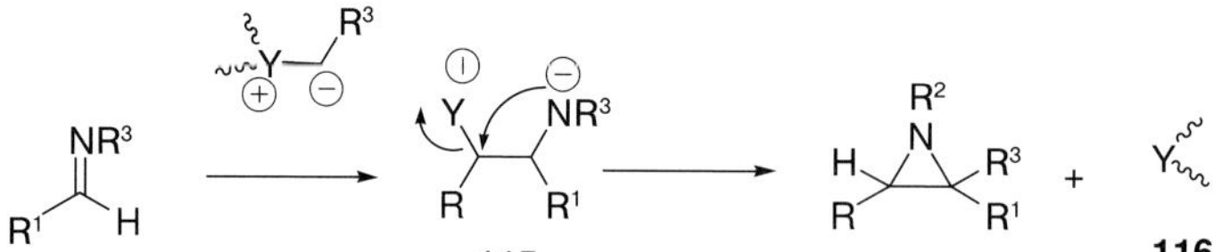

Scheme 1.37

1.3.2.1 Aryl, Vinyl, and Alkyl Aziridines: Stoichiometric Asymmetric Ylide-mediated Aziridination

Ruano has reported substrate-controlled asymmetric ylide aziridination by treatment of enantiopure sulfinyl imines **117** with dimethyloxosulfonium methylide **118** to form terminal aziridines [63]. The chiral *tert*-butylsulfinyl group was shown

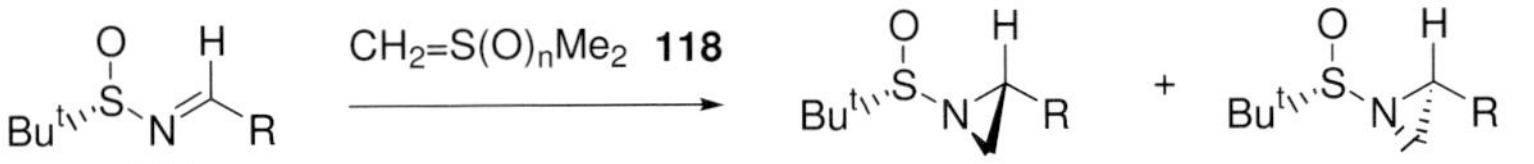

$CH_2{=}S(O)_nMe_2$ **118**

117

R = Ph	n = 0	70% yield, 15:85 ratio
R = Ph	n = 1	85% yield, 95:5 ratio
R = (*E*)PhCH	n = 0	72% yield, 18:82 ratio
R = (*E*)PhCH	n = 1	40% yield, 83:17 ratio

Scheme 1.38

to be the chiral auxiliary of choice, allowing the synthesis of aziridines in high yields and with good diastereoselectivities (Scheme 1.38).

The sense of asymmetric induction could be tuned in two ways: firstly through the chirality of the sufinyl group, and secondly through the use of dimethyloxosulfonium methylide ($n = 1$) or of dimethylsulfonium methylide ($n = 0$), which was found to provide aziridines with opposite diastereoselectivity. This was interpreted by assuming the process to be under thermodynamic control in the former

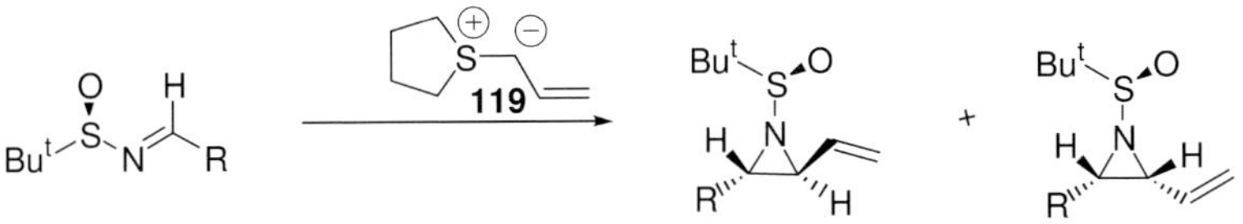

119

Scheme 1.39

Table 1.16 Chiral *tert*-butylsulfinylimines in asymmetric aziridine synthesis.

Entry	R	Yield of *trans* aziridines (%)	*trans:cis* aziridine
1	Ph	68	71:29
2	*p*-$OMeC_6H_4$	76	82:18
3	*p*-$NO_2C_6H_4$	74	59:41
4	1-naphthyl	64	80:20
5	ethyl	44	80:20
6	cyclopropyl	61	72:28
7	cyclohexyl	78	83:17
8	2-furyl	55	67:33
9	2-piperidine	54	88:12

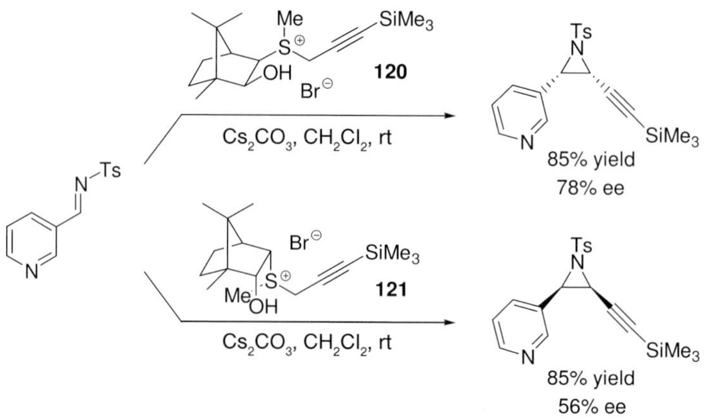

Scheme 1.40

case and under kinetic control in the latter case. When the reaction is under kinetic control the diastereoselectivity is determined in the attack of the ylide on the imine, whereas under thermodynamic control it is dependent on the relative stabilities of intermediate betaines and their relative rates of ring-closure.

Stockman has reported the preparation of alkyl-, aryl-, and vinyl-disubstituted aziridines with good diastereoselectivities and in good yields through treatment of *tert*-butylsulfinylimines with the ylide **119**, derived from *S*-allyl tetrahydrothiophenium bromide (Scheme 1.39) [64]. A range of substrates were tolerated, including heterocyclic, aromatic, and aliphatic substrates (Table 1.16).

Dai has also studied the synthesis of chiral vinyl aziridines through reactions between allylic ylides and *N*-tosyl imines [65], but did not examine an asymmetric variant because of the low diastereoselectivities. In contrast, propargyl-substituted ylides, generated *in situ* from the corresponding sulfonium salts in the presence of Cs_2CO_3 as the base, were found to afford aziridines in high yields and with exclusive *cis* diastereoselectivity (Scheme 1.40). When the camphor-derived chiral sulfonium salt **120** was employed, variable enantioselectivity (depending on substrate) was obtained, but with complete *cis* diastereoselection, whereas use of the diastereoisomeric sulfonium salt **121** resulted in opposite asymmetric induction. Aromatic, heteroaromatic, α,β-unsaturated, and aliphatic aldimines and ketimines could all be employed with high diastereoselectivity, and the chiral sulfide precursors of the sulfonium salts could usually be recovered in high yield. The enantioselectivities varied considerably (40–85% *ee*), however, depending on the substrate.

Saito has recently reported high yields and enantioselectivities in aziridine synthesis through reactions between aryl- or vinyl-substituted *N*-sulfonyl imines and aryl bromides in the presence of base and mediated by a chiral sulfide **122** (Scheme 1.41) [66]. Aryl substituents with electron-withdrawing and -donating groups gave modest *trans*:*cis* selectivities (around 3:1) with high enantioselectiv-

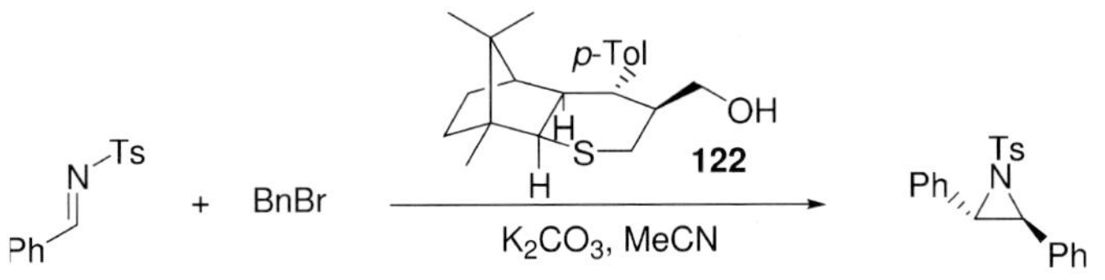

Scheme 1.41

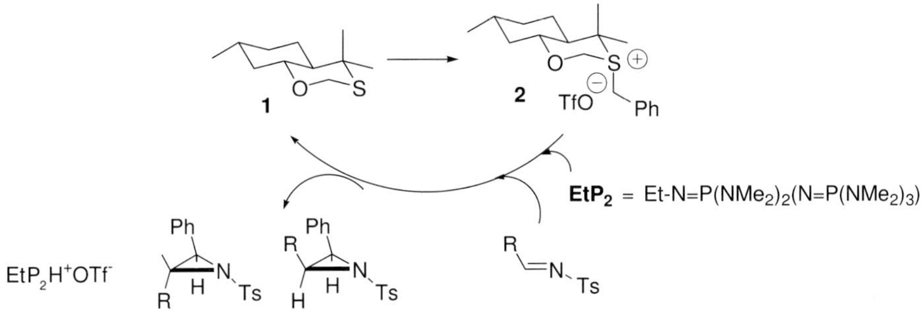

EtP₂ = Et-N=P(NMe₂)₂(N=P(NMe₂)₃)

Scheme 1.42

ities (85–99% *ee*). Vinyl-substituted imines gave similar enantioselectivities, but the diastereoselectivities were much lower.

Solladié-Cavallo has recently reported a two-step asymmetric synthesis of disubstituted *N*-tosylaziridines from (R,R,R,S_S)-(–)-sulfonium salt **2** (derived from Eliel's oxathiane; see Section 1.2.1.1) and *N*-tosyl imines with use of phosphazine base (EtP_2) to generate the ylide (Scheme 1.42) [67]. Although the diastereoselectivity was highly substrate-dependent, the enantioselectivities obtained were very high (98.7–99.9%). The chiral auxiliary, although used in stoichiometric quantities, could be isolated and reused, but the practicality and scope of this procedure is limited by the use of the strong – as well as expensive and sensitive – phosphazene base.

1.3.2.2 Aryl, Vinyl, and Alkyl Aziridines: Catalytic Asymmetric Ylide-mediated Aziridination

Of course, the key limitation of the ylide-mediated methods discussed so far is the use of stoichiometric amounts of the chiral reagent. Building on their success with catalytic asymmetric ylide-mediated epoxidation (see Section 1.2.1.2), Aggarwal and co-workers have reported an aza version that provides a highly efficient catalytic asymmetric synthesis of *trans*-aziridines from imines and diazo compounds or the corresponding tosylhydrazone salts (Scheme 1.43) [68–70].

A range of electron-withdrawing groups on the nitrogen – *N*-P(O)Ph_2, *N*-tosyl, and *N*-SES, for example – were tolerated. Imines derived from aromatic, heteroaromatic, unsaturated, and even aliphatic aldehydes and ketones were employed

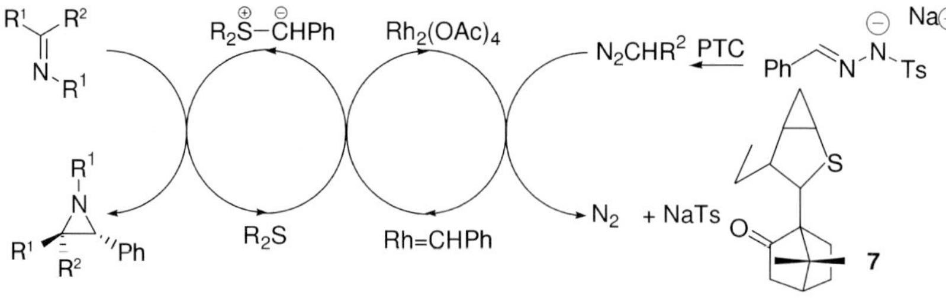

Scheme 1.43

Table 1.17 Catalytic asymmetric ylide-mediated aziridination.

Entry	R^1	R^2	R^3	Yield of *trans* aziridines (%)	*trans:cis* aziridine	*ee* (%) *trans*	*cis*
1	*p*-ClC_6H_4	H	TcBoc	56	6:1	94	90
2	*p*-ClC_6H_4	H	SES	82	2:1	98	81
3	C_6H_{11}	H	SES	50	2.5:1	98	89
4	*t*-Bu	H	Ts	53	2:1	73	95
5	(*E*)-PhCH=CH	H	SES	59	8:1	94	–
6	*p*-$MeOC_6H_4$	H	SES	60	2.5:1	92	78
7	3-furyl	H	Ts	72	8:1	95	–
8	Ph	Ph	$SO_2C_8H_7$	50	–	84	–

and good yields were obtained (Table 1.17). High enantioselectivities were obtained in all cases, but diastereoselectivities were dependent both on the nitrogen activating group and on the imine substituent, with carbamate groups giving better diastereoselectivities than the corresponding sulfonyl groups.

The variation of the diastereoselectivity with groups on the nitrogen can be explained by the model shown in Scheme 1.44. Large bulky groups on the nitrogen will increase the congestion in transition state **A**, resulting in reduced *trans* selectivity. However, small groups (e.g., alkoxycarbonyl) will be accommodated in this transition state more easily, resulting in increased amounts of the *trans* isomer. The high enantioselectivity observed is interpreted as in the epoxidation case

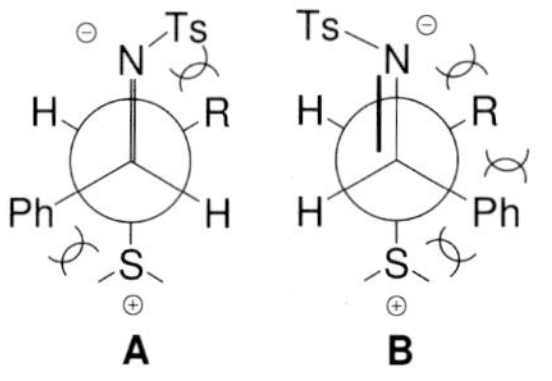

Scheme 1.44

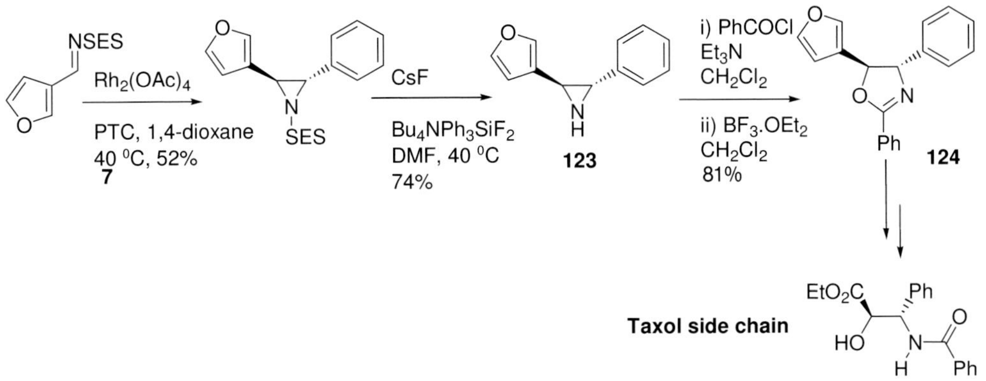

Scheme 1.45

(see Section 1.2.1.3), with the key difference being that the betaine formation is non-reversible [70], resulting in higher enantioselectivities and lower diastereoselectivities in general for aziridination than for epoxidation.

The main features of this process are: *i*) high convergency, *ii*) high enantioselectivity, *iii*) catalytic use of chiral sulfide and its quantitative reisolation, *iv*) ready availability of both enantiomers of the sulfide, and *v*) an efficient and user-friendly process. This methodology has been applied to construct the taxol side chain with a high degree of enantioselectivity via a *trans*-aziridine, followed by stereospecific rearrangement of the *trans*-benzoylaziridine **123** into the *trans*-oxazoline **124** (Scheme 1.45) [71].

1.4
Summary and Outlook

Two catalytic asymmetric sulfur ylide-mediated epoxidation processes have been developed. The method involving the reaction between a chiral sulfide and an alkyl halide and base in the presence of an aldehyde is generally limited to the synthesis of stilbene oxide derivatives. The method involving the reaction between a chiral sulfide and a diazo precursor in the presence of a PTC, metal catalyst, and aldehyde shows broader scope. Aromatic, heteroaromatic (but not pyridyl), aliphatic, and unsaturated aldehydes have been employed, together with a range of aromatic and heteroaromatic diazo precursors. Certain aldehydes and diazo precursors give rather low yields of epoxides, but in these cases an asymmetric stoichiometric process can be employed instead. The combined catalytic and stoichiometric processes allow access to a very broad range of epoxides, including glycidic amides and α,β-unsaturated epoxides, aziridines, and cyclopropanes, in many instances with control over both relative and absolute stereochemistry. This broad substrate scope of the process now allows the sulfur ylide disconnection to be applied with confidence in total synthesis. The synthesis of glycidic esters by a Darzens-type reaction remains challenging. Although stoichiometric processes employing chiral reagents or chiral auxiliaries have delivered high selectivity, no useful catalytic process has yet emerged.

Aziridination remains less well developed than epoxidation. Nevertheless, high selectivity in imine aziridination has been achieved through the use of chiral sulfinimines as auxiliaries. Highly successful catalytic asymmetric aziridination reactions employing either sulfur ylides or diazo esters and chiral Lewis acids have been developed, although their scope and potential applications in synthesis have yet to be established.

References

1 A. Solladié-Cavallo, A. Diep-Vohuule, V. Sunjic, V. Vinkovic, *Tetrahedron: Asymmetry* 1996, *7*, 1783.
2 A. Solladié-Cavallo, L. Bouérat, M. Roje, *Tetrahedron Lett.* 2000, *41*, 7309.
3 A. Solladié-Cavallo, M. Roje, T. Isarno, V. Sunjic, V. Vinkovic, *Eur. J. Org. Chem.* 2000, 1077.
4 V. K. Aggarwal, I. Bae, H.-Y. Lee, J. Richardson, D. T. Williams, *Angew. Chem. Int. Ed.* 2003, *42*, 3274.
5 N. Furukawa, Y. Sugihara, H. Fujihara, *J. Org. Chem.* 1989, *54*, 1222.
6 J. Zanardi, C. Leviverend, D. Aubert, K. Julienne, P. Metzner, *J. Org. Chem.* 2001, *66*, 5620.
7 J. Zanardi, D. Lamazure, S. Minière, V. Reboul, P. Metzner, *J. Org. Chem.* 2002, *67*, 9083.
8 H. Takada, P. Metzner, C. Philouze, *J. Chem. Soc., Chem. Commun.* 2001, 2350.
9 V. K. Aggarwal, E. Alonso, G. Hynd, K. M. Lydon, M. J. Palmer, M. Porcelloni, J. R. Studley, *Angew. Chem. Int. Ed.* 2001, *40*, 1430.
10 V. K. Aggarwal, E. Alonso, I. Bae, G. Hynd, K. M. Lydon, M. J. Palmer, M. Patel, M. Porcelloni, J. Richardson, R. A. Stenson, J. R. Studley, J.-L. Vasse, C. L. Winn, *J. Am. Chem. Soc.* 2003, *125*, 10926.
11 V. K. Aggarwal, S. Calamai, J. G. Ford, *J. Chem. Soc., Perkin Trans. 1* 1997, 593.
12 V. K. Aggarwal, J. N. Harvey, J. Richardson, *J. Am. Chem. Soc.* 2002, *124*, 5747.
13 V. K. Aggarwal, J. Richardson, *J. Chem. Soc., Chem. Commun.* 2003, 2644.
14 V. K. Aggarwal, S. Schade, B. Taylor, *J. Chem. Soc., Perkin Trans. 1* 1997, 2811.
15 A. Solladié-Cavallo, A. Diep-Vohuule, T. Isarno, *Angew. Chem. Int. Ed.* 1998, *37*, 1689.
16 K. Julienne, P. Metzner, V. Henryon, A. Greiner, *J. Org. Chem.* 1998, *63*, 4532.
17 M. A. Silva, B. R. Bellenie, J. M. Goodman, *Org. Lett.* 2004, *6*, 2559.
18 V. K. Aggarwal, J. Charmant, L. Dudin, M. Porcelloni, J. Richardson, *Proc. Nat. Acad. Sci.* 2004, *101*, 5467.
19 V. K. Aggarwal, A. Ali, M. P. Coogan, *J. Org. Chem.* 1997, *62*, 8628.
20 V. K. Aggarwal, M. P. Coogan, R. A. Stenson, R. V. H. Jones, R. Fieldhouse, J. Blacker, *Eur. J. Org. Chem.* 2002, 319.
21 B. R. Bellenie, J. M. Goodman, *J. Chem. Soc., Chem. Commun.* 2004, 1076.
22 A. Solladié-Cavallo, A. Diep-Vohuule, *J. Org. Chem.* 1995, *60*, 3494.
23 Y.-G. Zhou, X.-L. Hou, L.-X. Dai, L.-J. Xia, M.-H. Tang, *J. Chem. Soc., Perkin Trans. 1* 1999, 77.
24 V. K. Aggarwal, G. Hynd, W. Picoul, J.-L. Vasse, *J. Am. Chem. Soc.* 2002, *124*, 9964.
25 V. K. Aggarwal, R. Robiette, unpublished results.
26 C. Baldoli, P. Del Buttero, E. Licandro, S. Maiorana, A. Papagni, *J. Chem. Soc., Chem. Commun.* 1987, 762.
27 C. Baldoli, P. Del Buttero, S. Maiorana, *Tetrahedron* 1990, *46*, 7823.
28 L. N. Prigden, A. Abdel-Magid, I. Lantos, S. Shilcrat, D. S. Eggleston, *J. Org. Chem.* 1993, *58*.
29 A. Abdel-Magid, L. N. Prigden, D. S. Eggleston, I. Lantos, *J. Am. Chem. Soc.* 1986, *108*, 4595.

30 R. Takagi, J. Kimura, Y. Ohba, K. Takezono, Y. Hiraga, S. Kojima, K. Ohkata, *J. Chem. Soc., Perkin Trans. 1* 1998, 689.
31 K. Ohkata, J. Kimura, Y. Shinohara, R. Takagi, H. Yoshkazu, *J. Chem. Soc., Chem. Commun.* 1996, 2411.
32 Y.-C. Wang, C.-L. Li, H.-L. Tseng, S.-C. Chuang, T.-H. Yan, *Tetrahedron: Asymmetry* 1999, *10*, 3249.
33 Y.-C. Wang, D.-W. Su, C.-M. Lin, H. L. Tseng, C.-L. Li, T.-H. Yan, *J. Org. Chem.* 1999, *64*, 6495.
34 A. K. Ghosh, J.-H. Kim, *Org. Lett.* 2004, *6*, 2725.
35 E. J. Corey, S. Choi, *Tetrahedron Lett.* 1991, *32*, 2857.
36 E. J. Corey, S. S. Kim, *J. Am. Chem. Soc.* 1990, *112*, 4976.
37 S.-i. Kiyooka, K. A. Shahid, *Tetrahedron: Asymmetry* 2000, *11*, 1537.
38 J. C. Humelen, H. Wynberg, *Tetrahedron Lett.* 1978, *12*, 1089.
39 S. Colonna, R. Fornasier, U. Pfeiffer, *J. Chem. Soc., Perkin Trans. 1* 1978, 8.
40 P. Bakó, Á. Szöllõsy, P. Bombicz, L. Tõke, *Synlett* 1997, 291.
41 P. Bakó, K. Vizvárdi, Z. Bajor, L. Tõke, *J. Chem. Soc., Perkin Trans. 1* 1998, 1193.
42 P. Bakó, E. Czinege, T. Bakó, M. Czugler, L. Tõke, *Tetrahedron: Asymmetry* 1999, *10*, 4539.
43 S. Arai, Y. Shirai, T. Ishida, T. Shioiri, *J. Chem. Soc., Chem. Commun.* 1999, 49.
44 S. Arai, T. Shioiri, *Tetrahedron Lett.* 1998, *39*, 2145.
45 S. Arai, Y. Shirai, T. Ishida, T. Shioiri, *Tetrahedron* 1999, *55*, 6375.
46 S. Arai, T. Ishida, T. Shioiri, *Tetrahedron Lett.* 1998, *39*, 8299.
47 S. Arai, T. Shioiri, *Tetrahedron* 2002, *58*, 1407.
48 S. Arai, K. Tokumaru, T. Aoyama, *Tetrahedron Lett.* 2004, *45*, 1845.
49 J. Sweeney, *J. Chem. Soc., Chem. Rev.* 2002, *31*, 247.
50 D. Tanner, *Angew. Chem. Int. Ed.* 1994, *33*, 599.
51 L. Dai, *Pure and Applied Chemistry* 1999, *71*, 369.
52 F. A. Davis, H. Liu, P. Zhou, R. Fang, V. Reddy, Y. Zhang, *J. Org. Chem* 1999, *64*, 7559.
53 F. A. Davis, T. Ramachandar, Y. Wu, *J. Org. Chem.* **2003**, *68*, 6894.
54 F. A. Davis, Y. Wu, W. McCoull, K. Prasad, *J. Org. Chem.* **2003**, *68*, 2410.
55 J. B. Sweeney, A. B. McLaren, *Org. Lett.* 1999, *1*, 1339.
56 E. N. Jacobsen, K. B. Hansen, N. Finney, *Angew. Chem. Int. Ed.* **1995**, *34*, 676.
57 K. A. Jorgensen, K. Juhl, R. G. Hazel, *J. Chem. Soc., Perkin Trans. 1* 1999, 2293.
58 K. A. Jorgensen, K. G. Rasmussen, *J. Chem. Soc., Perkin Trans. 1* 1997, 1287.
59 W. D. Wulff, J. C. Antilla, *J. Am. Chem. Soc.* 1999, *121*, 5099.
60 W. D. Wulff, J. C. Antilla, *Angew. Chem. Int. Ed.* 2000, *39*, 4518.
61 W. D. Wulff, C. Loncaric, Org. *Lett. 2001*, 3, 3675.
62 T. Ishikawa, K. Hada, T. Watanabe, T. Isobe, J. *Am. Chem. Soc. 2001*, 123, *7707*.
63 J. L. G. Ruano, I. Fernandez, M. Catalina, A. A. Cruz, Tetr*ahedron: Asymmetry 1996*, 7, 3407.
64 R. A. Stockman, D. Morton, D. Pearson, R. A. Field, Org. *Lett. 2004*, 6, *2377*.
65 L. Dai, A. Li, Y. Zhou, X. Hou, L. Xia, L. Lin, Ange*w. Chem. Int. Ed. 1997*, 36, *1317*.
66 T. Saito, M. Sakairi, D. Akiba, *Tetrahedron Lett.* **2001**, *42*, 5451.
67 A. Solladie-Cavallo, M. Roje, R. Welter, V. Sunjic, J. Org. *Chem. 2004*, 69, *1409*.
68 V. K. Aggarwal, A. Thompson, R. V. H. Jones, M. C. H. Standen, J. Org. *Chem. 1996*, 61, *8368*.
69 V. K. Aggarwal, E. Alonso, G. Fang, M. Ferrara, G. Hynd, M. Porcelloni, Ange*w. Chem. Int. Ed. 2001*, 40, 1433.
70 V. K. Aggarwal, J. P. H. Charment, C. Ciampi, J. M. Hornby, C. J. O'Brian, G. Hynd, R. Parsons, J. Chem. *Soc., Perkin Trans. 1 2001, 1*, 3159.
71 V. K. Aggarwal, J. Vasse, Org. Lett. *2003, 5*, 3987.

2
Vinylaziridines in Organic Synthesis

Hiroaki Ohno

2.1
Introduction

Among the variously functionalized aziridines, vinylaziridines are increasingly being exploited as versatile building blocks for the stereoselective synthesis of biologically and synthetically important compounds, thanks to their high reactivity and ability to function as carbon electrophiles. Ring-opening reactions of vinylaziridines can be effected by various carbon and heteroatom nucleophiles to produce a variety of functionalized amine derivatives such as sphingosines, allyl amines, and (*E*)-alkene dipeptide isosteres. Elaboration through rearrangement, including ring-expansion, provides direct access to structural motifs in the synthesis of alkaloids and other biologically active substances such as β-lactams. This chapter serves as an introduction to vinylaziridines in organic chemistry. Synthesis, ring-opening reactions, isomerization including rearrangement, cycloadditions, and other important transformations of vinylaziridines are reviewed.

2.2
Direct Synthesis of Vinylaziridines [1]

2.2.1
Addition of Nitrene to Dienes

The most traditional method for the direct synthesis of vinylaziridines is by addition of nitrenes to conjugated dienes. In 1964, the synthesis of vinylaziridine **3** by 1,2-addition of alkoxycarbonylnitrene **2**, generated by photolysis of methoxycarbonyl azide (**1**), to butadiene was first reported by Hafner and co-authors (Scheme 2.1) [2]. The reactions between the nitrene **2** and (*E*)- or (*Z*)-but-2-ene proceeded in a stereospecific manner to give the corresponding *trans*- or *cis*-dimethylaziridines, respectively, suggesting that the nitrene reacts preferentially in a singlet state. Similar syntheses of vinylaziridines were also reported by other research groups [3]. The regioselectivity in the photochemical reaction of an azide-derived nitrene with

Aziridines and Epoxides in Organic Synthesis. Andrei K. Yudin

ISBN: 3-527-31213-7

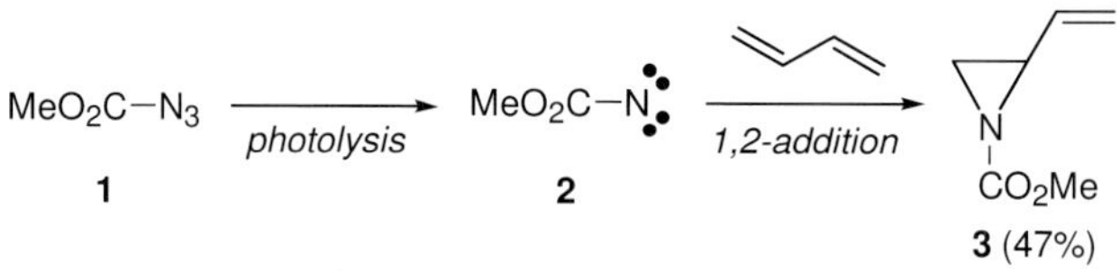

Scheme 2.1 1,2-Addition of methoxycarbonylnitrene (2) to 1,3-diene.

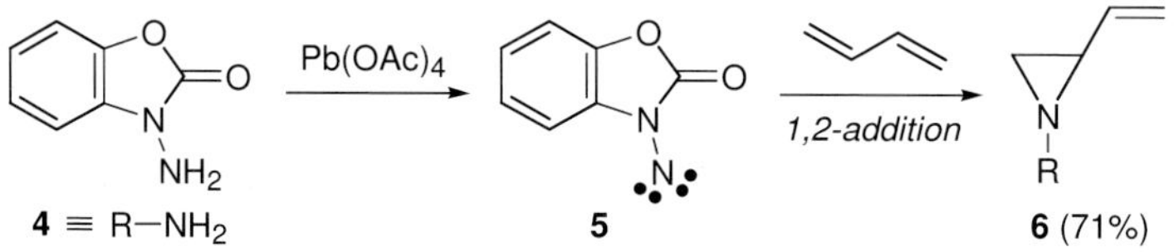

Scheme 2.2 1,2-Addition of amino nitrene 5 to 1,3-diene.

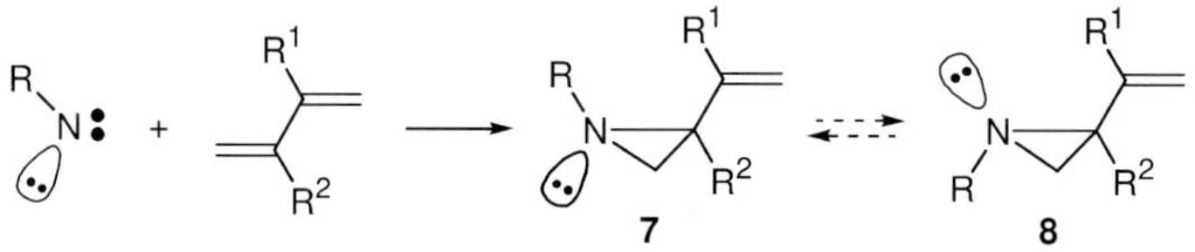

Scheme 2.3 Stereochemistry at the nitrogen on the 1,2-addition of nitrene.

isoprene depends on the diene concentration [4]. Reactions between vinyl azides and acrylic acid derivatives gave 1-vinylaziridines [5].

N-Aminobenzoxazolin-2-one (**4**), which was readily prepared by amination of benzoxazolin-2-one with hydroxylamine-*O*-sulfonic acid, is also a useful nitrene precursor (Scheme 2.2). Oxidation of **4** with lead(IV) acetate in the presence of a conjugated diene resulted in exclusive 1,2-addition of nitrene **5**, to yield vinylaziridine (**6**) in 71% yield [6]. The formation of vinylaziridines through 1,2-additions of methoxycarbonylnitrene (**2**) or amino nitrene **5** contrasts with the claimed 1,4-addition of nitrene itself to butadiene [7]. Since the reaction proceeded stereospecifically even at high dilution, the nitrene **5** appears to be generated in a resonance-stabilized singlet state, which is probably the ground state [8].

The stereochemistry at the aziridine nitrogen was also investigated by the same group, using the nitrenes derived from **4** or *N*-aminophthalimide. They found that the addition is stereospecific at low enough temperature (< -20 °C), giving the invertomers **7** with the vinyl group *cis* to the nitrogen substituent R (Scheme 2.3) [9].

In 1995, aziridination with 1,3-dienes **10** by treatment with PhI=NTs **9** was developed (Scheme 2.4) [10] on the foundation of pioneering works by Jacobsen and Evans on copper-catalyzed asymmetric aziridination of isolated alkenes [11].

Selective 1,2-addition of a nitridomanganese complex to conjugated dienes

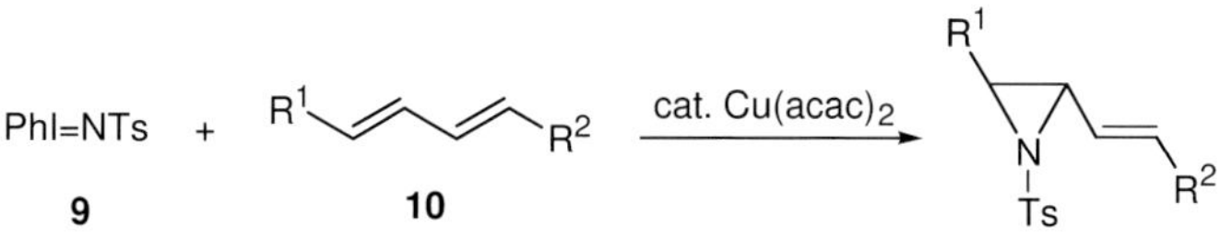

9 10 11 (48–77%)

Scheme 2.4 Aziridination of dienes with [(*N*-tosyl) imino]phenyliodinane **9**.

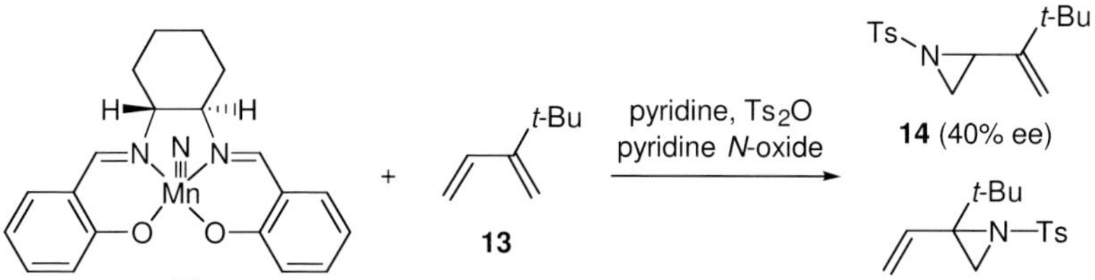

12 13 14 (40% ee) 15 (20% ee) (**14**:**15** = 62:38)

Scheme 2.5 First asymmetric synthesis of vinylaziridines with conjugated dienes.

(Scheme 2.5) was recently reported by Komatsu, Minakata, and coworkers [12]. The reaction with the (*R,R*)-complex **12** provided the first reagent-controlled asymmetric aziridination of conjugated dienes, although enantioselectivities were only low to moderate (20–40% *ee*).

2.2.2
Addition of Allylic Ylides and Related Reagents to Imines

The synthesis of vinylaziridines through reactions between allylic carbenoid reagents and imines (i.e., Darzen-type reactions) was first reported by Mauzé in 1980 [13]. Treatment of aldimines or ketimines **16** with *gem*-chloro(methyl)allyllithium (**17**) afforded *N*-substituted vinylaziridines **18** (Scheme 2.6). Similarly, 2,3-*trans*-*N*-diphenylphosphinyl-2-vinylaziridines **21** were prepared with good stereoselectivities (*trans:cis* = 10:1; Scheme 2.7) by treatment of α-bromoallyllithium (**20**) with *N*-diphenylphosphinyl aldimines **19** in the presence of zinc chloride [14].

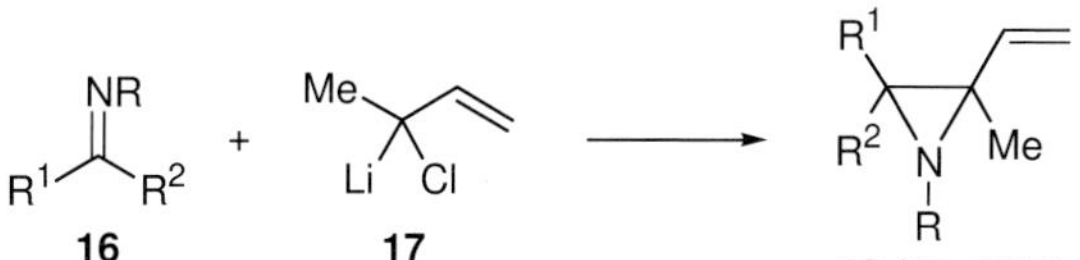

16 17 18 (28–65%)

Scheme 2.6 Aza-Darzen-type reaction with chloro(methyl) allyllithium (**17**).

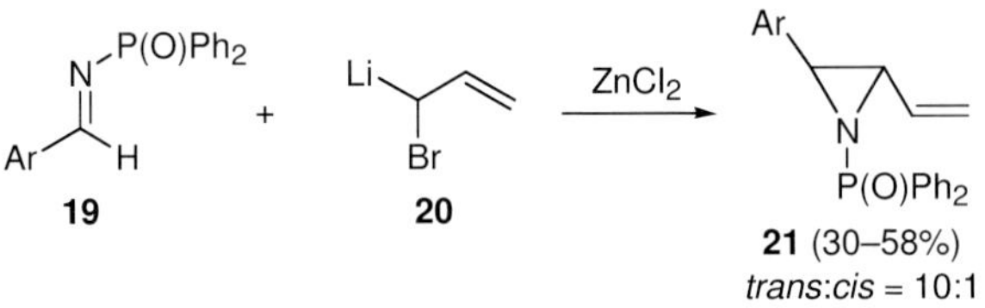

Scheme 2.7 Stereoselective aziridination of *N*-diphenylphosphinyl aldimines **19**.

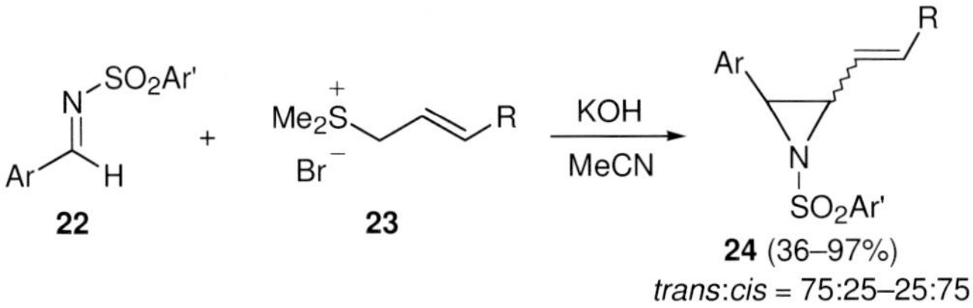

Scheme 2.8 Reactions between *N*-sulfonylimines and allylic ylides.

It is well known that aziridination with allylic ylides is difficult, due to the low reactivity of imines – relative to carbonyl compounds – towards ylide attack, although imines do react with highly reactive sulfur ylides such as Me_2S^+-CH_2-. Dai and coworkers found aziridination with allylic ylides to be possible when the activated imines **22** were treated with allylic sulfonium salts **23** under phase-transfer conditions (Scheme 2.8) [15]. Although the stereoselectivities of the reaction were low, this was the first example of efficient preparation of vinylaziridines by an ylide route. Similar results were obtained with use of arsonium or telluronium salts [16]. The stereoselectivity of aziridination was improved by use of imines activated by a phosphinoyl group [17]. The same group also reported a catalytic sulfonium ylide-mediated aziridination to produce (2-phenylvinyl)aziridines, by treatment of arylsulfonylimines with cinnamyl bromide in the presence of solid K_2CO_3 and catalytic dimethyl sulfide in MeCN [18]. Recently, the synthesis of 3-alkyl-2-vinylaziridines by extension of Dai's work was reported [19].

The stereochemistry of the sulfonium ylide-mediated aziridination is highly dependent on the reactivities of imines and ylides: imines and ylides of lower reactivity favor the formation of *trans*-aziridines **29**, whereas those of higher reactivity prefer *cis*-products **30** (Scheme 2.9). Mechanistic study of the ylide aziridination showed that the reaction proceeds in two steps: reversible addition of the sulfonium ylide to the imine to form intermediates **25** and **26**, which rotate to conformers **27** and **28**, followed by *anti*-elimination to yield the aziridines **29** and **30**, respectively. Steric repulsion between the aryl and R^1 groups would favor the intermediates **26** and **27**. Imines with a high reactivity and large k_1 promote the first addition step, and highly reactive ylides will provide a better leaving group to increase the rate of elimination (k_2) [20].

Syntheses of nonracemic vinylaziridines by reagent- or substrate-controlled

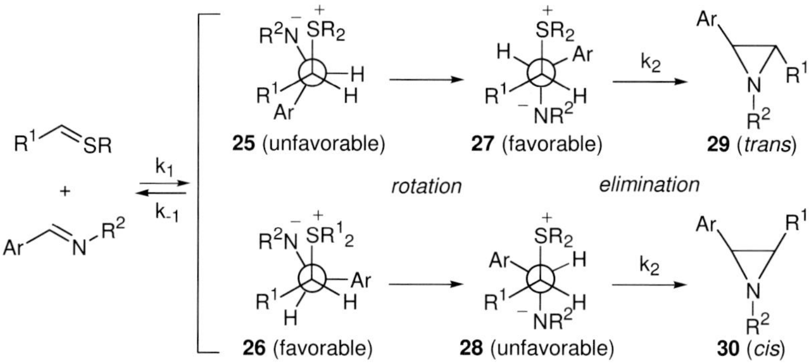

Scheme 2.9 Mechanism of aziridination with allylic ylides.

asymmetric induction have recently been reported. Addition of a chloroallyl phosphonamide anion derived from **31** to oximes **32** resulted in the formation of *N*-alkoxy-2-alkenylaziridines **33** in enantiomerically pure form (Scheme 2.10) [21]. Treatment of chiral *tert*-butylsulfinylimines **34** (R = aryl or alkyl) with the ylide derived from *S*-allyltetrahydrothiophenium bromide (**35**) yielded *trans*- and *cis*-2-vinylaziridines **36** and **37**, respectively, with moderate to good stereoselectivities (**36**/**37** = 59:41 to 88:12; 85–95% de for *trans* isomers **36**; Scheme 2.11) [22]. Although vinylaziridines are known to undergo acid-induced ring-opening reactions (see Section 2.3.3), the chiral sulfinyl group of **36** can be readily removed by treatment with anhydrous HCl in dioxane to give the corresponding *N*-unsubstituted vinylaziridines as hydrochloride salts in over 90% yields.

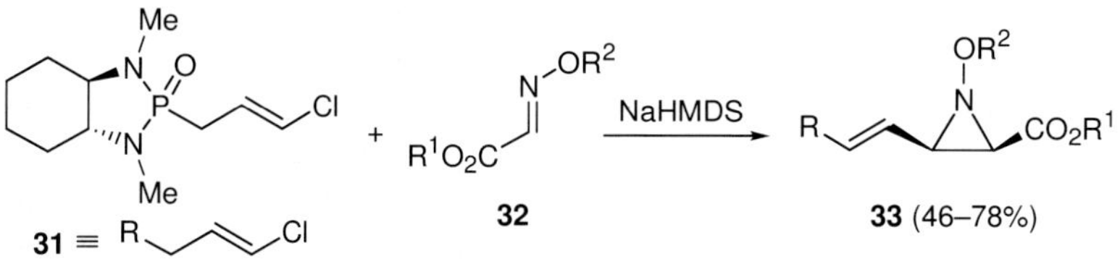

Scheme 2.10 Reagent-controlled synthesis of chiral vinylaziridines.

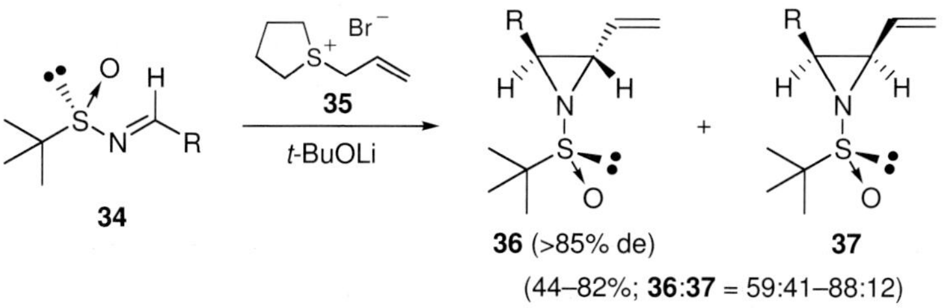

Scheme 2.11 Asymmetric synthesis of vinylaziridines by use of a chiral auxiliary.

2.2.3
Cyclization of Amino Alcohols and Related Compounds

Regioselective addition of bromine azide to dienes **38** at 25 °C gave the 1,4-adducts **39** or the 1,2-adducts **40** as thermodynamically favored products, their ratios depending on the substituent R on the terminal carbon (Scheme 2.12). These adducts were easily converted into vinylaziridines **41** on treatment with trimethylphosphite, although the stereochemistries of **39**, **40**, and **41** are unclear [23].

A related conversion based on regioselective ring-opening of alkenylepoxide **42** at the allylic position with sodium azide, followed by treatment of the resulting azido alcohols with PPh_3 to give *N*-unsubstituted vinylaziridine **43** with inversion of configuration, is also possible (Scheme 2.13) [24]. Somfai and coworkers reported direct aminolysis of vinylepoxides **44** with liquid ammonia in the presence of TsOH · H_2O (5 mol%), affording regioselective ring-opening of epoxides. Dehydration of the resulting amino alcohols **45** under Mitsunobu conditions gave the *N*-unsubstituted vinylaziridines **46** (Scheme 2.14) [25]. The main drawback of this synthetic route is the relatively low yields of the vinylaziridines **46** (0–54%); how-

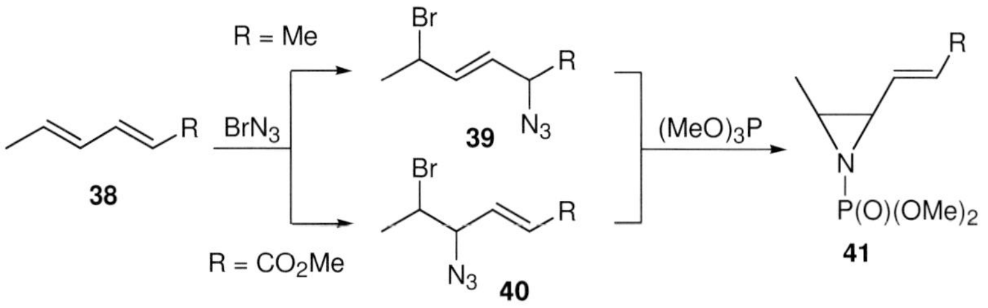

Scheme 2.12 Synthesis of vinylaziridines through addition of bromine azide to dienes.

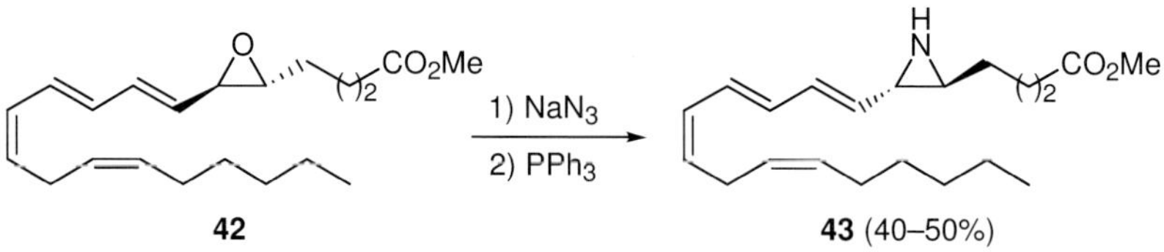

Scheme 2.13 Synthesis of vinylaziridines through sodium azide-mediated ring-opening of alkenylepoxides.

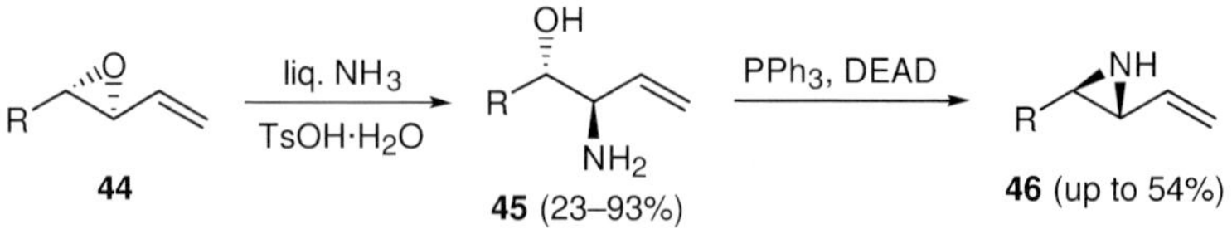

Scheme 2.14 Synthesis of vinylaziridines by aminolysis of vinylepoxides by liquid ammonia.

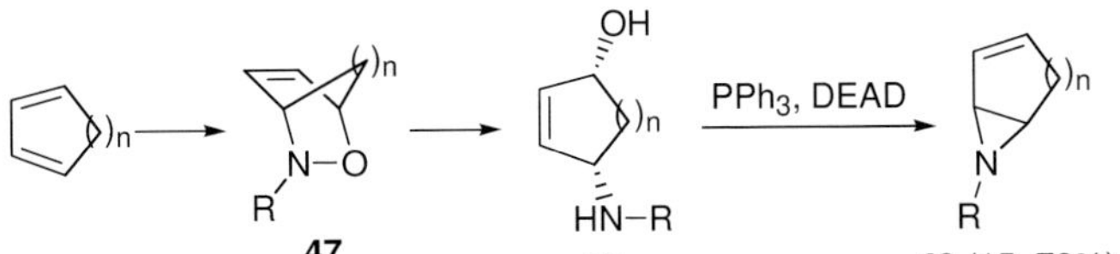

Scheme 2.15 Synthesis of vinylaziridines from 1,4-amino alcohols.

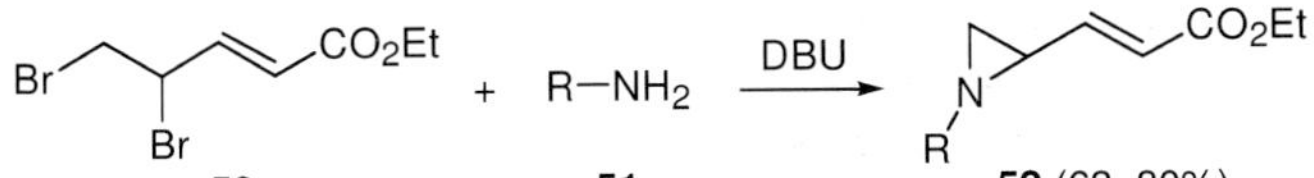

Scheme 2.16 Synthesis of vinylaziridines by the Gabriel-Cromwell reaction.

ever, a modified synthetic route involving protection of the nitrogen atom of **45** with a trityl group, followed by deprotection with TFA after ring-closure, gave *N*-unsubstituted vinylaziridines **46** in better yields (up to 77%) [26].

1,4-Amino alcohols are also good substrates for aziridination under Mitsunobu conditions. The *cis*-1,4-amino alcohols **48**, obtained by reductive cleavage of the nitrogen-oxygen bonds of the hetero Diels-Alder adducts **47**, underwent *syn*-S_N2'-type displacement on treatment with PPh_3 and DEAD to give cyclic vinylaziridines **49** (Scheme 2.15) [27].

The Gabriel-Cromwell reaction is a convenient strategy for the synthesis of vinylaziridines. Treatment of 4,5-dibromopent-2-enoate **50** with primary amines **51** in the presence of DBU afforded the corresponding conjugated aziridines **52** in 63–80% yields (Scheme 2.16) [28]. The use of DBU has proven to be essential for the successful conversion.

One of the most efficient methods for the stereoselective synthesis of vinylazir-

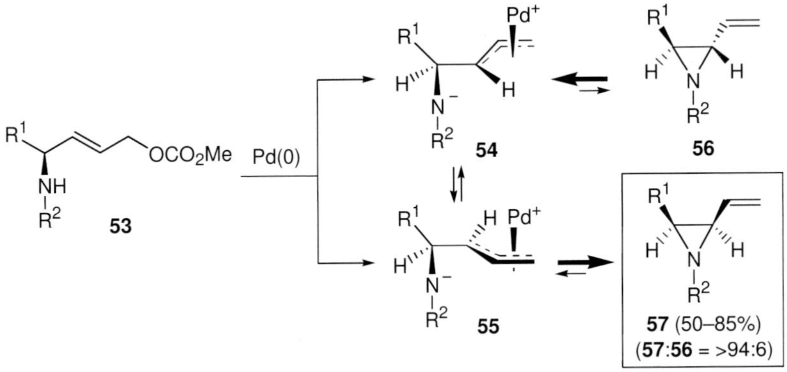

Scheme 2.17 Palladium(0)-catalyzed aziridination of allyl carbonates.

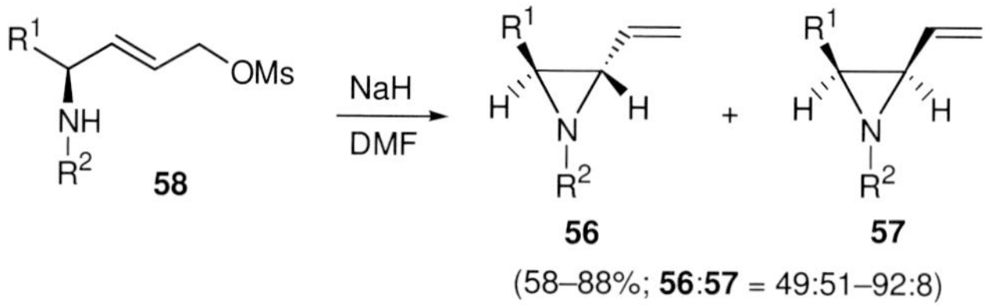

Scheme 2.18 NaH-mediated aziridination of allyl mesylates.

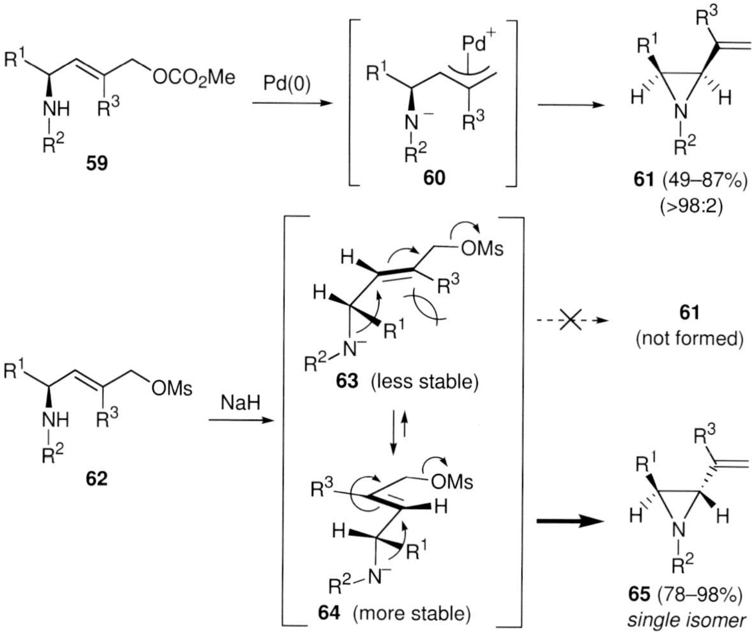

Scheme 2.19 A highly stereodivergent synthesis of *cis*- and *trans*-2-alkenylaziridines.

idines is palladium(0)-catalyzed aziridination of amino alcohol derivatives. In the context of the recent finding that the thermodynamically less stable *trans*-2-vinylaziridines **56** isomerize to the corresponding more stable *cis* isomers **57** via π-allylpalladium(II) intermediates **54** and **55** in the presence of palladium(0) (see Section 2.4.6) [29], Ohno, Ibuka, and coworkers demonstrated the first 2,3-*cis*-selective aziridination of allylic carbonates **53** to **57** via the same intermediates **54** and **55** (Scheme 2.17) [30]. Although sodium hydride-mediated aziridination of allylic mesylates **58** gave diastereomeric mixtures of vinylaziridines **56** and **57** (Scheme 2.18), reactions of **58** with branched R^1 groups resulted in predominant formation of *trans* isomers **56** (up to 92:8) [30, 31]. Interestingly, base-mediated cyclization of (*Z*)-allylic mesylates exclusively gives 3-pyrrolines in high yields, while the palladium(0)-catalyzed reactions of the corresponding methyl carbonates predominantly yield 2,3-*cis*-3-alkyl-2-vinylaziridines **57** [32].

We also reported a highly stereodivergent synthesis of sterically congested *cis*- and *trans*-alkenylaziridines **61** and **65** (Scheme 2.19) [33]. Whereas treatment of the methyl carbonates **59** with catalytic amounts of $Pd(PPh_3)_4$ in THF or 1,4-dioxane predominantly affords the corresponding thermodynamically more stable 2,3-*cis*-2-alkenylaziridines **61** (up to 99:1), treatment of the allylic mesylates **62** with sodium hydride in DMF exclusively yields the thermodynamically less stable *trans*-2-alkenylaziridines **65** (>99:1). The kinetically favored *trans*-selective aziridination could be attributable to unfavorable steric interaction between the R^1 and R^3 groups in aza-anionic intermediates **63** leading to the *cis*-vinylaziridines **61**. Similarly, treatment of mesylates **62** possessing bromine substituents (R^3 = Br) with sodium hydride gave *trans*-2-(1-bromovinyl)aziridines **65** (R^3 = Br) in a stereoselective manner; these are good precursors of 2-ethynylaziridines [34].

2.2.4 Cyclization of Amino Allenes

Transition metal-catalyzed cyclization of allenes bearing nucleophilic functionalities has attracted much attention in recent years. In particular, cyclization of amino allenes has become quite a useful methodology for the synthesis of five- or six-membered azacycles [35]. In contrast, ring-closure of amino allenes to yield aziridines was unknown until 1999, when Ohno, Ibuka, and coworkers found that the palladium-catalyzed reaction between the α-amino allene **66** and iodobenzene in the presence of K_2CO_3 at reflux in 1,4-dioxane exclusively yielded the corresponding 2,3-*cis*- and *trans*-2-alkenylaziridines **68** and **69**, respectively, each bearing a phenyl group on the double bond (Scheme 2.20) [36]. Interestingly, the reaction

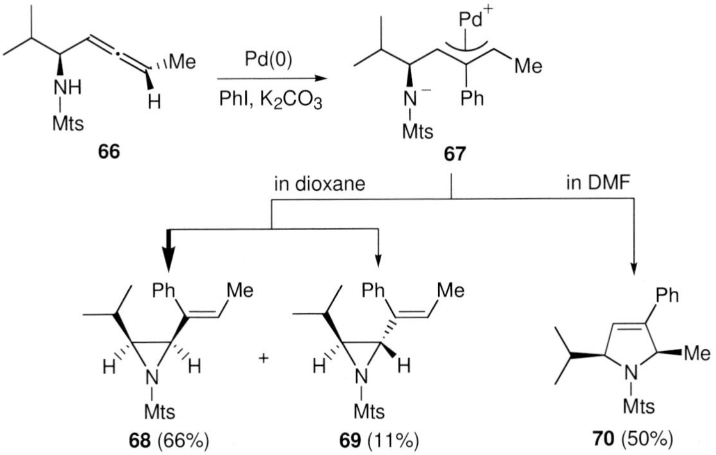

Scheme 2.20 Palladium(0)-catalyzed aziridination of α-amino allene **66**.

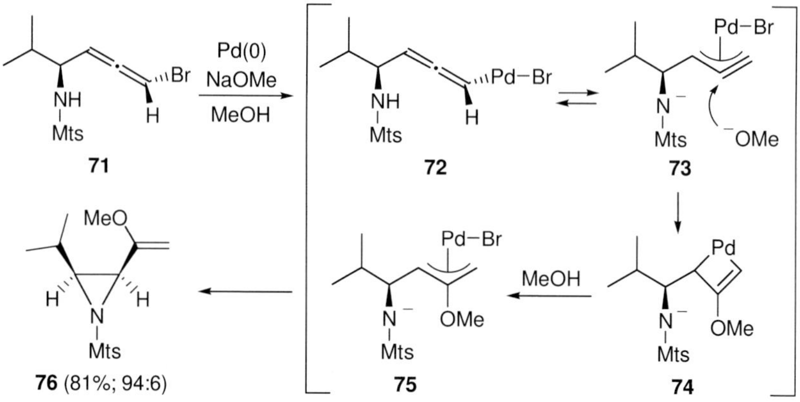

Scheme 2.21 Palladium(0)-catalyzed aziridination of bromoallene **71** in MeOH.

in DMF at around 70 °C afforded the corresponding 3-pyrroline derivative **70** as the sole isolable product. This aziridination reaction is extremely useful in that various aryl groups on the double bond can be introduced by use of readily available aryl halides in the final step of the synthetic route.

Ohno, Hamaguchi, Tanaka, and coworkers recently found that treatment of bromoallene **71** with $Pd(PPh_3)_4$ and NaOMe in MeOH provided the 2,3-*cis*-2-(1-methoxy)vinylaziridine **76** stereoselectively (Scheme 2.21) [37]. Since palladium(0) equilibrates vinylaziridines to give the thermodynamically more stable *cis* isomers [29], this result strongly suggests the formation of the π-allylpalladium complex **75** bearing a methoxy group on the central carbon. In other words, bromoallene **71** can act as allyl dication equivalent when treated with palladium(0) in an alcoholic solvent. This aziridination would proceed through oxidative addition of bromoallene **71** to palladium(0), reversible isomerization of **72** to π-propargylpalladium complex **73**, intramolecular nucleophilic addition by methoxide at the central carbon of **73** to produce a palladacyclobutene **74**, and finally, protonation and cyclization to give the vinylaziridine **76**.

2.2.5
Aziridination of α,β-unsaturated Oximes and Hydrazones

Unsaturated oximes are attractive substrates for aziridine synthesis. Treatment of oxime **77** with Red-Al yielded vinylaziridines **78**, **79**, and **80**, in various ratios depending on the *E*/*Z* ratio of the starting oxime **77** (Scheme 2.22) [38]. This reaction should proceed through abstraction of H_A, H_B, or H_C in the intermediate **81**, followed by hydride reduction of the resulting 2*H*-azirines **82-84**.

A related synthesis of vinylaziridines **88** by treatment of α,β-unsaturated hydrazones **85** with Grignard reagents has also been reported (Scheme 2.23) [39]. Nitrenes **86** were proposed as plausible intermediates, their cyclization to azirine **87**

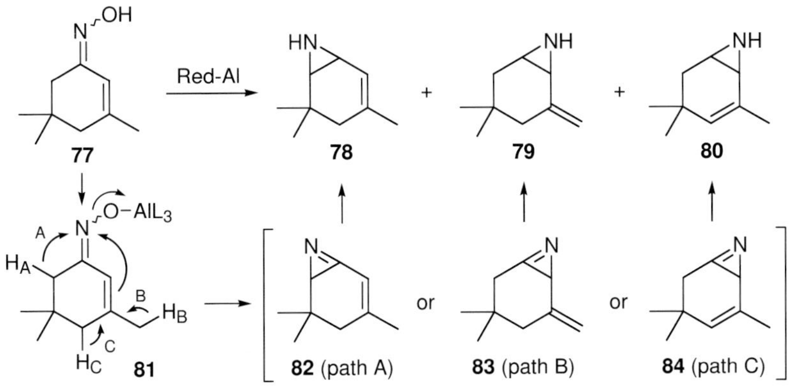

Scheme 2.22 Reductive aziridination of an α,β-unsaturated oxime.

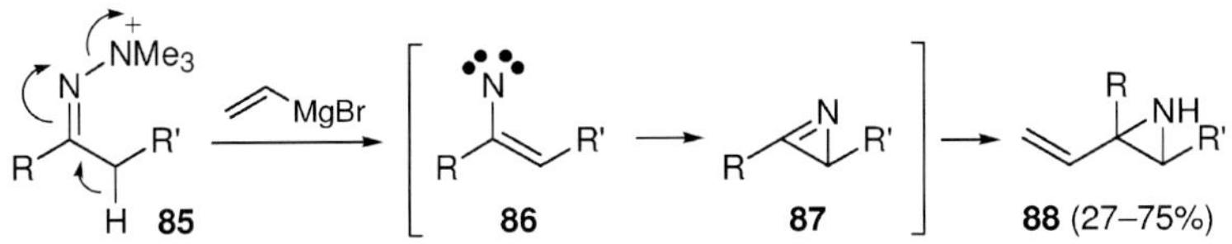

Scheme 2.23 Aziridination of α,β-unsaturated hydrazones with vinylmagnesium bromide.

and subsequent addition of vinylmagnesium bromide affording the vinylaziridines 88.

2.3 Ring-opening Reactions with Nucleophiles

2.3.1 Hydride Reduction

The first of the nucleophilic ring-opening reactions of vinylaziridines discussed in this section is diborane reduction, developed by Laurent and coworkers in 1976 (Scheme 2.24). Treatment of *N*-unsubstituted vinylaziridines **89** with B_2H_6 gives allyl amines **92** by S_N2' reduction via cyclic intermediates **90** [40]. In contrast, treatment with 9-BBN gives 2-(hydroxyethyl)aziridines **93** after oxidative workup (Scheme 2.25) [41].

In 1995, palladium-catalyzed reduction of vinylaziridines with formic acid was reported [42]. As shown in Scheme 2.26, 1,2-reduction products **95** and 1,4-products **96** were obtained in ratios depending on the reaction conditions, such as the additive, solvent, and catalyst employed. Both the *E*/*Z* selectivity of **95** and the

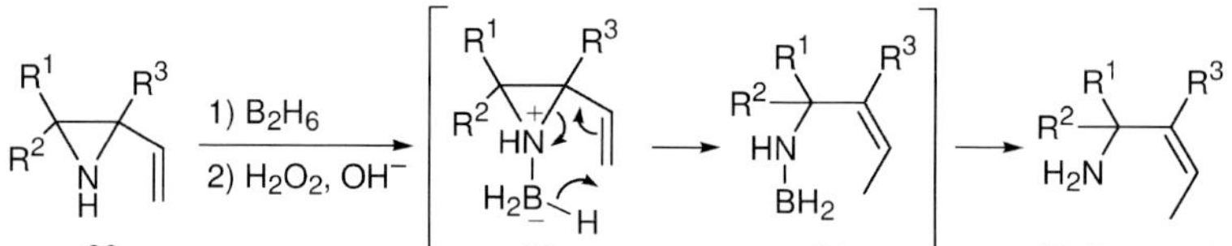

Scheme 2.24 Diborane-mediated S_N2' reduction of vinylaziridines.

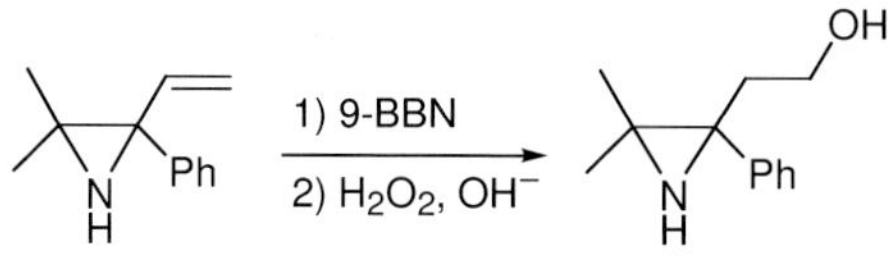

Scheme 2.25 Hydroboration of vinylaziridines with 9-BBN.

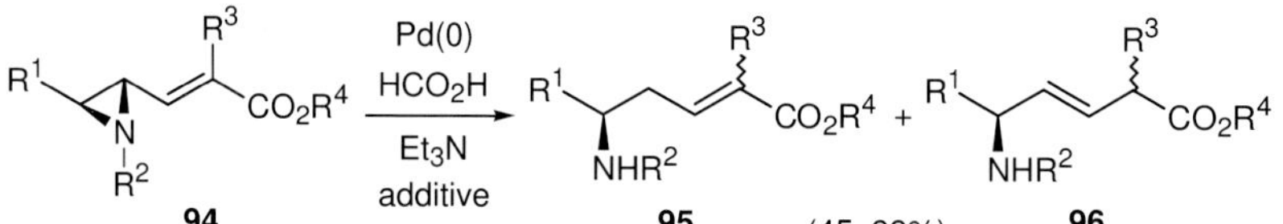

Scheme 2.26 Palladium(0)-catalyzed reduction with formic acid.

stereoselectivity at the α-carbon of **96** are strongly affected by palladium-catalyzed isomerization of β-aziridinyl-α,β-enoates [29, 43] (see Section 2.4.6).

2.3.2
Organocopper-mediated Alkylation

In 1994, two research groups independently reported organocopper-mediated ring-opening reactions of vinylaziridines. The original stereospecific S_N2' alkylation by Ibuka, Fujii, Yamamoto, and coworkers was carried out by treatment of β-aziridinyl-α,β-enoates **97**, **98**, **99**, or **100** with organocyanocuprate species such as RCu(CN)Li or $R_3ZnLi/CuCN$ (cat.) to give diastereomerically pure (*E*)-alkene dipeptide isosteres **101** or **102** in 82–98% yields, together with small amounts of S_N2 products (<6% yield, Scheme 2.27) [44]. A few months later, Wipf's group reported a closely related alkylative ring-opening reaction of β-aziridinyl-α,β-enoates **103**, possessing carbonyl substituents on their nitrogen atoms, with organocyanocuprate in the presence of BF_3 (Scheme 2.28) [45]. This reaction sometimes provides a considerable amount of 1,4-reduction products (<24%) or S_N2-alkylation products (<15% yield).

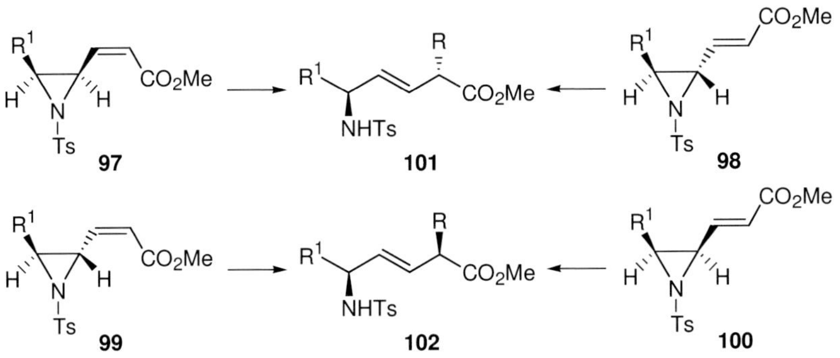

Reagent: R_3ZnLi/30 mol% CuCN or RCu(CN)Li.

Scheme 2.27 Ibuka's ring-opening reaction of β-aziridinyl-α,β-enoates with organocuprates.

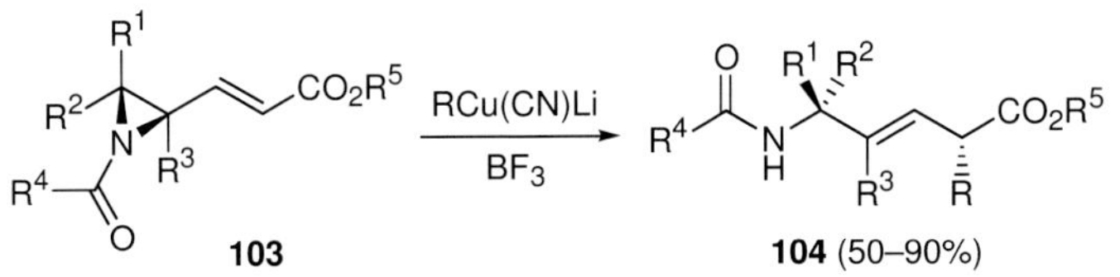

Scheme 2.28 Wipf's ring-opening reaction of β-aziridinyl-α,β-enoates with organocuprates.

Treatment of cyclic vinylaziridine **105** with organocuprates of the R_2CuLi type proceeds in a highly *syn*-selective manner (Scheme 2.29) [46]. The *syn* stereochemistry of the reaction reflects the effect of the acetonide group, which directs the nucleophilic attack to the less hindered α-face. The formation of S_N2 products **109** from the cyclic (chlorovinyl)aziridine **107** can be explained by assuming a *syn*-S_N2'

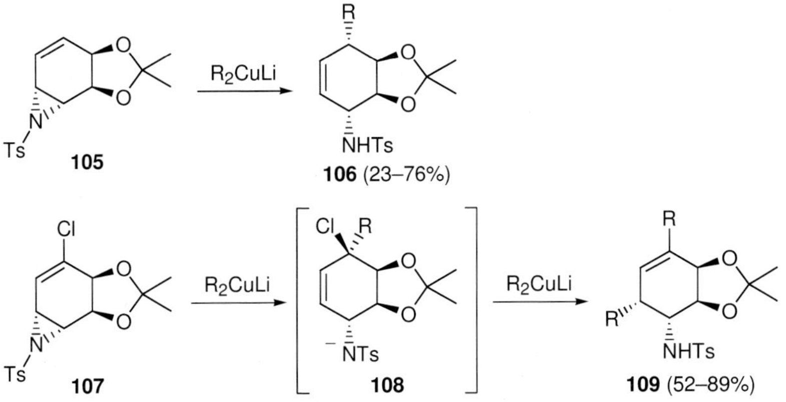

Scheme 2.29 Ring-opening reactions of cyclic vinylaziridines.

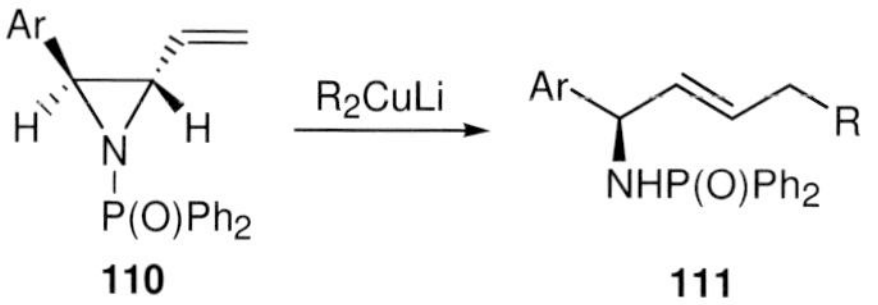

Ar, H, H, N, P(O)Ph_2 **110** $\xrightarrow{R_2CuLi}$ Ar, R, NHP(O)Ph_2 **111**

Scheme 2.30 Ring-opening reaction with organocopper reagents to form (*E*)-allyl amines.

ring-opening reaction of **107** followed by an *anti*-S_N2' reaction of the resulting allyl chloride intermediate **108** [47].

The final example of the organocopper-mediated ring-opening reaction involves the reactions of aziridines possessing terminal vinyl groups. Sweeney and coworkers reported that 2,3-*trans*-3-aryl-2-vinylaziridines **110** exclusively gave (*E*)-allyl amines **111** on treatment with Gilman-type organocopper reagents (Scheme 2.30) [14].

At almost the same time, Ibuka and coworkers reported the closely related reactions of various 3-alkyl-2-vinylaziridines and demonstrated the stereochemical outcomes of the reactions in detail (Scheme 2.31) [48]. Treatment of *cis*-aziridines **112** with organocopper reagents gave exclusively (*E*)-allyl amines **114** in high yields, while the *trans*-aziridines **113** afforded mixtures of (*E*)- and (*Z*)-allyl amines (**114**:**115** = >85:15). The exclusive formation of (*E*) isomers **114** from *cis*-aziridine **112** is presumably due to steric crowding of the disfavored conformer **117**, which gives rise to (*Z*)-allyl amine **115**. In contrast, allylic 1,3-strain may be minimized in conformer **116**. Therefore, organocopper reagents react with **112** by an *anti*-S_N2' reaction pathway from the favorable conformer **116** to yield the (*E*) isomer **114** exclusively.

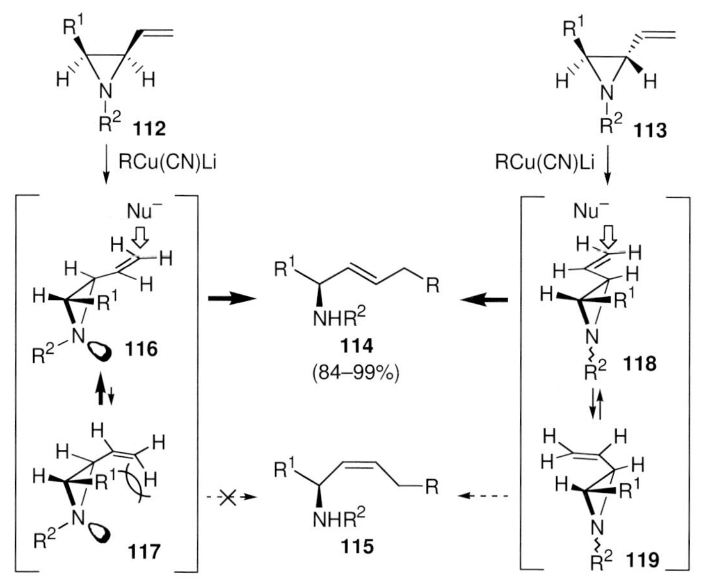

Scheme 2.31 Stereochemical course of ring-opening reactions with organocopper reagents.

2.3.3
Reactions with Oxygen Nucleophiles

Ring-opening reactions of vinylaziridines with oxygen nucleophiles are generally mediated by acid. This type of reaction is extremely useful for the stereodivergent synthesis of *vic*-amino alcohols. In 1996, Davis and coworkers demonstrated a highly stereoselective synthesis of *threo*- and *erythro*-sphingosines from the common intermediate *cis*-*N*-sulfinylaziridine **120** (Scheme 2.32) [49]. Treatment of **120** in acetone/TFA/H_2O gave β-hydroxy-α-amino acid ester **121** as a single isomer in 72% yield. Reduction of **121** with $LiBH_4$ gave L-*threo*-sphingosine **124** in 80% yield. In sharp contrast, treatment of **120** with trifluoroacetic anhydride (TFAA) in CH_2Cl_2 resulted in an 88:12 mixture of *erythro*- and *threo*-**123**. The unusual formation of *erythro*-**123** by nucleophilic attack from the same side as the departing amino group can be attributed to the Pummerer-type rearrangement of sulfoxide, the activated sulfoxide complex undergoing a stereospecific [3,3]-sigmatropic rearrangement involving the migration of the trifluoromethyl acetoxy group to a developing carbocation or ion pair **122**.

Tamamura, Fujii, and coworkers reported a stereodivergent synthesis of (*E*)-alkene dipeptide isosteres by a combination of regiospecific ring-opening reactions of vinylaziridines with an acid and organocuprate (Scheme 2.33) [50]. Thus, γ-aziridinyl-α,β-enoates **126**, readily obtained by palladium-catalyzed isomerization of diastereomeric mixtures **125** and subsequent recrystallization, were subjected to the organocopper-mediated ring-opening reaction to give L,L-type (*E*)-alkene dipeptide isosteres **127** in a highly stereoselective manner. In contrast, methanesulfonic acid-mediated ring-opening reactions of **126** yielded the corresponding allyl mesylates **128** with inversion of configuration, and these were converted into L,D-type **129** by *anti*-S_N2' substitution with organocopper reagents. By use of organozinc-derived copper reagents, (L-Xaa, L-Glu)- and (L-Xaa, D-Glu)-type dipeptide

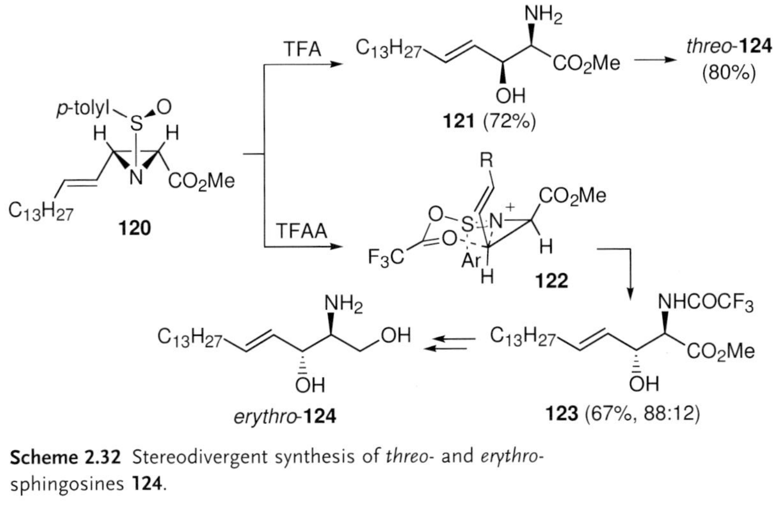

Scheme 2.32 Stereodivergent synthesis of *threo*- and *erythro*-sphingosines **124**.

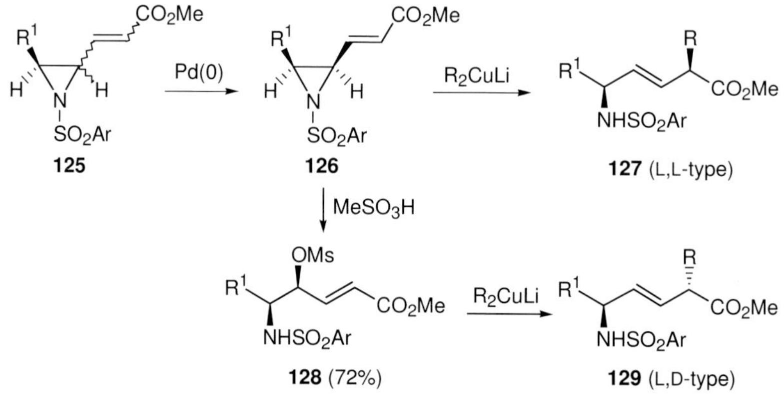

Scheme 2.33 Stereodivergent synthesis of (*E*)-alkene dipeptide isosteres.

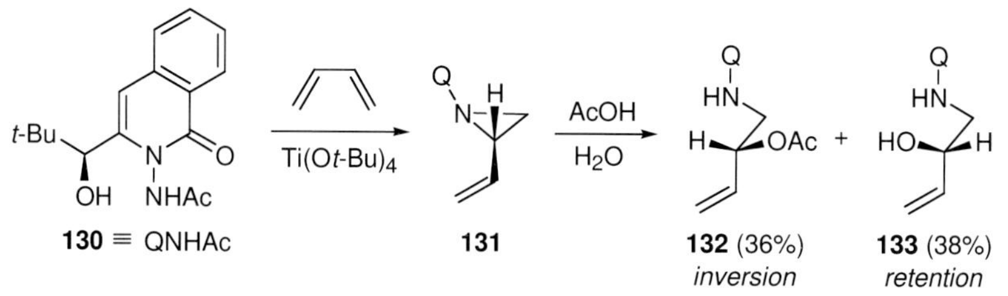

Scheme 2.34 Acetic acid-catalyzed ring-opening.

isosteres **127** and **129** (R = $CH_2CH_2CO_2R'$), respectively, can be synthesized in good to excellent yields [51].

Ring-opening of diastereomerically pure vinylaziridine **131**, prepared by aziridination of butadiene with 3-acetoxyaminoquinazolinone **130** [52], yielded acetate **132** with inversion of configuration, together with amino alcohol **133** with retention (Scheme 2.34) [53]. The formation of **133** can be explained by assuming participation by the quinazolinone carbonyl oxygen, which produces an intramolecular reaction with the aziridine carbon with retention of configuration.

A stereodivergent synthesis of *vic*-amino alcohols by ring-opening of *N*-unsubstituted aziridines **134** with an acid was reported by Somfai and coworkers (Scheme 2.35) [54]. Whereas aziridines **134** were regioselectively hydrolyzed into *anti*-amino alcohols **135** by treatment with $HClO_4$ in THF/H_2O, acetylation of **134** and subsequent rearrangement of the resulting acetates by treatment with $BF_3 \cdot OEt$ gave oxazolines **136** with retention of configuration, and these were hydrolyzed and deacetylated to give the corresponding *syn*-amino alcohols **137** in a stereospecific manner. This stereodivergent ring-opening reaction is useful for the synthesis of D-*erythro*- and L-*threo*-sphingosines [55].

An interesting oxidative ring-opening reaction is shown in Scheme 2.36. Oxidation of vinylaziridines **138** with *m*-CPBA gave **142** through [2,3] Meisenheimer

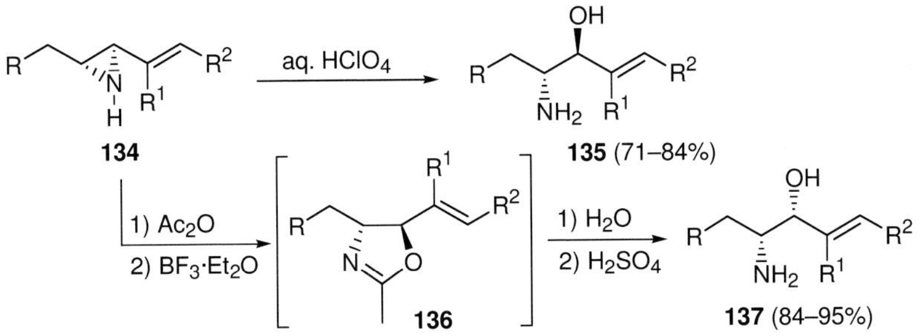

Scheme 2.35 Stereodivergent synthesis of *syn*- and *anti*-amino alcohols.

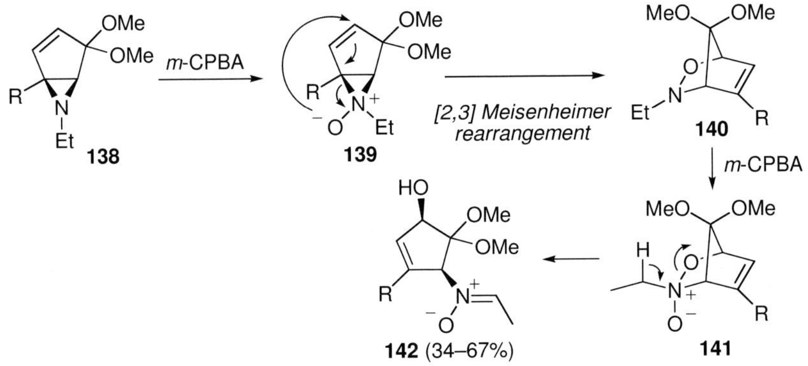

Scheme 2.36 Possible mechanism of the oxidative rearrangement of vinylaziridines.

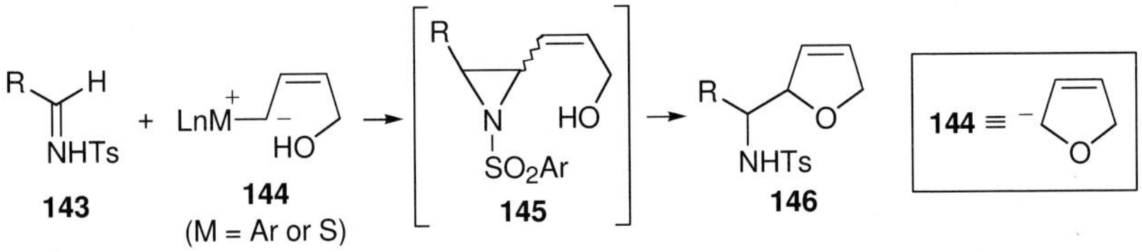

Scheme 2.37 Intramolecular reaction of an oxygen nucleophile to give dihydrofurans.

rearrangement of the initial *N*-oxide **139** followed by further oxidation of **140** and ring-cleavage [56].

Finally, an intramolecular reaction of an oxygen nucleophile to give 2,5-dihydrofuran derivatives **146** is shown in Scheme 2.37. Since the vinylaziridines were generated *in situ* by treatment of imines **143** with ylide **144**, this ylide is formally acting as an equivalent of the 2,5-dihydrofuran anion [57].

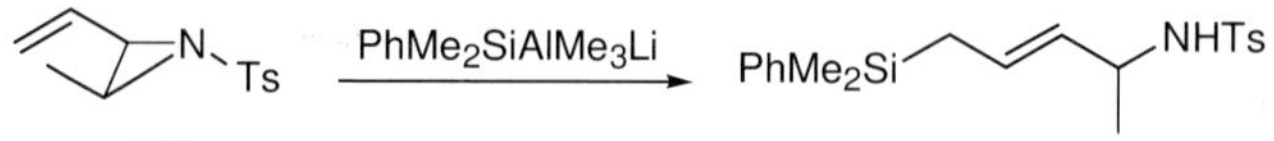

Scheme 2.38 Synthesis of allylsilane **148** by ring-opening of **147**.

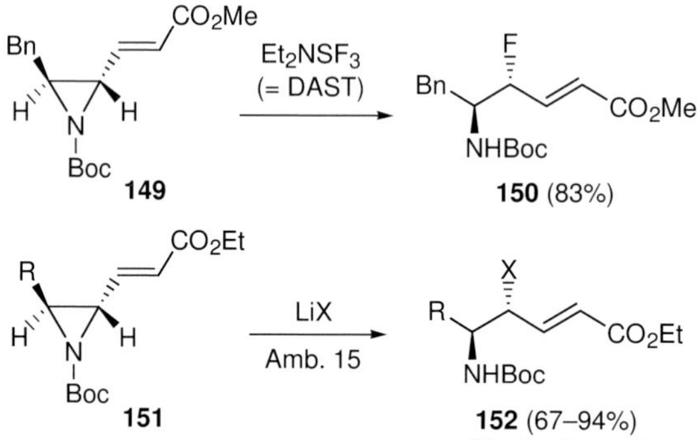

Scheme 2.39 Ring-opening reactions with halogen nucleophiles.

2.3.4 Reactions with Other Nucleophiles

Oshima and coworkers reported the preparation of allylsilanes such as **148** by ring-opening reactions of three-membered rings such as aziridines **147** with silylaluminium reagents as shown in Scheme 2.38 [58].

Halogen atoms can be stereoselectively introduced by ring-opening of γ-aziridinyl-α,β-enoates (Scheme 2.39). Treatment of **149** with diethylaminosulfur trifluoride (DAST) results in stereospecific ring-opening to yield fluorinated derivative **150** [59]. A related stereoselective conversion of γ-aziridinyl-α,β-enoates **151** into allyl halides **152** by use of lithium halide in the presence of Amberlyst 15 was also reported recently [60].

2.4 Isomerization Including Rearrangement

Isomerization of vinylaziridines is widely used in organic synthesis. Six types of isomerization of vinylaziridines are shown in Scheme 2.40. Outlined in this section are: *i*) azepine formation by aza-[3,3]-Claisen rearrangement of 1,2-divinyl- or 2,3-divinylaziridines **153** (Section 2.4.1), *ii*) pyrroline formation from **155** (Section 2.4.2), *iii*) aza-[2,3]-Wittig rearrangement of anionic species **157** (Section 2.4.3),

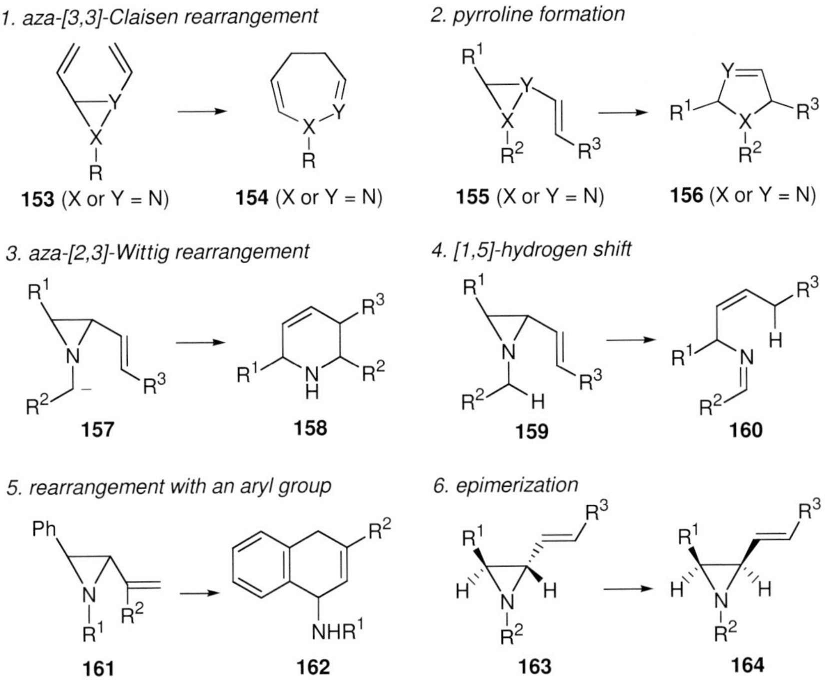

1. aza-[3,3]-Claisen rearrangement

153 (X or Y = N) → 154 (X or Y = N)

2. pyrroline formation

155 (X or Y = N) → 156 (X or Y = N)

3. aza-[2,3]-Wittig rearrangement

157 → 158

4. [1,5]-hydrogen shift

159 → 160

5. rearrangement with an aryl group

161 → 162

6. epimerization

163 → 164

Scheme 2.40 Isomerization of vinylaziridines.

iv) hydrogen shift from **159** (Section 2.4.4), *v*) rearrangement of vinylaziridines **161** with an aryl group (Section 2.4.5), and *vi*) epimerization of **163** at the allylic position (Section 2.4.6).

2.4.1 Aza-[3,3]-Claisen Rearrangement

Formation of azepine derivatives by aza-[3,3]-Claisen rearrangements was first reported by Stogryn and Brois, an isomeric mixture of 2,3-divinylaziridines **166**, generated by treatment of sulfonate **165** with NaOH, being converted into azepine **167** by steam distillation (Scheme 2.41) [61]. In this case, unchanged *trans*-azir-

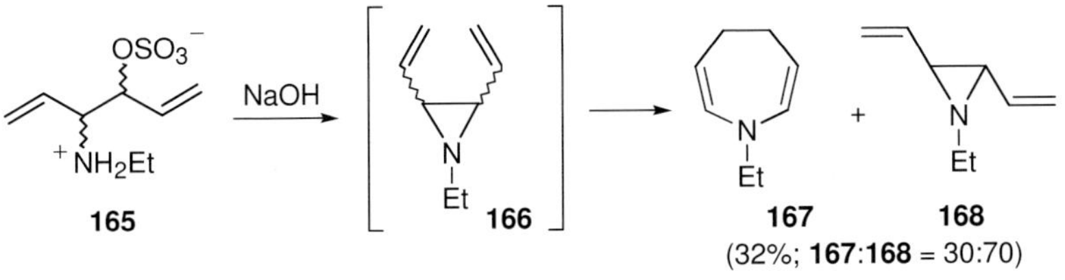

165 —NaOH→ [166] → 167 + 168

(32%; **167**:**168** = 30:70)

Scheme 2.41 Pioneering works on isomerization of divinylaziridines to azepines.

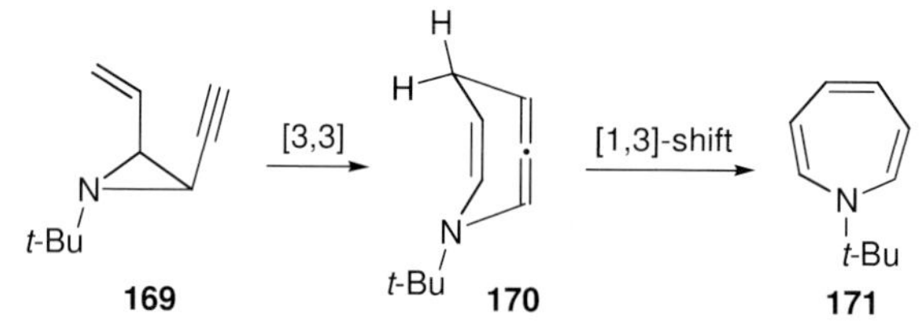

Scheme 2.42 Formation of 1*H*-azepine from **169**.

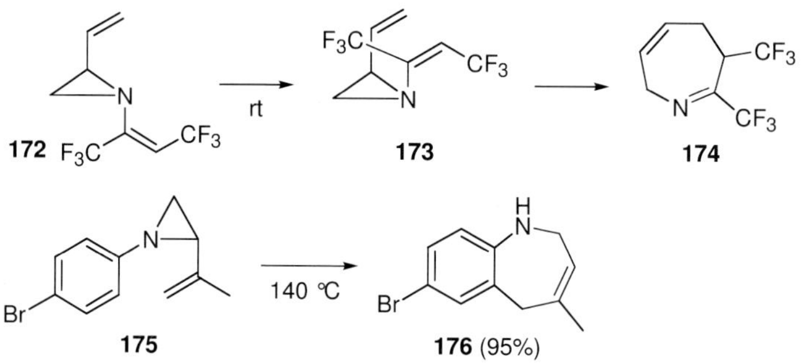

Scheme 2.43 Pioneering works on isomerization of 1,2-divinylaziridine derivatives to azepines.

idine **168** was also obtained (**167**:**168** = 30:70). A similar azepine formation using diastereomerically pure 2,3-*cis*-1-methyl-2,3-divinylaziridine has also been reported [62].

Thermal rearrangement of *cis*-2-ethynyl-3-vinylaziridine **169** forms 1*H*-azepine **171** (Scheme 2.42). This reaction proceeds through aza-[3,3]-Claisen rearrangement followed by a [1,3]-hydrogen shift of the cyclic allenene intermediate **170** [63]. The corresponding *trans*-aziridine also yielded the same azepine **171**, by thermal isomerization to the *cis* isomer **169** through carbon-carbon bond cleavage followed by reaziridination [64].

Azepine formation from 1,2-divinylaziridine derivatives was first reported by two independent research groups in 1967 (Scheme 2.43). Stogryn and Brois found that 1,2-divinylaziridine **172**, prepared by low-temperature addition of *N*-unsubstituted 2-vinylaziridines to hexafluoro-2-butyne, isomerized to azepine **174** via the 1,2-*cis* isomer **173** on standing at room temperature overnight [65]. At almost the same time, Scheiner reported that heating of vinylaziridine **175**, possessing a substituted phenyl group on the nitrogen atom, resulted in a clean rearrangement to a fused ring **176** (Scheme 2.43) [66]. Similarly, thermal isomerization of 1-heteroaromatic-2-vinylaziridines such as **177** yielded the corresponding pyrido-, isothiazolo-, and thienoazepines as shown in Scheme 2.44 [67]. Since the isomerization of *N*-pyridino-2-vinylaziridines to the corresponding azepine shows substituent effects completely analogous to those of the related benzenic *O*-Claisen rearrangement, the ring-expansion reaction of divinylaziridines should be described as a

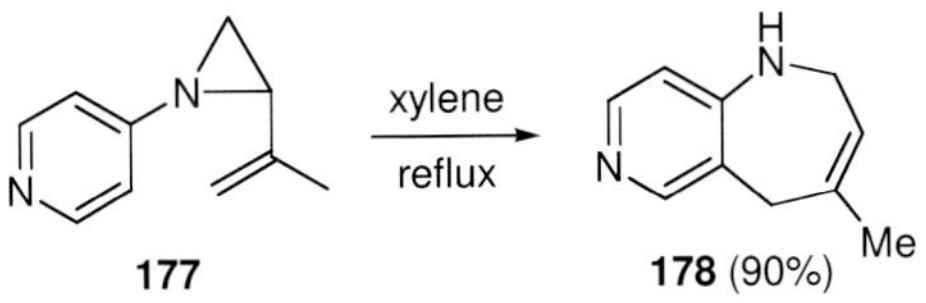

Scheme 2.44 Formation of pyrido-azepine **178**.

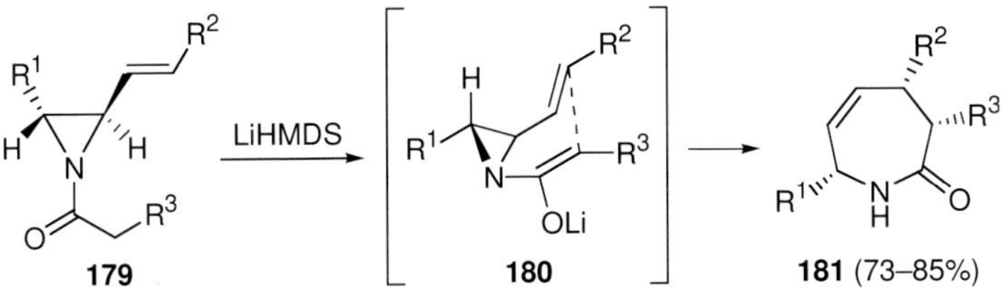

Scheme 2.45 Aza-[3,3]-Claisen enolate rearrangements of 1-acyl-2-vinylaziridines.

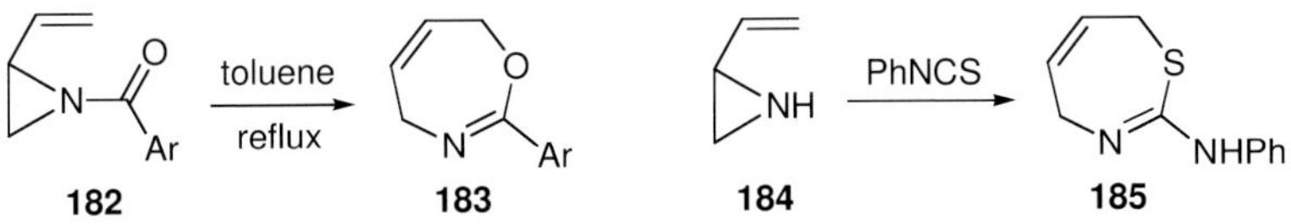

Scheme 2.46 Rearrangements of 2-vinylaziridines possessing carbonyl or thiocarbonyl groups.

concerted [3,3]-sigmatropic rearrangement [68]. A related isomerization of 4-(2-vinylaziridino)-5-methoxy-1,2-benzoquinones to bicyclic benzoquinones was also reported [69].

In 1997, Lindström and Somfai reported aza-[3,3]-Claisen enolate rearrangements of vinylaziridines (Scheme 2.45) [70]. Treatment of 1-acyl-2-vinylaziridines **179** with LHMDS resulted in the stereoselective formation of seven-membered lactams **181**, presumably through a boat-like transition state **180**.

Finally, formation of seven-membered rings by vinylaziridines possessing carbonyl or thiocarbonyl groups are shown in Scheme 2.46. Thermolysis of **182** in toluene at reflux yielded 4,7-dihydro-1,3-oxazepine **183** [71]. Similarly, treatment of *N*-unsubstituted vinylaziridine **184** with phenyl isothiocyanate in ether at 0 °C gave seven-membered ring **185**, via a 1-thiocarboxanilide-2-vinylaziridine intermediate [72].

2.4.2
Pyrroline Formation

Pyrroline formation through rearrangement of vinylaziridines was first reported with 1-vinylaziridine derivatives by Whitlock and Smith (Scheme 2.47) [73]. So-

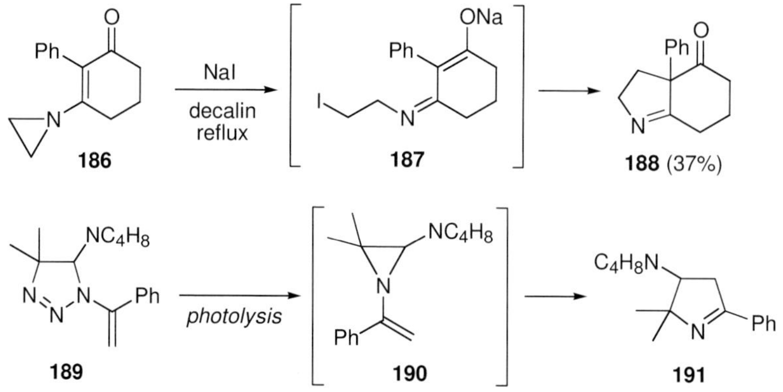

Scheme 2.47 Rearrangement of 1-vinylaziridines to 1-pyrroline derivatives.

dium iodide-mediated ring-opening of **186** and subsequent ring-formation afforded a bicyclic 1-pyrroline **188** in 37% yield, presumably through the intermediate **187**. A related conversion was also observed on the photolysis of 1-vinyl-1,2,3-triazole **189**, via the 1-vinylaziridine intermediate **190** [74].

Thermal rearrangement of 2-vinylaziridines yields 3-pyrrolines. When aziridines **192** were heated in decalin at 180 °C, 3-pyrrolines **193** were formed in 75–85% yields (Scheme 2.48) [75]. A similar rearrangement was also observed by Lwowski and coworkers [4]. This type of rearrangement can be effectively pro-

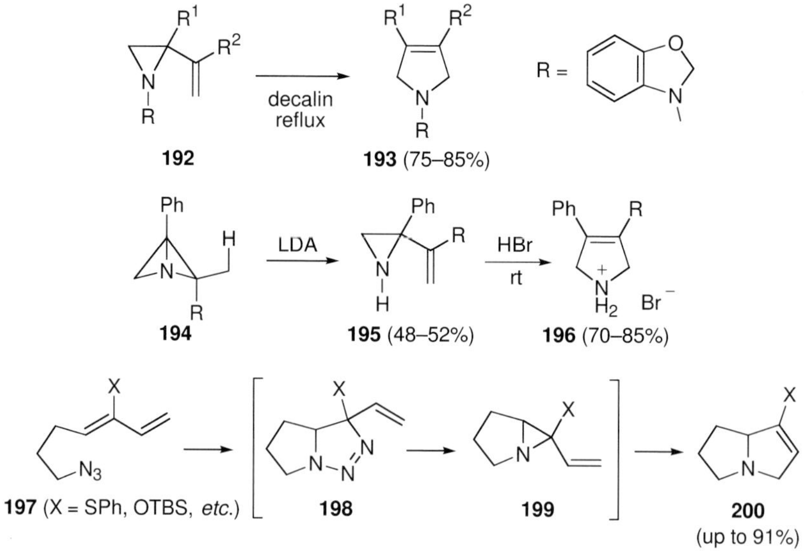

Scheme 2.48 Rearrangement of 2-vinylaziridines to 3-pyrroline derivatives.

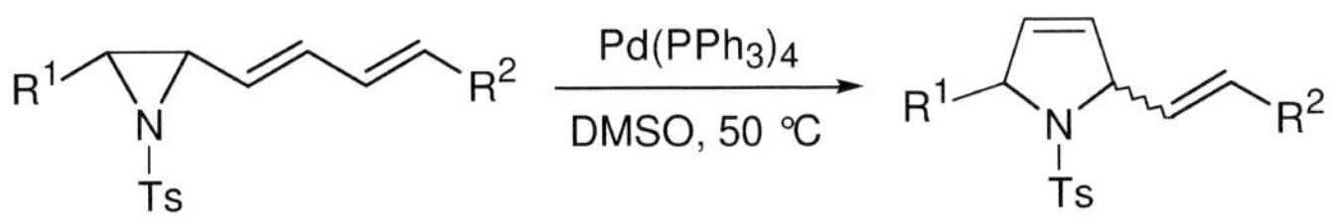

201 **202** (76–98%)

Scheme 2.49 Palladium(0)-catalyzed isomerization of 2-dienylaziridines to 3-pyrrolines.

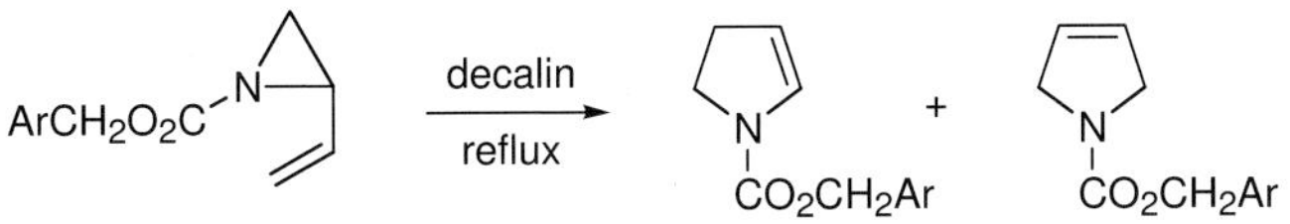

203 (Ar = 4-NO_2Ph) **204** **205**

(63–75%, **204**:**205** = 96:4)

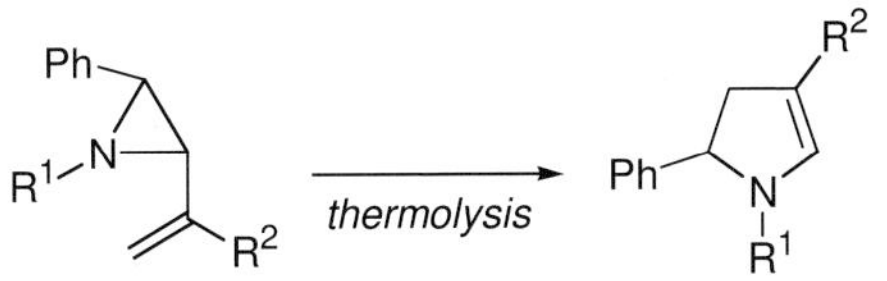

206 (R^1 = alkyl) **207** (63–75%)

Scheme 2.50 Thermal rearrangement to 2-pyrrolines.

moted by an acid: treatment of 2-vinylaziridines **195**, prepared by LDA-induced ring-opening of 1-azabicyclobutanes **194**, with 48% HBr gave 3-pyrrolines **196** as their hydrobromide salts, while heating of **195** in decalin at 175–180 °C resulted in decomposition of **195** without formation of the pyrrolines **196** [76]. Bicyclic vinylaziridines **199** are regarded as plausible intermediates in the intramolecular reactions of azides **197** with electron-rich 1,3-dienes to form bicyclic 3-pyrrolines **200** [77].

Palladium(0)-catalyzed isomerization of 2-dienylaziridines **201** to 3-pyrrolines **202** was reported in 1985 by Oshima, Nozaki, and coworkers (Scheme 2.49) [78]. This isomerization is in striking contrast to Ibuka's palladium-catalyzed isomerization of 2,3-*trans*-2-vinylaziridines to the corresponding 2,3-*cis* isomers (see Section 2.4.6) [29].

Thermal rearrangement of 2-vinylaziridine **203**, with an electron-withdrawing substituent on the nitrogen atom, at reflux in decalin predominantly yielded 2-pyrroline **204** along with some 3-pyrroline derivatives **205** (Scheme 2.50) [79]. A similar reaction was also observed with 1-alkyl-2-phenyl-3-vinylaziridines **206** [80]. From these observations, the nature of the products formed depends on the natures of the substituents on the aziridine ring: if the ring carbon carries a phenyl substituent, the thermolysis in most cases preferentially yields 2-pyrrolines.

Hudlicky and coworkers also reported a related 2-pyrroline formation from vinylaziridines [81], which are extremely useful for the synthesis of pyrrolizidine alkaloids such as the protected (+)-trihydroxyheliotridane **210** (Scheme 2.51). Since the pyrolysis of either diastereomer of **208** furnished the cyclized product **209** as a

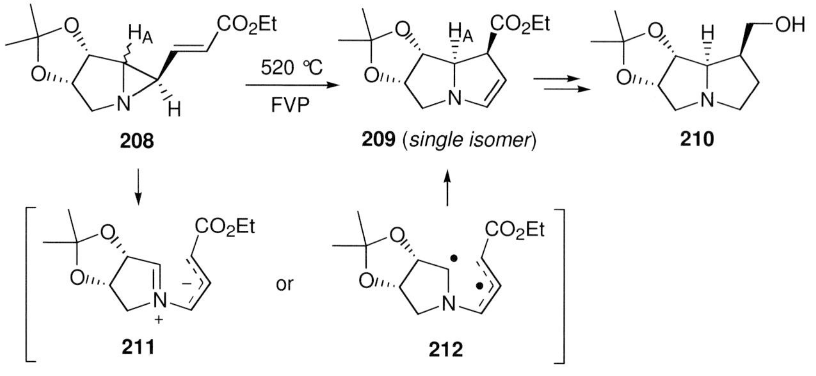

Scheme 2.51 Pyrolysis of vinylaziridine **208** to 2-pyrroline **209**.

single isomer, they proposed a diradical or zwitterionic closure of the intermediate species **211** and **212**.

2.4.3
Aza-[2,3]-Wittig Rearrangement

Aza-[2,3]-Wittig rearrangements of vinylaziridines were originally reported by Åhman and Somfai in 1994 (Scheme 2.52) [82]. Treatment of *N-tert*-butoxycarbonylmethyl vinylaziridines **213** with LDA afforded the corresponding *cis*-2,6-tetrahydropyridines **215** as single isomers. The stereochemical outcome in the rearrangement can be correctly accounted for by a concerted process through **214**. This type of reaction is also applicable to the 1-propargyl-2-vinylaziridines [83]. The same group also demonstrated the synthesis of some indolizidine alkaloids such as **216** and **217** by use of the aza-[2,3]-Wittig rearrangement [84]. Coldham and coworkers reported a similar rearrangement of 2-(1-phenylvinyl)aziridines possessing *tert*-butoxycarbonylmethyl moieties on their nitrogen atoms [85].

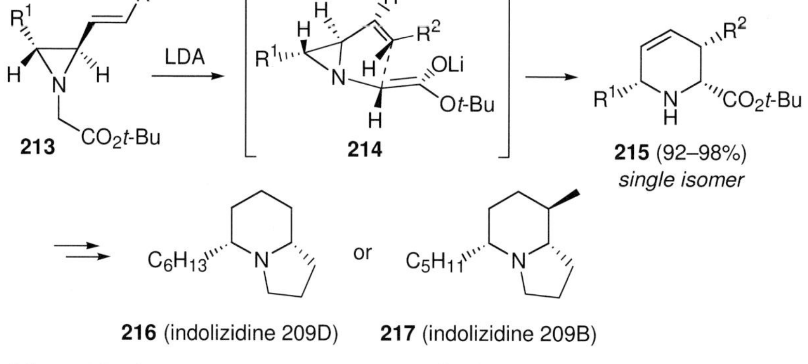

Scheme 2.52 Aza-[2,3]-Wittig rearrangements of vinylaziridines **213** and alkaloid syntheses.

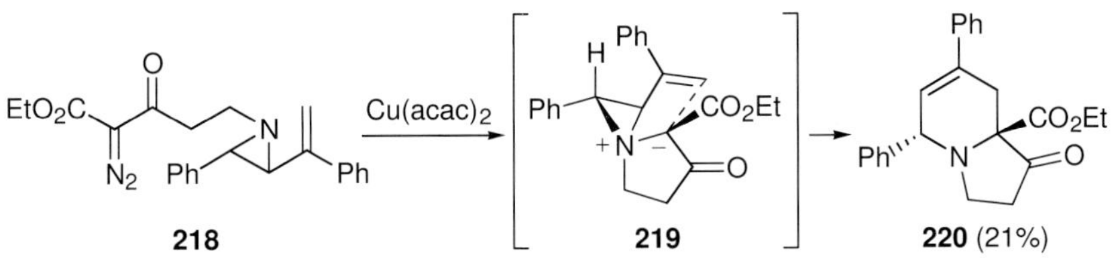

Scheme 2.53 [2,3]-Stevens rearrangement of a vinylaziridine-derived ammonium ylide **219**.

The aza-[2,3]-Wittig rearrangement of a vinylaziridine-derived quaternary aziridinium ylide (i.e., [2,3]-Stevens rearrangement) has recently been reported (Scheme 2.53) [86]. The aziridinium ylide **219**, generated by the intramolecular reaction of a copper carbenoid tethered to a vinylaziridine, underwent a [2,3]-Stevens rearrangement to furnish the bicyclic amine **220** with the indolizidine skeleton.

2.4.4 Hydrogen Shift

Hydrogen shifts are often observed in thermal isomerizations of vinylaziridines. Heating of compounds **221** at 180 °C produced mixture of 3-pyrrolines **222** and hydrazones **223** (Scheme 2.54) [87]. The formation of **223** can be explained in terms either of a concerted hydrogen shift as depicted in **224** or of diradical intermediates **225**, both of which would be followed by thermal isomerization of the (*Z*)-carbon-carbon double bonds to provide the (*E*) isomers **223**.

The shift of a hydrogen on the nitrogen substituent of the 1-benzyl-2-vinylaziridine **226** was observed on heating at 160 °C, selectively affording imines **227** through [1,5]-hydrogen shift (Scheme 2.55) [80b]. Similar [1,5]-hydrogen shifts were also reported by Pearson and coworkers [77] in their study on the thermal isomerization of bicyclic vinylaziridines **199** already described in Section 2.4.2 (Scheme 2.48) and by another research group [88].

Somfai and coworkers investigated the thermal isomerization of vinylaziridines **228** and found that while LDA-mediated isomerization of **228**, possessing a *tert*-butyl acetate moiety ($R^2 = CO_2t$-Bu), afforded tetrahydropyridines by aza-[2,3]-Wit-

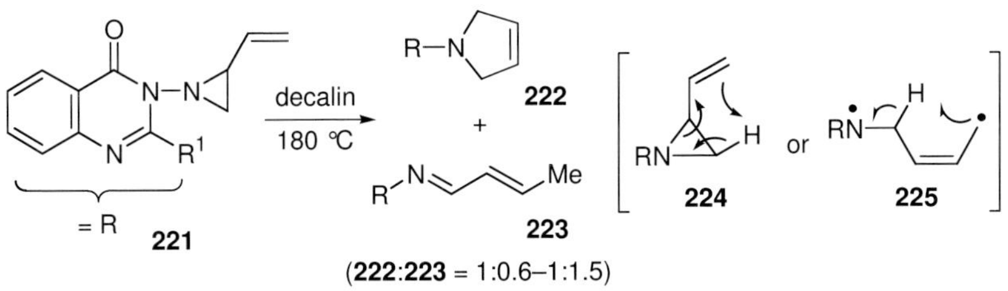

Scheme 2.54 Imine formation through hydrogen shifts.

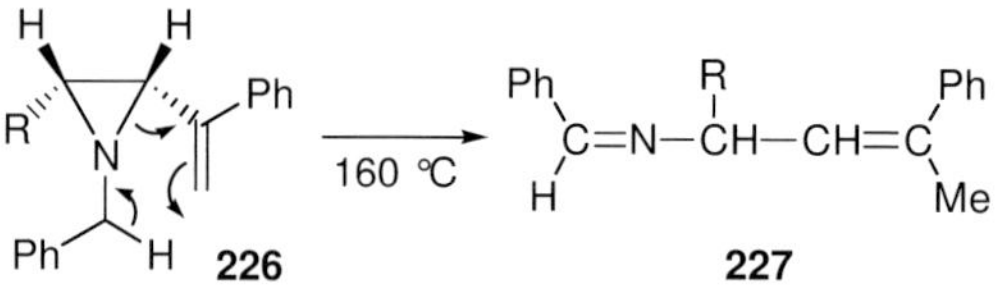

Scheme 2.55 [1,5]-Hydrogen shifts in 1-benzyl-2-vinylaziridines **226**.

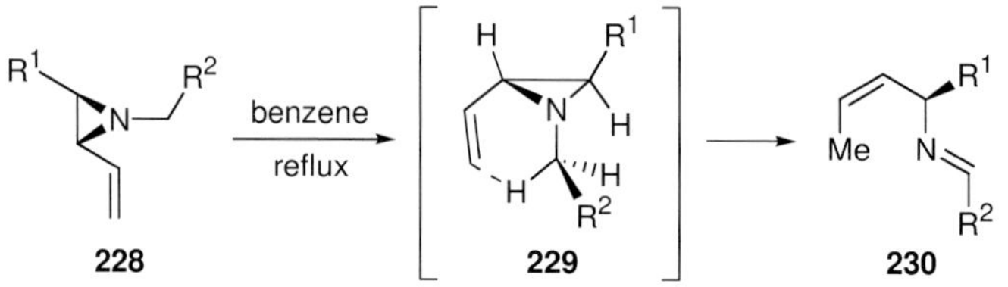

Scheme 2.56 Stereochemical course of the [1,5]-hydrogen shifts.

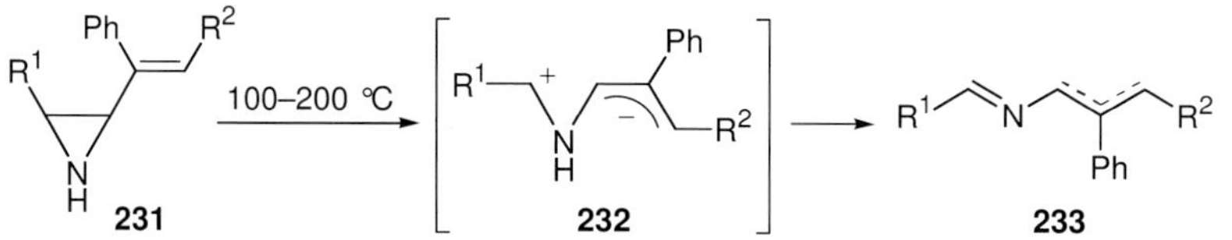

Scheme 2.57 Migration of hydrogen on the nitrogen atom.

tig rearrangement as already shown in Scheme 2.52, thermal rearrangement of **228** stereoselectively gave allylic imines **230** in quantitative yields through [1,5]-hydrogen shifts, presumably via the transition states **229** (Scheme 2.56) [89].

Finally, hydrogen migration in the *N*-unsubstituted 2-vinylaziridines **231** is shown in Scheme 2.57. Thermal isomerization of **231** gave imines **233** through heterolytic cleavage of the carbon-carbon bonds of the aziridine rings and migration of hydrogen on the nitrogen through intermediates such as **232**.

2.4.5
Rearrangement with an Aryl Group on the Aziridine Carbon

An interesting rearrangement of 1,3-divinyl-2-phenylaziridines was reported in 1979. As shown in Scheme 2.58, *trans* addition of *N*-unsubstituted aziridines to ethyl propiolate gave 1,3-divinylaziridines **234**, which could be rearranged to provide 1,4-dihydronaphthalenes **236** by heterolytic cleavage of the carbon-nitrogen bond via the intermediates **235**. To the best of my knowledge, this is the only example demonstrating rearrangement of 2-vinylaziridines possessing aryl groups on their aziridine carbons. Related reactions of *N*-aryl-2-vinylaziridines have already been mentioned in Section 2.4.1 (Schemes 4.43 and 4.44).

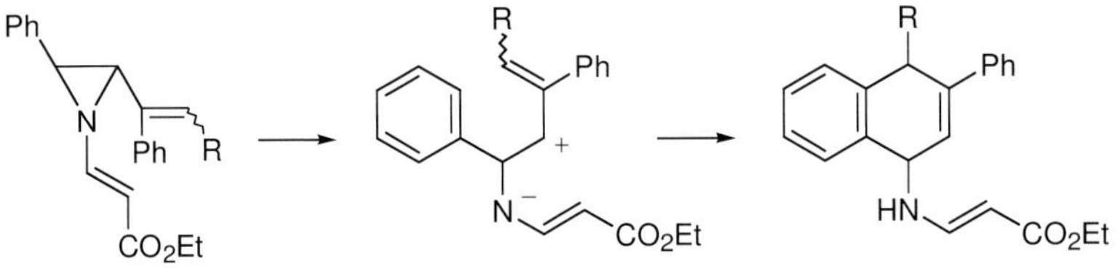

Scheme 2.58 Formation of 1,4-dihydronaphthalenes **236** through rearrangement of **234**.

2.4.6
Epimerization

As described in Section 2.3.2, vinylaziridines are versatile intermediates for the stereoselective synthesis of (*E*)-alkene dipeptide isosteres. One of the simplest methods for the synthesis of alkene isosteres such as **242** and **243** via aziridine derivatives of type **240** and **241** (Scheme 2.59) involves the use of chiral *anti*- and *syn*-amino alcohols **238** and **239**, synthesizable in turn from various chiral amino aldehydes **237**. However, when a chiral *N*-protected amino aldehyde derived from a natural α-amino acid is treated with an organometallic reagent such as vinylmagnesium bromide, a mixture of *anti*- and *syn*-amino alcohols **238** and **239** is always obtained. Highly stereoselective syntheses of either *anti*- or *syn*-amino alcohols **238** or **239**, and hence 2,3-*trans*- or 2,3-*cis*-3-alkyl-2-vinylaziridines **240** or **241**, from readily available amino aldehydes **237** had thus hitherto been difficult. Ibuka and coworkers overcame this difficulty by developing an extremely useful epimerization of vinylaziridines. Palladium(0)-catalyzed reactions of 2,3-*trans*-2-vinylaziridines **240** afforded the thermodynamically more stable 2,3-*cis* isomers **241** predominantly over **240** (**241**:**240** >94:6) through π-allylpalladium intermediates, in accordance with *ab initio* calculations [29]. This epimerization allowed a highly stereoselective synthesis of (*E*)-alkene dipeptide isosteres **243** with the desired L,L-

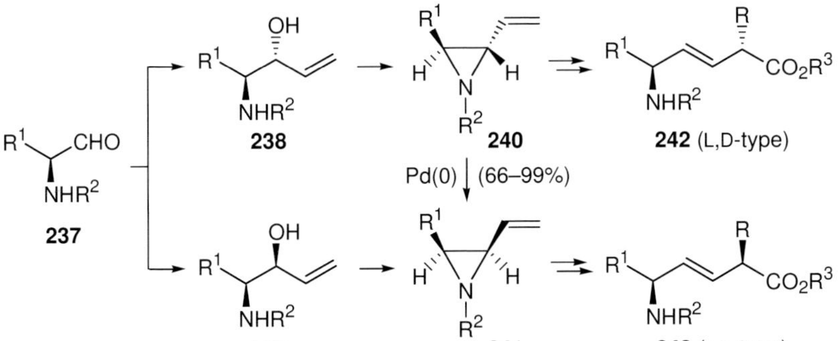

Scheme 2.59 Synthesis of L,L-type (*E*)-alkene dipeptide isosteres **243** through palladium(0)-catalyzed isomerization of vinylaziridines **240**.

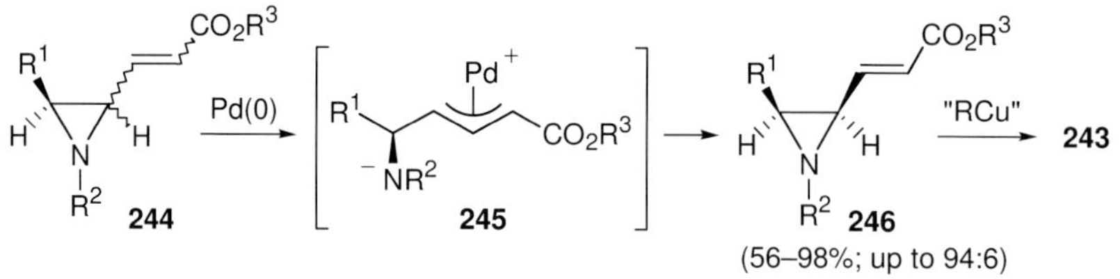

Scheme 2.60 Epimerization of γ-aziridinyl-α,β-enoates and synthesis of alkene isosteres.

stereochemistries through mixtures of *trans*- and *cis*-2-vinylaziridines **240** and **241**. Rhodium or iridium complexes also catalyze this epimerization [90].

A related palladium(0)-catalyzed epimerization of γ-aziridinyl-α,β-enoates **244** was also reported by Ibuka, Ohno, Fujii, and coworkers (Scheme 2.60) [43]. Treatment of either isomer of **244** with a catalytic amount of $Pd(PPh_3)_4$ in THF yielded an equilibrated mixture in which the isomer **246** with the desired configuration predominated (**246**:other isomers = 85:15 to 94:6). In most cases the isomer **246** could be easily separated from the diastereomeric mixture by a simple recrystallization, and the organocopper-mediated ring-opening reaction of **246** directly afforded L,L-type (*E*)-alkene dipeptide isosteres **243**.

2.5 Cycloaddition

2.5.1 Cycloadditions of Isocyanates and Related Compounds

Cycloadditions of isocyanates and their derivatives with vinylaziridines were first reported by Alper and coworkers. From their previous studies of cycloadditions to vinylepoxides or alkylaziridines, they investigated cycloadditions to vinylaziridines and found that such reactions with isocyanates, carbodiimides, or isothiocyanates in the presence of catalytic amounts of $Pd(OAc)_2$ (2 mol%) and PPh_3 (10 mol%) at room temperature afforded five-membered ring products **249** in 34–97% yields (Scheme 2.61) [91]. When an aziridine **247** possessing an alkyl substituent at the

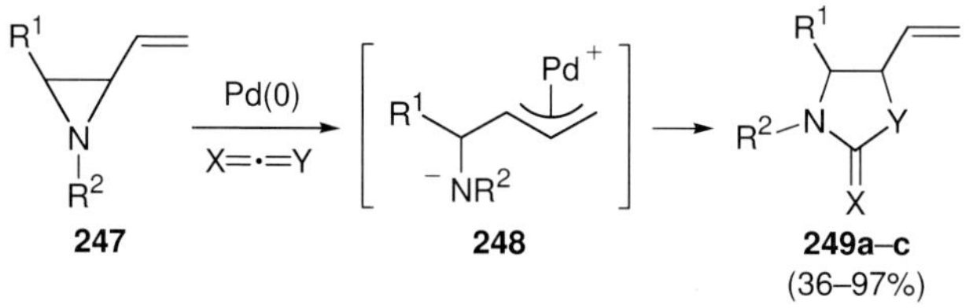

	X	Y
a	O	NAr
b	NAr	NAr
c	NAr	S

Scheme 2.61 Palladium(0)-catalyzed cycloaddition of heterocumulenes.

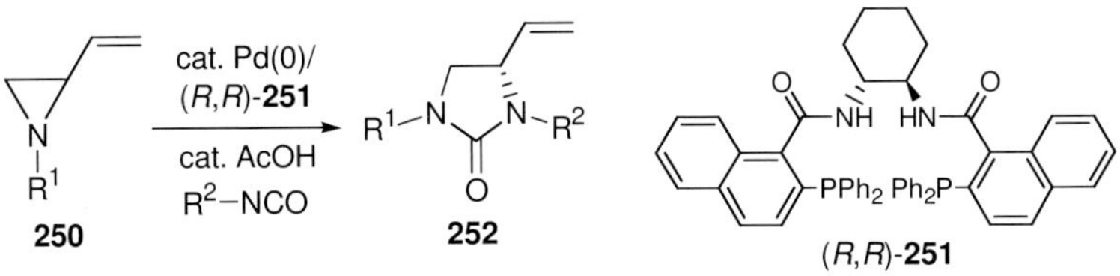

Scheme 2.62 Dynamic kinetic asymmetric cycloaddition of isocyanates.

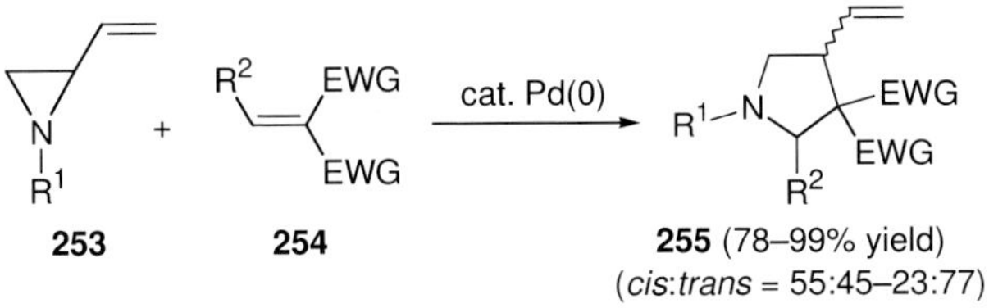

Scheme 2.63 Palladium(0)-catalyzed cycloadditions of activated olefins.

3-position (R^1 = alkyl) was used, an approximately 2:1 mixture of *cis* and *trans* five-membered ring **249** was obtained. The palladium catalyst activates the vinylaziridines by forming π-allylpalladium intermediates **248**, similarly to Ibuka's isomerization of vinylaziridines [29].

An asymmetric version of the cycloaddition of isocyanates to vinylaziridines was recently reported by Trost and Fandrick (Scheme 2.62) [92]. Treatment of racemic vinylaziridines **250** with isocyanates in the presence of $(\eta^3\text{-}C_3H_5PdCl)_2$ (2 mol%) and chiral ligand (*R*,*R*)-**251** (6 mol%) gave the corresponding chiral imidazolidinones **252** in up to 95% *ee* by dynamic kinetic asymmetric transformation through η^3–η^1–η^3 interconversion. The obtained imidazolidinones **252** were easily converted into the corresponding chiral diamines by successive treatment with $LiAlH_4$ and $H_2NOH \cdot H_2O$. A related $CeCl_3$-promoted asymmetric cycloaddition using (*S*)-BINAP and $Pd_2(dba)_3 \cdot CHCl_3$ giving up to 83% *ee* was recently reported by Alper's group [93].

Yamamoto and coworkers developed cycloadditions between activated olefins and vinylaziridines **253** with the aid of a palladium(0) catalyst (Scheme 2.63) [94], based on their three-component aminoallylation reaction. The corresponding 4-vinylpyrrolidines **255** were obtained as mixtures of diastereomers (*cis*:*trans* = 55:45 to 23:77).

2.5.2
Carbonylative Ring-expansion to Lactams

Ring expansions of vinylaziridines with carbon monoxide are a useful transformation with which to produce lactams. In 1974, a light-induced reaction between vinylaziridine **256** and $Fe(CO)_5$, forming the π-allyltricarbonyliron lactam complex

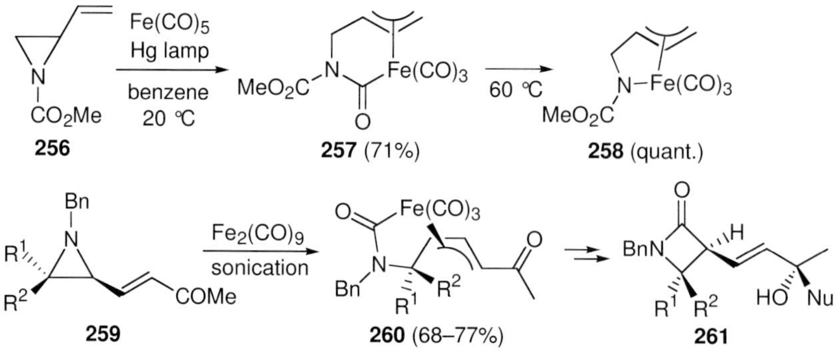

Scheme 2.64 Formation of π-allyltricarbonyliron lactam complexes.

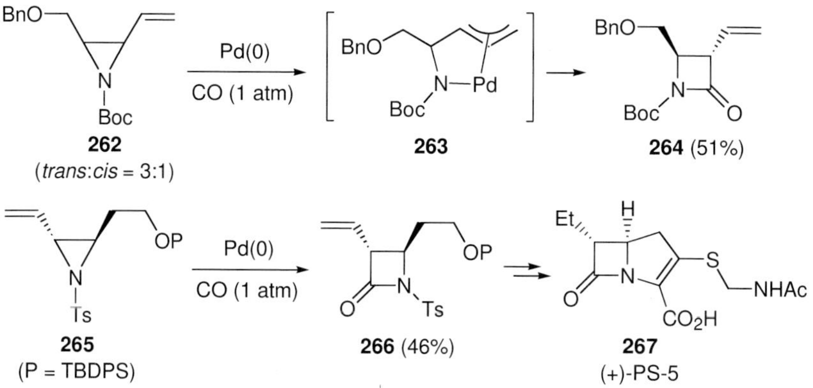

Scheme 2.65 Palladium(0)-catalyzed carbonylative ring-expansion.

257 by ring-opening and subsequent insertion of carbon monoxide into the Fe-N bond, was first reported by Aumann and coworkers (Scheme 2.64) [95]. At 60 °C, **257** decomposes quantitatively with a loss of one equivalent of CO to form **258**.

Ley and Middleton synthesized ketone-functionalized lactam complexes **260** (Scheme 2.64) by sonication of vinylaziridines **259** with $Fe_2(CO)_9$ in benzene. These complexes were easily converted into the corresponding β-lactams **261** by stereoselective addition of nucleophiles such as $NaBH_4$ or trialkylaluminium to the carbonyl group followed by decomplexation with Me_3NO [96].

In 1991, Ohfune and coworkers reported palladium(0)-catalyzed carbonylation of vinylaziridines **262** with carbon monoxide (1 atm.) in benzene (Scheme 2.65) [31]. Interestingly, 3,4-*trans*-azetidinone **264** was exclusively obtained from a diastereomeric mixture of *trans*- and *cis*-vinylaziridines **262** (3:1). Tanner and Somfai synthesized (+)-PS-5 (**267**) by use of palladium(0)-catalyzed *trans*-selective β-lactam formation in the presence of $Pd(dba)_3 \cdot CHCl_3$ (15 mol%) and excess PPh_3 in toluene.

2.6 Electron Transfer to Vinylaziridines

Electron transfer to vinylaziridines results in ring-opening reactions, yielding allyl amines. Treatment of **268** with SmI_2/DMEA (*N,N*-dimethylethanolamine) provided allyl amine **269** as a 2:1 mixture of olefinic isomers in 88% yield (Scheme 2.66) [97].

Another example is indium(0)-induced electron transfer to aziridines **270** incorporating allyl iodide moieties (Scheme 2.66). Treatment with indium(0) in MeOH at reflux gave the corresponding chiral (*E*)-dienylamines **271** in excellent yields [98]. It should be noted that indium was found to be more effective for this transformation than other metals such as zinc, samarium, and yttrium.

In 2001, an interesting umpolung of optically active vinylaziridines through palladium(0)-catalyzed metalation with indium(I) iodide and addition to aldehydes, resulting in the stereoselective formation of *syn,syn*-2-vinyl-1,3-amino alcohols **274**, each possessing three contiguous chiral centers, was reported by Takemoto and coworkers (Scheme 2.67) [99]. The ratio of the *syn,syn* isomer to the other three isomers was up to 93:7, being significantly affected both by the C-3 substituents on the aziridines (R^1) and by the R groups of the aldehydes. The stereoselectivities can be reasonably explained by considering the preferable cyclic transition states such as **273**. This is the only example that demonstrates the reactivity of vinylaziridines as nucleophiles for carbon-carbon bond-forming reactions. A related umpolung of chiral 2-ethynylaziridines by indium(I) and stereoselective addition of the resulting

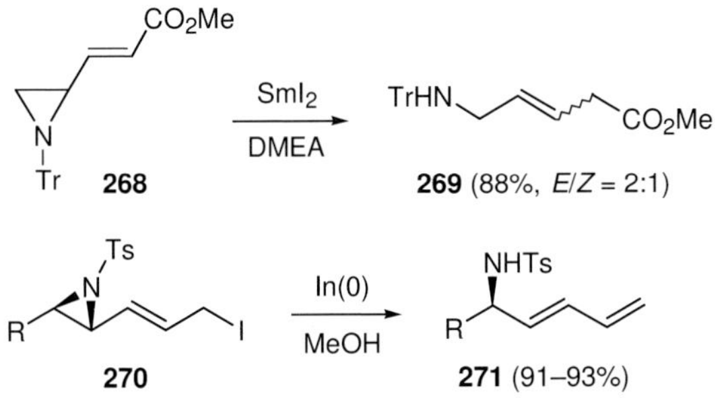

Scheme 2.66 Ring-opening reactions resulting from electron transfer.

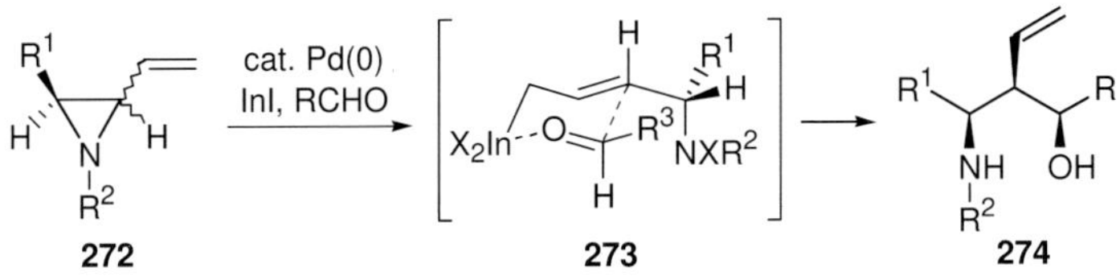

Scheme 2.67 Indium(I)-mediated umpolung and addition to aldehydes.

allenylindium reagents to aldehydes was reported by Ohno, Hamaguchi, and Tanaka in 2000 [100].

2.7 Conclusions

As described, vinylaziridines are versatile intermediates for many kinds of compounds containing nitrogen, thanks to the interesting chemical properties arising from strained aziridine rings at reactive allylic positions. Ethynylaziridines, acetylene analogues of vinylaziridines, have also attracted much attention in recent years as useful synthetic intermediates for such compounds as amino alcohols [100] and amino allenes [101], although restrictions of space do not allow detailed description of the synthesis [102], [103], [104], [105] and reactions of ethynylaziridines. Vinylaziridines and their related compounds are becoming a more useful and important class of compounds in organic synthesis and will continue to fascinate organic chemists.

References

1 General methods that can be widely applied for the synthesis of various aziridines and not only for vinylaziridines are not included in this section. Furthermore, although synthesis of vinylaziridines is often accomplished by formation of a carbon-carbon double bond after the aziridine ring has been formed, this type of synthetic route is not within the category of "direct synthesis of vinylaziridines". For examples, see: (a) D. Borel, Y. Gelas-Mialhe, R. Vessiére, *Can. J. Chem.* **1976**, *54*, 1582; (b) S. K. Nayak, L. Thijs, B. Zwanenburg, *Tetrahedron Lett.* **1999**, *40*, 981–984.

2 K. Hafner, W. Kaiser, R. Puttner, *Tetrahedron Lett.* **1964**, *5*, 3953–3956.

3 (a) M. P. Sammes, A. Rahman, *J. Chem. Soc., Perkin Trans. 1* **1972**, 344–346; (b) R. G. Sheppard, S. J. Norton, *J. Agric. Food Chem.* **1980**, *28*, 1300–1303.

4 A. Mishra, S. N. Rice, W. Lwowski, *J. Org. Chem.* **1968**, *33*, 481–486.

5 Y. Nomura, N. Hatanaka, Y. Takeuchi, *Chem. Lett.* **1976**, 901–904.

6 R. S. Atkinson, C. W. Rees, *Chem. Commun.* **1967**, 1230–1231.

7 R. Appel, O. Büchner, *Angew. Chem.* **1962**, *74*, 430–431.

8 R. S. Atkinson, C. W. Rees, *J. Chem. Soc., Perkin Trans. 1* **1969**, 772–778.

9 (a) R. S. Atkinson, J. R. Malpass, *J. Chem. Soc., Chem. Commun.* **1975**, 555–556; (b) R. S. Atkinson, J. R. Malpass, *J. Chem. Soc., Perkin Trans. 1* **1977**, 2242–2249.

10 J. G. Knight, M. P. Muldowney, *Synlett* **1995**, 949–951.

11 (a) Z. Li, K. R. Conser, E. N. Jacobsen, *J. Am. Chem. Soc.* **1993**, *115*, 5326–5327; (b) D. A. Evans, M. M. Faul, M. T. Bilodeau, B. A. Anderson, D. M. Barnes, *J. Am. Chem. Soc.* **1993**, *115*, 5328–5329.

12 M. Nishimura, S. Minakata, S. Thongchant, I. Ryu, M. Komatsu, *Tetrahedron Lett.* **2000**, *41*, 7089–7092.

13 B. Mauzé, *J. Organomet. Chem.* **1980**, *202*, 233–239.

14 A. A. Cantrill, A. N. Jarvis, H. M. I. Osborn, A. Ouadi, J. B. Sweeney, *Synlett* **1996**, 847–849.

15 (a) A.-H. Li, L.-X. Dai, X.-L. Hou, *Chem. Commun.* **1996**, 491–492; (b) A.-H. Li, L.-X. Dai, X.-L. Hou, *J. Chem. Soc., Perkin Trans. 1* **1996**, 2725–2729.

16 (a) A.-H. Li, L.-X. Dai, X.-L. Hou, M.-B. Chen, *J. Org. Chem.* **1996**, *61*, 4641–4648; telluronium allylides can react with *N*-Boc-aliphatic imines to afford *cis*-vinyla-

ziridines stereoselectively; see (b) W.-W. Liao, X.-M. Deng, Y. Tang, *Chem. Commun.* **2004**, 1516–1517.
17 X.-L. Hou, X.-F. Yang, L.-X. Dai, X.-F. Chen, *Chem. Commun.* **1998**, 747–748.
18 A.-H. Li, L.-X. Dai, X.-L. Hou, *J. Chem. Soc., Perkin Trans. 1* **1996**, 867–869.
19 L. G. Arini, A. Sinclair, P. Szeto, R. A. Stockman, *Tetrahedron Lett.* **2004**, *45*, 1589–1591.
20 X.-F. Yang, M.-J. Zhang, X.-L. Hou, L.-X. Dai, *J. Org. Chem.* **2002**, *67*, 8097–8103.
21 S. Hanessian, L.-D. Cantin, *Tetrahedron Lett.* **2000**, *41*, 787–790.
22 D. Morton, D. Pearson, R. A. Field, R. A. Stockman, *Org. Lett.* **2004**, *6*, 2377–2380.
23 A. Hassner, J. Keogh, *Tetrahedron Lett.* **1975**, *16*, 1575–1578.
24 R. Zamboni, J. Rokach, *Tetrahedron Lett.* **1983**, *24*, 331–334.
25 U. M. Lindström, P. Somfai, *Synthesis* **1998**, 109–117.
26 B. Olofsson, R. Wijtmans, P. Somfai, *Tetrahedron Lett.* **2002**, *58*, 5979–5982.
27 H. F. Olivo, M. S. Hemenway, A. C. Hartwig, R. Chan, *Synlett* **1998**, 247–248.
28 R. L. Weller, S. R. Rajski, *Tetrahedron Lett.* **2004**, *45*, 5807–5810.
29 (a) N. Mimura, T. Ibuka, M. Akaji, Y. Miwa, T. Taga, K. Nakai, H. Tamamura, N. Fujii, Y. Yamamoto, *Chem. Commun.* **1996**, 351–352; (b) T. Ibuka, N. Mimura, H. Aoyama, M. Akaji, H. Ohno, Y. Miwa, T. Taga, K. Nakai, H. Tamamura, N. Fujii, Y. Yamamoto, *J. Org. Chem.* **1997**, *62*, 999–1015.
30 H. Ohno, K. Ishii, A. Honda, H. Tamamura, N. Fujii, Y. Takemoto, T. Ibuka, *J. Chem. Soc., Perkin Trans. 1* **1998**, 3703–3716.
31 G. W. Spears, K. Nakanishi, Y. Ohfune, *Synlett* **1991**, 91–92.
32 (a) K. Ishii, H. Ohno, Y. Takemoto, T. Ibuka, *Synlett* **1999**, 228–230; (b) K. Ishii, H. Ohno, Y. Takemoto, E. Osawa, Y. Yamaoka, N. Fujii, T. Ibuka, *J. Chem. Soc., Perkin Trans. 1* **1999**, 2155–2163.
33 (a) H. Ohno, A. Toda, N. Fujii, Y. Miwa, T. Taga, Y. Yamaoka, E. Osawa, T. Ibuka, *Tetrahedron Lett.* **1999**, *40*, 1331–1334; (b) H. Ohno, Y. Takemoto, N. Fujii, T. Tanaka, T. Ibuka, *Chem. Pharm. Bull.* **2004**, *52*, 111–119.
34 (a) H. Ohno, A. Toda, N. Fujii, T. Ibuka, *Tetrahedron: Asymmetry* **1998**, *9*, 3929–3933; (b) H. Ohno, A. Toda, Y. Takemoto, N. Fujii, T. Ibuka, *J. Chem. Soc., Perkin Trans. 1* **1999**, 2949–2962.
35 H. Ohno, *Yakugaku Zasshi* **2001**, *121*, 733–741.
36 (a) H. Ohno, A. Toda, Y. Miwa, T. Taga, E. Osawa, Y. Yamaoka, N. Fujii, T. Ibuka, *J. Org. Chem.* **1999**, *64*, 2992–2993; (b) H. Ohno, M. Anzai, A. Toda, S. Ohishi, N. Fujii, T. Tanaka, Y. Takemoto, T. Ibuka, *J. Org. Chem.* **2001**, *66*, 4904–4914.
37 (a) H. Ohno, H. Hamaguchi, M. Ohata, T. Tanaka, *Angew. Chem., Int. Ed.* **2003**, *42*, 1749–1753; (b) H. Ohno, H. Hamaguchi, M. Ohata, S. Kosaka, T. Tanaka, *J. Am. Chem. Soc.* **2004**, *126*, 8744–8754.
38 L. Ferrero, M. Rouillard, M. Decouzon, M. Azzaro, *Tetrahedron Lett.* **1974**, *15*, 131.
39 R. Chaabouni, A. Laurent, *Synthesis* **1975**, 464–467.
40 R. Chaabouni, A. Laurent, *Tetrahedron Lett.* **1976**, *17*, 757–758.
41 R. Chaabouni, A. Laurent, B. Marquet, *Tetrahedron* **1980**, *36*, 877–885.
42 (a) A. Satake, I. Shimizu, A. Yamamoto, *Synlett* **1995**, 64–68; (b) H. Ohno, N. Mimura, A. Otaka, H. Tamamura, N. Fujii, T. Ibuka, I. Shimizu, A. Satake, Y. Yamamoto, *Tetrahedron* **1997**, *53*, 12933–12946.
43 (a) T. Ibuka, M. Akaji, N. Mimura, H. Habashita, K. Nakai, H. Tamamura, N. Fujii, *Tetrahedron Lett.* **1996**, *37*, 2849–2852; (b) T. Ibuka, N. Mimura, H. Ohno, K. Nakai, M. Akaji, H. Habashita, H. Tamamura, Y. Miwa, T. Taga, N. Fujii, Y. Yamamoto, *J. Org. Chem.* **1997**, *62*, 2982–2991.
44 (a) T. Ibuka, K. Nakai, H. Habashita, Y. Hotta, N. Fujii, N. Mimura, Y. Miwa, T. Taga, Y. Yamamoto, *Angew. Chem., Int. Ed. Engl.* **1994**, *33*, 652–654; (b) N. Fujii, K. Nakai, H. Tamamura, A. Otaka, N. Mimura, Y. Miwa, T. Taga, Y. Yamamoto, T. Ibuka, *J. Chem. Soc., Perkin Trans. 1* **1995**, 1359–1371.
45 P. Wipf, P. C. Fritch, *J. Org. Chem.* **1994**, *59*, 4875–4886; (b) P. Wipf, T. C. Henninger, *J. Org. Chem.* **1997**, *62*, 1586–1587; (c) P. Wipf, T. C. Henninger, S. J. Geib, *J. Org. Chem.* **1998**, *63*, 6088–6089.

46 T. Hudlicky, X. Tian, K. Königsberger, J. Rouden, *J. Org. Chem.* **1994**, *59*, 4037–4039.

47 T. Hudlicky, X. Tian, K. Königsberger, R. Maurya, J. Rouden, B. Fan, *J. Am. Chem. Soc.* **1996**, *118*, 10752–10765.

48 (a) H. Aoyama, N. Mimura, H. Ohno, K. Ishii, A. Toda, H. Tamamura, A. Otaka, N. Fujii, T. Ibuka, *Tetrahedron Lett.* **1997**, *38*, 7383–7386; (b) A. Toda, H. Aoyama, N. Mimura, H. Ohno, N. Fujii, T. Ibuka, *J. Org. Chem.* **1998**, *63*, 7053–7061.

49 F. A. Davis, G. V. Reddy, *Tetrahedron Lett.* **1996**, *37*, 4349–4352.

50 (a) H. Tamamura, M. Yamashita, H. Muramatsu, H. Ohno, T. Ibuka, A. Otaka, N. Fujii, *Chem. Commun.* **1997**, 2327–2328; (b) H. Tamamura, M. Yamashita, Y. Nakajima, K. Sakano, A. Otaka, H. Ohno, T. Ibuka, N. Fujii, *J. Chem. Soc., Perkin Trans. 1*, **1999**, 2983–2996.

51 S. Oishi, H. Tamamura, M. Yamashita, Y. Odagaki, N. Hamanaka, A. Otaka, N. Fujii, *J. Chem. Soc., Perkin Trans. 1* **2001**, 2445–2451.

52 R. S. Atkinson, W. T. Gattrell, A. P. Ayscough, T. M. Raynham, *Chem. Commun.* **1996**, 1935.

53 R. S. Atkinson, A. P. Ayscough, W. T. Gattrell, T. M. Raynham, *Tetrahedron Lett.* **1998**, *39*, 4377–4380.

54 (a) B. Olofsson, U. Khamrai, P. Somfai, *Org. Lett.* **2000**, *2*, 4087–4089; (b) B. Olofsson, P. Somfai, *J. Org. Chem.* **2002**, *67*, 8574–8583.

55 B. Olofsson, P. Somfai, *J. Org. Chem.* **2003**, *68*, 2514–2517.

56 C. S. Penkett, I. D. Simpson, *Tetrahedron Lett.* **2001**, *42*, 3029–3032.

57 W.-P. Deng, A.-H. Li, L.-X. Dai, X.-L. Hou, *Tetrahedron* **2000**, *56*, 2967–2974.

58 K. Fugami, K. Oshima, K. Utimoto, H. Nozaki, *Bull. Chem. Soc. Jpn.* **1987**, *60*, 2509–2515.

59 W. Berts, K. Luthman, *Tetrahedron* **1999**, *55*, 13819–13830.

60 G. Righi, C. Potini, P. Bovicelli, *Tetrahedron Lett.* **2002**, *43*, 5867–5869.

61 E. L. Stogryn, S. J. Brois, *J. Org. Chem.* **1965**, *30*, 88–91.

62 J. C. Pommelet, J. Chuche, *Tetrahedron Lett.* **1974**, *44*, 3897–3898.

63 N. Manisse, J. Chuche, *J. Am. Chem. Soc.* **1977**, *99*, 1272–1273.

64 N. Manisse, J. Chuche, *Tetrahedron* **1977**, *33*, 2399–2406.

65 (a) E. L. Stogryn, S. J. Brois, *J. Am. Chem. Soc.* **1967**, *89*, 605–609; a related reaction of [2-(aziridin-1-yl)alkenyl]triphenylphosphonium bromides was also reported; see (b) M. A. Calcagno, E. E. Schweizer, *J. Org. Chem.* **1978**, *43*, 4207–4215; a related azepine formation through addition of *N*-unsubstituted azidirines to electrophilic carbon-carbon multiple bond systems such as acrylonitrile followed by aza-[3,3]-Claisen rearrangement was reported by Hassner: (c) A. Hassner, R. D'Costa, A. T. McPhail, W. Butler, *Tetrahedron Lett.* **1981**, *22*, 3691–3694; (d) A. Hassner, W. Chau, *Tetrahedron Lett.* **1982**, *23*, 1989–1992.

66 P. Scheiner, *J. Org. Chem.* **1967**, *32*, 2628–2630.

67 H. P. Figeys, R. Jammar, *Tetrahedron Lett.* **1980**, *21*, 2995–2998.

68 H. P. Figeys, R. Jammar, *Tetrahedron Lett.* **1981**, *22*, 637–640.

69 L. Viallon, O. Reinaud, P. Capdevielle, M. Maumy, *Tetrahedron* **1996**, *52*, 13605–13614.

70 (a) U. M. Lindström, P. Somfai, *J. Am. Chem. Soc.* **1997**, *119*, 8385–8386; (b) U. M. Lindström, P. Somfai, *Chem. Eur. J.* **2001**, *7*, 94–98.

71 P. G. Mente, H. W. Heine, G. R. Scharoubim, *J. Org. Chem.* **1968**, *33*, 4547–4548.

72 P. G. Mente, H. W. Heine, *J. Org. Chem.* **1971**, *36*, 3076–3078.

73 H. W. Whitlock, Jr., G. L. Smith, *Tetrahedron Lett.* **1965**, *6*, 1389–1393.

74 M. M. Ito, Y. Nomura, Y. Takeuchi, S. Tomoda, *Chem. Lett.* **1981**, 1519–1522.

75 (a) R. S. Atkinson, C. W. Rees, *Chem. Commun.* **1967**, 1232–1232; (b) R. S. Atkinson, C. W. Rees, *J. Chem. Soc. (C)* **1969**, 778–782.

76 A. G. Hortmann, J.-Y. Koo, *J. Org. Chem.* **1974**, *39*, 3781–3783.

77 (a) W. H. Pearson, *Tetrahedron Lett.* **1985**, *26*, 3527–3530; (b) W. H. Pearson, S. C. Bergmeier, S. Degan, K.-C. Lin, Y.-F. Poon, J. M. Schkeryantz, *J. Org. Chem.* **1990**, *55*, 5719–5738.

78 (a) K. Fugami, Y. Morizawa, K. Oshima, H. Nozaki, *Tetrahedron Lett.* **1985**, *26*, 857–860; (b) K. Fugami, K. Miura, Y.

Morizawa, K. Oshima, K. Utimoto, H. Nozaki, *Tetrahedron* **1989**, *45*, 3089–3098.

79 P. G. Mente, H. W. Heine, *J. Org. Chem.* **1971**, *36*, 3076–3078.

80 (a) J. C. Pommelet, J. Chuche, *Can. J. Chem.* **1976**, *54*, 1571–1581; (b) D. Borel, Y. Gelas-Mialhe, R. Vessière, *Can. J. Chem.* **1976**, *54*, 1590–1598.

81 (a) T. Hudlicky, J. O. Frazier, G. Seoane, M. Tiedje, A. Seoane, L. D. Kwart, C. Beal, *J. Am. Chem. Soc.* **1986**, *108*, 3755–3762; (b) T. Hudlicky, G. Sinai-Zingde, G. Seoane, *Syn. Commun.* **1987**, *17*, 1155–1163; (c) T. Hudlicky, G. Seoane, T. C. Lovelace, *J. Org. Chem.* **1988**, *53*, 2094–2099; (d) T. Hudlicky, H. Luna, J. D. Price, F. Rulin, *J. Org. Chem.* **1990**, *55*, 4683–4387.

82 (a) J. Åhman, P. Somfai, *J. Am. Chem. Soc.* **1994**, *116*, 9781–9782; (b) J. Åhman, T. Jarevång, P. Somfai, *J. Org. Chem.* **1996**, *61*, 8148–8159.

83 J. Åhman, P. Somfai, *Tetrahedron Lett.* **1996**, *37*, 2495–2498.

84 (a) J. Åhman P. Somfai, *Tetrahedron Lett.* **1995**, *36*, 303–306; (b) J. Åhman, P. Somfai, *Tetrahedron* **1995**, *51*, 9747–9756; (c) P. Somfai, T. Jarevång, U. M. Lindström, A. Svensson, *Acta Chem. Scand.* **1997**, *51*, 1024–1029.

85 (a) I. Coldham, A. J. Collis, R. J. Mould, R. E. Rathmell, *Tetrahedron Lett.* **1995**, *36*, 3557–3560; (b) I. Coldham, A. J. Collis, R. J. Mould, R. E. Rathmell, *J. Chem. Soc., Perkin Trans. 1* **1995**, 2739–2745.

86 G. J. Rowlands, W. K. Barnes, *Tetrahedron Lett.* **2004**, *45*, 5347–5350.

87 T. L. Gilchrist, C. W. Rees, E. Stanton, *J. Chem. Soc. (C)* **1971**, 3036–3040.

88 W. Eberbach, J. C. Carré, *Chem. Ber.* **1983**, *116*, 563–586.

89 (a) J. Åhman, P. Somfai, D. Tanner, *J. Chem. Soc., Chem. Commun.* **1994**, 2785–2786; (b) J. Åhman, P. Somfai, *Tetrahedron Lett.* **1995**, *36*, 1953–1956; (c) J. Åhman, P. Somfai, *Tetrahedron* **1999**, *55*, 11 595–11600.

90 A. Honda, H. Ohno, N. Mimura, T. Ibuka, *Synlett* **1998**, 969–970.

91 D. C. D. Butler, G. A. Inman, H. Alper, *J. Org. Chem.* **2000**, *65*, 5887–5890.

92 B. M. Trost, D. R. Fandrick, *J. Am. Chem. Soc.* **2003**, *125*, 11 836–11837.

93 C. Dong, H. Alper, *Tetrahedron: Asymmetry* **2004**, *15*, 1537–1540.

94 K. Aoyagi, H. Nakamura, Y. Yamamoto, *J. Org. Chem.* **2002**, *67*, 5977–5980.

95 R. Aumann, K. Fröhlich, H. Ring, *Angew. Chem., Int. Ed.* **1974**, *13*, 275–276.

96 S. V. Ley, B. Middleton, *Chem. Commun.* **1998**, 1995–1996.

97 G. A. Molander, P. J. Stengel, *Tetrahedron* **1997**, *53*, 8887–8912.

98 J. S. Yadav, A. Bandyopadhyay, B. V. S. Reddy, *Synlett* **2001**, 1608–1610.

99 (a) Y. Takemoto, M. Anzai, R. Yanada, N. Fujii, H. Ohno, T. Ibuka, *Tetrahedron Lett.* **2001**, *42*, 1725–1728; (b) M. Anzai, R. Yanada, N. Fujii, H. Ohno, T. Ibuka, Y. Takemoto, *Tetrahedron* **2002**, *58*, 5231–5239.

100 (a) H. Ohno, H. Hamaguchi, T. Tanaka, *Org. Lett.* **2000**, *2*, 2161–2163; (b) H. Ohno, H. Hamaguchi, T. Tanaka, *J. Org. Chem.* **2001**, *66*, 1867–1875.

101 (a) H. Ohno, A. Toda, Y. Miwa, T. Taga, N. Fujii, T. Ibuka, *Tetrahedron Lett.* **1999**, *40*, 349–352; (b) H. Ohno, A. Toda, N. Fujii, Y. Takemoto, T. Tanaka, T. Ibuka, *Tetrahedron* **2000**, *56*, 2811–2820.

102 For synthesis through the reactions between *N*-tosylimines and sulfonium ylides, see: (a) A.-H. Li, Y.-G. Zhou, L.-X. Dai, X.-L. Hou, L.-J. Xia, L. Lin, *Angew. Chem., Int. Ed. Engl.* **1997**, *36*, 1317–1319; (b) D.-K. Wang, Li.-X. Dai, X.-L. Hou, *Chem. Commun.* **1997**, 1231–1232; (c) A.-H. Li, Y.-G. Zhou, L.-X. Dai, X.-L. Hou, L.-J. Xia, L. Lin, *J. Org. Chem.* **1998**, *63*, 4338–4348.

103 For synthesis by dehydrobromination of 2-(1-bromovinyl)aziridines, see: Ref. [34].

104 For synthesis by treatment of imines with nucleophiles, see: (a) F. Chemla, V. Hebbe, J. F. Normant, *Tetrahedron Lett.* **1999**, *40*, 8093–8096; (b) S. Florio, L. Troisi, V. Capriati, G. Suppa, *Eur. J. Org. Chem.* **2000**, 3793–3797; (c) F. Chemla, F. Ferreira, *Synlett* **2004**, 983–986.

105 For synthesis by intramolecular amination of bromoallenes, see: (a) H. Ohno, H. Hamaguchi, T. Tanaka, *Org. Lett.* **2001**, *3*, 2269–2271; (b) H. Ohno, K. Ando, H. Hamaguchi, Y. Takeoka, T. Tanaka, *J. Am. Chem. Soc.* **2002**, *124*, 15 255–15266.

3 Asymmetric Syntheses with Aziridinecarboxylate and Aziridinephosphonate Building Blocks

Ping Zhou, Bang-Chi Chen, and Franklin A. Davis

3.1 Introduction

Chiral nonracemic *trans*- and *cis*-aziridine-2-carboxylates **1** (Scheme 3.1) are special members of the vast and diverse family of aziridines. These three-membered nitrogen heterocyclic molecules are considered amino acids, not only because of the ease with which they are converted into acyclic α- and β-amino acids, but also because of the pronounced biological activities resulting from their incorporation into peptides as special amino acid units. Synthetically, aziridines have been widely used as versatile chiral building blocks for the synthesis of a variety of biologically and pharmaceutically important molecules. In addition, naturally occurring aziridine-2-carboxylic acid has also been reported. More recently, methods for the syntheses of analogous chiral nonracemic *trans*- and *cis*-aziridine-2-phosphonates **2** have been developed, and their uses as chiral building blocks have also emerged.

The chemistry of aziridine-2-carboxylates and phosphonates has been discussed in part in several reviews covering the literature through 1999 [1–3]. This chapter is intended to give an overview of asymmetric syntheses using chiral nonracemic aziridine-2-carboxylates and -phosphonates with particular emphasis on their applications as chiral building blocks in asymmetric synthesis since 2000. Some overlap with earlier reviews is necessary for the sake of continuity.

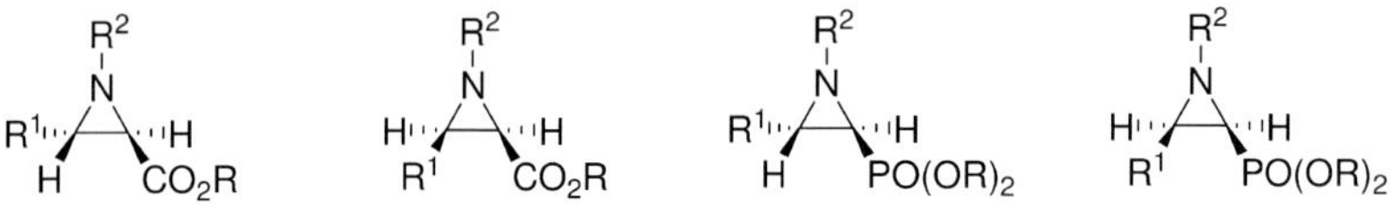

R^2, N, R^1, H, H, CO_2R — *trans*-**1**
R^2, N, H, H, R^1, CO_2R — *cis*-**1**
R^2, N, R^1, H, H, $PO(OR)_2$ — *trans*-**2**
R^2, N, H, H, R^1, $PO(OR)_2$ — *cis*-**2**

Scheme 3.1

Aziridines and Epoxides in Organic Synthesis. Andrei K. Yudin

ISBN: 3-527-31213-7

3.2
Preparation of Aziridine-2-carboxylates and Aziridine-2-phosphonates

3.2.1
Preparation of Aziridine-2-carboxylates

A variety of methods for the asymmetric syntheses of aziridine-2-carboxylates have been developed. They can be generally classified into eight categories based on the key ring-forming transformation and starting materials employed: (*i*) cyclization of hydroxy amino esters, (*ii*) cyclization of hydroxy azido esters, (*iii*) cyclization of α-halo- and α-sulfonyloxy-β-amino esters, (*iv*) aziridination of α,β-unsaturated esters, (*v*) aziridination of imines, (*vi*) aziridination of aldehydes, (*vii*) 2-carboxylation of aziridines, and (*viii*) resolution of racemic aziridine-2-carboxylates.

3.2.1.1 Cyclization of Hydroxy Amino Esters

The earliest method developed for the preparation of nonracemic aziridine-2-carboxylates was the cyclization of naturally occurring β-hydroxy-α-amino acid derivatives (serine or threonine) [4]. The β-hydroxy group is normally activated as a tosyl or mesyl group, which is ideal for an intramolecular S_N2 displacement. The cyclization has been developed in both one-pot and stepwise fashion [4–9]. As an example, serine ester **3** (Scheme 3.2) was treated with tosyl chloride in the presence of triethylamine to afford aziridine-2-carboxylate **4** in 71% yield [9]. Cyclization of α-hydroxy-β-amino esters to aziridine-2-carboxylates under similar conditions has also been described [10].

The aziridine ring-formation is sometimes sensitive to the *N*-protecting group, as well as to other factors. Treatment of **5** (Scheme 3.3) with mesyl chloride (MsCl) and DMAP, for example, afforded bis-mesylate **6** in 97% yield, but attempted cyclization by use of Et_3N in THF failed to give the desired aziridine **7** [11]. Use of a combination of $LiClO_4/Et_3N$ in THF gave rise to the desired product **7** in 71% yield along with a small amount of azetidine **8** in a ratio of 25:1. Use of $LiClO_4$/*s*-collidine in dioxane afforded exclusively **7** in 91% isolated yield.

Aziridine-2-carboxylates **12** (Scheme 3.4) have also been prepared from β-hydroxy-α-amino esters **9** by treatment with sulfuryl chloride in place of tosyl or mesyl chloride. Treatment of **9** with thionyl chloride in the presence of triethylamine, followed by oxidation of **10** with sodium periodate and a catalytic amount of

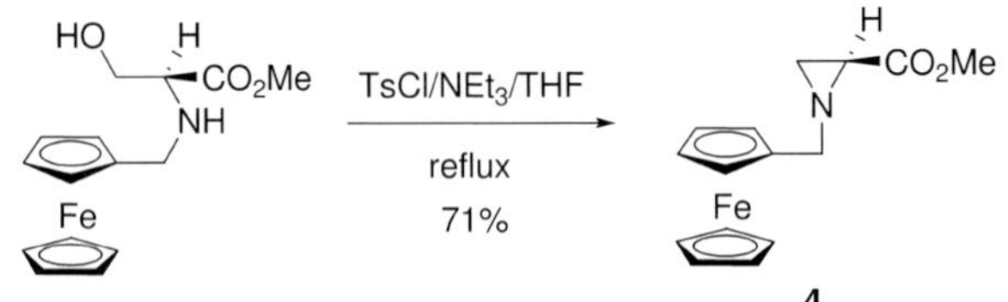

HO H CO$_2$Me NH Fe **3** — TsCl/NEt$_3$/THF, reflux, 71% → H CO$_2$Me N Fe **4**

Scheme 3.2

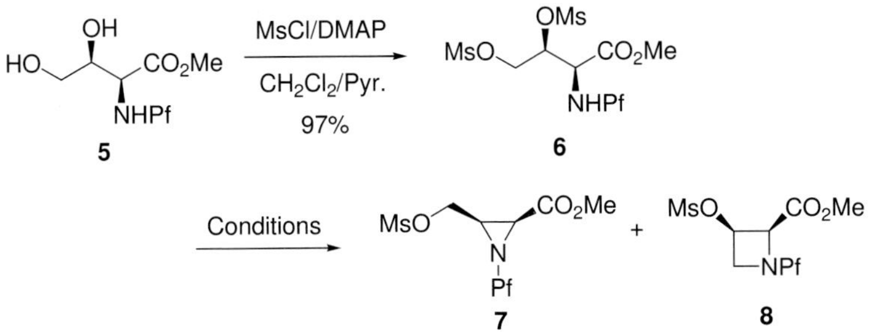

Conditions	Yield (%) **7**	ratio **7/8**
Et_3N/THF, reflux	no reaction	--
$LiClO_4/Et_3N$/THF, reflux	71	25/1
$LiClO_4$/s-collidine/dioxane, 75°C	91	100/0

Scheme 3.3

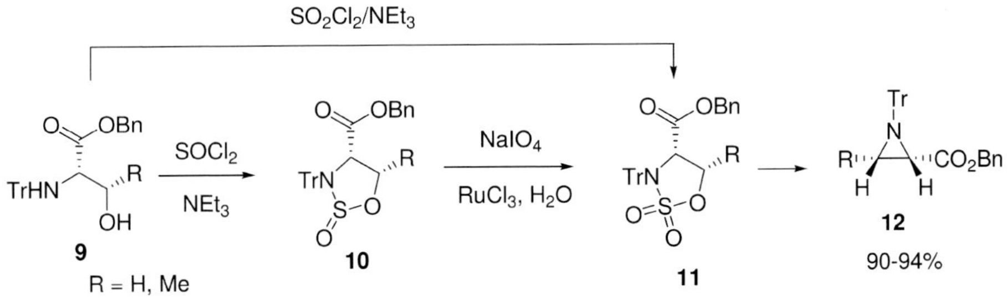

Scheme 3.4

ruthenium(III) chloride, afforded aziridine **12**. The same compound was obtained on treatment of **9** with sulfuryl chloride and excess triethylamine. Both reactions were believed to proceed through the intermediate **11**, although it was not detected [12].

Cyclization of β-hydroxy-α-amino esters under Mitsunobu reaction conditions is an alternative approach to aziridine-2-carboxylic esters [6b, 13–16]. In this case the β-hydroxy group is activated by a phosphorus reagent. Treatment of Boc-α-Me-D-Ser-OMe **13** (Scheme 3.5) with triphenylphosphine and diethyl azodicarboxylate (DEAD), for example, gave α-methyl aziridinecarboxylic acid methyl ester **14** in 85% yield [15]. In addition to PPh_3/DEAD [13b, 15], several other reagent combi-

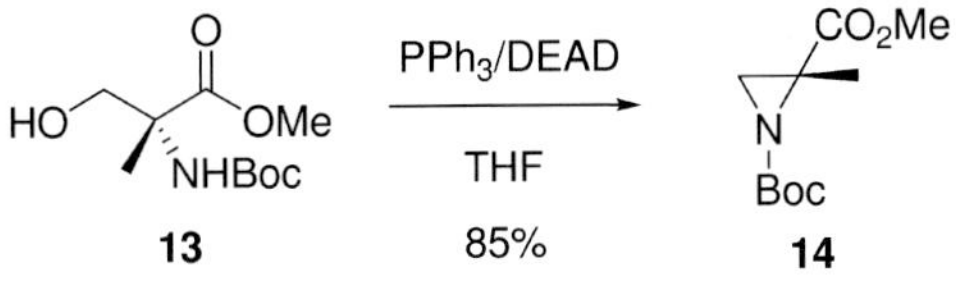

Scheme 3.5

nations have also been used, including $PPh_3/CCl_4/NEt_3$ [13a], $PPh_3/CBr_4/NEt_3$ [13a], PPh_3/DIAD [14a], and $Ph_3P(OEt)_2$ [16].

3.2.1.2 Cyclization of Hydroxy Azido Esters

A general method for the synthesis of *N*-unsubstituted aziridine-2-carboxylates involves a triphenylphosphine-mediated reductive cyclization of hydroxy azido esters [17–22]. A recent example involves the treatment of β-hydroxy-α-azido ester **15** (Scheme 3.6) with PPh_3 to give aziridine **16** in 90% yield [19]. α-Hydroxy-β-azido esters undergo similar reactions to give aziridine-2-carboxylates [20–22].

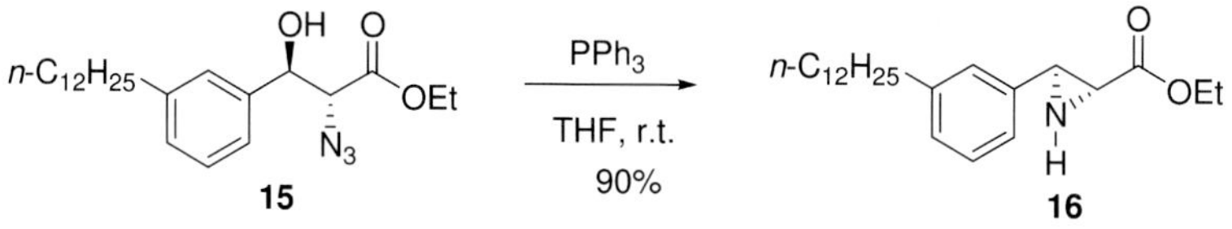

15 → (PPh_3, THF, r.t., 90%) → **16**

Scheme 3.6

3.2.1.3 Cyclization of α-Halo- and α-Sulfonyloxy-β-amino Esters and Amides

Michael addition of amines to α-halo-α,β-unsaturated esters and derivatives gives rise to α-halo-β-amino esters, which often undergo spontaneous cyclization to aziridine-2-carboxylates [23–28]. The diastereoselectivity of this process is not usually satisfactory, but high asymmetric induction could be achieved through careful choice of chiral directing group and reaction conditions. Treatment of α-bromo acrylamide **17** (Scheme 3.7) with benzylamine, for example, produced aziridine **20** in 60–86% yield and with 100% de [26]. The key stereodifferentiating step involved a face-selective protonation of enolate **18** to give α-bromo-β-amino amides **19**, which then cyclized under the reaction conditions to give aziridines **20**.

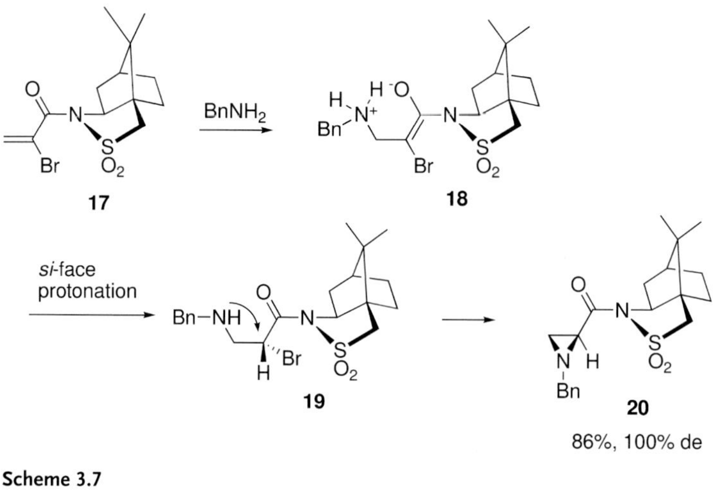

17 → ($BnNH_2$) → **18** → (*si*-face protonation) → **19** → **20** (86%, 100% de)

Scheme 3.7

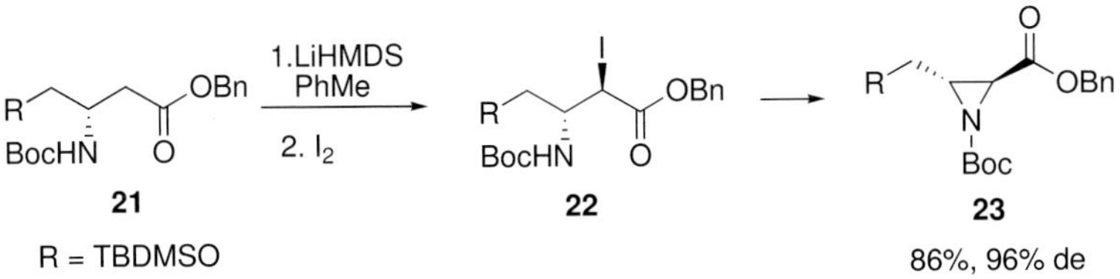

Scheme 3.8

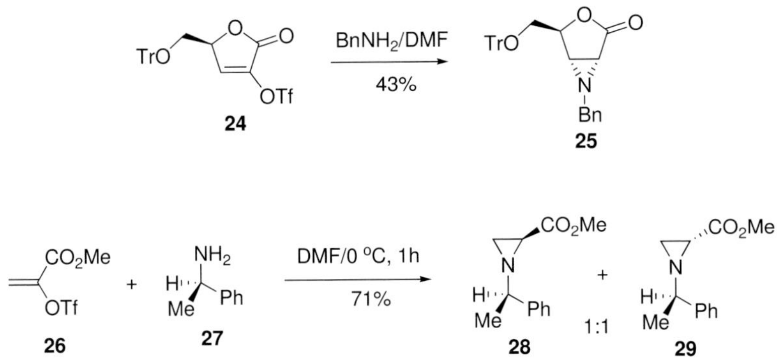

Scheme 3.9

Amines also react with α,β-dibromo esters to give aziridine-2-carboxylates [23–25, 29–30]. Here it has been suggested that initial formation of an α-bromo-α,β-unsaturated ester through base-promoted elimination, followed by Michael addition of the amine, gives rise to an α-bromo-β-amino ester, followed by aziridine ring-formation [24]. Like the direct aziridination of α-bromo-α,β-unsaturated esters, this approach also suffers from generally low diastereoselectivity. The β-amino-α-halo carboxylate intermediates involved in this approach to aziridine-2-carboxlate could alternatively be generated by halogenation of β-amino esters: treatment of β-amino ester **21** (Scheme 3.8) with LiHMDS in toluene, followed by addition of iodine, for example, afforded aziridine **23** in 86% yield and 96% de via the α-iodo intermediate **22** [31].

More recently, Dodd and co-workers reported that treatment of triflate **24** (Scheme 3.9) with benzylamine in DMF afforded aziridine lactone **25** in 43% yield [32]. Treatment of vinyl triflate **26** with (*R*)-phenylethylamine **27** gave a mixture of diastereoisomers **28** and **29** in 71% yield and a ratio of 1:1 [33].

3.2.1.4 Aziridination of α,β-unsaturated Esters

Asymmetric aziridination of α,β-unsaturated esters by use of *N*-nitrenes was studied in great detail by Atkinson and co-workers [34, 35]. Here, lead tetraacetate-mediated oxidative addition of *N*-aminoquinazolone **30** (Scheme 3.10) to α-methylene-γ-butyrolactone **32** was reported to proceed with complete asymmetric induc-

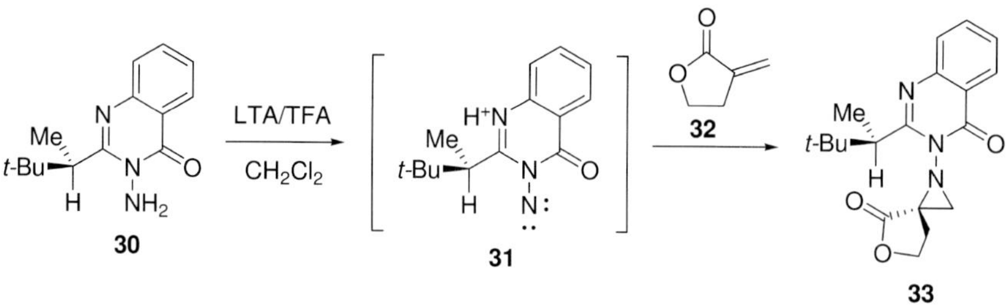

Scheme 3.10

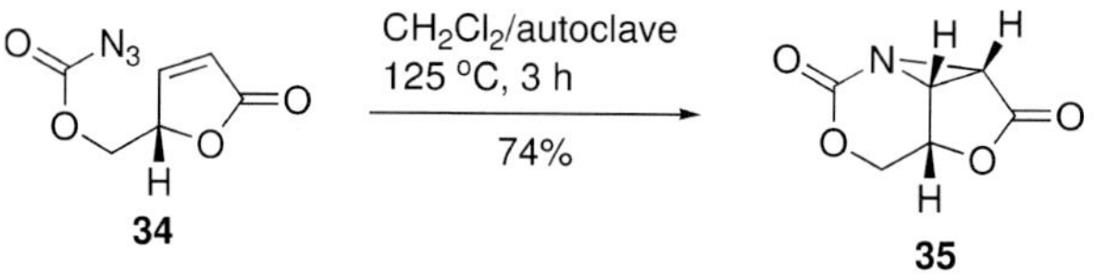

Scheme 3.11

tion in the presence of trifluoroacetic acid. The only aziridine stereoisomer obtained was shown by X-ray crystallography to be compound **33**. It should be pointed out that evidence against the involvement of *N*-nitrene **31** has also been reported [36]. Use of the nonracemic oxidant lead tetra-(*S*)-2-methylbutanoate [37] and chiral camphor-based auxiliaries [38] in this type of transformation has also been described.

Thermolysis of acyl azides and subsequent cyclization to give aziridines has been reported by Egli and Dreiding [39]. Heating of acylazide **34** (Scheme 3.11) in an autoclave afforded aziridine **35** in 74% yield [39].

More recently, Tardella and co-workers reported that treatment of 2-trifluoromethyl acrylate **36** (Scheme 3.12) with the anion generated from nosyloxycarbamate **37** gave rise to aziridine-2-carboxylate **38** in 96% yield and 72% de with undetermined stereochemistry [40]. Aza-MIRC (Michael-initiated ring closure) was used to account for this transformation. A number of other hydroxylamine derivatives have been employed successfully in this type of aziridination reaction, includ-

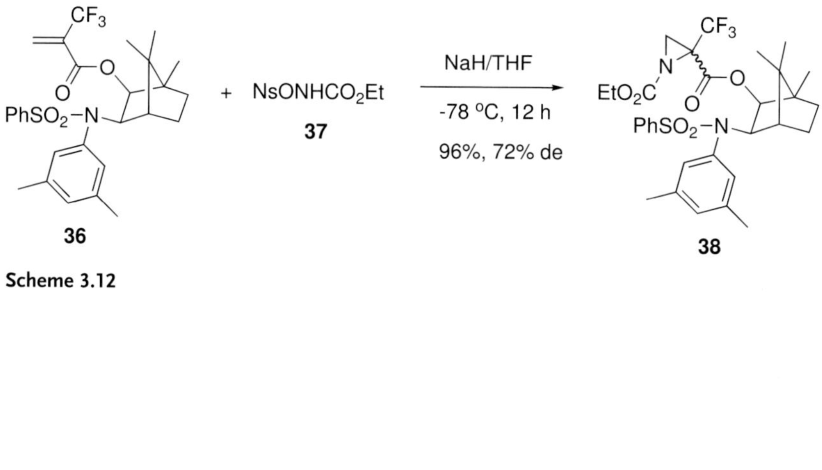

Scheme 3.12

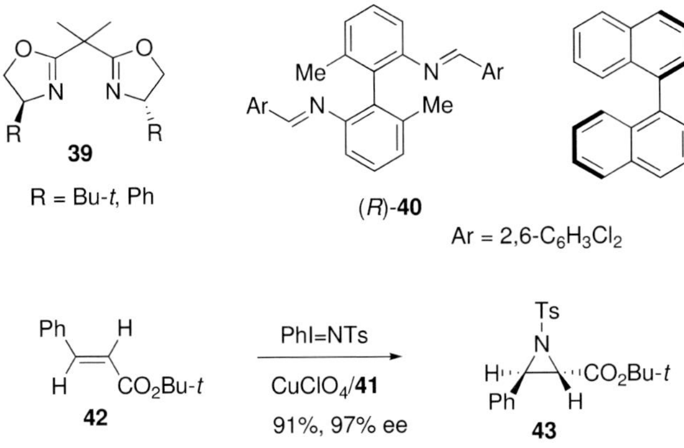

39

R = Bu-*t*, Ph

(*R*)-40

41

Ar = 2,6-$C_6H_3Cl_2$

42 → 43: PhI=NTs, $CuClO_4$/41, 91%, 97% ee

Scheme 3.13

ing *N,O*-bistrimethylsilylhydroxyamine [41–43], *N*-arylhydroxamates [44], and *N*-benzylhydroxyamine [45]. The stereoselectivity, however, is generally poor to modest.

A much more versatile and useful asymmetric synthesis of aziridine-2-carboxylates involves reactions between α,β-unsaturated esters and [*N*-(*p*-toluenesulfonyl)imino]phenyliodinane (PhI=NTs) in the presence of catalytic amounts of Cu(I) salts, such as Cu(OTf) and $CuClO_4$, together with a chiral ligand. A number of chiral ligands have been developed and studied for this purpose [46–54]: some examples include bis(oxazoline) **39** (Scheme 3.13) [46, 47], biphenylimine **40** [48, 49], and binaphthylimine **41** [50]. Treatment of **42** with PhI=NTs in the presence of $CuClO_4$ and ligand **41**, for example, gave **43** in 91% yield and 97% *ee* [50].

3.2.1.5 Aziridination of Imines

Reactions between imines and α-diazo carboxylates afford aziridine-2-carboxylates [55]. An asymmetric version of this reaction using chiral nonracemic catalysts has been described [53, 56–58]. As an example, catalytic aziridination of imine **44** (Scheme 3.14) with ethyl diazoacetate in the presence of 10% catalyst generated

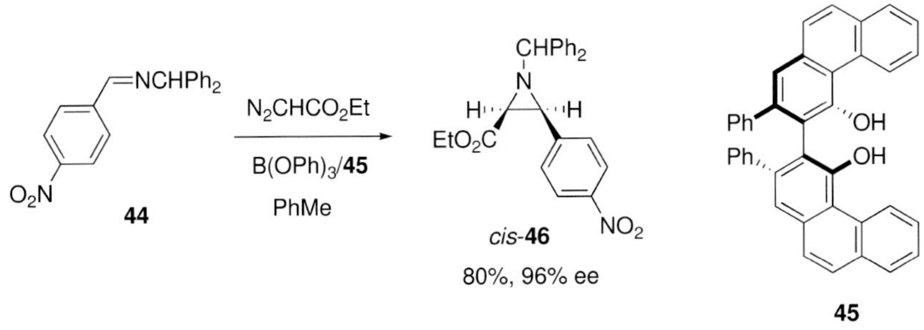

44 → *cis*-46: N_2CHCO_2Et, $B(OPh)_3$/45, PhMe; 80%, 96% ee

45

Scheme 3.14

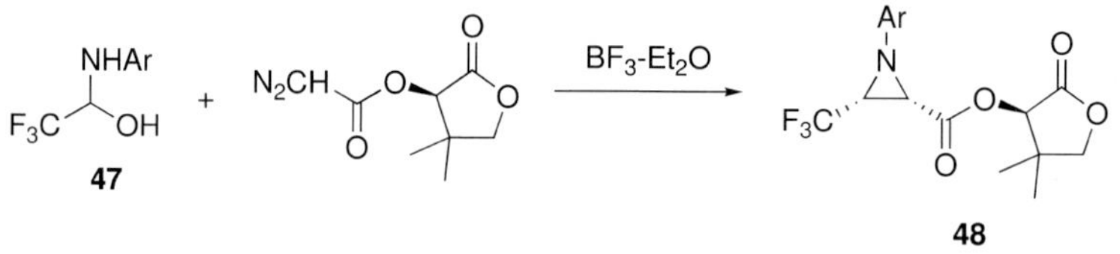

Ar	Yield (%)	cis/trans	de%
C_6H_4OMe-*p*	81	>95/<5	94
C_6H_4Br-*p*	83	>99/<1	92

Scheme 3.15

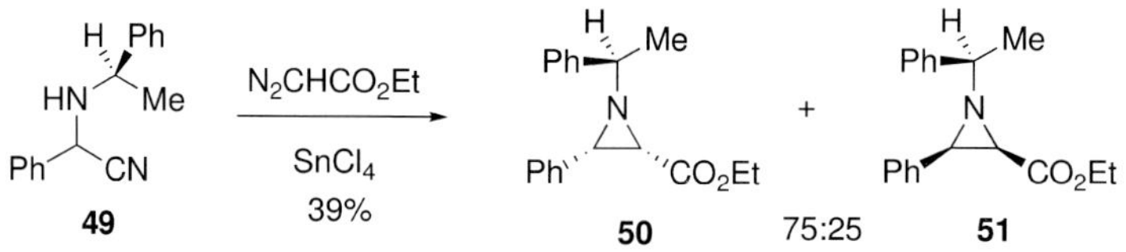

Scheme 3.16

from triphenylborate and ligand **45** afforded aziridine-2-carboxylate **46** in 80% yield [57]. This reaction is highly stereoselective, giving the *cis*-aziridine **46** in 96% *ee*.

The $BF_3 \cdot Et_2O$-catalyzed aziridination of compounds **47** (Scheme 3.15) with a diazo ester derived from (*R*)-pantolacetone gave aziridine-2-carboxylates **48** [59]. The reaction exhibited both high *cis* selectivity (>95:<5) and excellent diastereoselectivity. Treatment of α-amino nitrile **49** (Scheme 3.16) with ethyl diazoacetate in the presence of 0.5 equivalent of $SnCl_4$ afforded aziridines **50** and **51** in 39% yield in a ratio of 75:25 [60].

Darzens reactions between the chiral imine **52** and α-halo enolates **53** for the preparation of nonracemic aziridine-2-carboxylic esters **54** (Scheme 3.17) were studied by Fujisawa and co-workers [61]. It is interesting to note that the lithium enolate afforded (2*R*,3*S*)-aziridine (2*R*,3*S*)-**54** as the sole product, whereas the zinc enolate give rise to the isomer (2*S*,3*R*)-**54**. The α-halogen did not seem to affect the stereoselectivity.

More recently, Davis and co-workers developed a new method for the asymmetric syntheses of aziridine-2-carboxylates through the use of an aza-Darzens-type reaction between sulfinimines (*N*-sulfinyl imines) and α-haloenolates [62–66]. The reaction is highly efficient, affording *cis*-*N*-sulfinylaziridine-2-carboxylic esters in high yield and diastereoselectivity. This method has been used to prepare a variety of aziridines with diverse ring and nitrogen substituents. As an example, treatment of sulfinimine (S_S)-**55** (Scheme 3.18) with the lithium enolate of *tert*-butyl bromoacetate gave aziridine **56** in 82% isolated yield [66].

2*H*-Azirine-2-carboxylates such as **57** (Scheme 3.19) are a special class of imines that undergo additions to their C=N double bonds to give aziridine-2-carboxylates

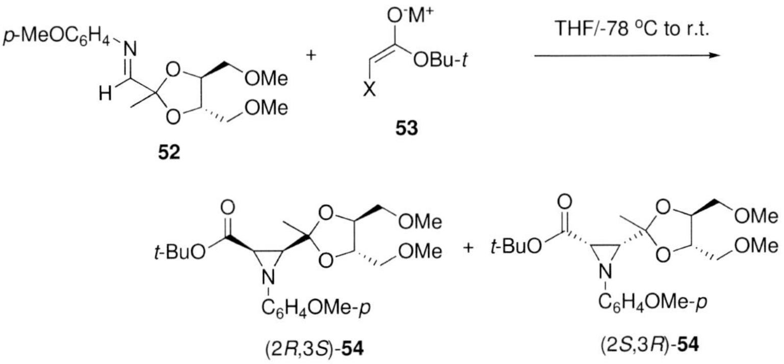

X	M	Enolate **53** Equiv.	Yield (%)	Ratio **54** (2*R*,3*S*)/(2*S*,3*R*)
Cl	Li	3	44	91/9
Br	Li	10	49	100/0
Cl	Zn	3	59	0/100
Br	Zn	3	59	0/100

Scheme 3.17

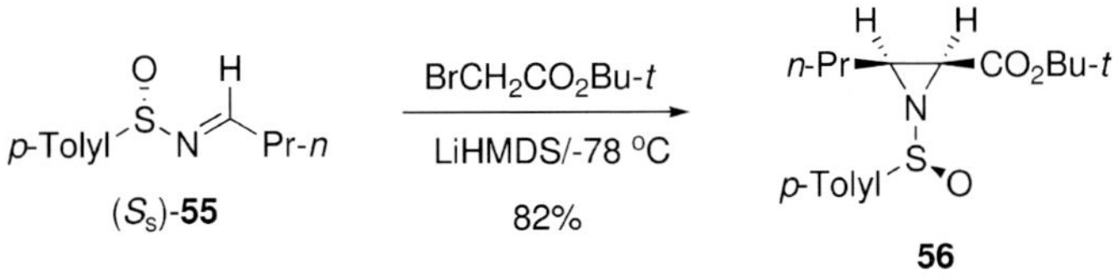

Scheme 3.18

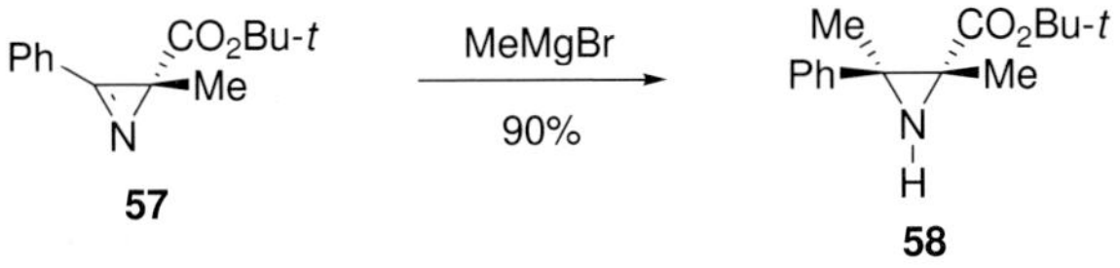

Scheme 3.19

[65–67]. Indeed, treatment of azirine **57** with MeMgBr afforded aziridine-2-carboxylate **58** in 90% yield [65].

Lewis acid-mediated asymmetric Diels-Alder reactions between 2*H*-azirines **59**, bearing chiral auxiliaries, with enophiles such as **60** afforded mixtures of bicyclic aziridine-2-carboxylates **61** (Scheme 3.20) [68]. 8-Phenylmenthol appeared to be the auxiliary of choice in this reaction in terms of yield and diastereoselectivity.

An enantioselective Diels-Alder reaction between the prochiral 2*H*-aziridine **62** and cyclopentadiene **63** (Scheme 3.21) has also been reported [68]. Through the use of (*S*)-BINOL **64** together with $AlMe_3$, aziridine-2-carboxylate **65** was obtained in 50% *ee* and 41% isolated yield [68].

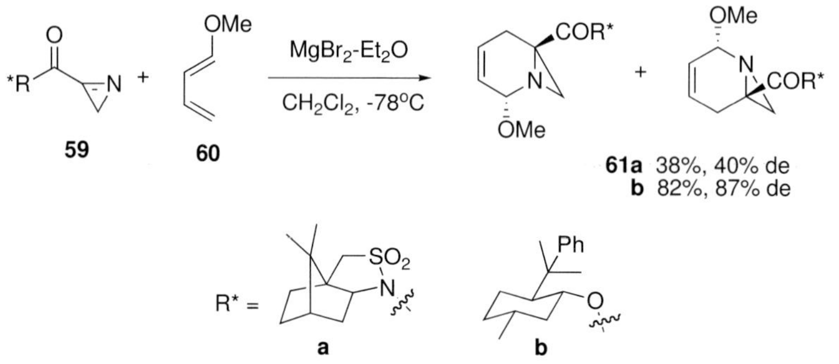

Scheme 3.20

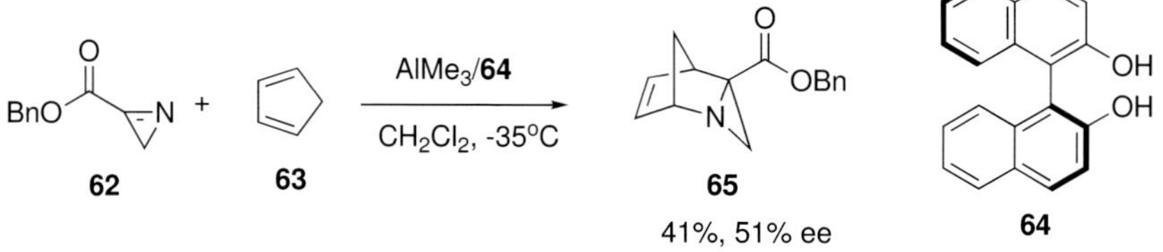

Scheme 3.21

3.2.1.6 Aziridination of Aldehydes

Ishikawa and co-workers reported that treatment of chiral nonracemic guanidinium bromide **66** (Scheme 3.22) with aldehyde **67**, in the presence of tetramethyl-

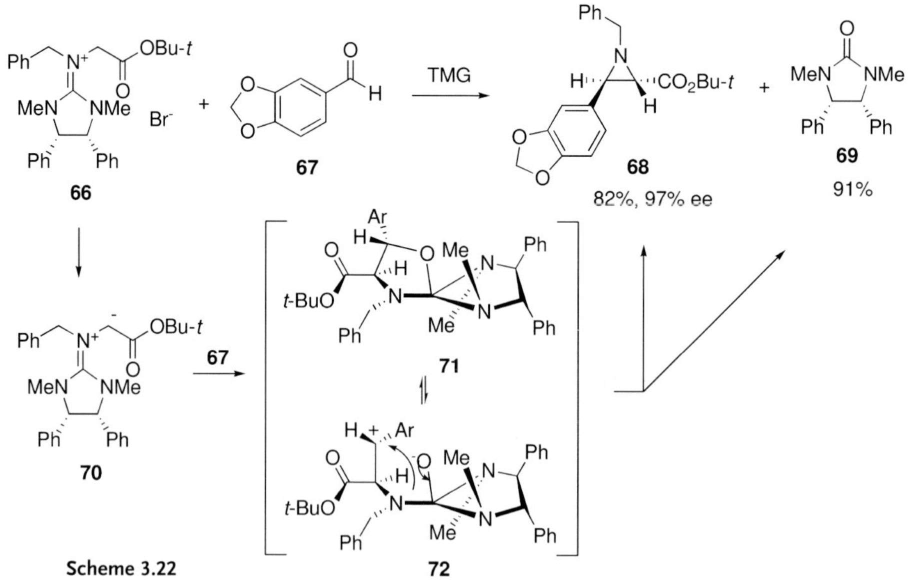

Scheme 3.22

guanidine as the base, resulted in the formation of *trans*-aziridine-2-carboxylate **68** in 82% yield and 97% *ee* [69]. The reaction was believed to proceed through the guanidinium ylide **70**, with the *re*-face of the ylide being attacked by the aldehyde function in the C–C bond formation to afford a spiro oxazolidine **71**. The enantioselectivity for the *trans*-aziridine can be explained by assuming minimized steric repulsion between the substituents of each component in the benzylic cation/hydroxide ion-pair intermediate **72** derived from **71**.

3.2.1.7 2-Carboxylation of Aziridines

Treatment of sulfinylaziridine **73** (Scheme 3.23) with MeMgBr and then with *tert*-butyllithium gave aziridinyllithium **74**, which reacted with ethyl chloroformate to afford aziridine-2-carboxylate **75** in 64% yield [70]. The reaction was stereospecific, giving **75** as a single diastereomer.

Recently, Uneyama reported that treatment of (*R*)-1-tosyl-2-trifluoromethylaziridine **76** (Scheme 3.24) with *n*-BuLi at –100 °C and subsequent trapping of the anion with electrophiles such as chloroformates produced aziridine-2-carboxylates **77** in good to excellent yields [71]. The retention of the configuration of the trifluoromethylated quaternary carbon center in the course of the reaction was confirmed by derivatization of the product and by X-ray studies.

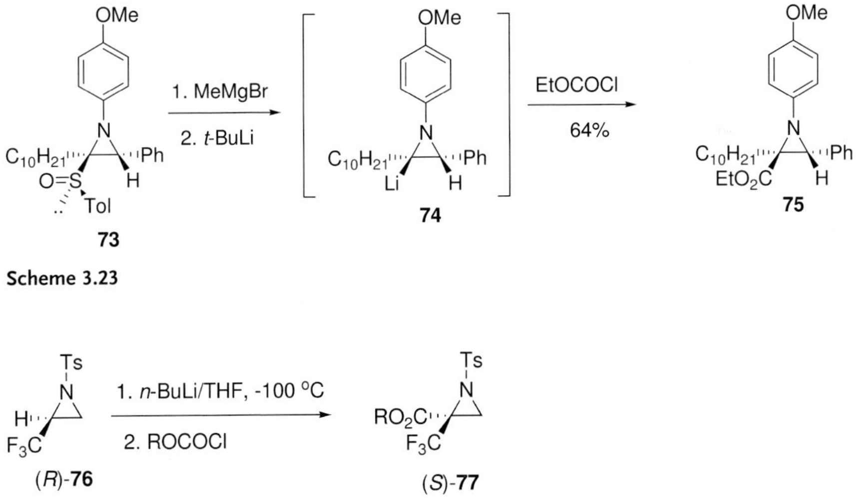

OMe N $C_{10}H_{21}$ Ph O=S H Tol **73** — 1. MeMgBr, 2. *t*-BuLi → [OMe N $C_{10}H_{21}$ Ph Li H **74**] — EtOCOCl, 64% → OMe N $C_{10}H_{21}$ Ph EtO_2C H **75**

Scheme 3.23

Ts N H F_3C (*R*)-**76** — 1. *n*-BuLi/THF, -100 °C, 2. ROCOCl → Ts N RO_2C F_3C (*S*)-**77**

R	Yield (%)
Me	85
Et	89
Bn	95

Scheme 3.24

3.2.1.8 Resolution of Racemic Aziridine-2-carboxylates

Several early reports dealt with the resolution of racemic aziridine-2-carboxylic acids [72, 73]. Treatment of (±)-**78** (Scheme 3.25) with (–)-*trans*-2,3-bis(hydroxydiphenylmethyl)-1,4-dioxaspiro[5.4]decane (**79**), for example, afforded the 1:1 ratio inclusion compound **80**. Upon distillation, the inclusion compound **80** gave enantiomerically pure (–)-**78** in 33% yield.

Kinetic resolution of the racemic aziridine-2-carboxylate **82** (Scheme 3.26) was reported by Iqbal and co-workers [74]. When **82** was allowed to react with *N*-cinnamoyl-L-proline (**81**) under mixed anhydride coupling conditions, the *N*-acyl aziridine **83** was obtained in optically pure form along with aziridine **84**.

Kinetic asymmetric amidation of the racemic *N*-methoxyaziridinedicarboxylic ester **85** (Scheme 3.27) with 0.5 equivalent of L-ephedrine (**86**) in absolute ethanol with a catalytic amount of EtONa afforded (–)-(*S*)-**85** in 59.2% yield [75]. The configurational stability of (–)-(*S*)-**85** was estimated to be $t_{1/2}$ = 323 years at 20 °C. On the other hand, (+)-(*S*)-**87** was obtained in 46% yield and 50% *ee* by the enzymatic hydrolysis of racemic **87** [76, 77]. It is interesting to note that the *N*-chloro aziridine (+)-(*S*)-**87** has a much shorter half life ($t_{1/2}$ = 32 h at 20 °C) than its *N*-methoxy counterpart (–)-(*S*)-**85**.

Kumar and co-workers reported that *N*-arylaziridine-2-carboxylates could be resolved with the aid of lipase from *Candida rugosa* [78]. Moderate to high enantiose-

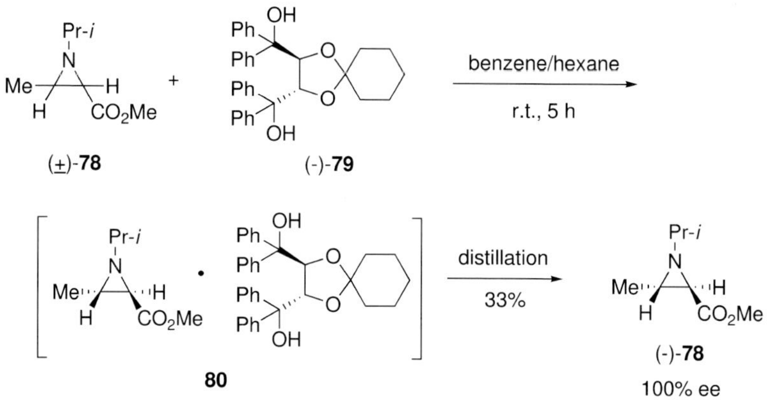

Scheme 3.25

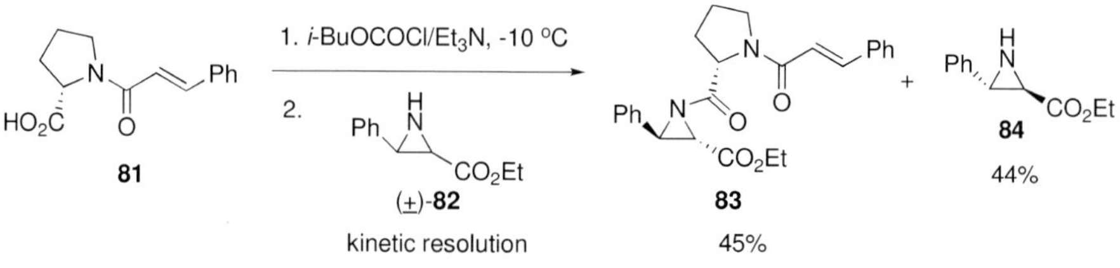

Scheme 3.26

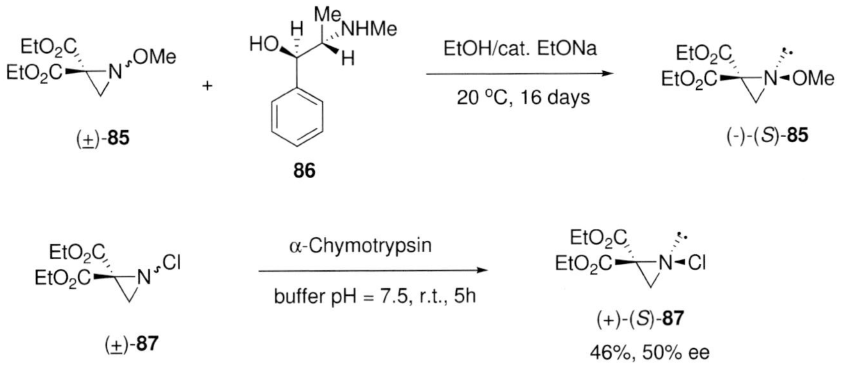

Scheme 3.27

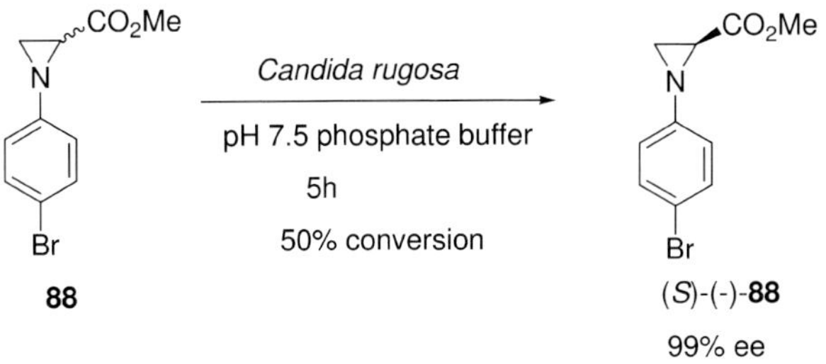

Scheme 3.28

lectivity was observed; treatment of racemic **88** (Scheme 3.28) with *Candida rugosa* for 5 hours, for example, afforded (*S*)-**88** in 99% *ee* at 50% conversion.

3.2.2
Preparation of Aziridine-2-phosphonates

Methodology for the cyclization of α-hydroxy-β-amino phosphonates has also been developed and employed in synthesis of aziridine-2-phosphonates [79, 80]. Mesylation of α-hydroxy-β-amino phosphonates **89** (Scheme 3.29), for example, gave α-mesyloxy-β-amino phosphonates **90**. Treatment of **90** with K_2CO_3 afforded aziridine-2-phosphonates **91** in 93–95% yield [79].

An aza-Darzens reaction, involving the addition of chloromethylphosphonate anions to enantiopure *N*-sulfinimines, has also been developed by Davis and others for the asymmetric synthesis of aziridine-2-phosphonates [81–84]. As an example, treatment of the lithium anion generated from dimethyl chloromethylphosphonate (**93**; Scheme 3.30) with *N*-sulfinimine (S_S)-**92** gave the α-chloro-β-amino phosphonate **94**, which could be isolated in 51% yield. Cyclization of **94** with *n*-BuLi gave *cis*-*N*-sulfinylaziridine-2-phosphonate **95** in 82% yield [81].

More recently, an improved method for the asymmetric synthesis of aziridine-

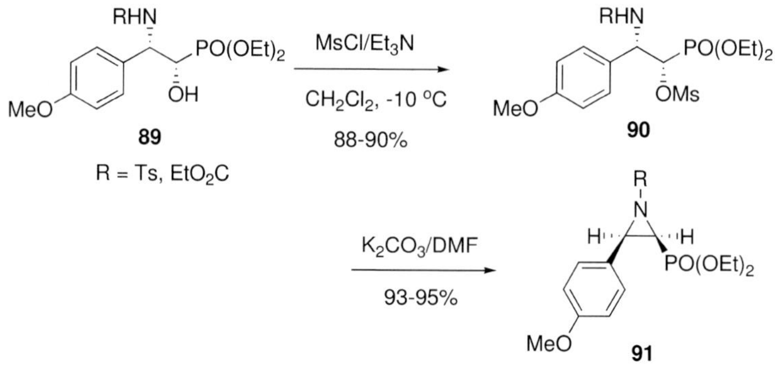

Scheme 3.29

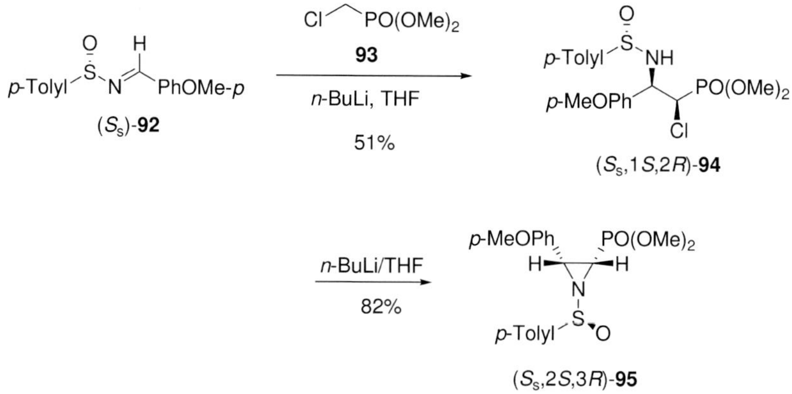

Scheme 3.30

2-phosphonates through the use of the enantiopure *N*-(2,4,6-trimethylphenylsulfinyl) imine **96** (Scheme 3.31) was reported by Davis and co-workers [85]. Treatment of a mixture of *N*-sulfinimine **96** and iodomethylphosphonate **97** with LiHMDS in THF at −78 °C gave aziridine **98** directly in 78% isolated yield [85].

Reduction of azirine-2-phosphonates **99** (Scheme 3.32) with $NaBH_4$ in ethanol exclusively gave *cis*-aziridine-2-phosphonates **100** in 81–82% yield [86, 87]. A Diels-Alder reaction between azirine-2-phosphonate **101** and *trans*-piperylene **102**

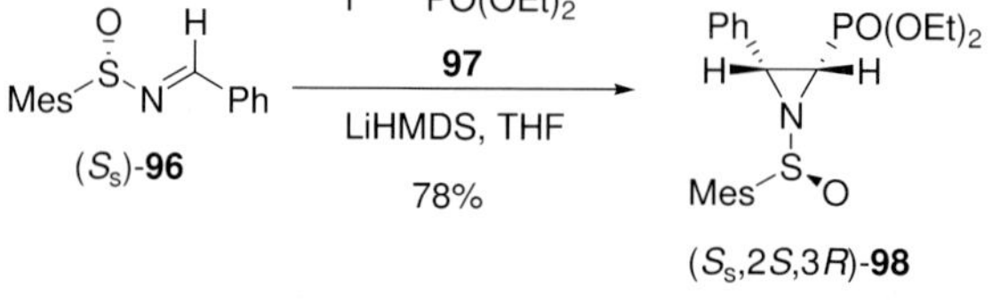

Scheme 3.31

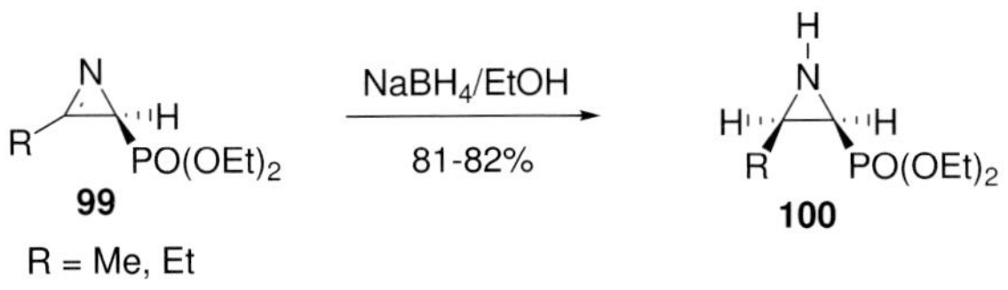

NaBH$_4$/EtOH, 81-82%; 99 → 100; R = Me, Et

Scheme 3.32

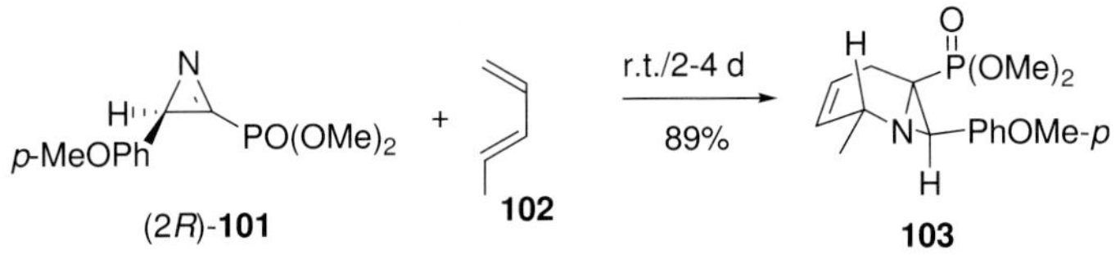

r.t./2-4 d, 89%; (2R)-101 + 102 → 103

Scheme 3.33

MeMgBr/LiBu-t, THF; (EtO)$_2$POCl, 43%; 104 → [105] → 106

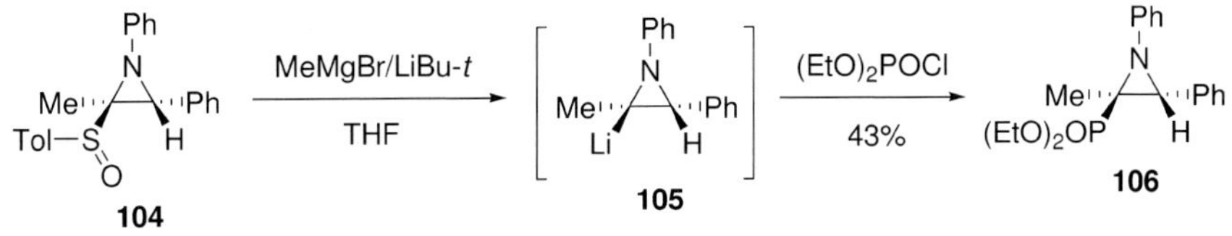

Scheme 3.34

(Scheme 3.33) afforded the bicyclic aziridine-2-phosphonate **103** in 89% yield [83].

Treatment of sulfinylaziridine **104** (Scheme 3.34) with base, followed by trapping of the intermediate aziridinyllithium **105** with (EtO)$_2$POCl, produced **105** in 43% yield [88].

3.3 Reactions of Aziridine-2-carboxylates and Aziridine-2-phosphonates

3.3.1 Reactions of Aziridine-2-carboxylates

Aziridine-2-carboxylates **1** and **2** are versatile synthetic intermediates. In addition to the normally expected reactions at the N-l nitrogen and the C-2 carboxylic group, the C-3 and C-2 positions can undergo a variety of nucleophilic ring-opening reactions (Figure 3.1). The C-2 proton, thanks to its proximity to a carbonyl group, can be abstracted, making it possible to carry out electrophilic substitution reactions at this position. Aziridine-2-carboxylates **1** are also convenient sources of azomethine ylides upon thermolysis. Only those reactions involving the C-2 and C-3 ring atoms are presented in this section.

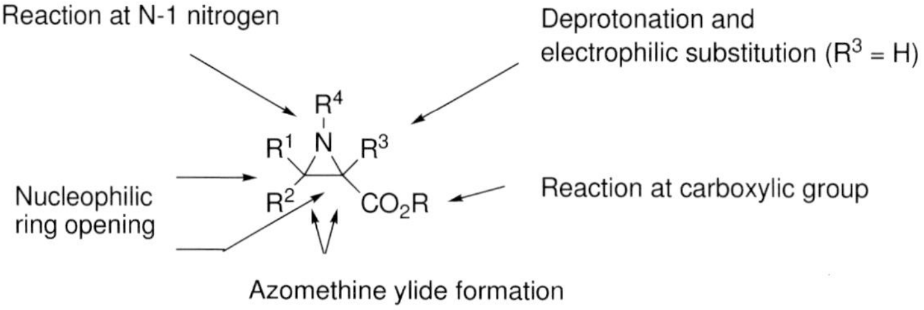

Figure 3.1 Reactions of aziridinecarboxylates.

3.3.1.1 **Reductive Ring-opening**

Hydrogenation of aziridine-2-carboxylates results in aziridine ring-opening and formation of amino acid derivatives. Aziridines with C-3 aryl substituents usually undergo ring-opening at the benzylic position, which gives rise to α-amino esters [64, 70, 89]. Hydrogenation of aziridine **107** (Scheme 3.35) with 10% Pd/C in methanol, for example, afforded (*S*)-phenylalanine **108** in 90% yield [21]. When aliphatic groups are present at the C-3 position, hydrogenolysis of the aziridine ring occurs at the C-2 carbon to give β-amino esters, such as in the transformation of **109** into **110** (Scheme 3.36) [66, 90]. Transfer hydrogenation using formic acid [46, 65] and Raney-Ni-mediated reductive ring-opening [66] have also been reported.

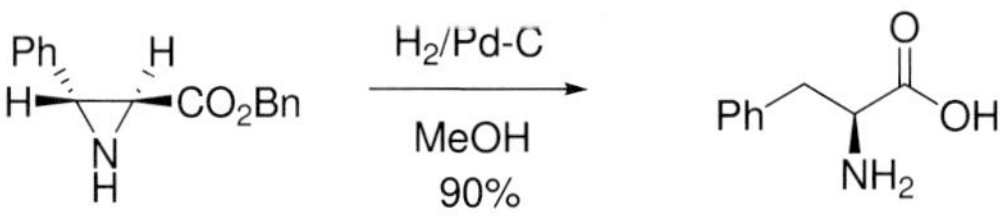

107 (*S*)-**108**

Scheme 3.35

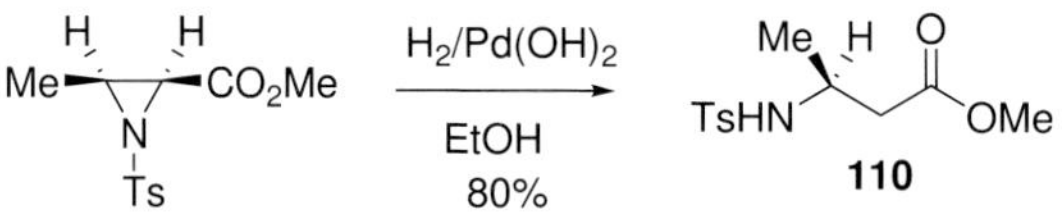

109 **110**

Scheme 3.36

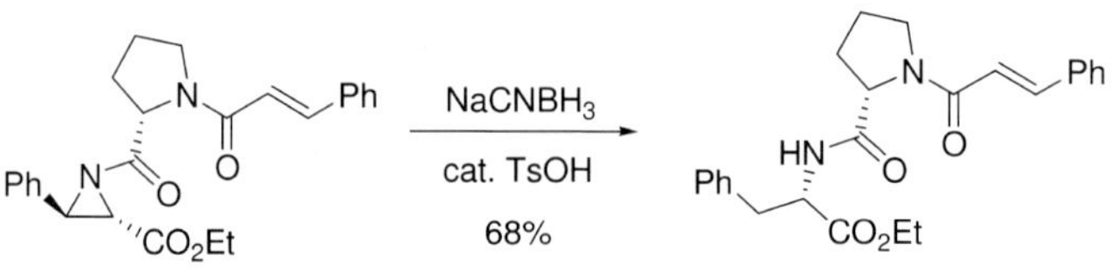

111 **112**

Scheme 3.37

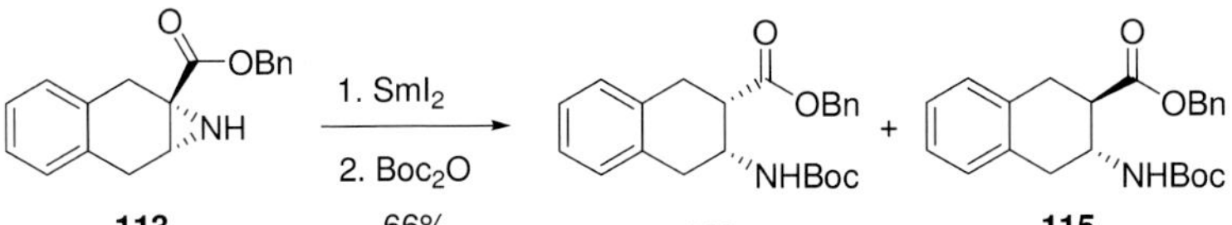

1. SmI_2
2. Boc_2O
66%

113 → 114 + 115 (1.6:1)

Scheme 3.38

Reductive aziridine ring-opening with sodium cyanoborohydride has been described [74, 91]. In the presence of a catalytic amount of TsOH, compound **111** (Scheme 3.37) gave **112** in 68% yield on treatment with sodium cyanoborohydride [74, 91].

Samarium(II) iodide has also been used as a reducing agent for the opening of the aziridine ring [69, 92]. The reaction takes place at the 2-position, giving β-amino ester derivatives. More recently, it was found that SmI_2-mediated reductive aziridine ring-opening of **113** (Scheme 3.38) resulted in the formation of **114** and **115** in 66% yield in a ratio of 1.6:1 [93].

3.3.1.2 Base-promoted Ring-opening

The C-2 proton in an aziridine-2-carboxylate is acidic, due to the adjacent carboxylic group. Upon treatment with base, such aziridines may undergo ring-opening reactions to give α-amino-α,β-unsaturated carboxylates [74, 94, 95]. As an example, treatment of **111** (Scheme 3.39) with $TMSI/Et_3N$ gave **116** in 64% yield [74].

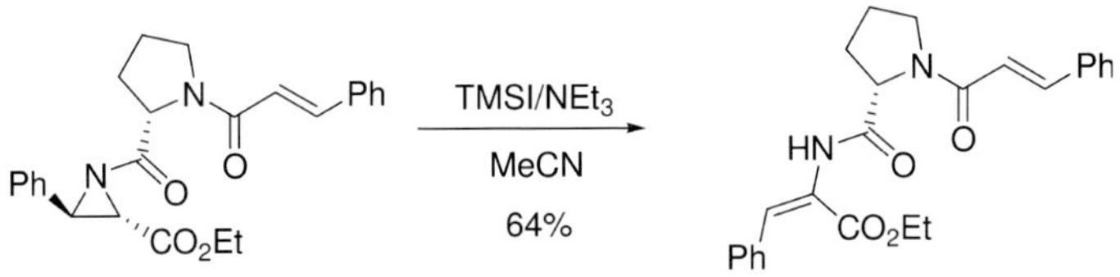

$TMSI/NEt_3$
MeCN
64%

111 → 116

Scheme 3.39

3.3.1.3 Nucleophilic Ring-opening

With Halogen Nucleophiles Aziridine-2-carboxylates react with halogen nucleophiles to give α- or β-halo-substituted amino acid derivatives. A number of halogen nucleophiles have been reported to effect aziridine ring-opening, including HCl [96–98], HI [96], NaBr/Amberlyst [99], $MgBr_2$ [99, 100], and MgI_2 [99, 100]. While the regio- and stereochemical outcomes of the ring-opening are dependent on the nucleophilic reagent as well as the aziridine substitution pattern, high selectivity could be obtained with appropriate choice of reaction conditions. Treatment of *N*-acylaziridine peptide **117** (Scheme 3.40) with 1 M hydrochloric acid, for example,

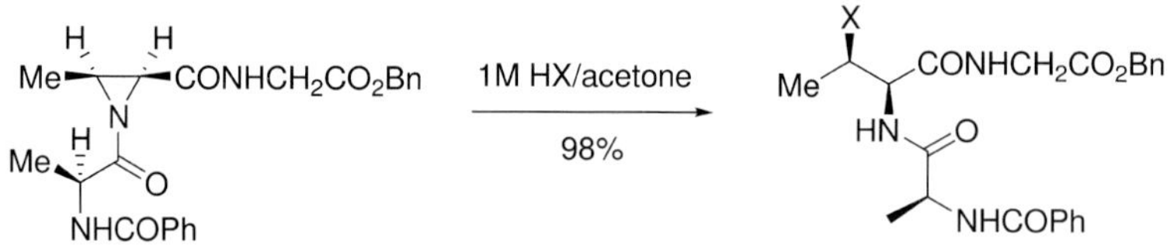

Scheme 3.40

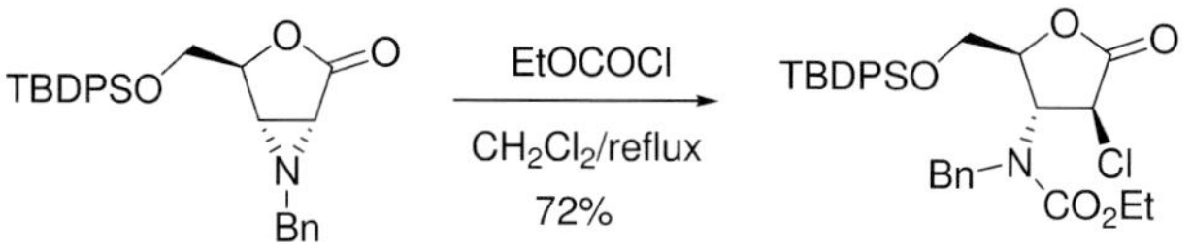

Scheme 3.41

afforded **118a** in 98% yield [96]. Similarly, **117** reacted with 1 M hydriodic acid to give compound **118b**, also in 98% yield. In each case the reaction was highly regio- and stereoselective, giving *syn*-118 as a single product.

In an attempt to effect a chloroformate-mediated dealkylation of aziridine lactone **119** (Scheme 3.41), Dodd and co-workers observed aziridine ring-opening. α-Chloro lactone **120** was obtained in 72% yield [32].

With Oxygen Nucleophiles Aziridine ring-opening of **111** (Scheme 3.42) with water in the presence of a catalytic amount of TsOH gave the corresponding β-hydroxyphenylalanine derivative **121** in 72% yield as the major isomer [74]. Treatment of *N*-(*p*-tolylsulfinyl) aziridine-2-carboxylates with TFA and subsequent aqueous workup resulted in the formation of β-substituted serine derivatives [62, 63, 101]. Under these reaction conditions, not only was the aziridine ring opened, but also the *N*-sulfinyl group was removed; treatment of **122** (Scheme 3.43) with TFA at 73 °C, for example, afforded **123** in 75% yield [101].

When aziridine-2-carboxylic ester **124** (Scheme 3.44) was treated with 1 equivalent of TFA, **125** was formed as a single diastereomer in 80% yield [57]. Use of 10 equivalents of TFA resulted in the formation of *N*-trifluoroacetyl product **126**. This

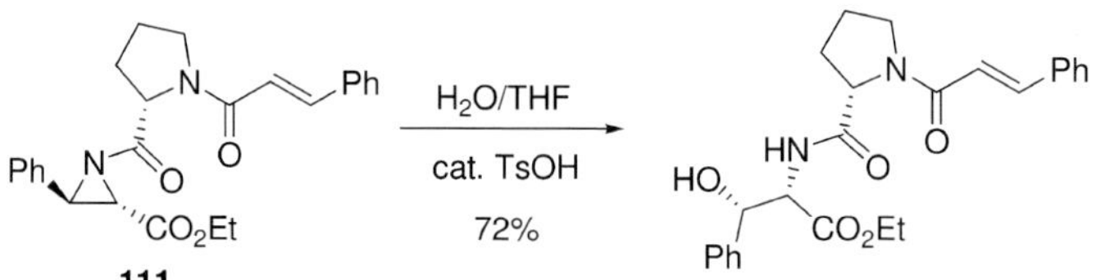

Scheme 3.42

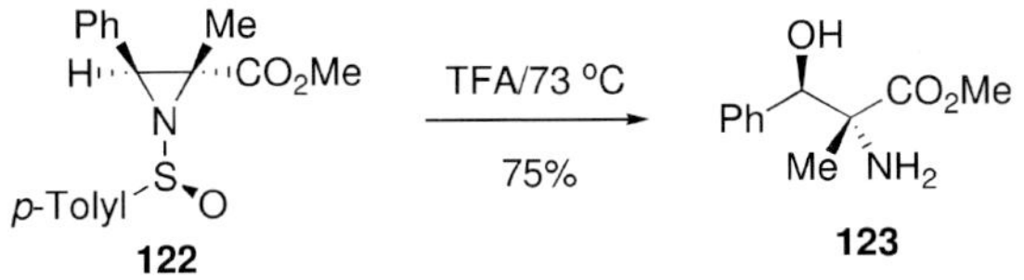

Scheme 3.43

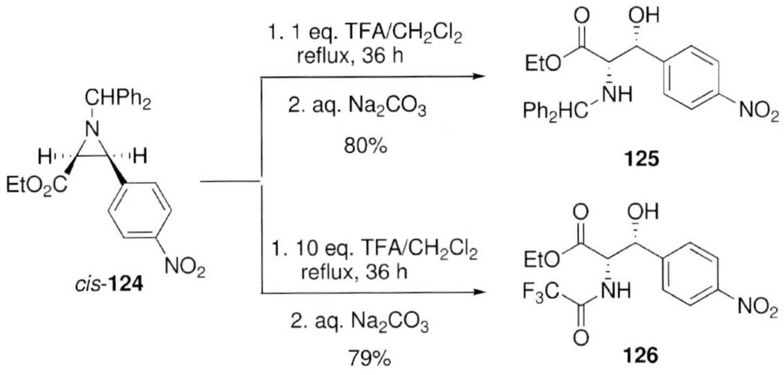

Scheme 3.44

transformation was explained in terms of a sequence of events including protonation of the aziridine nitrogen, attack of trifluoroacetate at the 3-benzylic carbon, debenzhydrylation, and migration of the trifluoroacetyl group from oxygen to nitrogen.

Ring-opening of aziridine-2-carboxylates with alcohols has been reported to give β-alkoxy-α-amino esters [16, 102]. Treatment of *cis*-aziridine **127** (Scheme 3.45) with alcohol in the presence of a catalytic amount of boron trifluoride etherate afforded β-alkoxy-α-amino esters **128** in 57–100% yields [16, 102a]. The reaction is both regio- and stereoselective, affording **128** as the only product.

Carboxylic acids have also been investigated as nucleophiles in the ring-opening of aziridine-2-carboxylates [103]. Solvation of compounds **129** (Scheme 3.46) in acetic acid, for example, gave **130a** and **130b** in 83% and 89% yields, respectively.

Interestingly, treatment of methyl *trans*-1*H*-3-phenylaziridine-2-carboxylate (**131**) with acetic acid gave the *N*-acylated product **133** (Scheme 3.47). It was proposed

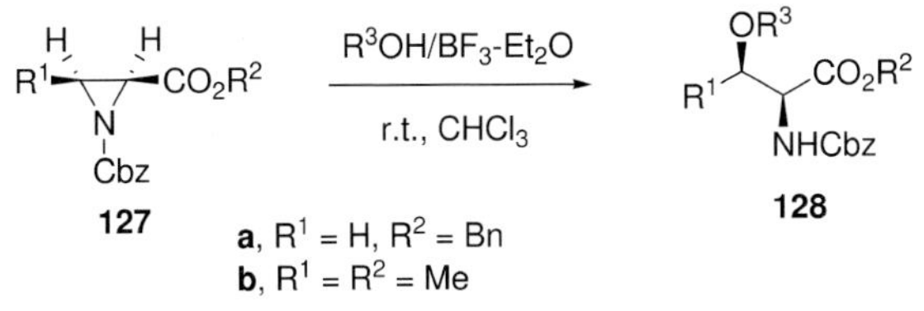

Scheme 3.45

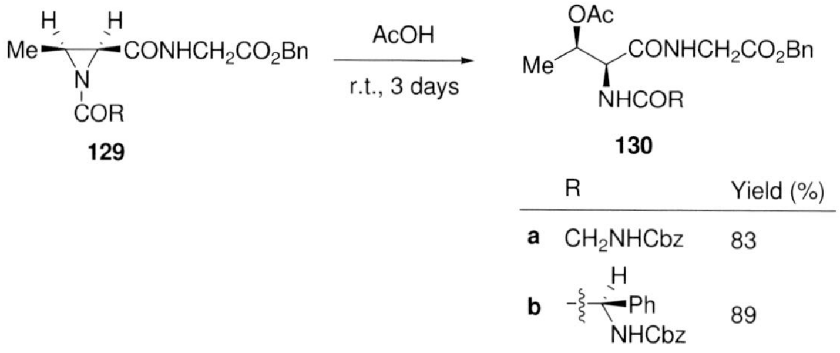

Scheme 3.46

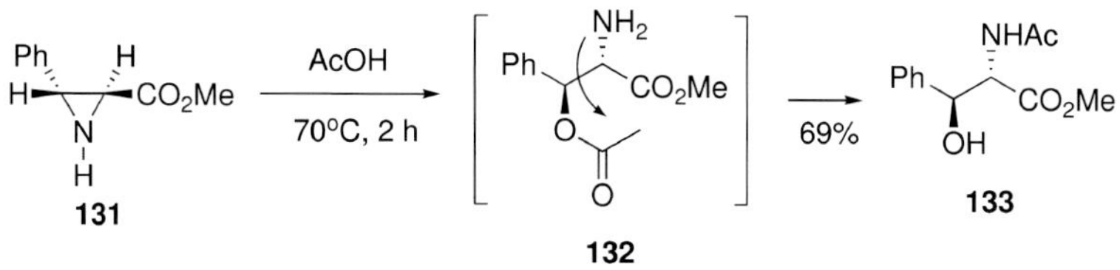

Scheme 3.47

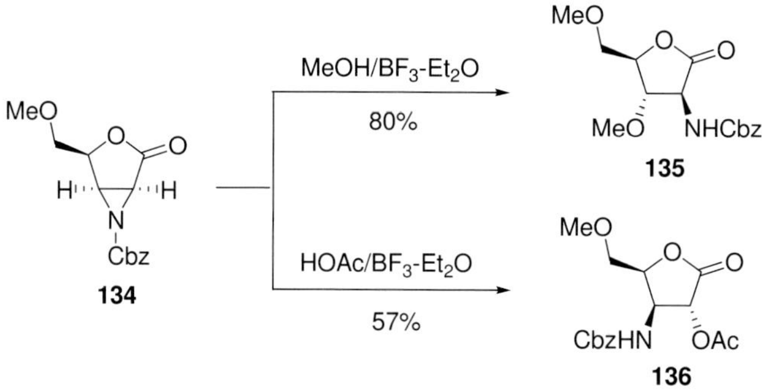

Scheme 3.48

that aziridine was first opened by attack by acetate anion at C-3 of **131**, followed by an acyl migration from oxygen to nitrogen via the intermediate **132** [97].

With methanol in the presence of $BF_3 \cdot Et_2O$, aziridine lactone **134** (Scheme 3.48) gave **135** in 80% yield [104, 105]. Use of HOAc afforded **136** in 57% yield [104, 105].

Treatment of *N*-acetyl aziridinecarboxylate with acetic anhydride and heating resulted in the aziridine ring-opened product [106], whereas treatment of the aziridine **137** (Scheme 3.49) with acetic anhydride in the presence of pyridine and DMAP as base similarly resulted in the formation of acetate **138** in 90% yield [45].

Reaction of aziridine peptides **139** (Scheme 3.50) and dibenzylphosphoric acid

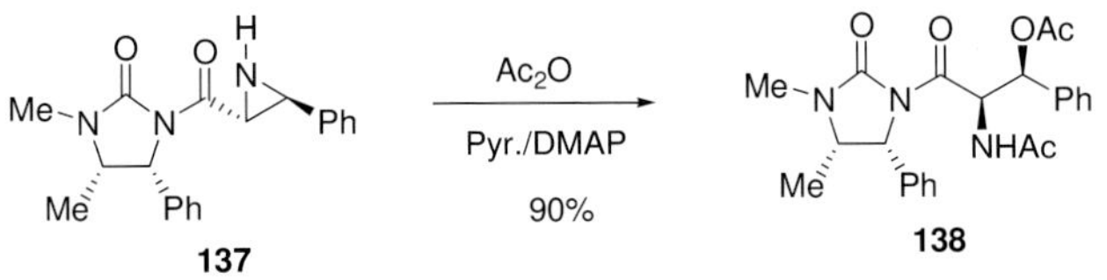

Scheme 3.49

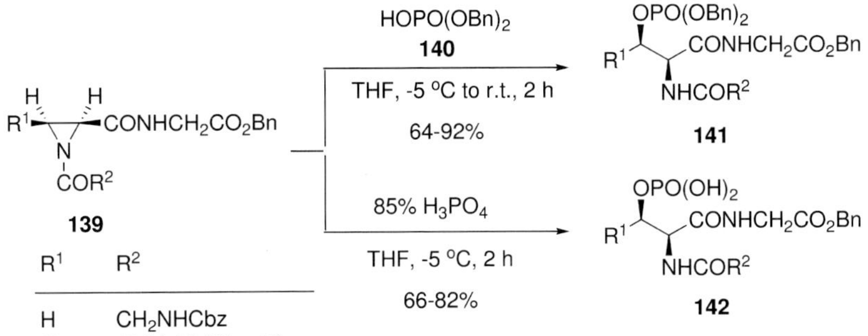

R^1	R^2
H	CH_2NHCbz
H	CHMeNHCbz-(*S*)
Me	CHMeNHCbz-(*S*)
Me	CH_2NHCbz

Scheme 3.50

(**140**) in THF afforded compounds **141** in 64–92% yields. Similar treatment of **139** with 85% phosphoric acid gave **142** in 66–82% yields [107].

With Sulfur Nucleophiles *N*-Carboxy-protected aziridine-2-carboxylates react with thiols to give β-mercapto-α-amino acid derivatives. The reaction is usually catalyzed by BF_3 and the yields range from fair to excellent [15, 16, 108–111]. With *N*-unprotected 3-substituted aziridine-2-carboxylates, the ring-opening with thiols usually takes place with *anti* stereoselectivity, especially in the case of the C-3 aliphatic substituted substrates. In cases in which C-3 is aromatic, however, the stereoselectivity has been found to be a function of the substitution pattern on the aromatic ring: 3-*p*-methoxyphenyl-substituted aziridines **143a** (Scheme 3.51) and

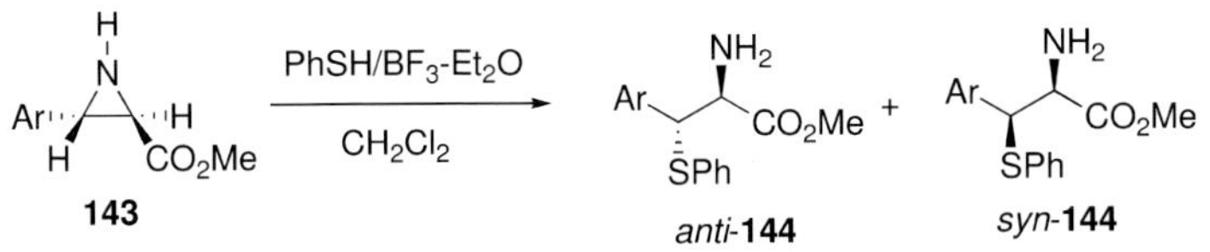

	Ar	Conditions	anti/syn	Yield (%)
a	C_6H_4OMe-*p*	r.t., < 15min	60/40	65
b	Ph	r.t., 22 h	100/0	80
c	$C_6H_4NO_2$-*p*	reflux, 9 h	100/0	65

Scheme 3.51

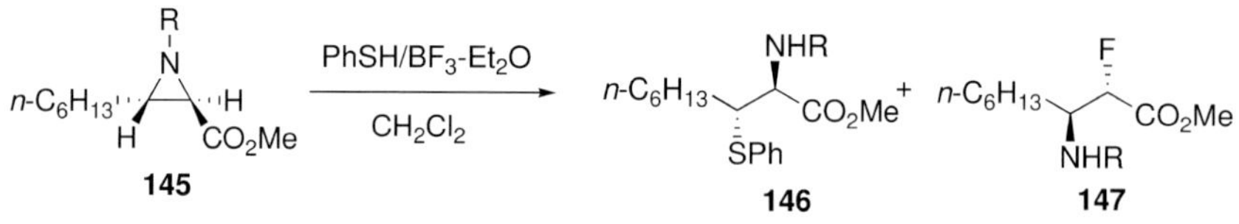

	R	146/147	Yield (%)
a	TEOC	60/40	70
b	Ac	100/0	38
c	Cbz	100/0	95
d	Ms	100/0	40
e	Ts	100/0	63
f	$BnSO_2$	100/0	63

Scheme 3.52

thiophenol, for example, gave mixtures of *anti/syn*-144a whereas the phenyl and *p*-nitrophenyl aziridines **143b** and **143c** exclusively afforded *anti*-**144b** and *anti*-**144c**. The *p*-methoxy group apparently assists the C3-N bond cleavage to such an extent that the transition state of the nucleophilic reaction has considerable S_N1 character [97].

Zwanenburg studied the reactions between thiophenol and various *N*-protected aziridine-2-carboxylates **145** (Scheme 3.52) [98]. While the ring-opening usually took place at the C-3 position, it was observed that the *N*-TEOC-protected (TEOC = trimethylsilylethoxycarbonyl) aziridine **145a** gave a mixture of C-3 and C-2 ring-opening products **146a** and **147a**, with **147a** incorporating a fluorine atom at thc α-position. The other protected aziridines, **145b-f**, proceeded normally, affording the C-3 ring-opening products **146b-f** exclusively [98].

Reactions between aziridine-2-carboxylic acids and thiols in aqueous solution have been explored by Hata and Watanabe [112]. The reactions occurred predominantly at C-2 instead of C-3 to afford β-amino acids, with the reaction between **148** (Scheme 3.53) and thiophenol in 0.2 M sodium phosphate buffer at room tem-

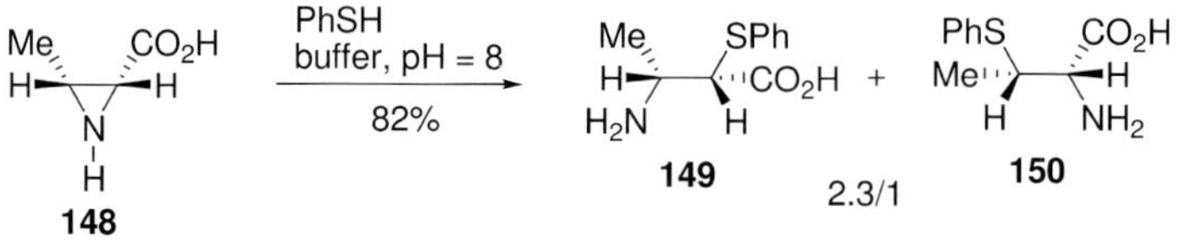

Scheme 3.53

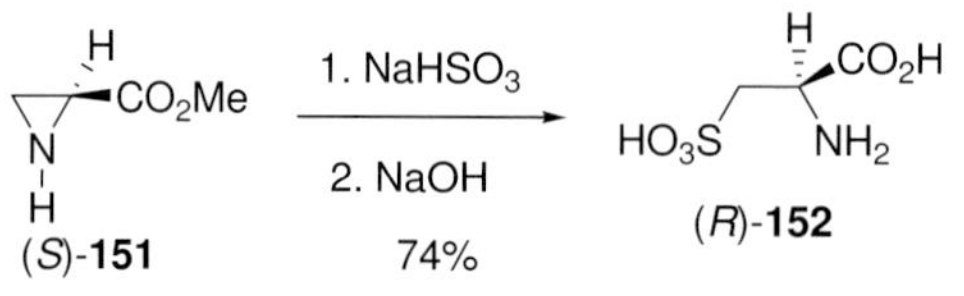

Scheme 3.54

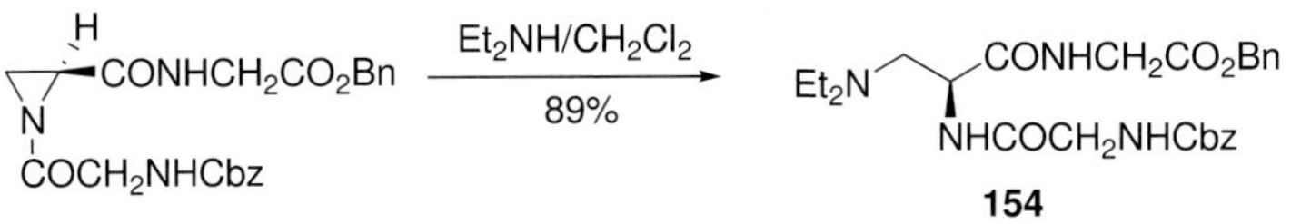

Scheme 3.55

perature, for example, affording **149** and **150** in 82% yield and in a ratio of 2.3/1.

The opening of *N*-unsubstituted aziridine-2-carboxlic esters with sodium bisulfite was recently described by Xu [113]. Thus, treatment of (*S*)-**151** (Scheme 3.54) with $NaHSO_3$ followed by NaOH afforded (*R*)-cysteic acid **152** in 74% yield. (*S*)-Cysteic acid was prepared similarly in 72% yield.

With Nitrogen Nucleophiles Aziridine-2-carboxylates react with primary and secondary amines, including anilines, to give α,β-bisamino carboxylates [71, 113]. As an example, treatment of aziridine **153** (Scheme 3.55) with diethylamine in methylene chloride afforded compound **154** in 89% yield after chromatographic separation.

Aziridine-2-carboxylates also react with azides to give ring-opened products [100, 114, 115]. Treatment of *N*-acetylaziridine **155a** (Scheme 3.56) with NaN_3 in DMF in the presence of $BF_3 \cdot Et_2O$ afforded a mixture of C-3 and C-2 ring-opening products **156a/157a** in 32% yield and in a ratio of 93:7. However, treatment of *N*-sulfonylaziridine-2-carboxylic esters **155b** and **155c** with azidotrimethylsilane in DMF containing one equivalent of ethanol occurred with completely opposite regioselectivity, giving **157b** and **157c** as single isomers [98].

With Carbon Nucleophiles The reactions between organometallic reagents and aziridine-2-carboxylates have also been explored [116–118]. Compound **158** (Scheme 3.57) reacts with Gilman cuprate **159a**, for example, to give the α-amino ester **160a** in 68% yield. It is interesting to note that the Lipshutz higher-order cyanocuprate **159c** was completely unreactive [100]. Aziridine-2-carboxylic acid **161**

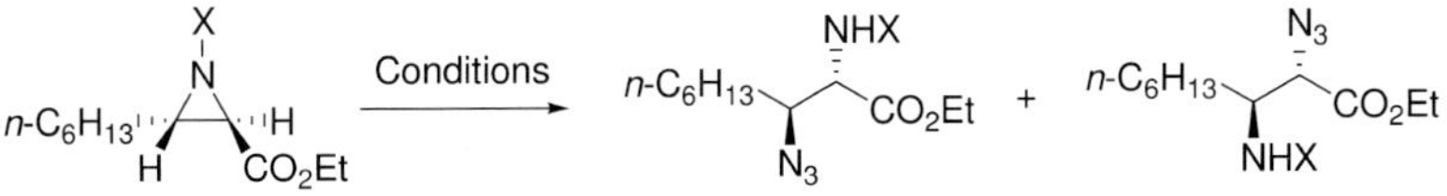

	X	Conditions	156/157	Yield (%)
a	Ac	NaN_3, BF_3-Et_2O, DMF	93/7	32
b	Ms	$TMSN_3$, EtOH, DMF	0/100	88
c	Ts	$TMSN_3$, EtOH, DMF	0/100	56

Scheme 3.56

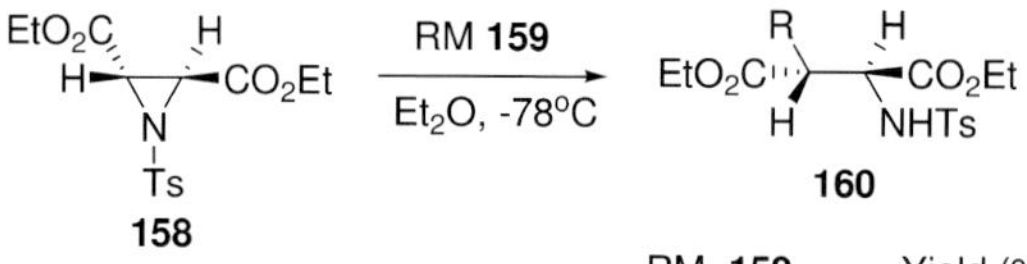

	RM **159**	Yield (%)
a	$LiMe_2Cu$	68
b	$LiBu_2Cu$	54
c	Li_2Bu_2CuCN	0

Scheme 3.57

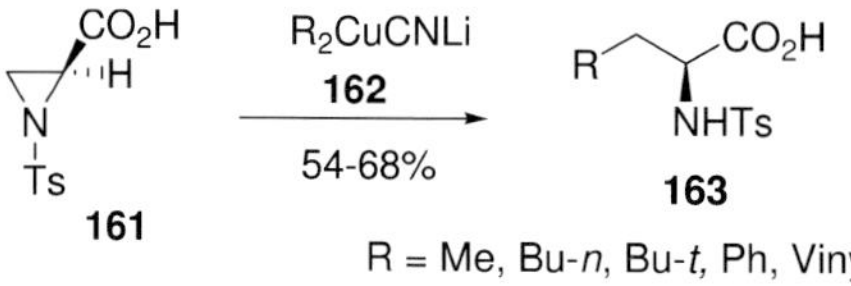

R = Me, Bu-*n*, Bu-*t*, Ph, Vinyl

Scheme 3.58

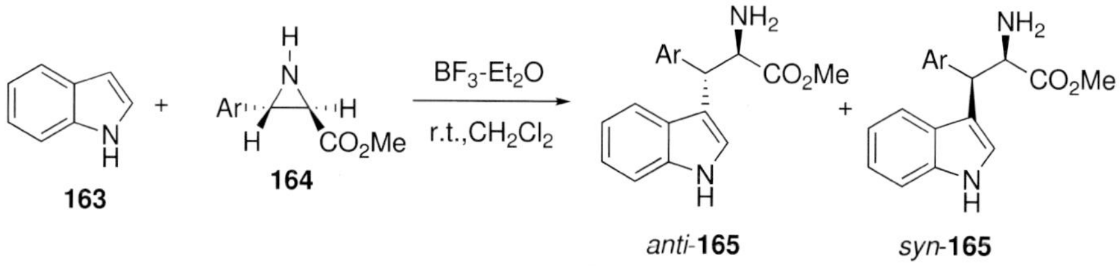

	Ar	Time	anti/syn	Yield (%)
a	C_6H_4OMe-*p*	1 h	55/45	68
b	Ph	2 h	100/0	53
c	$C_6H_4NO_2$-*p*	5 d	100/0	60

Scheme 3.59

(Scheme 3.58) and organocuprates **162** afforded α-amino acids **163** in 54–68% yields [92].

Tryptophan derivatives such as **165** (Scheme 3.59) can be prepared by nucleophilic ring-opening of aziridine-2-carboxylates **164** with indole [97, 98]. The reaction was C-3 regioselective. Unlike in the case of substrate **164a**, where S_N1-type substitution is possible, aziridines **163b** and **163c** exclusively afforded the *anti* products **165b** and **165c**. Treatment of 3,3-dimethylaziridine-2-carboxylate with *N*-methylindole under similar conditions failed to give the desired product, however [119], with formation of a fluorinated valine derivative being observed instead. Attempted ring-opening of aziridine-2-carboxylate with indole by use of scandium triflate also failed [120].

In contrast to aziridine-2-carboxylic esters, treatment of aziridine lactone **166** (Scheme 3.60) with *N*-methylindole (**167**) in the presence of $BF_3 \cdot Et_2O$ proceeded

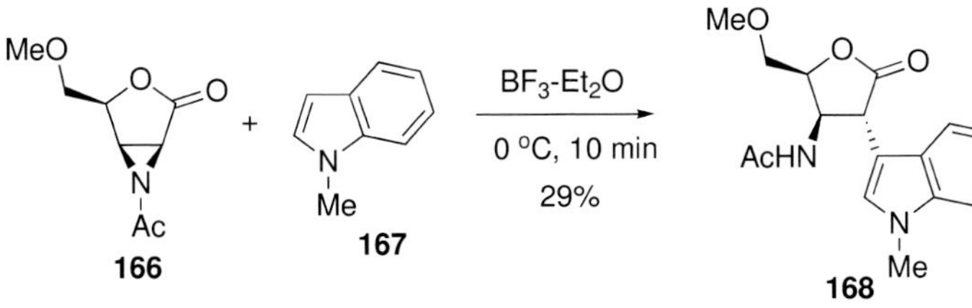

Scheme 3.60

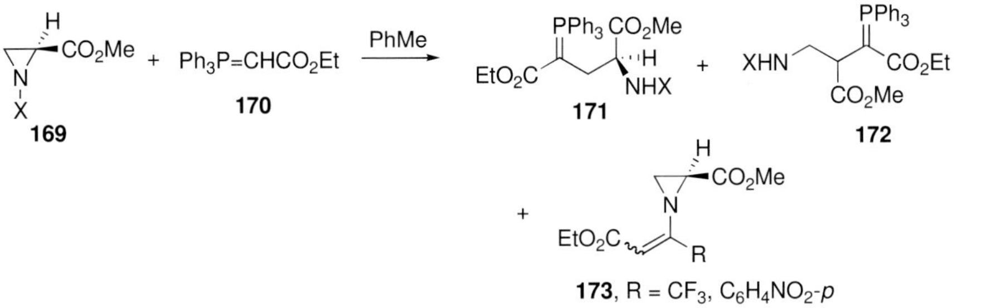

X	**170** (eq.)	Conditions	**171** (%)	**172** (%)	**173** (%)
$COC_6H_4NO_2$-*p*	1.1	reflux, 2 h	49	12	16 (R = $C_6H_4NO_2$-*p*)
$CO_2C_6H_4NO_2$-*p*	2.0	reflux, 24 h	30		
CO_2Bn	2.0	reflux, 24 h	30		
$COCF_3$	1.0	50 °C, 6 h			95 (R = CF_3)
$SO_2C_6H_4Me$-*p*	1.1	reflux, 2 h	66	13	

Scheme 3.61

exclusively at the C-2 position, affording **168** [104]. MNDO calculations suggest that the regioselectivity differences between aziridine lactones and aziridine-2-carboxylic esters is related to the variation in the LUMO coefficients at C-2 and C-3 in these aziridines.

Treatment of aziridine **169** (Scheme 3.61) with ylide **170** in toluene gave a mixture of products **171–173** [121]. The product distribution was highly dependent on the protecting group on the aziridine nitrogen atom. It was interesting to note that product **173** was only observed when the X substituent on aziridine nitrogen was an acyl group.

3.3.1.4 Electrophilic Substitutions at the C-2 Carbon Atom

Unlike regular aziridine-2-carboxylic esters, aziridine-2-carboxylic thioester **174** (Scheme 3.62) forms stable carbanions at the 2-position upon treatment with base [13b, 122]. Thus, electrophilic alkylations of aziridine **174** afforded products **175**. The reactions were highly diastereoselective, affording **175** in moderate to good

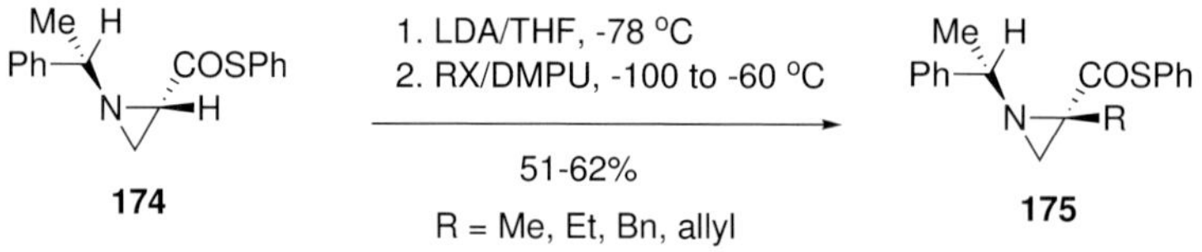

1. LDA/THF, -78 °C
2. RX/DMPU, -100 to -60 °C
51-62%
R = Me, Et, Bn, allyl

174 → 175

Scheme 3.62

yields as single isomers [122]. β-Nitrostyrene was also found to react with the anion generated from **174** to give the Michael addition product [122].

3.3.1.5 Ring-expansion Reactions

N-Acylaziridine-2-carboxylates readily rearrange to oxazolines under thermal, acidic, or nucleophilic conditions [91, 123–127]. Treatment of *trans*-aziridine-2-carboxylate **176** (Scheme 3.63) with NaI in acetonitrile, for example, resulted in ring-expansion product **177** through the so-called Heine reaction. The reaction involves initial opening of the aziridine ring by iodide and subsequent oxazoline ring-closure by S_N2 displacement of the resultant iodide intermediate [127].

Treatment of aziridine **137** (Scheme 3.64) with benzoyl chloride in the presence of pyridine and DMAP afforded ring-expansion product **179** in 90% yield [45]. The reaction was believed to proceed through the intermediate *N*-acylaziridine **178**.

Aziridines **180** (Scheme 3.65) react with acetonitrile and $BF_3 \cdot Et_2O$, a Lewis catalyst, to give imidazolines **181** in 65–95% yield [98]. Under the same reaction conditions, however, the C-3 phenyl-substituted aziridine **182** (Scheme 3.66) afforded oxazoline **183** in 59% yield [97].

Treatment of *N*-Boc-2-aziridine-2-carboxylate **184** (Scheme 3.67) with a catalytic

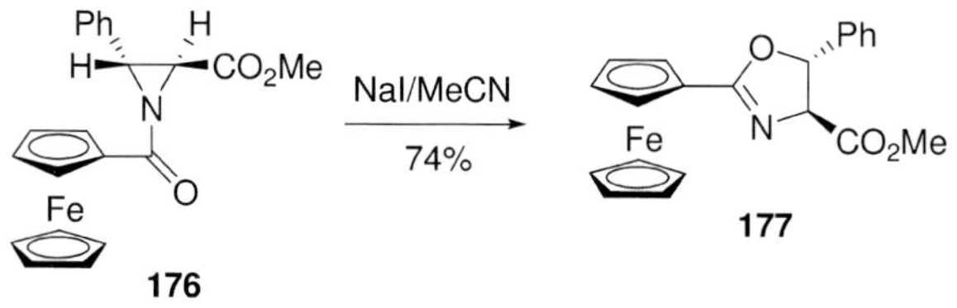

NaI/MeCN, 74%

176 → 177

Scheme 3.63

PhCOCl, Pyr./DMAP; 90%

137 → [178] → 179

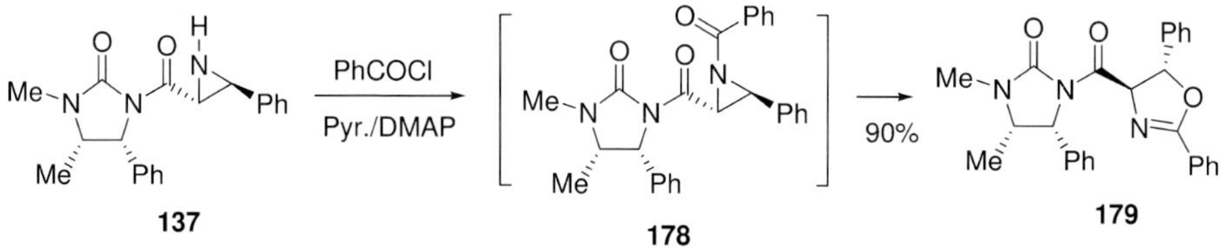

Scheme 3.64

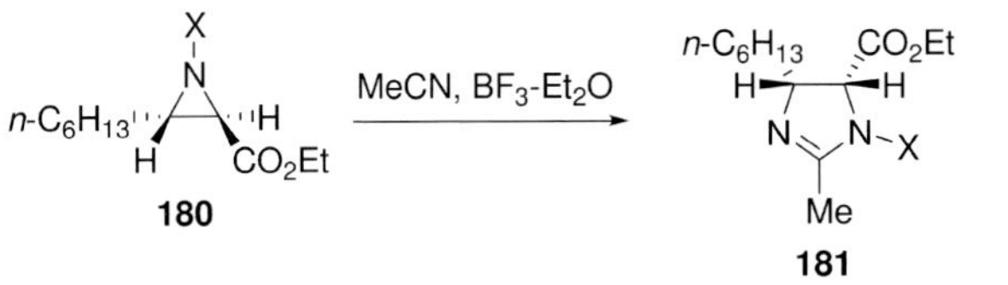

180 → (MeCN, BF_3-Et_2O) → **181**

	X	Yield (%)
a	Ac	91
b	Ms	92
c	Ts	65
d	Cbz	85
e	TEOC	95

Scheme 3.65

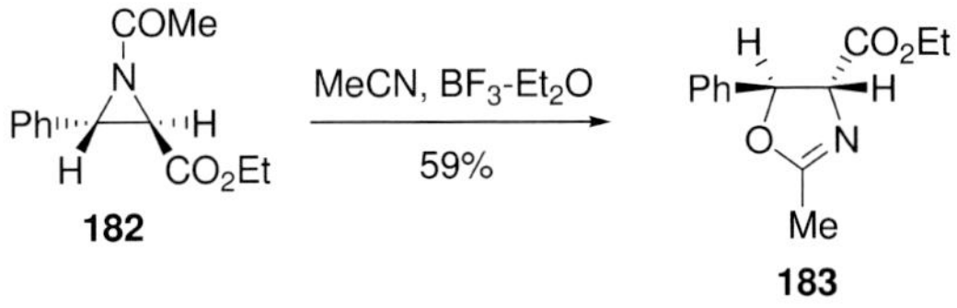

182 → (MeCN, BF_3-Et_2O, 59%) → **183**

Scheme 3.66

184 → (1. 0.2 eq. $Sn(OTf)_2/CH_2Cl_2$; 2. silica gel chromatography; 80%) → **185**

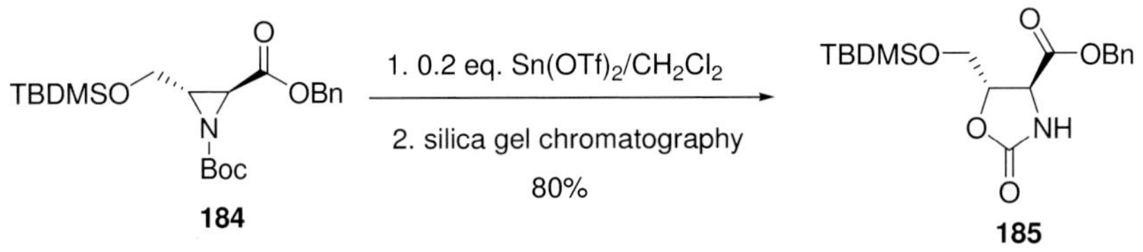

Scheme 3.67

amount of $Sn(OTf)_2$ in methylene chloride resulted in the formation of oxazolidin-2-one **185** [105]. Compound **185** was obtained in 80% yield [31]. $BF_3 \cdot Et_2O$ has also been found to be an effective Lewis acid for this transformation.

Recently, Lee and co-workers reported an efficient method for the preparation of enantiomerically pure oxazolidin-2-ones from aziridine-2-carboxylates **186** (Scheme 3.68) [128]. This one-pot aziridine ring-opening and subsequent intramolecular cyclization process was highly regio- and stereoselective, affording **187** in high yield.

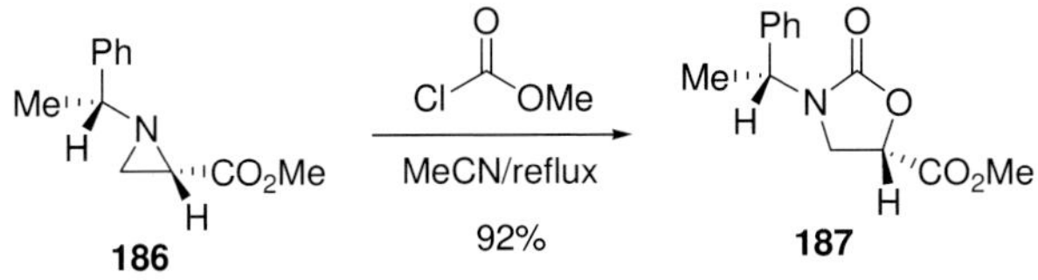

186 → ($ClCO_2Me$, MeCN/reflux, 92%) → **187**

Scheme 3.68

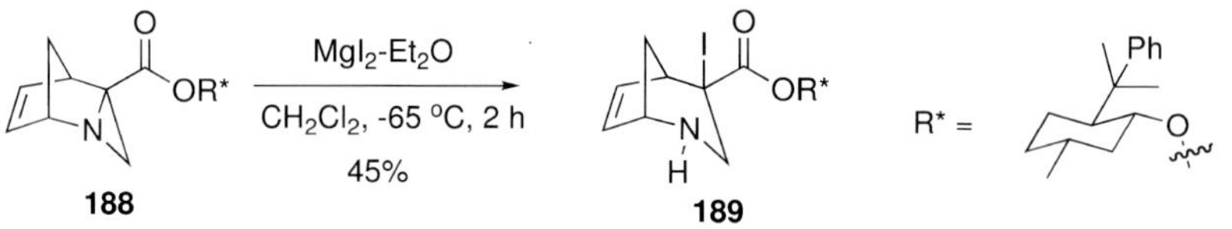

Scheme 3.69

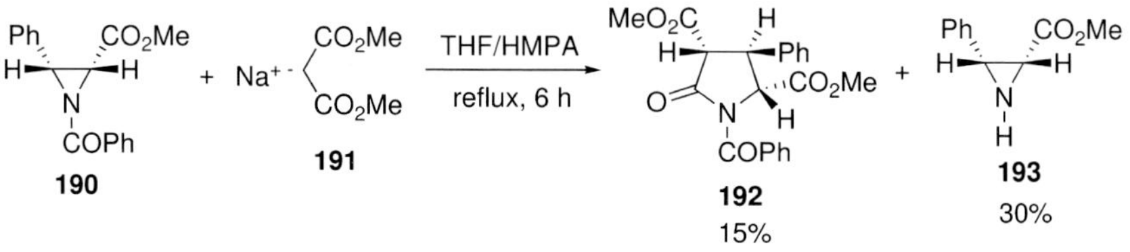

Scheme 3.70

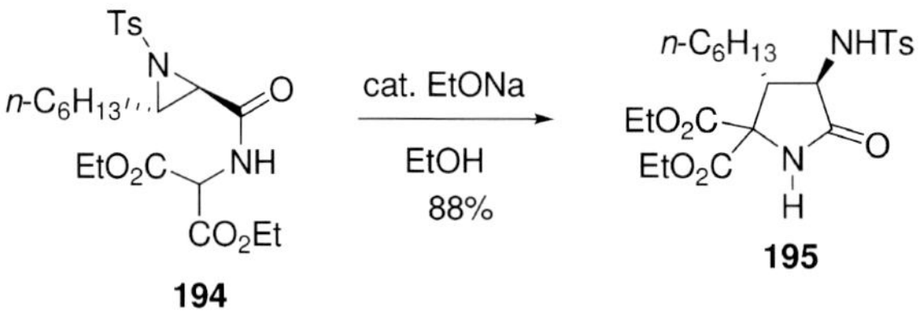

Scheme 3.71

With $MgI_2 \cdot Et_2O$, **188** (Scheme 3.69) reacts in methylene chloride at –65 °C to give **189** in 45% yield after 2 hours [68]. The corresponding α-bromo and -chloro esters could be prepared similarly by using $MgBr_2 \cdot Et_2O$ and $MgCl_2 \cdot Et_2O$ instead of $MgI_2 \cdot Et_2O$.

Carbanions of active methylene compounds also react with aziridine-2-carboxylic esters to give ring-opened products [129]. The ring-opened intermediates usually cyclize spontaneously to pyrrolidones. Treatment of **190** (Scheme 3.70) with the sodium enolate of dimethyl malonate **191**, for example, afforded pyrrolidone **192** in 15% isolated yield, together with 30% of the debenzoylated product **193**.

When aziridine **194** (Scheme 3.71) was treated with a catalytic amount of NaOEt in ethanol it underwent an intramolecular ring-expansion to pyrrolidinone **195** in 88% yield [130]. The ring-opening took place via an internal S_N2 reaction, which was confirmed by an X-ray analysis of the product **195**. It is interesting to note that under similar reaction conditions **196** (Scheme 3.72) afforded β-lactam product **197** [130].

Intramolecular and intermolecular 1,3-dipolar cycloadditions of aziridine-2-carboxylic esters with alkenes and alkynes have been investigated [131, 132]. Upon heating, aziridine-2-carboxylates undergo C-2–C-3 bond cleavage to form azome-

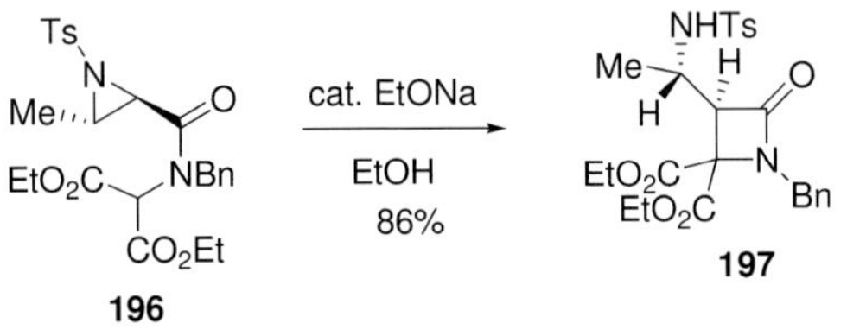

Scheme 3.72

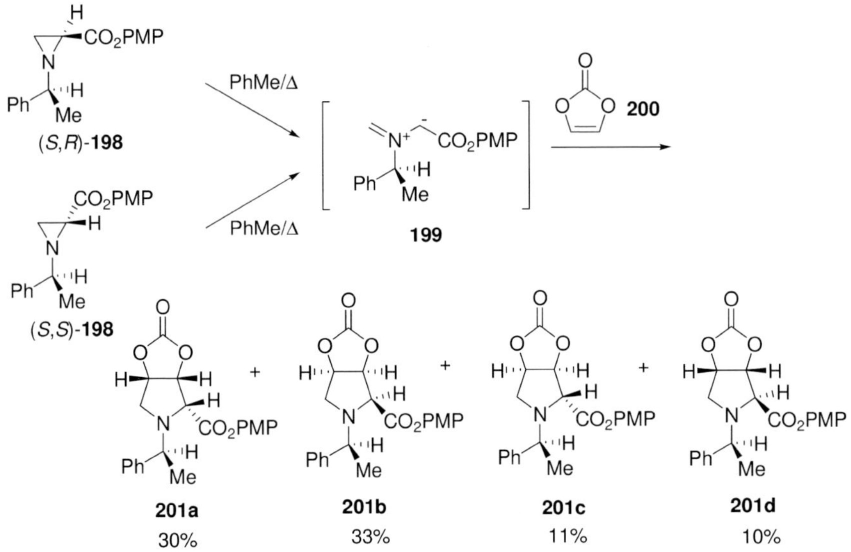

Scheme 3.73

thine ylide intermediates, which can be trapped by alkenes to give pyrrolidine-2-carboxylic acids [133]. In this context, heating of **198** with three equivalents of vinyl carbonate **200** (Scheme 3.73) in toluene in a sealed tube at 280 °C produced four chromatographically separable products **201** in 88% yield [134, 135]. The fact that both (*S*,*R*)-**198** and (*S*,*S*)-**198** afforded the same product distribution suggested the involvement of the azomethine ylide **199** in the process.

The spiro aziridine **202** (Scheme 3.74) reacts with a catalytic amount of boron trifluoride etherate to give rise to ring-expanded compound **205** in almost quantitative yield [55]. A possible mechanism involving intermediates **203** and **204** was proposed.

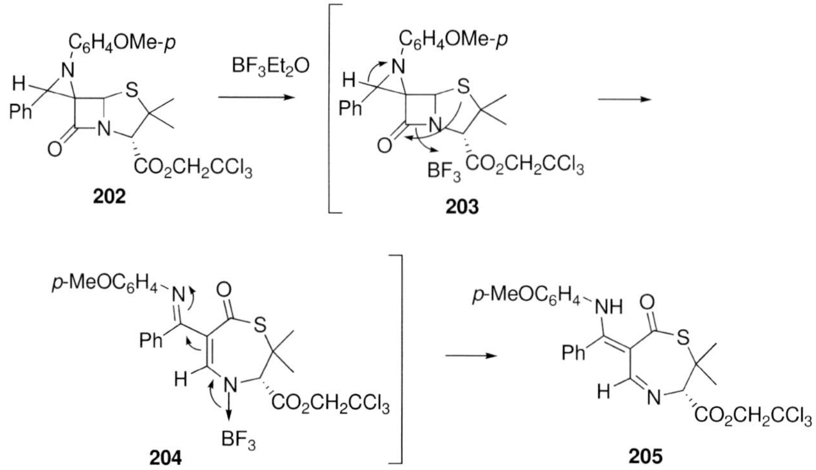

Scheme 3.74

3.3.1.6 Conversion to Azirine-2-carboxylates

Treatment of *N*-sulfinyl-*cis*-aziridine-2-carboxylate **206** (Scheme 3.75) with LDA at −78 °C afforded 2*H*-azirine-2-carboxylate **207** in 47% yield. The *trans*-aziridine **206** gave only a 9% yield of the desired product **207** [95]. Treatment of *N*-tosyl 2-substituted aziridines **208** (Scheme 3.76) with LDA resulted in the formation of azirines **209** in 61–87% yield [95].

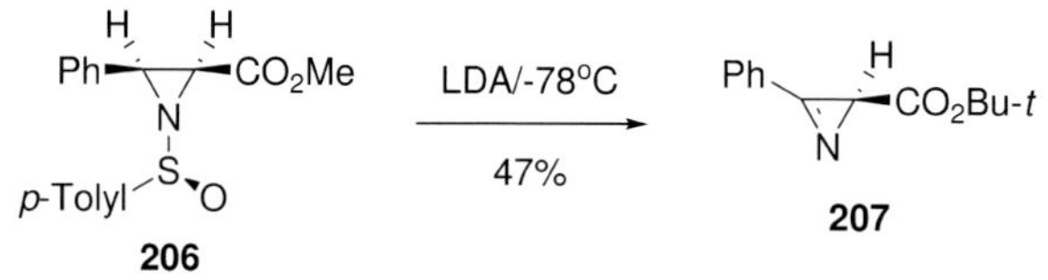

Scheme 3.75

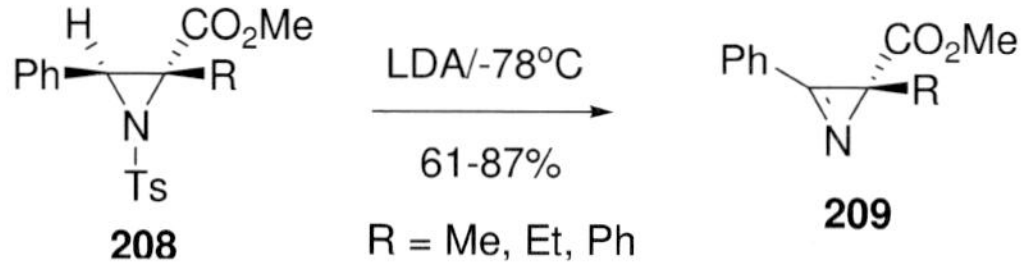

Scheme 3.76

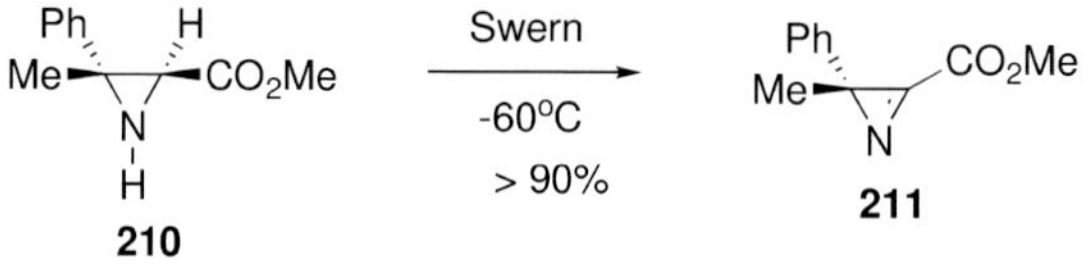

Scheme 3.77

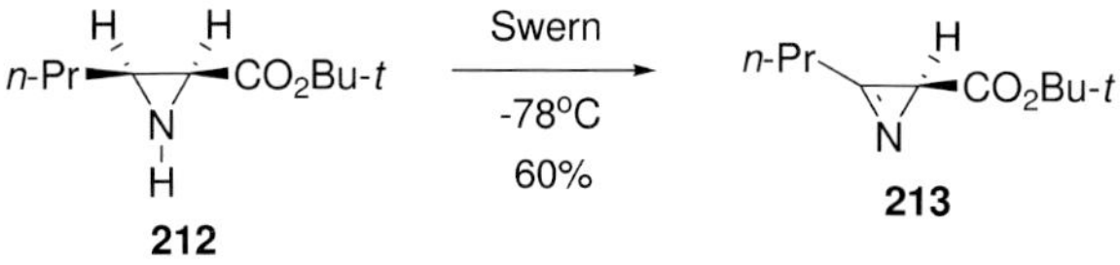

Scheme 3.78

Swern oxidation of *N*-unsubstituted aziridine-2-carboxylate **210** (Scheme 3.77) resulted in the formation of 2*H*-azirine **211** in >90% yield [95] Similar oxidation of **212** (Scheme 3.78) afforded azirine **213** in 60% yield [66].

3.3.2
Reactions of Aziridine-2-phosphonates

Reductive ring-opening of aziridine-2-phosphonates to give α-amino phosphonates has been reported by Davis and others [80–82, 84, 86, 87, 136]. With 3-aryl aziridine **214** (Scheme 3.79), hydrogenation afforded α-amino phosphonate **215** [87], while the 3-alkyl derivative **216** (Scheme 3.80) gave β-amino phosphonates **217** [87]. Reductive ring-opening of *N*-tosyl aziridine-2-phosphonate **218** (Scheme 3.81) with $NaBH_4$ resulted in the formation of mixture of α- and β-aminophosphonates **219** and **220** in a ratio of approximately 1:1 [87].

Hydrogenation of the bicyclic aziridine-2-phosphonate **221** (Scheme 3.82) in THF resulted in the formation of piperidinephosphonate **222** in 81% yield [83].

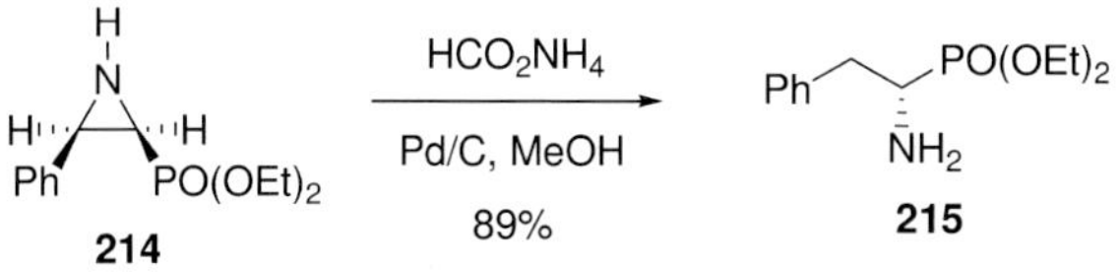

Scheme 3.79

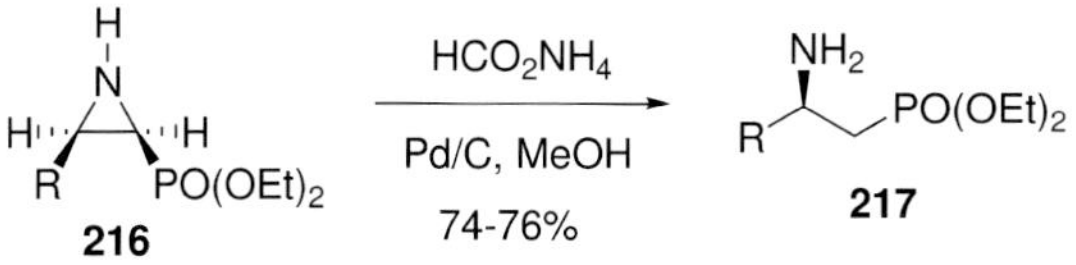

Scheme 3.80

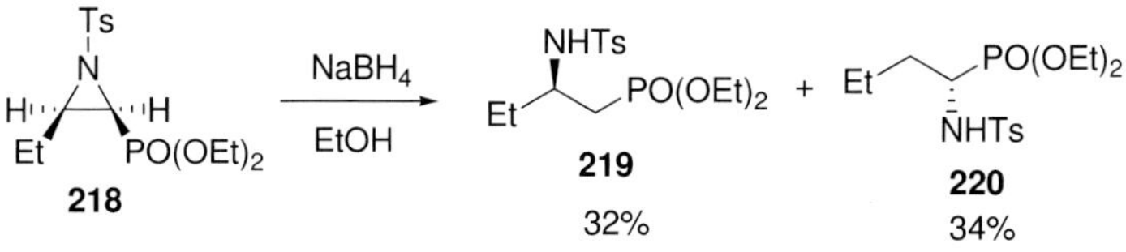

Scheme 3.81

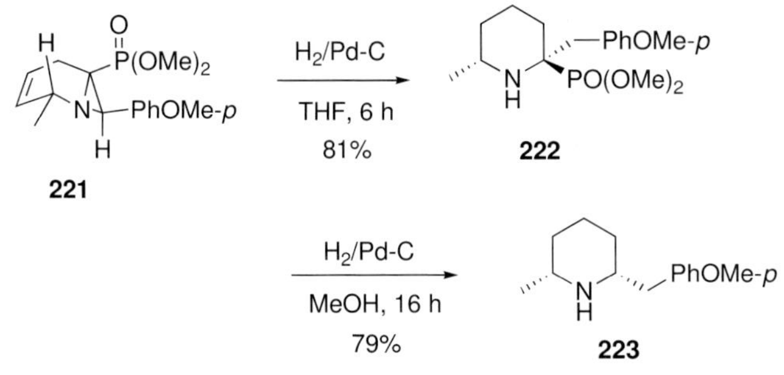

Scheme 3.82

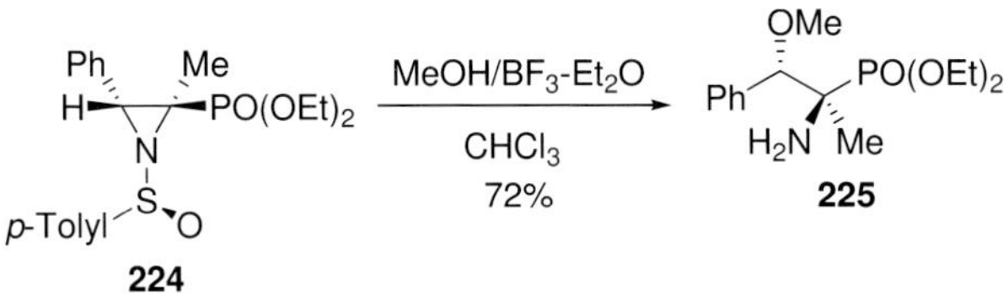

Scheme 3.83

When methanol was employed as the solvent for a long period of time, further reduction took place to give *cis*-piperidine **223** in 79% yield [83].

Treatment of *N*-sulfinylaziridine **224** (Scheme 3.83) with $BF_3 \cdot Et_2O$ in the presence of methanol afforded **225** in 72% yield [136].

Swern oxidation of *N*-unsubstituted aziridine-2-phosphonates resulted in the formation of both 2*H*-azirine-2-phosphonates and 2*H*-azirine-3-phosphonates [81, 83]. Treatment of *cis*-aziridine-2-phosphonate **226** (Scheme 3.84) with DMSO/$(COCl)_2/Et_3N$ afforded azirine-phosphonates **227** and **228**, with the former predominating [81]. Under similar conditions, however, *trans*-aziridine-2-phospho-

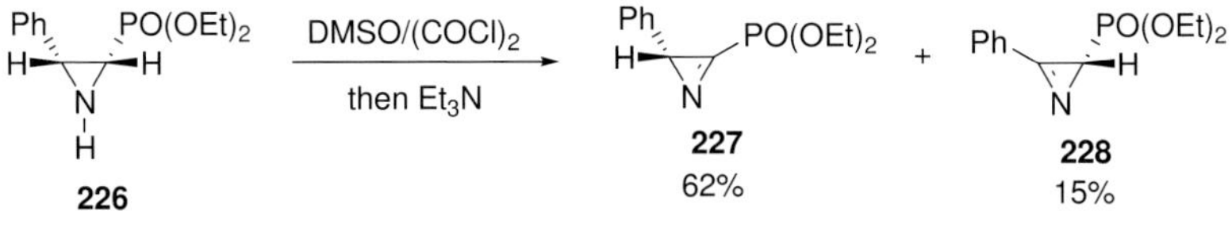

Scheme 3.84

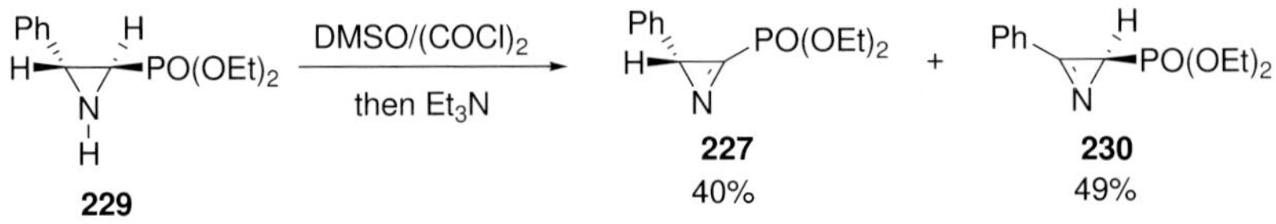

Scheme 3.85

nate **229** (Scheme 3.85) gave **227** and **230** in nearly equal amounts [81]. Arguments concerning the acidities of the C-2 and C-3 protons were advanced to explain these results [81].

3.4
Applications in Natural Product Syntheses

Aziridine-2-carboxylates are playing important roles in the synthesis of natural products and pharmaceutically useful molecules. In this section, applications of chiral nonracemic aziridine-2-carboxylates in the synthesis of natural products are discussed.

(2*S*,3*S*)-(+)-Aziridine-2,3-dicarboxylic acid (**234**; Scheme 3.86), an example of a naturally occurring aziridinecarboxylic acid, is a metabolite of *Streptomyces* MD398-Al. This aziridine was prepared by treatment of diethyl (2*R*,3*R*)-(–)-oxirane-2,3-dicarboxylate (**231**) with trimethylsilyl azide in EtOH/DMF to produced azido alcohol **232** [137], and treatment of this alcohol with triphenylphosphine afforded the aziridine dicarboxylate **233** in 71% yield. Hydrolysis of **233** afforded the natural product **234** in 69% yield.

Aziridine lactone **235** (Scheme 3.87) underwent ring-opening with allyl alcohol to give a 53% yield of α-amino lactone **236**, which was successfully transformed to the unnatural enantiomer of polyoxamic acid (–)-**237** [32].

One of the key steps in the synthesis of actinomycin D (**241**; Scheme 3.88) and its serine analogue involved the regio- and stereoselective ring-opening of aziridine **238** with the acid **239** [138, 139]. This transformation took place in methylene

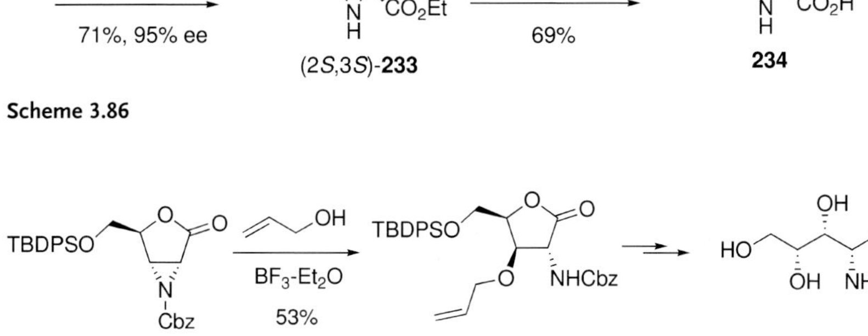

(2*R*,3*R*)-**231** → [1. $TMSN_3$, EtOH/DMF; 2. NH_4Cl, EtOH/H_2O; 72%] → (2*R*,3*S*)-**232** → [1. PPh_3, DMF; 2. heat, 3 h; 71%, 95% ee] → (2*S*,3*S*)-**233** → [1. LiOH, EtOH; 2. H^+; 69%] → **234**

Scheme 3.86

235 → [allyl alcohol, BF_3-Et_2O, 53%] → **236** → (-)-**237**

Scheme 3.87

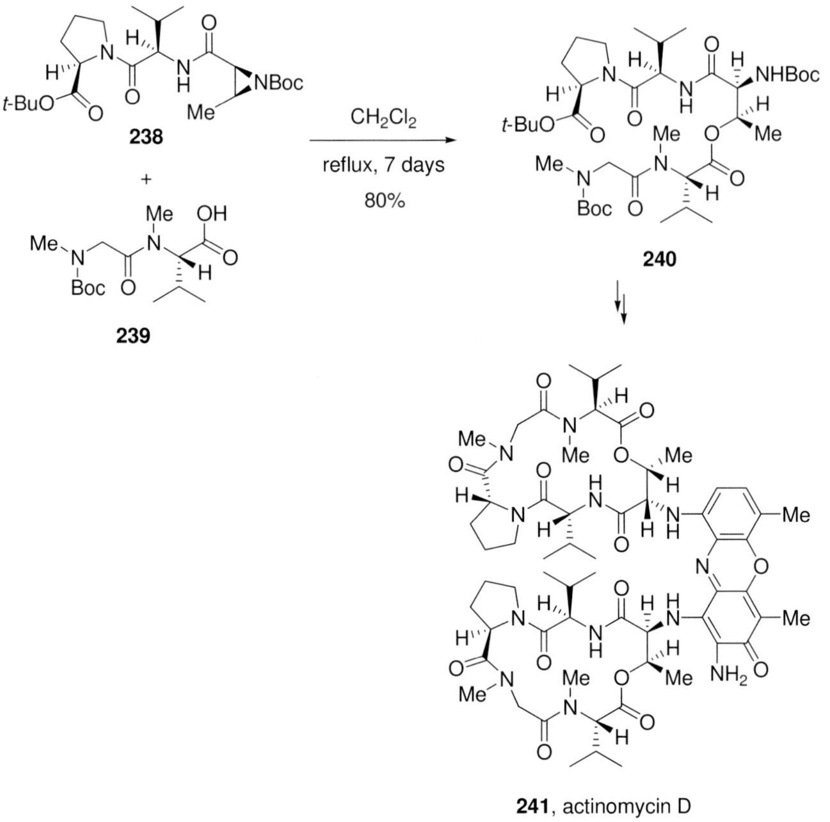

Scheme 3.88

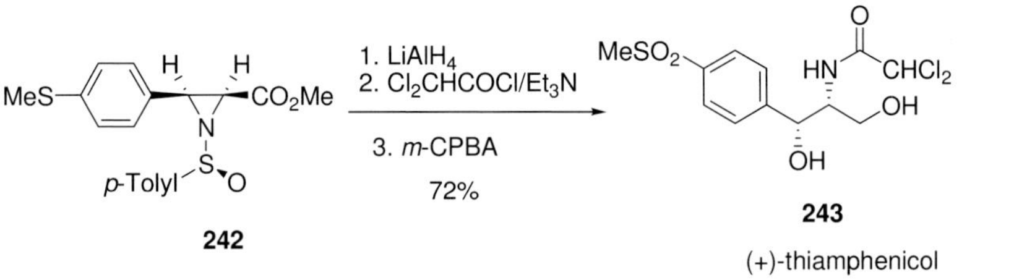

Scheme 3.89

chloride at reflux and gave the ester **240** in 80% isolated yield. Compound **240** has been successfully converted into the naturally occurring cyclic peptide lactone antibiotic actinomycin D (**241**).

When *N*-sulfinylaziridine-2-carboxylate **242** (Scheme 3.89) was treated with LAH, followed by *N*-acetylation and sulfide oxidation, (+)-thiamphenicol (**243**) was isolated in 72% yield [63]. Thiamphenicol is a broad-spectrum synthetic antibacter-

n-$C_{12}H_{25}$ CO_2Me H H N S p-Tolyl O **244**

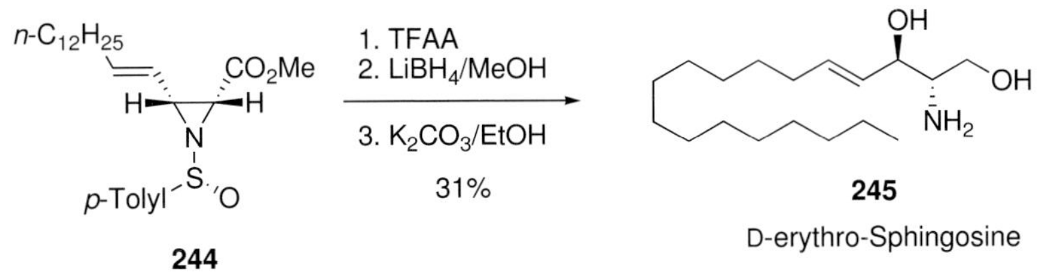

1. TFAA
2. $LiBH_4$/MeOH
3. K_2CO_3/EtOH
31%

OH OH NH_2 **245** D-erythro-Sphingosine

Scheme 3.90

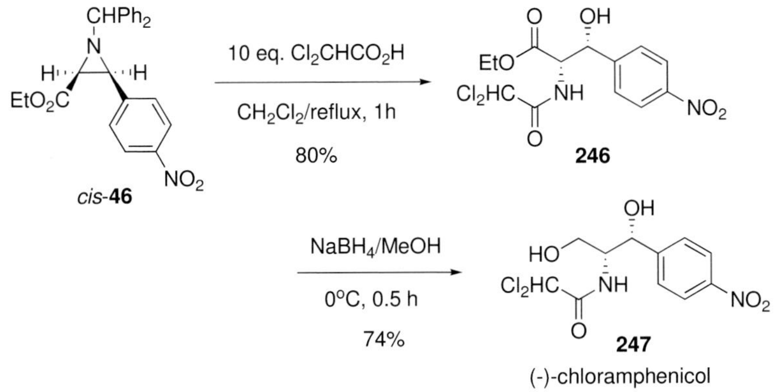

$CHPh_2$ N H H EtO_2C NO_2 *cis*-**46**

10 eq. Cl_2CHCO_2H
CH_2Cl_2/reflux, 1h
80%

O OH EtO Cl_2HC NH O NO_2 **246**

$NaBH_4$/MeOH
0°C, 0.5 h
74%

OH HO Cl_2HC NH O NO_2 **247** (-)-chloramphenicol

Scheme 3.91

ial agent used in the animal health industry. D-erythro-Sphingosine (**245**; Scheme 3.90), which is the major constituent of the lipid backbone of sphingolipids and plays an important role in cell recognition events, was prepared from *N*-sulfinylaziridine **244** in a procedure including a novel trifluoroacetic anhydride-induced Pummerer-type rearrangement of the aziridine [140]. Ring-opening of aziridine **46** with 10 equivalents of dichloroacetic acid afforded **246** in 80% yield (Scheme 3.91) [57]. Reduction of **246** with $NaBH_4$ gave (–)-chloramphenicol **247**, one of the oldest antibacterial agents, originally isolated from *Streptomyces venezuelae*.

Nakajima and Okawa prepared cysteine (**250**; Scheme 3.92) by treating aziridine **127a** with dry H_2S gas in the presence of a catalytic amount of $BF_3 \cdot Et_2O$ [141]. The resulting thiol **248** was then oxidized with iodine to afford disulfide **249** in 70% overall yield. Removal of the Cbz protecting group afforded cysteine (**250**).

Treatment of aziridine-2-carboxylic ester **251** (Scheme 3.93) with benzyl mercaptan in the presence of boron trifluoride etherate afforded **252** in 71% yield [142]. Compound **252** has been transformed into peptide **253**, an analogue of a penicillin precursor.

Protected 3-methyl-D-cystein (**257**; Scheme 3.94), a structural unit of the peptide antibiotics nisin and subtilin, has been synthesized through the ring-opening of the aziridinecarbamide **254** with thiobenzoic acid (**255**) [143, 144]. The reaction took place overnight at room temperature and in methylene chloride to give **256** in greater than 95% yield.

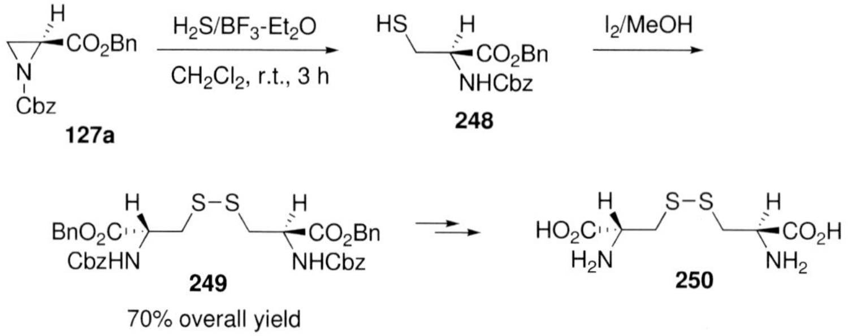

Scheme 3.92

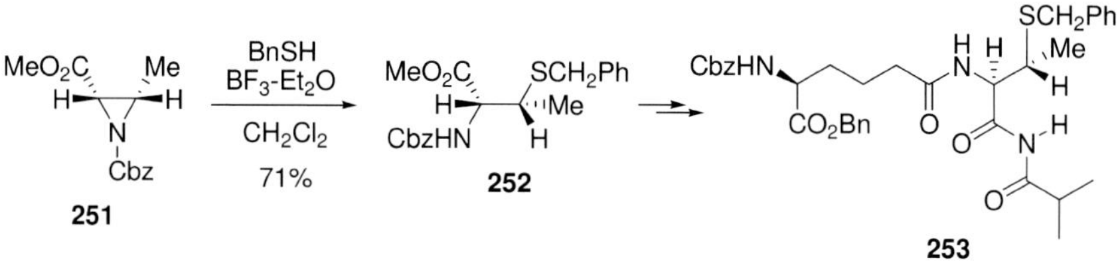

Scheme 3.93

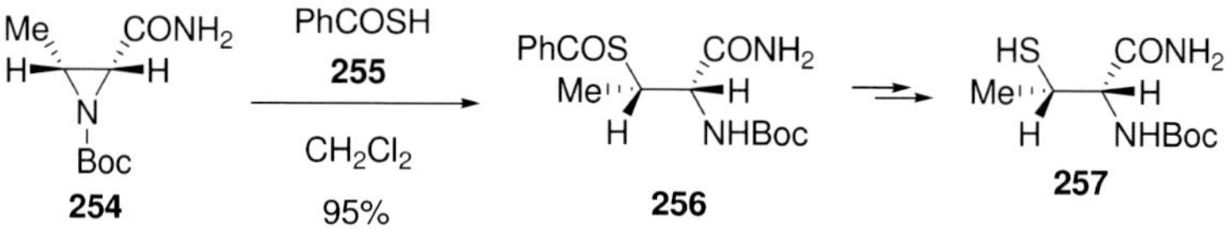

Scheme 3.94

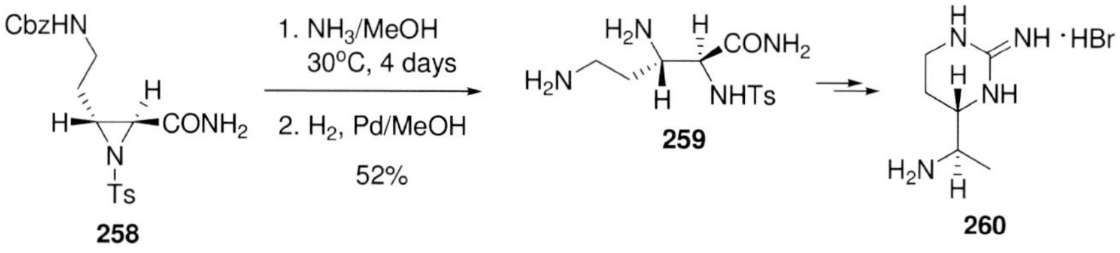

Scheme 3.95

(2*S*,3*R*)-Aziridine-2-carboxylic amide **258** (Scheme 3.95) has been used in the synthesis of the cyclic guanidino amino acid, L-epicapreomycidine (**260**) [145]. Treatment of **258** with saturated ammonia in methanol at 30 °C for 4 days in a pressure bottle resulted in the aziridine ring-opening product, which afforded **259** in 52% yield after removal of the Cbz protecting group.

FR900490 (**264**; Scheme 3.96), a new immunomodulator, has been prepared from aziridine-2-carboxylic ester **261** [146, 147]. This aziridine reacts with (*S*)-histidine (**262**) in the presence of 1 N sodium hydroxide, giving **263** in 31–44% yield

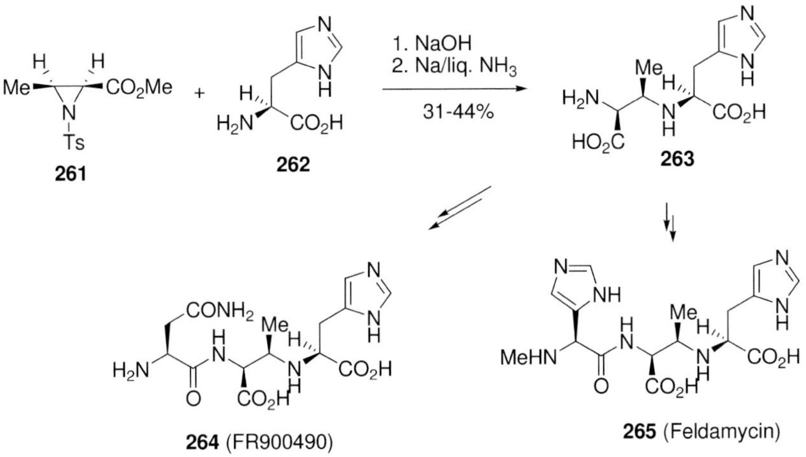

Scheme 3.96

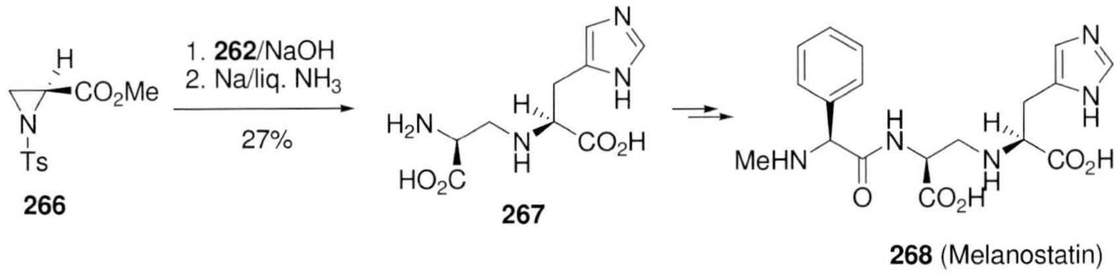

Scheme 3.97

after removal of the tosyl group with sodium in liquid ammonia. Feldamycin (**265**), an inhibitor of melanin synthesis in the growing cells of *S. bikiniensis* and B16 melanoma, was also prepared from aziridine **261**. On the other hand, melanostatin (**268**; Scheme 3.97) has been synthesized in a manner similar to **266** via **267** [147].

Aziridine **269** (Scheme 3.98), on treatment with pyrazole (**270**) and subsequent TFA deprotection, gave β-pyrazolylalanine **271**, first isolated from *Citrullus vulgaris* (a water melon) and isosteric with histidine [148]. Treatment of **269** with 1,2,4-oxadiazolidine-3,5-dione **272** gave (*S*)-quisqualic acid (**273**), an active ingredient of the ancient Chinese drug Shihchuntze, an anthelmintic made from the seeds of *Quisqualis indica* [148].

(α*R*,β*R*)-l,β-Dimethyltryptophan (**275**; Scheme 3.99), an intermediate in the syntheses of analogues of the potent PAF (platelet-activating factor) inhibitor FR900452 (**276**), has been prepared from aziridine **274** [149]. Treatment of aziridine **274** with *N*-methylindole (**167**) in the presence of boron trifluoride etherate gave **275** as single isomer in 60% yield. Other tryptophan derivatives have been synthesized similarly [104, 150].

[3+2] Cycloaddition reactions of aziridine-2-carboxylic esters have also been used

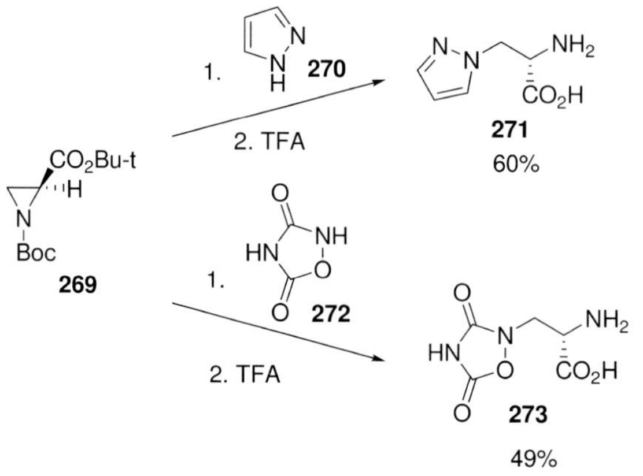

Scheme 3.98

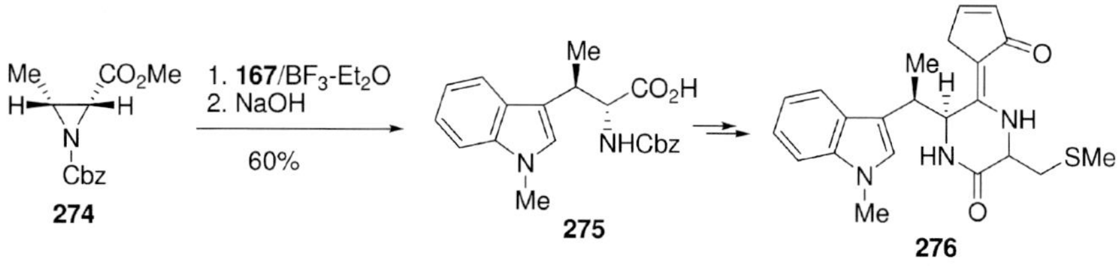

Scheme 3.99

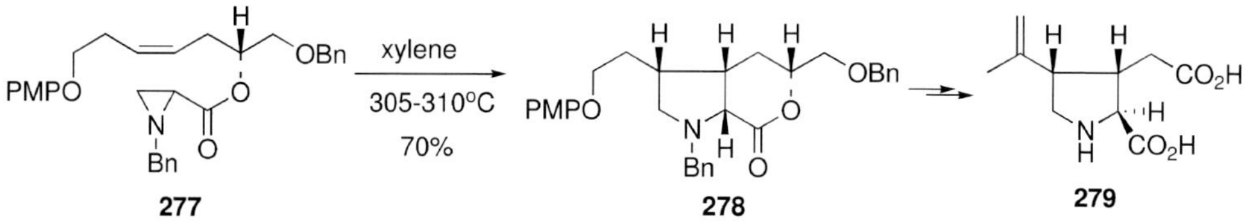

Scheme 3.100

in the synthesis of natural products. Thermolysis of aziridine **277** (Scheme 3.100) in xylene at 305–310 °C in a sealed tube afforded pyrrolidine lactone **278** in 70% yield [151]. This intramolecular cycloaddition was highly diastereoselective, affording **278** as a single isomer. Compound **278** has been further transformed into the natural product (–)-kainic acid (**279**), which possesses neuroexcitant properties as well as anthelmintic and insecticidal activities. A similar strategy has been used for the synthesis of acromelic acid B (**282**; Scheme 3.101) [152] and (–)-mesembrine (**285**; Scheme 3.102) from aziridines **280** and **283**, respectively [132c].

The marine cytotoxic antibiotic (*R*)-(–)-dysidazirine (**286**; Scheme 3.102) was prepared for the first time by Davis and co-workers from *N*-sulfinylaziridine **244** [95]. Treatment of *N*-sulfinylaziridine **244** with LDA at –95 °C in the presence of TMSCl afforded **286** in 62% yield [95].

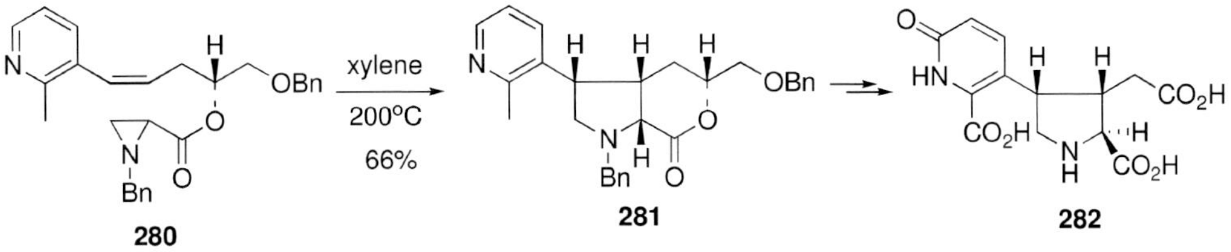

Scheme 3.101

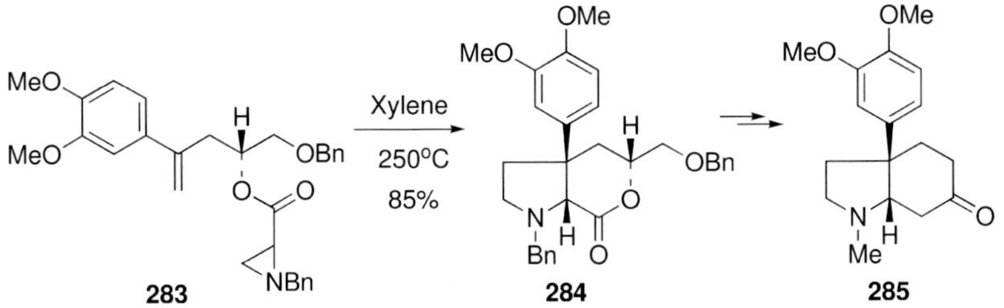

Scheme 3.102

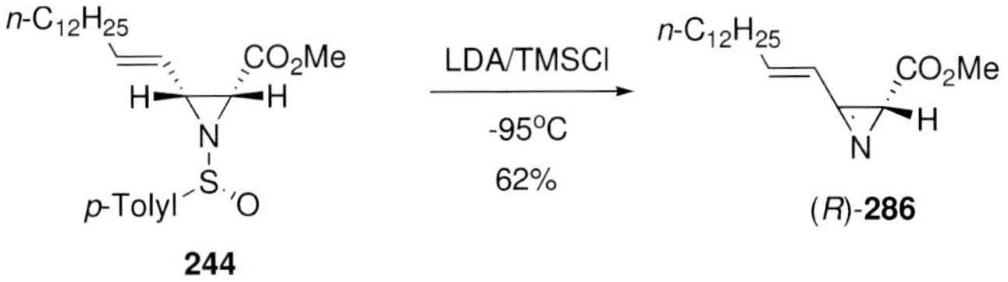

Scheme 3.103

3.5 Summary and Conclusions

The predictable regio- and stereoselective ring-opening reactions of aziridine-2-carboxylates with various nucleophiles serve as valuable sources of structurally diverse α- and β-amino acids. Activation of the aziridine nitrogen by an electron-withdrawing group (acyl, sulfonyl), by protonation, or by Lewis acids promotes either C-2 attack to give β-amino acids or C-3 attack to give α-amino acids. Since ring-opening does not affect the stereochemistry at C-2 or C-3, the aziridine stereochemistry is maintained in the product. The ring substituents and the reaction conditions determine the stereo- and regioselectivity of the ring-opening. Oxidation of *N*-unsubstituted aziridine-2-carboxylates produces 2*H*-azirine-2-carboxylates, providing additional opportunities for aziridinecarboxylate synthesis. While less is known concerning the properties of chiral aziridine-2-phosphonates, their ring-opening reactions to afford α- and β-amino phosphonates, important amino acids surrogates, appear to parallel those of the aziridine-2-carboxylates. The exception is that oxidation of *N*-unsubstituted aziridinephosphonates produces 2*H*-azirine-3-phosphonates, which are new chiral imino dienophiles. The availability of efficient methods for the asymmetric synthesis of aziridinecarboxylates and aziridinephosphonates, coupled with their varied chemistry, suggests that their utility as building blocks for asymmetric synthesis will greatly increase in the future.

References

1 (a) J. A. Deyrup in *Small Ring Heterocycles*; Hassner, A. Ed;, Wiley: New York, 1983; Vol. 1, p. 1. (b) W. H. Pearson, B. W. Lian, S. C. Bergmeier, *Aziridines and Azirines: Monocyclic*, in *Comprehensive Heterocyclic Chemistry II*; Padwa, A., Ed.; Pergamon: Oxford, 1996, p. 1.
2 B. Zwanenburg, *Pure & Appl. Chem.* **1999**, *71*, 423.
3 W. McCoull, F. A. Davis, *Synthesis* **2000**, 1347.
4 (a) K. Okawa, T. Kinutani, K. Sakai, *Bull. Chem. Soc. Jpn.* **1968**, *41*, 1353. (b) K. Okawa, K. Nakajima, T. Tanaka, Y. Kawana, *Chem. Lett.* **1975**, 591.
5 M. R. Paleo, N. Aurrecoechea, K.-Y. Jung, H, Rapoport, *J. Org. Chem.* **2003**, *68*, 130.
6 (a) K. Nakajima, F. Takai, T. Tamaka, K. Okawa, *Bull. Chem. Soc. Jpn.* **1978**, *51*, 1577. (b) M. A. Krook, M. J. Miller, *Org. Chem.* **1985**, *50*, 1126.
7 (a) F. Roeehiecioli, F.-X. Jarreau, M. Pais, *Tetrahedron Lett.* **1978**, *34*, 2917. (b) K. J. Shaw, J. R. Luly, H. Rapoport, *J. Org. Chem.* **1985**, *50*, 4515.
8 P. C. B. Page, S. M. Allin, S. J. Maddocks, M. R. J. Elsegood, *J. Chem. Soc. Perkin Trans. 1*, **2002**, 2827.
9 M.-C. Wang, D.-K. Wang, Y. Zhu, L.-T. Liu, Y.-F. Guo, *Tetrahedron: Asymmetry* **2004**, *15*, 1289.
10 A. S. Axelsson, K. J. O'Toole, P. A. Spencer, D. W. Young, *J. Chem. Soc. Chem. Commun.* **1991**, 1085.
11 E. Fernandez-Megia, M. A. Montaos, F. J. Sardina, *J. Org. Chem.* **2000**, *65*, 6780.
12 E. Kuyl-Yeheskiely, M. Lodder, G. A. van der Marel, J. H. van Boom, *Tetrahedron Lett.* **1992**, *33*, 3013.
13 (a) R. Haner, B. Olano, D. Seebach, *Helv. Chim. Acta.* **1987**, *70*, 1676. (b) M. M. Midland, R. L. Halterman, *J. Org. Chem.* **1981**, *46*, 1229.
14 (a) P. Wipf, C. P. Miller, *Tetrahedron Lett.* **1992**, *33*, 6267. (b) K. Nakajima, H. Sasaki, M. Neya, M. Morishita, S. Sakai, K. Okawa, *Pept. Chem.* **1982**, *19*; *Chem. Abstr.* **1983**, *99*, 122863y.
15 N. D. Smith, M. Goodman, *Org. Lett.* **2003**, *5*, 1035.
16 E. Kuyl-Yeheskiely, C. M. Dreef-Tromp, G. A. van der Marel, J. H. van Boom, *Reel. Trav. Chim. Pays-Bas*, **1989**, *108*, 314.
17 J. Legters, L. Thijs, B. Zwanenburg, *Tetrahedron Lett.* **1989**, *30*, 4881.
18 J. Legters, L. Thijs, B. Zwanenburg, *Recl. Trav. Chim. Pays-Bas* **1992**, *111*, 1.
19 J. Chun, L. He, B. Hoe-Sup; B. Robert. *J. Org. Chem.* **2000**, *65*, 7634.
20 L. Thijs, J. J. M. Porskamp, A. A. W. M. van Loon, M. P. W. Derks, R. W. Feenstra, J. Legters, B. Zwanenburg, *Tetrahedron* **1990**, *46*, 2611.
21 C. Xiong, W. Wang, C. Cai, V. J. Hruby, *J. Org. Chem.* **2002**, *67*, 1399.
22 C. Xiong, W. Wang, V. J. Hruby, *J. Org. Chem.* **2002**, *67*, 3514.
23 J. W. Lawn, T. Itoh, N. Ono, *Can. J. Chem.* **1973**, *51*, 856.
24 O. Ploux, M. Caruso, G. Chassaing, A. Marquet, *J. Org. Chem.* **1988**, *53*, 3154.
25 F. Polyak, T. Dorofeeva, R. Sturkovich, Y. Goldberg, *Synth. Commun.* **1991**, *21*, 239.
26 P. Garner, O. Dogan, S. Pillai, *Tetrahedron Lett.* **1994**, *35*, 1653.
27 J. M. Trochet, A. M. Massoud, *Heterocycle* **1989**, *29*, 419.
28 K. Jahnisch, F. Tittelbach, E. Grundemann, M. Schneider, *Eur. J. Org. Chem.* **2000**, 3957.
29 (a) K. Harada, I. Nakamura, *J. Chem. Soc. Chem. Commun.* 1978, 522. (b) I. Nakamura, K. Harada, *Chem. Lett.* **1979**, 313.
30 S. Farooq, W. Swain, Jr., R. Daeppen, G. Rihs, *Tetrahedron: Asymmetry* **1992**, *3*, 51.
31 G. Luppi, C. Tomasini, *Synlett.* **2003**, 797.

32 A. Tarrade, P. Dauban, R. H. Dodd, *J. Org. Chem.* **2003**, *68*, 9521.
33 M.-J. Tranchant, V. Dalla, I. Jabin, B. Decroix, *Tetrahedron* **2002**, *58*, 8425.
34 (a) R. S. Atkinson, G. Tughan, *J. Chem. Soc. Perkin Trans. 1*, **1987**, 2787. (b) R. S. Atkinson, G. Tughan, *J. Chem. Soc. Perkin Trans. 1*, **1987**, 2803. (c) R. S. Atkinson, J. Fawcett, D. R. Russell, G. Tughan, *J. Chem. Soc. Chem. Commun.* **1986**, 832.
35 (a) R. S. Atkinson, J. Fawcett, D. R. Russell, G. Tughan, *J. Chem. Soc. Chem. Commun.* **1987**, 458. (b) R. S. Atkinson, G. Tughan, *J. Chem. Soc. Chem. Commun.* **1987**, 456.
36 R. S. Atkinson, B. J. Kelly, *J. Chem. Soc. Chem. Commun.* **1987**, 1362.
37 G. V. Zhalnina, M. A. Kuznetsov, V. V. Semenovskii, G. V. Shustov, G. V. *Vestn. Leningr. Univ., Ser. 4: Fiz., Khim.* **1990**, 72; *Chem. Abstr.* **1991**, *114*, 163 892d.
38 K.-S. Yang, K. Chen, *J. Org. Chem.* **2001**, 66, 1676.
39 M. Egli, A. S. Dreiding, *Helv. Chim. Acta* **1986**, *69*, 1442.
40 D. Colantoni, S. Fioravanti, L. Pellacani, P. A. Tardella, *Org. Lett.* **2004**, *6*, 197.
41 G. Cardillo, S. Casolari, L. Gentilucci, C. Tomasini, *Angew. Chem. Int. Ed. Engl.* **1996**, *35*, 1848.
42 G. Cardillo, S. Fabbroni, L. Gentilucci, M. Gianotti, R. Perciaccante, S. Selva, A. Tolomelli, *Tetrahedron: Asymmetry* **2002**, *13*, 1411.
43 G. Cardillo, S. Fabbroni, L. Gentilucci, M. Gianotti, R. Percacciante, A. Tolomelli, *Tetrahedron: Asymmetry* **2002**, *13*, 1407.
44 J. Aires-de-Sousa, S. Prabhakar, A. M. Lobo, A. M. Rosa, M. J. S. Gomes, M. C. Corvo, D. J. Williams, A. J. P. White, *Tetrahedron: Asymmetry* **2002**, *12*, 3349.
45 G. Cardillo, L. Gentilucci, A. Tolomelli, *Tetrahedron Lett.* **1999**, *40*, 8261.
46 D. A. Evans, M. M. Faul, M. T. Bilodeau, B. A. Anderson, D. M. Barnes, *J. Am. Chem. Soc.* **1993**, *115*, 5328.
47 C. Langham, S. Taylor, D. Bethell, P. McMorn, P. C. B. Page, D. J. Willock, C. Sly, F. E. Hancock, F. King, G. J. Hutchings, *J. Chem. Soc. Perkin Trans. 2*, **1999**, 1043.
48 K. M. Gillespie, C. J. Sanders, P. O'Shaughnessy, I. Westmoreland, C. P. Thickitt, P. Scott, *J. Org. Chem.* **2002**, *67*, 3450.
49 K. M. Gillespie, E. J. Crust, R. J. Deeth, P. Scott, *J. Chem. Soc. Chem. Commun.* **2001**, 785.
50 M. Shi, C.-J. Wang, A. S. C. Chan, *Tetrahedron: Asymmetry* **2001**, *12*, 3105.
51 H. Suga, A. Kakehi, S. Ito, T. Ibata, T. Fudo, Y. Watanabe, Y. Kinoshita, *Bull. Chem. Soc. Jpn*, **2003**, *76*, 189.
52 J. C. Antilla, W. D. Wulff, *J. Am. Chem. Soc.* **1999**, *121*, 5099.
53 V. K. Aggarwal, M. Ferrara, C. J. O'Brien, A. Thompson, R. V. H. Jones, R. Fieldhouse, *J. Chem. Soc. Perkin Trans. 1*, **2001**, 1635.
54 M. Shi, C.-J. Wang, *Chirality* **2002**, *14*, 412.
55 J. C. Sheehan, K. Nakajima, E. Chacko, *Heterocycles* **1979**, *13*, 227.
56 J. R. Krumper, M. Gerisch, J. M. Suh, R. G. Bergman, T. D. Tilley, *J. Org. Chem.* **2003**, *68*, 9705.
57 C. Loncaric, W. D. Wulff, *Org. Lett.* **2001**, *3*, 3675.
58 J. C. Antilla, W. D. Wulff, *Angew. Chem. Int. Ed. Engl.* **2000**, *39*, 4518.
59 T. Akiyama, S. Ogi, K. Fuchibe, K. *Tetrahedron Lett.* **2003**, *44*, 4011.
60 K.-D. Lee, J.-M. Suh, J.-H. Park, H.-J. Ha, H. G. Choi, C. S. Park, J. W. Chang, W. K. Lee, Y. Dong, H. Yun, *Tetrahedron* **2001**, *57*, 8267.
61 T. Fujisawa, R. Hayakawa, M. Shimizu, *Tetrahedron Lett.* **1992**, *33*, 7903.
62 F. A. Davis, P. Zhou, G. V. Reddy, *J. Org. Chem.* **1994**, *59*, 3243.
63 F. A. Davis, P. Zhou, *Tetrahedron Lett.* **1994**, *35*, 7525.
64 F. A. Davis, H. Liu, P. Zhou, T. Fang, G. V. Reddy, Y. Zhang, *J. Org. Chem.* **1999**, *64*, 7559.
65 F. A. Davis, Y. Zhang, A. Rao, Z. Zhang, Z. *Tetrahedron* **2001**, *57*, 6345.
66 F. A. Davis, J. Deng, Y. Zhang, R. C. Haltiwanger, *Tetrahedron* **2002**, *58*, 7135.
67 G. A. Molander, P. J. Stengel, P. J. *Tetrahedron* **1997**, *53*, 8887.
68 S. Timen, P. Somfai, P. *J. Org. Chem.* **2003**, *68*, 9958.
69 K. Hada, T. Watanabe, T. Isobe, T. Ishikawa, *J. Am. Chem. Soc.* **2001**, *123*, 7705.
70 T. Satoh, Y. Fukuda, *Tetrahedron* **2003**, *59*, 9803.

71 Y. Yamauchi, T. Kawate, T. Katagiri, K. Uneyama, *Tetrahedron* **2003**, *59*, 9839.

72 E. Francotte, R. M. Wolf, *J. Chromatography* **1992**, *595*, 63.

73 K. Mori, F. Toda, *Tetrahedron: Asymmetry* **1990**, *1*, 281.

74 B. Saha, J. P. Nandy, S. Shukla, I. Siddiqui, J. Iqbal, *J. Org. Chem.* **2002**, *67*, 7858.

75 V. F. Fudchenko, O. A. D'yachenko, A. B. Zolotoi, L. O. Atovmyan, I. I. Chervin, R. G. Kostyanovsky, *Tetrahedron* **1982**, *38*, 961.

76 M. Bucciarelli, A. Forni, I. Moretti, F. Prati, *Tetrahedron: Asymmetry* **1990**, *1*, 5.

77 L. Antolini, M. Bucciarelli, A. Forni, I. Moretti, F. Pratl, G. Torre, *J. Chem. Soc. Perkin Trans. 2* **1992**, 959.

78 H. M. S. Kumar, M. S. Rao, P. P. Chakravarthy, J. S. Yadav, J. S. *Tetrahedron: Asymmetry* **2004**, *15*, 127.

79 A. A. Thomas, K. B. Sharpless, *J. Org. Chem.* **1999**, *64*, 8379.

80 C. Pousset, M. Larcheveque, *Tetrahedron Lett.* **2002**, *43*, 5257.

81 F. A. Davis, W. McCoull, *Tetrahedron Lett.* **1999**, *40*, 249.

82 F. A. Davis, Y. Wu, H. Yan, W. McCoull, K. R. Prasad, *J. Org. Chem.* **2003**, *68*, 2410.

83 F. A. Davis, Y. Wu, H. Yan, K. R. Prasad, W. McCoull, *Org. Lett.* **2002**, *4*, 655.

84 D. Y. Kim, K. H. Suh, J. S. Choi, J. Y. Mang, S. K. Chang, *Synth. Commun.* **2000**, *30*, 87.

85 F. A. Davis, T. Ramachandar, Y. Wu, *J. Org. Chem.* **2003**, *68*, 6894.

86 F. Palacios, A. M. Ochoa de Retana, J. I. Gil, *Tetrahedron Lett.* **2000**, *41*, 5363.

87 F. Palacios, D. Aparicio, M. A. Ochoa de Retana, J. M. de los Santos, J. I. Gil, R. Lopez de Munain, *Tetrahedron: Asymmetry* **2003**, *14*, 689.

88 T. Satoh, M. Ozawa, K. Takano, M. Kudo, *Tetrahedron Lett.* **1998**, *39*, 2345.

89 F. A. Davis, C.-H. Liang, H. Liu, *J. Org. Chem.* **1997**, *62*, 3796.

90 Y. Lim, W. K. Lee, *Tetrahedron Lett.* **1995**, *36*, 8431.

91 E. N. Prabhakaran, J. P. Nandy, S. Shukla, A. Tewari, S. Kumar Das, J. Iqbal, *Tetrahedron Lett.* **2002**, *43*, 6461.

92 N. J. Church, D. W. Young, *Tetrahedron Lett.* **1995**, *36*, 151.

93 N. H. Kawahata, M. Goodman, *Tetrahedron Lett.* **1999**, *40*, 2271.

94 K. Nakajima, H. Oda, K. Okawa, *Bull. Chem. Soc. Jpn.* **1982**, *55*, 3232.

95 F. A. Davis, H. Liu, C.-H. Liang, G. V. Reddy, Y. Zhang, T. Fang, D. D. Titus, *J. Org. Chem.* **1999**, *64*, 8929.

96 K. Okawa, K. Nakajima, T. Tanaka, M. Neya. *Bull. Chem. Soc. Jpn.* **1982**, *55*, 174.

97 J. Legters, L. Thijs, B. Zwanenburg, *Recl. Trav. Chim. Pays-Bas* **1992**, *111*, 16.

98 J. Legters, J. E. H. Willems, L. Thijs, B. Zwanenburg, *Recl. Trav. Chim. Pays-Bas* **1992**, *111*, 59.

99 G. Righi, R. DÀchille, C. Bonini, *Tetrahedron Lett.* **1996**, *37*, 6893.

100 D. Tanner, C. Birgersson, H. K. Dhaliwal, *Tetrahedron Lett.* 1990, *31*, 1903.

101 F. A. Davis, H. Liu, G. V. Reddy, *Tetrahedron Lett.* **1996**, *37*, 5473.

102 (a) K. Nakajima, M. Neya, S. Yamada, K. Okawa, K. *Bull. Chem. Soc. Jpn.* **1982**, *55*, 3049. (b) M. Ho, W. Wang, M. Douvlos, T. Pham, T. Klock, *Tetrahedron Lett.* **1991**, *32*, 1283.

103 N. Hayashi, K. Okawa, *Bull. Chem. Soc. Jpn.* **1979**, *52*, 3579.

104 L. Dubois, A. Mehta, E. Tourette, R. H. Dodd, *J. Org. Chem.* **1994**, *59*, 434.

105 C. Tomasini, A. Vecchione, *Org. Lett.* **1999**, *1*, 2153.

106 G. Cardillo, L. Gentilucci, A. Tolomelli, C. Tomasini, *J. Org. Chem.* **1998**, *63*, 3458.

107 K. Okawa, M. Yuki, T. Tanaka, *Chem. Lett.* **1979**, 1085.

108 R. J. Parry, A. E. Mizusawa, I. C. Chiu, M. V. Naidu, M. Riceiardone, *J. Am. Chem. Soc.* **1985**, *107*, 2512.

109 R. J. Parry, M. V. Naidu, *Tetrahedron Lett.* **1983**, *24*, 1133.

110 K. Nakajima, H. Oda, K. Okawa, *Bull. Chem. Soc. Jpn.* **1983**, *56*, 520.

111 H. Shao, Q. Zhu, M. Goodman, *J. Org. Chem.* **1995**, *60*, 790.

112 Y. Hata, M. Watanabe, *Tetrahedron* **1987**, *43*, 3881.

113 J. Xu, *Tetrahedron: Asymmetry* **2002**, *13*, 1129.

114 (a) H. Han, J. Yoon, K. D. Janda, *J. Org. Chem.* **1998**, *63*, 2045. (b) H. Harada, T. Morie, T. Suzuki, T. Yoshida, S. Kato, *Tetrahedron* **1998**, *54*, 10671. (c) S. Kato, H. Harada, T. Morie, *J. Chem. Soc. Chem. Commun.* **1997**, 3219.

115 L. Antolini, M. Bucciarelli, E. Caselli, P. Davoli, A. Forni, I. Moretti, F. Prati, G. Torre, *J. Org. Chem.* **1997**, *62*, 8784.

116 J. E. Baldwin, R. M. Adlington, I. A. O'Neil, C. Schofield, A. C. Spivey, J. B. Sweeney, *J. Chem. Soc. Chem. Commun.* **1989**, 1852.

117 B. C. M. Burgaud, D. C. Horwell, A. Padova, M. C. Pritchard, *Tetrahedron* **1996**, *52*, 13035.

118 J. E. Baldwin, A. C. Spivey, C. J. Schofield, J. B. Sweeney, *Tetrahedron Lett.* **1993**, *49*, 6309.

119 R. Reddy, J. B. Jaquith, W. R. Neelagiri, S. Saleh-Hanna, T. Durst, *Org. Lett.* **2002**, *4*, 695.

120 F. Yokokawa, H. Sugiyama, T. Aoyama, T. Shioiri, *Synthesis* **2004**, 1476.

121 J. E. Baldwin, R. M. Adlington, N. G. Robinson, *J. Chem. Soc. Chem. Commun.* **1987**, 153.

122 D. Seebach, R. Haner, R. *Chem. Lett.* **1987**, 49.

123 G. Cardillo, L. Gentilucci, A. Tolomelli, C. Tomasini, *Tetrahedron Lett.* **1997**, *38*, 6953.

124 G. Cardillo, L. Gentilucci, A. Tolomelli, *J. Chem. Soc. Chem. Commun.* **1999**, 167.

125 G. Cardillo, L. Gentilucci, A. Tolomelli, *Tetrahedron* **1999**, *55*, 15151.

126 (a) F. W. Eastwood, P. Perlmutter, Q. Yang, *Tetrahedron Lett.* **1994**, *35*, 2039. (b) F. W. Eastwood, P. Perlmutter, Q. Yang, *J. Chem. Soc. Perkin Trans. 1*, **1997**, 35.

127 B. F. Bonini, M. Fochi, M. Comes-Franchini, A. Ricci, L. Thijs, B. Zwanenburg, *Tetrahedron: Asymmetry* **2003**, *14*, 3321.

128 T. B. Sim, S. H. Kang, K. S. Lee, W. K. Lee, H. Yun, Y. Dong, H.-J. Ha, *J. Org. Chem.* **2003**, *68*, 104.

129 Z. Bouayad, J. Chanet-Ray, S. Ducher, R. Vessiere, *J. Heterocycl. Chem.* **1991**, *28*, 1757.

130 I. Funaki, L. Thijs, B. Zwanenburg, *Tetrahedron* **1996**, *52*, 9909.

131 Y. Gelas-Mialhe, T. Touraud et Roger, R. Yessiere, *Can. J. Chem.* **1982**, 60, 2830.

132 (a) S. Takano, T. Iwabuchi, K. Ogasawara, *J. Chem. Soc. Chem. Commun.* **1988**, 1204. (b) S. Takano, S. Tomita, T, Iwabuchi, K. Ogasawara, *Heterocycles* **1989**, *29*, 1473. (c) S. Takano, K. Samizu, K. Ogasawara, *Chem. Lett.* **1990**, 1239.

133 J. Vebrel, D. Gree, R. Carrie, *Can. J. Chem.* **1984**, *62*, 939.

134 S. Takano, M. Moriya, K. Ogasawara, *Tetrahedron: Asymmetry* **1992**, *3*, 681.

135 A. Derdour, F. Texier, *Can. J. Chem.* **1985**, *63*, 2245.

136 F. A. Davis, W. McCoull, D. D. Titus, *Org. Lett.* **1999**, *1*, 1053.

137 J. Legters, L. Thijs, B. Zwanenburg, *Tetrahedron* **1991**, *47*, 5287.

138 (a) T. Tanaka, K. Nakajima, K. Okawa, *Bull. Chem. Soc. Jpn.* **1980**, *53*, 1352. (b) K. Okawa, K. Nakajima, T. Tanaka, *J. Heterocycl. Chem.* **1980**, *17*, 1815.

139 (a) K. Nakajima, T. Tanaka, M. Neya, K. Okawa, *Bull. Chem. Soc. Jpn.* **1982**, *55*, 3237. (b) K. Nakajima, T. Tanaka, M. Neya, K. Okawa, *Pept. Chem.* **1981**, *19*, 143..

140 F. A. Davis, G. V. Reddy, *Tetrahedron Lett.* **1996**, *37*, 4349.

141 K. Nakajima, K. Okawa, *Bull. Chem. Soc. Jpn.* 1983, *56*, 1565.

142 J. E. Baldwin, R. M. Adlington, N. Moss, N. G. Robinson, *J. Chem. Soc. Chem. Commun.* **1987**, 1664.

143 (a) T. Wakamiya, K. Shimbo, T. Shiba, K. Nakajima, M. Neya, K. Okawa, *Bull. Chem. Soc. Jpn.* **1982**, *55*, 3878. (b) T. Wakamiya, K. Fukase, K. Shimbo, T. Shiba, *Bull. Chem. Soc. Jpn.* **1983**, *56*, 1559.

144 K. Nakajima, K. Nakashio, K. Okawa, *Pept. Chem.* **1983**, *21*, 19.

145 T. Teshima, K. Konishi, T. Shiba, *Bull. Chem. Soc. Jpn.* **1980**, *53*, 508.

146 N. Shigemaatsu, H. Setoi, I. Uchida, *Tetrahedron Lett.* **1988**, *29*, 5147.

147 K. Imae, H. Kamachi, H. Yamashita, T. Okita, S. Okuyama, T. Tsuno, T. Yamasaki, Y. Sawada, M. Ohbayashi, T. Naito, T. Oki, *J. Antibiotics* **1991**, *44*, 76.

148 C. N. Farthing, J. E. Baldwin, A. T. Russell, C. J. Schofield, A. C. Spivey, *Tetrahedron Lett.* **1996**, *37*, 5225.

149 I. Shima, N. Shimazaki, K. Imai, K. Hemmi, M. Hashimoto, *Chem. Pharm. Bull.* **1990**, *38*, 564.

150 K. Sato, A. P. Lozikowski, *Tetrahedron Lett.* **1989**, *30*, 4073.

151 S. Takano, Y. Iwabuchi, K, Ogasawara, *J. Chem. Soc. Chem. Commun.* **1988**, 1204.

152 S. Takano, S. Tomita, Y. Iwabuchi, K. Ogasawara, *Heterocycles* **1989**, *29*, 1473.

4
Synthesis of Aziridines

Dedicated, with respect, to Professor Sir Charles Rees, FRS

Joseph B. Sweeney

4.1
Introduction

In 1999, Bob Atkinson wrote [1] that aziridination reactions were "<qs>epoxidation's poor relation<qe>", and this was undoubtedly true at that time: the scope of the synthetic methods available for preparation of aziridines was rather narrow when compared to the diversity of the procedures used for the preparation of the analogous oxygenated heterocycles. The preparation of aziridines has formed the basis of several reviews [2] and the reader is directed towards those works for a comprehensive analysis of the area: this chapter presents a concise overview of classical methods and focuses on modern advances in the area of aziridine synthesis, with particular attention to stereoselective reactions between nitrenes and alkenes on the one hand, and carbenes and imines on the other.

4.2
Overview and General Features

Those reactions that have found general use for the preparation of aziridines can be grouped into two broad classes: *addition* and *cyclization* processes, and each of these categories can be further divided. Addition processes can be classified as being C_2+N_1 reactions (addition of nitrenes, or nitrene equivalents ["nitrenoids"], to alkenes; Scheme 4.1) or C_1N_1+C_1 reactions (addition of carbenes or carbenoids to imines; Scheme 4.2).

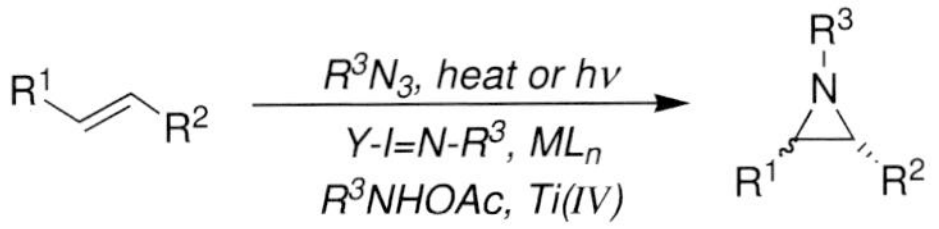

R^3N_3, heat or hν
Y-I=N-R^3, ML_n
R^3NHOAc, Ti(IV)

C_2+N_1 reactions: addition of nitrenes, or nitrenoids, to alkenes

Scheme 4.1

Aziridines and Epoxides in Organic Synthesis. Andrei K. Yudin

ISBN: 3-527-31213-7

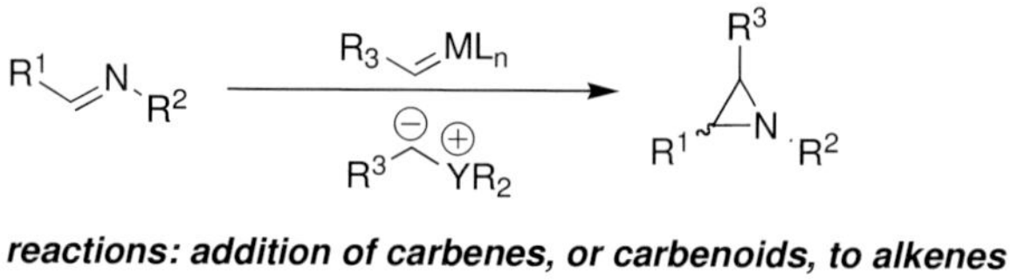

C_1N_1 + C_1 reactions: addition of carbenes, or carbenoids, to alkenes

Scheme 4.2

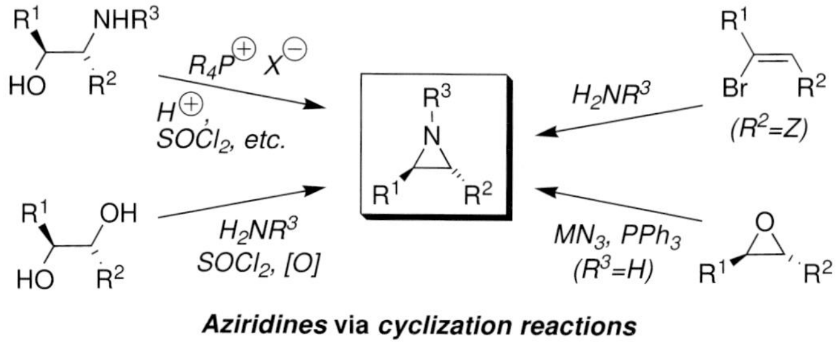

Aziridines* via *cyclization reactions

Scheme 4.3

The carbene-imine route to aziridines has attracted increasing attention of late, though there have also been notable recent advances in addition processes involving alkenes and nitrene equivalents.

Cyclization processes involve ring-closure of amino alcohols (Scheme 4.3) or equivalents, and as the aziridine precursors are often available as single stereoisomers, the method is an attractive one if the aziridine target is a key synthetic intermediate.

4.2.1 Addition to Alkenes

The addition of oxygen to alkenes by use of "active" oxygen donors is the most widely used generally applicable method for epoxide synthesis, but (in contrast with the rapid reactions of alkenes with peroxyacids and alkylhydroperoxides) the analogous nitrogenous reaction has not been widely used. Alkenes are often inert to the influence of the aza-analogues, *O*-acylhydroxylamines, and *O*-alkylhydroxylamines; *O*-sulfonyl and *O*-alkylhydroxylamines have been used as aziridinating reagents, but the scope of such reactions is limited [3].

The analogy with epoxidation of alkenes by use of dioxiranes is better maintained, and the analogous compounds (diaziridines and oxaziridines) have been shown to act as promising aziridination agents (Schemes 4.4 [4] and 4.5 [5]); Houk and Armstrong [6] have carried out modeling studies of the requirements for efficient aziridination through the use of oxaziridines and diaziridines, and have identified *N*-silyl, *N*-trifluoroacetyl, and *N*-alkyl oxaziridines and diaziridiniums as potentially useful alkene aziridinators, but the area is far from mature.

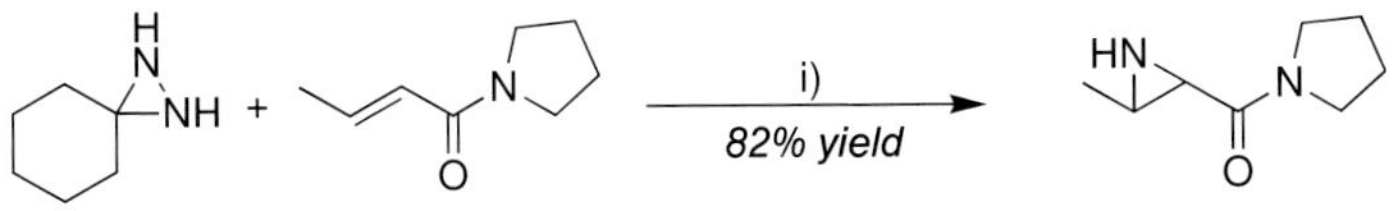

i) BuLi, THF, 78 °C→rt

82% yield

Scheme 4.4

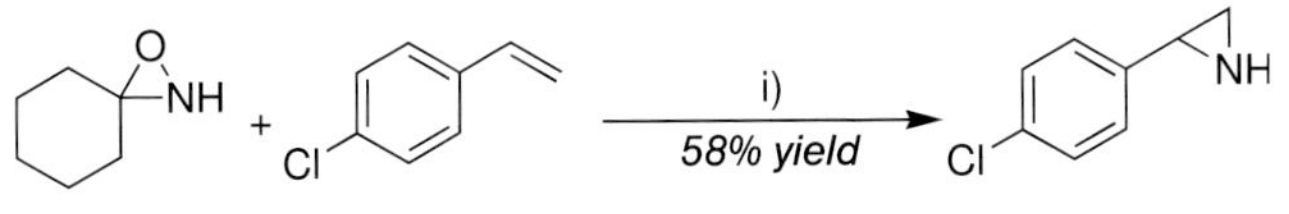

i) $PhCH_3$, 100 °C

58% yield

Scheme 4.5

Thus, the key reason for the paucity of methods available by analogy with epoxidation methods is the comparative inertness of N–O and N–N bonds relative to the peroxide bond. This means that the synthetic methods that have been developed for preparation of aziridines are distinct from those that have evolved for epoxide synthesis.

4.2.1.1 **Addition of Nitrenes and Nitrenoids to Alkenes**

There are two general methods within this subcategory, involving one- or two-step mechanisms. Nitrenes and metalonitrenes thus add to alkenes by a direct aziridination reaction, whereas nonmetallic nitrenoids usually react through an addition-elimination process (Scheme 4.6).

Notwithstanding the drawbacks to the method, the addition of nitrenes to alkenes is a well studied classical method for direct aziridination. The original reactions (often involving alkoxycarbonylnitrenes) employed harsh conditions, resulting in nonstereoselective transformations. In these pioneering reports, the requi-

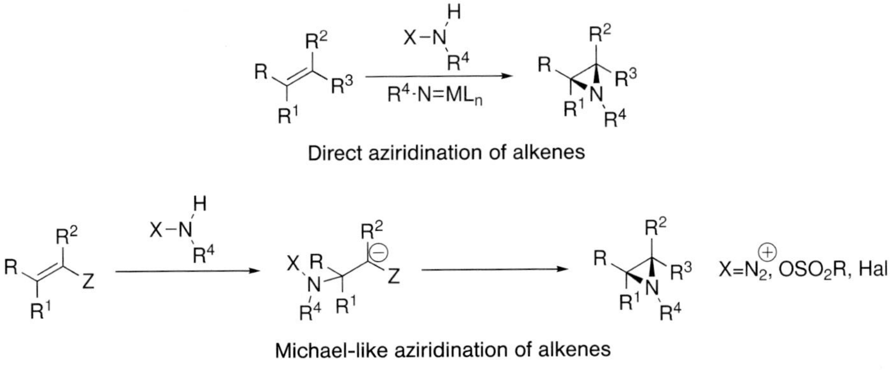

R^4·N=ML_n

Direct aziridination of alkenes

X=N_2^+, OSO_2R, Hal

Michael-like aziridination of alkenes

Scheme 4.6

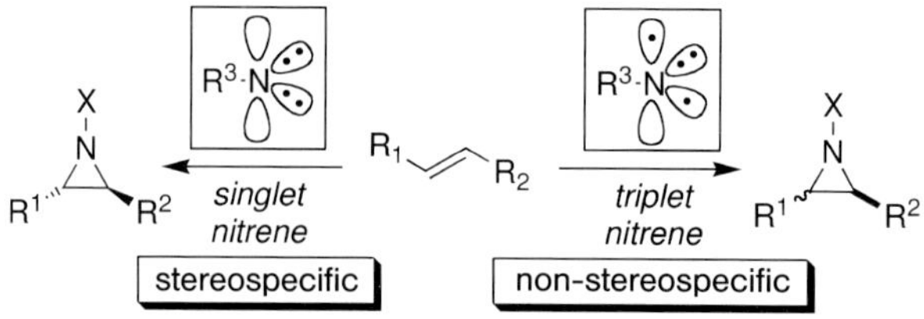

Scheme 4.7

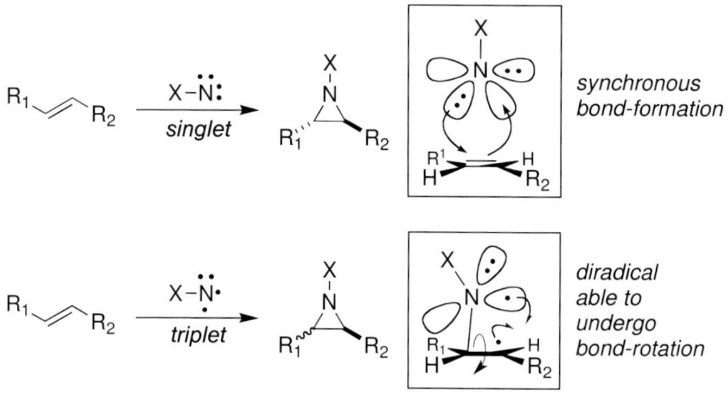

Scheme 4.8

site acylnitrenes were typically generated by thermal or photochemical decomposition of the corresponding azides, and these methods intrinsically give mixtures of (more reactive) singlet and (more stable) triplet nitrenes; only singlet nitrenes react stereospecifically with 1,2-disubstituted alkenes, while triplet nitrenes react unselectively (Scheme 4.7).

"Free" nitrenes exist as mixtures of singlet and triplet forms and, since the triplet is simultaneously more thermodynamically stable and unable to react by a one-step mechanism, mixtures of *cis* and *trans* isomers invariably resulted from the early methods (Scheme 4.8).

When unacylated azides are used as nitrene precursors, the first reaction with an alkene is a cycloaddition, generating the corresponding 1,2,3-triazoline, which often eliminates N_2 under the fierce reaction conditions to give an aziridine product (Scheme 4.9).

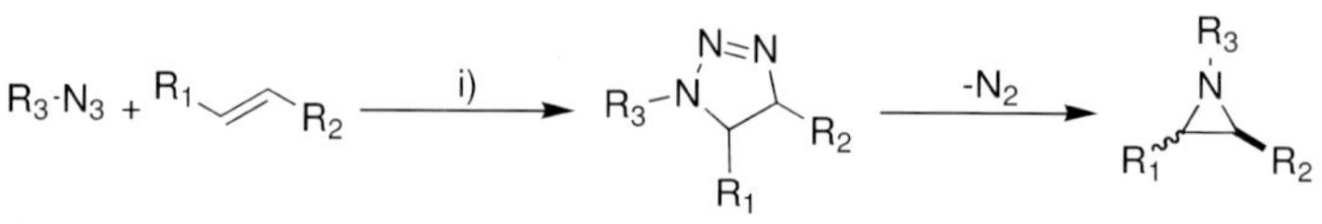

i) hν or Δ

Scheme 4.9

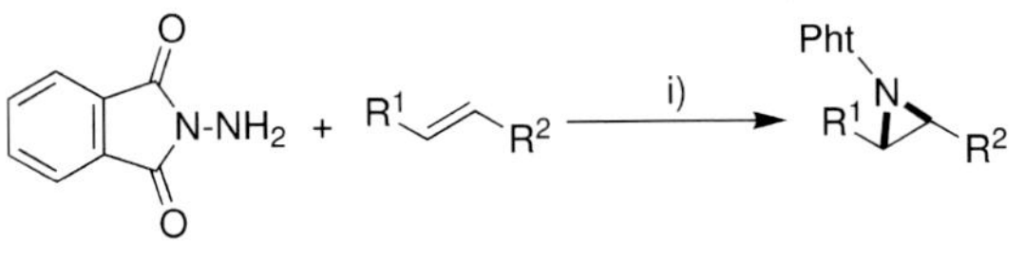

i) $Pb(OAc)_4$

Scheme 4.10

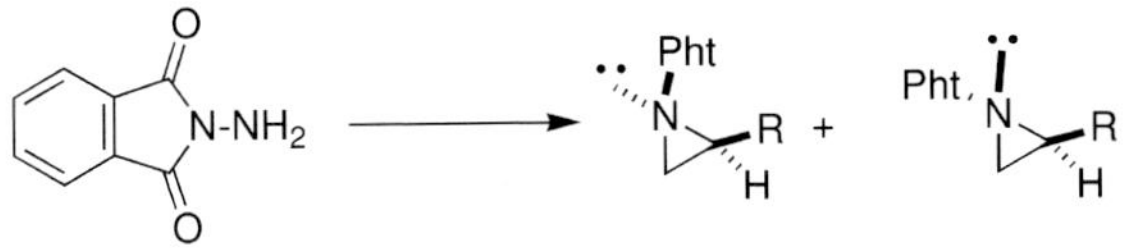

-20 °C 100:0 (R=Ph, CO_2Me)

0 °C 0:100 (R=Ph)
14:86 (R=CO_2Me)

i) $RCH{=}CH_2$, $Pb(OAc)_4$

Scheme 4.11

In many instances, however, the intermediate triazoline can be isolated and separately converted into the aziridine, often with poor stereoselectivity. The first practical modification to the original reaction conditions generated the (presumed) nitrenes by *in situ* oxidation of hydrazine derivatives. Thus, Atkinson and Rees prepared a range of *N*-amino aziridine derivatives by treatment of *N*-aminophthalimides (and other *N*-aminoheterocycles) with alkenes in the presence of lead tetraacetate (Scheme 4.10) [7].

These reactions proved highly selective, and considerably more stereoselective than those of nitrenes generated by the methods mentioned previously, seemingly implying that the nitrene intermediate was formed in the singlet state. The reactions are highly diastereoselective even for monosubstituted alkenes, the selectivity arising from the fact that substituted aziridines are chiral at nitrogen: styrene and methyl acrylate thus react with *N*-aminophthalimide at low temperature (–20 °C) to give exclusively *cis*-*N*-phthalimidoaziridines (Scheme 4.11) [8].

This unusually high stereoselectivity is due to the high barrier to *N*-inversion and the presence of the electron-withdrawing phthalimido group. It can be demonstrated that the reaction is under kinetic control when the reaction mixtures described above are allowed to warm to temperatures near or above 0 °C, whereupon a partial or complete inversion of configuration at nitrogen can be seen (by NMR spectroscopy).

It has since transpired that the active aziridinating species in these reactions are the *N*-acetoxylated hydrazine derivatives, rather than nitrenes: exposure of ethylquinazolinone ("Q^1-NH_2") to lead tetraacetate at low temperature generates *N*-(acetoxyamino)-quinazolinone ("Q^1-NHOAc"), which is a relatively stable sub-

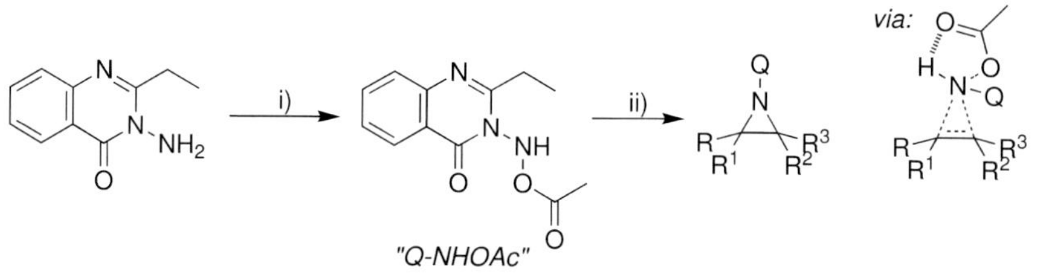

"Q-NHOAc"

i)Pb(OAc)$_4$, CH$_2$Cl$_2$, -20 °C ii) RR^{1}C=CR^{2}R^3

Scheme 4.12

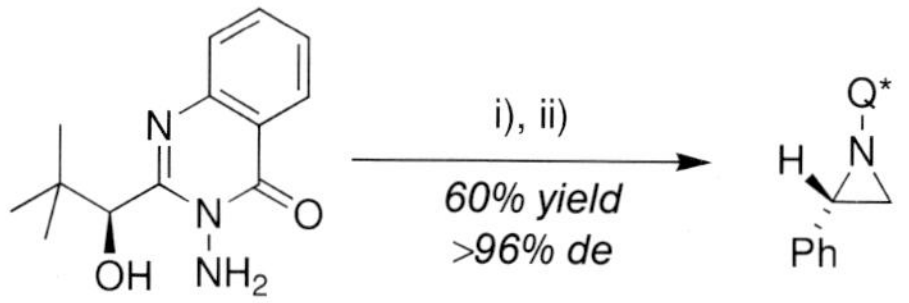

i), ii) 60% yield >96% de

i) Pb(OAc)$_4$, -20ûC; ii) styrene, Ti(OBut)$_4$

Scheme 4.13

stance, capable of aziridinating alkenes in good yields to give the corresponding *N*-quinazolinonylaziridines (Scheme 4.12) [9].

These reagents may be considered to be one of the elusive aza-analogues of peroxyacids, and there are significant mechanistic similarities between the Rees-Atkinson reaction and the Bartlett epoxidation. Chiral Q-reagents have been used to effect highly stereoselective aziridination of alkenes (Scheme 4.13) [1].

Although an efficient reaction, the Rees-Atkinson aziridination method suffers from two drawbacks: the necessity for an *N*-phthalimido or *N*-quinazolinonyl substituent and the use of a highly toxic oxidant. Thus, recent efforts (especially in these green times) have focussed upon more benign methods for generation of the key nitrenoids. Yudin demonstrated the power of electrochemistry with a novel method that removes the need for an added metal oxidant, demonstrating an unusually and impressively broad substrate tolerance compared to many alkene aziridination reactions (Scheme 4.14) [10].

Che et al. have reported the use of iodobenzene diacetate as an alternative to lead tetraacetate in the original Rees-Atkinson reactions of a relatively narrow range of olefin substrates (primarily styrenes) [11].

Since the mid-1990s, synthetic attention has been directed more towards the use of metal-stabilized nitrenes as synthetic effectors of alkene aziridination. In 1969 it was reported that Cu(I) salts were capable of mediating alkene aziridination when treated with tosyl azide, but the method was limited in scope and was not adopted as a general method for the synthesis of aziridines [12]. Metaloporphyrins [13] were shown to be catalysts for the aziridination of alkenes in the presence of the nitrene precursor *N*-tosyliminophenyliodinane [14] in the early 1980s, but the reaction did

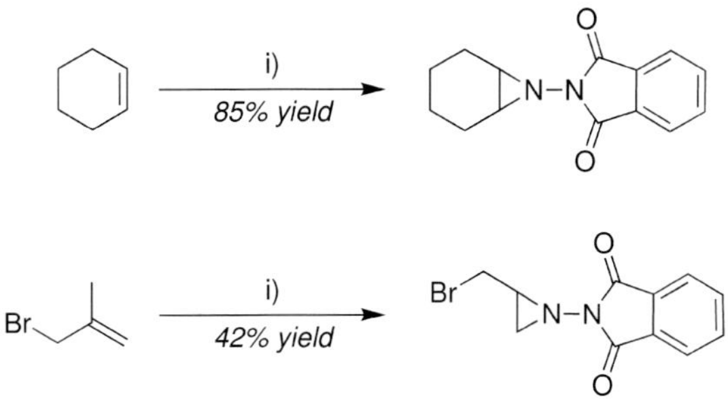

i) *N*-Aminophthalimide, + 1.80 V (*vs* Ag wire), MeCN, Et_3NHOAc, 25 °C

Scheme 4.14

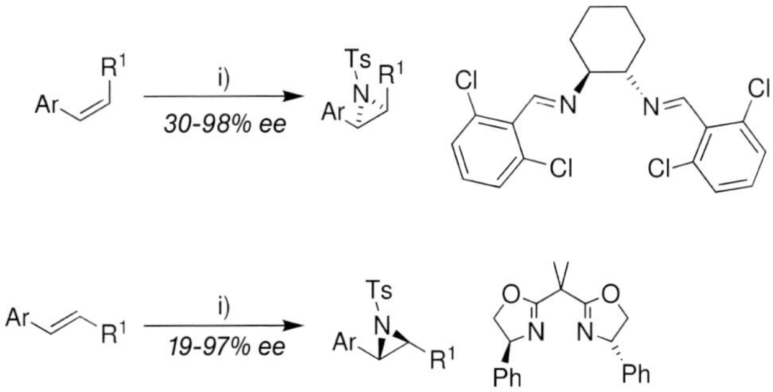

i) PhINTs, CuOTf, ligand

Scheme 4.15

not attract significant interest for some time. These pioneering studies provided the inspiration for a new generation of copper-based catalytic processes. In particular, Evans, Jacobsen, and Katsuki have described the utility of chiral bisoxazolines and 1,2-diamines in *enantioselective* aziridination of a range of alkenes (Scheme 4.15) [15].

Jacobsen et al. have made a convincing argument that these types of reaction proceed by way of discrete copper-nitrene complexes, rather than by some sort of single-electron process (Scheme 4.16) [13c].

The use of the bisoxazolinylanthracene ligand AnBOX in an Evans-like aziridination of chalcones has thrown up the interesting observation that the sense of enantiodiscrimination is inverted in relation to the reaction carried out with a traditional BOX ligand with similar stereogenicity (Scheme 4.17) [16].

These reactions, though representing a major achievement in the synthesis of enantioenriched aziridines, still retain some drawbacks, not least of which is the frequent requirement for the reactions to be conducted with a large excess (often five or more equivalents) of alkene, coupled with the explosive nature of the iodinane component. A further limitation to the methodology is the variable enantio-

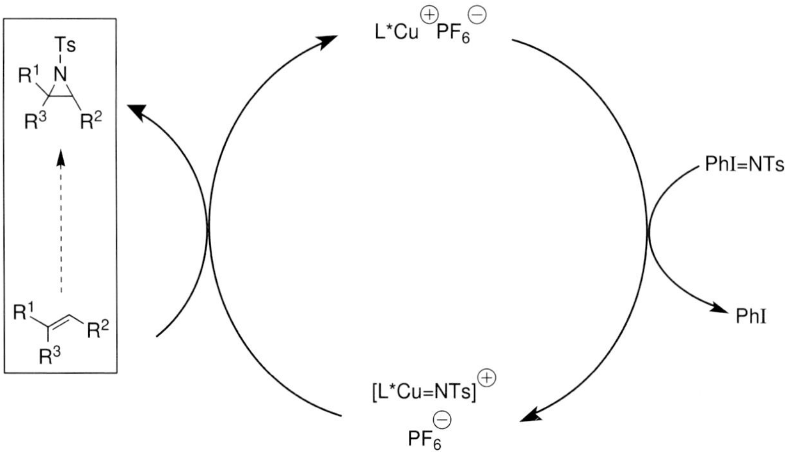

Scheme 4.16

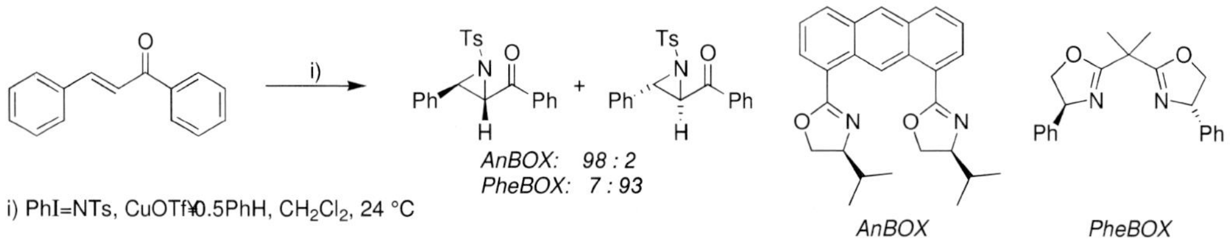

i) PhI=NTs, CuOTf·0.5PhH, CH_2Cl_2, 24 °C

Scheme 4.17

control exerted during the processes. Subsequent efforts inspired by the seminal work of these groups have ameliorated some, though by no means all, of the blemishes impairing the elegance of the method. Thus, the iodinane reagents may be prepared *in situ* rather than being isolated, from PhI=O [17] or $PhI(OAc)_2$ [18] and a sulfonamide. Copper(II) salts may be used in iodinane aziridination, and such reactions have been carried out in ionic liquids with the catalytic system proving to be recyclable up to five times with no loss of activity [19]. Particularly active copper catalysts are generated from copper complexes and "homoscorpionate" (trispyrazolylborate) ligands (Figure 4.1) [20]. The activity of these catalysts means that excess olefin is not required in the reaction, a most useful improve-

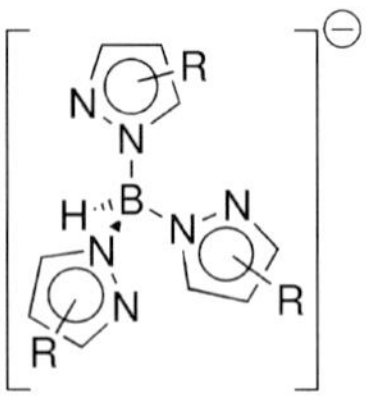

Figure 4.1 "Homoscorpionate" ligands.

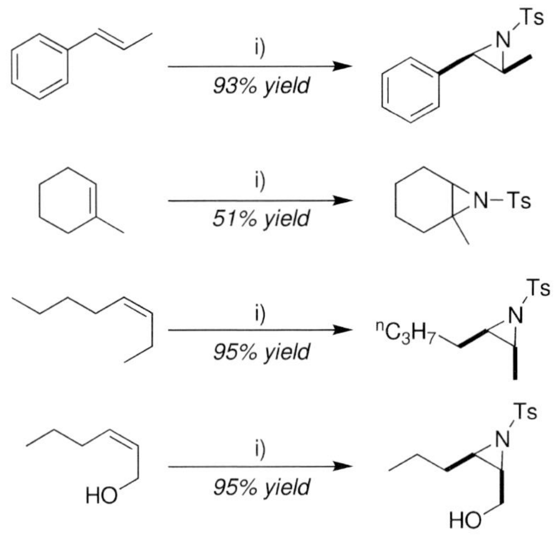

i) 93% yield

i) 51% yield

i) 95% yield

i) 95% yield

i) TsNNaCl, PTAB (10 mol%), MeCN, 25 °C

Scheme 4.18

ment on the existing methods. Disilver(I) catalysts may also be used in the reaction, though there is still the need for an excess of olefin [21].

The nature of the arylsulfonamides used in one-pot asymmetric aziridinations is important, with electron-deficient sulfonamides giving the best yields and *ees* [22]. The reaction can be enhanced if the sulfonamide also bears a ligating group: pyridylsulfonamides have been shown to exert a powerful influence upon the progress of copper-catalyzed aziridinations, acting both as reagents and as good ligands. In these reactions, the relatively mild cleavage procedure for the pyridylsulfonamide group (treatment with magnesium in methanol) is advantageous, as is the ability of the ligating sp^2 nitrogen to invert the regiochemistry of subsequent ring-opening by organocuprates (relative to the corresponding reactions of tolylsulfonylaziridines) [23].

The metal catalyst is not absolutely required for the aziridination reaction, and other positive nitrogen sources may also be used. After some years of optimization of the reactions of alkenes with positive nitrogen sources in the presence of bromine equivalents, Sharpless et al. reported the utility of chloramine-T in alkene aziridinations [24]. Electron-rich or electron-neutral alkenes react with the anhydrous chloramines and phenyltrimethylammonium tribromide in acetonitrile at ambient temperature, with allylic alcohols being particularly good substrates for the reaction (Schemes 4.18 and 4.19).

Other oxidants may be used in place of PTAB: chloramine-T efficiently aziridinates a range of alkenes in the presence of H_2O_2 (Scheme 4.20) [25].

Chloramine-T also functions as a nitrene source in the presence of heteropoly acids (HPAs) such as phosphomolybdic and phosphotungstic acids. The aziridination of alkenes by treatment with the combination of HPA and chloramine-T is

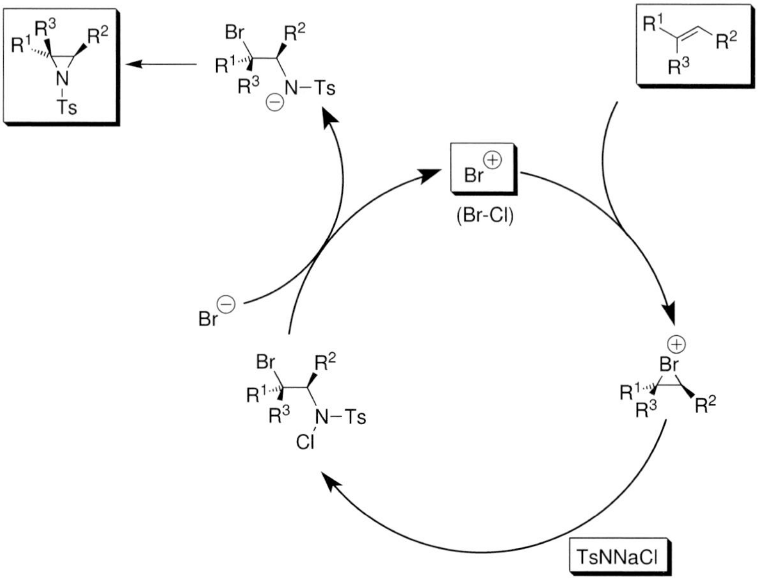

Scheme 4.19

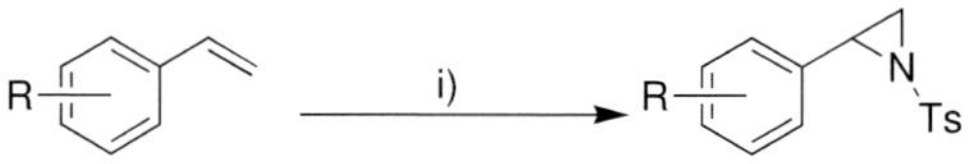

i) TsNClNa, H_2O_2, HBr, CH_3CN, $MgSO_4$, 25 °C

R	*Yield/%*
H	*75*
4-CH_3	*86*
3-CH_3	*80*

Scheme 4.20

instantaneous at room temperature and substoichiometric amounts of catalyst are needed, though the reaction often requires an excess of olefin, as in the traditional (i.e., metal-catalyzed, iodinane) process. The authors postulate a metalo-oxaziridine intermediate as the key reactive component (Scheme 4.21) [26].

Bromamine-T can also be utilized as a nitrene source, as reported by Zhang et al. [27]. Fe(III) porphyrins such as Fe(TPPC)Cl (Figure 4.2) thus catalyze the aziridination of alkenes when bromamine-T is used, whereas chloramine-T was inactive and iodinanes were inefficient reagents.

One of the limitations to the methods using addition of nitrenoids to alkenes is the prevalence of sulfonyl substitution at nitrogen, with the concomitant difficulty in subsequently removing the sulfonyl unit. Guthikonda and Du Bois' report of the use of trichloroethylsulfamate (which may be cleaved with metallic zinc under mild conditions) in Rh(I)-catalyzed alkene aziridinations thus represents a useful addition to the methodology (Scheme 4.22). In addition to providing a practical alternative to the tosyl group, the Du Bois process is unique in that the alkene does

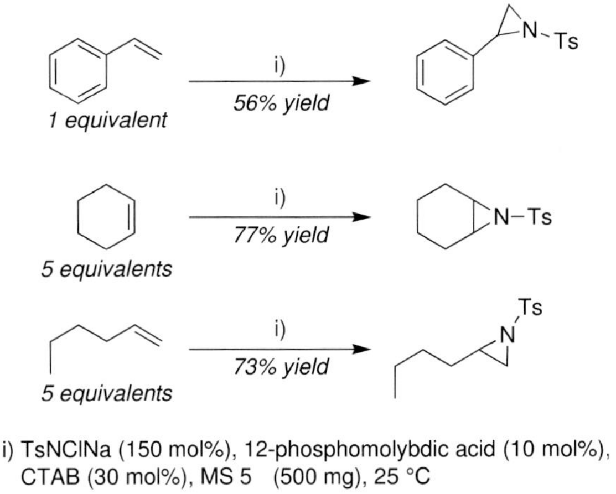

1 equivalent — i) *56% yield* — N–Ts

5 equivalents — i) *77% yield* — N–Ts

5 equivalents — i) *73% yield* — Ts, N

i) TsNClNa (150 mol%), 12-phosphomolybdic acid (10 mol%), CTAB (30 mol%), MS 5 (500 mg), 25 °C

Scheme 4.21

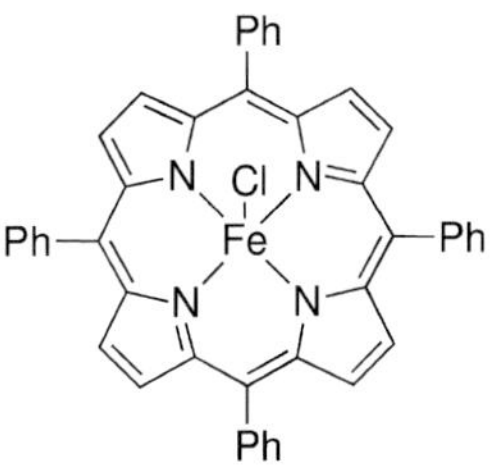

Ph, Ph, Ph, Ph, N, N, N, N, Cl, Fe

Figure 4.2 Fe(TPPC)Cl.

i) *85% yield* — $SO_3CH_2CCl_3$, N

i) *82% yield* — $N-SO_3CH_2CCl_3$

i) *72% yield* — $^nC_6H_{11}$, $SO_3CH_2CCl_3$, N

i) $Rh_2(tfacam)_4$(1 mol%), $H_2NSO_3CH_2CCl_3$ (1.1 eq.), PhH, 0 °C

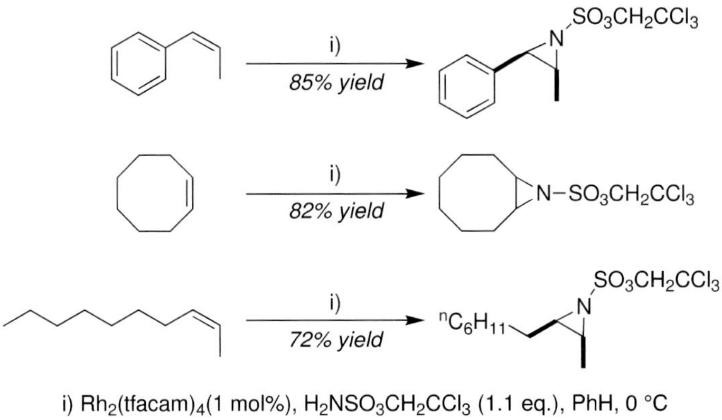

Scheme 4.22

not have to be present in excess (not normally the case in metal-nitrenoid aziridinations; *vide infra*) [28].

4.2.1.2 **Aziridines by Addition-elimination Processes**

The Gabriel-Cromwell aziridine synthesis involves nucleophilic addition of a formal nitrene equivalent to a 2-haloacrylate or similar reagent [29]. Thus, there is an initial Michael addition, followed by protonation and 3-*exo-tet* ring-closure. Asymmetric variants of the reaction have been reported. *N*-(2-Bromo)acryloyl camphorsultam, for example, reacts with a range of amines to give *N*-substituted (aziridinyl)acylsultams (Scheme 4.23) [30].

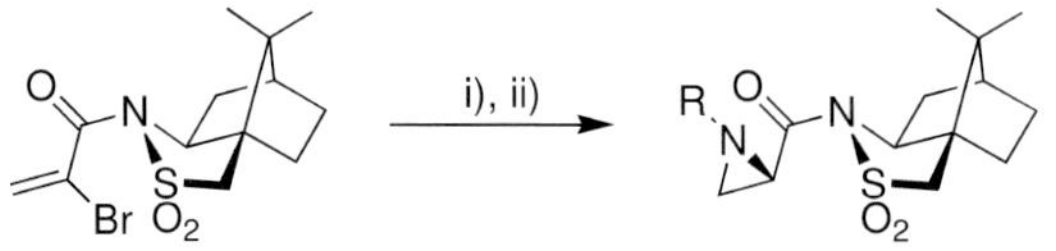

i), ii)

i) RNH_2 ii) Et_3N

Scheme 4.23

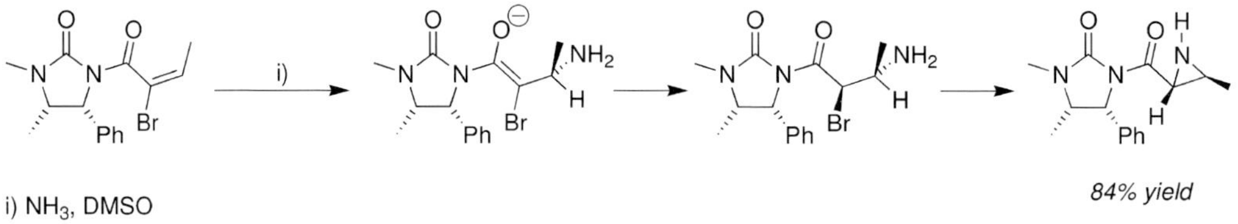

i)

84% yield

i) NH_3, DMSO

Scheme 4.24

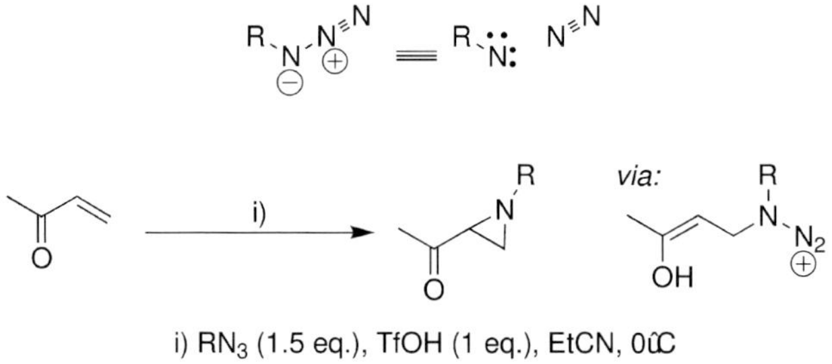

i)

via:

i) RN_3 (1.5 eq.), TfOH (1 eq.), EtCN, 0ûC

R	*Yield/%*
Bn	92
Ph_2CH	88
1-Ad	93
PMP	43
$^tBuO_2CCH_2$	66
Me_2C=$CHCH_2$	68

Scheme 4.25

Chiral *N*-(2-bromo)crotonoyl imidazolidinones react with ammonia to give *N*-unsubstituted aziridines (Scheme 4.24) [31].

Another conceptually unique approach in alkene aziridination has come from Johnston's labs. These workers shrewdly identified organic azides as nitrene equivalents when these compounds are in the amide anion/diazonium resonance form. Thus, when a range of azides were treated with triflic acid and methyl vinyl ketone at 0 °C, the corresponding aziridines were obtained, in synthetically useful yields. In the absence of the Brønsted acid catalyst, cycloaddition is observed, producing triazolines. The method may also be adapted, through the use of unsaturated imides as substrates, to give *anti*-aminooxazolidinones (Scheme 4.25) [32].

4.2.2
Addition to Imines

Since alkenes are relatively impotent precursors to aziridines, especially with regard to stereoselective reactions, substantially greater advances have been made in this field by means of the addition reactions between imines and a range of carbene equivalents.

4.2.2.1 Carbene Methodology

The synthesis of aziridines through reactions between nitrenes or nitrenoids and alkenes involves the simultaneous (though often asynchronous; *vide supra*) formation of two new C–N bonds. The most obvious other alternative synthetic analysis would be simultaneous formation of *one* C–N bond and *one* C–C bond (Scheme 4.26). Thus, reactions between *carbenes* or carbene equivalents and *imines* comprise an increasingly useful method for aziridination. In addition to carbenes and carbenoids, ylides have also been used to effect aziridinations of imines; in all classes of this reaction type the mechanism frequently involves a stepwise, addition-elimination process, rather than a synchronous bond-forming event.

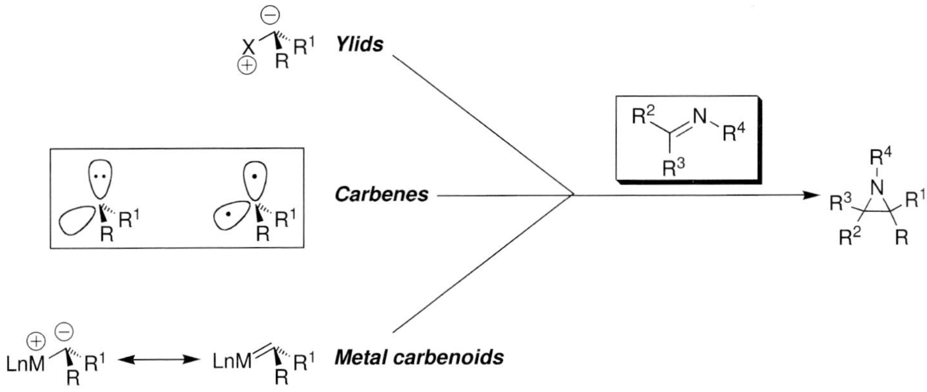

Scheme 4.26

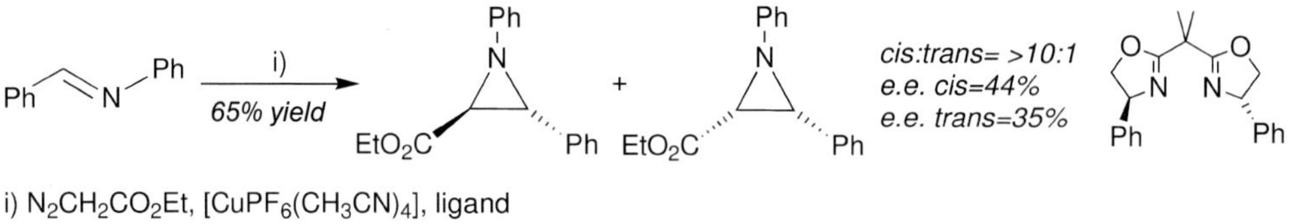

65% yield

cis:trans= >10:1
e.e. cis=44%
e.e. trans=35%

i) $N_2CH_2CO_2Et$, $[CuPF_6(CH_3CN)_4]$, ligand

Scheme 4.27

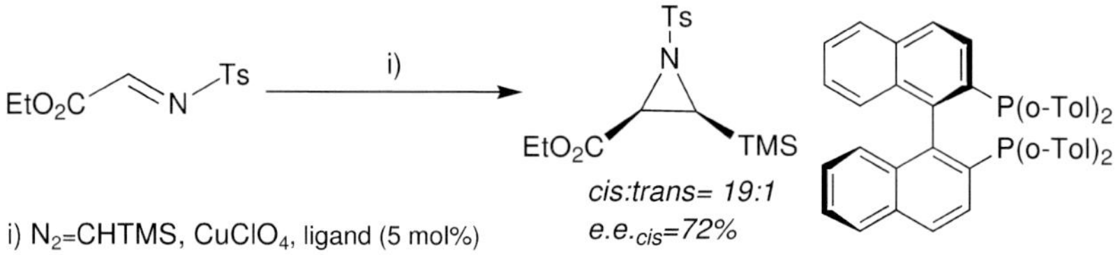

cis:trans= 19:1
$e.e._{cis}$=72%

i) N_2=CHTMS, $CuClO_4$, ligand (5 mol%)

Scheme 4.28

This area of research has only recently attracted the attention of synthetic organic chemists, but there has been a flurry of impressive activity in the area. Simple (i. e., unstabilized) carbenes suffer from many of the problems of nitrenes (*vide infra*) and most reported synthetically useful procedures use carbenoids: the majority of recent reports have focussed upon reactions between α-diazoesters and imines in the presence of a range of catalysts. In one of the earliest reports of enantioselective carbene-imine reactions, for instance, Jacobsen and Finney reported that ethyl diazoacetate reacts with *N*-arylaldimines in the presence of copper(I) hexafluorophosphate with mediocre stereoselectivity to give *N*-arylaziridine carboxylates. Though the diastereoselectivities of the reaction were often acceptable (usually >10:1, in favor of the *cis* isomers) the observed enantioselectivity was low (no more than 44% *ee*; Scheme 4.27) [33].

Trimethylsilyldiazomethane reacts under similar conditions with *N*-tosylimines in the presence of (*R*)-Tol-BINAP, with better enantiocontrol, but the process does not live up to the standards expected of modern asymmetric transformations (Scheme 4.28) [34].

Probably the most widely applicable asymmetric imine aziridination reaction reported to date is that of Wulff et al. These workers approached the reaction from a different perspective, utilizing the so-called "vaulted", axially chiral boron Lewis acids VANOL and VAPOL [35] to mediate reactions between ethyl diazoacetate and *N*-benzhydrylimines (Scheme 4.29) [36]. The reactions proceed with impressive enantiocontrol, but there is a requirement that the benzhydryl substituent be present; since this group is not an aziridine activator there is, therefore, a need for deprotection and attachment of a suitable activating group. Nonetheless, this method is a powerful one, with great potential for synthesis, as shown by the rapid synthesis of chloroamphenicol by the methodology [37].

Williams and Johnston have reported the first use of proton catalysis in the aziridination of imines by diazoesters (Scheme 4.30) [38]. A range of aryl and ali-

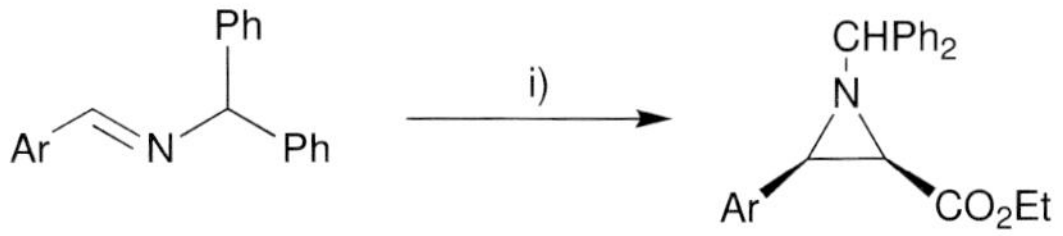

i) $N_2CH_2CO_2Et$, VOPOL-Lewis Acid

Ar	cis:trans	*Yield/%*	*ee/%*
Ph	*>50:1*	*77*	*97*
2-Naphthyl	*30:1*	*70*	*97*
p-AcOC$_6$H$_4$	*40:1*	*67*	*96*
p-NO$_2$C$_6$H$_4$	*11:1*	*68*	*91*
p-BrC$_6$H$_4$	*16:1*	*64*	*97*
o-MeC$_6$H$_4$	*3:1*	*51*	*98*
2-furyl	*16:1*	*55*	*95*
n-propyl	*>50:1*	*54*	*91*
cyclohexyl	*35:1*	*72*	*96*

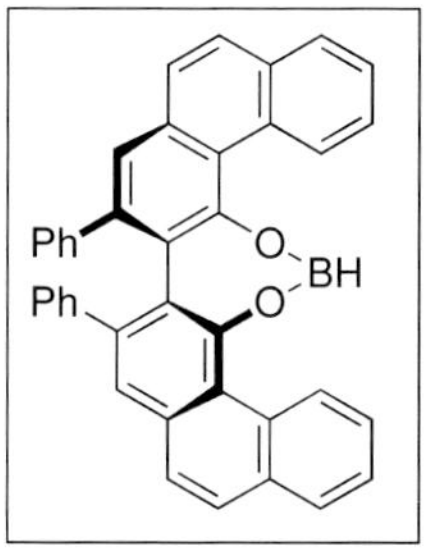

Scheme 4.29

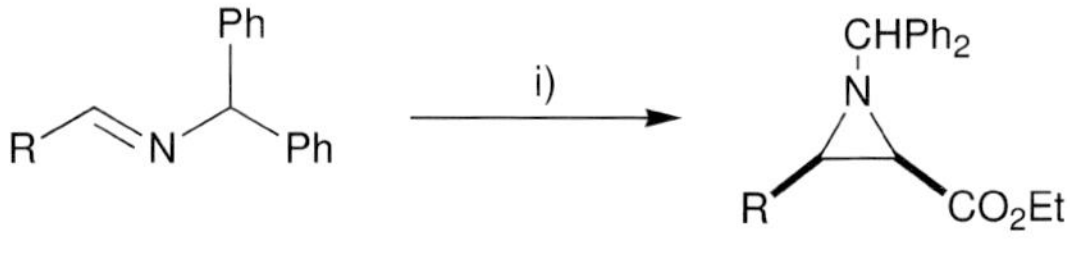

i) $N_2CH_2CO_2Et$, TfOH (25 mol%), EtCN, -78 °C→ 25 °C

R	cis:trans	*Yield/%*
Ph	*82:18*	*42*
MeO$_2$C	*>95:5*	*86*
2-pyridyl	*90:10*	*73*
EtO$_2$C–	*>95:5*	*53*

Scheme 4.30

phatic benzhydryl imines react with ethyl diazoacetate in the presence of triflic acid to give generally good yields of the corresponding aziridines. The reaction proceeds at low temperature, acid-sensitive functionalities (such as acetals) are left unscathed, and the reaction is capable of high diastereoselectivity: strongly electron-withdrawing imine substituents induce excellent *cis* selectivity in the reaction.

Diazoester aziridinations may be carried out in ionic liquids [39]. Other carbene equivalents have been investigated in aziridination reactions, though not to the same extent as diazocarbonyl compounds. Dibromo(*tert*-butyldimethylsilyl)methyllithium, for example, aziridinates *N*-arylimines to give 1-bromo-2-aryl-3-silylaziridines; these compounds function as useful synthetic intermediates, reacting

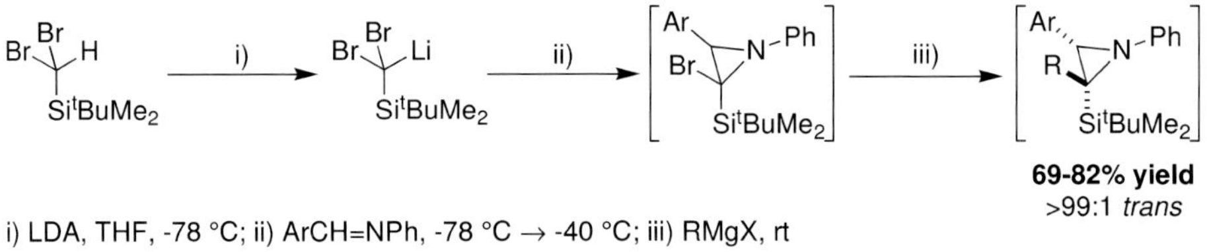

69-82% yield
>99:1 *trans*

i) LDA, THF, -78 °C; ii) ArCH=NPh, -78 °C → -40 °C; iii) RMgX, rt

Scheme 4.31

with nucleophiles to give a range of *trans*-1-substituted 2-aryl-3-silylaziridines in good yields, and in a highly stereoselective manner (Scheme 4.31) [40].

4.2.2.2 Aza-Darzens and Analogous Reactions

The reaction of sulfur (and analogous iodine [41] ylides) with imines to form aziridines is a Darzens-like reaction, analogous in gross terms to the carbene processes described above. Such ylides react to give β-sulfonium or β-iodonium amide anions, which are not isolated, but are instead allowed to react by ring-closure to give aziridines directly; in particular, this method is most often used for the synthesis of vinyl aziridines. Thus, Corey-Chakovsky sulfonium ylides have been used in asymmetric aziridination reactions of chiral sulfinyl imines. Stockman et al. extrapolated the initial findings of Garcia-Ruano [42] and Davis [43] and found that a range of aryl, aliphatic, and vinyl aziridines could be prepared in good yield and with high levels of stereoselectivity (Scheme 4.32) [44].

The Stockman group has also studied reactions between the same imines and allyl sulfonium ylides (first reported by Hou and Dai [45]) [46]. *N*-Sulfinyl vinylazir-

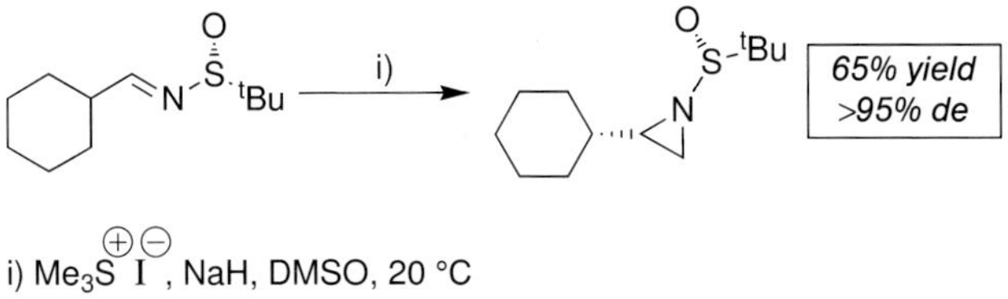

65% yield
>95% de

i) $Me_3S^{\oplus}I^{\ominus}$, NaH, DMSO, 20 °C

Scheme 4.32

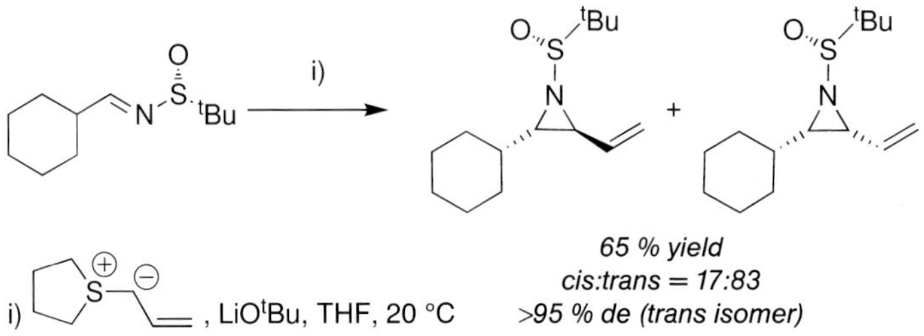

65 % yield
cis:trans = 17:83
>95 % de (trans isomer)

i) (tetrahydrothiophenium allyl ylide), LiO^{t}Bu, THF, 20 °C

Scheme 4.33

idines were thus prepared with moderate *cis/trans* selectivity, but the major stereoisomers were obtained with high diastereoexcess, with respect to the newly-formed asymmetric centers (Scheme 4.33).

In addition to their seminal contributions to epoxidation by such methodologies, Aggarwal et al. have also made significant contributions to the study of chiral sulfonium aza-Darzens reactions (Scheme 4.34). Terpene-derived sulfides react with diazo compounds to give aziridines in generally good yields and with useful levels of enantiocontrol. In addition to being highly selective, the reaction is operationally simplified by the cunning *in situ* generation of the requisite diazo compounds, by decomposition of the corresponding sodium salts of the analogous tosylhydrazones. Finally, the method is one of the few to furnish aziridines bearing *N*-substituents (tosyl, SES, Dpp, and carbamoyl, *inter alia*) amenable to subsequent synthetic manipulation under conditions compatible with a range of other functionality [47]. Despite recent advances in deprotection techniques [48], the widely used methodology to prepare *N*-sulfonyl aziridines can be flawed, the inherent stability of the sulfonamide group retarding sulfonylation. Aggarwal's method is therefore of considerable practical utility. The chiral additive responsible for the enantioselectivity is easily prepared from camphorsulfonic acid.

The classical aza-Darzens reaction (between bromoenolates and imines) has been investigated by several groups in recent years, especially with respect to the design and execution of asymmetric variants. Both stoichiometric and catalytic methods have been studied: thus, the reactions between *N*-Dpp imines and chiral α-bromoenolates [49] (derived from Oppolzer's sultams; Scheme 4.35) and between *S*-chiral sulfinylimines and achiral bromoenolates [50] (Scheme 4.36) have been reported.

Both of these transformations proceed in good yield and with high levels of enantio- and diastereoselectivity. The *N*-Dpp aziridines produced by the former reaction had already been shown by Sweeney et al. to be useful reagents for ring-opening (the Dpp group is a good activator to ring-cleavage and is removed by treatment with either Brønsted or Lewis acid at low or ambient temperature), whilst the products of the latter reaction are also activated aziridines. Thus, Davis et al. found that these chiral sulfinyl aziridines undergo hydrolytic and hydrogenolytic ring-cleavage efficiently, but must be oxidized to the corresponding sulfonyl

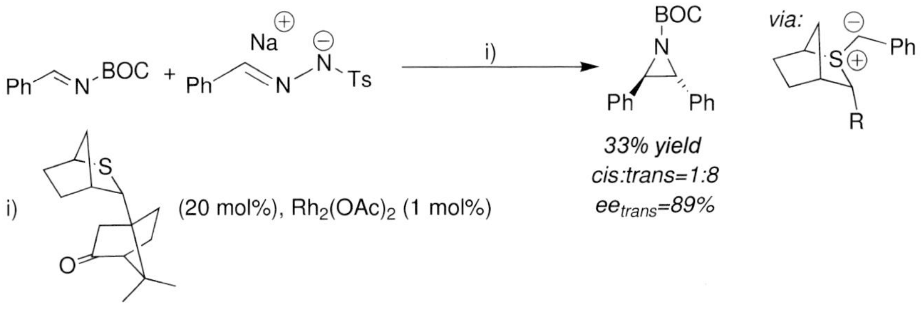

Ph–CH=N–BOC + Ph–CH=N–N⁻(Na⁺)–Ts →(i) N-BOC aziridine (Ph, Ph)

33% yield
cis:trans=1:8
ee_{trans}=89%

via: sulfonium ylide (S⊕, Ph, R)

i) chiral sulfide (20 mol%), $Rh_2(OAc)_2$ (1 mol%)

Scheme 4.34

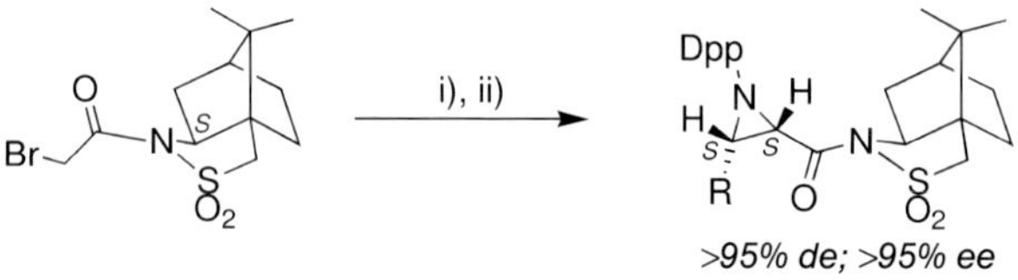

i) LHMDS, THF, -78 °C; ii) RCH=NDpp, THF, -78 °C

>95% de; >95% ee

R	Yield/%
Ph	71
p-MeOC$_6$H$_4$	60
o-NO$_2$C$_6$H$_4$	72
p-FC$_6$H$_4$	57
m-BrC$_6$H$_4$	60
p-BrC$_6$H$_4$	65
2-pyridyl	71

Scheme 4.35

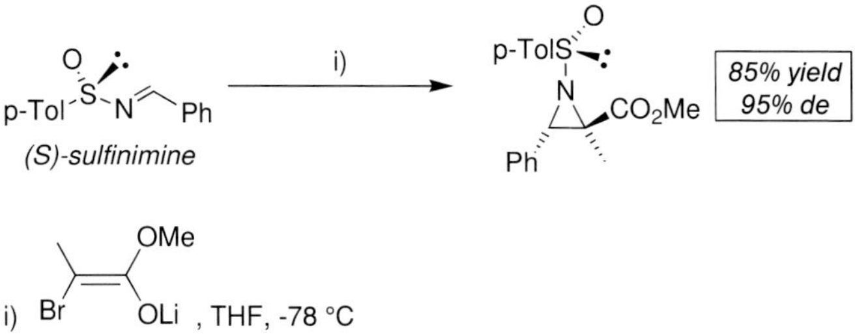

(S)-sulfinimine

85% yield
95% de

i) Br OMe OLi, THF, -78 °C

Scheme 4.36

analogues to be able to undergo ring-opening reactions with carbon nucleophiles efficiently. The same group have also reported the use of sulfonium ylides with similar substrates (*vide infra*).

Other reactions adapted from asymmetric aldol reactions suffer in comparison from the fact that (probably due to the strength of the boron-nitrogen bond) boron-mediated processes generally yield the intermediate 2-halo-3-aminoester products rather than aziridine products directly [51].

4.2.3
Addition to Azirines

Azirines (three-membered cyclic imines) are related to aziridines by a single redox step, and these reagents can therefore function as precursors to aziridines by way of addition reactions. The addition of carbon nucleophiles has been known for some time [52], but has recently undergone a renaissance, attracting the interest of several research groups. The cyclization of 2-(*O*-tosyl)oximino carbonyl compounds – the Neber reaction [53] – is the oldest known azirine synthesis, and asymmetric variants have been reported. Zwanenburg et al., for example, prepared nonracemic chiral azirines from oximes of 3-ketoesters, using cinchona alkaloids as catalysts (Scheme 4.37) [54].

Azirines may also be prepared by thermolysis of vinyl azides (Scheme 4.38) [55].

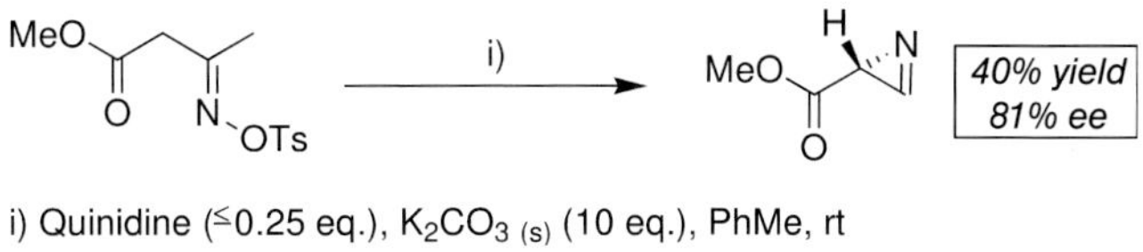

i) Quinidine (≤0.25 eq.), K_2CO_3 (s) (10 eq.), PhMe, rt

40% yield 81% ee

Scheme 4.37

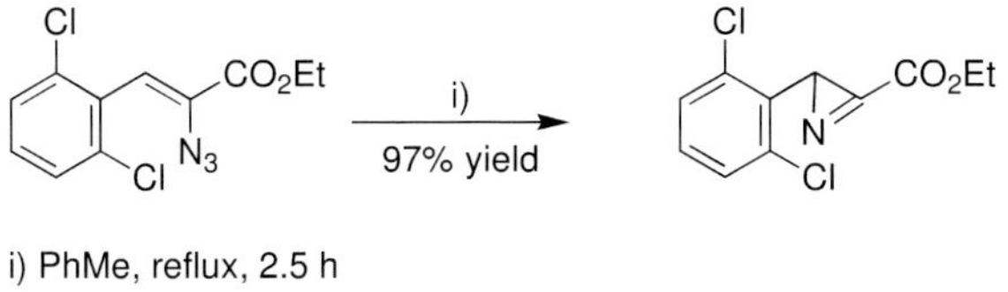

i) PhMe, reflux, 2.5 h

97% yield

Scheme 4.38

i) 1. LDA, -78 °C; 2. MeI; 3. H_2O ... OLi, THF, -78 °C

ii) 1. LDA, -78 °C; 2. MeI; 3. H_2O

70% yield 96% de; 42% yield; Dysidazirine

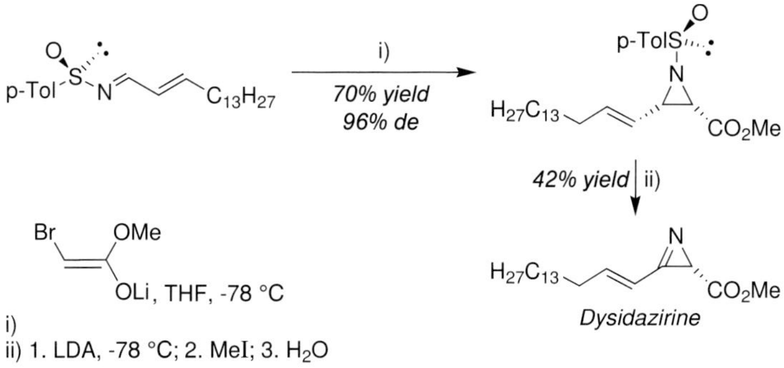

Scheme 4.39

Several groups have now reported alternative methods for the preparation of chiral azirines [56], and this has subsequently resulted in a renewed interest in the use of these intermediates for aziridine synthesis. The natural product antibiotic (–)-dysidazirine was synthesized by Davis et al. through the use of an unsaturated chiral *N*-sulfinyl aziridine prepared by the group's asymmetric aza-Darzens methodology (*vide infra*), revolving around the elimination of methyl tolylsulfinite as the key azirine-forming reaction (Scheme 4.39) [57]. The regiochemistry of the elimination is intriguing, proceeding by loss of the (apparently) *less* acidic proton.

The same group found that the yield of azirine was in some cases enhanced if the elimination was of a sulfenate, rather than a sulfinite, leaving group; treatment of the thus produced azirines with Grignard reagents was highly stereoselective, providing the tetrasubstituted aziridine products in good yield (Scheme 4.40). This report was the first concerning reactions between 2*H*-azirine-2-carboxylic esters and organometallic reagents [58]. It was subsequently found that the efficiency of elimination of sulfinyl aziridines is also enhanced when the key reaction is carried out at very low temperature in the presence of TMSCl [59].

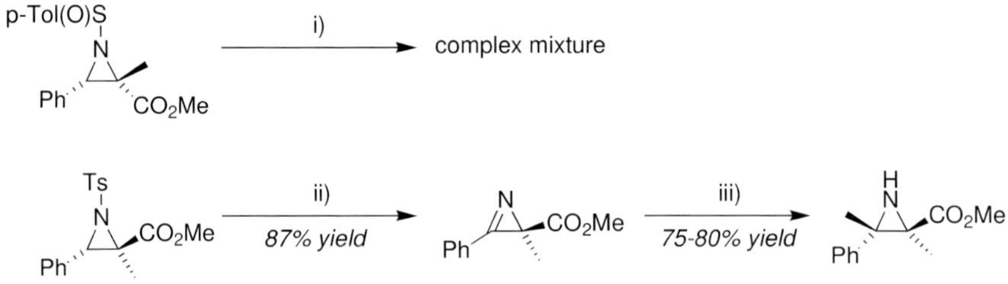

i) 1. LDA, -78 °C; 2. MeI; 3. H_2O; ii) 1. LDA, -78 °C; 2. H_2O; iii) MeMgBr, THF, -78 °C

Scheme 4.40

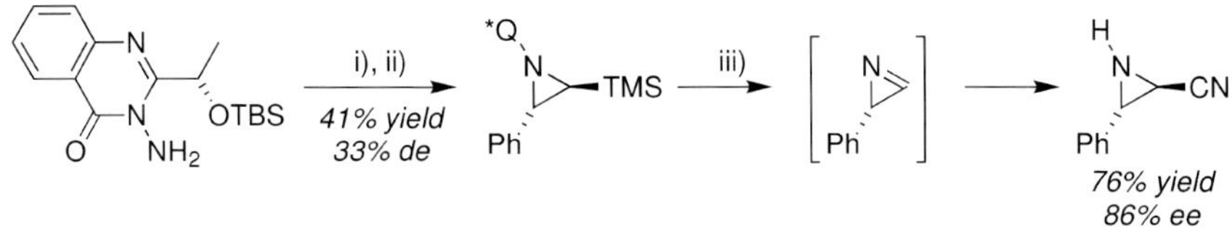

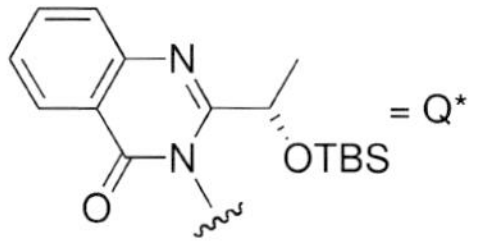

i) $Pb(OAc)_4$, CH_2Cl_2, -20 °C
ii) (*E*)-TMSCH=CHPh, HMDS
iii) CsF, KCN, DMF

Scheme 4.41

Atkinson Q-aziridines bearing silyl ring substituents undergo elimination in the presence of CsF to give chiral azirines, which react with cyanide ion to give cyanoaziridines in good yields [60]. The approach of the incoming nucleophile is *anti* to the preexisting asymmetric center, as would be predicted (Scheme 4.41).

Reactions between organolithium reagents and 3-(2-naphthyl)-2*H*-azirine in the presence of a range of chiral ligands have been studied. The product aziridines are obtained in at best moderate *ee*s [61].

Asymmetric transfer hydrogenation reducing agents react with 2-aryl azirines to give aziridines in good yields but with moderate enantiomeric excesses (Scheme 4.42) [62].

Carbon-centered radicals have been shown to undergo addition reactions with azirine-3-carboxylates. Methyl 2-(2,6-dichlorophenyl)azirine-2-carboxylate thus reacts with alkyl and aryl iodides in the presence of triethylborane to give aziridines in good yields. The radical approaches from the opposite face to the aryl substituent, giving the *cis* products as single diastereoisomers (Scheme 4.43) [63].

A similar but asymmetric variant of the reaction, involving the radical addition of alkyl iodides and trialkylboranes to chiral azirine esters derived from 8-phenylmenthol and camphorsultam, in the presence of a Cu(I) catalyst, has subsequently been reported [64]. The diastereoselectivity of the addition is variable (0–92% de)

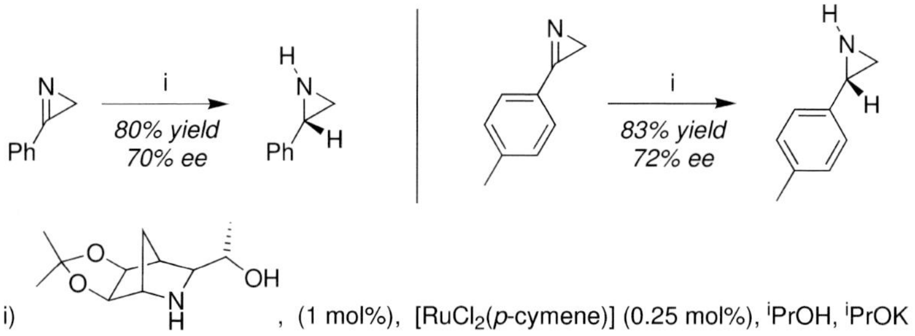

80% yield
70% ee

83% yield
72% ee

i) OH, (1 mol%), $[RuCl_2(p\text{-cymene})]$ (0.25 mol%), ^iPrOH, ^iPrOK

Scheme 4.42

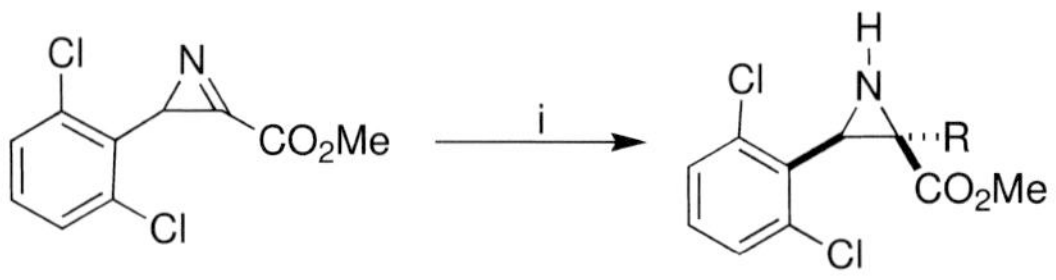

R	Yield/%
Et	71
^iPr	82
^tBu	89
$PhCH_2CH_2$	45
$^cC_6H_{11}$	79

i) Et_3B, R-I, CH_2Cl_2, -40 °C

Scheme 4.43

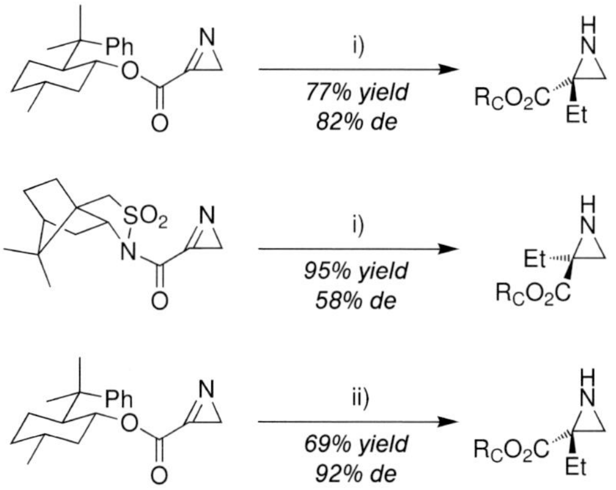

77% yield
82% de

95% yield
58% de

69% yield
92% de

i) EtI (10 eq.), Et_3B (5 eq.), O_2, -105 °C; ii) Et_3B (3 eq.), CuCl (0.1 eq), O_2, -105 °C

Scheme 4.44

and the process must be conducted at very low temperature for best results (Scheme 4.44).

Aziridines bearing heteroatom substituents are best prepared through treatment of the corresponding azirines with heteroatom nucleophiles. Thus, azirine carboxylates (in this case prepared by thermal decomposition of the corresponding vinyl

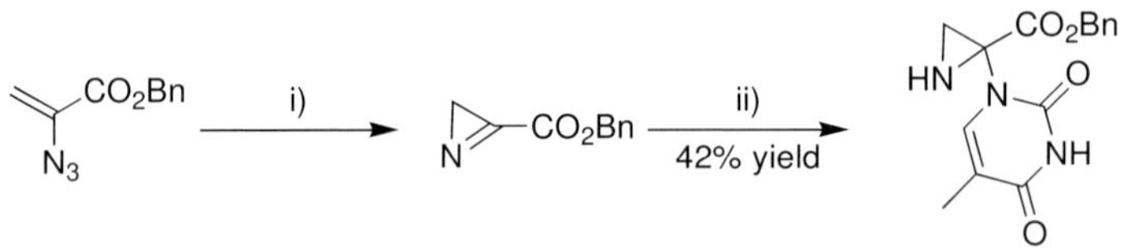

i) PhMe, 105 °C, 10 h; ii) thymine, K_2CO_3, rt, 1.5 h

Scheme 4.45

azides; *vide infra*) react with pyrimidine and purine bases in mediocre yield (Scheme 4.45) [65]. Enantiomerically enriched ethyl 3-methyl-2*H*-azirine-2-carboxylate acts as an efficient alkylating agent for a variety of five-membered aromatic nitrogen heterocycles [66].

Azirines are also the reagents of choice for the preparation of polycyclic aziridines in which the nitrogen atom is located at a ring junction, by means of heterocycloaddition reactions [67], though the C=N bond is sometimes rather unreactive in these process. Racemic arylazirinecarboxylates thus react relatively rapidly with 2,3-dimethylbutadiene to give bicyclo[4.1.0]heptene carboxylates (Scheme 4.46) [68], whereas chiral azirine-3-phosphonates undergo slow Diels-Alder cycloaddition with simple dienes, but the reactions proceed with essentially complete diastereoselectivity (Scheme 4.47) [69]

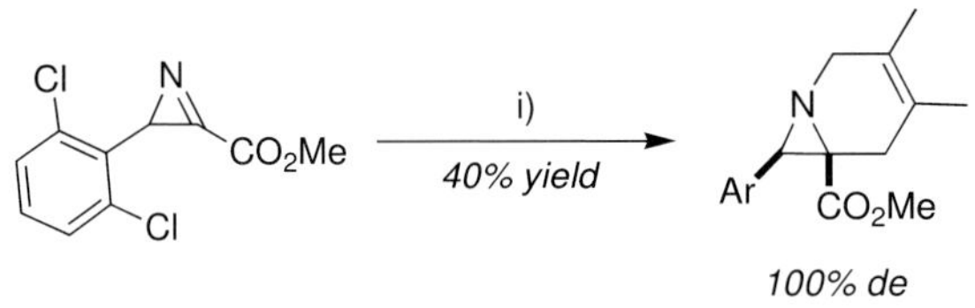

i) 2,3-Dimethylbutadiene (large excess), rt, 24h

Scheme 4.46

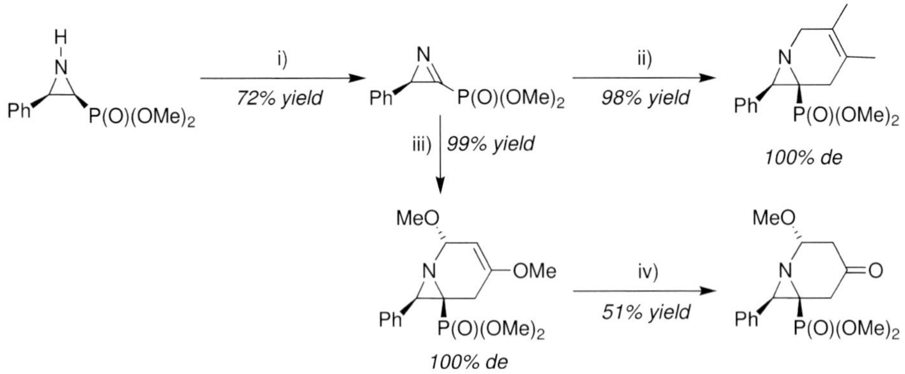

i) DMSO, $(COCl)_2$, Et_3N; ii) 2,3-dimethylbutadiene (100 eq.), rt, 4d; iii) Danishefsky's diene (5 eq., rt, 8h); iv) SiO_2

Scheme 4.47

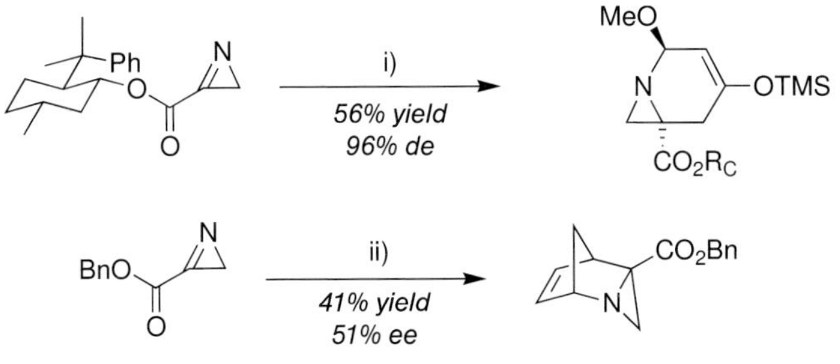

i) 56% yield 96% de

ii) 41% yield 51% ee

i) Danishefsky's diene, $MgBr_2 \cdot OEt_2$, CH_2Cl_2, -100 °C; ii) Cyclopentadiene, (*S*)-BINOL, $AlMe_3$, -40 °C, CH_2Cl_2

Scheme 4.48

There has been some investigation of auxiliary-controlled cycloadditions of azirines. Thus, camphor-derived azirine esters undergo cycloaddition with dienes, with poor diastereoselectivity [70]. The same azirines were also observed to react unselectively with phenylmagnesium bromide. Better selectivities were obtained when Lewis acids were used in the corresponding cycloaddition reactions of 8-phenylmenthyl esters of azirine 2-carboxylates (Scheme 4.48) [71]. The same report also describes the use of asymmetric Lewis acids in similar cycloadditions, but mediocre *ees* were observed.

4.2.4 Aziridines through Cyclization

4.2.4.1 From Epoxides

Jacobsen has utilized [(salen)Co]-catalyzed kinetic resolutions of terminal epoxides to prepare *N*-nosyl aziridines with high levels of enantioselectivity [72]. A range of racemic aryl and aliphatic epoxides are thus converted into aziridines in a four-step process, by sequential treatment with water (0.55 equivalents), Ns-NH-BOC, TFA, Ms_2O, and carbonate (Scheme 4.49). Despite the apparently lengthy procedure, overall yields of the product aziridines are excellent and only one chromatographic purification is required in the entire sequence.

Guanidines have been shown to act as nitrogen sources in the conversion of

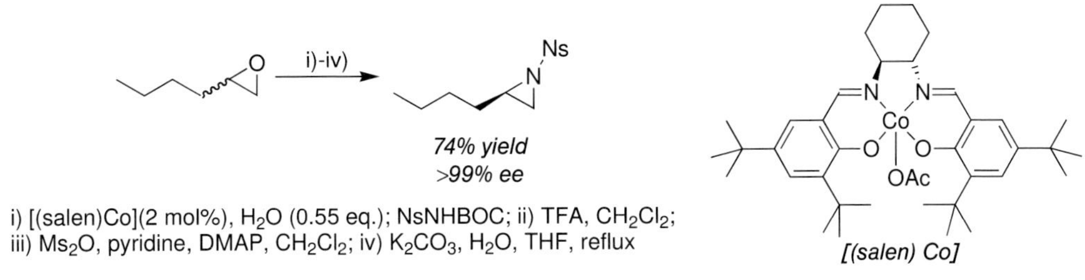

i)-iv) 74% yield >99% ee

i) [(salen)Co](2 mol%), H_2O (0.55 eq.); NsNHBOC; ii) TFA, CH_2Cl_2; iii) Ms_2O, pyridine, DMAP, CH_2Cl_2; iv) K_2CO_3, H_2O, THF, reflux

[(salen) Co]

Scheme 4.49

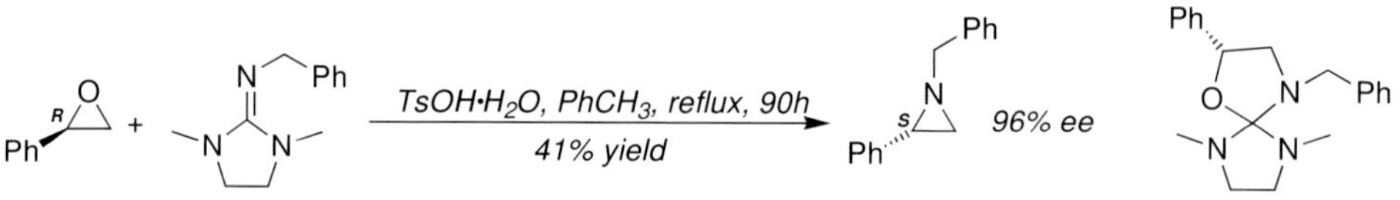

TsOH•H_2O, $PhCH_3$, reflux, 90h

41% yield

96% ee

Scheme 4.50

epoxides to aziridines (Scheme 4.50) [73]. The method utilizes the cyclic guanidine shown to mediate the aziridinations of aryl aldehydes in the presence of bromoesters [74]; the process involves ring-opening of the epoxide by the exocyclic nitrogen atom of the guanidine and subsequent oxygen-nitrogen exchange. Chiral epoxides may be converted into aziridines of opposite configuration by the process.

4.2.4.2 From 1,2-Aminoalcohols and 1,2-Aminohalides

This class of aziridine-forming reaction includes the first reaction reported to afford aziridines. In 1888 Gabriel reported that aziridines could be prepared in a two-step process, by chlorination of ethanolamines with thionyl chloride, followed by alkali-induced cyclization [75]. Wenker subsequently reported that heating of 600 g of ethanolamine with more than 1 kg of 96% sulfuric acid at high temperature produced "<qs>β-aminoethyl sulphuric acid<qe>"; 282 g of it was distilled from aqueous base to give 23 g of aziridine itself, the first preparation of the parent compound in a pure condition [76]. Though there is no evidence to substantiate the hypothesis, the intermediate in these reactions is perhaps a cyclic sulfamidate (Scheme 4.51).

Not surprisingly, these harsh conditions (Wenker commented about the first stage of the preparation that "<qs>at about 250 °C charring begins, necessitating the end of the operation<qe>") were not generally suitable, producing mixtures of cyclized and elimination product when any substitution was present α to the hydroxy moiety. This area has attracted considerable interest in the intervening years, with a wide range of conditions for activation of the hydroxy group having evolved, enabling the preparation of a wide range of achiral and enantiomerically pure aziridines for use in synthesis, and the reader is directed to review sources for fuller details of this type of aziridine-forming process [77]. In particular, Mitsunobu-like oxyphosphonium activation has been used extensively to execute the transformation.

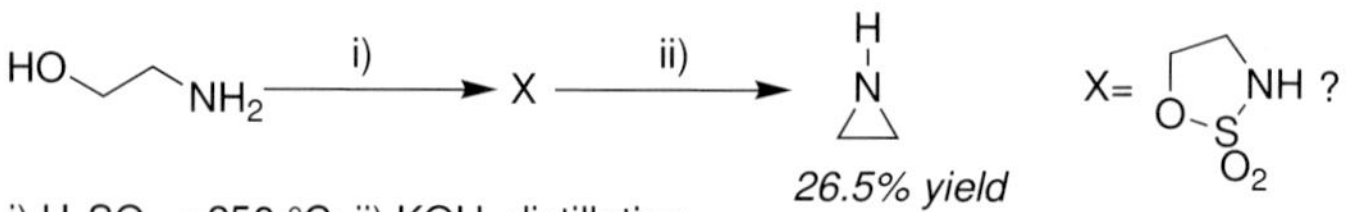

26.5% yield

i) H_2SO_4, >250 °C; ii) KOH, distillation

Scheme 4.51

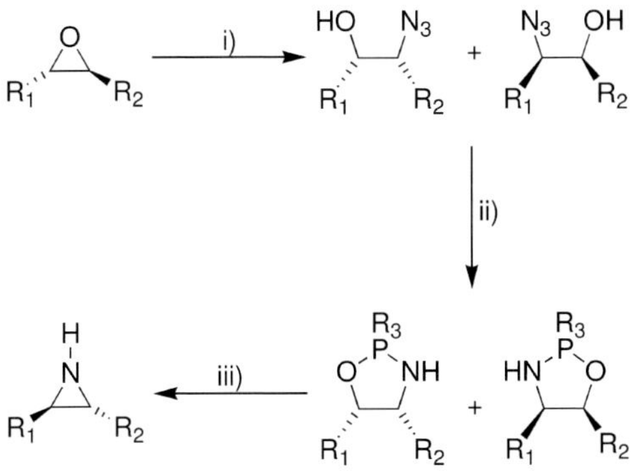

i) $N_3^{\ominus}$; ii) PR_3, MeCN, rt; iii) MeCN, Δ

Scheme 4.52

4.2.4.3 **From 1,2-Azidoalcohols [2, 3]**

Since the development of new asymmetric epoxide syntheses there has been a ready supply of enantiomerically pure epoxides, and there have been abundant reports of multistep preparation of aziridines from these precursors. In particular, phosphine-mediated ring-closures of azidoalcohols (one of the reactions to bear Staudinger's name), obtained from chiral epoxides by ring-opening reactions through the use of a range of nucleophilic azide sources, has been widely investigated. The key, aziridine-forming reaction in these strategies involves the treatment of hydroxyazides with trialkyl- or triarylphosphine, producing diastereomeric oxazaphospholidines, which are rapidly formed but slowly converted into *N*-unsubstituted aziridines upon heating in acetonitrile (Scheme 4.52). The reaction has been demonstrated in a wide range of chiral and achiral epoxides and there is no issue of regiochemistry: both asymmetric centers are cleanly and predictably inverted during the process.

4.3 Conclusions

Aziridines were amongst the first heterocycles to be prepared by synthesis, and they occupy a special place in organic synthesis. This chapter has, we hope, indicated the rich chemistry associated with this area of synthetic endeavor in both the distant and the recent past; it is clear that the recent exciting developments in asymmetric aziridine synthesis indicate a most promising future for the practitioners in the field and the greater synthetic community at large.

References

1 R. S. Atkinson, *Tetrahedron*, **1999**, *55*, 1519.
2 P. Müller, C. Fruit, *Chem. Rev.* **2003**, *103*, 2905; J. B. Sweeney, *Chem. Soc. Rev.* **2002**, *31*, 247; E. N. Jacobsen in *Comprehensive Asymmetric Catalysis*, E. N. Jacobsen, A. Pfaltz, H. Yamamoto, Eds. Springer-Verlag, Berlin, **1999**, *2*, 607
3 (a) J. E. G. Kemp in *Comprehensive Organic Synthesis*; B. M. Trost, I. Fleming, Eds.; Oxford: Pergamon, **1991**; Vol. 7, p. 467; (b) A. Padwa, S. S. Murphree, in *Progress in Heterocyclic Chemistry*, G. W. Gribble, T. L. Gilchrist, Eds.; Oxford: Pergamon Elsevier Science, **2000**; Vol. 12, Chapter 4.1, p. 57.
4 K. Hori, H. Sugahira, Y. N. Ito, T. Katsuki, *Tetrahedron Lett.* **1999**, *40*, 5207.
5 S. Andreae, E. Schmitz, *Synthesis* **1991**, 327.
6 I. Washington, K. N. Houk, A. Armstrong, *J. Org. Chem.* **2003**, *68*, 6497.
7 R. S. Atkinson, C. W. Rees, *J. Chem. Soc. (C)* **1969**, 772; D. J. Anderson, T. L. Gilchrist, D. C. Horwell, C. W. Rees, *J. Chem. Soc. (C)* **1970**, 576.
8 R. S. Atkinson, R. Martin, *J. Chem. Soc., Chem. Commun*, **1974**, 386; R. S. Atkinson, J. R. Malpass, *J. Chem. Soc., Perkin Trans. 1* 1977, 2242.
9 R. S. Atkinson, M. J. Grimshire, B. J. Kelly, *Tetrahedron* **1989**, *45*, 2875.
10 T. Siu, A. K. Yudin, *J. Am. Chem. Soc.* **2002**, *124*, 530.
11 J. Li, J.-L. Liang, P. W. H. Chan, C.-M. Che, *Tetrahedron Lett.* **2004**, *45*, 2685.
12 H. Kwart, A. A. Khan, *J. Am. Chem. Soc.* **1967**, *89*, 1950.
13 For other reports of metaloporphyrin catalysis of aziridination, see: R. Breslow, S. H. Gellman, *J. Chem. Soc., Chem. Commun.* **1982**, 1400; J. T. Groves, T. Takahashi, *J. Am. Chem. Soc.* **1983**, *105*, 2073; R. Breslow, S. Gellman, *J. Am. Chem. Soc.*, **1983**, *105*, 6729; D. Mansuy, J.-P. Mahy, A. Dureault, G. Bedi, G. Battioni, *J. Chem. Soc., Chem. Commun.* **1984**, 1161; J.-P. Mahy, P. Battioni, D. Mansuy, *J. Am. Chem. Soc.* **1986**, *108*, 1079; J.-P. Mahy, G. Bedi, P. Battioni, D. Mansuy, *J. Chem. Soc., Perkin Trans. 2* **1988**, 1517; T.-S. Lai, H.-L. Kwong, C.-M. Che, S.-M. Peng, *Chem. Commun.* **1997**, 2373; J.-P. Simonato, J. Pecaut, W. R. Scheidt, J.-C. Marchon, *Chem. Commun.* **1997**, 989; S.-M. Au, J.-S. Huang, W.-Y. Yu, W.-H. Fung, C.-M. Che, *J. Am. Chem. Soc.* **1999**, *121*, 9120; J.-L. Liang, J.-S. Huang, X.-Q. Yu, N. Zhu, C.-M. Che, *Chem. Eur. J.* **2002**, *8*, 1563.
14 (a) R. A. Abramovitch, T. D. Bailey, T. Takaya, V. Uma, *J. Org. Chem.* **1974**, *39*, 340; (b) A. Yamada, T. Yamamoto, M. Okawara, *Chem. Lett.* **1975**, 361.
15 (a) D. A. Evans, M. M. Faul, M. T. Bilodeau, *J. Am. Chem. Soc.*, **1994**, *116*, 2742; (b) R. W. Quan, Z. Li, E. N. Jacobsen, *ibid.*, **1993**, *114*, 8156; (c) Z. Li, R. W. Quan, E. N. Jacobsen, *ibid.*, **1995**, *117*, 5889; (d) H. Nishikori, T. Katsuki, *Tetrahedron Lett.*, **1996**, *37*, 9245.
16 J. Xu, L. Ma, P. Jiao, *Chem. Comm.* **2004**, 1616.
17 P. Dauban, L. Saniere, A. Tarrade, R. H. Dodd, *J. Am. Chem. Soc.* **2001**, *123*, 7707; F. Duran, L. Leman, A. Ghini, G. Burton, P. Dauban, R. H. Dodd, *Org. Lett.* **2002**, *4*, 2481.
18 J.-L. Hiang, J.-S. Huang, X.-Q. Yu, N. Zhu, C.-M. Che, *Chem. Eur. J.* **2002**, *8*, 1563.
19 M. L. Kantam, V. Neeraja, B. Kavita, Y. Haritha, *Synlett* **2004**, 525.
20 M. A. Mairena, M. M. Díaz-Requejo, T. R. Belderraín, M. C. Nicasio, S. Trofimenko, P. J. Pérez, *Organometallics* **2004**, *23*, 253.
21 Y. Cui, C. He, *J. Am. Chem. Soc.* **2003**, *125*, 16202.
22 H.-L. Kwong, D. Liu, K.-Y. Chan, C.-S. Lee, K.-H. Huang, C.-M. Che, *Tetrahedron Lett.* **2004**, *45*, 3965.
23 H. Han, I. Bae, E. J. Yoo, J. Lee, Y. Do, S. Chang, *Org. Lett.* **2004**, *6*, 4109.
24 J. U. Jeong, B. Tao, I. Sagasser, H. Henniges, K. B. Sharpless, *J. Am. Chem. Soc.* **1998**, *120*, 6844.
25 S. L. Jain, V. B. Sharma, B. Sain, *Tetrahedron Lett.* **2004**, *45*, 8731.
26 G. D. K. Kumar, S. Baskaran, *Chem. Comm.* **2004**, 1027.
27 R. Vyas, G.-Y. Gao, J. D. Harden, X. P. Zhang, *Org. Lett.* **2004**, *6*, 1907, see also L. Simkhovich, Z. Gross *Tetrahedron Lett.* **2001**, *42*, 8089.
28 K. Guthikonda, J. Du Bois, *J. Am. Chem. Soc.* **2002**, *124*, 13672.

29 G. V. Shustov, O. N. Krutius, V. N. Voznesensky, I. I. Chervin, A. V. Eremeev, R. G. Krostyanovsky, F. D. Polyak, *Tetrahedron* **1990**, *46*, 6741.

30 (a) O. Ploux, M. Caruso, G. Chassaing, A. Marquet, *J. Org. Chem.* **1988**, *53*, 3154; (b) P. Garner, O. Dogan, S. Pillai, *Tetrahedron Lett.* **1994**, *35*, 1653.

31 a) G. Cardillo, S. Casolari, L. Gentiluca, C. Tomasini, *Angew. Chem. Int. Ed. Engl.* **1996**, *35*, 1848; b) G. Cardillo, L. Gentilucci, C. Tomasini, M. P. V. Castejan-Bordas, *Tetrahedron: Asymmetry* **1996**, *7*, 755.

32 J. M. Mahoney, C. R. Smith, J. N. Johnston, *J. Am. Chem. Soc.* **2005**, *127*, 1354.

33 K. B. Hansen, N. S. Finney, E. N. Jacobsen, *Angew. Chem., Int. Ed. Engl.* **1995**, *34*, 676.

34 K. Juhl, R. G. Hazell, K. A. Jørgensen, *J. Chem. Soc., Perkin Trans. 1* **1999**, 2293.

35 Y. Su, C. Rabalakos, W. D. Mitchell, W. D. Wulff, *Org. Lett.* **2005**, *7*, 367.

36 W. D. Wulff, J. C. Antilla, *J. Am. Chem. Soc.* **1999**, *121*, 5099; J. C. Antilla, W. D. Wulff, *Angew. Chem., Int. Edn. Engl.* **2000**, *39*, 4518.

37 C. Loncaric, W. D. Wulff, *Org. Lett.* **2001**, *3*, 3675.

38 A. L. Williams, J. N. Johnston, *J. Am. Chem. Soc.* **2004**, *126*, 1613.

39 W. Sun, C. G. Xia, H. W. Wang, *Tetrahedron Lett.* **2003**, *44*, 2409.

40 K. Yagi, H. Shinokubo, K. Oshima, *Org Lett.* **2004**, *6*, 4339.

41 M. Ochiai, Y. Kitagawa, *Tetrahedron Lett.* **1998**, *39*, 5569.

42 J. L. García Ruano, I. Férnandez, C. Hamdouchi, *Tetrahedron Lett.* **1995**, *36*, 295; J. L. García Ruano, I. Férnandez, M. del Prado Catalina, A. A. Cruz, *Tetrahedron: Asymmetry* **1996**, *7*, 4307.

43 F. A. Davis, P. Zhou, C.-H. Liang, R. E. Reddy, *Tetrahedron: Asymmetry* **1995**, 6, 1511.

44 D. Morton, D. Pearson, R. A. Field, R. A. Stockman, *Synlett* **2003**, 1985.

45 X.-L. Hou, X.-F Yang, L.-X. Dai, X.-F. Chen, *Chem. Comm.* **1998**, 747, and references therein.

46 D. Morton, D. Pearson, R. A. Field, R. A. Stockman, *Org Lett.* **2004**, *6*, 2377.

47 V. K. Aggarwal, E. Alonso, G. Fang, M. Ferrera, G. Hynd, M. Porcelloni, *Angew. Chem., Int. Edn. Engl.* **2001**, *40*, 1433; V. K. Aggarwal, M. Ferrera, C. J. O'Brien, A. Thompson, R. V. H. Jones, R. Fieldhouse, *J. Chem. Soc., Perkin Trans. 1* **2001**, 1635.; V. K. Aggarwal, *Chem. Rev.* **1997**, *97*, 2341.

48 D. R. Alonso, P. G. Andersson, *J. Org. Chem.* **1998**, *63*, 9455.

49 A. A. Cantrill, Lee D. Hall, A. N. Jarvis, H. M. I. Osborn, J. Raphy, J. B. Sweeney, *J. Chem. Soc., Chem. Commun.* **1996**, 2631.

50 F. A. Davis, H. Liu, P. Zhou, T. N. Fang, G. V. Reddy, Y. L. Zhang, *J. Org. Chem.*, **1999**, *64*, 7559, and references therein.

51 See, for instance, C. Gennari, G. Pain, *Tetrahedron Lett.* **1996**, *37*, 3747.

52 R. M. Carlson, S. Y. Lee, *Tetrahedron Lett.* **1969**, 4001.

53 P. W. Neber, A. Friedolsheim, *Justus Liebigs Ann. Chem.* **1926**, *449*, 109; C. O'Brien, *Chem. Rev.* **1964**, *64*, 81; J. A. Hyatt, *J. Org. Chem.* **1981**, *46*, 3953.

54 M. M. H. Verstappen, G. J. A. Ariaans, B. Zwanenberg, *J. Am. Chem. Soc.* **1995**, *118*, 8491. See also: T. Ooi, M. Takahashi, K. Doda, K. Maruoka, *J. Am. Chem. Soc.* **2002**, *124*, 7640.

55 L. Henn, D. M. B. Hickey, C. J. Moody, C. W. Rees, *J. Chem. Soc., Perkin Trans. 1*, **1984**, 2189

56 See: M. Alajarín, R.-Á. Orenes, Á. Vidal, A. Pastor, *Synthesis* **2003**, 49 for a novel synthesis of non-chiral azirines.

57 F. A. Davis, G. V. Reddy, H. Liu, *J. Am. Chem. Soc.* **1995**, *117*, 3651.

58 F. A. Davis, C.-H. Liang, H. Liu, *J. Org. Chem.* **1997**, *62*, 3796; F. A. Davis, J. H. Deng, Y. L. Zhang, R. C. Haltiwanger *Tetrahedron* **2002**, *58*, 7135.

59 F. A. Davis, H. Liu, C.-H. Liang, G. V. Reddy, Y. Zhang, T. Fang, D. D. Titus, *J. Org. Chem.* **1999**, *64*, 8929.

60 R. S. Atkinson, M. P. Coogan, I. S. T. Lochrie, *J. Chem. Soc., Perkin Trans. 1* **1997**, 897.

61 E. Risberg, P. Somfai *Tetrahedron: Asymmetry* **2002**, *13*, 1957.

62 P. Roth, P. G. Andersson, P. Somfai, *Chem. Comm.* **2002**, 1752.

63 M. J. Alves, G. Fortes, E. Guimarães, A. Lemos, *Synlett* **2003**, 1403.

64 E. Risberg, A. Fischer, P. Somfai, *Chem. Comm.* **2004**, 2088.

65 T. L. Gilchrist, R. Mendonça, *Synlett* **2000**, 1843.

66 M. J. Alves, A. G. Fortes, L. F. Goncalves *Tetrahedron Lett.* **2003**, *44*, 6277.

67 For a review of the chemistry of azirines, see: T. L. Gilchrist, *Aldrichimica Acta* **2001**, *34*, 51.

68 M. J. Alves, T. L. Gilchrist, *J. Chem. Soc., Perkin Trans. 1* **1998**, 299.

69 F. A. Davis, Y. Wu, H. Yan, K. R. Prasad, W. McCoull, *Org. Lett.* **2002**, *4*, 655.

70 Y. S. P. Álvares, M. J. Álves, N. G. Azoia, J. F. Bickley, T. L. Gilchrist, *J. Chem. Soc., Perkin Trans. 1* **2002**, 1911.

71 A. S. Timén, P. Somfai, *J. Org. Chem.* **2003**, *68*, 9958.

72 S. Y. Kim, E. N. Jacobsen, *Angew. Chem., Int. Edn. Engl.* **2004**, *43*, 3952.

73 Y. Tsuchiya, T. Kumamoto, T. Ishikawa, *J. Org. Chem.* **2004**, *69*, 8504.

74 T. Ishikawa, *J. Am. Chem. Soc.* **2001**, *123*, 7707.

75 S. Gabriel, *Chem. Ber.*,**1888**, *21*, 1049; S. Gabriel, *Chem. Ber.* **1888**, *21*, 2664.

76 H. Wenker, *J. Am. Chem. Soc.* **1935**, *57*, 2328.

77 H. M. I. Osborn, J. Sweeney, *Tetrahedron: Asymmetry* **1997**, *8*, 1693.

5
Metalated Epoxides and Aziridines in Synthesis

David M. Hodgson and Christopher D. Bray

5.1
Introduction

The chemistry of epoxides and aziridines is dominated by reactions that involve cleavage of the strained heterocyclic ring by nucleophiles (see Chapters 7 and 12). However, a less well known aspect of the chemistry of epoxides and aziridines is that which occurs upon reaction with a strong base. Abstraction of a β-proton may occur, resulting in the formation of allylic alcohols or amines, a process termed β-elimination [1a]. However, the β-hydrogens are not usually the most acidic; due to the electron-withdrawing effect of the oxygen/nitrogen and the acidifying nature of the three-membered ring, removal of an α-proton can occur to give an α-metalated epoxide (oxiranyl anion) or aziridine (aziridinyl anion) **1** (Scheme 5.1).

The chemistry of α-metalated epoxides and aziridines (the α prefix will from now on not be included but should be assumed) has been reviewed previously [1], but in this chapter it is our intention to focus on those reactions involving them that are useful in synthesis, rather than just of pedagogical interest. Beginning with metalated epoxides, since the greater amount of work has involved them, we intend to present carefully chosen examples of their behavior that delineate the diverse nature of their chemistry. We will then move on to metalated aziridines, the chemistry of which, it will become apparent, closely mirrors that of their epoxide cousins.

1: α-metalated epoxide (X = O)
α-metalated aziridine (X = NR)

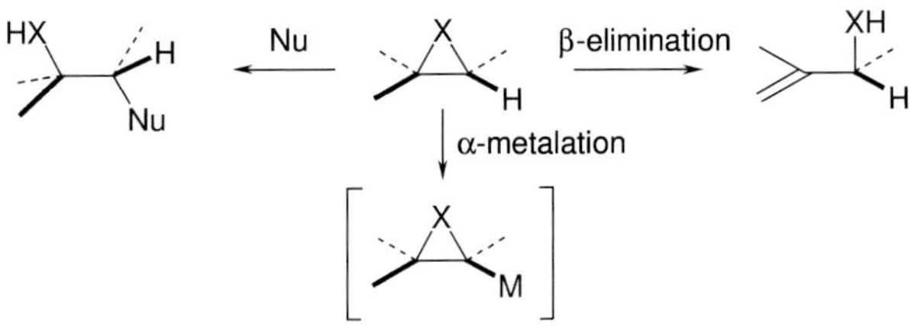

Scheme 5.1

Aziridines and Epoxides in Organic Synthesis. Andrei K. Yudin

ISBN: 3-527-31213-7

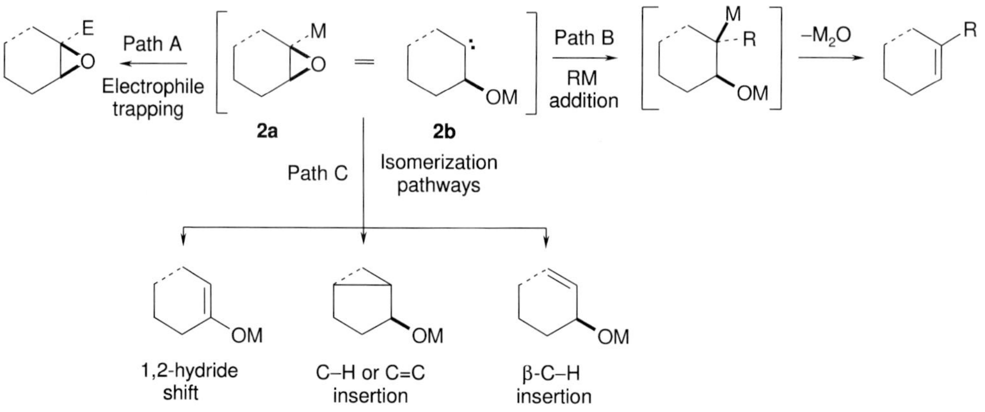

Scheme 5.2

5.2
Metalated Epoxides

Metalated epoxides undergo a variety of transformations; most simply, they can act as classical nucleophiles and react with a range of electrophiles to give more highly substituted epoxides, which constitutes an attractive route to such compounds (Scheme 5.2, Path A).

When lithiated, the ring strain of the three-membered heterocycle remains important, and this strain, combined with a weakening of the α-C–O bond, due to its greater polarization, make metalated epoxides highly electrophilic species [2]. They react with strong nucleophiles (often the base that was used to perform the α-deprotonation) to give olefins following the elimination of M_2O (Scheme 5.2, Path B), a process often referred to as "reductive alkylation".

Finally, metalated epoxides undergo isomerization processes characteristic of traditional carbenoids (Scheme 5.2, Path C). The structure of a metalated epoxide is intermediate in nature between the structures **2a** and **2b** (Scheme 5.2). The existence of this intermediacy is supported by computational studies, which have shown that the α-C–O bond of oxirane elongates by ~12% on α-lithiation [2]. Furthermore, experimentally, the α-lithiooxycarbene **4a** (Scheme 5.3) returned cyclopentene oxide **7** among its decomposition products; indeed, computational studies of singlet **4a** suggest it possesses a structure in the gas phase that is intermediate in nature between an α-lithiocarbene and the lithiated epoxide **4b** [3].

The existence of a metalated epoxide was first proposed by Cope and Tiffany, to explain the rearrangement of cyclooctatetraene oxide (**8**) to cycloocta-1,3,5-trien-7-one (**11**) on treatment with lithium diethylamide. They suggested that lithiated epoxide **9** rearranged to enolate **10**, which gave ketone **11** on protic workup (Scheme 5.4) [4].

Following this, House and Ro presented the first experimental evidence for the existence of metalated epoxides. Treatment of *cis*- and *trans*-α-methyl-β-(phenyl-

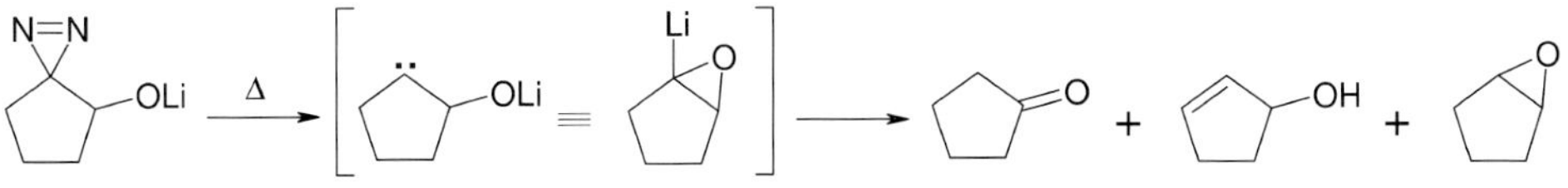

Scheme 5.3

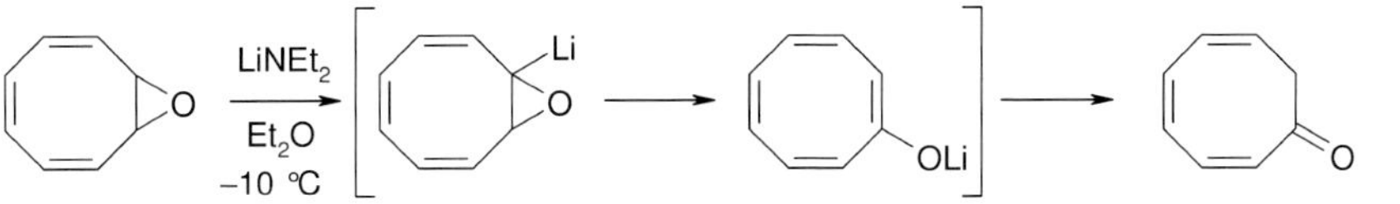

Scheme 5.4

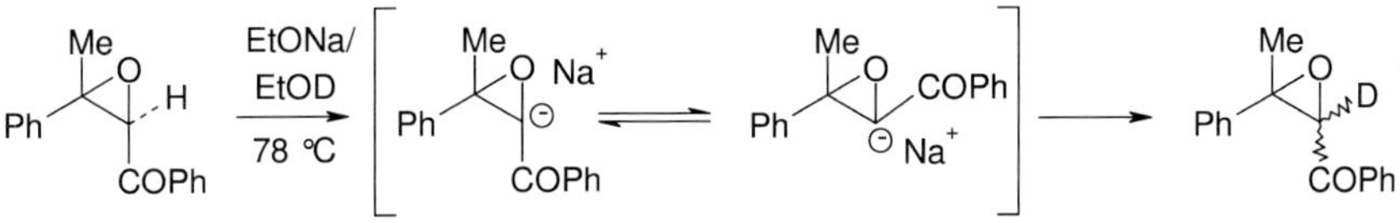

Scheme 5.5

carbonyl)acetophenone oxide (**12**) with sodium ethoxide in deuterioethanol resulted in a geometric mixture of **14**, in which one deuterium had been incorporated onto the epoxide ring (Scheme 5.5) [5]. This experiment not only indicates the intermediacy of anion **13a**, but also highlights the issue of oxiranyl anion configurational stability (**13a** ⇌ **13b**), a topic we shall return to later.

5.2.1
C–H Insertions

5.2.1.1 Transannular C–H Insertions in Epoxides of Medium-sized Cycloalkenes

It was the early observation by Cope et al. that treatment of epoxides of medium-sized cycloalkenes with strong base results in the generation of new carbon skeletons, which prompted the initial interest in the chemistry of metalated epoxides [6].

Treatment of *cis*-cyclooctene oxide (**15**) with lithium diethylamide in ether at reflux gave *endo-cis*-bicyclo[3.3.0]octan-2-ol (**16**) in 70% yield. Under identical conditions, *trans*-cyclooctene oxide (**17**) gave *exo-cis*-bicyclo[3.3.0]octan-2-ol (**18**) in 55–60% yield (Scheme 5.6) [7]. In each case, the reaction is completely stereospecific, with neither of the epimeric alcohols being observed, which suggests that the reactions proceed through insertion of a carbenoid (rather than the same α-li-

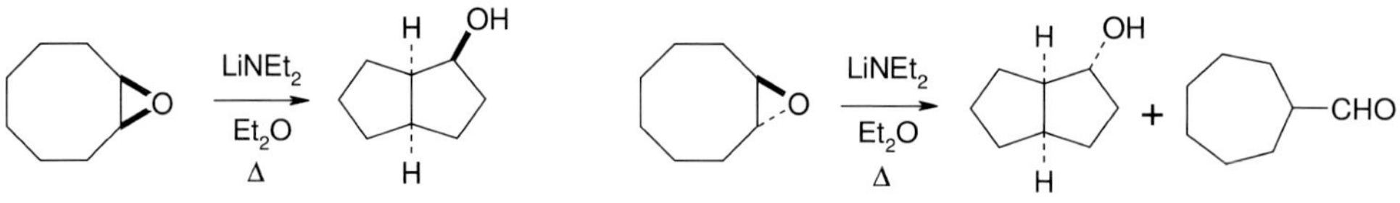

15 → LiNEt$_2$, Et$_2$O, Δ → 16: 70%

17 → LiNEt$_2$, Et$_2$O, Δ → 18: 55-60% + 19: 32%

Scheme 5.6

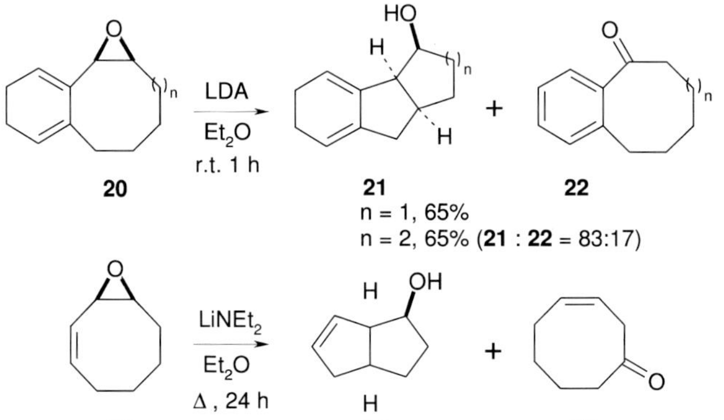

20 → LDA, Et$_2$O, r.t. 1 h → 21 + 22

n = 1, 65%

n = 2, 65% (**21** : **22** = 83:17)

23 → LiNEt$_2$, Et$_2$O, Δ, 24 h → 24 + 25

73% (**24** : **25** = 92 : 8)

Scheme 5.7

thiooxy carbene) into the transannular C–H bond. The synthesis of alcohol **16** was later reported to proceed in improved yield through the use of LDA in ether/hexane (80%) [8]. Cope et al. also showed that *cis-* and *trans*-cyclodecene oxide underwent similar transformations [9].

Benzo-fused epoxides **20** ($n = 1$ or 2) underwent facile rearrangement to alcohols **21** ($n = 1$ or 2, respectively) upon treatment with LDA at room temperature in only 1 h, owing to the acidic nature of their benzylic protons (Scheme 5.7) [10]. Similarly, 3,4-epoxycyclooctene **23** gave alcohol **24** [11].

n-BuLi can be used in place of lithium amides for the rearrangement of epoxides of eight- and ten-membered cycloalkenes [12]. Epoxides of larger and smaller cycloalkenes were found to give a variety of products (Table 5.1). This dependence of product profile on ring size demonstrates that close proximity between the metalated epoxide and the target C–H bond [as is the case for epoxides of eight-, (nine-; see below) and ten-membered cycloalkenes] is important for efficient transannular insertion to occur.

In recent years, enantioselective variants of the above transannular C–H insertions have been extensively studied. The enantiodetermining step involves discrimination between the enantiotopic protons of a *meso*-epoxide by a homochiral base, typically an organolithium in combination with a chiral diamine ligand, to generate a chiral nonracemic lithiated epoxide (e.g., **26**; Scheme 5.8). Hodgson

Table 5.1 Ring size and product distribution.

n-BuLi; Et_2O / hexane; −78 °C → 25 °C; 60-90%

n	Ratio of products			
1	29	56	15	0
2	0	84	3	13 (a = 2, b = 1)
3	0	0	trace	99 (a = 1, b = 3)
5	trace	0	trace	97.5 (a = 2, b = 4)
7	80	10	10	0

and Lee reported the first such example in 1996 [13]. Treatment of *cis*-cyclooctene oxide with *s*-BuLi/(–)-sparteine gave alcohol **16** in 79% yield and 73% *ee*. The use of *i*-PrLi (which does not contain a stereocenter) in place of commercial *s*-BuLi gave improved results (86% yield, 84% *ee*); these conditions were also successfully applied to cyclonon- and cyclododecene oxides (**27** and **29**). Interestingly, the use of substoichiometric amounts (20 mol%) of (–)-α-isosparteine also resulted in good yield (86%) and *ee* (84%) of alcohol **16** with *i*-PrLi as base; this indicates potential for asymmetric catalysis (Scheme 5.8) [14, 15].

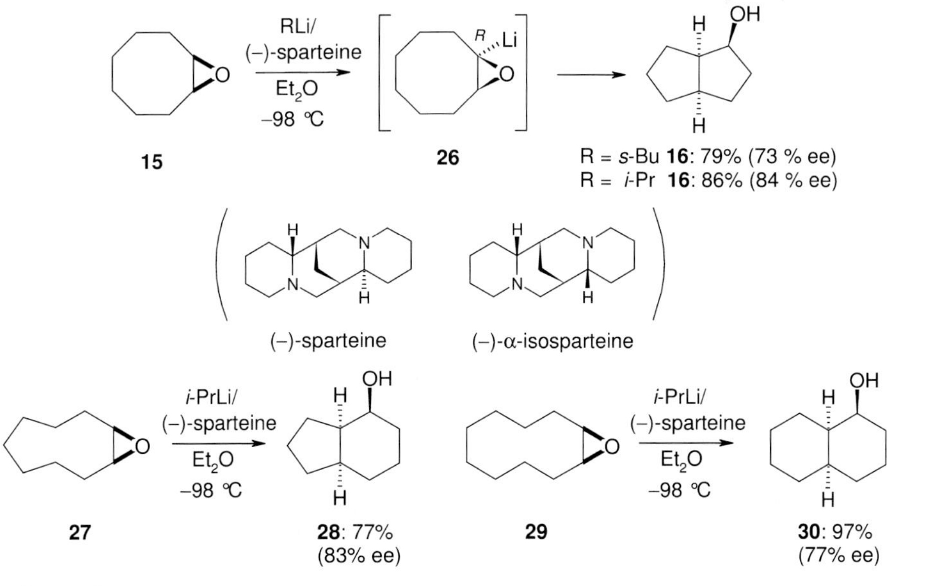

Scheme 5.8

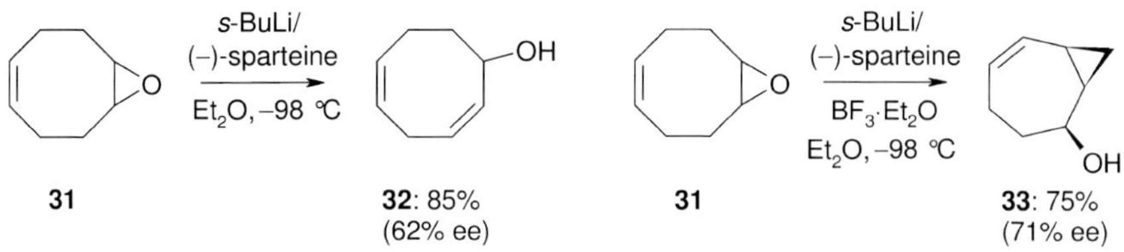

Scheme 5.9

A significant effect of Lewis acids on such transannular C-H insertion reactions has been demonstrated. Treatment of 5,6-epoxycyclooctene (**31**) with *s*-BuLi/(−)-sparteine gave allylic alcohol **32**, formally the product of β-elimination, in good yield (and *ee*) (Scheme 5.9). In the presence of $BF_3 \cdot Et_2O$, however, alcohol **33** was produced as a result of α-lithiation, in 75% yield and 71% *ee* [16].

To extend the scope of asymmetric transannular C–H insertions, more highly functionalized medium-sized cyclic epoxides have been investigated. A triad of cyclooctene oxides **34**, **36**, and **38**, possessing protected diol units, gave the expected alcohols **35**, **37**, and **39** (Scheme 5.10) [17, 18]; an asymmetric synthesis of (−)-xialenon A has been achieved starting from alcohol **39** [19]. In comparison,

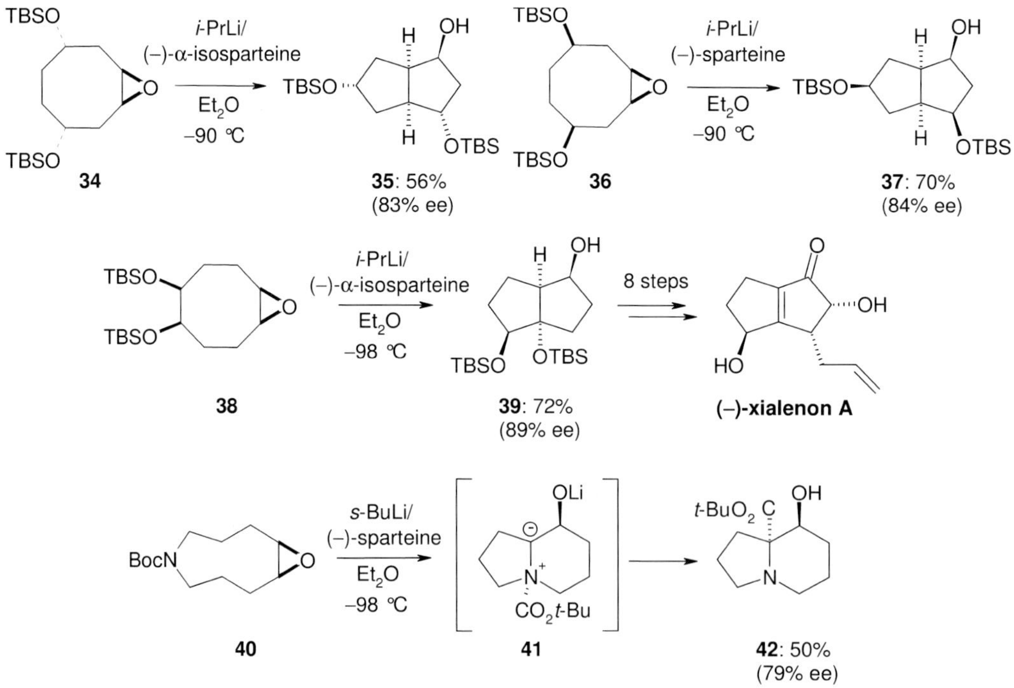

Scheme 5.10

azacyclic epoxide **40** gave indolizidine **42**, probably through trapping of the lithiated epoxide by the carbamate nitrogen and subsequent Stevens rearrangement [18, 20].

5.2.1.2 Transannular C–H Insertions in Epoxides of Polycyclic Alkenes

Transannular C–H insertions are not limited to epoxides of medium-sized cycloalkenes. Lithium amide-induced rearrangement of *exo*-norbornene oxide **43** gave nortricyclanol **44** in up to 90% yield, depending on the base employed (Scheme 5.11) [15, 21, 22]. Similarly, epoxides **45** and **47** gave alcohols **46** and **48**, respectively [21, 23]. Under similar conditions, chloronorbornene oxide **49** gave predominantly nortricyclanone **50** [24]. An intriguing reaction was found to occur with *exo,exo*-norbornadiene dioxide **51**; presumably epoxide **52**, the direct result of transannular insertion, was highly strained and so was reduced by the excess of LDA present to give *meso*-diol **53** [15, 22].

The extension of this methodology into asymmetric processes has also been explored: the use of lithium (*S,S*)-bis(1-phenyl)ethylamide as base allowed access to nortricyclanol **44** (73% yield, 49% *ee*), and the use of *s*-BuLi/(–)-sparteine gave similar results (73% yield, 52% *ee*) [25]. During a targeted natural product synthesis, base-induced transannular C-H insertion of *N*-Boc-7-azanorbornene oxide **54** was examined. The use of LDA as base gave azanortricyclanol **55** in 52% yield (Scheme 5.12); application of aryllithiums in combination with diamine ligands provided the best yields and *ees*. A subsequent radical rearrangement of the methyl xanthate of **55** allowed access to a biologically active analogue of epibatidine [26].

In all examples of enantioselective deprotonation of *meso*-epoxides with organo-

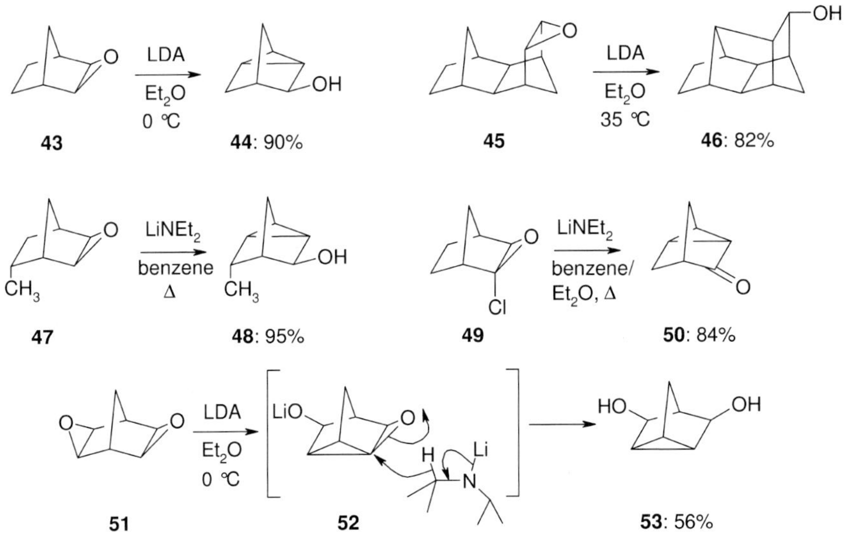

Scheme 5.11

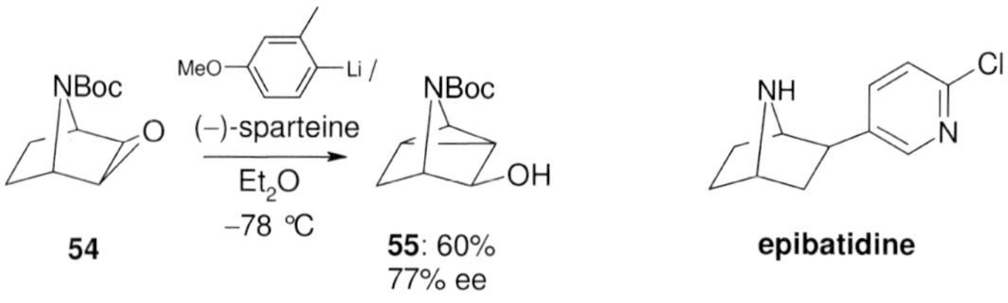

Scheme 5.12

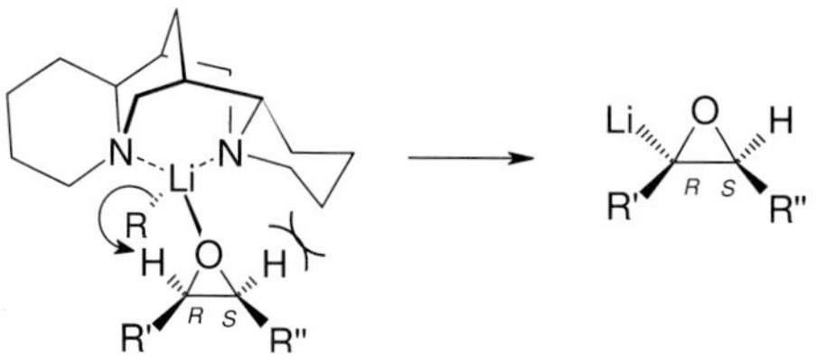

Figure 5.1 Ternary organolithium(RLi) · sparteine · epoxide complex in which the proton on the (*R*)-epoxide stereocenter is held closer to the organolithium.

lithiums in combination with the sparteines reported to date, removal of the proton from the (*R*)-epoxide stereocenter is preferred. This enantioselectivity may be explained in terms of a ternary organolithium(RLi) · sparteine · epoxide complex, in which the proton on the (*R*)-epoxide stereocenter is held closer to the organolithium (Figure 5.1).

5.2.1.3 Nontransannular Examples of C–H Insertion

Limited examples of nontransannular C–H insertions occurring in metalated epoxides exist. Treatment of *trans*-di-*tert*-butylethylene oxide **56** with *t*-BuLi predominantly gave the diastereomeric alcohols **58** and **59** (Scheme 5.13) [27]. Mioskowski

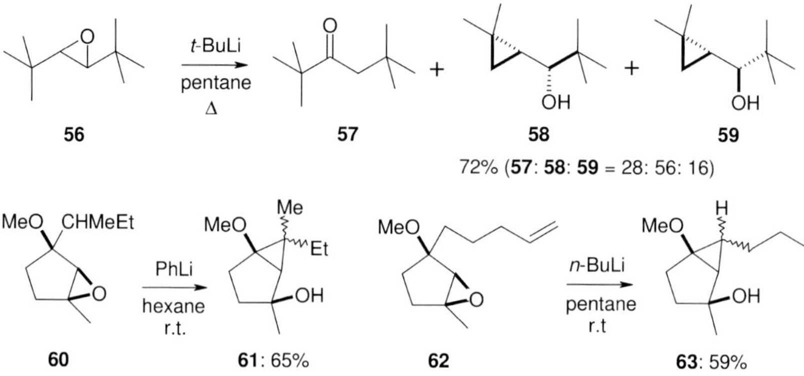

Scheme 5.13

et al. have also demonstrated that epoxides such as **60** and **62** provide cyclopropanes **61** and **63** as mixtures of diastereomers in good yields [28, 29]. In all three cases, one can still imagine the lithiated epoxides and the target C–H bonds oriented close to one another, though in the last two examples the base needed to be added slowly to a dilute solution of the epoxide to minimize the formation of products derived from "reductive alkylation".

5.2.1.4 Isomerization of Epoxides to Ketones

Whereas *exo*-norbornene oxide rearranges to nortricyclanol on treatment with strong base through transannular C–H insertion (Scheme 5.11), *endo*-norbornene oxide **64** gives norcamphor **65** as the major product (Scheme 5.14) [15, 22]. This product arises from 1,2-hydrogen migration; very little transannular rearrangement is observed. These two reaction pathways are often found to be in competition with one another, and subtle differences in substrate structure, and even in the base employed, can have a profound influence on product distribution.

Base-induced rearrangement of bicyclo[2.2.2]octane oxide **67** gives predominantly bicyclo[2.2.2]octanone **68** (Scheme 5.15), which once again indicates that close proximity between the carbenoid center and the C-H bond into which it may insert is important if such an insertion is to occur [30]. In comparison, the sense of product distribution is reversed for the related substrate bicyclo[2.2.2]octadiene oxide **70** on treatment with LDA [15, 22], alcohol **72** being the favored product.

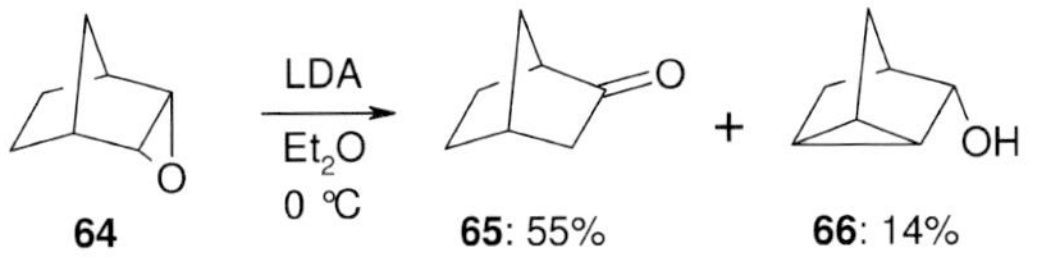

64 → LDA, Et_2O, 0 °C → **65**: 55% + **66**: 14%

Scheme 5.14

67 → LDA, Et_2O, Δ → **68** + **69**; 95% **71**: **72** = 19: 1

70 → base, Et_2O, Δ → **71** + **72**
LDA: **71**: 29%, **72**: 55%
(chiral lithium amide, Ph/N/Li/Ph): **71**: 50%, **72**:15%

73 → $LiNEt_2$, Et_2O, Δ → **74**: 95%

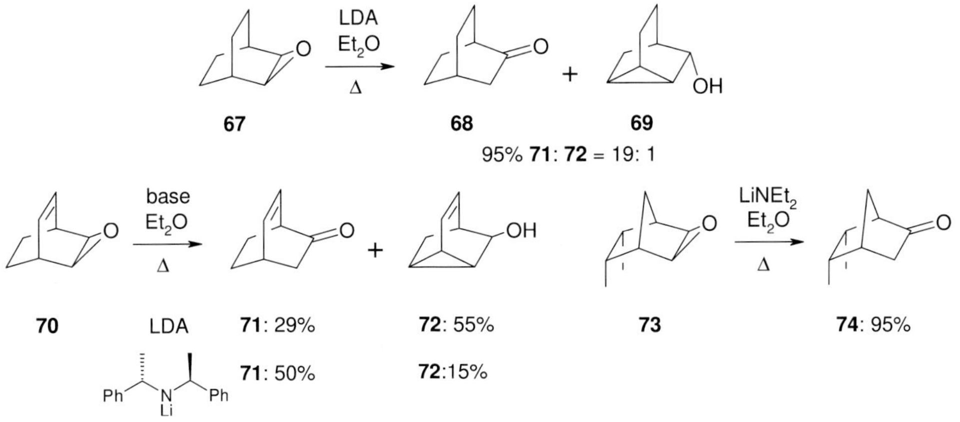

Scheme 5.15

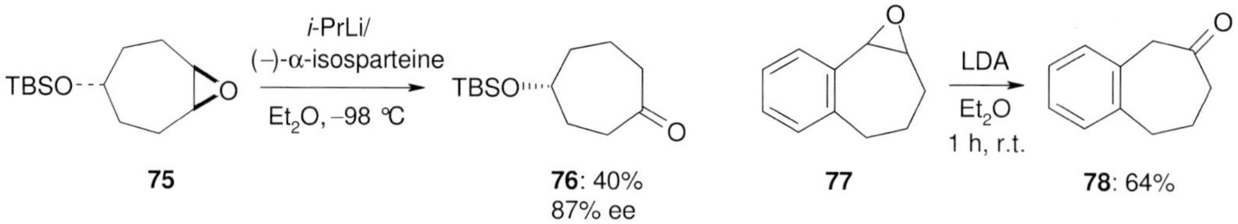

Scheme 5.16

Interestingly, the use of (*S*,*S*)-bis(1-phenyl)ethylamide as base with epoxide **70** predominantly yields ketone **71**. Where the possibility for competing C–H insertion is removed (e.g., with epoxide **73**), isomerization to ketone **74** occurs in excellent yield.

Epoxides of eight-, nine-, and ten-membered cycloalkenes undergo facile transannular rearrangement (see above); in contrast, cycloheptene oxides predominantly undergo rearrangement to give ketones. Boeckmann found that treatment of cycloheptene oxide itself with *n*-BuLi gave cycloheptanone among other products (Table 5.1, Entry 2) [12]. This type of rearrangement was applied to substituted cycloheptene oxide **75** (Scheme 5.16): enantioselective lithiation/rearrangement gave ketone **76**, a known intermediate in an asymmetric synthesis of (*S*)-physoperuvine [31]. Benzo-fused cycloheptene oxide **77** underwent facile rearrangement to ketone **78** [10].

The effect of substrate structure on product profile is further illustrated by the reactions of *cis*- and *trans*-stilbene oxides **79** and **83** with lithium diethylamide (Scheme 5.17) [32]. Lithiated *cis*-stilbene oxide **80** rearranges to enolate **81**, which gives ketone **82** after protic workup, whereas with lithiated *trans*-stilbene oxide **84**, phenyl group migration results in enolate **85** and hence aldehyde **86** on workup. Triphenylethylene oxide **87** underwent efficient isomerization to ketone **90** [32].

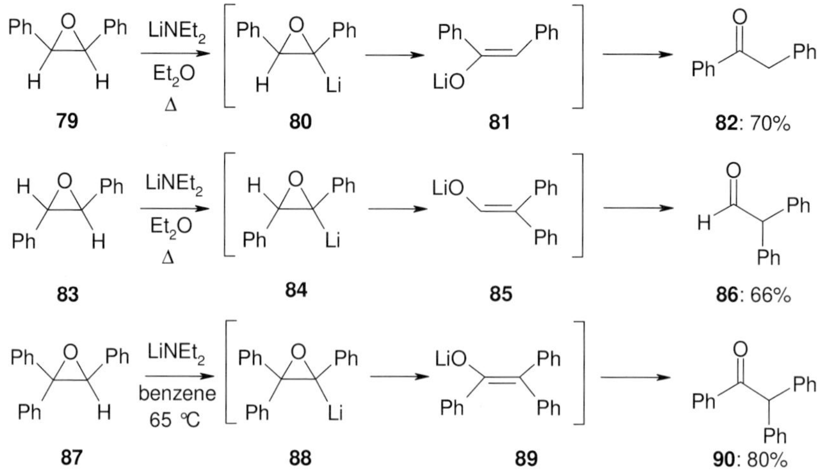

Scheme 5.17

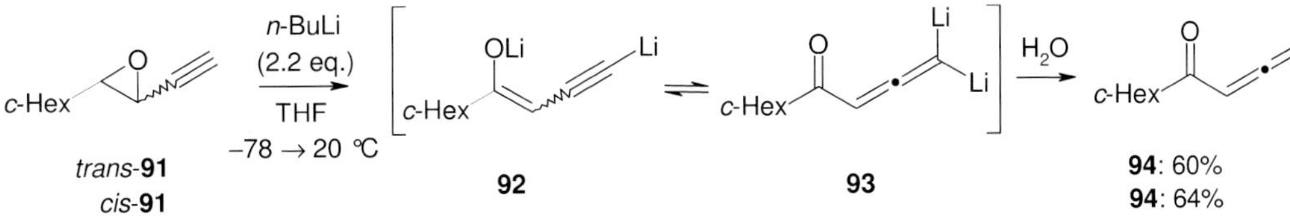

Scheme 5.18

The isomerization of acetylenic oxiranes *cis*- and *trans*-**91** to allenic ketone **94** has recently been described (Scheme 5.18). It is proposed that the rearrangement proceeds via a dilithium ynenolate [33].

5.2.2
Cyclopropanations

Crandall and Lin first demonstrated the reaction between a metalated epoxide and an alkene to generate a cyclopropane, a process typical of a carbenoid, in 1967. Treatment of 1,2-epoxyhex-5-ene with *t*-BuLi gave products predominantly derived from incorporation of the organolithium, but a small amount (9%) of *trans*-bicyclo[3.1.0]hexan-2-ol (**97**) was also observed (Scheme 5.19) [27]. The reaction was completely stereoselective, with none of the epimeric alcohol being observed. Hodgson et al. subsequently demonstrated that the use of lithium 2,2,6,6-tetramethylpiperidide (LTMP) as base, rather than *t*-BuLi, allowed access to alcohol **97** in much improved yield (79%) [34a]. The stereoselective nature of the reaction was attributed to an initial *trans*-lithiation, followed by trapping of the carbenoidic center by the tethered olefin occurring at the stage of the lithium carbenoid rather than an α-lithiooxy carbene (compare this to the formation of alcohols **16** and **18**; Scheme 5.6), in a chair-like transition state **96** (Scheme 5.19). This behavior should be contrasted with that observed with use of the superbase LIDAKOR with 1,2-epoxyhex-5-ene, which gives an isomeric cyclobutanol through allylic deprotonation and subsequent epoxide ring-opening [34b].

Hodgson et al. showed that a series of bis- and tris-homoallylic terminal epoxides underwent intramolecular cyclopropanation to give a range of bicyclic alcohols. A short asymmetric synthesis of sabina ketone based on this chemistry was demonstrated (Scheme 5.20). A practical advantage with this process is that the volatile epoxides can be replaced with readily available chlorohydrins, an extra

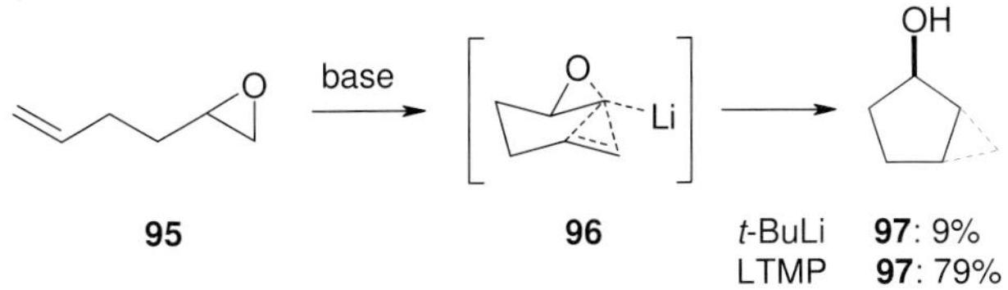

Scheme 5.19

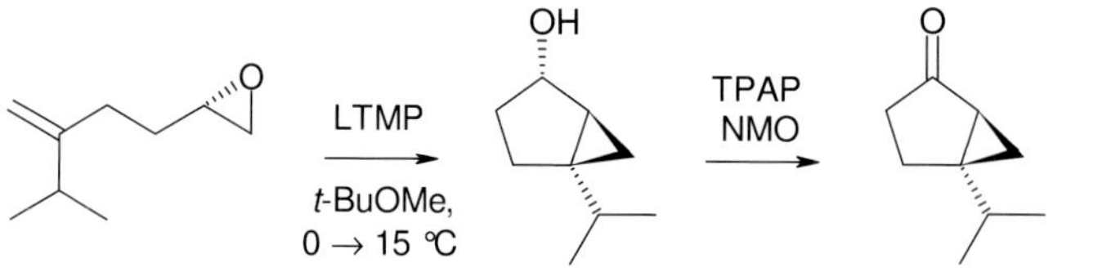

98 99: 80% sabina ketone: 95%

Scheme 5.20

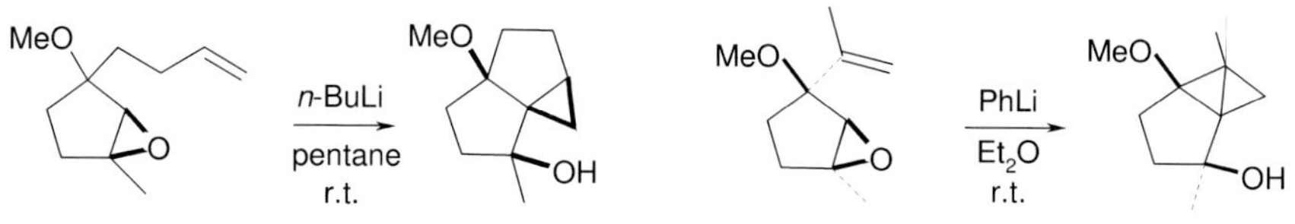

100 101: 47% 102 103: 74%

Scheme 5.21

equivalent of base simply being employed in the reaction to generate the epoxide *in situ* [34c].

Mioskowski et al. have demonstrated a route to spirocyclopropanes. As an example, treatment of epoxide **100** with *n*-BuLi in pentane stereoselectively gave tricyclic alcohol **101**, albeit in only 47% yield (Scheme 5.21) [29]. With a related substrate, epoxide **102** stereoselectively gave dicyclopropane **103** on treatment with PhLi; uniquely, the product was isolable after column chromatography in 74% yield [35]. As was also seen with attempts to perform C–H insertion reactions in a non-transannular sense, one should note that steps were taken to minimize the formation of olefin products, either by the use of a base with low nucleophilicity (LTMP) and/or by slow addition of the base to a dilute solution (10^{-3} M in the case of **102**) of the epoxide.

Other such dicyclopropanes underwent efficient acid-catalyzed (SiO_2) hydrolysis to give α-ketocyclopropanes (e.g., **104**), which could further be elaborated to 2,3-disubstituted-2-cyclopentenones (e.g., **105**) in good to excellent yield through acid-assisted nucleophilic ring-opening (Scheme 5.22) [35].

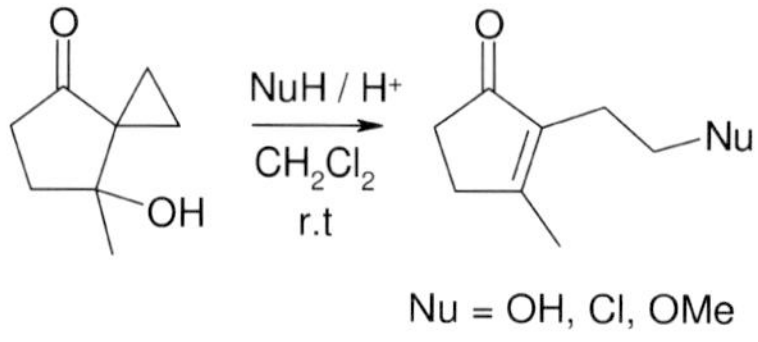

104 105: 81-95%

Scheme 5.22

5.2.3 Olefin Formation

Metalated epoxides can react with organometallics to give olefins after elimination of dimetal oxide, a process often referred to as "reductive alkylation" (Path B, Scheme 5.2). Crandall and Lin first described this reaction in their seminal paper in 1967; treatment of *tert*-butyloxirane **106** with 3 equiv. of *tert*-butyllithium, for example, gave *trans*-di-*tert*-butylethylene **110** in 64% yield (Scheme 5.23), Stating that this reaction "*should have some synthetic potential*", [36] they proposed a reaction pathway in which *tert*-butyllithium reacted with α-lithiooxycarbene **108** to generate dianion **109** and thence olefin **110** upon elimination of dilithium oxide. The epoxide has, in effect, acted as a vinyl cation equivalent.

Despite its obvious potential, this reaction received little further attention until Eisch and Galle reported that the treatment of α,β-epoxysilanes **111** and *cis*- or *trans*-**113** with excess PhLi gave alkenes **112** and **114**; the isolation of Ph_4Si as a byproduct from these reactions indicates that they proceeded via metalated epoxides generated through silicon-lithium exchange (Scheme 5.24) [37]. In this case, it is not clear whether the formation of *trans*-stilbene from *cis*-**113** is due to configurational instability of the intermediate lithiated epoxide, whether it is simply due to elimination of dilithium oxide occurring to give the thermodynamically more stable olefin, or perhaps both. For the control of olefin geometry in synthesis, these are important considerations.

Different substrate geometries can even result in alternate reaction pathways operating. The reactions between *trans*-α,β-epoxytrimethylsilane **115** and organometals (metal = Li, Ce, or La) give predominantly *trans*-alkene **116** in high yields (Scheme 5.25) [38]. In contrast, treatment of *cis*-**115** with some of the same organometals produces (*Z*)-vinylsilanes. The use of a bulkier substituent on silicon (e.g., **118**) resulted in an almost complete reversal of stereoselectivity to give (*E*)-vinylsilane **119** [39]; the factors concerning olefin geometry discussed above are obviously equally valid here.

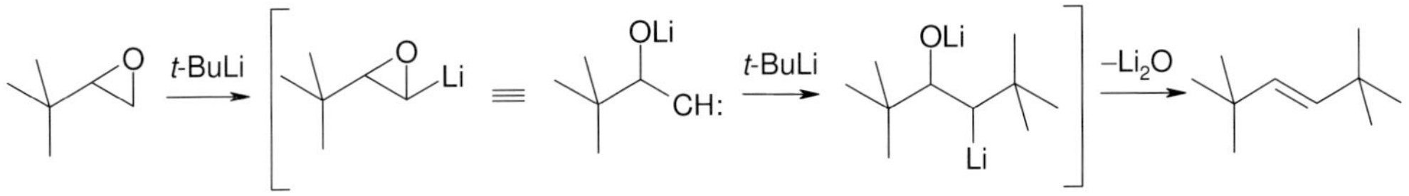

t-BuLi; OLi, Li, CH:; *t*-BuLi; $-Li_2O$

106 **107** **108** **109** **110**: 64%

Scheme 5.23

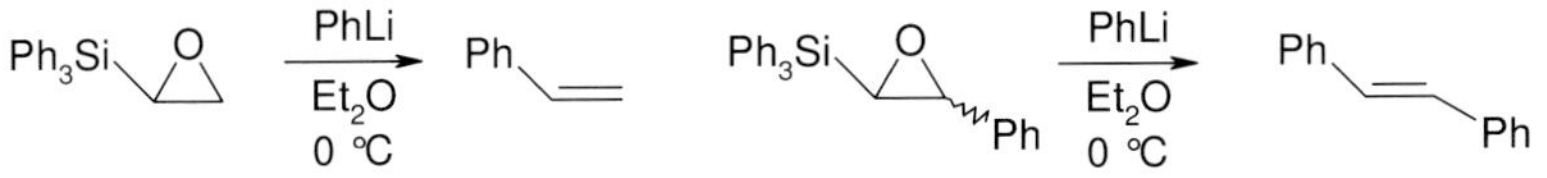

Ph_3Si; PhLi, Et_2O, 0 °C; Ph; PhLi, Et_2O, 0 °C; Ph, Ph

111 **112** *cis*- or *trans*-**113** **114** (*trans* : *cis* > 95:5)

Scheme 5.24

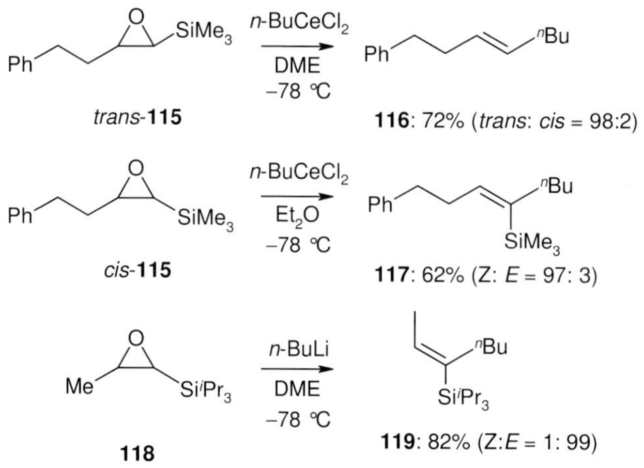

trans-**115** → n-BuCeCl$_2$, DME, −78 °C → **116**: 72% (*trans*: *cis* = 98:2)

cis-**115** → n-BuCeCl$_2$, Et$_2$O, −78 °C → **117**: 62% (Z: *E* = 97: 3)

118 → n-BuLi, DME, −78 °C → **119**: 82% (Z:*E* = 1: 99)

Scheme 5.25

120 → Me$_3$SiCH$_2$Li, Et$_2$O, −78 °C → [**121**] → Me$_3$SiCH$_2$Li, −Li$_2$O → **122**: 68% (*E*:*Z* = 97:3)

123 → Me$_3$SiCH$_2$Li, hexane, r.t. → **124**: 75%

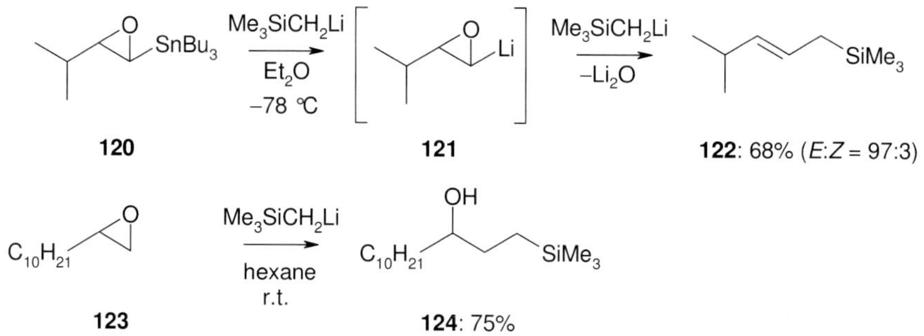

Scheme 5.26

trans-α,β-Epoxystannanes (e.g., **120**) also undergo "reductive alkylation" (Scheme 5.26). On treatment with excess organolithium (Me$_3$SiCH$_2$Li, for example), tin-lithium exchange furnishes lithiated epoxide **121** and hence ultimately *trans*-allylsilane **122** [40]. It should be noted that simple terminal epoxides such as **123** do not undergo "reductive alkylation" on treatment with Me$_3$SiCH$_2$Li (probably because it is insufficiently basic to abstract one of the ring protons), but instead undergo direct nucleophilic ring-opening to give secondary alcohols (e.g., **124**) [41].

Mioskowski et al. examined the "reductive alkylation" of simple epoxides by organolithiums in THF in considerable detail, and found that the best yields and stereoselectivities were obtained with secondary and tertiary alkyllithiums (Table 5.2, Entries 1–5) [42]. n-BuLi gave a mixture of olefins (Entry 6), whereas PhLi and MeLi (Entries 7 and 8) gave very poor yields. Similar transformations have been reported with the use of lithium tetraalkylcerate reagents (Entries 9 and 10) [43].

Table 5.2 "Reductive alkylation" of simple epoxides by organolithiums.

Entry	Substrate	Organometal	Product	Yield % (ratio)
1	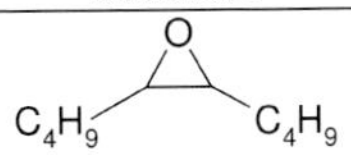 O C_4H_9 C_4H_9	*s*-BuLi	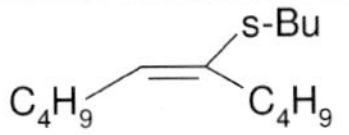 s-Bu C_4H_9 C_4H_9	76
2	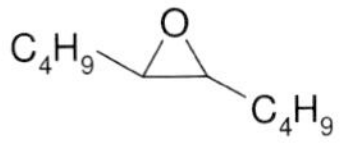 C_4H_9 O C_4H_9	*s*-BuLi	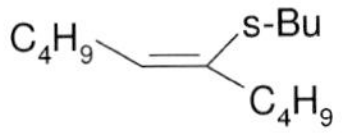 C_4H_9 s-Bu C_4H_9	65
3	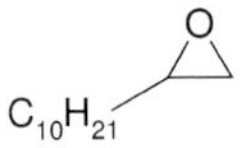 O $C_{10}H_{21}$	*t*-BuLi	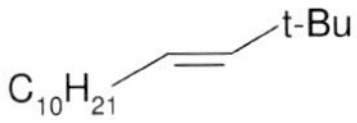 t-Bu $C_{10}H_{21}$	98
4	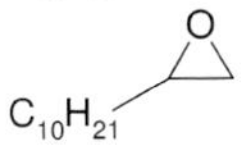 O $C_{10}H_{21}$	*s*-BuLi	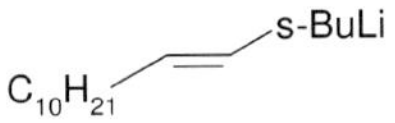 s-BuLi $C_{10}H_{21}$	96
5	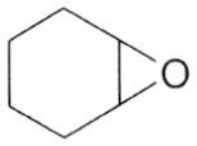 O	*t*-BuLi	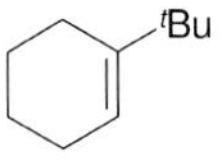 ^{t}Bu	56
6	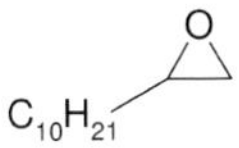 O $C_{10}H_{21}$	*n*-BuLi	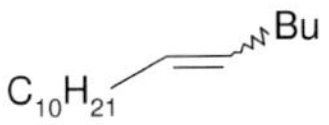 Bu $C_{10}H_{21}$	91 (74:26; *E*:*Z*)
7	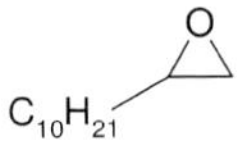 O $C_{10}H_{21}$	PhLi	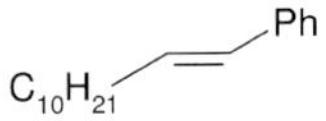 Ph $C_{10}H_{21}$	20
8	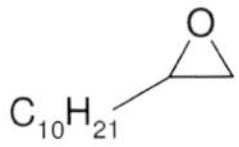 O $C_{10}H_{21}$	MeLi	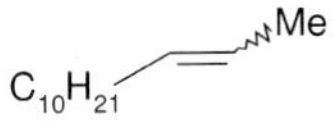 Me $C_{10}H_{21}$	48 (66:34; *E*:*Z*)
9	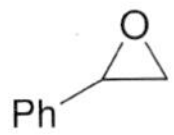 O Ph	*n*-Bu_4CeLi	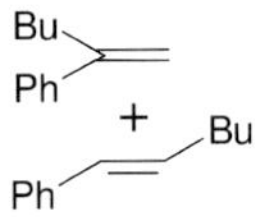 Bu Ph (93) + Bu Ph (7)	84
10	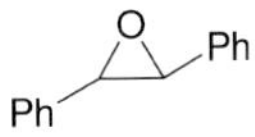 O Ph Ph	*n*-Bu_4CeLi	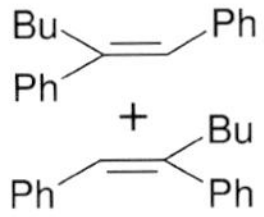 Bu Ph Ph (84) + Bu Ph Ph (16)	98

It was demonstrated that when a better leaving group than lithium oxide (Li_2O) is present at the α-position (e.g., epoxide **125**; Scheme 5.27), alkene formation occurs with retention of the alcohol moiety [44].

Hodgson and coworkers extended this concept to epoxides of unsaturated cyclic ethers **128** [5] and amines **130** [46, 47] (Scheme 5.28). It is interesting that the use of trimethylsilylmethyllithium as the organolithium in this case resulted in substituted allylsilanes **129** and **131** (R = CH_2SiMe_3); presumably the epoxide ring protons of **128** and **130** are more acidic than those of a simple terminal epoxide (see Scheme 5.26).

These processes could be performed in an enantioselective manner by addition of (–)-sparteine or bisoxazoline ligands; in general the yields and *ee*s were found to

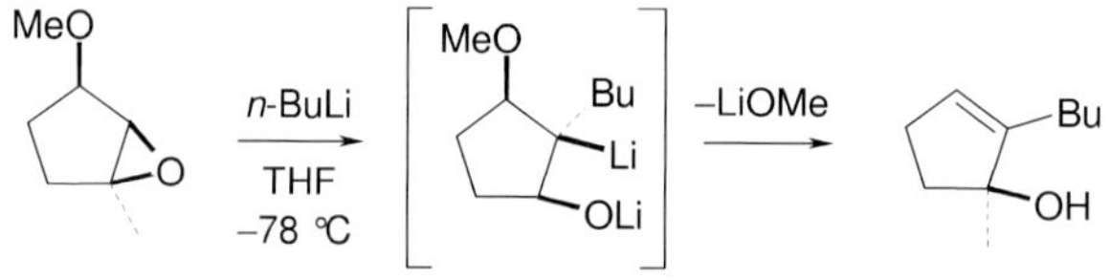

MeO

n-BuLi

THF

–78 °C

MeO Bu Li OLi

–LiOMe

Bu OH

125 **126** **127**: 95%

Scheme 5.27

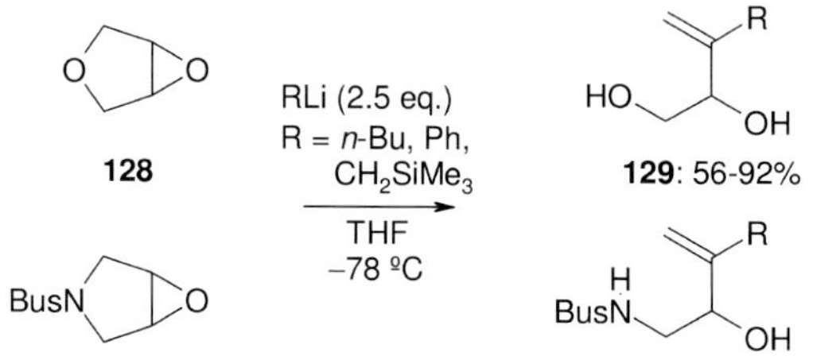

O O

128

RLi (2.5 eq.)
R = *n*-Bu, Ph, CH_2SiMe_3

THF
–78 °C

HO R OH

129: 56-92%

BusN O

130

H BusN R OH

131: 44-72%

Scheme 5.28

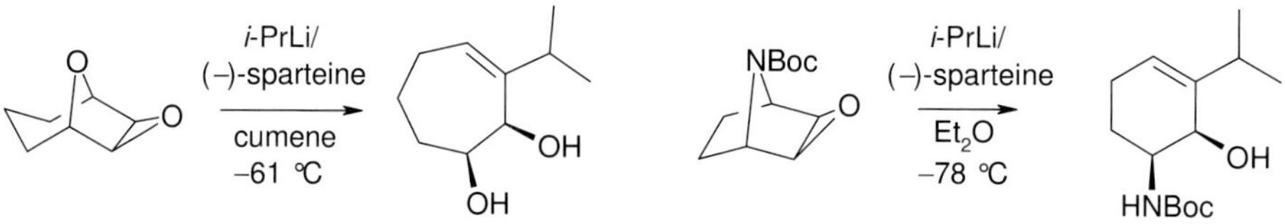

O O

i-PrLi/
(–)-sparteine

cumene
–61 °C

OH OH

132 **133**: 58%
73% ee

NBoc O

i-PrLi/
(–)-sparteine

Et_2O
–78 °C

OH HNBoc

134 **135**: 78%
87% ee

Scheme 5.29

be highly substrate- and ligand-dependant [47, 48], although in certain cases the results were very good (Scheme 5.29).

There are some synthetic drawbacks with using the methodology described above to access alkenes: *i*) the base and the nucleophile are always the same species, and *ii*) typically only simple alkyl/aryl units can be incorporated. Steps have been taken to address these issues. Utimoto et al. demonstrated that epoxysilanes could be deprotonated with *s*-BuLi and that the lithiated epoxides thus generated could then be treated with organoaluminiums or organozincs to give a range of triphenylsilyl-substituted olefins (Table 5.3). The use of diethylhaloaluminiums to generate vinylhalides is also notable [49].

Whitby and Kasatkin have extended this idea to insertion of stabilized metalated epoxides into zirconacycles (Scheme 5.30) [50].

Hodgson et al. have demonstrated that arylalkenes **139** and dienes **140** can readily be prepared from simple terminal epoxides in a highly stereoselective manner by employing LTMP as base in combination with aryl and vinyllithiums as nucleophiles at 0 °C (Scheme 5.31) [41]. Without addition of LTMP, secondary alcohols

Table 5.3 Deprotonation with *s*-BuLi and treatment with organoaluminiums or organozincs to give a range of triphenylsilyl-substituted olefins.

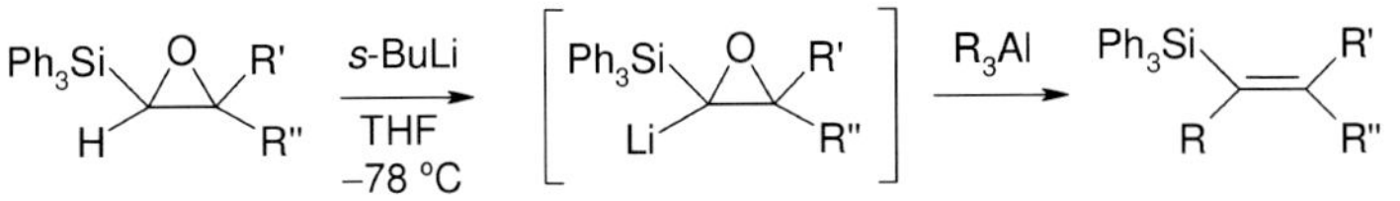

Entry	R'	R''	R_3Al	Yield (%)
1	H	H	Me_3Al	86
2			Et_3Al	83
3			Et_2AlCl	78 (R = Cl), 8 (R = Et)
4			Et_2AlI	77 (R = I), 3 (R = Et)
5			(*E*)-*n*-BuCH=CHAl(*i*-Bu)$_2$	80 (R = (*E*)-*n*-BuCH=CH)
6			$Me_3SiC{\equiv}CAl(i\text{-}Bu)_2$	75 (R = $C{\equiv}CSiMe_3$), 9 (R = *i*-Bu)
7			Me_2AlSBu	52 (R = SBu), 17 (R= Me)
8			Et_2Zn	87
9	Me	H	Me_3Al	92
10			Et_3Al	94
11			Et_2AlCl	60 (R = Cl), 23(R = Et)
12			Et_2AlI	67 (R = I), 13 (R = Et)
13			(*E*)-*n*-BuCH=CHAl(*i*-Bu)$_2$	80 (R = (*E*)-*n*-BuCH=CH), 4 (R = *i*-Bu)
14			$Me_3SiC{\equiv}CAl(i\text{-}Bu)_2$	65 (R = $C{\equiv}CSiMe_3$), 14 (R = *i*-Bu)
15			Et_2Zn	80
16	H	Me	Me_3Al	93
17			Et_3Al	92
18	Me	Me	Me_3Al	83
19			(*E*)-*n*-BuCH=CHAl(*i*-Bu)$_2$	66 (R = (*E*)-*n*-BuCH=CH), 6 (R = *i*-Bu)

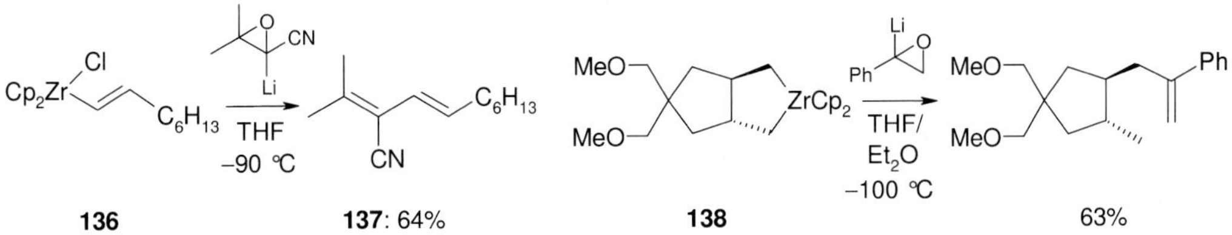

Scheme 5.30

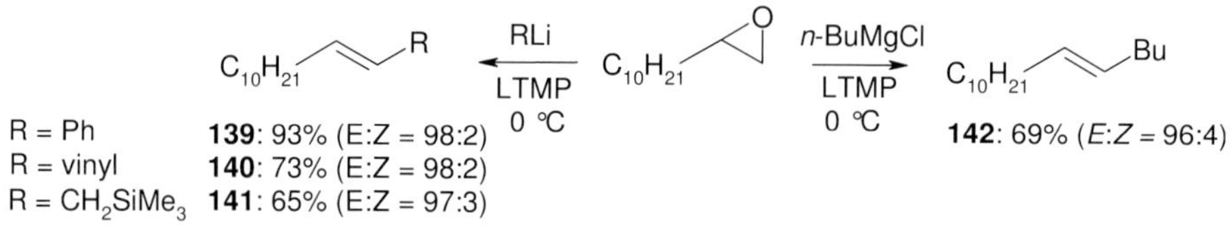

R = Ph — **139**: 93% (E:Z = 98:2)
R = vinyl — **140**: 73% (E:Z = 98:2)
R = CH_2SiMe_3 — **141**: 65% (E:Z = 97:3)

(RLi, LTMP, 0 °C; n-BuMgCl, LTMP, 0 °C) **142**: 69% (*E*:*Z* = 96:4)

Scheme 5.31

were the predominant products, the result of direct ring opening (see above). By using commercially available Me_3SiCH_2Li as the nucleophile one can also access allylsilanes (e. g., **141**) in a regio- and stereocontrolled manner. As was shown by Mioskowski et al., "reductive alkylation" of simple terminal epoxides with primary alkyllithiums (e. g., *n*-BuLi) occurs in good yield but with poor stereoselectivity (Table 5.2, Entry 6), this being attributable to the initial α-deprotonation by *n*-BuLi, occurring unselectively. The use of a simple alkyl Grignard nucleophile (e. g., *n*-BuMgCl) in combination with LTMP allows access to isolated olefins **142** in a highly stereocontrolled manner. This is due to exclusive *trans*-α-deprotonation taking place with LTMP.

A new reactivity mode of terminal epoxides with hindered lithium amides, providing hindered aldehyde enamines such as **145**, has recently been uncovered (Scheme 5.32). The reaction is believed to involve the addition of a lithium amide to a lithiated epoxide to give a dianion (e. g., **144**) [51]. The aldehyde enamines (e. g., **145**) thus generated provided the first examples of enamines reacting with unactivated primary and secondary alkyl iodides in synthetically useful yields and gave a range of α-substituted aldehydes (e. g., **146**). Acid-catalyzed (SiO_2) hydrolysis of the highly hindered aldehyde enamines obtained by treatment of terminal epoxides with LTMP (2.5 equiv.), provides a method for their efficient regioselective "isomerization" to aldehydes [52].

Finally, the nucleophile to a lithiated epoxide need not be the base originally used to generate it, or even one that has been externally added, but can be another lithiated epoxide. This disproportionation/carbenoid dimerization of (enantiopure) lithiated epoxides provides 2-ene-1,4-diols (Scheme 5.33) [53]. Syntheses of D-mannitol and D-iditol in three steps from (*S*)-tritylglycidyl ether were achieved with this method.

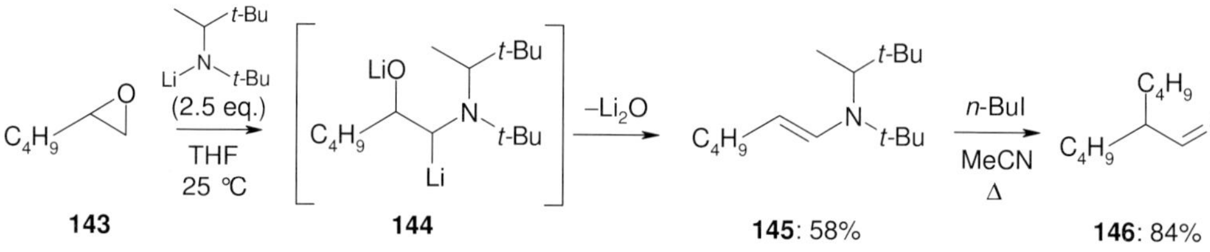

143 (2.5 eq., THF, 25 °C) **144** (−Li_2O) **145**: 58% (*n*-BuI, MeCN, Δ) **146**: 84%

Scheme 5.32

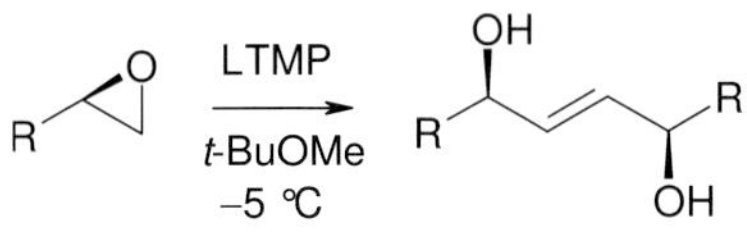

R	Yield (%)	E:Z
t-Bu	86	100:0
Cy	77	51:26
n-C_5H_{11}	62	43:19

Scheme 5.33

5.2.4 Electrophile Trapping

5.2.4.1 Introduction

An attractive route to more highly substituted epoxides is treatment of a metalated epoxide with an electrophile. The majority of work in this area has concentrated on electrophile trapping of metalated epoxides containing an anion-stabilizing group, since such metalated epoxides have been found to exist at low temperature sometimes for a few minutes, and in some cases for several hours. Eisch and Galle were the pioneers in this area, reporting the direct deprotonation of a range of epoxides possessing anion-stabilizing groups, and their subsequent reactions with electrophiles (Table 5.4) [54]. A plethora of other examples using a wide range of anion-stabilizing groups have since followed.

Table 5.4 Electrophile trapping of lithiated epoxides containing anion-stabilizing groups.

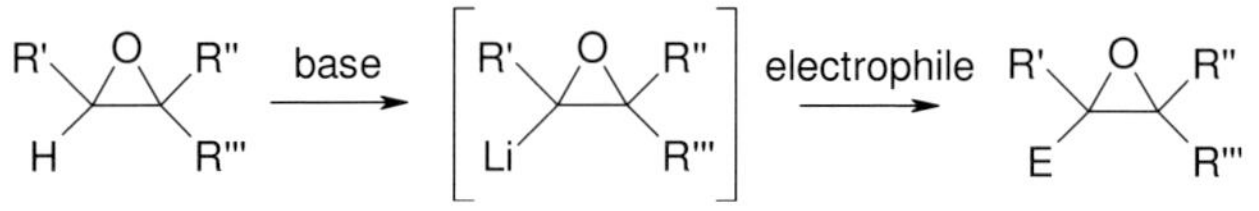

Entry	R'	R''	R'''	Base	Electrophile (EX)	Yield (%)
1	$SiPh_3$	H	H	n-BuLi	MeI	91
2	$SiPh_3$	H	H	n-BuLi	$PhC(O)N(CH_2)_2$	78
3	$SiPh_3$	H	H	n-BuLi	Me_3SnCl	55
4	$SiPh_3$	H	H	n-BuLi	Me_3SiCl	79
5	$SiMe_3$	n-C_6H_{13}	H	t-BuLi/TMEDA	MeI	80
6	SO_2Ph	H	Ph	n-BuLi	MeI	96
7	$PO(OEt)_2$	H	H	LDA	Me_3SiCl	84
8	CN	Me	Me	n-BuLi	Me_3SiCl	74
9	Ph	H	H	t-BuLi	Me_3SiCl	78

5.2.4.2 Silyl-stabilized Lithiated Epoxides

Commonly employed anion-stabilizing groups are those containing silicon (Table 5.4, Entries 1–5). Magnus et al. reported that epoxysilane **147** could be deprotonated with *t*-BuLi, and that the lithiated epoxide **148** thus generated could be trapped with allyl bromide to give epoxysilane **149** in a synthetically useful yield (Scheme 5.34) [55]. Iodomethane (88 %) and chlorotrimethylsilane (60 %) could also be trapped.

Molander and Mautner demonstrated that deprotonation of *cis*-α,β-epoxysilane **150** with *s*-BuLi/TMEDA was complete in 10 minutes, whereas the corresponding *trans*-isomer **150** required 4 hours [56]. Similarly, treatment with butyraldehyde was more efficient with *cis*-**151** (Scheme 5.35), which could also be trapped with a wide variety of other carbonyl-containing electrophiles. The results demonstrated that lithiated epoxides *cis*- and *trans*-**151** were configurationally stable at −116 °C for periods of up to 4 hours. Only in the case of *cis*-**151** (*t*-butyl = *n*-octyl) was the lithiated epoxysilane found to be configurationally unstable.

During studies directed towards the synthesis of spatol, Salomon and Murthi studied epoxysilane **153** bearing a temporary tether (Scheme 5.36), which circumvented the potential problem of configurational instability described above [57].

In their synthesis of (+)-cerulenin, Mani and Townsend employed lithiated epoxysilane **157**, which they trapped with (4*E*,7*E*)-nonadienal to give a 77 % yield of **158**, which was further manipulated to give the natural product (Scheme 5.37) [58].

cis-α,β-Epoxy-γ,δ-vinylsilanes **159** are regioselectively lithiated at the α-silyl position, and can subsequently be stereoselectively trapped with a range of electrophiles to give α-substituted epoxyvinylsilanes **160**, which can in turn be isomerized to α-silyl-β-vinylketones **161** (Scheme 5.38) [59].

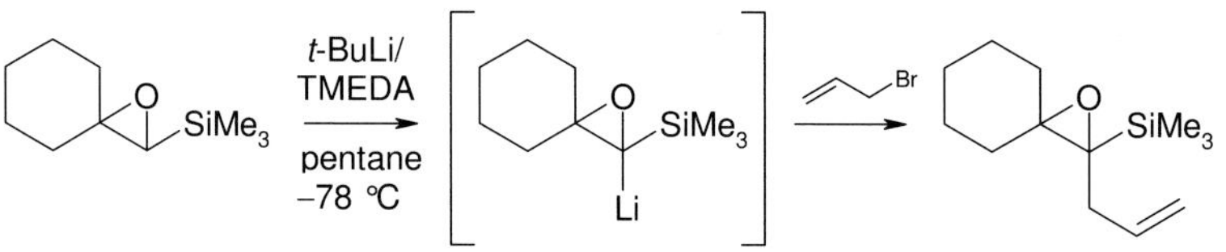

t-BuLi/TMEDA, pentane, −78 °C; allyl bromide

147 → [**148**] → **149**: 70%

Scheme 5.34

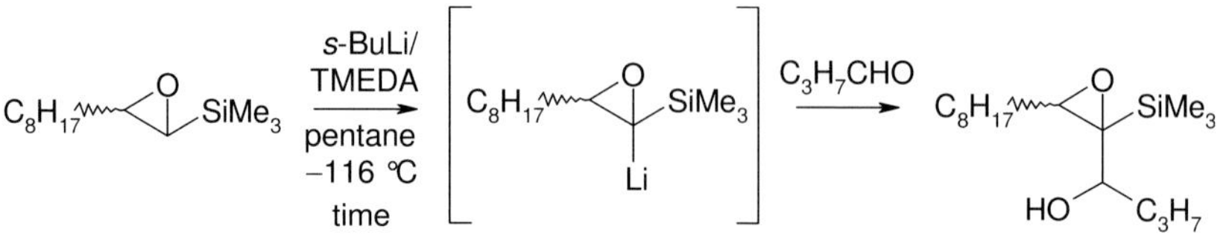

s-BuLi/TMEDA, pentane, −116 °C, time; C_3H_7CHO

cis-**150**, 10 min. → *cis*-**151** → *cis*-**152**: 83%

trans-**150**, 4 h → *trans*-**151** → trans-**152**: 63%

Scheme 5.35

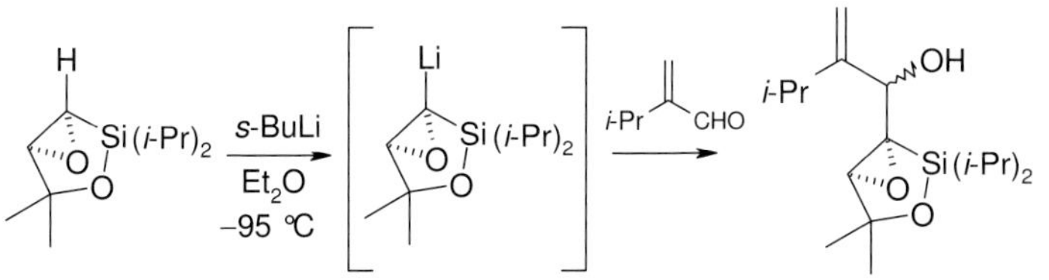

s-BuLi, Et_2O, –95 °C; *i*-Pr(C=CH$_2$)CHO

153 → [154] → 155: β-OH 31%, α-OH 69%

Scheme 5.36

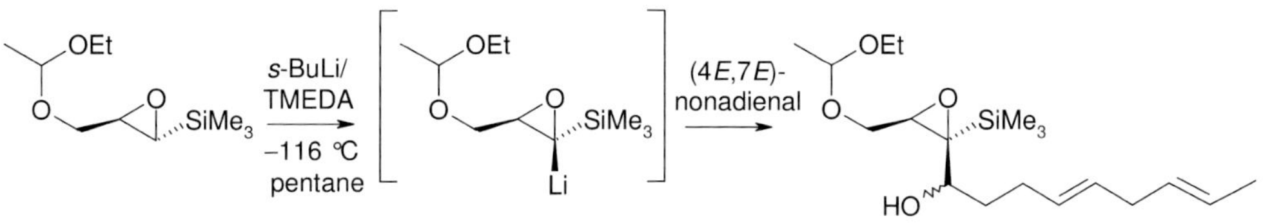

s-BuLi/TMEDA, –116 °C, pentane; (4*E*,7*E*)-nonadienal

156 → [157] → 158: 77%

Scheme 5.37

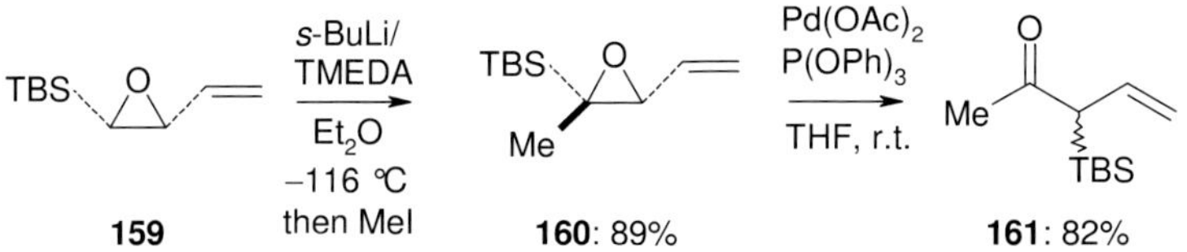

s-BuLi/TMEDA, Et_2O, –116 °C, then MeI; $Pd(OAc)_2$, $P(OPh)_3$, THF, r.t.

159 → 160: 89% → 161: 82%

Scheme 5.38

5.2.4.3 Sulfonyl-stabilized Lithiated Epoxides

Along with Eisch and Galle (Table 5.4, Entry 6), Jackson and coworkers have also developed the use of sulfonyl-stabilized metalated epoxides in synthesis. Phenylsulfonyloxirane can be deprotonated with *n*-BuLi and then trapped with electrophiles (Scheme 5.39) [60]. Trapping with carbonyl-containing compounds alone did not give the desired products, but addition of the Lewis acid $MgBr_2$ to **163** prior to addition of the electrophile gave good results. More highly substituted phenylsulfonyloxiranes could also be further elaborated by this method, though they displayed markedly different reactivities depending on their substitution patterns. A synthetically useful route to α,β-epoxyketones **169** from **166** was thus developed [61].

Mori et al. have demonstrated the most dramatic uses of lithiated epoxides in natural product synthesis [62]. By employing the chemistry developed by Jackson, and subsequently performing a Lewis acid-catalyzed ($BF_3 \cdot OEt_2$) cyclisation, tetrahydrofuran, tetrahydropyran, and oxepane rings are readily accessed; this strategy is demonstrated by the synthesis of the marine epoxy lipid **173** (Scheme 5.40) [63].

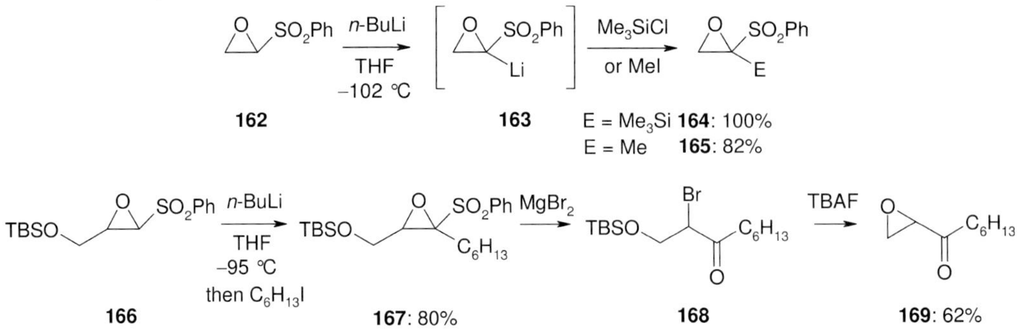

Scheme 5.39

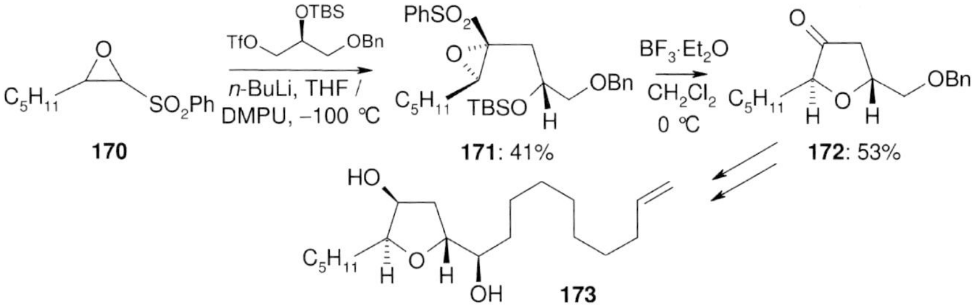

Scheme 5.40

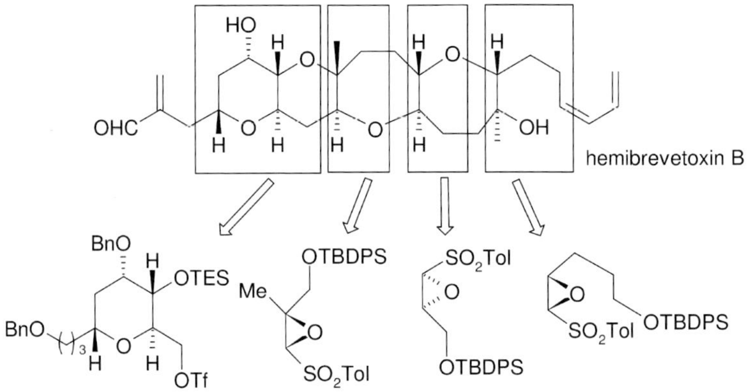

Scheme 5.41

This lithiated epoxysulfone cyclisation strategy has been iteratively applied in the total synthesis of hemibrevitoxin B, a polycyclic ether marine toxin from the red tide organism *Gymnodinium breve* (Scheme 5.41) [64].

5.2.4.4 Organyl-stabilized Lithiated Epoxides

Since Eisch and Galle first introduced organyl substituents as anion-stabilizing groups for lithiated epoxides (Table 5.4, Entries 8–9) they have examined them extensively (Table 5.5) [54, 65].

Florio et al. demonstrated that the lithiation/electrophile trapping of enantiopure styrene oxide, as well as the β-substituted styrene oxides **180** and **182**, is totally stereoselective (Scheme 5.42) [66]. These results demonstrate that the intermediate benzylic anions are configurationally stable within the timescale of deprotonation/electrophile trapping.

Shimizu et al. have introduced an indirect route to stabilized lithiated epoxides. Treatment of dichlorohydrin **184** with 3 equiv. of vinyllithium in the presence of 3 equiv. of LTMP gave lithiated epoxide **185**, which could be trapped with a range of electrophiles (e.g., Me_3SnCl) to give 1,2-divinyl epoxides **186**; these in turn underwent Cope rearrangements on heating to give oxepanes **187** (Scheme 5.43) [67].

Table 5.5 Lithiation-electrophile trapping of α-organyl-substituted epoxides.

Entry	Epoxide	Base	Organyl group	Electrophile	Yield (%)
1	**174**	*t*-BuLi · TMEDA	phenyl	$SiMe_3$	96
2	**174**	*t*-BuLi · TMEDA	phenyl	$PhCONEt_2$	65
3	**79**	*t*-BuLi · TMEDA	phenyl	$SiMe_3$	94
4	**79**	*t*-BuLi · TMEDA	phenyl	$PhCONEt_2$	91
5	**83**	*t*-BuLi · TMEDA	phenyl	$SiMe_3$	70
6	**23**	*t*-BuLi · TMEDA	vinyl	$SiMe_3$	95
7	**23**	*t*-BuLi · TMEDA	vinyl	$PhCONEt_2$	81
8	**175**	*t*-BuLi · TMEDA	vinyl	$SiMe_3$	45
9	**176**	*t*-BuLi · TMEDA	alkynyl	$SiMe_3$	65
10	**177**	LDA	cyano	$SiMe_3$	95
11	**178**	LDA	ethoxycarbonyl	$SiMe_3$	70
12	**179**	*t*-BuLi · TMEDA	phenyl	$SiMe_3$	81

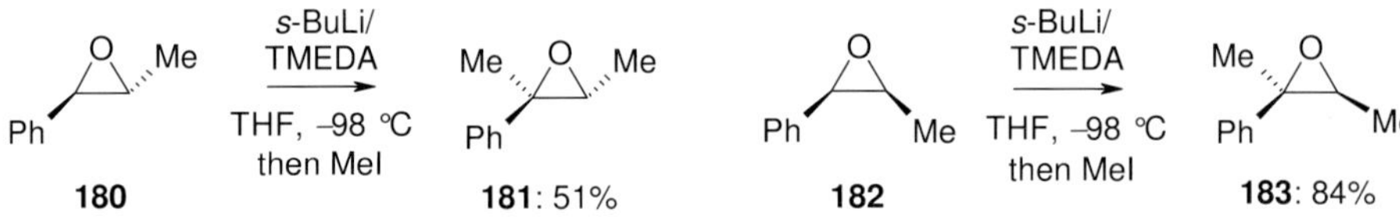

Scheme 5.42

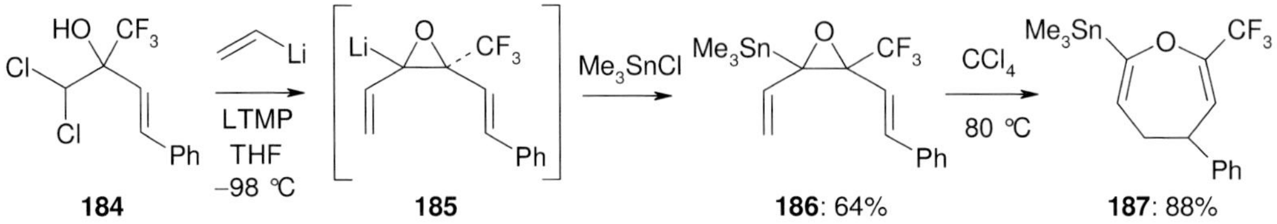

Scheme 5.43

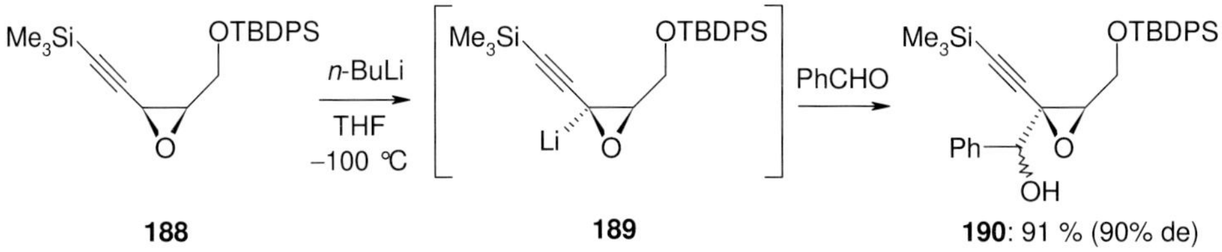

Scheme 5.44

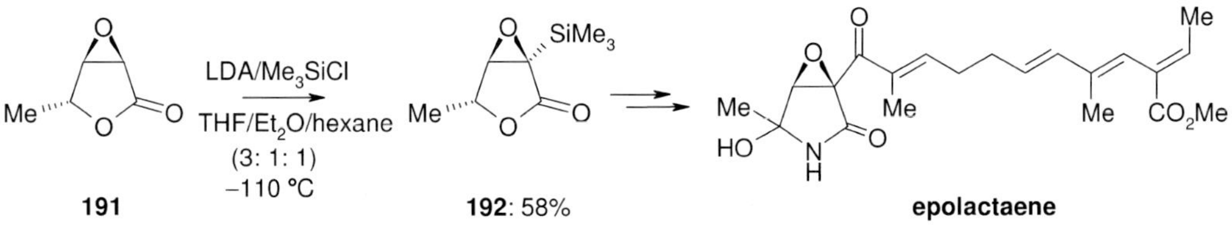

Scheme 5.45

Palladium-catalyzed cross-coupling with **187** gave a range of synthetically useful oxepanes.

Pale et al. have reported that the stereoselective electrophile trapping of alkynyl-stabilized lithiated epoxides **189**, generated from the parent epoxide **188** and *n*-BuLi, gives substituted epoxides such as **190**, in good yield and de (Scheme 5.44) [67].

The use of an ester as an anion-stabilizing group for a lithiated epoxide was demonstrated by Eisch and Galle (Table 5.5, Entry 11). This strategy has been extended to α,β-epoxy-γ-butyrolactone **191**, which could be deprotonated with LDA and trapped *in situ* with chlorotrimethylsilane to give **192**, which was used in a total synthesis of epolactaene (Scheme 5.45) [69]. The use of a lactone rather than a

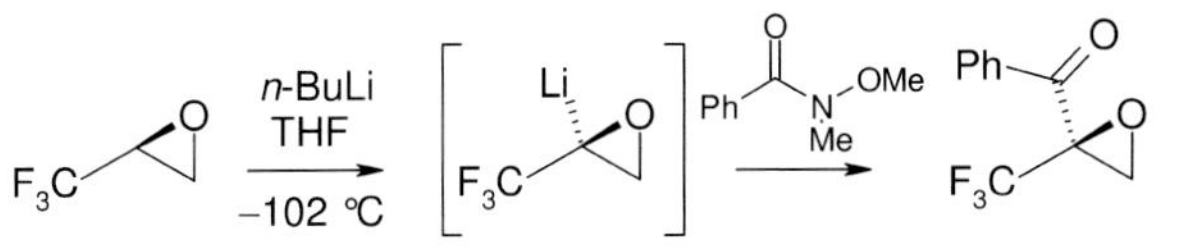

Scheme 5.46

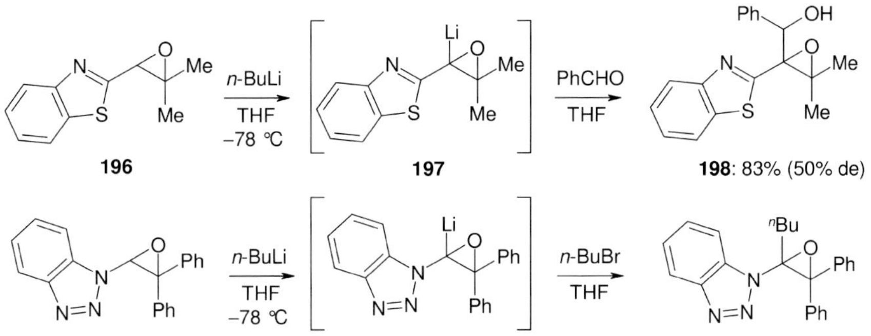

Scheme 5.47

simple ester as the anion-stabilizing group not only localizes the anion at the bridgehead, which has implications for its reactivity, but also means that configurational stability is not an issue.

Uneyama et al. have shown that enantiopure trifluoromethyloxirane (**193**) can be lithiated and stereoselectively trapped with a variety of electrophiles to give substituted trifluoromethyloxiranes such as **195** (Scheme 5.46) [70]; the use of a Weinreb amide as the electrophile is unusual.

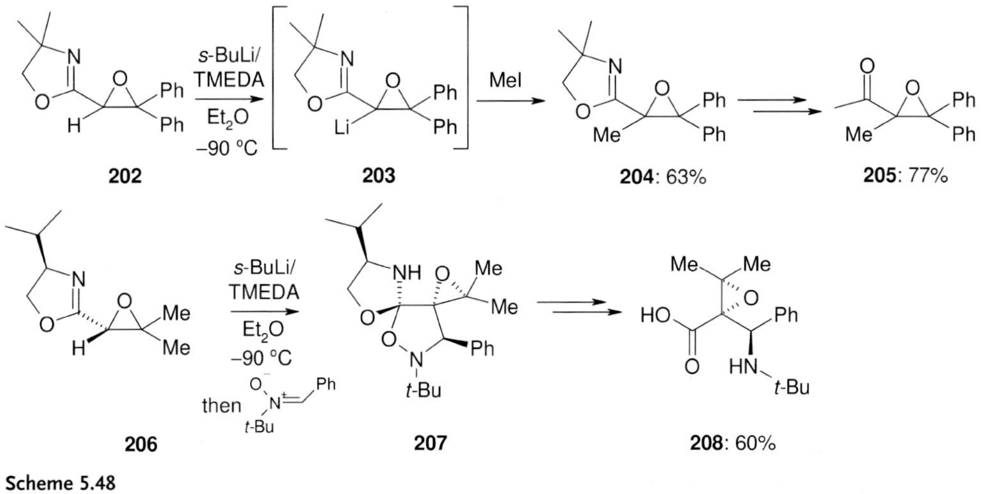

Scheme 5.48

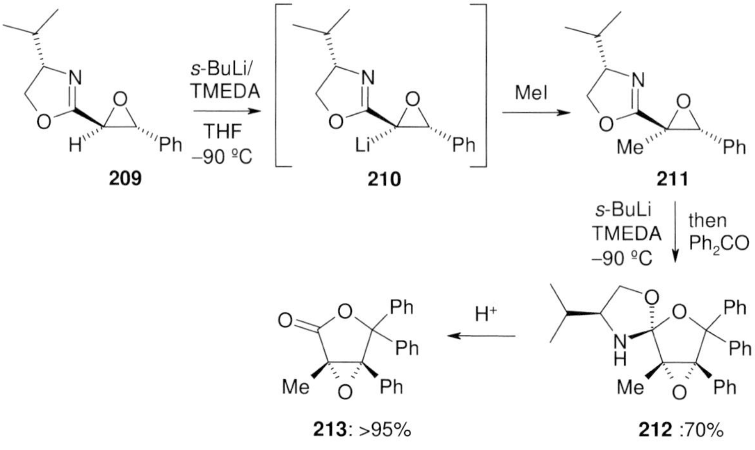

Scheme 5.49

Both benzothiazolyl and benzotriazolyl units have been employed as heteroaromatic anion-stabilizing groups for metalated epoxides (Scheme 5.47) [71]. The successful use of a simple alkyl bromide as electrophile with **200** is notable.

Florio and coworkers have also reported the use of oxazolinyl groups as anion-stabilizing substituents. Lithiation/electrophile trapping of oxazolinylepoxide **202** provided access to acyloxiranes **205** [72], while deprotonation/electrophile trapping of oxazolinylepoxide **206** with nitrones gave access to enantiopure α-epoxy-β-amino acids **208** (Scheme 5.48) [73].

Lithiation/electrophile trapping of enantiopure epoxide **209** stereoselectively gave epoxide **211**; further elaboration via a metalated epoxide gave spirocyclic epoxide **212**, which after treatment with acid gave epoxylactone **213** as a single diastereomer (Scheme 5.49) [74].

5.2.4.5 Remotely Stabilized Lithiated Epoxides

A new method of stabilizing metalated epoxides through "remote" coordination has recently been introduced, allowing stereoselective access to a range of epoxylactones. Epoxylactone **216** was converted into the phytotoxin xylobovide (Scheme 5.50) [75]

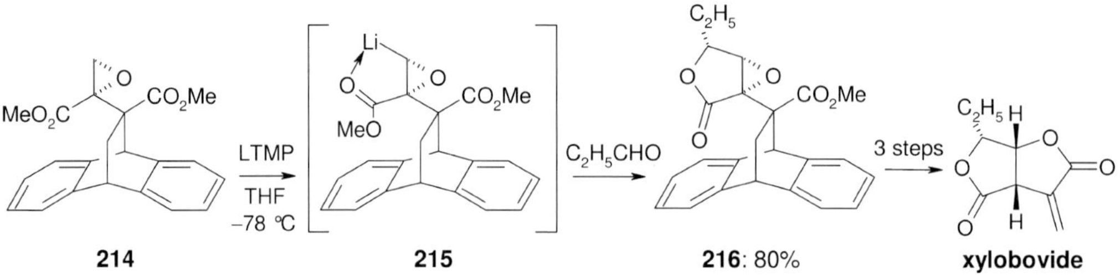

Scheme 5.50

5.2.4.6 Simple Metalated Epoxides

Electrophile trapping of simple metalated epoxides (i.e., those not possessing an anion-stabilizing group) is possible. Treatment of epoxystannane **217** with *n*-BuLi (1 equiv.) in the presence of TMEDA gave epoxy alcohol **218** in 77% yield after trapping with acetone (Scheme 5.51) [76]. In the absence of TMEDA, the non-stabilized epoxides underwent dimerization to give mixtures of enediols.

Satoh and Horiguchi introduced a desulfinylation method for the formation of simple lithiated epoxides, which could be trapped with a variety of electrophiles, such as Me_3SiCl (Scheme 5.52) [77].

When *n*-BuLi is used instead of *t*-BuLi, the byproduct after desulfinylation (*n*-BuS(O)Ph) possesses an acidic proton, which is abstracted by the metalated epoxide. Hence, overall, a stereoselective protodesulfinylation is achieved. This can be used for the asymmetric synthesis of epoxides, such as that of (–)-disparlure from enantiopure sulfoxide **222** (Scheme 5.53) [78].

Hodgson and coworkers have demonstrated that the use of diamine ligands in combination with *s*-BuLi allows the direct deprotonation/electrophile trapping of

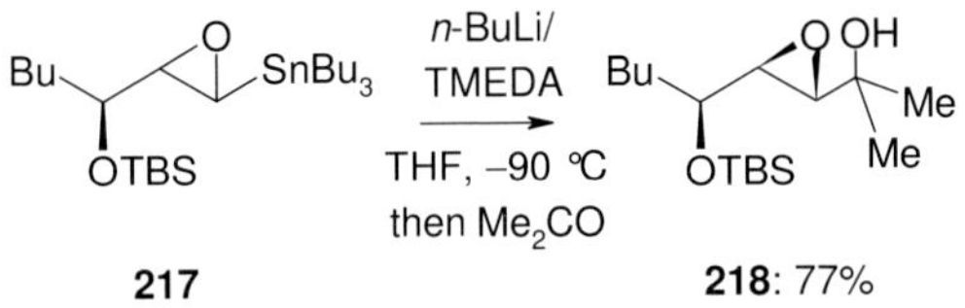

Scheme 5.51

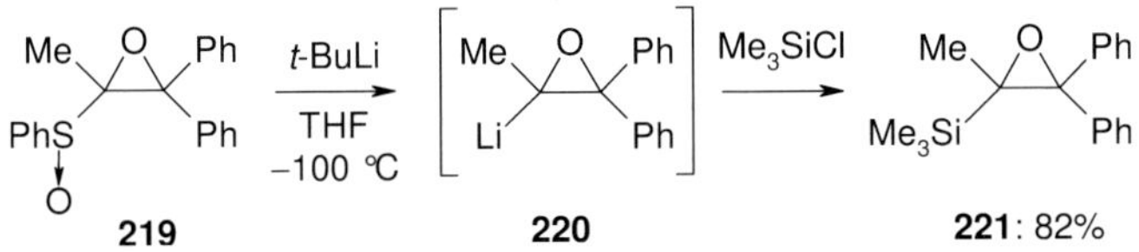

Scheme 5.52

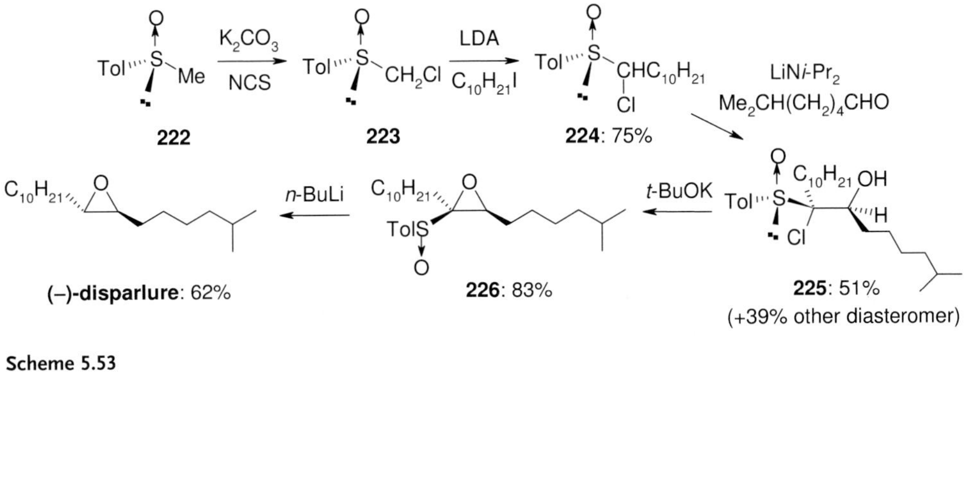

Scheme 5.53

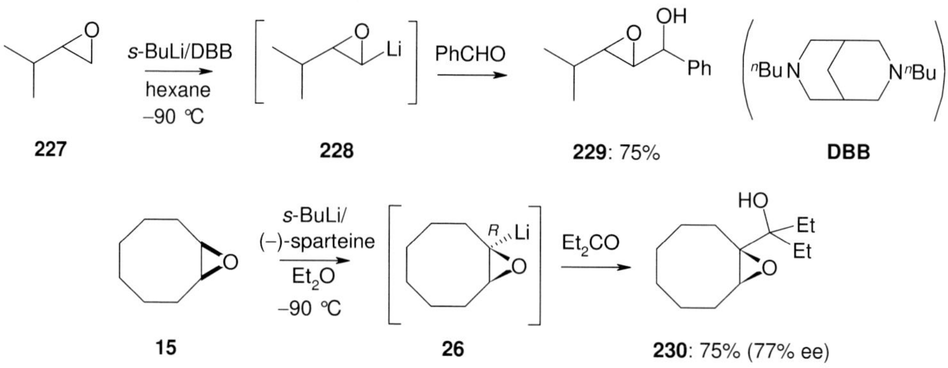

227 → (s-BuLi/DBB, hexane, −90 °C) → [228] → (PhCHO) → 229: 75%; DBB

15 → (s-BuLi/(−)-sparteine, Et_2O, −90 °C) → [26] → (Et_2CO) → 230: 75% (77% ee)

Scheme 5.54

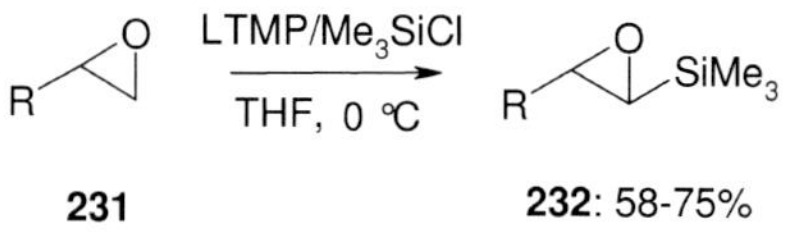

231 → (LTMP/Me_3SiCl, THF, 0 °C) → 232: 58-75%

Scheme 5.55

simple terminal epoxides **227**, which are readily available as single enantiomers, to give substituted epoxides **229** (Scheme 5.54) [79]. Enantioselective deprotonation/electrophile trapping of *meso*-epoxides (e.g., cyclooctene oxide **15**) gave enantioenriched substituted epoxides **230** [80].

Use of LTMP as base [52] *in situ* with Me_3SiCl allows straightforward access to a variety of synthetically useful α,β-epoxysilanes **232** at near ambient temperature directly from (enantiopure) terminal epoxides **231** (Scheme 5.55) [81]. This reaction relies on the fact that the hindered lithium amide LTMP is compatible with Me_3SiCl under the reaction conditions and that the electrophile trapping of the nonstabilized lithiated epoxide intermediate must be very rapid, since the latter are usually thermally very labile.

5.3 Metalated Aziridines

The chemistry of metalated aziridines is far less developed than the chemistry of metalated epoxides, although from what is known [1b], it is obvious that their chemistry is similar. Like metalated epoxides, metalated aziridines can act as classical nucleophiles with a variety of electrophiles to give more highly substituted aziridines (Scheme 5.56, Path A). A small amount is known about how they can act as electrophiles with strong nucleophiles to undergo "reductive alkylation" (Path B), and undergo C–H insertion reactions (Path C).

The first example of a metalated aziridine appeared in 1972. Treatment of azir-

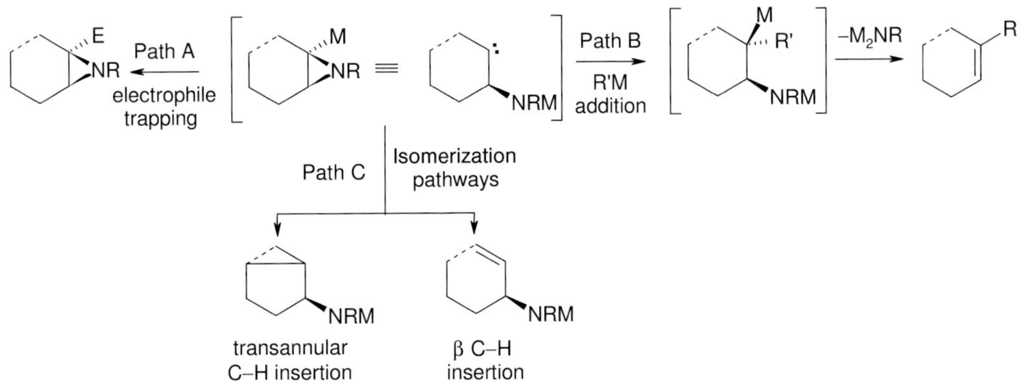

Scheme 5.56

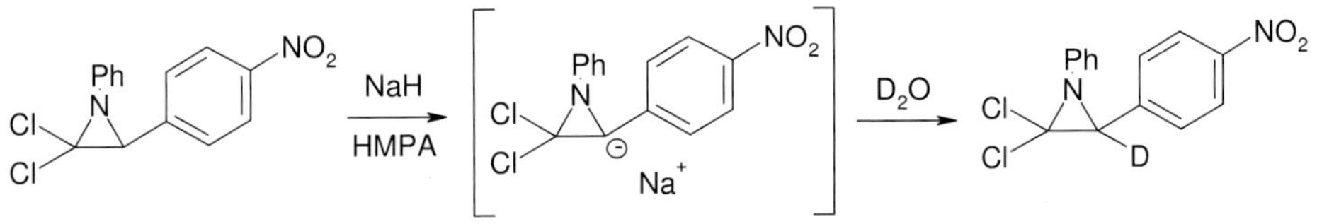

233 234 235: 80%

Scheme 5.57

idine **233** with NaH was found to give anion **234**, which was quenched with D_2O to give deuterated aziridine **235** in 80% yield (Scheme 5.57) [82].

5.3.1
Electrophile Trapping

5.3.1.1 Stabilized Metalated Aziridines

Seebach and coworkers examined the deprotonation/electrophile trapping of phenylthioaziridine carboxylates **236** (Scheme 5.58). These thioesters were found to be more stable than their oxy-ester congeners when lithiated; treatment of **236** with LDA at −78 °C, followed by trapping with MeI at −100 °C, stereoselectively afforded aziridine **237** [83].

A *tert*-butyl ester serves as an efficient organyl-stabilizing group for a lithiated aziridine when the *N*-protecting group is a chelating moiety. Deprotonation/elec-

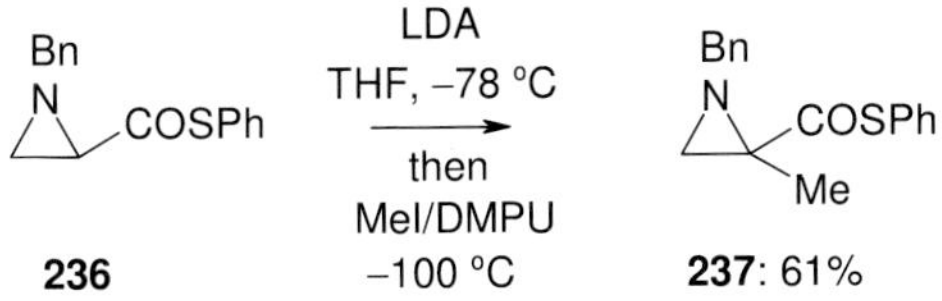

236 237: 61%

Scheme 5.58

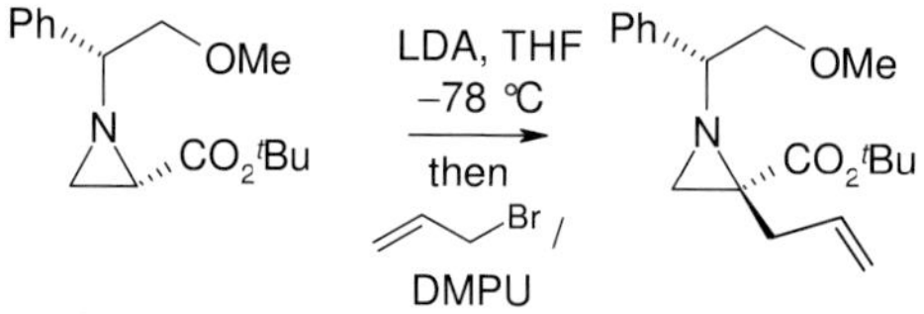

238 **239**: 64%

Scheme 5.59

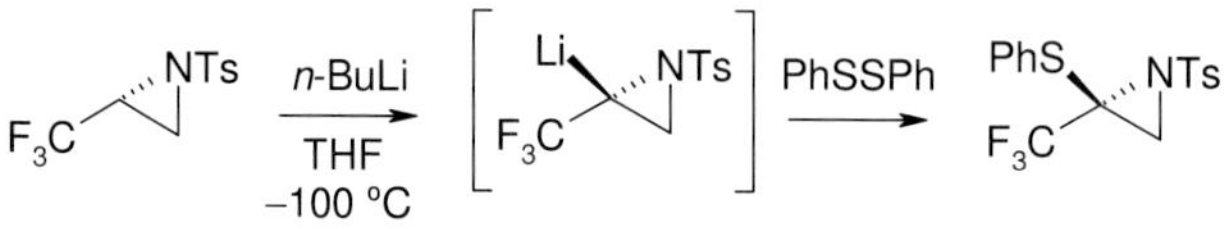

240 **241** **242**: 87%

Scheme 5.60

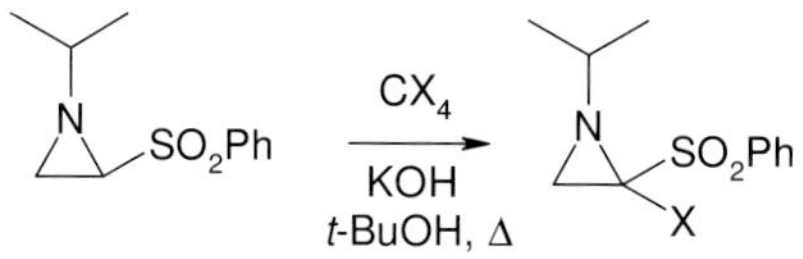

243

X = Cl **244**: 70%
X = Br **245**: 74%

Scheme 5.61

trophile trapping of aziridine **238** proceeds with retention of configuration (Scheme 5.59) [84].

As well as for metalated epoxides, the trifluoromethyl moiety also proved an effective organyl-stabilizing group for metalated aziridines. Lithiated aziridine **241** reacted stereoselectively with carbonyl-containing electrophiles, and phenyl disulfide and chlorotrimethylsilane were also trapped in good yield (Scheme 5.60) [70b, 85].

Sulfonylaziridine **243** was halogenated in carbon tetrahalides in the presence of KOH as base [86] (Scheme 5.61). Although other examples of electrophile trapping of sulfonyl- and phosphonyl-stabilized metalated aziridines exist, the reactions were not stereoselective [87].

Florio et al. have employed heteroaromatic rings as organyl-stabilizing groups for metalated aziridines as well as for metalated epoxides. Regioselective deprotonation of aziridine **246** with *n*-BuLi, followed by addition of MeI, gave aziridine **247** (Scheme 5.62) [88].

Enantiopure methyleneaziridine **248** could be lithiated and subsequently trapped with a variety of electrophiles in a highly diastereoselective manner (Scheme 5.63) [89].

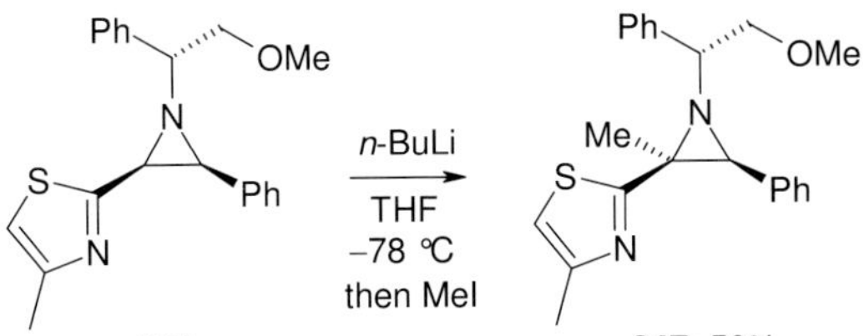

Scheme 5.62

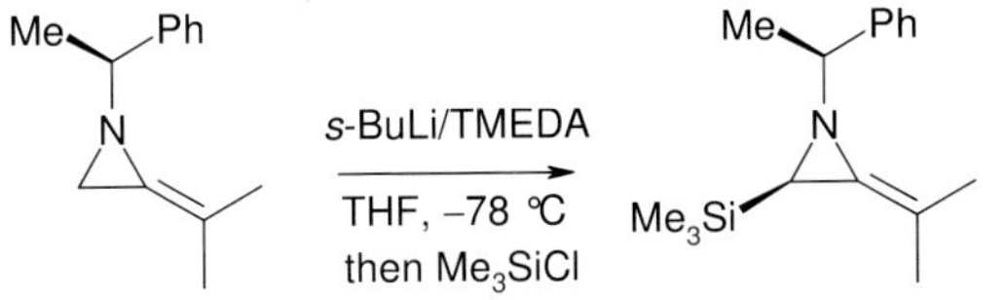

Scheme 5.63

5.3.1.2 Nonstabilized Metalated Aziridines

Reports of the generation and subsequent electrophile trapping of nonstabilized metalated aziridines appeared before those for metalated epoxides. Desulfinylation of sulfinylaziridine **250** with EtMgBr gave metalated aziridine **251**, which, remarkably, could be kept at 0 °C for 1 h before quenching with D_2O (Scheme 5.64). The deuterated aziridine **252** (E = D) was obtained in excellent yield, but acetaldehyde was the only other electrophile found to be trapped efficiently [90].

Desilylation of α,β-aziridinylsilanes is a route to nonstabilized metalated aziridines. Treatment of **254** with CsF in the presence of PhCHO furnished substituted aziridine **255** in good yield (Scheme 5.65). Interestingly, the isomeric aziridine **256** gave azirine **257** on treatment with CsF, presumably via a phenyl-stabilized metalated aziridine [91].

Vedejs et al. demonstrated that lithium-tin exchange is a feasible route to nonstabilized lithiated aziridines. Treatment of stannylaziridine **258** with *n*-BuLi gave lithiated aziridine **259**, which could be trapped with a variety of electrophiles such as benzaldehyde (Scheme 5.66) [92]. Furthermore, simple *N*-alkyl aziridine **261** can be activated by Lewis acid precomplexation to BH_3 (to give **262**) and subsequently

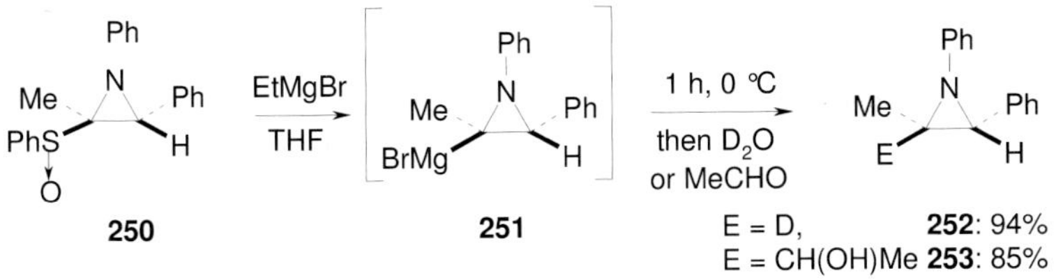

Scheme 5.64

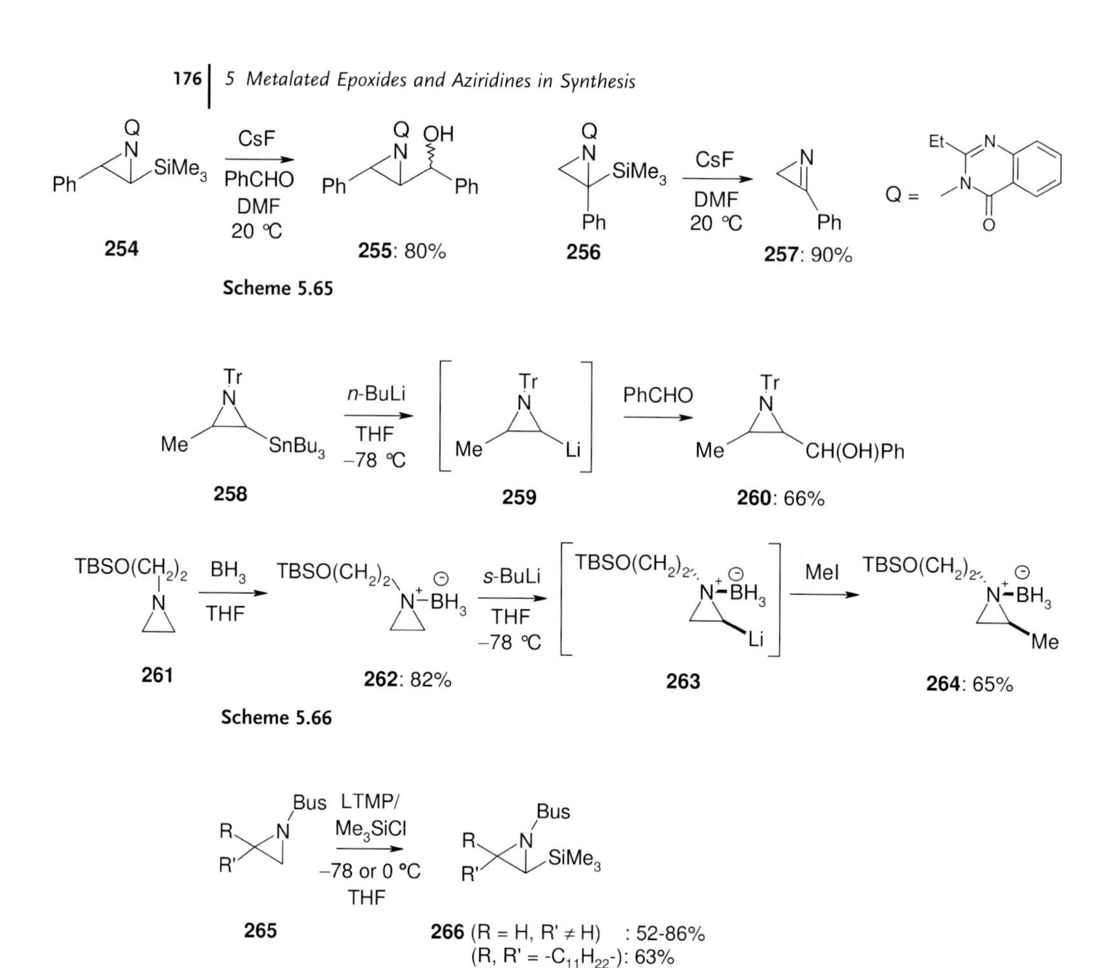

Scheme 5.65

Scheme 5.66

Scheme 5.67

lithiated with *s*-BuLi. Lithiation was found to occur almost exclusively *syn* to boron, and the resulting anion **263** could be trapped with methyl iodide to give aziridine **264**. Removal of the borane was achieved in >95% yield by heating at reflux in ethanol [93]. This methodology has been applied to the synthesis of functionalized sugar-derived aziridines [94] and to the stereoselective functionalization of 2-(1-aminoalkyl)aziridines [95].

Direct deprotonation/electrophile trapping of simple aziridines is also possible. Treatment of a range of *N*-Bus-protected terminal aziridines **265** with LTMP in the presence of Me_3SiCl in THF at −78 °C stereospecifically gave *trans*-α,β-aziridinylsilanes **266** (Scheme 5.67) [96]. By increasing the reaction temperature (to 0 °C) it was also possible to α-silylate a β-disubstituted aziridine; one should note that attempted silylation of the analogous epoxide did not provide any of the desired product [81].

In direct contrast with the chemistry of epoxides [81], trapping of the intermediate lithiated aziridine was possible with external electrophiles (which would

Table 5.6 Trapping of the intermediate lithiated aziridine with external electrophiles.

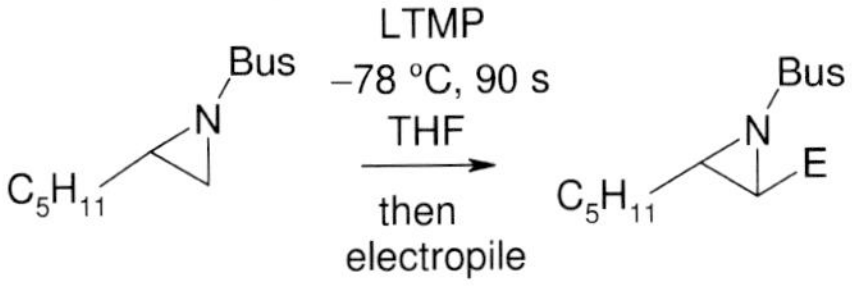

Electrophile	E	Isolated yield (%)
Bu_3SnCl	$SnBu_3$	87
Et_2CO	$C(OH)Et_2$	76
$PhCONMe_2$	C(O)Ph	57
DMF	CHO	63
$PhSO_2F$	SO_2Ph	92
CO_2	CO_2Me	63[a]

[a] Following treatment with CH_2N_2

otherwise be incompatible *in situ* with LTMP). A range of *trans*-α,β-disubstituted aziridines were thus synthesized (Table 5.6). The increased scope of reactivity observed here for terminal aziridines relative to terminal epoxides could be attributed to the *trans*-α-proton of the former having a lower pK_a.

5.3.2 Olefin Formation

O'Brien et al. provided the first examples of olefin formation by "reductive alkylation" of aziridines [97]. Treatment of aziridine **267** with *s*-BuLi gave olefin **270** in 76% yield (Scheme 5.68). For the formation of olefin **270** they suggest a reaction pathway that proceeds in a manner analogous to that proposed for epoxides [36]; namely, nucleophilic attack of *s*-BuLi on lithiated aziridine **268** to form dilithiated species **269**, which eliminates Li_2NTs ($TsNH_2$ was observed as a product of this reaction) to yield olefin **270**.

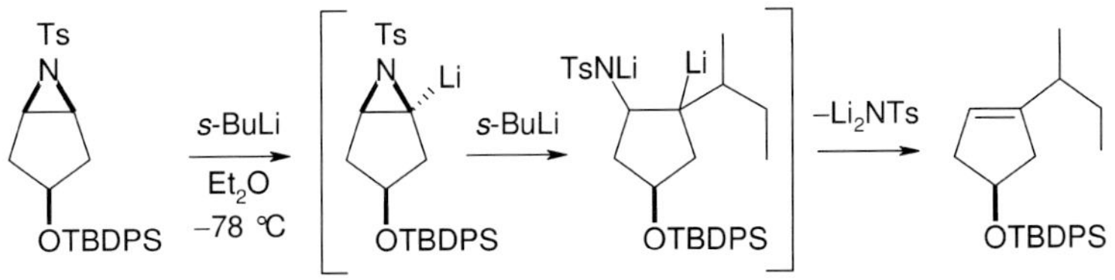

Scheme 5.68

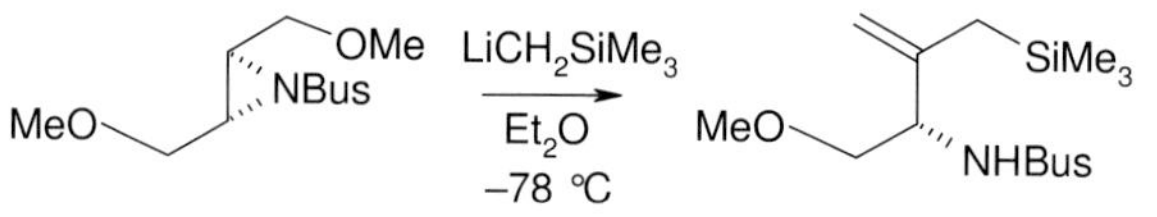

MeO / OMe / NBus — $LiCH_2SiMe_3$, Et_2O, −78 °C → MeO / $SiMe_3$ / NHBus

271 **272**: 75%

Scheme 5.69

When a better leaving group than $LiNSO_2R$ (e.g., OMe) is present at the α-position, retention of the potentially useful sulfonamide moiety occurs (e.g., in the conversion of aziridine **271** into the highly functionalized amino ether **272**; Scheme 5.69) [98]. It should be noted that the analogous chemistry with epoxides of allylic diethers failed; this could again (see above) be possibly due to the higher pK_a of the epoxide proton relative to the aziridine proton.

5.3.3
C–H Insertions

Only recently have C–H insertions reactions of metalated aziridines demonstrating character typical of traditional carbenoids been uncovered. The first of these involved treatment of aziridine **273** with LDA (Scheme 5.70) [99]. Upon lithiation, transannular C–H insertion gave tricyclic amine **274** as the major product. Norcamphor **65** was also produced in 15% yield, presumably through a 1,2-hydride shift to give an imine, which underwent subsequent hydrolysis to give the ketone. Müller et al. demonstrated that the use of *s*-BuLi/(–)-sparteine as base with aziridine **273** gave tricyclic amine **274** in 72% yield and 43% *ee* [100]. Furthermore, they showed that aziridine **275** underwent transannular rearrangement in a manner analogous to that observed with cyclooctene oxide, to give bicyclic amines (Table 5.7, Entry 1), although the yields and *ee*s for larger ring sizes were less comparable (Entries 2 and 3).

The β-elimination of epoxides to allylic alcohols on treatment with strong base is a well studied reaction [1a]. Metalated epoxides can also rearrange to allylic alcohols via β-C–H insertion, but this is not a synthetically useful process since it is usually accompanied by competing α-C–H insertion, resulting in ketone enolates. In contrast, aziridine **277** gave allylic amine **279** on treatment with *s*-BuLi/(–)-sparteine (Scheme 5.71) [97]. By analogy with what is known about reactions of epoxides with organolithiums, this presumably proceeds via the α-metalated aziridine **278** [101].

It should be noted that the sense of asymmetric induction in the lithiation/rearrangement of aziridines **274**, **276**, and **279** by treatment with *s*-butyllithium/(–)-sparteine is opposite to that observed for the corresponding epoxides (i.e. removal of the proton occurs at the (*S*)-stereocenter) [102]. If one accepts the proposed model to explain the selective abstraction of the proton at the (*R*)-stereocenter of an epoxide (Figure 5.1), then, from the large difference in steric bulk (and Lewis basicity) between an oxygen atom and a tosyl-protected nitrogen atom, it is obvious that this model cannot be applied to the analogous aziridines.

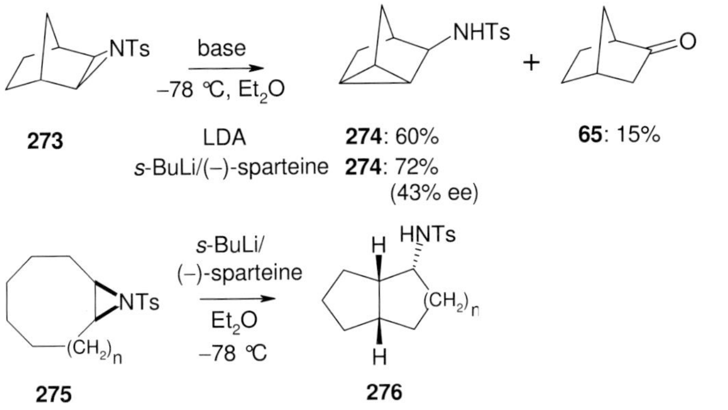

NTs
base
–78 °C, Et_2O
NHTs
+
O
273
LDA
s-BuLi/(–)-sparteine
274: 60%
274: 72%
(43% ee)
65: 15%
s-BuLi/
(–)-sparteine
Et_2O
–78 °C
NTs
$(CH_2)_n$
H
HNTs
H
275
276

Scheme 5.70

Table 5.7 Transannular rearrangement: dependence on ring size.

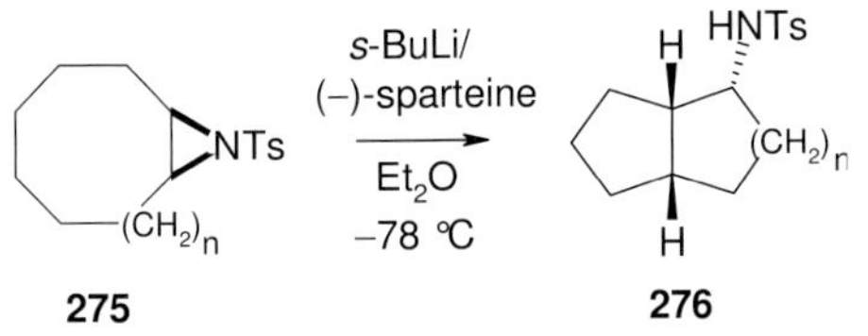

s-BuLi/
(–)-sparteine
Et_2O
–78 °C
NTs
$(CH_2)_n$
H
HNTs
H
275
276

Entry/*n*	Yield[a] (%)	*Ee*[a] (%)
1	69 (79)	75 (73)
2	35 (77)	68 (83)
3	15 (97)	47 (77)

[a] Results for epoxides in brackets for comparison

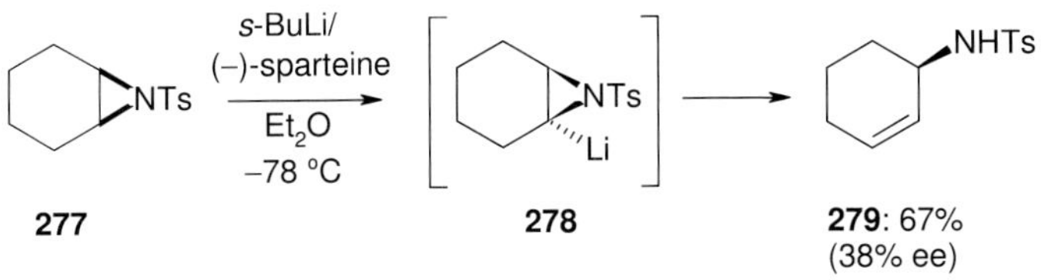

s-BuLi/
(–)-sparteine
Et_2O
–78 °C
NTs
NTs
Li
NHTs
277
278
279: 67%
(38% ee)

Scheme 5.71

5.4 Outlook

In the fifty or so years since the discovery of α-metalated epoxides, our understanding of their reactivity has advanced to such a level that their use in routine organic synthesis is now possible. Many research groups continue to examine their unusual reaction pathways and to develop these into synthetically useful processes. In contrast, the chemistry of α-metalated aziridines is still in its infancy and there are undoubtedly many interesting facets of their nature still to be explored and applied in organic synthesis.

References

1 (a) Crandall, J. K.; Apparu, M. *Org. React. (N. Y.)* **1983**, *29*, 345–443; (b) Satoh, T. *Chem. Rev.* **1996**, *96*, 3303–3325; (c) Mori, Y. *Rev. Heteroat. Chem.* **1997**, *17*, 183–210; (d) Doris, E.; Dechoux, L.; Mioskowski, C. *Synlett* **1998**, 337–343; (e) Hodgson, D. M.; Gras, E. *Synthesis* **2002**, 1625–1642; (f) Hodgson, D. M.; Tomooka, K.; Gras, E. in *Topics in Organometallic Chemistry*; Hodgson, D. M., Ed.; Springer: Berlin, **2003**, pp. 217–250; (g) Chemla, F.; Vrancken, E. in *The Chemistry of Organolithium Reagents*; Rappoport, Z., Marek, I., Eds.; J. Wiley and Sons: New York, **2004**, pp. 1165–1242.

2 Boche, G.; Lohrenz, J. C. W. *Chem. Rev.* **2001**, *101*, 697–756.

3 Morgan, K. M.; O'Connor, M. J.; Humphrey, J. L.; Buschman, K. E. *J. Org. Chem.* **2001**, *66*, 1600–1606.

4 Cope, A. C.; Tiffany, B. D. *J. Am. Chem. Soc.* **1951**, *73*, 4158–4161.

5 House, H. O.; Ro, R. S. *J. Am. Chem. Soc.* **1958**, *80*, 2428–2433.

6 Cope, A. C.; Martin, M. M.; McKervey, M. A. *Quart. Rev. (London)* **1966**, *20*, 119–152.

7 Cope, A. C.; Lee, H.-H.; Petree, H. E. *J. Am. Chem. Soc.* **1958**, *80*, 2849–2852.

8 Whitesell, J. K.; White, P. D. *Synthesis* **1975**, 602–603.

9 Cope, A. C.; Brown, M.; Lee, H.-H. *J. Am. Chem. Soc.* **1958**, *80*, 2855–2859.

10 Thies, R. W.; Chiarello, R. H. *J. Org. Chem.* **1979**, *44*, 1342–1344.

11 Crandall, J. K.; Chang, L.-H. *J. Org. Chem.* **1967**, *32*, 532–536.

12 Boeckman, R. K., Jr. *Tetrahedron Lett.* **1977**, 4281–4284.

13 Hodgson, D. M.; Lee, G. P. *Chem. Commun.* **1996**, 1015–1016.

14 Hodgson, D. M.; Lee, G. P. *Tetrahedron: Asymmetry* **1997**, *8*, 2303–2306.

15 Hodgson, D. M.; Lee, G. P.; Marriott, R. E.; Thompson, A. J.; Wisedale, R.; Witherington, J. *J. Chem. Soc., Perkin Trans. 1* **1998**, 2151–2162.

16 Alexakis, A.; Vrancken, E.; Mangeney, P. *J. Chem. Soc., Perkin Trans. 1* **2000**, 3354–3355.

17 Hodgson, D. M.; Cameron, I. D. *Org. Lett.* **2001**, *3*, 441–444.

18 Hodgson, D. M.; Cameron, I. D.; Christlieb, M.; Green, R.; Lee, G. P.; Robinson, L. A. *J. Chem. Soc., Perkin Trans. 1* **2001**, 2161–2174.

19 (a) Hodgson, D. M.; Galano, J.-M.; Christlieb, M. *Chem. Commun.* **2002**, 2436–2437; (b) Hodgson, D. M.; Galano, J.-M.; Christlieb, M. *Tetrahedron* **2003**, *59*, 9719–9728.

20 Hodgson, D. M.; Robinson, L. A. *Chem. Commun.* **1999**, 309–310.

21 Crandall, J. K. *J. Org. Chem.* **1964**, *29*, 2830–2833.

22 Hodgson, D. M.; Marriott, R. E. *Tetrahedron Lett.* **1997**, *38*, 887–888.

23 Neff, J. R.; Nordlander, J. E. *Tetrahedron Lett.* **1977**, 499–500.

24 (a) McDonald, R. N.; Steppel, R. N.; Cousins, R. C. *J. Org. Chem.* **1975**, *40*, 1694–1698; see also: (b) Arjona, O.; Menchaca, R.; Plumet, J. *Tetrahedron* **2000**, *56*, 3901–3907.

25 Hodgson, D. M.; Wisedale, R. *Tetrahedron: Asymmetry* **1996**, *7*, 1275–1276.

26 (a) Hodgson, D. M.; Maxwell, C. R.; Matthews, I. R. *Synlett* **1998**, 1349–1350; (b) Hodgson, D. M.; Maxwell, C. R.; Matthews, I. R. *Tetrahedron: Asymmetry* **1999**, *10*, 1847–1850; (c) Wisedale, R.; Matthews, I. R.; Carpenter, K. J.; Dickenson, A. H.; Wonnacott, S. *J. Chem. Soc., Perkin Trans. 1* **2001**, 3150–3158.

27 (a) Crandall, J. K.; Lin, L.-H. C. *J. Am. Chem. Soc.* **1967**, *89*, 4526–4527; (b) Lin, L.-H. C. Ph. D. Thesis, Indiana University, USA, **1967**.

28 Dechoux, L.; Agami, C.; Doris, E.; Mioskowski, C. *J. Org. Chem.* **1999**, *64*, 9279–9281.

29 Dechoux, L.; Agami, C.; Doris, E.; Mioskowski, C. *Tetrahedron* **2003**, *59*, 9701–9706.

30 Crandall, J. K.; Crawley, L. C.; Banks, D. B.; Lin, L. C. *J. Org. Chem.* **1971**, *36*, 510–513.

31 Hodgson, D. M.; Robinson, L. A.; Jones, M. L. *Tetrahedron Lett.* **1999**, *40*, 8637–8640.

32 Cope, A. C.; Trumbull, P. A.; Trumbull, E. R. *J. Am. Chem. Soc.* **1958**, *80*, 2844–2849.

33 Denichoux, A.; Ferreira, F.; Chemla, F. *Org. Lett.* **2004**, *6*, 3509–3512.

34 (a) Hodgson, D. M.; Chung, Y. K.; Paris, J.-M. *J. Am. Chem. Soc.* **2004**, *126*, 8664–8665; (b) Mordini, A.; Peruzzi, D.; Russo, F.; Valacchi, M.; Reginato, G.; Brandi, A. *Tetrahedron* **2005**, *61*, 3349–3360; (c) Hodgson, D. M.; Chung, Y. K.; Paris, J.-M. *Synthesis* **2005**, 2264–2266.

35 Dechoux, L.; Agami, C.; Doris, E.; Mioskowski, C. *Eur. J. Org. Chem.* **2001**, 4107–4110.

36 Crandall, J. K.; Lin, L.-H. C. *J. Am. Chem. Soc.* **1967**, *89*, 4527–4528.

37 Eisch, J. J.; Galle, J. E. *J. Org. Chem.* **1976**, *41*, 2615–2621.

38 Ukaji, Y.; Yoshida, A.; Fujisawa, T. *Chem. Lett.* **1990**, 157–160.

39 Santiago, B.; Lopez, C.; Soderquist, J. A. *Tetrahedron Lett.* **1991**, *32*, 3457–3460.

40 Soderquist, J. A.; Lopez, C. *Tetrahedron Lett.* **1991**, *32*, 6305–6306.

41 Hodgson, D. M.; Fleming, M. J.; Stanway, S. J. *J. Am. Chem. Soc.* **2004**, *126*, 12250–12251.

42 Doris, E.; Dechoux, L.; Mioskowski, C. *Tetrahedron Lett.* **1994**, *35*, 7943–7946.

43 Ukaji, Y.; Fujisawa, T. *Tetrahedron Lett.* **1988**, *29*, 5165–5168.

44 Dechoux, L.; Doris, E.; Mioskowski, C. *Chem. Commun.* **1996**, 549–550.

45 (a) Hodgson, D. M.; Stent, M. A. H.; Wilson, F. X. *Org. Lett.* **2001**, *3*, 3401–3403; (b) Hodgson, D. M.; Stent, M. A. H.; Wilson, F. X. *Synthesis* **2002**, 1445–1453.

46 (a) Hodgson, D. M.; Miles, T. J.; Witherington, J. *Synlett* **2002**, 310–312; (b) Hodgson, D. M.; Miles, T. J.; Witherington, J. *Tetrahedron* **2003**, *59*, 9729–9742.

47 Hodgson, D. M.; Paruch, E. *Tetrahedron* **2004**, *60*, 5185–5199.

48 (a) Hodgson, D. M.; Maxwell, C. R.; Miles, T. J.; Paruch, E.; Stent, M. A. H.; Matthews, I. R.; Wilson, F. X.; Witherington, J. *Angew. Chem., Int. Ed.* **2002**, *41*, 4313–4316; (b) Hodgson, D. M.; Stent, M. A. H.; Stefane, B.; Wilson, F. X. *Org. Biomol. Chem.* **2003**, *1*, 1139–1150; (c) Hodgson, D. M.; Maxwell, C. R.; Miles, T. J.; Paruch, E.; Matthews, I. R.; Witherington, J. *Tetrahedron* **2004**, *60*, 3611–3624.

49 Taniguchi, M.; Oshima, K.; Utimoto, K. *Tetrahedron Lett.* **1991**, *32*, 2783–2786.

50 (a) Kasatkin, A. N.; Whitby, R. J. *Tetrahedron Lett.* **2000**, *41*, 5275–5280; (b) Kasatkin, A. N.; Whitby, R. J. *Tetrahedron Lett.* **2000**, *41*, 6201–6205; (c) Kasatkin, A.; Whitby, R. J. *Tetrahedron* **2003**, *59*, 9857–9864.

51 Hodgson, D. M.; Bray, C. D.; Kindon, N. D. *J. Am. Chem. Soc.* **2004**, *126*, 6870–6871.

52 Yanagisawa, A.; Yasue, K.; Yamamoto, H. *J. Chem. Soc.; Chem. Commun.* **1994**, 2103–2104.

53 Hodgson, D. M.; Bray, C. D.; Kindon, N. D. *Org. Lett.* **2005**, *7*, 2305–2308.

54 (a) Eisch, J. J.; Galle, J. E. *J. Am. Chem. Soc.* **1976**, *98*, 4646–4648; (b) Eisch, J. J.; Galle, J. E. *J. Organomet. Chem.* **1976**, *121*, C10-C14; (c) Eisch, J. J.; Galle, J. E. *J. Organomet. Chem.* **1988**, *341*, 293–313.

55 Burford, C.; Cooke, F.; Roy, G.; Magnus, P. *Tetrahedron* **1983**, *39*, 867–876.

56 Molander, G. A.; Mautner, K. *J. Org. Chem.* **1989**, *54*, 4042–4050.

57 Murthi, K. K.; Salomon, R. G. *Tetrahedron Lett.* **1994**, *35*, 517–520.

58 Mani, N. S.; Townsend, C. A. *J. Org. Chem.* **1997**, *62*, 636–640.

59 (a) Marie, J.-C.; Courillon, C.; Malacria, M. *Synlett* **2002**, 553–556; (b) Courillon, C.; Marie, J.-C.; Malacria, M. *Tetrahedron* **2003**, *59*, 9759–9766.

60 (a) Ashwell, M.; Jackson, R. F. W. *J. Chem. Soc.; Chem. Commun.* **1988**, 645–647; (b) Ashwell, M.; Clegg, W.; Jackson, R. F. W. *J. Chem. Soc., Perkin Trans. 1* **1991**, 897–908; (c) Jackson, R. F. W.; Dunn, S. F. C.; McCamley, A.; Clegg, W. *Org. Biomol. Chem.* **2003**, *1*, 2527–2530.

61 (a) Hewkin, C. T.; Jackson, R. F. W.; Clegg, W. *Tetrahedron Lett.* **1988**, *29*, 4889–4892; (b) Ashwell, M.; Jackson, R. F. W. *J. Chem. Soc., Perkin Trans.1* **1989**, 835–837; (c) Hewkin, C. T.; Jackson, R. F. W. *Tetrahedron Lett.* **1990**, *31*, 1877–1880; (d) Hewkin, C. T.; Jackson, R. F. W. *J. Chem. Soc., Perkin Trans. 1* **1991**, 3103–3111; (e) Hewkin, C. T.; Jackson, R. F. W.; Clegg, W. *J. Chem. Soc., Perkin Trans.1* **1991**, 3091–3101; (f) Dunn, S. F. C.; Jackson, R. F. W. *J. Chem. Soc., Perkin Trans.1* **1992**, 2863–2870.

62 For a recent review see: (a) Mori, Y.; Yaegashi, K. *Rec. Res. Dev. Org. Chem.* **1999**, *3*, 207–217. For recent applications see: (b) Mori, Y.; Furuta, H.; Takase, T.; Mitsuoka, S.; Furukawa, H. *Tetrahedron Lett.* **1999**, *40*, 8019–8022; (c) Mori, Y.; Yaegashi, K.; Furukawa, H. *Tetrahedron Lett.* **1999**, *40*, 7239–7242; (d) Mori, Y.; Mitsuoka, S.; Furukawa, H. *Tetrahedron Lett.* **2000**, *41*, 4161–4164; (e) Mori, Y.; Hayashi, H. *Tetrahedron* **2002**, *58*, 1789–1797; (f) Furuta, H.; Takase, T.; Hayashi, H.; Noyori, R.; Mori, Y. *Tetrahedron* **2003**, *59*, 9767–9777; (g) Mori, Y.; Nogami, K.; Hayashi, H.; Noyori, R. *J. Org. Chem.* **2003**, *68*, 9050–9060; (h) Mori, Y.; Takase, T.; Noyori, R. *Tetrahedron Lett.* **2003**, *44*, 2319–2322; (i) Inoue, M.; Yamashita, S.; Tatami, A.; Miyazaki, K.; Hirama, M. *J. Org. Chem.* **2004**, *69*, 2797–2804; (j) Furuta, H.; Hase, M.; Noyori, R.; Mori, Y. *Org. Lett.* **2005**, *7*, 4061–4064.

63 Mori, Y.; Sawada, T.; Furukawa, H. *Tetrahedron Lett.* **1999**, *40*, 731–734.

64 (a) Mori, Y.; Yaegashi, K.; Furukawa, H. *J. Am. Chem. Soc.* **1997**, *119*, 4557–4558; (b) Mori, Y.; Yaegashi, K.; Furukawa, H. *J. Org. Chem.* **1998**, *63*, 6200–6209.

65 Eisch, J. J.; Galle, J. E. *J. Org. Chem.* **1990**, *55*, 4835–4840.

66 (a) Capriati, V.; Florio, S.; Luisi, R.; Salomone, A. *Org. Lett.* **2002**, *4*, 2445–2448; (b) Capriati, V.; Florio, S.; Luisi, R.; Nuzzo, I. *J. Org. Chem.* **2004**, *69*, 3330–3335; (c) Florio, S.; Aggarwal, V.; Salomone, A. *Org. Lett.* **2004**, *6*, 4191–4194; (d) For a recent review, see: Capriati, V.; Florio, S.; Luisi, R. *Synlett* **2005**, 1359–1369.

67 (a) Shimizu, M.; Fujimoto, T.; Minezaki, H.; Hata, T.; Hiyama, T. *J. Am. Chem. Soc.* **2001**, *123*, 6947–6948; (b) Shimizu, M.; Fujimoto, T.; Liu, X.; Minezaki, H.; Hata, T.; Hiyama, T. *Tetrahedron* **2003**, *59*, 9811–9823; (c) Shimizu, M.; Fujimoto, T.; Liu, X.; Hiyama, T. *Chem. Lett.* **2004**, *33*, 438–439.

68 (a) Grandjean, D.; Pale, P.; Chuche, J. *Tetrahedron: Asymmetry* **1993**, *4*, 1991–1994; (b) Klein, S.; Zhang, J. H.; Holler, M.; Weibel, J.-M.; Pale, P. *Tetrahedron* **2003**, *59*, 9793–9802.

69 (a) Kuramochi, K.; Itaya, H.; Nagata, S.; Takao, K.-I.; Kobayashi, S. *Tetrahedron Lett.* **1999**, *40*, 7367–7370; (b) Kuramochi, K.; Nagata, S.; Itaya, H.; Takao, K.-I.; Kobayashi, S. *Tetrahedron Lett.* **1999**, *40*, 7371–7374; (c) Kuramochi, K.; Nagata, S.; Itaya, H.; Matsubara, Y.; Sunoki, T.; Uchiro, H.; Takao, K.-i.; Kobayashi, S. *Tetrahedron* **2003**, *59*, 9743–9758.

70 (a) Yamauchi, Y.; Katagiri, T.; Uneyama, K. *Org. Lett.* **2002**, *4*, 173–176; (b) Yamauchi, Y.; Kawate, T.; Katagiri, T.; Uneyama, K. *Tetrahedron* **2003**, *59*, 9839–9847.

71 (a) Florio, S.; Ingrosso, G.; Troisi, L.; Lucchini, V. *Tetrahedron Lett.* **1993**, *34*, 1363–1366. (b) Katritzky, A. R.; Manju, K.; Steel, P. J. *J. Org. Chem.* **2003**, *68*, 407–411.

72 (a) Florio, S.; Capriati, V.; Di Martino, S.; Abbotto, A. *Eur. J. Org. Chem.* **1999**, 409–417; (b) Luisi, R.; Capriati, V.; Carlucci, C.; Degennaro, L.; Florio, S. *Tetrahedron* **2003**, *59*, 9707–9712.

73 Luisi, R.; Capriati, V.; Degennaro, L.; Florio, S. *Org. Lett.* **2003**, *5*, 2723–2726.

74 Capriati, V.; Degennaro, L.; Favia, R.; Florio, S.; Luisi, R. *Org. Lett.* **2002**, *4*, 1551–1554.

75 (a) Lertvorachon, J.; Thebtaranonth, Y.; Thongpanchang, T.; Thongyoo, P. *J. Org. Chem.* **2001**, *66*, 4692–4694; (b) Chaiyanurakkul, A.; Jitchati, R.; Kaewpet, M.; Rajviroongit, S.; Thebtaranonth, Y.; Thongyoo, P.; Watcharin, W. *Tetrahedron* **2003**, *59*, 9825–9837.

76 Lohse, P.; Loner, H.; Acklin, P.; Sternfeld, F.; Pfaltz, A. *Tetrahedron Lett.* **1991**, *32*, 615–618.

77 (a) Satoh, T.; Horiguchi, K. *Tetrahedron Lett.* **1995**, *36*, 8235–8238; (b) Satoh, T.; Kobayashi, S.; Nakanishi, S.; Horiguchi, K.; Irisa, S. *Tetrahedron* **1999**, *55*, 2515–2528.

78 (a) Satoh, T.; Oohara, T.; Ueda, Y.; Yamakawa, K. *Tetrahedron Lett.* **1988**, *29*, 313–316; (b) Satoh, T.; Oohara, T.; Ueda, Y.; Yamakawa, K. *J. Org. Chem.* **1989**, *54*, 3130–3136.

79 (a) Hodgson, D. M.; Norsikian, S. L. M. *Org. Lett.* **2001**, *3*, 461–463; (b) Hodgson, D. M.; Kirton, E. H. M. *Synlett* **2004**, 1610–1612; (c) Hodgson, D. M.; Reynolds, N. J.; Coote, S. J. *Org. Lett.* **2004**, *6*, 4187–4189; (d) Hodgson, D. M.; Kirton, E. H. M.; Miles, S. M.; Norsikian, S. L. M.; Reynolds, N. J.; Coote, S. J. *Org. Biomol. Chem.* **2005**, *3*, 1893–1904.

80 (a) Hodgson, D. M.; Gras, E. *Angew. Chem., Int. Ed.* **2002**, *41*, 2376–2378; (b) Hodgson, D. M.; Buxton, T. J.; Cameron, I. D.; Gras, E.; Kirton, E. H. M. *Org. Biomol. Chem.* **2003**, *1*, 4293–4301.

81 Hodgson, D. M.; Reynolds, N. J.; Coote, S. J. *Tetrahedron Lett.* **2002**, *43*, 7895–7897.

82 Rubottom, G. M.; Stevenson, G. R.; Chabala, J. C.; Pascucci, V. L. *Tetrahedron Lett.* **1972**, 3591–3594.

83 (a) Seebach, D.; Haener, R. *Chem. Lett.* **1987**, 49–52; (b) Haener, R.; Olano, B.; Seeback, D. *Helv. Chem. Acta* **1987**, *70*, 1676–1693;

84 (a) Alezra, V.; Bonin, M.; Micouin, L.; Husson, H.-P. *Tetrahedron Lett.* **2000**, *41*, 651–654; (b) Alezra, V.; Bonin, M.; Micouin, L.; Policar, C.; Husson, H.-P. *Eur. J. Org. Chem.* **2001**, 2589–2594; (c) See also: Patwardhan, A. P.; Pulgam, V. R.; Zhang, Y.; Wulff, W. D. *Angew. Chem., Int. Ed.* **2005**, *44*, 6169–6172.

85 Yamauchi, Y.; Kawate, T.; Itahashi, H.; Katagiri, T.; Uneyama, K. *Tetrahedron Lett.* **2003**, *44*, 6319–6322.

86 Gaillot, J. M.; Gelas-Mialhe, Y.; Vessiere, R. *Can. J. Chem.* **1979**, *57*, 1958–1966

87 (a) Reutrakul, V.; Prapansiri, V.; Panyachotipun, C. *Tetrahedron Lett.* **1984**, *25*, 1949–1952; (b) Coutrot, P.; Elgadi, A.; Grison, C. *Heterocycles* **1989**, *28*, 1179–1192.

88 (a) Luisi, R.; Capriati, V.; Florio, S.; Ranaldo, R. *Tetrahedron Lett.* **2003**, *44*, 2677–2681; (b) De Vitis, L.; Florio, S.; Granito, C.; Ronzini, L.; Troisi, L.; Capriati, V.; Luisi, R.; Pilati, T. *Tetrahedron* **2004**, *60*, 1175–1182; (c) Luisi, R.; Capriati, V.; Florio, S.; Di Cunto, P.; Musio, B. *Tetrahedron* **2005**, *61*, 3251–3260.

89 Hayes, J. F.; Prevost, N.; Prokes, I.; Shipman, M.; Slawin, A. M. Z.; Twin, H. *Chem. Commun.* **2003**, 1344–1345.

90 (a) Satoh, T.; Oohara, T.; Yamakawa, K. *Tetrahedron Lett.* **1988**, *29*, 4093–4096. See also: (b) Satoh, T.; Sato, T.; Oohara, T.; Yamakawa, K. *J. Org. Chem.* **1989**, *54*, 3973–3978; (c) Satoh, T.; Ozawa, M.; Takano, K.; Chyouma, T.; Okawa, A. *Tetrahedron* **2000**, *56*, 4415–4425.

91 Atkinson, R. S.; Kelly, B. J. *Tetrahedron Lett.* **1989**, *30*, 2703–2704.

92 Vedejs, E.; Moss, W. O. *J. Am. Chem. Soc.* **1993**, *115*, 1607–1608.

93 (a) Vedejs, E.; Kendall, J. T. *J. Am. Chem. Soc.* **1997**, *119*, 6941–6942; (b) Vedejs, E.; Bhanu Prasad, A. S.; Kendall, J. T.; Russel, J. S. *Tetrahedron* **2003**, *59*, 9849–9856.

94 Bisseret, P.; Bouix-Peter, C.; Jacques, O.; Henriot, S.; Eustache, J. *Org. Lett.* **1999**, *1*, 1181–1182.

95 (a) Concellon, J. M.; Suarez, J. R.; Garcia-Granda, S.; Diaz, M. R. *Angew. Chem., Int. Ed.* **2004**, *43*, 4333–4336; (b) Concellon, J. M.; Bernad, P. L.; Suarez, J. R. *Chem. Eur. J.* **2005**, *11*, 4492–4501.

96 Hodgson, D. M.; Humphreys, P. G.; Ward, J. G. *Org. Lett.* **2005**, *7*, 1153–1156. See also: Beak, P.; Wu, S.; Yum, E. K.; Jun, Y. M. *J. Org. Chem.* **1994**, *59*, 276–277.

97 O'Brien, P.; Rosser, C. M.; Caine, D. *Tetrahedron* **2003**, *59*, 9779–9791.

98 Hodgson, D. M.; Stefane, B.; Miles, T. J.; Witherington, J. *Chem. Commun.* **2004**, 2234–2235.

99 Arjona, O.; Menchaca, R.; Plumet, J. *Heterocycles* **2001**, *55*, 5–7.

100 (a) Müller, P.; Nury, P. *Helv. Chem. Acta* **2001**, *84*, 662–677; (b) Müller, P.; Riegert, D.; Bernardinelli, G. *Helv. Chem. Acta* **2004**, *87*, 227–239.

101 See also: Mordini, A.; Russo, F.; Valacchi, M.; Zani, L.; Degl'Innocenti, A.; Reginato, G. *Tetrahedron* **2002**, *58*, 7153–7163.

102 O'Brien, P.; Rosser, C. M.; Caine, D. *Tetrahedron Lett.* **2003**, *44*, 6613–6615.

6
Metal-catalyzed Synthesis of Epoxides

Hans Adolfsson and Daniela Balan

6.1
Introduction

Epoxides are highly versatile intermediates used in the formation of more complex organic compounds. This is well reflected in the substantial use of the simple compound propylene oxide, approximately 4 million tons of which are produced annually, which is employed as a building block in the industrial production of wide variety of organic materials. One of the industrial methods used for the formation of simple epoxides such as propylene oxide is the *chlorohydrin* process. In this process, which currently accounts for about 50% of the world production of propylene oxide, propylene reacts with chlorine in the presence of sodium hydroxide, initially to generate an intermediate 1,2-chloro alcohol (chlorohydrin). In a second step, this intermediate undergoes dehydrochlorination to yield the final epoxide product. (Scheme 6.1a) [1]. The major disadvantage with this process is the huge amount of byproducts (2.01 ton NaCl and 0.102 ton 1,2-dichloropropane) produced per ton of propylene oxide. These significant amounts of waste are not acceptable in the long run, and efforts in replacing such chemical plants with "greener" epoxidation processes are certainly necessary. The central topic for the development of future oxidation systems both in industry and in academia needs to be focused on the use of environmentally benign oxidants as well as on highly

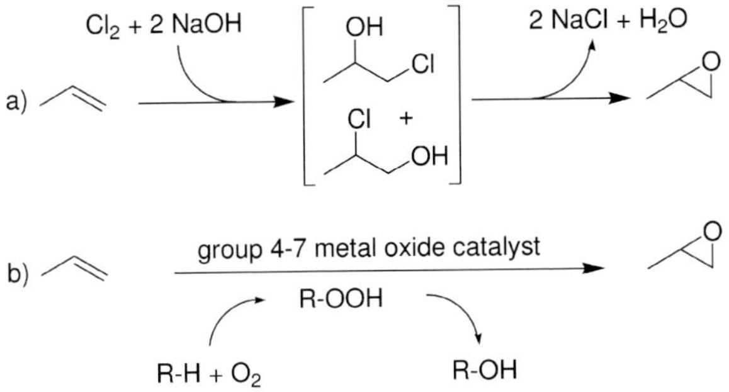

a) Cl_2 + 2 NaOH; OH, Cl + Cl, OH; 2 NaCl + H_2O

b) group 4-7 metal oxide catalyst; R-OOH; R-H + O_2; R-OH

Scheme 6.1

Aziridines and Epoxides in Organic Synthesis. Andrei K. Yudin

ISBN: 3-527-31213-7

atom-economic processes. Of course, traditional aspects such as broad substrate scope and high selectivity must come into play, and if all these facets are taken into account the use of catalysis becomes a natural choice. In this context, the Halcon-Arco process (Scheme 6.1b), which uses transition metal oxides and organic hydroperoxides as reagents, represents a significantly more eco-compatible route for the production of propylene oxide [2]. When it comes to the production of fine chemicals, uncatalyzed processes with traditional oxidants (e.g., peroxyacetic acid and *meta*-chloroperoxybenzoic acid) are often used. However, the direct formation of epoxides by olefin oxidation with use either of transition metal catalysts or of simple organic ketone-based catalysts is once again considerably more elegant and environmentally friendly [3, 4, 5]. There are several advantages displayed by transition metal-based systems, and the major issues in the development of novel methods are the choice of terminal oxidant along with the compatibility of the metal catalyst with this particular oxidant. This is of major importance, as the conservation and management of resources should be of primary interest when novel chemical processes are developed. Innovation in and improvements to catalytic epoxidation methods should therefore be directed towards systems employing molecular oxygen or hydrogen peroxide as terminal oxidants. This chapter covers the development of efficient and selective metal-catalyzed epoxidation methods, starting with early transition metals (Ti, V) and ending with group VIII metals (Fe, Ru). We are not presenting a complete review of the latest developments in transition metal-catalyzed epoxidations, but rather highlighting the most relevant catalysts from each group of metals. The main focus is on methods for fine chemical production, including asymmetric epoxidations, although recent developments of metal-catalyzed epoxidations suitable for bulk production are also described. In particular, we emphasize modern, efficient, and reliable methods for selective transition metal-catalyzed epoxidation by olefin oxidation with environmentally benign oxidants (e.g., aqueous hydrogen peroxide).

6.2
Oxidants Available for Selective Transition Metal-catalyzed Epoxidation

There are several available terminal oxidants for the transition metal-catalyzed epoxidation of olefins (Table 6.1). Typical oxidants compatible with most metal-based epoxidation systems are various alkyl hydroperoxides, hypochlorite, or iodosylbenzene. A problem associated with these oxidants is their low active oxygen content (Table 6.1), while there are further drawbacks with these oxidants from the point of view of the nature of the waste produced. Thus, from an environmental and economical perspective, molecular oxygen should be the preferred oxidant, because of its high active oxygen content and since no waste (or only water) is formed as a byproduct. One of the major limitations of the use of molecular oxygen as terminal oxidant for the formation of epoxides, however, is the poor product selectivity obtained in these processes [6]. Aerobic oxidations are often difficult to control and can sometimes result in combustion or in substrate overoxidation. In

Table 6.1 Oxidants used in transition metal-catalyzed epoxidations, and their active oxygen content.

Oxidant	Active oxygen content (wt. %)	Waste product
Oxygen (O_2)	100	–
Oxygen (O_2)/reductor	50	H_2O
H_2O_2	47	H_2O
NaOCl	21.6	NaCl
CH_3CO_3H	21.1	CH_3CO_2H
t-BuOOH (TBHP)	17.8	*t*-BuOH
$KHSO_5$	10.5	$KHSO_4$
BTSP[a]	9	hexamethyldisiloxane
PhIO	7.3	PhI

[a] Bistrimethylsilyl peroxide

combination with the limited number of catalysts available for direct activation of molecular oxygen, this strongly confines the use of this oxidant. On the other hand, hydrogen peroxide displays much better properties as terminal oxidant. The active oxygen content of H_2O_2 is about as high as for typical applications of molecular oxygen in epoxidations (since a reductor is required in almost all cases), and the waste produced by employing this oxidant is plain water. As in the case of molecular oxygen, the epoxide selectivity of H_2O_2 can sometimes be relatively poor, although recent developments have provided transition metal-based methods in which excellent reactivity and epoxide selectivity can be obtained [7]. From a safety point of view, catalytic systems based on hydrogen peroxide as the terminal oxidant should be carried out with H_2O_2 concentrations of less than 60%, since higher concentrations can result in hazardous use, storage, and transportation problems. As well as molecular oxygen and hydrogen peroxide, alkyl hydroperoxides also serve as readily available and inexpensive oxidants for the epoxidation of olefins. There are several advantages in using hydroperoxides rather than molecular oxygen or hydrogen peroxide in organic oxidations: alkyl hydroperoxides show higher stability towards decomposition by metal contaminants, they display high thermal stability in dilute organic solvents, and they are compatible with a large number of functional groups. The obvious disadvantage, which makes oxidation system based on alkyl hydroperoxides less atom-economical, is the formation of alcohol byproducts, which have to be separated from the product. The following sections present various available oxidation systems for the selective epoxidation of olefins with transition metal catalysts and, in particular, with hydrogen peroxide as the terminal oxidant.

6.3
Epoxidations of Olefins Catalyzed by Early Transition Metals

High-valent early transition metals such as titanium(IV) and vanadium(V) have been shown to catalyze the epoxidation of olefins efficiently. The preferred oxidants used with these catalysts are various alkyl hydroperoxides, typically *tert*-butyl hydroperoxide (TBHP) or ethylbenzene hydroperoxide (EBHP). One of the routes for the industrial production of propylene oxide is based on a heterogeneous Ti^{IV}/SiO_2 catalyst, employing EBHP as the terminal oxidant [8]. On the other hand, the use of hydrogen peroxide as oxidant in combination with early transition metal catalysts (Ti and V) is rather limited. The reason for the poor reactivity can be traced to severe inhibition of the metal complexes by strongly coordinating ligands such as alcohols and, in particular, water.

6.3.1
Titanium-catalyzed Epoxidations

The Sharpless-Katsuki asymmetric epoxidation (AE) procedure for the enantioselective formation of epoxides from allylic alcohols is a milestone in asymmetric catalysis [9]. This classical asymmetric transformation uses TBHP as the terminal oxidant, and the reaction has been widely used in various synthetic applications. There are several excellent reviews covering the scope and utility of the AE reaction [10]. This asymmetric transformation has – besides its outstanding historical value [11] – the extraordinary features of being easily accessible and performable, highly predictable in its outcome, and applicable to a wide range of substrates. The titanium tartrate catalyst (**1**; Figure 6.1) is readily accessible from dialkyl tartrate and titanium alkoxide (most commonly diethyl tartrate (DET) and titanium isopropoxide), and the alkyl hydroperoxide used as oxidant is in general commercially available.

The reason for the efficient epoxidation of explicitly allylic alcohols with this system can be found in the strong associative interactions occurring between the substrate and the catalyst. The $[Ti(tartrate)(OR)_2]_2$ dimer **1**, which is considered to be the active catalyst in the reaction, will generate structure **2** after the addition of

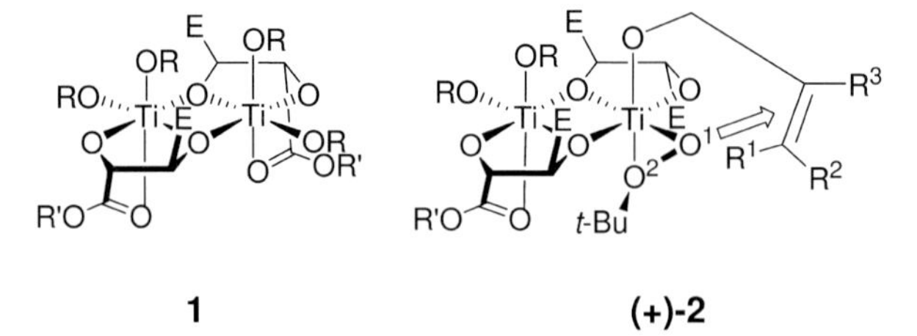

Figure 6.1 Proposed structure for the titanium tartrate complex (**1**) and its transformation after addition of reagents (e.g., TBHP and olefin), forming **(+)-2**.

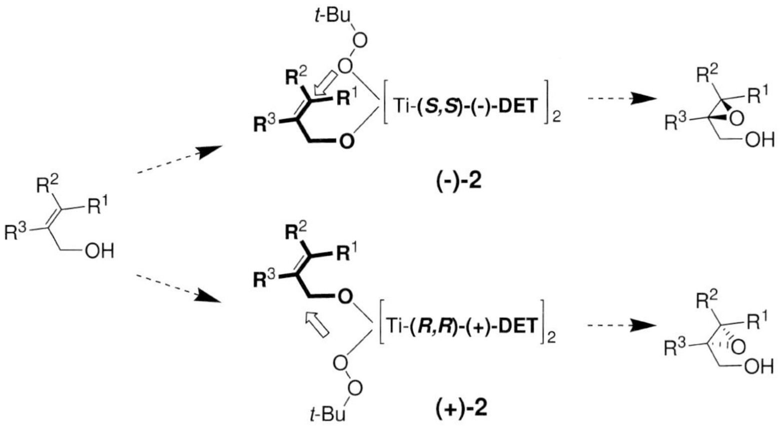

Figure 6.2 Enantiofacial differentiation in AE, depending on the configuration of the diethyl tartrate ligand in the titanium complex **2**.

the allylic alcohol and the oxidant (TBHP) [12, 13, 14]. The coordination of the distal oxygen atom (O^2) from TBHP to the titanium ion (complex **2**; Figure 6.2) activates the peroxide and facilitates the oxygen transfer. The allylic alcohol, which coordinates to the same titanium center through its alcohol functionality, will be activated and close enough for oxygen delivery in an intramolecular fashion. In this way, complete regioselectivity can be obtained when the substrate is an allylic alcohol containing several double bonds. Oxidation at the C2-olefin is entropically more favored than at any other double bond situated at a more advanced position along the carbon chain [10]. With regard to enantioselectivity, it is generally the case that oxygen transfer from the coordinated alkyl peroxide to a primary allylic alcohol will take place from "above" when (*S,S*)-(−)-DET is employed as chiral ligand, and from "below" in when (*R,R*)-(+)-DET is used for generating the chiral titanium complex **1** (Figure 6.2) [9]. The stereochemical predictability of the AE reaction is a consequence of the positioning of the two reacting partners – the oxidant and the allylic alcohol – in complex **2**, as well as of the inherent C_2-symmetry of complex **1**.

From the structure of the active catalyst (**1**; Figure 6.1) it is obvious that the titanium and the alkyl tartrate are present in a 1:1 stoichiometry. More thorough studies show that a 20% excess of tartrate is desirable, in order to avoid the formation of complexes containing substoichiometric amounts of ligand relative to titanium. These complexes are active catalysts but induce low or no enantioselectivity in the epoxidation of allylic alcohols. The use of the correct stoichiometry will result in the predominant formation of complex **1**, which has superior catalytic activity and gives the epoxide products with high enantioselectivity. This scenario is perhaps one of the best illustrative examples of the subject of *ligand acceleration catalysis* (LAC) [15]. A high excess of tartrate is undesirable as well, since this will

Table 6.2 Examples of epoxides generated by AE, applied to different primary allylic alcohols showing all eight basic substitution patterns.

Substitution pattern	Epoxidation results		
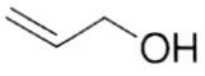	65% (90% *ee*) [16]		
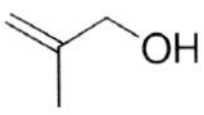	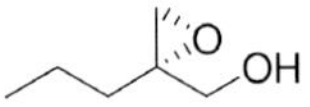 88% (95% *ee*) [16]	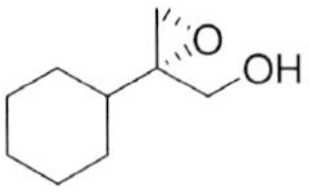 81% (>95% *ee*) [9]	 74% (>95% *ee*) [17]
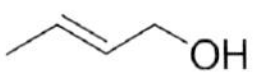	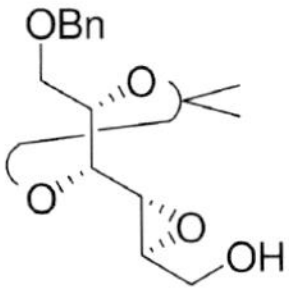 70% (>95% *ee*) [18]	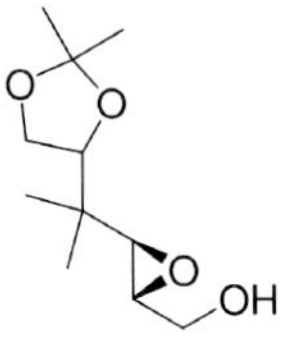 70% (>95% *ee*) [19]	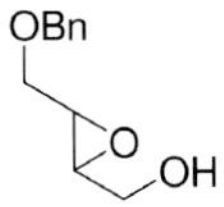 85% (>98% *ee*) [18]
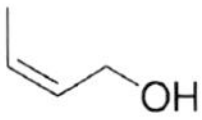	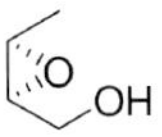 68% (92% *ee*) [16]	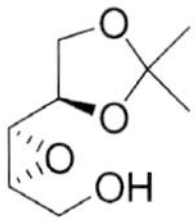 55% (93% *ee*) [18]	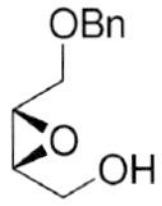 84% (92% *ee*) [18]
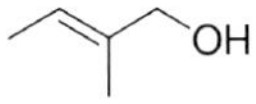	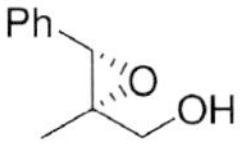 79% (>98% *ee*) [16]	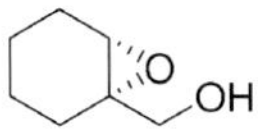 77% (93% *ee*) [16]	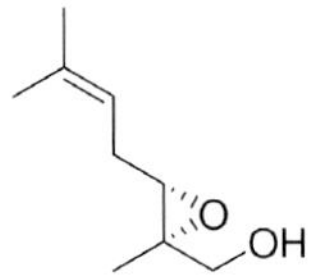 64% (>90% *ee*) [20]
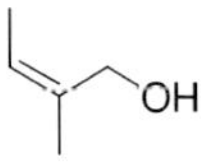	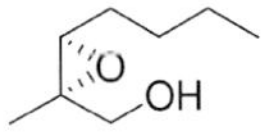 80% (89% *ee*) [9]	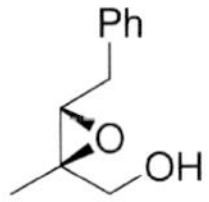 90% (91% *ee*) [21]	
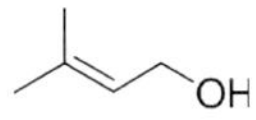	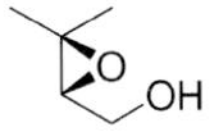 25% (>90% *ee*) [22]	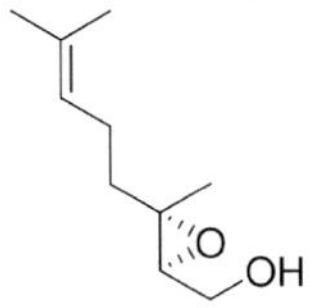 95% (91% *ee*) [16]	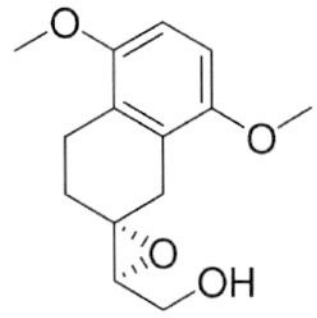 97% (93% *ee*) [23]
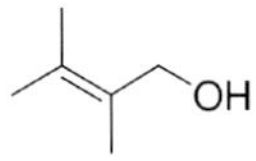	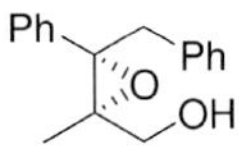 90% (94% *ee*) [24]	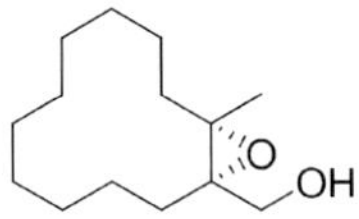 72% (94% *ee*) [25]	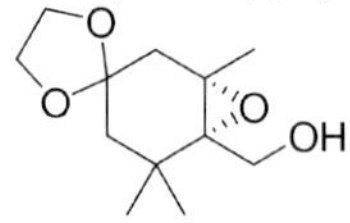 95% (95% *ee*) [26]

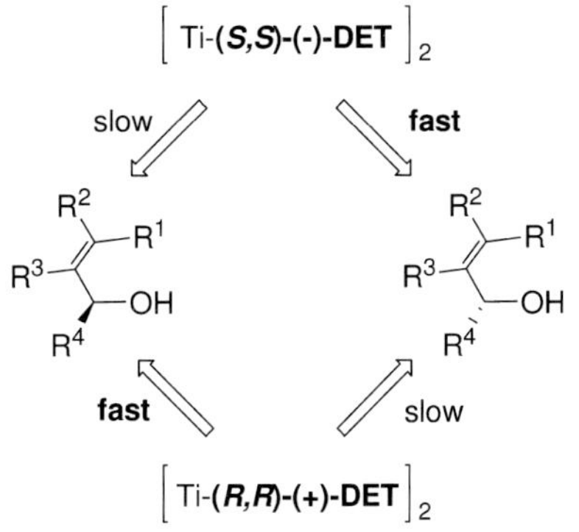

Figure 6.3 Diastereofacial differentiation in titanium-catalyzed AE of secondary allylic alcohols.

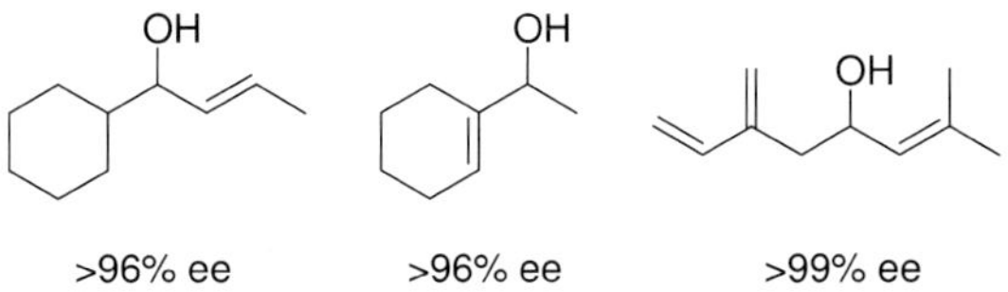

Figure 6.4 Some successful examples of kinetic resolution of allylic alcohols by enantioselective epoxidation [21, 27].

generate a new complex with the stoichiometry $[Ti(tartrate)_2]_x$, known to be inactive in the epoxidation reaction [13].

The AE reaction catalyzed by titanium tartrate **1** and with alkyl hydroperoxide as terminal oxidant has been applied to a large variety of primary allylic alcohols containing all eight basic substitution patterns. A few examples are presented in Table 6.2.

The empirical rule described above for the enantiofacial differentiation in AE of primary allylic alcohols also applies to secondary allylic alcohols. The new aspect that needs to be taken into consideration in this case is the steric hindrance arising from the presence of a substituent (R^4) at the carbon bearing the hydroxy group (Figure 6.3). This substituent will interfere in the process of oxygen delivery, making the oxidation of one enantiomer much faster than the reaction of the other one. The phenomenon is so acute that in practice kinetic resolution is often achieved (Figure 6.4) [27].

The titanium-catalyzed AE reaction is a fairly robust system and it can be performed on substrates containing a wide range of different functional groups (FGs) (Table 6.3) [13]. However, it is important to point out that an intramolecular reaction with the formed epoxide is possible whenever the FG present in the molecule has a favorable position to facilitate such a transformation. An illustration of this phenomenon is presented in Eq. (1) [28].

Table 6.3 Functional group tolerance in the titanium-catalyzed AE reaction.

Compatible FG			Incompatible FG
acetal, ketal	carboxylic ester	nitro	amine (except pyridines)
acetylene	epoxide	olefin	carboxylic acid
alcohol	ether	silyl	mercaptan
aldehyde	hydrazide	sulfone	phenol
amide	ketone	sulfoxide	phosphine
azide	nitrile	urea, urethane	

Ti(OPr^i)$_4$, (+)-DET, TBHP

-25°C

OH O

O O OH

73% (95% ee)

(1)

When it comes to reactions employing aqueous hydrogen peroxide as the terminal oxidant together with titanium catalysts, the only properly working systems are based on heterogeneous catalysts. The development of the titanium(IV) silicate catalyst (TS-1) by chemists at Enichem represented a breakthrough for reactions performed with hydrogen peroxide [29]. This hydrophobic molecular sieve demonstrated excellent properties (i. e., high catalytic activity and selectivity) for the epoxidation of small linear olefins in methanol. The substrates are adsorbed into the micropores of the TS-1 catalyst, which efficiently prevents inhibition by water [8]. After the epoxidation reaction, the TS-1 catalyst can easily be separated and recycled. To broaden the scope of this epoxidation method and thereby allow for the oxidation of a wider range of substrates, several different titanium-containing silicate zeolites have been prepared. Its range has consequently been improved somewhat, although the best epoxidation results with titanium silicates as catalysts are obtained with smaller, unbranched substrates.

6.3.2
Vanadium-catalyzed Epoxidations

The phenomenon that early transition metals in combination with alkyl hydroperoxides could participate in olefin epoxidation was discovered in the early 1970s [30, 31]. While *m*-CPBA was known to oxidize more reactive isolated olefins, it was discovered that allylic alcohols were oxidized to the corresponding epoxides at the same rate or even faster than a simple double bond when V^V or Mo^{VI} catalysts were employed in the reaction [Eq. (2)] [30].

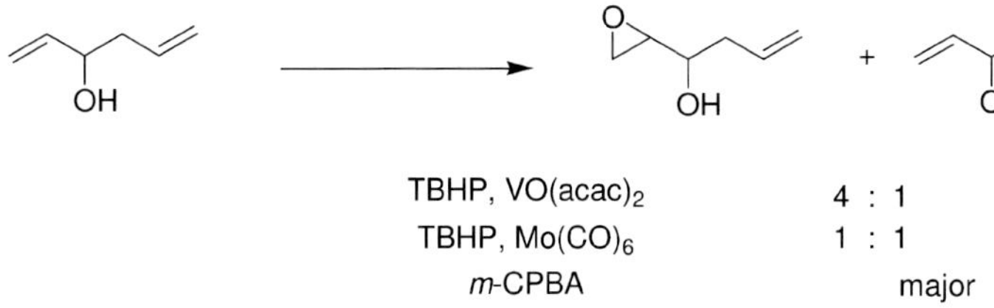

(2)

TBHP, $VO(acac)_2$ 4 : 1
TBHP, $Mo(CO)_6$ 1 : 1
m-CPBA major

The mechanism for such a process was explained in terms of a structure as depicted in Figure 6.5. The allylic alcohol and the alkyl hydroperoxide are incorporated into the vanadium coordination sphere and the oxygen transfer from the peroxide to the olefin takes place in an intramolecular fashion (as described above for titanium tartrate catalyst) [30, 32].

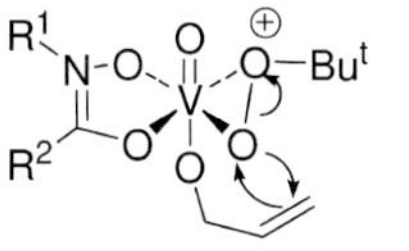

Figure 6.5 Proposed structure for the vanadium complex prior to the oxygen transfer from peroxide to the allylic olefin.

Vanadium-catalyzed asymmetric epoxidation was reported as early as 1977. The study performed showed the hydroxamic acid **3** to be the most promising ligand in the reaction and further revealed the importance of using a specific ratio between the vanadium catalyst precursor and the chiral ligand [Eq. (3)] [32]. As shown in Eq. (3), use of a fivefold excess of ligand **3** gave considerably better enantioselectivity than when only two ligands per vanadium were employed. The yield, however, was much better in the latter case, which indicates a reaction scenario displaying a *ligand decelerating effect* (LDC). This is in direct contradiction to the titanium-catalyzed AE discussed above, and it highlights one of the difficulties that may be encountered during the development of methods for asymmetric reactions.

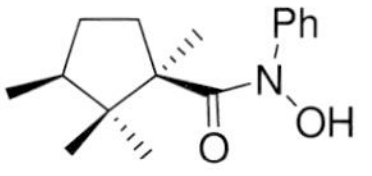

3

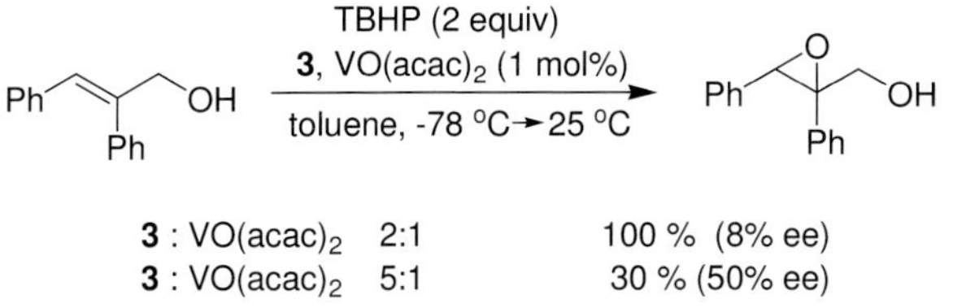

TBHP (2 equiv)
3, $VO(acac)_2$ (1 mol%)
toluene, -78 °C→25 °C

(3)

3 : $VO(acac)_2$ 2:1 100 % (8% ee)
3 : $VO(acac)_2$ 5:1 30 % (50% ee)

While the titanium-catalyzed epoxidation reaction was flourishing, the pioneer version of the reaction remained undeveloped and it was not until 1999 that new

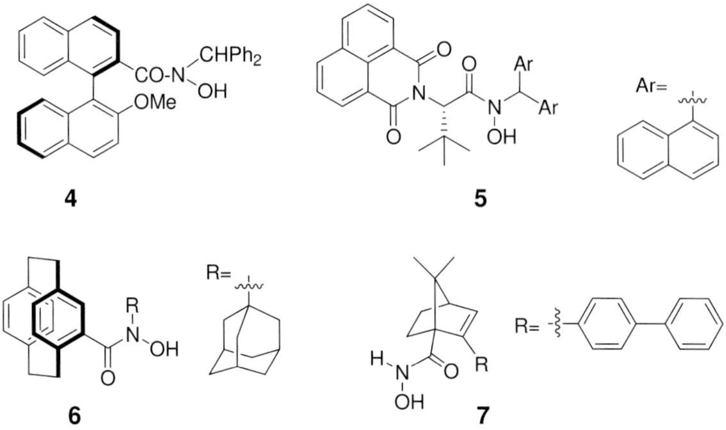

Figure 6.6 Chiral ligands employed in the vanadium-catalyzed AE of allylic alcohols.

Table 6.4 Comparison of the performances showed by the vanadium complexes with ligands **4–7** in the AE reaction generating the epoxides **A** and **B**, respectively.

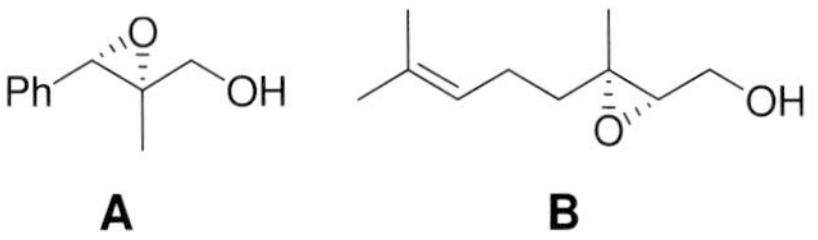

$VO(O^iPr)_3$ (mol %)	Ligand (mol %)	Oxidant	Reaction time	Temp. (°C)	A Yield (%)	*Ee* (%)	B Yield (%)	*Ee* (%)	Ref.
5	**4** (7.5)	TrOOH[a]	2–3 d	−20	96	91	80	66	34
1	**5** (1.5)	TBHP	3–6 h	0	97	95	95	81	36
5	**6** (7.5)	TBHP	3 d	−20	85	71	89	45	37
5	**7** (7.5)	TBHP	1 d	0	86	55	83	46	38

[a] Trityl hydroperoxide.

discoveries in vanadium-catalyzed AE were revealed [33]. In the epoxidation of various allylic alcohols, the hydroxamic acid of structure **4** (Figure 6.6) was presented as the best ligand for vanadium, allowing enantiomeric excesses of up to 94% [33, 34].

Great interest in vanadium-catalyzed AE has been shown in recent years, and many promising results continue to appear [35]. Some of the most successful complexes to date are based on the ligands shown in Figure 6.6. A comparison of their catalytic activities is presented in Table 6.4.

6.4 Chromium-, Molybdenum-, and Tungsten-catalyzed Epoxidations

Group VI metals display quite diverse characteristics when used as reagents or catalysts for the formation of epoxides. Chromium(v) complexes generated from chromyl nitrate [$CrO_2(NO_3)_2$] are known to epoxidize olefins, although these reactions are strictly stoichiometric with respect to the metal reagent [4, 39]. The use of terminal oxidants such as hydrogen peroxide or alkyl hydroperoxides results only in decomposition of the oxidant. However, chromium salen complexes (exemplified by **8**) can be used in catalytic amounts, although the preferred oxidant in these reactions is iodosyl benzene [40, 41]. This system, with use of complex **9**, has recently generated some interest in asymmetric epoxidations [42, 43]. Unlike the corresponding manganese-based epoxidation systems (*vide infra*), chromium oxo complexes such as **9** are isolable and have therefore been studied in great detail in both stoichiometric and catalytic epoxidations. These chiral chromium(v) catalysts give good enantioselectivity in the epoxidation of *trans*-disubstituted olefins, as is demonstrated by the epoxidation of *trans*-β-methylstyrene by use of a stoichiometric quantity of **8** ($X^- = NO_3$-, L = Ph_3PO) and 1.2 equivalents of iodosyl benzene, in which the corresponding epoxide was obtained in 45% yield and in 92% *ee*. Changing the terminal oxidant to hydrogen peroxide resulted in a poor yield (2%) and no *ee*. Under catalytic conditions (10 mol% of **9**) in the presence of PhIO, the epoxide was formed in higher yield (71%) but with a slightly lower enantiomeric excess (82%) [42].

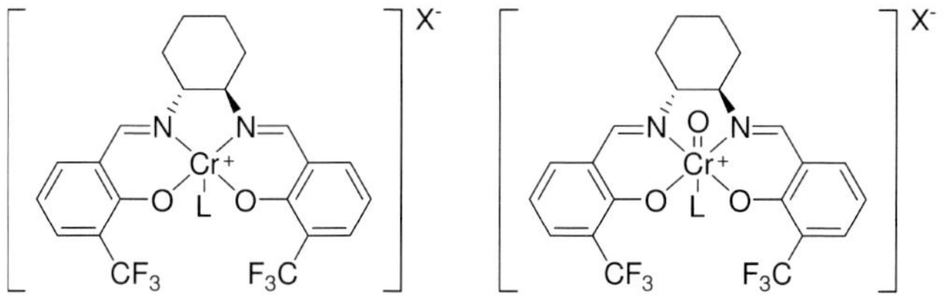

8 **9**

Epoxidation systems based on molybdenum and tungsten catalysts have been extensively studied for more than 40 years. The typical catalysts – Mo^{VI}-oxo or W^{VI}-oxo species – do, however, behave rather differently, depending on whether anionic or neutral complexes are employed. Whereas the anionic catalysts, especially the use of tungstates under phase-transfer conditions, are able to activate aqueous hydrogen peroxide efficiently for the formation of epoxides, neutral molybdenum or tungsten complexes do react with hydrogen peroxide, but better selectivities are often achieved with organic hydroperoxides (e.g., TBHP) as terminal oxidants [44, 45].

6.4.1
Homogeneous Systems Using Molybdenum and Tungsten Catalysts and Alkyl Hydroperoxides or Hydrogen Peroxide as the Terminal Oxidant

Molybdenum hexacarbonyl [$Mo(CO)_6$] has been used in combination with TBHP for the epoxidation of terminal olefins [44]. Good yields and selectivity for the epoxide products were obtained when reactions were performed under anhydrous conditions in hydrocarbon solvents such as benzene. The inexpensive and considerably less toxic $MoO_2(acac)_2$ is a robust alternative to $Mo(CO)_6$ [2]. A number of different substrates ranging from simple α-olefins to more complex terpenes have been oxidized with very low catalytic loadings of this particular molybdenum complex (Scheme 6.2). The epoxidations were carried out with use of dry TBHP (~70%) in toluene.

The use of molybdenum catalysts in combination with hydrogen peroxide is not so common. Nevertheless, there are a number of systems in which molybdates have been employed for the activation of hydrogen peroxide. A catalytic amount of sodium molybdate in combination with monodentate ligands (e.g., hexaalkyl phosphorus triamides or pyridine-*N*-oxides), and sulfuric acid allowed the epoxidation of simple linear or cyclic olefins [46]. The selectivity obtained by this method was quite low, and significant amounts of diol were formed, even though highly concentrated hydrogen peroxide (>70%) was employed.

More recently, Sundermeyer and coworkers reported on the use of long-chain trialkylamine oxides, trialkylphosphane oxides, or trialkylarsane oxides as monodentate ligands for neutral molybdenum peroxo complexes [47]. These compounds were employed as catalysts for the epoxidation of oct-1-ene and cyclooctene with aqueous hydrogen peroxide (30%) under biphasic conditions ($CHCl_3$). The epoxide products were obtained in high yields and with good selectivity. The high selectivity achieved by this method was ascribed to high solubility of the product in the organic phase, thus protecting the epoxide from deleterious hydrolysis. This procedure has not been employed for the formation of hydrolytically sensitive epoxides, so the generality of the method can be questioned.

With regard to the use of tungsten-based systems, Payne and Williams reported on the selective epoxidation of maleic, fumaric, and crotonic acids with a catalytic amount of sodium tungstate (2 mol%) in combination with aqueous hydrogen

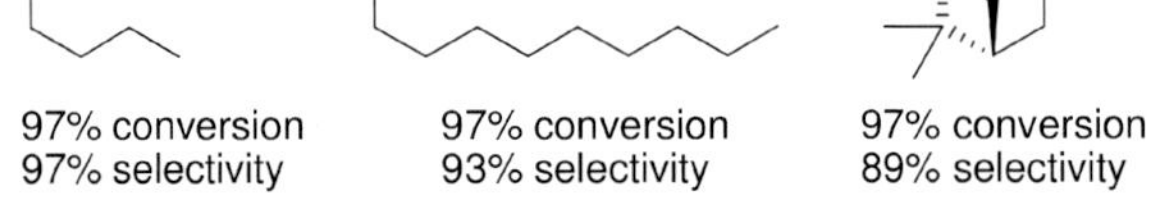

R3
R1 R2
TBHP (~ 70% in toluene)
$MoO_2(acac)_2$ (0.01 mol%)
100 - 125 °C, 1 - 4h
R3 O
R1 R2

97% conversion
97% selectivity

97% conversion
93% selectivity

97% conversion
89% selectivity

Scheme 6.2

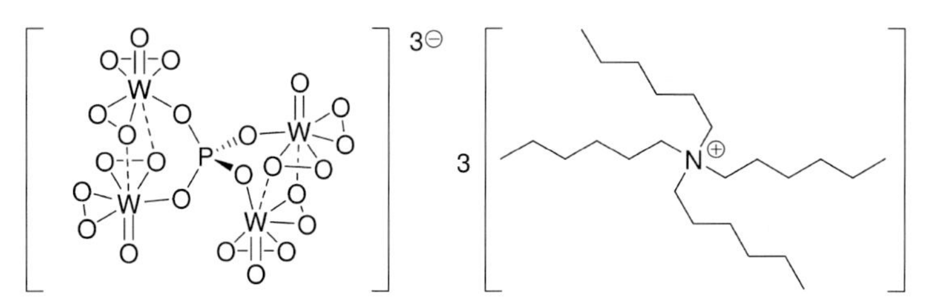

Figure 6.7 The Venturello $(n\text{-hexyl}_4N)_3[PO_4(W(O)(O_2)_2)_4]$ catalyst.

peroxide as the terminal oxidant back in 1959 [48]. The key to success was careful control of the pH (4–5.5) in the reaction media. These electron-deficient substrates were notoriously difficult to oxidize selectively by the standard techniques (peroxy acid reagents) available at the time. Previous attempts to use sodium tungstate and hydrogen peroxide had resulted in the isolation of the corresponding diols, due to rapid hydrolysis of the intermediate epoxides. Significant improvements in this catalytic system were introduced by Venturello and coworkers [49, 50], who found that the addition of phosphoric acid and the introduction of quaternary ammonium salts as PTC reagents considerably increased the scope of the reaction. The active tungstate catalysts are often generated *in situ*, although catalytically active peroxo complexes such as $(n\text{-hexyl}_4N)_3[PO_4(W(O)(O_2)_2)_4]$ have been isolated and characterized (Figure 6.7) [51].

In recent work, Noyori and coworkers found conditions for the selective epoxidation of aliphatic α-olefins either in toluene or in a completely solvent-free reaction setup [52, 53, 54]. One of the disadvantages with the previous systems was the use of chlorinated solvents. The conditions established by Noyori, however, provided an overall "greener" epoxidation process, since the reactions were efficiently performed in nonchlorinated solvents. In this reaction, sodium tungstate (2 mol%), (aminomethyl)phosphonic acid, and methyltri-*n*-octylammonium bisulfate (1 mol% of each) were employed as catalysts for epoxidation, with aqueous hydrogen peroxide (30%) as the terminal oxidant. The epoxidation of various terminal olefins under these conditions (90 °C, no solvent added) gave high yields for a number of substrates (Table 6.5). The workup procedure was exceptionally simple, since the product epoxides could be distilled directly from the reaction mixture. The use of appropriate additives turned out to be crucial for successful outcomes of these epoxide-forming reactions, significantly lower conversions being obtained when the (aminomethyl)phosphonic acid was replaced by other phosphonic acids or simply by phosphoric acid. The nature of the phase-transfer reagent was further established as an important parameter. Use of ammonium bisulfate (HSO_4^-) was superior to that of the corresponding chloride or hydroxide salts. The size, and hence the lipophilicity, of the ammonium ion was important, since tetra-*n*-butyl- or tetra-*n*-hexylammonium bisulfate were inferior to phase-transfer agents containing larger alkyl groups. The epoxidation system was later extended to encompass

Table 6.5 Epoxidation of terminal olefins by the Noyori system.

R–CH=CH$_2$ → (1.5 equiv. H_2O_2 (30% aq); $Na_2WO_4 \times 2H_2O$ (2 mol%); $[CH_3(n\text{-}C_8H_{17})_3N]HSO_4$ (1 mol%); $NH_2CH_2PO_3H$ (1 mol%)) → R-epoxide

Entry	Olefin	Time (h)	Conversion (%)	Yield (%)
1	oct-1-ene	2	89	86
2	dec-1-ene	2	94	93
3[a]	dec-1-ene	4	99	99
4[a]	allyl octyl ether	2	81	64
5[a]	styrene	3	70	2

[a] 20 mmol olefin in 4 mL toluene

other substrates, such as simple olefins with different substitution patterns, and to olefins containing various functionalities (alcohols, ethers, ketones, and esters).

A major limitation of this method is the low pH at which the reactions are performed, which resulted in substantially lower yields in reactions with substrate progenitors of acid-sensitive epoxides, in which competing ring-opening processes effectively reduced the usefulness of the method. As an example, the oxidation of styrene had proceeded with 70% conversion after 3 h at 70 °C, but the observed yield of styrene oxide was only 2% (Table 6.5, Entry 5).

The epoxidation method developed by Noyori was subsequently applied to the direct formation of dicarboxylic acids from olefins [55]. Cyclohexene was oxidized to adipic acid in 93% yield with the tungstate/ammonium bisulfate system and 4 equivalents of hydrogen peroxide. The selectivity problem associated with the Noyori method was circumvented to a certain degree by the improvements introduced by Jacobs and coworkers [56]. Additional amounts of (aminomethyl)phosphonic acid and Na_2WO_4 were introduced into the standard catalytic mixture, and the pH of the reaction media was adjusted to 4.2–5 with aqueous NaOH. These changes allowed for the formation of epoxides from α-pinene, 1-phenyl-1-cyclohexene, and indene, with high levels of conversion and good selectivity (Scheme 6.3).

Another highly efficient tungsten-based system for the epoxidation of olefins was recently introduced by Mizuno and coworkers [57]. The tetrabutylammonium salt of a Keggin-type silicodecatungstate $[\gamma\text{-SiW}_{10}O_{34}(H_2O)_2]^{4-}$ (Scheme 6.4) was found to catalyze the epoxidation of various olefin substrates with aqueous hydrogen peroxide as the terminal oxidant. The characteristics of this system are very high epoxide selectivity (99%) and excellent efficiency in the use of the terminal oxidant (99%). Terminal and di-and trisubstituted olefins were all epoxidized in high yields within reasonably short reaction times with use of 0.16 mol% catalyst (1.6 mol% in tungsten, Scheme 6.4). The X-ray structure of the catalyst precursor

R^3, R^1, R^2 → R^3, O, R^1, R^2

1.5 equiv. H_2O_2 (35% aq)

$PW_4O_{24}[CH_3(n\text{-}C_8H_{17})_3N]$ (2-2.6 mol%)

$Na_2WO_4 \times 2H_2O$ (0-7 mol%)

$NH_2CH_2PO_3H$ (2-7 mol%)

pH = 4.2-5 (NaOH addition)

60 °C, 2-4h

97% conversion
83% selectivity

96% conversion
68% selectivity

83% conversion
92% selectivity

Scheme 6.3

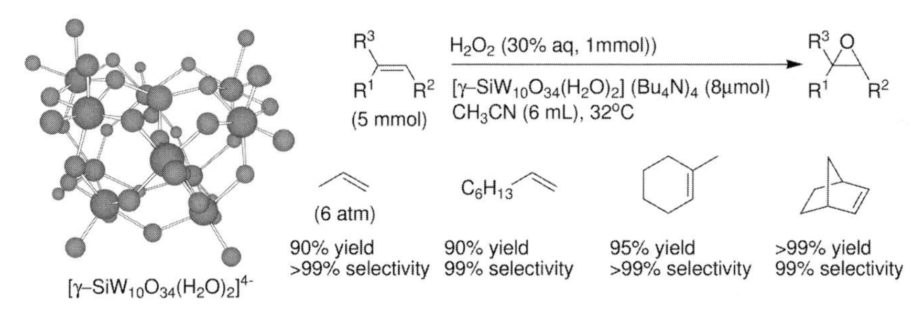

Scheme 6.4

revealed ten tungsten atoms connected to a central SiO_4 unit. *In situ* infrared spectroscopy of the reaction mixture during the epoxidation reaction indicated high structural stability of the catalyst. Furthermore, it was demonstrated that the catalyst could be recovered and recycled up to five times without loss of activity or selectivity (epoxidation of cyclooctene). Interestingly, the often encountered problem with hydrogen peroxide decomposition was negligible with this catalyst. The efficient use of hydrogen peroxide (99%), combined with the high selectivity and productivity in propylene epoxidation, may represent potential for industrial applications.

6.4.2
Heterogeneous Catalysts

One problem associated with the peroxotungstate-catalyzed epoxidation system described above is the separation of the catalyst after the completed reaction. To overcome this obstacle, efforts to prepare heterogeneous tungstate catalysts have been conducted. De Vos and coworkers employed W catalysts derived from sodium tungstate and layered double hydroxides (LDH – coprecipitated $MgCl_2$, $AlCl_3$, and NaOH) for the epoxidation of simple olefins and allyl alcohols with aqueous hydrogen peroxide [58]. They found that, depending on the nature of the catalyst (both hydrophilic and hydrophobic catalysts were used), different reactivity and selectivity were obtained for nonpolar and polar olefins. The hydrophilic LDH-WO_4 catalyst was particularly effective for the epoxidation of allyl and homoallyl alcohols, whereas the hydrophobic catalyst (containing *p*-toluenesulfonate) showed better reactivity with unfunctionalized substrates.

Gelbard and coworkers have reported on the immobilization of tungsten catalysts with polymer-supported phosphine oxide, phosphonamide, phosphoramide, and phosphotriamide ligands [59]. Use of these heterogeneous catalysts together with hydrogen peroxide for the epoxidation of cyclohexene resulted in moderate to good conversion of the substrate, although low epoxide selectivity was observed in most cases. A significantly more selective heterogeneous catalyst was obtained by Jacobs and coworkers upon treatment of the macroreticular ion-exchange resin Amberlite IRA-900 with an ammonium salt of the Venturello anion $\{PO_4[WO(O_2)_2]_4\}^{3-}$ [56, 60]. The formed catalyst was used for the epoxidation of a number of terpenes, and high yields and good selectivity of the corresponding epoxides were achieved.

In a different strategy, siliceous mesoporous MCM-41-based catalysts were prepared [60]. Quaternary ammonium salts and alkyl phosphoramides were grafted on MCM-41 and the material obtained was treated with tungstic acid for the preparation of heterogeneous tungstate catalysts. The catalysts were employed in the epoxidation of simple cyclic olefins with aqueous hydrogen peroxide (35%) as terminal oxidant, but levels of conversion and selectivity for the formed epoxide were rather low. In the case of cyclohexene, the selectivity could be improved by the addition of pyridine. The low level of tungsten leaching (< 2%) seen with these catalysts is certainly advantageous.

A particularly interesting system for the epoxidation of propylene to propylene oxide, working under pseudo-heterogeneous conditions, was reported by Zuwei and coworkers [61]. The catalyst, which was based on the Venturello anion combined with long-chained alkylpyridinium cations, showed unique solubility properties. In the presence of hydrogen peroxide the catalyst was fully soluble in the solvent, a 4:3 mixture of toluene and tributyl phosphate, but when no more oxidant was left, the tungsten catalyst precipitated and could simply be removed from the

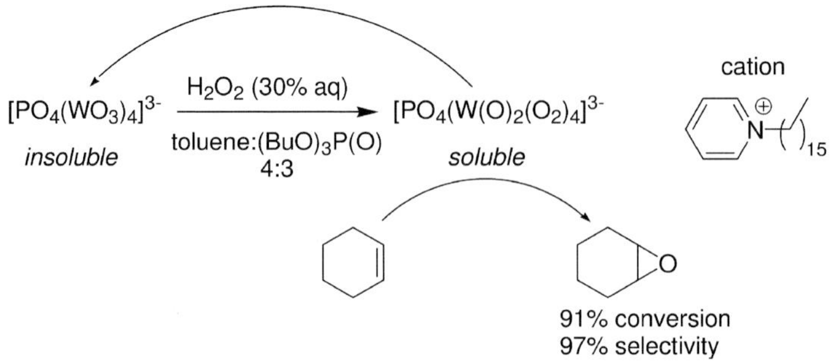

$$[PO_4(WO_3)_4]^{3-} \xrightarrow[\text{toluene:(BuO)}_3\text{P(O) 4:3}]{H_2O_2\ (30\%\ aq)} [PO_4(W(O)_2(O_2)_4]^{3-}$$

insoluble *soluble*

cation

91% conversion
97% selectivity

Scheme 6.5

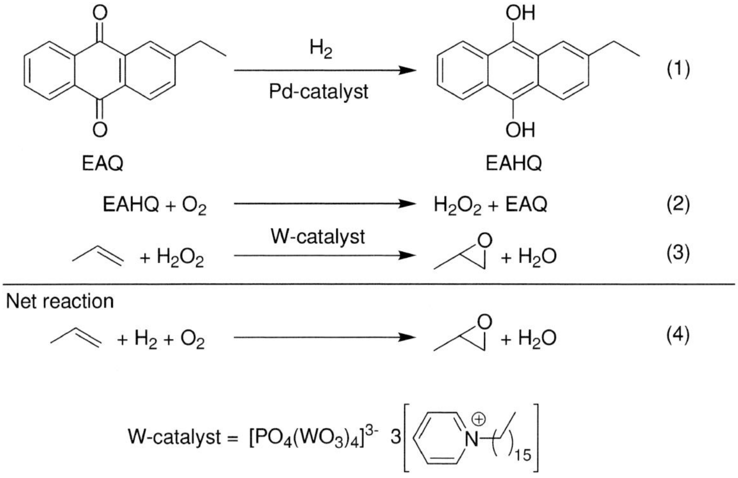

EAQ $\xrightarrow[\text{Pd-catalyst}]{H_2}$ EAHQ (1)

EAHQ + O_2 ⟶ H_2O_2 + EAQ (2)

propylene + H_2O_2 $\xrightarrow{\text{W-catalyst}}$ propylene oxide + H_2O (3)

Net reaction

propylene + H_2 + O_2 ⟶ propylene oxide + H_2O (4)

W-catalyst = $[PO_4(WO_3)_4]^{3-}$ 3[N-hexadecylpyridinium^{+}]

Scheme 6.6

reaction mixture (Scheme 6.5). Furthermore, this epoxidation system could be combined with the 2-ethylanthraquinone (EAQ)/2-ethylanthrahydroquinone (EAHQ) process for hydrogen peroxide formation (Scheme 6.6), and good conversion and selectivity for propylene oxide were obtained in three consecutive cycles. The catalyst was recovered by centrifugation after each cycle and was used directly in the next reaction.

6.5 Manganese-catalyzed Epoxidations

Historically, the interest of using manganese complexes as catalysts for the epoxidation of olefins comes from biologically relevant oxidative manganese porphyrins. The terminal oxidants compatible with manganese porphyrins were initially restricted to iodosyl benzene, sodium hypochlorite, alkyl peroxides and hydroperoxides, *N*-oxides, $KHSO_5$, and oxaziridines. Molecular oxygen can also be used in the presence of an electron source. The use of hydrogen peroxide often results in oxidative decomposition of the catalyst, due to the potency of this oxidant.

6.5.1 Hydrogen Peroxide as Terminal Oxidant

The introduction of chlorinated porphyrins (**10**) allowed for hydrogen peroxide to be used as terminal oxidant [62]. These catalysts, discovered by Mansuy and coworkers, were demonstrated to resist decomposition, and efficient epoxidations of olefins were achieved when they were used together with imidazole or imidazolium carboxylates as additives, (Table 6.6, Entries 1 and 2).

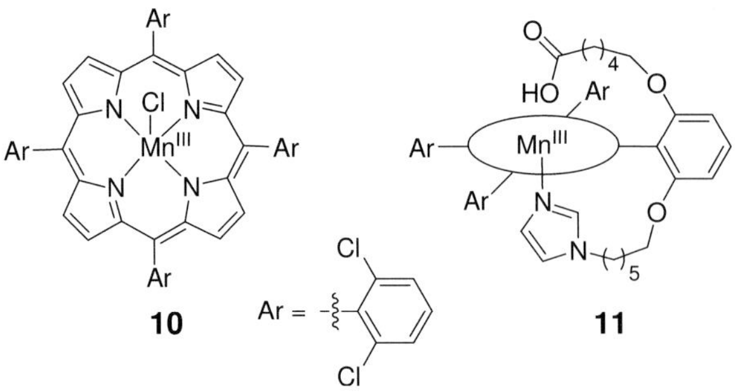

Table 6.6 Manganese/porphyrin-catalyzed epoxidation of *cis*-cyclooctene with aqueous H_2O_2 (30%).

Entry	Catalyst	Additive	T (°C)	Time (min)	Yield (%)
1	**10** 2.5 mol%	imidazole (0.6 equiv.)	20	45	90
2	**10** 0.5 mol%	*N*-hexyl-imidazole (0.5 mol%) benzoic acid (0.5 mol%)	0	15	100
3	**11** 0.1 mol%	–	0	3	100

The observation that addition of imidazoles and carboxylic acids significantly improved the epoxidation reaction resulted in the development of Mn-porphyrin complexes containing these groups covalently linked to the porphyrin platform as attached pendant arms (**11**) [63]. When these catalysts were employed in the epoxidation of simple olefins with hydrogen peroxide, enhanced oxidation rates were obtained in combination with perfect product selectivity (Table 6.6, Entry 3). In contrast with epoxidations catalyzed by other metals, the Mn-porphyrin system yields products with scrambled stereochemistry: the epoxidation of *cis*-stilbene with Mn(TPP)Cl (TPP = tetraphenylporphyrin) and iodosylbenzene, for example, generated *cis*- and *trans*-stilbene oxide in a ratio of 35:65. The low stereospecificity was improved by use of heterocyclic additives such as pyridines or imidazoles. The epoxidation system, with hydrogen peroxide as terminal oxidant, was reported to be stereospecific for *cis*-olefins, whereas *trans*-olefins are poor substrates with these catalysts.

Apart from the porphyrin-based catalysts, manganese complexes of *N*-alkylated 1,4,7-triazacyclononane (TMTACN, **12**) have also been found to catalyze the epoxidation of olefins efficiently in the presence of acid additives (typically oxalic, ascorbic, or squaric acid) and hydrogen peroxide [64, 65, 66]. Reactions performed without acid present required huge excess (ca. 100 equivalents) of hydrogen peroxide for efficient epoxidation. The rather difficult preparation of the TACN ligands has given rise to increased activity with the goal of finding alternative ligands with similar coordinating properties. In this respect, pyridylamine ligands represent an interesting alternative. Feringa and coworkers found that the dinuclear manganese complex **14**, prepared from the tetra-pyridyl ligand **13**, was an efficient catalyst for the epoxidation of simple olefins [67]. Only 0.1 mol% of catalyst **14** was required for high levels of conversion (87%) of cyclohexene into its corresponding epoxide. An excess of aqueous hydrogen peroxide (8 equivalents) was used, due to the usual problem of peroxide decomposition in the presence of manganese complexes.

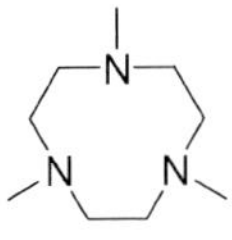

12 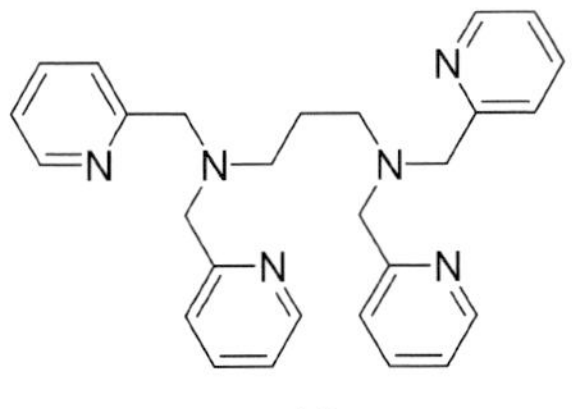 **13** 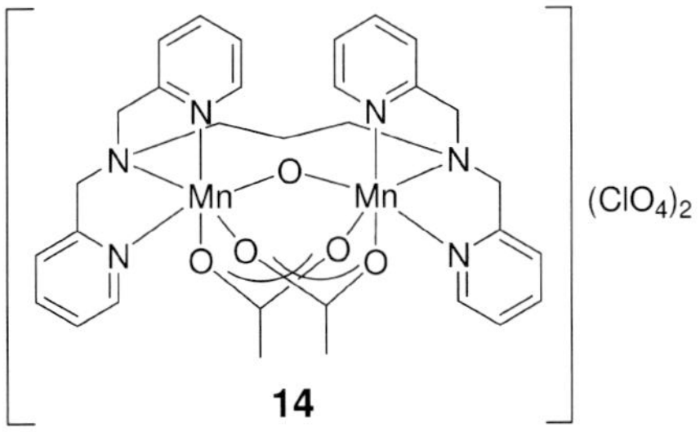 **14** $(ClO_4)_2$

In a recent screening of different metal salts, Lane and Burgess found that simple manganese(II) and (III) salts catalyzed the formation of epoxides in DMF or *t*-BuOH in the presence of aqueous hydrogen peroxide (Scheme 6.7) [68]. It was further established that the addition of bicarbonate was of importance for the epoxidation reaction.

It was established by spectroscopic methods that peroxymonocarbonate (HCO_4^-) is formed upon the mixing of hydrogen peroxide and bicarbonate [69]. In the absence of the metal catalyst, the oxidizing power of the peroxymonocarbonate formed *in situ* with respect to its reaction with olefins was demonstrated to be moderate. In the initial reaction setup, this $MnSO_4$-catalyzed epoxidation required a considerable excess of hydrogen peroxide (10 equiv.) for efficient epoxide formation. With regard to the scope of the reaction, it was found that electron-rich substrates such as di-, tri-, and tetrasubstituted olefins gave moderate to good yields of their corresponding epoxides. Styrene and styrene derivatives were also demonstrated to react smoothly, whereas monoalkyl-substituted substrates were completely unreactive under these conditions. The basic reaction medium used was highly beneficial for product protection, with acid-sensitive epoxides being formed in good yields. Different additives were screened in order to improve this epoxidation system, and it was found that the addition of sodium acetate was beneficial for reactions performed in *t*-BuOH. Similarly, the addition of salicylic acid improved the outcome of the reaction performed in DMF. The use of these additives efficiently reduced the number of hydrogen peroxide equivalents necessary for pro-

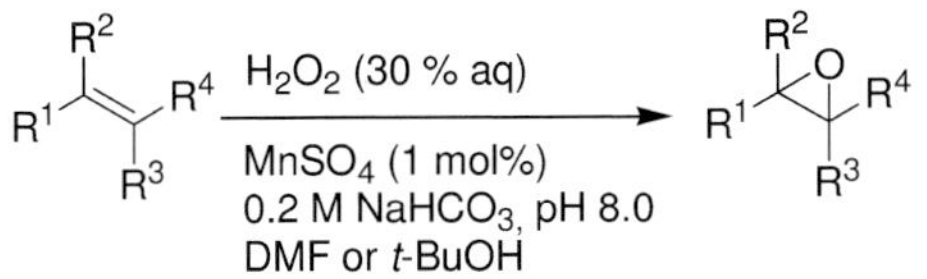

R^1, R^2, R^3, R^4 — H_2O_2 (30 % aq); $MnSO_4$ (1 mol%); 0.2 M $NaHCO_3$, pH 8.0; DMF or *t*-BuOH

Scheme 6.7

Table 6.7 Manganese sulfate-catalyzed epoxidation of olefins using aqueous H_2O_2 (30%).[a]

Olefin	No additive		Salicylic acid (4 mol%)	
	equiv. H_2O_2	yield	equiv. H_2O_2	yield %
	10	99	2.8	96
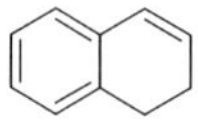	10	87	5	97[b]
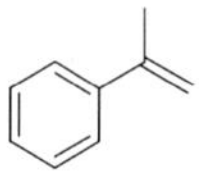	10	96	5	95[b]
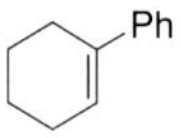 Ph	10	95	5	95[b]
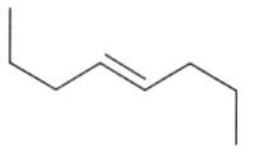	25	60	25	75
	25	54	25	75
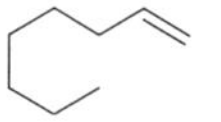	25	0	25	0

[a] Conditions as in Scheme 6.7. [b] Isolated yields.

ductive epoxidation (Table 6.7). The reaction is not completely stereospecific, since the epoxidation of *cis*-oct-4-ene yielded a *cis/trans* mixture of the product (1:1.45 without additive and 1:1.1 in the presence of 4 mol% salicylic acid).

The use of the ionic liquid [bmim][BF_4] further improved the Burgess epoxidation system [70]. Chan and coworkers found that replacement of sodium bicarbonate for tetramethylammonium bicarbonate and performing the reaction in [bmim][BF_4] allowed for efficient epoxidation of a number of different olefins, including substrates affording acid-labile epoxides (such as dihydronaphthalene (99% yield) and 1-phenylcyclohexene (80% yield)).

6.5.2 Manganese-catalyzed Asymmetric Epoxidations

A breakthrough in the area of asymmetric epoxidation came at the beginning of the 1990s, when the groups of Jacobsen and Katsuki more or less simultaneously discovered that chiral Mn-salen complexes (**15**) catalyzed the enantioselective formation of epoxides [71, 72, 73]. The discovery that simple achiral Mn-salen complexes could be used as catalysts for olefin epoxidation had already been made about five years earlier, and the typical terminal oxidants used with these catalysts closely resemble those of the porphyrin systems [74]. Unlike the titanium-catalyzed asymmetric epoxidation discovered by Sharpless, the Mn-salen system does not require precoordination of the olefin substrate to the catalyst, so unfunctionalized olefins could be oxidized efficiently and selectively. The enantioselectivity was shown to be highly sensitive towards the substitution pattern of the olefin substrate. Excellent selectivity (> 90% *ee*) was obtained for aryl- or alkynyl-substituted terminal-, *cis*-disubstituted-, and trisubstituted olefins, especially with the second-generation catalysts (here represented by **16** [75]), whereas *trans*-disubstituted olefins were epoxidized with low rates and low *ees* (<40%) (Scheme 6.8 and Table 6.8). In this respect, the analogous chromium-salen system described above serves as good complement to its manganese counterpart, since good enantioselectivity was obtained in the epoxidation of the "difficult" *trans*-disubstituted olefins. Nevertheless, the use of catalyst **15** in asymmetric epoxidations is now widespread, and one example of its efficiency in generating chiral epoxides can be found in the industrial production of the HIV-protease inhibitor Indinavir [76]. One of the key steps in this process is the enantioselective epoxidation of indene to indene oxide (Scheme 6.9).

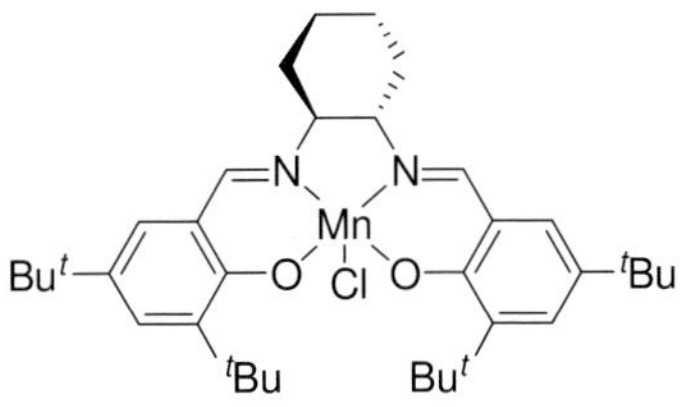

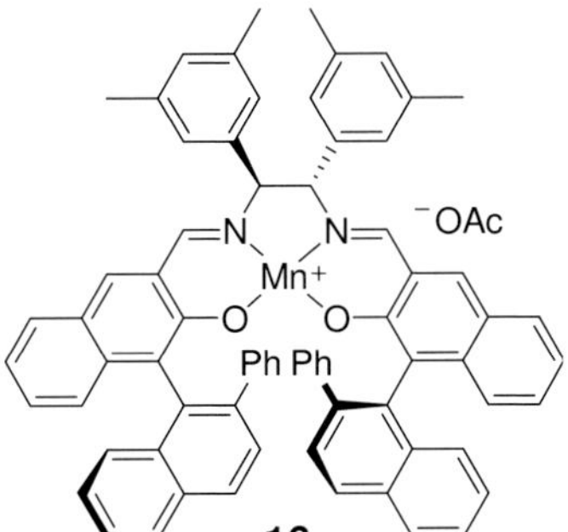

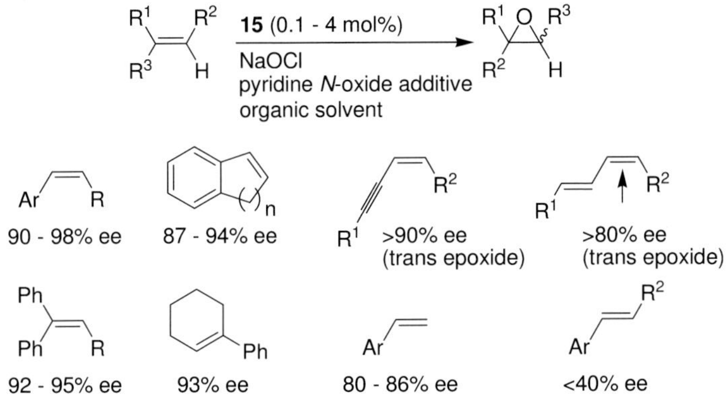

Scheme 6.8

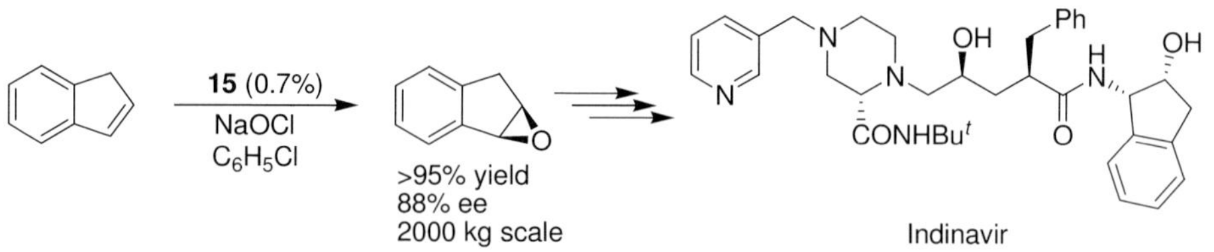

15 (0.7%)
NaOCl
C_6H_5Cl

>95% yield
88% ee
2000 kg scale

Indinavir

Scheme 6.9

The mechanism of the enantioselective epoxidation of olefins with Mn-salen complexes as catalysts has been the object of intense discussion. The most widely accepted suggestion involves a two-step reaction, in which the initial step is a transfer of oxygen from the terminal oxidant to the Mn(III)-salen catalyst. The existence of a Mn(V)-oxo intermediate was first postulated by Kochi and coworkers [74], and later verified by mass spectrometry techniques by Feichtinger and Plattner [77]. In the second step, oxygen is transferred from the transition metal center to the olefin, generating the epoxide. The details regarding this second step are keenly debated, and two different catalytic systems – incorporating either neutral or cationic manganese intermediates responsible for the oxygen transfer – have been proposed [78]. The enantioselection was initially believed to origin from an approach of the olefin according to Figure 6.8a. According to this model, the only trajectory accessible for the incoming olefin towards the oxo group is the one that passes over the chiral diimine backbone of the Mn complex. The other possible routes were believed to be sterically hindered by the bulky *tert*-butyl groups in the 3- and 5-positions of the arenes [73]. However, recent density functional theory calculations in combination with molecular mechanics methods suggest that the origin of enantioselectivity is an olefin approach from the side of the aromatic ring, as depicted in Figure 6.8b [79]. According to this study, the specific factors governing high selectivity in the reaction can be summarized in terms of the stereocenters positioned close to the diimine bridge inducing a folding of the ligand and thereby creating a chiral pocket. The low-energy path into this chiral pocket is a side-on approach over one of the aromatic rings. Attack of the olefin from the front

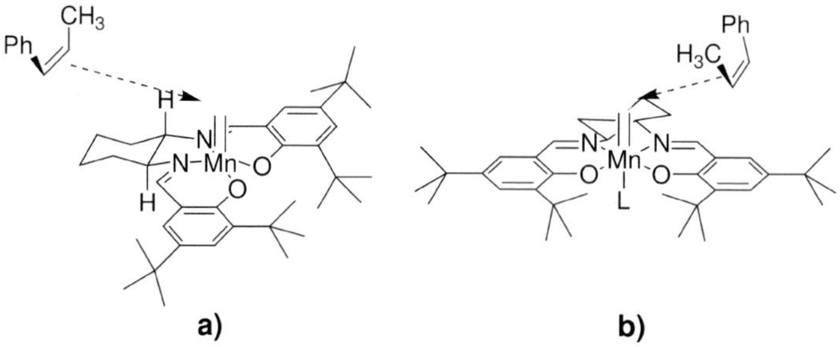

a)

b)

Figure 6.8 Rationale for enantioselection in Mn-salen epoxidation. a) Initially proposed model (the axial ligand L is omitted for clarity). b) Recently proposed model based on calculation.

is efficiently hindered by bulky groups in the 3,3'-positions of the ligand, and an approach according to Figure 6.8a was calculated to be disfavored by ca. 4 kcal · mol^{-1}.

The typical oxidant used in Mn-salen asymmetric epoxidations is NaOCl, although recent work by the groups of Berkessel and Katsuki have opened the way for hydrogen peroxide to be employed [80, 81]. Berkessel found that imidazole additives were crucial for the formation of the active oxo-manganese intermediates, and a Mn catalyst (**17**) based on a salen ligand incorporating a pendant imidazole was used for asymmetric epoxidation with aqueous H_2O_2 (Table 6.8). Yields and enantioselectivity, however, did not reach the levels obtained when other oxidants were used. In Katsuki's work, *N*-methylimidazole was used as an additive in the epoxidation of a chromene derivative with use of the sterically hindered Mn-salen catalyst **19** (Scheme 6.10). At low substrate concentrations (0.1 M) the yield was only 17%, but use of higher concentrations gave substantially more product formation. Unfortunately, other substrates were not oxidized with the same yield and degree of enantioselectivity.

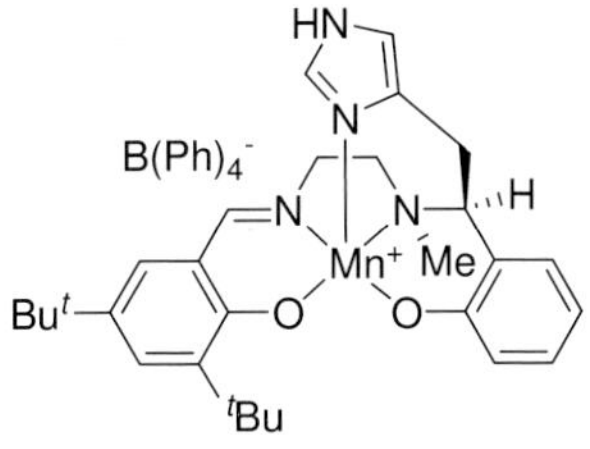

17 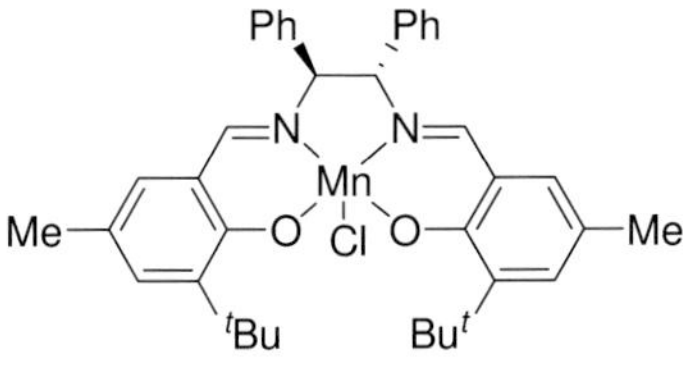 **18**

Pietikäinen reported on the use of ammonium acetate (20 mol%) as an additive together with the Jacobsen catalyst (**15**) or the manganese salen **18** for the epoxidation of a number of olefins with aqueous hydrogen peroxide as terminal oxidant [82]. In general, this method generated epoxides in 50–90% yields with enantioselectivities of up to 96%, although only a narrow range of substrates has been examined. A major problem with the use of aqueous hydrogen peroxide in the Mn-salen-catalyzed reactions is associated with Mn-catalyzed oxidant decomposition

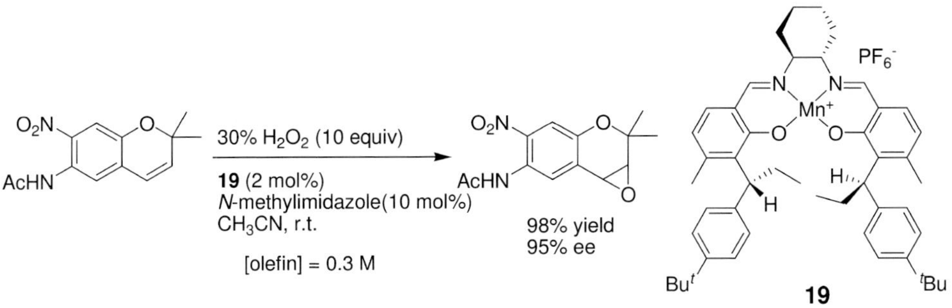

30% H_2O_2 (10 equiv)
19 (2 mol%)
N-methylimidazole(10 mol%)
CH_3CN, r.t.
[olefin] = 0.3 M

98% yield
95% ee

19

Scheme 6.10

Table 6.8 Enantioselective epoxidation of 1,2-dihydronaphthalene with Mn^{III} salen complexes.

Oxidant
Mn-salen

Entry	Mn-salen (amount)	Oxidant	Additive[a]	Solvent	Temp (°C)	Yield (%)	*Ee* (%)	Ref.
1	**15** (0.5 mol%)	NaOCl	4-PPNO (10 mol%)	chlorobenzene	0	67	86	86
2	**16** (2.5 mol%)	NaOCl (5 equiv..)	4-PPNO (25 mol%)	CH_2Cl_2	0	78	98	75
3	**17** (10 mol%)	NaOCl (4 equiv..)	–	CH_2Cl_2	r.t	76	56	80
4	**17** (10 mol%)	30% H_2O_2 (10 equiv.)	–	CH_2Cl_2 : H_2O (1:1)	0	73	45	80
5	**17** (10 mol%)	1% H_2O_2 (10 equiv.)	–	CH_2Cl_2 : H_2O (1:1)	0	72	64	80
6	**18** (5 mol%)	30% H_2O_2 (2.5–3 equiv.)	NMO (0.4 equiv..)	CH_2Cl_2 : MeOH (1:1)	2	74	69	82
6[b]	**18** (4.8–6.4 mol%)	UHP (1.5 equiv..)	MA (2 equiv..)	CH_2Cl_2 : DMF (23:2)	−18	70	73	83
7	**19** (2 mol%)	H_2O_2 (10 equiv.)	*N*-methyl-imidazole	CH_3CN	r.t	[c]	60	81

[a] 4-PPNO: 4-phenylpyridine *N*-oxide. NMO: *N*-methylmorpholine *N*-oxide. MA: maleic anhydride. [b] In this case the active oxidant is peroxymaleic acid formed *in situ* prior to the epoxidation reaction. [c] Not reported.

and catalyst deactivation due to the presence of water. Anhydrous hydrogen peroxide, in the forms either of the urea/H_2O_2 peroxide adduct (UHP) or of the triphenylphosphine oxide/H_2O_2 adduct, has been employed to circumvent this problem [83, 84, 85]. Although epoxide yield and enantioselectivity are in the range of what can be obtained with NaOCl, the catalyst loading is often significantly higher, and the removal of urea or Ph_3PO constitute an additional problem.

6.6 Rhenium-catalyzed Epoxidations

The use of rhenium-based systems for the epoxidation of olefins has increased considerably during the last ten years [87]. In 1989, Jørgensen stated, "the catalytic activity of rhenium in epoxidation reactions is low" [4]. The very same year, a few patents describing the use of porphyrin complexes containing rhenium as catalysts for the production of epoxides were released. The first major breakthrough, however, came in 1991, when Herrmann introduced methyltrioxorhenium (MTO, **20**) as a powerful catalyst for olefin epoxidation with hydrogen peroxide as the terminal oxidant [88]. This organometallic rhenium compound, formed in tiny amounts in the reaction between $(CH_3)_4ReO$ and air, was first detected by Beattie and Jones in 1979 [89].

$H_3C{-}Re(=O)_3$

20

A more reliable method for the preparation of MTO was introduced by Herrmann and coworkers in 1988 [90]. In this process, dirhenium heptoxide (Re_2O_7) was allowed to react with tetramethyltin, forming MTO and an equimolar amount of tin perrhenate. The maximum yield in this reaction was only 50% relative to the initial rhenium, and to improve this procedure a more efficient route towards MTO was developed. Treatment of Re_2O_7 with trifluoroacetic anhydride in acetonitrile quantitatively generated $CF_3CO_2ReO_3$, which upon further reaction with $MeSn(Bu)_3$ gave MTO in high yield (95%) [91, 92]. The main advantage of this route, apart from the efficient use of the rhenium source, is the replacement of the rather unpleasant tetramethyltin reagent with the more easily accessible alkyl-$Sn(Bu)_3$. This procedure is also compatible with the formation of other $RReO_3$ compounds, such as ethyltrioxorhenium (ETO). Additional routes towards MTO involve the treatment of perrhenates with trialkylsilyl chlorides to generate $ClReO_3$, followed by treatment with $(CH_3)_4Sn$ to form MTO in almost quantitative yield [93]. Today there is a whole range of organorhenium oxides available, and they can be regarded as one of the best examined classes of organometallic compounds [94, 95]. From a catalytic point of view, though, MTO is one of few organorhenium oxides that have been shown to act effectively as a catalyst in epoxidation reactions. With regard to the physical properties of organorhenium oxides, MTO shows the greatest thermal stability (decomposes >300 °C), apart from the catalytically inert 18e $(\eta^5\text{-}C_5Me_5)ReO_3$ complex. Furthermore, the high solubility of MTO in virtually any solvent from pentane to water makes this compound particularly attractive for catalytic applications.

The 14e compound MTO readily forms coordination complexes of the type MTO-L and MTO-L_2 with anionic and uncharged Lewis bases [96]. These yellow adducts are typically five- or six-coordinate complexes, and the Re-L system is highly labile. Apart from their fast hydrolysis in wet solvents, MTO-L adducts are much less thermally stable then MTO itself. The pyridine adduct of MTO, for instance, decomposes even at room temperature. In solution, methyltrioxorhenium displays high stability in acidic aqueous media, although its decomposition is strongly accelerated at increased hydroxide concentrations [97, 98]. Thus, under basic aqueous conditions MTO decomposes as shown in Equation (4).

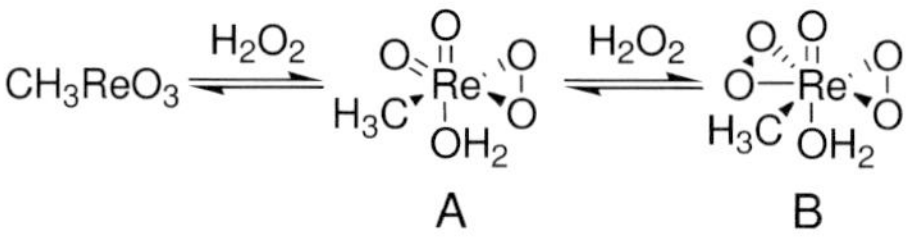

$$CH_3ReO_3 \underset{}{\overset{H_2O_2}{\rightleftharpoons}} \mathbf{A}\ [CH_3Re(O)(O)(\eta^2\text{-}O_2)(OH_2)] \underset{}{\overset{H_2O_2}{\rightleftharpoons}} \mathbf{B}\ [CH_3Re(O)(\eta^2\text{-}O_2)_2(OH_2)]$$

Scheme 6.11

$$CH_3ReO_3 + H_2O \longrightarrow CH_4 + ReO_4^- + H^+ \tag{4}$$

For catalytic epoxidation applications, perhaps the most important feature of MTO is its ability to activate hydrogen peroxide (5–85%) without decomposing the oxidant. The halflife of hydrogen peroxide is 20000 times higher in the presence of MTO than in that of $RuCl_3$, and 50 times higher than in the presence of MnO_2 [99]. Upon treatment of MTO with hydrogen peroxide a rapid equilibrium takes place as shown in Scheme 6.11.

MTO reacts with hydrogen peroxide to form a monoperoxo complex (**A**), which undergoes further reaction to yield a bis-peroxorhenium complex (**B**). The formation of the peroxo complexes is evident from the appearance of an intensive yellow color in the solution. Both peroxo complexes (**A** and **B**) have been detected through their methyl resonances by ^1H and ^{13}C NMR spectroscopy [97]. Furthermore, the structure of the bis-peroxo complex **B** has also been determined crystallographically [100]. In solution, **B** is the most abundant species in the equilibrium, suggesting that this is the more thermodynamically stable peroxo complex. The coordination of a water molecule to **B** has been established by NMR spectroscopy, but no such coordination has been observed for **A**, indicating either no coordinated water or high lability of such a ligand. The protons of the coordinated water molecule in **B** are highly acidic, and this has important implications for the epoxidation reaction (*vide infra*). With regard to catalytic activity, however, it has been demonstrated that both complexes are active as oxygen-transfer species. While decomposition of the MTO catalyst under basic conditions is often negligible, the presence of hydrogen peroxide completely changes the situation. The combination of basic media and H_2O_2 rapidly induces an irreversible decomposition of MTO as in Equation (5), and this deleterious side reaction is usually a great problem in the catalytic system [97].

$$CH_3ReO_3 \xrightarrow[\text{Pyridine}]{H_2O_2} HOReO_3 \cdot 2py + CH_3OH \tag{5}$$

In this oxidative degradation, MTO decomposes into catalytically inert perrhenate and methanol. The decomposition reaction is accelerated at higher pH, presumably through the reaction between the more potent nucleophile HO_2- and MTO. The decomposition of MTO under basic conditions is rather problematic, since the selectivity for epoxide formation certainly profits from the use of nonacidic conditions.

6.6.1 MTO as Epoxidation Catalyst – Original Findings

The rapid formation of peroxo complexes in the reaction between MTO and hydrogen peroxide makes this organometallic compound useful as an oxidation catalyst. In the original report on olefin epoxidation with MTO, Herrmann and coworkers employed a preformed solution of hydrogen peroxide in *tert*-butanol as the terminal oxidant [88]. This solution was prepared by mixing *tert*-butanol and aqueous hydrogen peroxide, followed by the addition of anhydrous $MgSO_4$. After filtration, this essentially water-free solution of hydrogen peroxide was used in the epoxidation reactions. It was further reported that MTO, or rather its peroxo complexes, were stable for weeks in this solution if kept at low temperatures (below 0 °C). As seen above, later studies by Espenson revealed the instability of MTO in hydrogen peroxide solutions [97]. Epoxidation of various olefins by use of 0.1–1 mol% of MTO and the H_2O_2/t-BuOH solution generally resulted in high levels of conversion to epoxide, but a significant amount of *trans*-1,2-diol was often formed by ring-opening of the epoxide. The reason for using "anhydrous" hydrogen peroxide was of course an attempt to avoid this side-reaction, but since hydrogen peroxide generates water upon reaction with MTO it was impossible to work under strictly water-free conditions. The ring-opening process may be catalyzed either directly by MTO, due to the intrinsic metal Lewis acidity, or simply by protonation of the epoxide. To overcome this problem, Herrmann used excesses of amines (e.g., 4,4'-dimethyl-2,2'-bipyridine, quinine, and cinchonine), to coordinate to the metal and thus suppress the ring-opening process [101]. This resulted in better selectivity for the epoxide, at the expense of decreased, or in some cases completely inhibited, catalytic activity. In an attempt to overcome the problems of low selectivity for epoxide formation and the decreased catalytic activity obtained with amine additives, Adam introduced the urea/hydrogen peroxide (UHP) adduct as the terminal oxidant for the MTO-catalyzed system [102]. This resulted in substantially better selectivity for several olefins, although substrates giving highly acid-sensitive epoxides still suffered from deleterious ring-opening reactions.

6.6.2 The Influence of Heterocyclic Additives

The second major discovery regarding the use of MTO as an epoxidation catalyst came in 1996, when Sharpless and coworkers reported on the use of substoichiometric amounts of pyridine as a co-catalyst in the system [103]. A change of solvent from *tert*-butanol to dichloromethane and the introduction of 12 mol% of pyridine even allowed the synthesis of very sensitive epoxides with aqueous hydrogen peroxide as the terminal oxidant. A significant rate acceleration was also observed for the epoxidation reaction performed in the presence of pyridine. This discovery was the first example of an efficient MTO-based system for epoxidation under neutral to basic conditions. Under these conditions the detrimental acid-induced decomposition of the epoxide is effectively avoided. With this novel system, a variety of

Table 6.9 MTO-catalyzed epoxidation of olefins with H_2O_2.

Olefin	No additive[a]	Pyridine[b]	3-Cyanoipyridine[b]	Pyrazole[b]
	90 (5)	96 (6)		
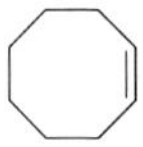	100 (2)[b]	99 (2)		89 (0.02)
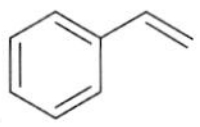		84 (16)	96 (5)[c]	96 (5)
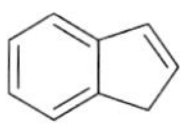	48 (37)	96 (5)		
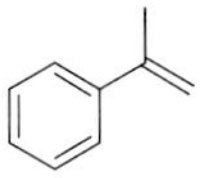		82 (6)	74 (1.5)[c]	93 (1.5)
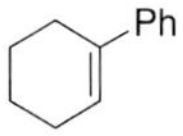 Ph		98 (1)	96 (1)[c]	95 (1)
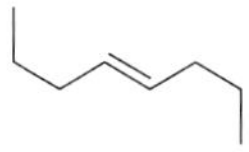	95 (2)	91 (24)	97 (12)	
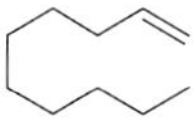	75 (12)	82 (48)	99 (14)	99 (14)

Yield % (reaction time h) [a] Anhydrous H_2O_2 in *t*-BuOH. [b] Aqueous H_2O_2 (30%), [c] Pyridine and 3-cyanopyridine (6 mol% of each).

olefin substrates were converted into their corresponding epoxides in high yields and with high epoxide selectivities (Scheme 6.12 and Table 6.9).

The increased rate of epoxidation observed with pyridine as an additive has been studied by Espenson and Wang and was explained to a certain degree in terms of an accelerated formation of peroxorhenium species [104]. Stabilization of the rhenium catalyst through pyridine coordination was also detected, although the excess of pyridine needed in the procedure unfortunately resulted in increased catalyst deactivation. As can be seen above, MTO is stable under acidic conditions, but accelerated decomposition of the catalyst into perrhenate and methanol occurs at high pH. The Brønsted basicity of pyridine produces increased amounts of HO_2^-, which speeds up the formation of the peroxo complexes and the decomposition of the catalyst. The addition of pyridine to the epoxidation system hence gave certain improvements regarding rate and selectivity for epoxide formation, but at the expense of catalyst lifetime. This turned out to be a minor problem for highly reactive substrates such as tetra-, tri- and *cis*-disubstituted olefins, since these com-

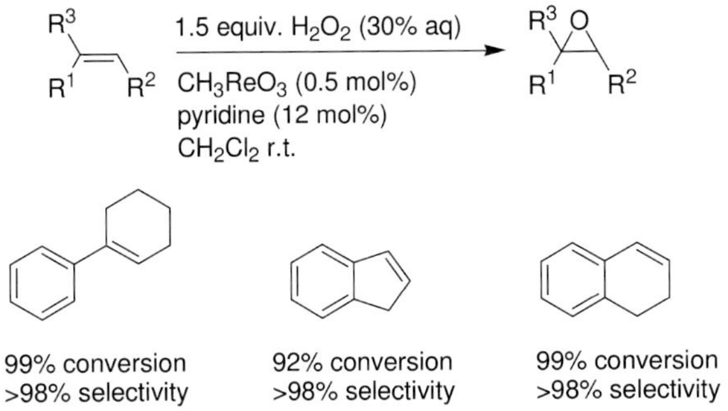

R^3, R^1, R^2 olefin → 1.5 equiv. H_2O_2 (30% aq), CH_3ReO_3 (0.5 mol%), pyridine (12 mol%), CH_2Cl_2 r.t. → epoxide

99% conversion
>98% selectivity

92% conversion
>98% selectivity

99% conversion
>98% selectivity

Scheme 6.12

pounds are converted into epoxides at a rate significantly higher then the rate for catalyst decomposition. Less electron-rich substrates such as α-olefins, however, react more slowly with electrophilic oxygen-transfer agents, and require longer reaction times to reach acceptable conversions. Under the pyridine (12 mol%) conditions, neither dec-1-ene nor styrene fully converted, even after prolonged reaction times.

A major improvement regarding epoxidation of terminal olefins was achieved upon exchanging pyridine for its less basic analogue 3-cyanopyridine (pK_a pyridine = 5.4; pK_a 3-cyanopyridine = 1.9) [105]. This improvement turned out to be general for a number of different terminal olefins, irrespective of the existence of steric hindrance at the α-position of the olefin or the presence of other functional groups in the substrate (Scheme 6.13 and Table 6.9).

Terminal olefins giving acid-labile epoxides, however, were not efficiently protected by this procedure. This problem was solved by use of a cocktail consisting of 3-cyanopyridine and pyridine (5–6 mol% of each additive) in the epoxidation reaction. The additive 3-cyanopyridine was also successfully employed in epoxidation of *trans*-disubstituted olefins, a problematic substance class when using the parent pyridine system [106]. For these reactions, the amount of the MTO catalyst could be reduced to 0.2–0.3 mol%, with only 1–2 mol% of 3-cyanopyridine added. Again, acid-sensitive epoxides were obtained with use of a mixture of 3-cyanopyridine and the parent pyridine. It should be pointed out that the pyridine additives undergo oxidation reactions, forming the corresponding pyridine-*N*-oxides [107],

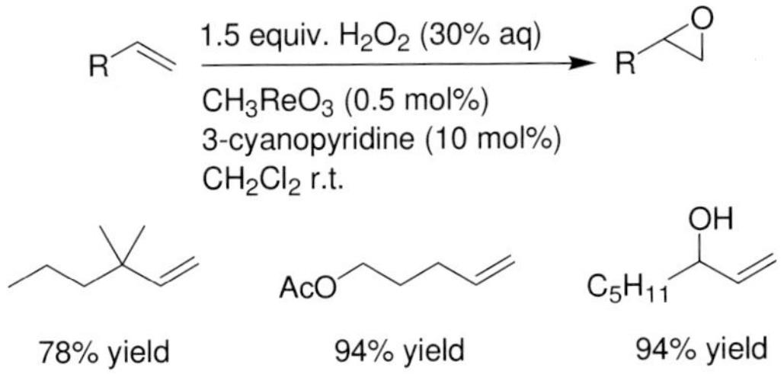

R-olefin → 1.5 equiv. H_2O_2 (30% aq), CH_3ReO_3 (0.5 mol%), 3-cyanopyridine (10 mol%), CH_2Cl_2 r.t. → R-epoxide

78% yield

AcO (substrate): 94% yield

C_5H_{11}, OH (substrate): 94% yield

Scheme 6.13

which will of course effectively decrease the amount of additive present in the reaction mixture. In fact, as pointed out by Espenson, the use of a pyridinium salt (a mixture of pyridine and, i. e., acetic acid) can be more effective in protecting the additive from *N*-oxidation [104, 108]. This can be beneficial for slowly reacting substrates, where *N*-oxidation would compete with olefin epoxidation. The Herrmann group introduced an improvement to the Sharpless system by employing pyrazole as an additive [109]. Pyrazole is a less basic heterocycle ($pK_a = 2.5$) than pyridine and does not undergo *N*-oxidation by the MTO/H_2O_2 system. Furthermore, use of pyrazole as the additive allowed for the formation of certain acid-sensitive epoxides. In terms of which additive to choose, pyrazole is perhaps the most effective for the majority of olefins, although pyridine would be the preferred additive for certain acid-labile compounds (Table 6.9) [110].

6.6.3
The Role of the Additive

The use of various heterocyclic additives in the MTO-catalyzed epoxidation has been demonstrated to be of great importance for substrate conversion, as well as for the product selectivity. With regard to selectivity, the role of the additive is obviously to protect the product epoxides from deleterious, acid-catalyzed (Brønsted or Lewis acid) ring-opening reactions. This can be achieved by direct coordination of the heterocyclic additive to the rhenium metal, thereby significantly decreasing its Lewis acidity. In addition, the basic nature of the additives will increase the pH of the reaction media.

There are a number of different interpretations available of the accelerating effects observed when pyridine or pyrazole is added to the MTO system. One likely explanation is that the additives serve as phase-transfer agents. Upon addition of MTO to an aqueous H_2O_2 solution, immediate formation of the peroxo complexes **A** and **B** (cf. Scheme 6.11) occurs, as is seen in the intense bright yellow color of the solution. If an immiscible organic solvent is added, the yellow color remains present in the aqueous layer, but addition of pyridine to this mixture results in an instantaneous transfer of the peroxo complexes into the organic phase. The transportation of the active oxidants into the organic layer should thus favor the epoxidation reaction, since the olefin concentration is significantly higher in this phase (Scheme 6.14). Additionally, the rate at which MTO is converted into **A** and **B** is accelerated when basic heterocycles are added. This has been attributed to the Brønsted basicity of the additives, which increases the amount of peroxide anion present in the reaction mixture. A higher concentration of HO_2-, however, is detrimental to the MTO catalyst, but the coordination of a Lewis base to the metal seems to have a positive effect in protecting the catalyst from decomposition.

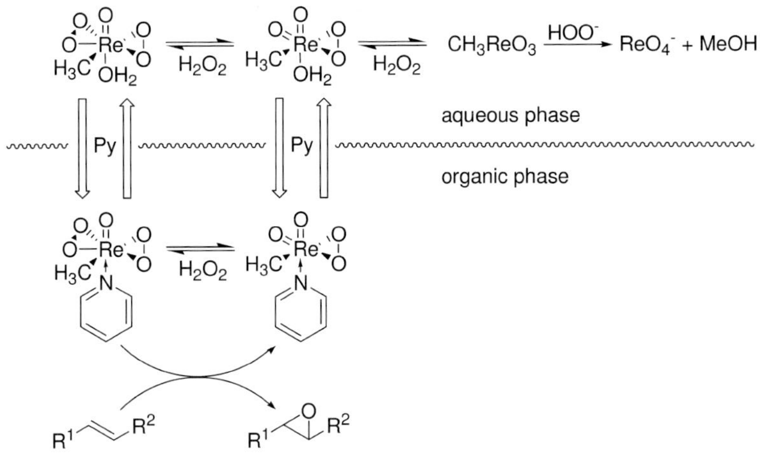

Scheme 6.14

6.6.4
Other Oxidants

While aqueous hydrogen peroxide is certainly the most practical oxidant for MTO-catalyzed epoxidations, the use of other terminal oxidants can sometimes be advantageous. As mentioned above, the urea-hydrogen peroxide adduct has been employed in olefin epoxidations. The anhydrous conditions obtained with UHP improved the system by decreasing the amount of diol formed in the reaction. The absence of significant amounts of water further helped in preserving the active catalyst from decomposition. A disadvantage, however, is the poor solubility of UHP in many organic solvents, which makes these reactions heterogeneous.

Another interesting terminal oxidant that has been applied in MTO-catalyzed epoxidations is sodium percarbonate (SPC) [111]. The fundamental structure of SPC involves hydrogen peroxide encapsulated through hydrogen bonding in a sodium carbonate matrix [112]. SPC slowly decomposes in water, and in organic solvents, to release hydrogen peroxide. The safety of this oxidant is reflected in its common use as an additive in household washing detergents and toothpaste. When this "solid form" of hydrogen peroxide was employed in MTO-catalyzed (1 mol%) oxidation of a wide range of olefins, good yields of the corresponding epoxides were obtained. An essential requirement for a successful outcome of the reaction was the addition of an equimolar amount (relative to the oxidant) of trifluoroacetic acid (TFA). In the absence of acid, or with acetic acid added, no or poor reactivity was observed. The role of the acid in this heterogeneous system is to facilitate slow release of hydrogen peroxide. Despite the presence of acid, even hydrolytically sensitive epoxides were formed in high yields. This can be explained in terms of efficient buffering of the system by $NaHCO_3$ and CO_2, formed in the reaction between TFA and SPC. The initial pH was measured as 2.5, but after 15

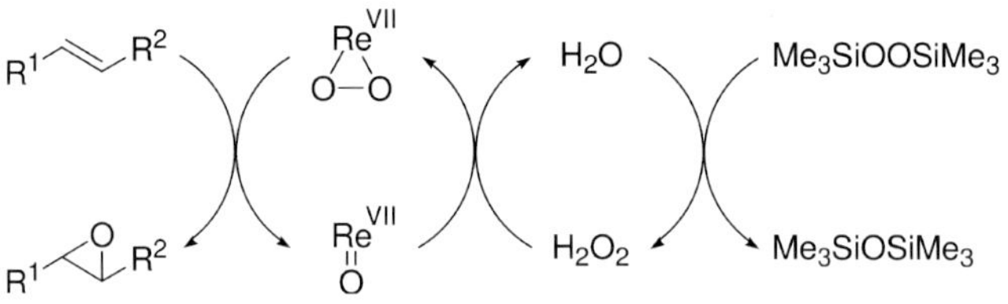

Re^{VII} (peroxo) H_2O $Me_3SiOOSiMe_3$

Re^{VII} (oxo) H_2O_2 $Me_3SiOSiMe_3$

Scheme 6.15

min a constant pH of 10.5 was established, ensuring protection of acid-sensitive products.

Bis-trimethylsilyl peroxide (BTSP) represents another form of "anhydrous" hydrogen peroxide [113]. The use of strictly anhydrous conditions in MTO-catalyzed olefin epoxidations would efficiently eliminate problems with catalyst deactivation and product decomposition due to ring-opening reactions. BTSP, which is the disilylated form of hydrogen peroxide, has been used in various organic transformations [114]. Upon reaction, BTSP is converted into hexamethyldisiloxane, thereby assuring anhydrous conditions. In initial experiments, MTO showed little or no reactivity towards BTSP under stoichiometric conditions [115]. This was very surprising, in view of the high reactivity observed for BTSP relative to hydrogen peroxide in oxidation of sulfides to sulfoxides [116]. The addition of one equivalent of water to the MTO/BTSP mixture rapidly facilitated the generation of the active peroxo complexes, however, this being explained in terms of hydrolytic formation of H_2O_2 from BTSP in the presence of MTO (Scheme 6.15). In fact, other proton sources proved to be equally effective in promoting this hydrolysis. Thus, no epoxidation occurred when the MTO/BTSP system was used under strictly aprotic conditions, whilst the addition of trace amounts of a proton source triggered the activation of BTSP and the formation of epoxides was observed.

Under optimum reaction conditions, MTO (0.5 mol%), water (5 mol%), and 1.5 equiv. of BTSP were used for efficient epoxide formation. The discovery of these essentially water-free epoxidation conditions resulted in another interesting breakthrough: the use of simple oxorhenium compounds as catalyst precursors [115, 117]. The catalytic activity of rhenium compounds such as Re_2O_7, $ReO_3(OH)$, and ReO_3 in oxidation reactions with aqueous hydrogen peroxide as the terminal oxidant is typically very poor. Attempts to form epoxides by use of catalytic Re_2O_7 in 1,4-dioxane with H_2O_2 (60%) at elevated temperatures (90 °C) mainly yielded 1,2-diols [118]. When hydrogen peroxide was replaced by BTSP in the presence of a catalytic amount of a proton source, however, each of the rhenium oxides Re_2O_7, $ReO_3(OH)$, or ReO_3 was equally effective as MTO in olefin epoxidations. In fact, the use of ReO_3 proved to be highly practical, since this compound, unlike Re_2O_7, is hydrolytically stable. There are several benefits associated with these epoxidation conditions. The amount of BTSP used in the reaction can easily be monitored by gas chromatography, and, furthermore, the simple workup procedure associated with this protocol is very appealing, since evaporation of the solvent (typically dichloromethane) and the formed hexamethyldisiloxane yields the epoxide.

Table 6.10 MTO-catalyzed epoxidation of olefins with anhydrous H_2O_2 or with aqueous H_2O_2 in fluorous solvents.[a]

Olefin	UHP[b]	SPC[c]	BTSP[d]	UHP[e] ionic liquid	H_2O_2[f] CF_3CH_2OH	H_2O_2[g] $(CF_3)_2CHOH$
	97 (18)			99 (8)	99 (0.5)	
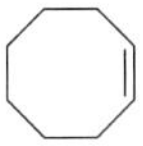		94 (2)		95 (8)	99 (1)	93 (1)
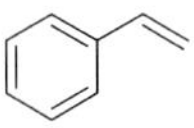	44 (19)	96 (12)		95 (8)	82 (2)	
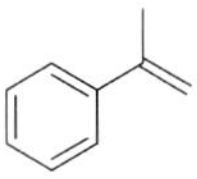	55 (21)[h]	91 (3)				
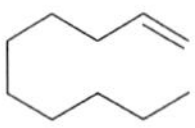		94 (15)	94 (14)	46 (72)	97 (21)	88 (24)[i]

[a] Yield % (reaction time h). [b] 1 mol% MTO. [c] 1 mol% MTO, 12 mol% pyrazole. [d] 0.5 mol% Re_2O_7. [e] 2 mol% MTO. [f] 0.1 mol% MTO, 10 mol% pyrazole, 60% H_2O_2. [g] 0.1 mol% MTO, 10 mol% pyrazole, 30% H_2O_2. [h] An additional 26% of the diol was formed. [i] Dodec-1-ene was used as substrate.

6.6.5
Solvents/Media

The high solubility of the MTO catalyst in almost any solvent opens up a broad spectrum of reaction media to choose from when performing epoxidations. The most commonly used solvent, however, is still dichloromethane. From an environmental point of view this is certainly not the most appropriate solvent in large-scale epoxidations. Interesting solvent effects for the MTO-catalyzed epoxidation was reported by Sheldon and coworkers, who performed the reaction in trifluoroethanol [119]. The change from dichloromethane to the fluorinated alcohol allowed a further reduction of the catalyst loading down to 0.1 mol%, even for terminal olefin substrates (Table 6.10). It should be pointed out that this method does require 60% aqueous hydrogen peroxide for efficient epoxidations.

Bégué and coworkers recently achieved an improvement in this method by performing the epoxidation reaction in hexafluoro-2-propanol [120]. They found that the activity of hydrogen peroxide was significantly increased in this fluorous alcohol, in relation to trifluoroethanol, which allowed for the use of 30% aqueous H_2O_2. Interestingly, the nature of the substrate and the choice of additive turned out to have important consequences for the lifetime of the catalyst. Cyclic disubstituted olefins were efficiently epoxidized with 0.1 mol% of MTO and 10 mol%

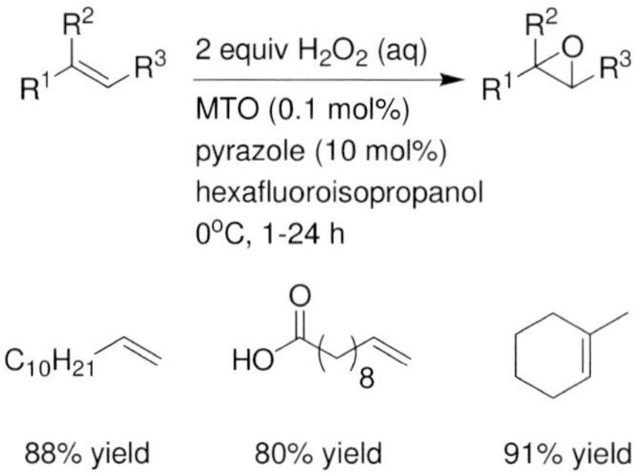

R^1, R^2, R^3 alkene → epoxide: 2 equiv H_2O_2 (aq), MTO (0.1 mol%), pyrazole (10 mol%), hexafluoroisopropanol, 0°C, 1-24 h

$C_{10}H_{21}$ (alkene): 88% yield; HO(C=O)-(CH₂)₈-alkene: 80% yield; 1-methylcyclohexene: 91% yield

Scheme 6.16

pyrazole as the catalytic mixture, but for trisubstituted substrates the use of the additive 2,2'-bipyridine turned out to be crucial for high conversion (Scheme 6.16). The use of pyrazole in this case proved to be highly deleterious for the catalyst, as indicated by the loss of the yellow color of the reaction solution. This observation is certainly counter intuitive, since more basic additives normally decrease the catalyst lifetime. The fact that full conversion of long-chained α-olefins was achieved after 24 h with pyrazole as the additive, and the observation that the catalyst was still active after this period of time, is very surprising in view of the outcome with more highly functionalized substrates. To increase conversion for substrates displaying poor solubility in hexafluoro-2-propanol, trifluoromethylbenzene was added as a co-solvent. In this way, dodec-1-ene was converted into its corresponding epoxide in high yield.

The use of involatile ionic liquids as environmentally benign solvents has received significant attention in recent years. Abu-Omar and coworkers developed an efficient MTO-catalyzed epoxidation procedure with 1-ethyl-3-methylimidazolium tetrafluoroborate ([emim]BF_4) as solvent and urea/hydrogen peroxide (UHP) as the terminal oxidant [121, 122]. A major advantage of this system is the high solubility of UHP, MTO, and its peroxo complexes, making the reaction media completely homogeneous. Under these essentially water-free conditions, high levels of conversion and good epoxide selectivity were achieved for the epoxidation of variously substituted olefins. Replacement of UHP with aqueous hydrogen peroxide for the epoxidation of 1-phenylcyclohexene resulted in a poor yield of this acid-sensitive epoxide, the formation of the corresponding diol being observed instead. A disadvantage of the UHP/ionic liquid system, in relation to other MTO procedures, is the higher catalyst loading (2 mol%) required for efficient epoxide formation.

6.6.6
Asymmetric Epoxidations with MTO

The MTO-based epoxidation system offers a particularly effective and practical route for the formation of racemic epoxides. Attempts to prepare chiral MTO complexes and to employ them in catalytic epoxidation have so far been scarce, and the few existing reports are unfortunately quite discouraging. In the epoxidation of *cis*-β-methylstyrene with MTO and hydrogen peroxide in the presence of the additive (*S*)-1-(*N,N*-dimethyl)phenylethylamine, an enantiomeric excess of 86% in the product has been claimed [123]. The epoxides from other substrates such as styrene and oct-1-ene were obtained with significantly lower enantioselectivities (13% *ee*). Furthermore, the MTO-catalyzed epoxidation of 1-methylcyclohexene with L-prolineamide, (+)-2-aminomethylpyrrolidine, or (*R*)-1-phenylethylamine as additives was reported to yield the product in low yield and with poor enantioselectivity (20% *ee* or less) [124]. A significant amount of the corresponding diol was formed in these reactions. More recently, Herrmann and coworkers reported the use of chiral pyrazole derivatives as ligands in the MTO-catalyzed epoxidation of *cis*-β-methylstyrene [125]. Levels of conversion and enantiomeric excess were unfortunately also rather poor, with a maximum of 27% *ee*. These workers have also reported on the use of chiral diols as ligands for the epoxidation of the same substrate (*cis*-β-methylstyrene), the best enantioselectivity obtained being 41% *ee*, although in this particular case the reaction was only completed to 5% conversion. A general method for the enantioselective formation of epoxides through the use of rhenium catalysts is thus still lacking. It would certainly be a breakthrough if such a system could be developed, in view of the efficiency of MTO-catalyzed epoxidation reactions with hydrogen peroxide as terminal oxidant.

6.7 Iron-catalyzed Epoxidations

The use of iron salts and complexes for olefin epoxidation is in many respects similar to that of manganese catalysts. Iron porphyrins can be used as epoxidation catalysts, but conversion and selectivity are often inferior to what is obtained with their manganese counterparts. The potential to use hydrogen peroxide as terminal oxidant efficiently are limited, due to the rapid iron-catalyzed decomposition of the oxidant. Traylor and coworkers, however, found conditions under which a polyfluorinated Fe(TPP)-catalyst (**21**) could be employed in the epoxidation of cyclooctene to yield the corresponding epoxide in high yield (Scheme 6.17) [126]. High catalyst loading (5 mol%) and slow addition of the oxidant were required, certainly limiting the usefulness of this procedure.

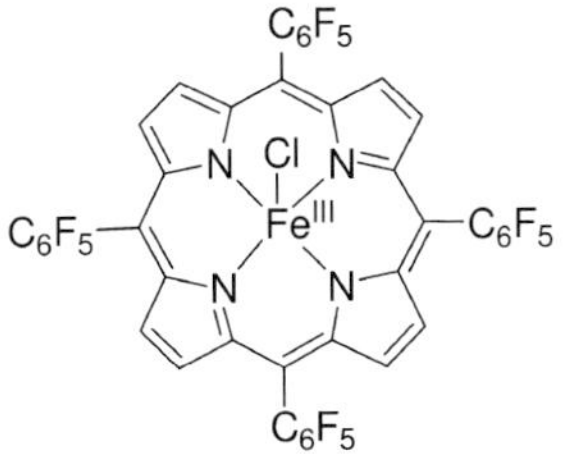

21 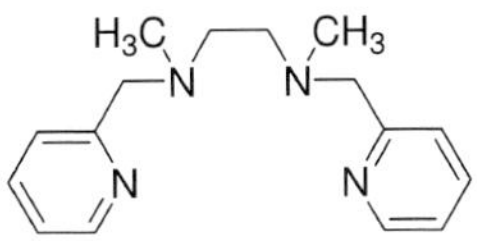 **22** 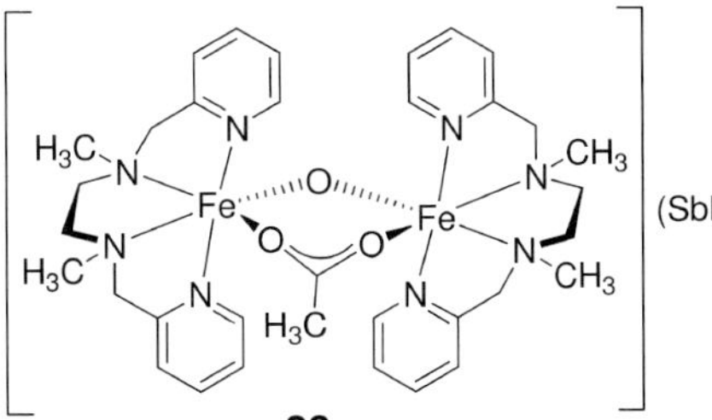 **23** $(SbF_6)_2$

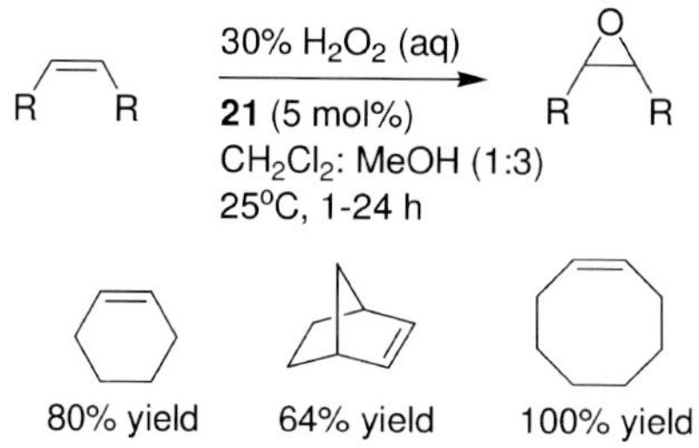

30% H_2O_2 (aq), **21** (5 mol%), CH_2Cl_2: MeOH (1:3), 25°C, 1-24 h

80% yield 64% yield 100% yield

Scheme 6.17

A number of iron complexes with biomimetic non-heme ligands have recently been introduced as catalysts for alkane hydroxylation and olefin epoxidation and dihydroxylation. These complexes have been demonstrated to activate hydrogen peroxide without the formation of free hydroxyl radicals, a feature commonly observed in iron oxidation chemistry. A particularly efficient catalytic system for selective epoxidation of olefins was developed by Jacobsen and coworkers [127]. In this procedure, a tetradentate ligand (BPMEN = *N,N'*-dimethyl-*N,N'*-bis(2-pyridylmethyl)-diaminoethane, **22**) was combined with an iron(II) precursor and acetic acid to yield a self-assembled μ-oxo, carboxylate-bridged diiron(III) complex (**23**). This dimeric iron complex, resembling the active site found in the hydroxylase methane monooxygenase (MMO), was demonstrated to epoxidize olefins efficiently in the presence of aqueous hydrogen peroxide (50%). This catalyst turned out to be particularly active for the epoxidation of terminal olefins, normally the most difficult substrates to oxidize. Thus, dodec-1-ene was transformed into its corresponding epoxide in 90% yield after 5 min. with use of 3 mol% of the catalyst. This system was also effective for the epoxidation of other simple nonterminal olefins such as cyclooctene and *trans*-dec-5-ene (Scheme 6.18).

Que and coworkers reported on a similar monomeric iron complex, formed with the BPMEN ligand but without acetic acid [128]. This complex was able to epoxidize cyclooctene in reasonably good yield (75%), but at the same time a small amount of the *cis*-diol (9%) was formed. This feature observed with this class of complexes has been further studied and more selective catalysts have been prepared. Even though poor levels of conversion are often obtained with the current

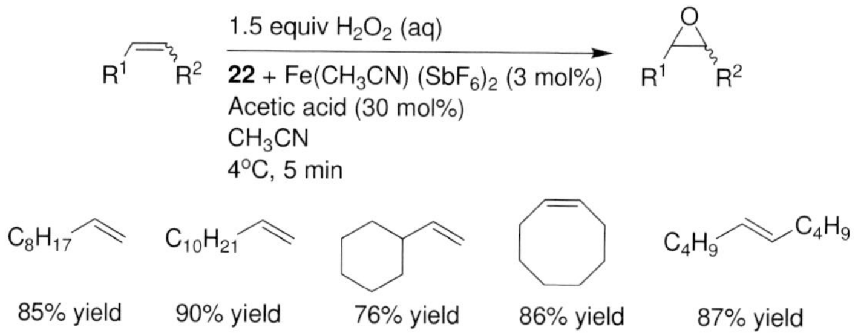

1.5 equiv H_2O_2 (aq), **22** + $Fe(CH_3CN)(SbF_6)_2$ (3 mol%), Acetic acid (30 mol%), CH_3CN, 4°C, 5 min

C_8H_{17}: 85% yield; $C_{10}H_{21}$: 90% yield; 76% yield; 86% yield; C_4H_9/C_4H_9: 87% yield

Scheme 6.18

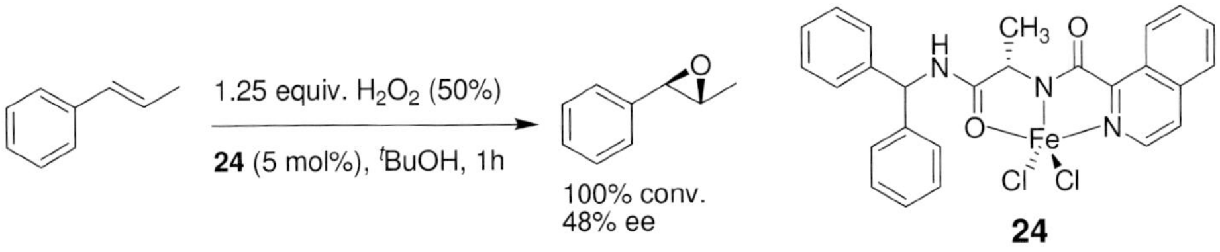

1.25 equiv. H_2O_2 (50%)

24 (5 mol%), ^{t}BuOH, 1h

100% conv.
48% ee

24

Scheme 6.19

catalysts, this method represents an interesting alternative to other *cis*-dihydroxylation systems [129, 130]. Use of similar chiral ligands based on 1,2-diaminocyclohexane resulted in complexes that were able to catalyze the formation of epoxides in low yields and with low enantioselectivities (0–12% *ee*). The simultaneous formation of *cis*-diols occurred with significantly better enantioselectivity (up to 82% *ee*), but these products were also obtained in low yields.

Using high-throughput screening techniques, Francis and Jacobsen discovered a novel iron-based method for the preparation of enantiomerically enriched epoxides [131]. In this system, chiral complexes prepared from polymer-supported peptide-like ligands and iron(II) chloride were evaluated as catalysts for the epoxidation of *trans*-β-methylstyrene with aqueous hydrogen peroxide (30%) as terminal oxidant. The best polymer-supported catalysts yielded the corresponding epoxide at up to 78% conversion with enantioselectivities ranging from 15 to 20% *ee*. With a homogeneous catalyst derived from this combinatorial study, *trans*-β-methylstyrene was epoxidized in 48% *ee* after 1 h (100% conversion, 5 mol% catalyst, 1.25 equiv. 50% hydrogen peroxide in *t*-BuOH) (Scheme 6.19).

Even though only moderate enantiomeric excess was obtained with catalyst **24**, this example is of particular interest since it demonstrates how novel catalysts based on "nonclassic" combinations of ligands and transition metals can be discovered.

6.8
Ruthenium-catalyzed Epoxidations

High-valent ruthenium oxides (e.g., RuO_4) are powerful oxidants and react readily with olefins, mostly resulting in cleavage of the double bond [132]. If reactions are performed with very short reaction times (0.5 min.) at 0 °C it is possible to control the reactivity better and thereby to obtain *cis*-diols. On the other hand, the use of less reactive, low-valent ruthenium complexes in combination with various terminal oxidants for the preparation of epoxides from simple olefins has been described [133]. In the more successful earlier cases, ruthenium porphyrins were used as catalysts, especially in combination with *N*-oxides as terminal oxidants [134, 135, 136]. Two examples are shown in Scheme 6.20, terminal olefins being oxidized in the presence of catalytic amounts of Ru-porphyrins **25** and **26** with the sterically hindered 2,6-dichloropyridine *N*-oxide (2,6-DCPNO) as oxidant. The use

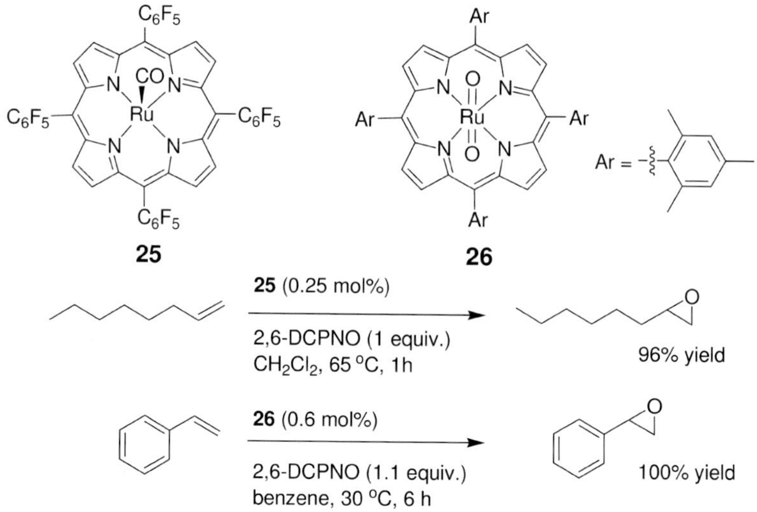

25 (0.25 mol%)
2,6-DCPNO (1 equiv.)
CH_2Cl_2, 65 °C, 1h
96% yield

26 (0.6 mol%)
2,6-DCPNO (1.1 equiv.)
benzene, 30 °C, 6 h
100% yield

Ar =

Scheme 6.20

of this particular oxidant was crucial for high epoxide yield, since simple pyridine, generated from pyridine *N*-oxide, significantly lowered the epoxidation rate by coordination to the ruthenium catalyst. Interestingly, aerobic epoxidation of olefins under 1 atm. of molecular oxygen without any additional reducing agent has been reported with catalyst **26** [137]. Recently, Beller and coworkers reported on the use of $RuCl_3$ and pyridine-2,6-dicarboxylic acid as an efficient catalyst for the epoxidation of olefins with 3% aqueous hydrogen peroxide as terminal oxidant [138]. With $RuCl_3$ loadings ranging from 0.01 to 5 mol% and slow addition of the oxidant, to avoid ruthenium-catalyzed H_2O_2 decomposition, a number of different olefins were successfully converted into epoxides in yields of up to 99%.

Asymmetric epoxidation of olefins with ruthenium catalysts based either on chiral porphyrins or on pyridine-2,6-bisoxazoline (pybox) ligands has been reported (Scheme 6.21). Berkessel et al. reported that catalysts **27** and **28** were efficient catalysts for the enantioselective epoxidation of aryl-substituted olefins (Table 6.10) [139]. Enantioselectivities of up to 83% were obtained in the epoxidation of 1,2-dihydronaphthalene with catalyst **28** and 2,6-DCPNO. Simple olefins such as oct-1-ene reacted poorly and gave epoxides with low enantioselectivity. The use of pybox ligands in ruthenium-catalyzed asymmetric epoxidations was first reported by Nishiyama et al., who used catalyst **30** in combination with iodosyl benzene, bisacetoxyiodo benzene [PhI(OAc)$_2$], or TBHP for the oxidation of *trans*-stilbene [140]. In their best result, with PhI(OAc)$_2$ as oxidant, they obtained *trans*-stilbene oxide in 80% yield and with 63% *ee*. More recently, Beller and coworkers have reexamined this catalytic system, finding that asymmetric epoxidations could be performed with ruthenium catalysts **29** and **30** and 30% aqueous hydrogen peroxide (Table 6.11) [141]. Development of the pybox ligand provided ruthenium complex **31**, which turned out to be the most efficient catalyst for asymmetric

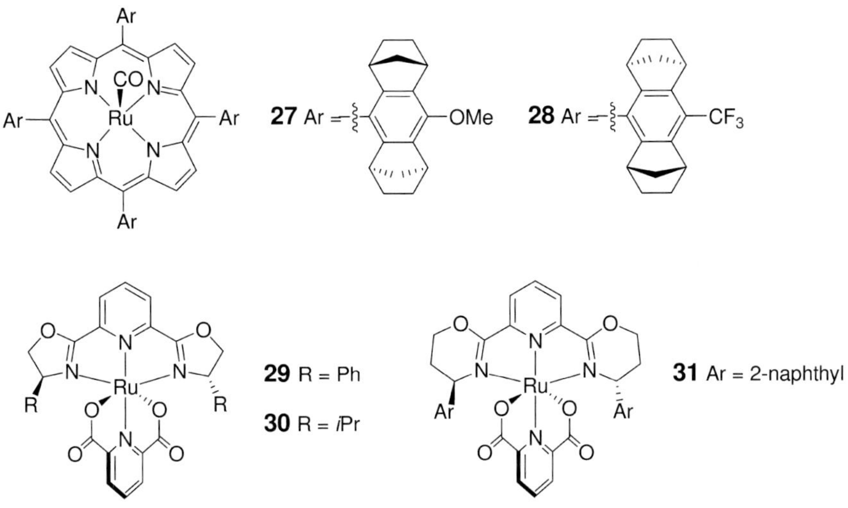

Ar
N CO N
Ar Ru Ar
N N
Ar

27 Ar = OMe **28** Ar = CF_3

29 R = Ph

30 R = *i*Pr

31 Ar = 2-naphthyl

Scheme 6.21

olefin epoxidation, use of 5 mol% of **31** and slow addition of hydrogen peroxide allowing a number of aryl-substituted olefins to be epoxidized in yields of up >99% and with enantioselectivities of up to 84%. A comparison of different chiral ruthenium catalysts used in the epoxidation of styrene is summarized in Table 6.11.

Table 6.11 Ruthenium-catalyzed epoxidation of styrene.

oxidant / Ru-cat

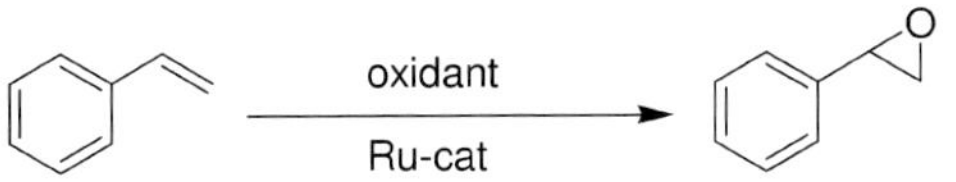

Entry	Ru-cat (amount)	Oxidant	Solvent	Temp (°C)	Yield (%)	*Ee* (%)	Ref.
1	**27** (0.1 mol%)	2,6-DCPN 1.1 equiv.	benzene	r. t.	82	76	139
2	**28** (0.1 mol%)	2,6-DCPN 1.1 equiv.	benzene	r. t.	96	79	139
3[a]	**29** (5 mol%)	TBHP 1.5 equiv.	*t*-BuOH	r. t.	56	30	142
4[a]	**29** (5 mol%)	30% H_2O_2 3 equiv.	AmOH[b]	r. t.	70	31	141
5	**30** (5 mol%)	$PhI(OAc)_2$ 3 equiv.	AmOH[b]	0	48	5	143
6[a]	**31** (5 mol%)	30% H_2O_2 3 equiv.	*t*-BuOH	r. t.	85	59	141

[a] Slow addition of oxidant over 12 h. [b] Amyl alcohol.

6.9
Concluding Remarks

The epoxidation of olefins with the aid of transition metal-based catalysts is certainly a well studied reaction. There are, however, only a few really good and general systems efficiently compatible with environmentally benign oxidants such as aqueous hydrogen peroxide. An important feature for selective epoxidations with H_2O_2 as terminal oxidant is the catalyst's capability to activate hydrogen peroxide by heterolytic cleavage of the O–O bond, hence avoiding radical formation. The generation of free radicals by homolytic cleavage usually results in uncontrolled side reactions such as substrate and/or catalyst degradation and poorly selective product formation. A comparison of the efficiencies achieved with the best catalysts described in this chapter for racemic epoxidations is presented in Table 6.12.

From comparison of the different methods presented in Table 6.12 it is evident that there are advantages and limitations with almost all available epoxidation systems. In terms of the scope of the reactions, some methods are better for linear terminal olefins, whereas others show enhanced reactivity with more functionalized substrates. The environmentally attractive TS-1 system is highly efficient, but restricted to linear substrates. The various available tungsten systems efficiently produce epoxides from simple substrates, but acid-sensitive products undergo further reactions, thus effectively reducing the selectivity of the process. MTO is a highly active epoxidation catalyst and, when it is combined with heterocyclic additives, even hydrolytically sensitive products are obtained in good yields and with good selectivities. However, the fact that most MTO-catalyzed epoxidations are performed in chlorinated or fluorinated solvents is a limitation with

Table 6.12 Transition metal-catalyzed epoxidation of olefins with H_2O_2 as terminal oxidant.

Catalyst	S/C	Solvent	Temp (°C)	1-Olefin[g] yield (%)/TOF (h^{-1})	Cyclooctene yield (%)/TOF (h^{-1})	Ref.
Ti[a]	100	MeOH	25	74/108[h]	–	29
W[b]	50/500	toluene	90	91/12	98/122	53
Mo[c]	100/200	CH_2Cl_2	60	96/4	100/100	47
Mn[d]	100	DMF	25	–	67/4	69
Re[e]	200	CH_2Cl_2	25	99/14	89/8900	109
Re[e]	1000	CF_3CH_2OH	25	97/48	99/990	119
Fe[f]	33	CH_3CN	4	85/337	86/341	127

[a] TS-1. [b] Na_2WO_4, $NH_2CH_2PO_3H_2$, $(C_8H_{17})_3NCH_3^+HSO_4^-$. [c] $MoO_5(OAs(C_{12}H_{25}))_3$. [d] $MnSO_4$. [e] MTO, pyrazole. [f] **23**. [g] Dec-1-ene. [h] Oct-1-ene.

regards to their eco-compatibility. Nevertheless, from a synthetic perspective, the use of MTO in combination with pyridine or pyrazole is, in our opinion, the current method of choice for lab-scale epoxidations, in view of its broad substrate scope and its operational simplicity. With regard to asymmetric processes employing hydrogen peroxide as terminal oxidant, there are only a few reported systems that produce epoxides with moderate levels of enantioselectivity. The iron-based catalyst developed by Francis and Jacobsen is a promising candidate, whilst the ruthenium catalysts containing pybox ligands are other interesting systems. However, this field is currently quite premature, and further developments along these lines may result in more selective and efficient epoxidation techniques. In addition to solving problems associated with substrate scope and selectivity, perhaps the most important challenge is to create more efficient epoxidation systems. Most epoxidation methods, asymmetric or non-asymmetric, display rather poor substrate to catalyst ratios. In the best cases, the catalyst loading is as low as 0.1 mol%, but most often 0.5–1 mol% of the catalyst needs to be added to ensure full conversion of the starting olefin. In comparison to the corresponding highly efficient existing catalysts for the reduction of olefins with molecular hydrogen, there is a long way to go before oxidation catalysts reach the same level of efficiency.

In conclusion, the above summary of oxidation methods shows that there is still room for further improvements in the field of selective olefin epoxidation. The development of active and selective catalysts capable of oxidizing a broad range of olefin substrates with aqueous hydrogen peroxide as terminal oxidant in inexpensive and environmentally benign solvents remains a continuing challenge.

References

1 D. Kahlich, K. Wiechern, J. Lindner in *Ullmann's Encyclopedia of Industrial Chemistry*, 5th ed., *Vol. A22* (Eds.: B. Elvers, S. Hawkins, W. Russey, G. Schultz), VCH, Weinheim, **1993**, pp. 239–260.

2 T. Kratz, W. Zeiss, in *Peroxide Chemistry* (Ed.: W. Adam), Wiley-VCH, Weinheim, **2000**, 41–59.

3 R. A. Sheldon, in *Applied Homogeneous Catalysis with Organometallic Compounds*, 2nd ed., *1* (Eds.: B. Cornils, W. A. Herrmann), Wiley-VCH, Weinheim, **2002**, pp. 412–427.

4 K. A. Jørgensen, *Chem. Rev.*, **1989**, *89*, 431.

5 Y. Shi, in *Modern Oxidation Methods* (Ed.: J.-E. Bäckvall), Wiley-VCH, Weinheim, **2004**, pp. 51–78.

6 J. R. Monnier, *Applied Catalysis A: General*, **2001**, *221*, 73.

7 B. S. Lane, K. Burgess, *Chem. Rev.*, **2003**, *103*, 2457.

8 R. A. Sheldon in *Aspects of Homogeneous Catalysis, Vol. 4* (Ed.: R. Ugo), Reidel, Dordrecht, **1981**, pp. 3–70.

9 T. Katsuki, K. B. Sharpless, *J. Am. Chem. Soc.*, **1980**, *102*, 5974.

10 For a recent comprehensive review, see: T. Katsuki in *Comprehensive Asymmetric Catalysis II* (Eds.: E. N. Jacobsen, A. Pfaltz, H. Yamamoto), Springer, Heidelberg, **1999**, pp. 621–648.

11 http://nobelprize.org/chemistry/laureates/2001/index.html.

12 S. S. Woodard, M. G. Finn, K. B. Sharpless, *J. Am. Chem. Soc.*, **1991**, *113*, 106.

13 M. G. Finn, K. B. Sharpless, *J. Am. Chem. Soc.*, **1991**, *113*, 113.

14 R. A. Johnson, K. B. Sharpless, in *Catalytic Asymmetric Synthesis* (Ed.: I. Ojima), 2nd edition, Wiley-VCH, New York, **2000**, pp. 231–280.

15 D. J. Berrisford, C. Bolm, K. B. Sharpless,

Angew. Chem. Int. Ed. Engl., **1995**, *34*, 1059.
16 Y. Gao, R.M. Hanson, J.M. Klunder, S.J. Ko, H, Masamune, K.B. Sharpless, *J. Am. Chem. Soc.*, **1987**, *109*, 5765.
17 D. Tanner, P. Somfai, *Tetrahedron*, **1986**, *42*, 5985.
18 T. Katsuki, A.W.M. Lee, P. Ma, V.S. Martin, S. Masamune, K.B. Sharpless, D. Tuddenham, F.J. Walker, *J. Org. Chem.*, **1982**, *47*, 1373.
19 R.E. Dolle, K.C. Nicolaou, *J. Am. Chem. Soc.*, **1985**, *107*, 1691.
20 K. Mori, H. Ueda, *Tetrahedron*, **1981**, *37*, 2581.
21 K.B. Sharpless, C.H. Behrens, T. Katsuki, A.W.M. Lee, V.S. Martin, M. Takatani, S.M. Viti, F.J. Walker, S.W. Woodard, *Pure Appl. Chem.*, **1983**, *55*, 589.
22 S. Yamada, M. Shiraishi, M. Ohmori, H. Takayama, *Tetrahedron Lett.*, **1984**, *25*, 3347.
23 M. Sodeoka, T. Iimori, M. Shibasaki, *Tetrahedron Lett.*, **1985**, *26*, 6497.
24 T. Erickson, *J. Org. Chem.*, **1986**, *51*, 934.
25 J.A. Marshall, T.M. Jenson, *J. Org. Chem.*, **1984**, *49*, 1707.
26 M. Acemoglu, P. Uebelhart, M. Rey, C.H. Eugster, *Helv. Chim. Acta*, **1988**, *71*, 931.
27 V. S. Martin, S. S. Woodard, T. Katsuki, Y. Yamada, M. Ikeda, K. B. Sharpless, *J. Am. Chem. Soc.*, **1981**, *103*, 6237.
28 B. D. Johnston, A. C. Oehlschlager, *J. Org. Chem.*, **1982**, *47*, 5384.
29 B. Notari, *Catal. Today*, **1993**, *18*, 163.
30 M. N. Sheng, J. G. Zajacek, *J. Org. Chem.*, **1970**, *35*, 1839.
31 K. B. Sharpless, R. C. Michaelson, *J. Am. Chem. Soc.*, **1973**, *95*, 6136.
32 R. C. Michaelson, R. E. Palermo, K. B. Sharpless, *J. Am. Chem. Soc.*, **1977**, *99*, 1992.
33 N. Murase, Y. Hoshino, M. Oishi, H. Yamamoto, *J. Org. Chem.*, **1999**, *64*, 338.
34 N. Murase, Y. Hoshino, M. Oishi, H. Yamamoto, *Bull. Chem. Soc. Jpn.*, **2000**, *73*, 1653.
35 For a recent review, see C. Bolm, *Coord. Chem. Rev.*, **2003**, *237*, 245.
36 Y. Hoshino, H. Yamamoto, *J. Am. Chem. Soc.*, **2000**, *122*, 10452.
37 C. Bolm, T. Kühn, *Synlett*, **2000**, 899.
38 H.-L. Wu, B.-J. Uang, *Tetrahedron: Asymmetry*, **2002**, *13*, 2625.
39 M. Miyaura, J. K. Kochi, *J. Am. Chem. Soc.*, **1983**, *105*, 2368.
40 E. G. Samsel, K. Srinivasan, J. K. Kochi, *J. Am. Chem. Soc.*, **1985**, *107*, 7606.
41 K. Srinivasan, P. Michaud, J. K. Kochi, *J. Am. Chem. Soc.*, **1986**, *108*, 2309.
42 A. M. Daly, M. F. Renehan, D. G. Gilheany, *Org. Lett.*, **2001**, *3*, 663.
43 For a recent mechanistic discussion, see: P. Brandt, P.-O. Norrby, A. M. Daly, D. G. Gilheany, *Chem. Eur. J.*, **2002**, *8*, 4299.
44 K. B. Sharpless, T. R. Verhoeven, *Aldrichmica Acta*, **1979**, *12*, 63.
45 W. R. Thiel, in *Transition Metals for Organic Synthesis 2* (Eds.: M. Beller, C. Bolm), Wiley-VCH, Weinheim, **1998**, pp. 290–300.
46 O. Bortolini, F. Di Furia, G. Modena, R. Seraglia, *J. Org. Chem.*, **1985**, *50*, 2688.
47 G. Wahl, D. Kleinhenz, A. Schorm, J. Sundermeyer, R. Stowasser, C. Rummey, G. Bringmann, C. Fickert, W. Kiefer. *Chem. Eur. J.*, **1999**, *5*, 3237.
48 G. B. Payne, P. H. Williams, *J. Org. Chem.*, **1959**, *24*, 54.
49 C. Venturello, E. Alneri, M. Ricci, *J. Org. Chem.*, **1983**, *48*, 3831.
50 C. Venturello, R. D'Aloisio, *J. Org. Chem.*, **1988**, *53*, 1553.
51 C. Venturello, R. D'Aloisio, J. C. J. Bart, M. Ricci, *J. Mol. Catal.*, **1985**, *32*, 107.
52 K. Sato, M. Aoki, M. Ogawa, T. Hashimoto, R. Noyori, *J. Org. Chem.*, **1996**, *61*, 8310.
53 K. Sato, M. Aoki, M. Ogawa, T. Hashimoto, D. Paynella, R. Noyori, *Bull. Chem. Soc. Jpn.*, **1997**, *70*, 905.
54 R. Noyori, M. Aoki, K. Sato, *Chem. Commun.*, **2003**, 1977.
55 K. Sato, M. Aoki, R. Noyori, *Science*, **1998**, *281*, 1646.
56 A. L. Villa de P., B. F. Sels, D. E. De Vos, P. A. Jacobs, *J. Org. Chem.*, **1999**, *64*, 7267.
57 K. Kamata, K. Yonehara, Y. Sumida, K. Yamaguchi, S. Hikichi, N. Mizuno, *Science*, **2003**, *300*, 964.
58 B. F. Sels, D. E. De Vos, P. A. Jacobs, *Tetrahedron Lett.*, **1996**, *37*, 8557.
59 For a comprehensive summary, see: G. Gelbard *C. R. Chim.*, **2000**, *3*, 757.
60 D. Hoegaerts, B. F. Sels, D. E. De Vos, F. Verpoort, P. A. Jacobs, *Catal. Today*, **2000**, *60*, 209.

61 X. Zuwei, Z. Ning, S. Yu, L. Kunlan, *Science*, **2001**, *292*, 1139.
62 P. Battioni, J.-P. Renaud, J. F. Bartoli, M. Reina-Artiles, M. Fort, D. Mansuy, *J. Am. Chem. Soc.*, **1988**, *110*, 8462.
63 P. L. Anelli, L. Banfi, F. Legramandi, F. Montanari, G. Pozzi, S. Quici, *J. Chem. Soc., Perkin Trans. 1*, **1993**, 1345.
64 R. Hage, J. E. Iburg, J. Kerschner, J. H. Koek, E. L. M. Lempers, R. J. Martens, U. S. Racheria, S. W. Russell, T. Swarthoff, M. R. P. van Vliet, J. B. Warnaar, L. van der Wolf, B. Krijnen, *Nature*, **1994**, *369*, 637.
65 D. E. De Vos, B. F. Sels, M. Reynaers, Y. V. S. Rao, P. A. Jacobs, *Tetrahedron Lett.*, **1998**, *39*, 3221.
66 A. Berkessel, C. A. Sklorz, *Tetrahedron Lett.*, **1999**, *40*, 7965.
67 J. Brinksma, R. Hage, J. Kerschner, B. L. Feringa, *Chem. Commun.*, **2000**, 537.
68 B. S. Lane, K. Burgess, *J. Am. Chem. Soc.*, **2001**, *123*, 2933.
69 B. S. Lane, M. Vogt, V. J. DeRose, K. Burgess, *J. Am. Chem. Soc.*, **2002**, *124*, 11946.
70 K.-H. Tong, K.-Y. Wong, T. H. Chan, *Org. Lett.*, **2003**, *5*, 3423.
71 W. Zhang, J. L. Loebach, S. R. Wilson, E. N. Jacobsen, *J. Am. Chem. Soc.*, **1990**, *112*, 2801.
72 R. Irie, K. Noda, Y. Ito, N. Matsumoto, T. Katsuki, *Tetrahedron Lett.*, **1990**, *31*, 7345.
73 E. N. Jacobsen, M. H. Wu in *Comprehensive Asymmetric Catalysis II* (Eds.: E. N. Jacobsen, A. Pfaltz, H. Yamamoto), Springer, Heidelberg, **1999**, pp. 649–677.
74 K. Srinivasan, P. Michaud, J. K. Kochi, *J. Am. Chem. Soc.*, **1986**, *108*, 2309.
75 H. Sasaki, R. Irie, T. Hamada, K. Suzuki, T. Katsuki, *Tetrahedron*, **1994**, *50*, 11827.
76 J. Vacca, B. Dorsey, W. A. Schlief, R. Levin, S. McDaniel, P. Darke, J. Zugay, J. Quintero, O. Blahy, E. Roth, V. Sardena, A. Schlabach, P. Graham, J. Condra, L. Gotlib, M. Holloway, J. Lin, I. Chen, K. Vastag, D. Ostovic, P. S. Anderson, E. A. Emini, J. R. Huff, *Proc. Nat. Acad, Sci.*, **1994**, *91*, 4096.
77 D. Feichtinger, D. A. Plattner, *Angew. Chem. Int. Ed. Engl.*, **1997**, *36*, 1718.
78 For recent mechanistic discussions, see; a) C. Linde, N. Koliaï, P.-O. Norrby, B. Åkermark, *Chem. Eur. J.*, **2002**, *8*, 2568; b) Y. G. Abashkin, S. K. Burt, *Org. Lett.*, **2004**, *6*, 59; c) H. Jacobsen, L. Cavallo, *Phys. Chem. Chem. Phys.*, **2004**, *6*, 3747.
79 H. Jacobsen, L. Cavallo, *Chem. Eur. J.*, **2001**, *7*, 800.
80 A. Berkessel, M. Frauenkron, T. Schwenkreis, A. Steinmetz, G. Baum, D. Fenske, *J. Mol. Catal. A: Chem.*, **1996**, *113*, 321.
81 R. Irie, N. Hosoya, T. Katsuki, *Synlett*, **1994**, 255.
82 P. Pietikäinen, *Tetrahedron*, **1998**, *54*, 4319.
83 P. Pietikäinen, *J. Mol. Catal. A: Chem.*, **2001**, *165*, 73.
84 R. I. Kureshy, N. H. Khan, S. H. R. Abdi, S. T. Patel, R. V. Jasra, *Tetrahedron: Asymmetry*, **2001**, *12*, 433.
85 R. I. Kureshy, N. H. Kahn, S. H. R. Abdi, S. Singh, I. Ahmed, R. S. Shukla, R. V. Jasra, *J. Catal.*, **2003**, *219*, 1.
86 J. F. Larrow, E. N. Jacobsen, *J. Am. Chem. Soc.*, **1994**, *116*, 12129.
87 F. E. Kühn, W. A. Herrmann, *Chemtracts – Org. Chem.*, **2001**, *14*, 59.
88 W. A: Herrmann, R. W. Fischer, D. W. Marz, *Angew, Chem. Int. Ed. Engl.*, **1991**, *30*, 1638.
89 I. R. Beattie, P. J. Jones, *Inorg. Chem.*, **1979**, *18*, 2318.
90 W. A. Herrmann, J. G. Kuchler, J. K. Felixberger, E. Herdtweck, W. Wagner, *Angew. Chem. Int. Ed. Engl.*, **1988**, *27*, 394.
91 W. A. Herrmann, F. E. Kühn, R. W. Fischer, W. R. Thiel, C. C. Romão, *Inorg. Chem.*, **1992**, *31*, 4431.
92 W. A. Herrmann, W. R. Thiel, F. E. Kühn, R. W. Fischer, M. Kleine, E. Herdtweck, W. Scherer, J. Mink, *Inorg. Chem.*, **1993**, *32*, 5188.
93 W. A. Herrmann, R. Kratzer, R. W. Fischer, *Angew. Chem. Int. Ed. Engl.*, **1997**, *36*, 2652.
94 W. A. Herrmann, F. E. Kühn, *Acc. Chem. Res.*, **1997**, *30*, 169.
95 C. C. Romão, F. E. Kühn, W. A. Herrmann, *Chem. Rev.*, **1997**, *97*, 3197.
96 J. H. Espenson, *Chem. Commun.*, **1999**, 479.
97 M. M. Abu-Omar, P. J. Hansen, J. H. Espenson, *J. Am. Chem. Soc.*, **1996**, *118*, 4966.
98 G. Laurenczy, F. Lukács, R. Roulet, W. A. Herrmann, R. W. Fischer, *Organometallics*, **1996**, *15*, 848.

99 F. E. Kühn, A. Scherbaum, W. A. Herrmann, *J. Organomet. Chem.*, **2004**, *689*, 4149.
100 W. A. Herrmann, R. W. Fischer, W. Scherer, M. U. Rauch, *Angew. Chem. Int. Ed. Engl.*, **1993**, *32*, 1157.
101 W. A. Herrmann, R. W. Fischer, M. U. Rauch, W. Scherer, *J. Mol. Catal.*, **1994**, *86*, 243.
102 W. Adam, C. M. Mitchell, *Angew. Chem. Int. Ed. Engl.*, **1996**, *35*, 533.
103 J. Rudolph, K. L. Reddy, J. P. Chiang, K. B. Sharpless, *J. Am. Chem. Soc.*, **1997**, *119*, 6189.
104 W.-D. Wang, J. H. Espenson, *J. Am. Chem. Soc.*, **1998**, *120*, 11 335.
105 C. Copéret, H. Adolfsson, K. B. Sharpless, *Chem. Commun.*, **1997**, 1565.
106 H. Adolfsson, C. Copéret, J. P. Chiang, A. K. Yudin, *J. Org. Chem.*, **2000**, *65*, 8651.
107 C. Copéret, H. Adolfsson, T.-A. V. Khuong, A. K. Yudin, K. B. Sharpless, *J. Org. Chem.*, **1998**, *63*, 1740.
108 H. Adolfsson, K. B. Sharpless, *unpublished results.*
109 W. A. Herrmann, R. M. Kratzer, H. Ding, W. R. Thiel, H. Glas, *J. Organomet. Chem.*, **1998**, *555*, 293.
110 H. Adolfsson, A. Converso, K. B. Sharpless, *Tetrahedron Lett.*, **1999**, *40*, 3991.
111 A. R. Vaino, *J. Org. Chem.*, **2000**, *65*, 4210.
112 A. McKillop, W. R. Sanderson, *Tetrahedron*, **1995**, *51*, 6145.
113 W. P. Jackson, *Synlett*, **1990**, 536.
114 C. Jost, G. Wahl, D. Kleinhenz, J. Sundermeyer, in *Peroxide Chemistry* (Ed.: W. Adam), Wiley-VCH, Weinheim, **2000**, 341–364.
115 A. K. Yudin, K. B. Sharpless, *J. Am. Chem. Soc.*, **1997**, *119*, 11 536.
116 R. Curci, R. Mello, L. Troisi, *Tetrahedron*, **1986**, *42*, 877.
117 A. K. Yudin, J. P. Chiang, H. Adolfsson, C. Copéret, *J. Org. Chem.*, **2001**, *66*, 4713.
118 S. Warwel, M. Rüsch den Klaas, M. Sojka, *J. Chem. Soc., Chem. Commun.*, **1991**, 1578.
119 M. C. A. van Vliet, I. W. C. E. Arends, R. A. Sheldon, *Chem · Commun.*, **1999**, 821.
120 J. Iskra, D. Bonnet-Delpon, J.-P. Bégué, *Tetrahedron Lett.*, **2002**, *43*, 1001.
121 G. S. Owens, M. M. Abu-Omar, *Chem. Commun.*, **2000**, 1165.
122 G. S. Owens, A. Durazo, M. M. Abu-Omar, *Chem. Eur. J.*, **2002**, *8*, 3053.
123 C. E. Tucker, K. G. Davenport, **1997**, Hoechst Celanese Corporation US Patent 5618958.
124 M. J. Sabater, M. E. Domine, A. Corma, *J. Catal.*, **2002**, *210*, 192.
125 J. J. Haider, R. M. Kratzer, W. A. Herrmann, J. Zhao, F. E. Kühn *J. Organomet. Chem.*, **2004**, *689*, 3735.
126 T. G. Traylor, S. Tsuchiya,Y.-S. Byun, C. Kim, *J. Am. Chem. Soc.*, **1993**, *115*, 2775.
127 M. C. White, A. G. Doyle, E. N. Jacobsen, *J. Am. Chem. Soc.*, **2001**, *123*, 7194.
128 K. Chen, L. Que, Jr., *Chem. Commun.*, **1999**, 1375.
129 M. Costas, A. K. Tipton, K. Chen, D.-H. Jo, L. Que, Jr., *J. Am. Chem. Soc.*, **2001**, *123*, 6722.
130 K. Chen, M. Costas, J. Kim, A. K. Tipton, L. Que, Jr., *J. Am. Chem. Soc.*, **2002**, *124*, 3026.
131 M. B. Francis, E. N. Jacobsen, *Angew. Chem. Int. Ed. Engl.*, **1999**, *38*, 937.
132 S.-I. Murahashi, N. Komiya in *Ruthenium in Organic Synthesis* (Ed. S.-I. Murahashi), Wiley-VCH, Weinheim, **2004**, 53–87.
133 G. A. Barf, R. A. Sheldon, *J. Mol. Catal, A: Chem.*, **1995**, *102*, 23.
134 J. T. Groves, M. Bonchio, T. Carofiglio, K. Shalyaev, *J. Am. Chem. Soc.*, **1996**, *118*, 8961.
135 T. Higuchi, H. Othake, M. Hirobe, *Tetrahedron Lett.*, **1989**, *30*, 6545.
136 H. Ohtake, T. Higuchi, M. Hirobe, *Tetrahedron Lett.*, **1992**, *33*, 2521.
137 J. T. Groves, R. Quinn, *J. Am. Chem. Soc.*, **1985**, *107*, 5790.
138 M. Klawonn, M. K. Tse, S. Bhor, C. Döbler, M. Beller, *J. Mol. Catal. A: Chem.*, **2004**, *218*, 13.
139 A. Berkessel, P. Kaiser, J. Lex, *Chem. Eur. J.*, **2003**, *9*, 4746.
140 H. Nishiyama, T. Shimida, H. Itoh, H. Sugiyama, Y. Motoyama, *Chem. Commun.*, **1997**, 1863.
141 M. K. Tse, C. Döbler, S. Bhor, M. Klawonn, W. Mägerlein, H. Hugl, M. Beller, *Angew. Chem. Int. Ed.*, **2004**, *43*, 5255.
142 S. Bohr, M. K. Tse, M. Klawonn, C. Döbler, W. Mägerlein, M. Beller, *Adv. Synh. Catal.*, **2004**, *346*, 263.
143 M. K. Tse, S. Bhor, M. Klawonn, C. Döbler, M. Beller, *Tetrahedron Lett.*, **2003**, *44*, 7479.

7
Catalytic Asymmetric Epoxide Ring-opening Chemistry

Lars P. C. Nielsen and Eric N. Jacobsen

7.1
Introduction

Asymmetric catalysis of epoxide ring-opening reactions may be divided simply into two categories: desymmetrization of *meso*-epoxides and kinetic resolution of racemic epoxides (Scheme 7.1). Over the past few years, both have been the focus of substantial interest, and important advances have been made in the development of useful catalysts for these purposes. This chapter is concerned specifically with nonenzymatic [1] catalytic asymmetric ring-opening (ARO) reactions of epoxides. Nucleophilic addition reactions are discussed according to the identity of the reactive nucleophilic center (nitrogen, sulfur, oxygen, carbon, halide, and hydride). Enantioselective catalytic rearrangements of both *meso*-epoxides and chiral epoxides are treated in a separate section.

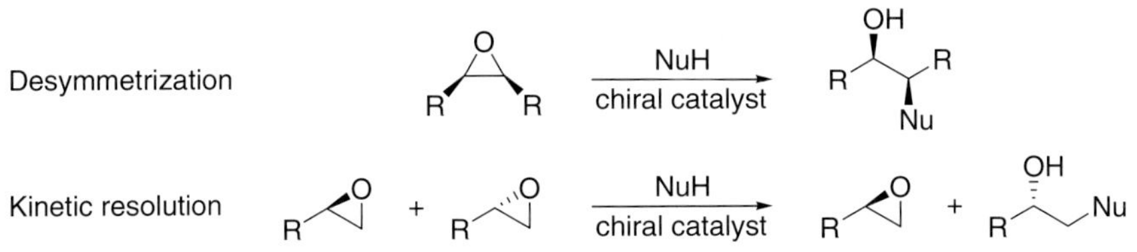

Desymmetrization
NuH
chiral catalyst

Kinetic resolution
NuH
chiral catalyst

Scheme 7.1

7.2
Enantioselective Nucleophilic Addition to *Meso*-Epoxides

7.2.1
Nitrogen-centered Nucleophiles

Azide is widely useful as a surrogate for ammonia in nucleophilic substitution reactions, due to its high nucleophilicity, low basicity, and stability towards a variety of conditions for subsequent transformations. In particular, the azidolysis of

Aziridines and Epoxides in Organic Synthesis. Andrei K. Yudin

ISBN: 3-527-31213-7

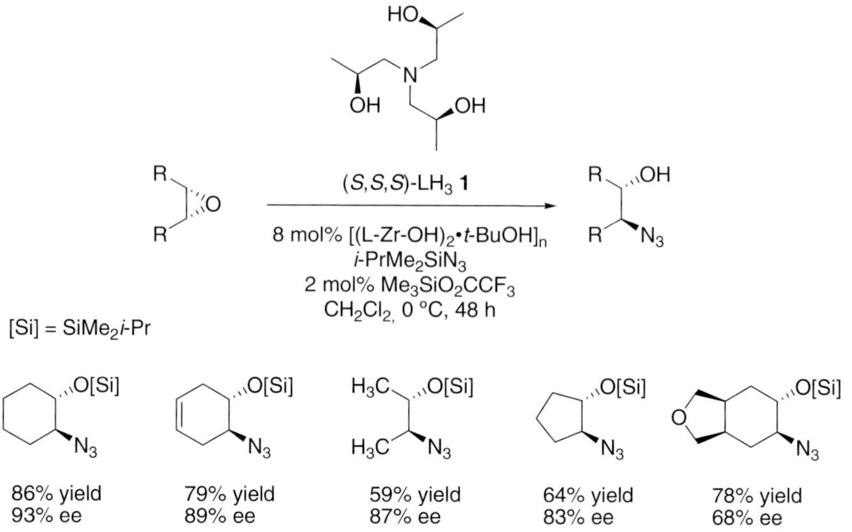

(S,S,S)-LH_3 **1**

8 mol% $[(L\text{-}Zr\text{-}OH)_2\cdot t\text{-}BuOH]_n$
$i\text{-}PrMe_2SiN_3$
2 mol% $Me_3SiO_2CCF_3$
CH_2Cl_2, 0 °C, 48 h

[Si] = $SiMe_2i\text{-}Pr$

86% yield, 93% ee; 79% yield, 89% ee; 59% yield, 87% ee; 64% yield, 83% ee; 78% yield, 68% ee

Scheme 7.2

meso-epoxides provides an attractive approach to chiral 1,2-amino alcohols. Building on the work of Yamashita and Oguni, who used stoichiometric amounts of chiral titanium alkoxides to effect asymmetric addition to *meso*-epoxides [2], Nugent developed the first catalytic and highly efficient asymmetric ring-opening of epoxides (Scheme 7.2) [3]. Nugent's result with a zirconium trialkanolamine complex **1** and $TMSN_3$ validated the ARO strategy as a worthy goal in asymmetric catalysis.

Jacobsen subsequently reported a practical and efficient method for promoting the highly enantioselective addition of $TMSN_3$ to *meso*-epoxides (Scheme 7.3) [4]. The chiral (salen)Cl–Cl catalyst **2** is available commercially and is bench-stable. Other practical advantages of the system include the mild reaction conditions, tolerance of some Lewis basic functional groups, catalyst recyclability (up to 10 times at 1 mol% with no loss in activity or enantioselectivity), and amenability to use under solvent-free conditions. Song later demonstrated that the reaction could be performed in room temperature ionic liquids, such as 1-butyl-3-methylimidazolium salts. Extraction of the product mixture with hexane allowed catalyst recycling and product isolation without recourse to distillation (Scheme 7.4) [5].

Detailed investigations indicated an interesting mechanism for azide openings catalyzed by **2** [6]. Chloride-epoxide addition products were observed in the initial stages of the ARO reaction with **2** in amounts commensurate with the catalyst loading. Azide complex **3**, characterized as the THF adduct, was isolated from the reaction mixture and proved to be an active and recyclable catalyst for the ARO, pointing to the role of **2** as that of a precatalyst. Kinetic experiments revealed a second-order dependence on the concentration of **3**, a zero-order dependence on azide source, and inverse-order dependence on epoxide concentration. The suc-

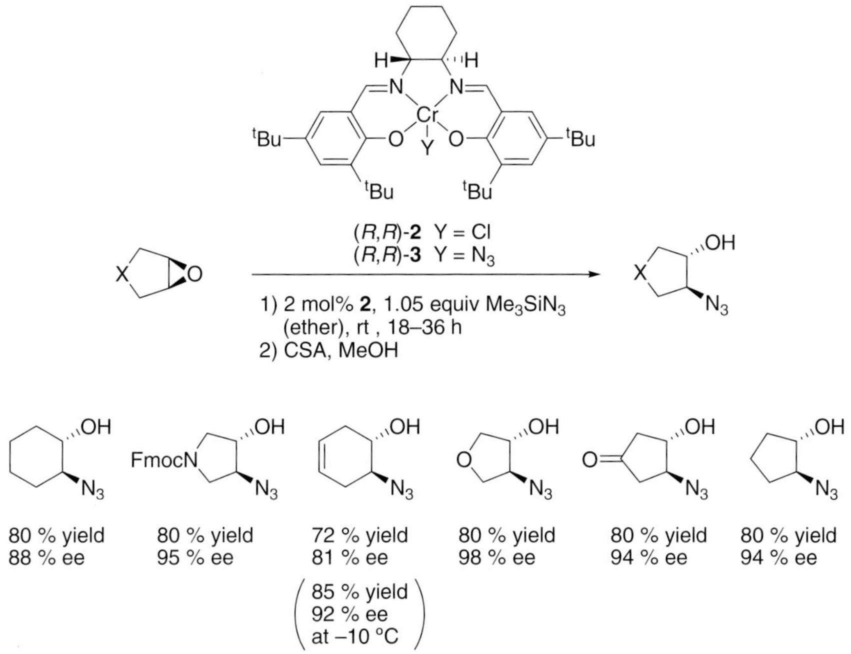

(R,R)-**2** Y = Cl
(R,R)-**3** Y = N_3

1) 2 mol% **2**, 1.05 equiv Me_3SiN_3 (ether), rt , 18–36 h
2) CSA, MeOH

80 % yield, 88 % ee; 80 % yield, 95 % ee; 72 % yield, 81 % ee (85 % yield, 92 % ee at –10 °C); 80 % yield, 98 % ee; 80 % yield, 94 % ee; 80 % yield, 94 % ee

Scheme 7.3

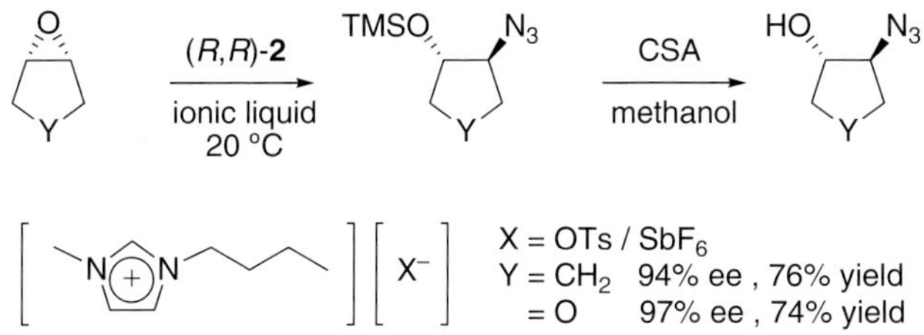

(R,R)-**2**, ionic liquid, 20 °C; CSA, methanol

X = OTs / SbF_6
Y = CH_2 94% ee , 76% yield
= O 97% ee , 74% yield

Scheme 7.4

cessful use of $TMSN_3$ required a small amount of water to form a catalytic amount of HN_3, the species that mediates turnover. Using these and other data, the authors proposed a catalytic cycle involving cooperative activation of both nucleophile and electrophile by distinct catalyst complexes (Scheme 7.5).

The relative orientation of the (salen)metal units in the enantioselectivity-determining transition state stands as an important question underlying the cooperative mechanism of the ARO. "Head to head" and "head to tail" limiting geometries for the bimetallic ring-opening are shown in Figure 7.1 [7]. Dimeric complexes containing various tethering motifs were prepared, and only catalysts capable of achieving "head to tail" geometry (Figure 7.1) were found to display *ees* and enhanced rates comparable to those seen with the monomeric catalyst **3**, which

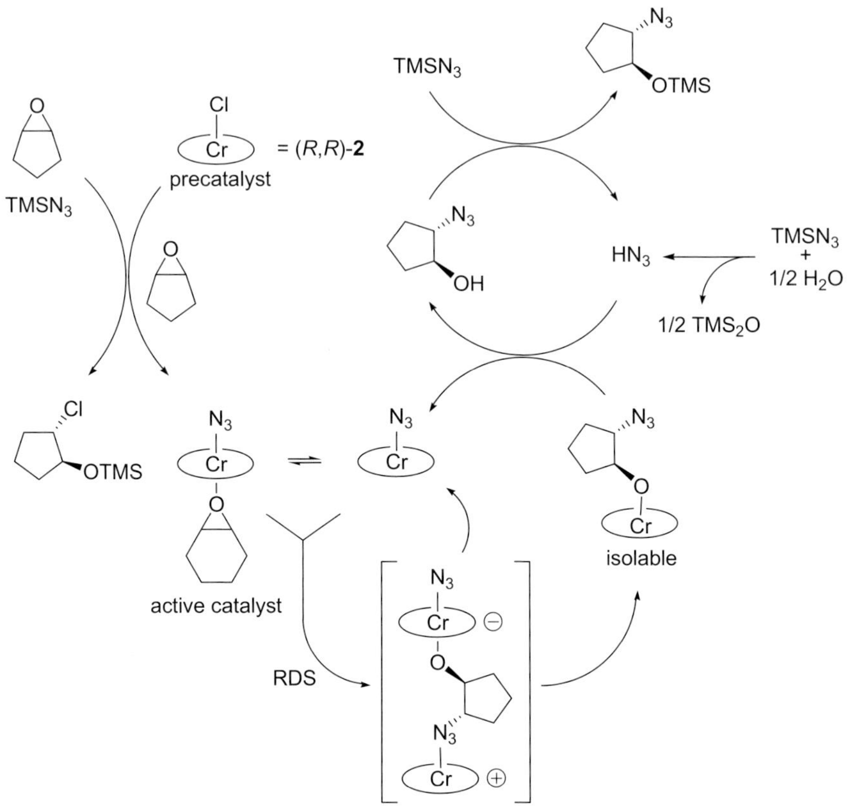

Scheme 7.5

suggested a transition structure approaching the "head to tail" geometry. Such mechanistic insight foreshadowed other mechanism-driven catalyst improvements in several ARO reactions (*vide infra*) [8].

The (salen)chromium-catalyzed ARO of epoxides with azide provides an efficient route to a range of chiral building blocks useful for synthesis. Several biologically interesting molecules have been prepared by this methodology, including the protein kinase C inhibitor balanol [9], carbocyclic nucleoside analogues such as aristeromycin and carbovir [10], and prostaglandin derivatives [11] (Figure 7.2). Ganem and co-workers provided an elegant application of this ARO reaction in the desymmetrization of a highly substituted epoxycyclopentane as part of a formal synthesis of (–)-allosamidin [12]. Catalyst **2** can also be used to desymmetrize resin-bound epoxides, as was demonstrated in the solid-phase synthesis of a library of cyclic RGD pharmacophores for platelet receptors [13].

Despite these significant results in azide additions, only limited success has been obtained in enantioselective addition of other sp^2-hybridized nitrogen-centered nucleophiles to *meso*-epoxides. Bartoli et al. demonstrated that aniline was a

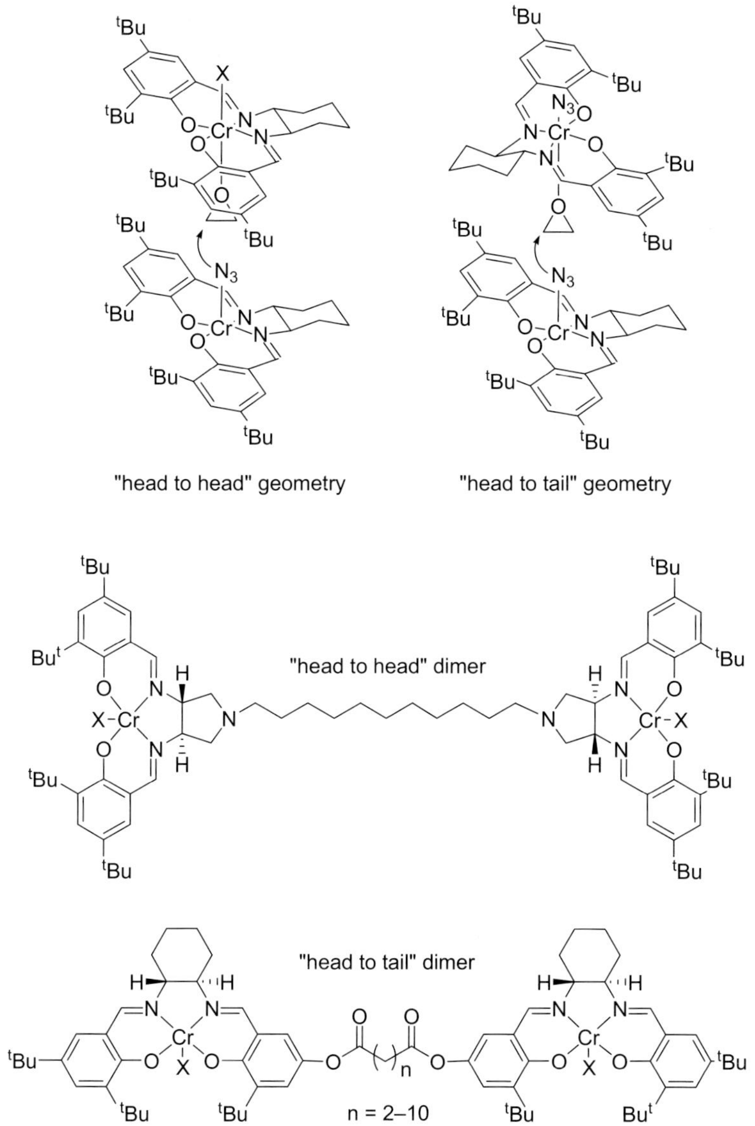

Figure 7.1 Head to head and head to tail limiting geometries for the bimetallic ring-opening.

competent nucleophile for the ARO of *cis*-stilbene oxide with catalyst **2** and triethylamine as an additive (Scheme 7.6) [14]. The reaction is highly enantioselective and oxidative dearylation to the primary amine can be effected without erosion of enantiomeric excess. However, the scope of this reaction has yet to be described.

Hou and Shibasaki independently reported the application of lanthanide-BI-

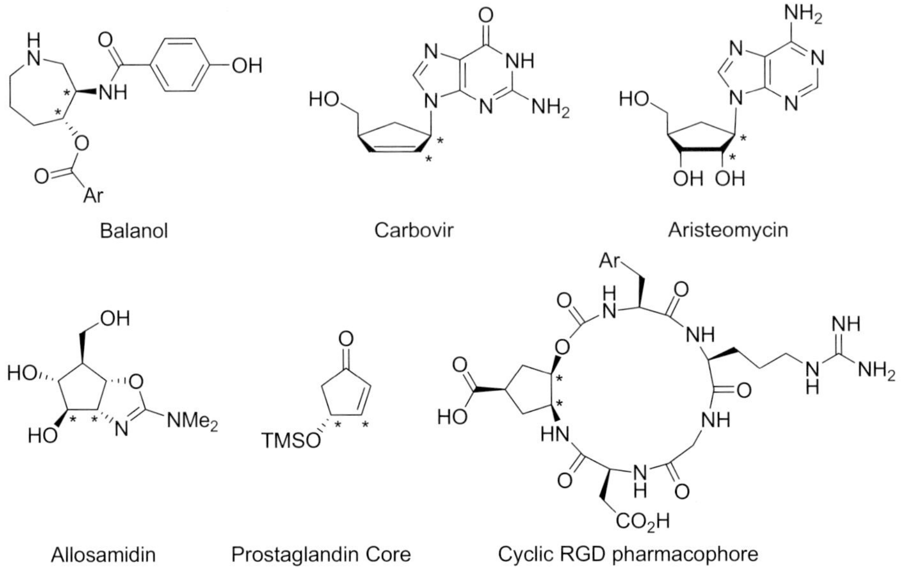

Balanol   Carbovir   Aristeomycin

Allosamidin   Prostaglandin Core   Cyclic RGD pharmacophore

Figure 7.2 Biologically interesting molecules prepared by (salen)chromium-catalyzed ARO of epoxides with azide.

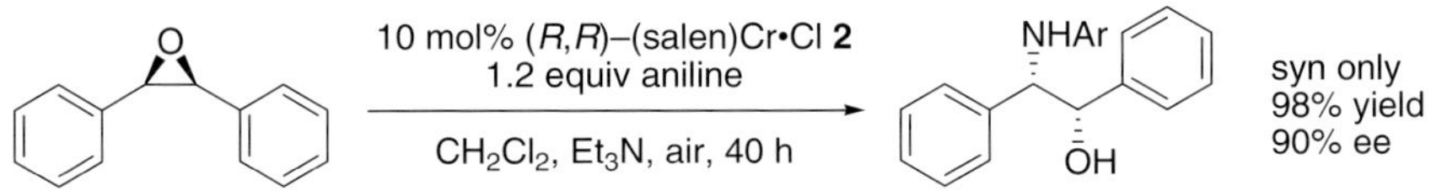

10 mol% (*R,R*)–(salen)Cr•Cl **2**, 1.2 equiv aniline, CH_2Cl_2, Et_3N, air, 40 h → syn only, 98% yield, 90% ee

Scheme 7.6

NOL-based systems for addition of (substituted) anilines to *meso* epoxides. Hou found that a ytterbium-BINOL complex catalyzed desymmetrization of cyclohexene oxide in up to 80% *ee* [15]. Shibasaki demonstrated that a praseodymium-BINOL complex could promote addition of *p*-anisidine to several epoxides in moderate yields with modest enantioselectivities (Scheme 7.7) [16].

The catalytic addition of alkylamines to *meso*-epoxides is generally difficult, presumably due to inhibition of the Lewis acid catalyst by the strongly Lewis basic nucleophile. Inaba et al., however, have shown that 3,5,8-trioxabicyclo[5.1.0]octanes are useful substrates with a 1:1 Ti-BINOL complex in ring-openings with primary alkylamines such as benzylamine. The resulting enantioselective addition reaction provided access to optically pure 2-amino-1,3,4-butanetriols-versatile chiral four-carbon building blocks (Scheme 7.8) [17]. Use of racemic or enantiomerically enriched α-methylbenzylamines as nucleophiles gave increased selectivity (>97% de). However, the method as reported was very limited in scope, as no other epoxides investigated were found to undergo ring-opening.

Relatively little has been published regarding asymmetric catalysis of ring-open-

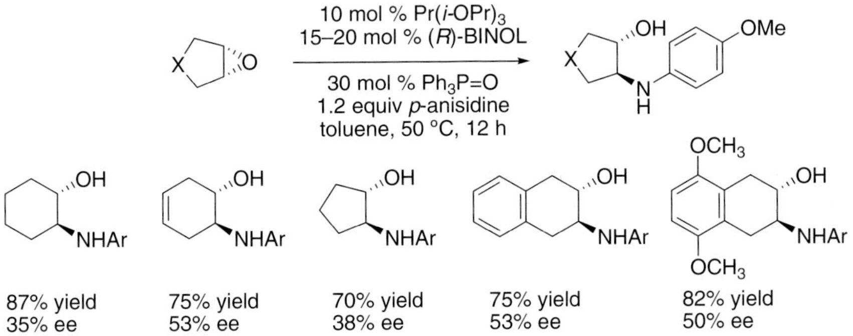

10 mol % Pr(i-OPr)3
15–20 mol % (R)-BINOL
30 mol % Ph3P=O
1.2 equiv p-anisidine
toluene, 50 °C, 12 h
OH
OMe
N
H
X
O
OH
NHAr
OCH3
87% yield
35% ee
75% yield
53% ee
70% yield
38% ee
75% yield
53% ee
82% yield
50% ee

Scheme 7.7

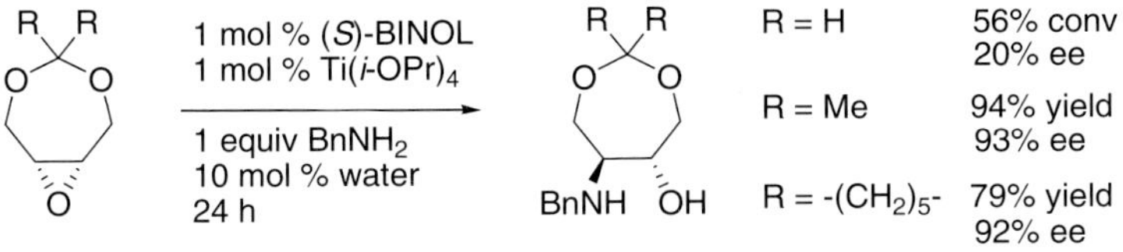

1 mol % (S)-BINOL
1 mol % Ti(i-OPr)4
1 equiv BnNH2
10 mol % water
24 h
BnNH OH
R = H 56% conv 20% ee
R = Me 94% yield 93% ee
R = -(CH2)5- 79% yield 92% ee

Scheme 7.8

Bn
CH3
Ga
CH3
(R)-4
1) 10 mol % (R)-4 , 1.2 equiv TMSCN
CH2Cl2, –78 °C to rt, 10 h
2) KF / MeOH
OH
NC
70 % yield
95 % ee
61 % yield
69 % ee
58 % yield
72 % ee

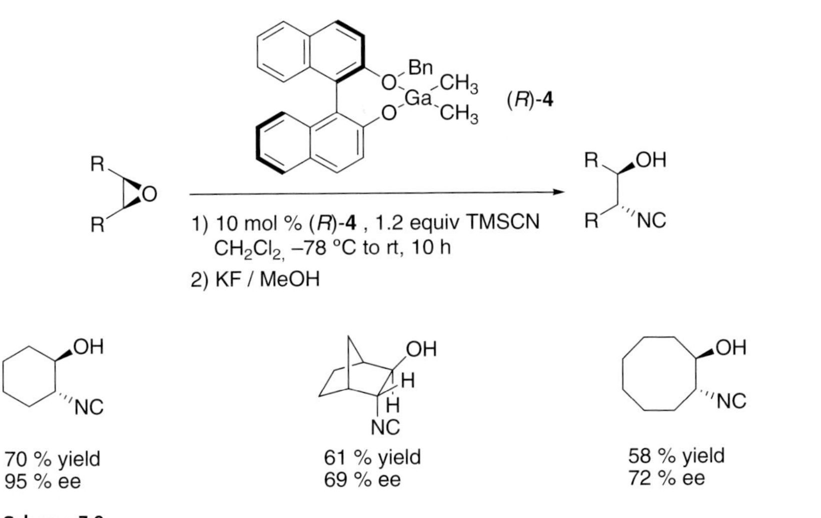

Scheme 7.9

ing reactions employing nucleophiles containing sp-hybridized nitrogen [18]. Whereas certain aluminium complexes are moderately effective for the asymmetric addition of cyanide to epoxides through the C terminus of the nucleophile (e.g., Scheme 7.23, *vide infra*), Pan and Zhu discovered that a Ga-BINOL complex **4** catalyzed the isocyanosilylation of *meso*-epoxides with TMSCN with good enantioselectivity (Scheme 7.9) [19]. The different regioselectivities promoted by metals in the same column of the periodic table was interpreted by these authors on the basis of the softer nature of gallium relative to the more electropositive aluminium.

7.2.2
Sulfur-centered Nucleophiles

One of the earliest useful methods for asymmetric opening of *meso*-epoxides with sulfur-centered nucleophiles was reported by Yamashita and Mukaiyama, who employed a heterogeneous zinc tartrate catalyst (Scheme 7.10) [20]. Epoxides other than cyclohexene oxide were not investigated, and the enantioselectivity depended strongly on the identity of the thiol.

Without question, the most significant advance in the use of sulfur-centered nucleophiles was made by Shibasaki, who discovered that 10 mol% of a novel gallium-lithium-bis(binaphthoxide) complex **5** could catalyze the addition of *tert*-butylthiol to various cyclic and acyclic *meso*-epoxides with excellent enantioselectivities and in good yields (Scheme 7.11) [21]. This work builds on Shibasaki's broader studies of heterobimetallic complexes, in which dual activation of both the electrophile and the nucleophile is invoked [22]. This method has been applied to an efficient asymmetric synthesis of the prostaglandin core through an oxidation/elimination sequence (Scheme 7.12).

Jacobsen demonstrated that the (salen)Cr system used to effect intermolecular, cooperative asymmetric azidolysis of *meso*-epoxides (Schemes 7.3 and 7.5) could be applied to sulfur-centered nucleophiles (Scheme 7.13). In order to overcome moderate enantioselectivity (<60% *ee*), a dithiol nucleophile was employed as part of a double resolution strategy in which the minor enantiomer of the monoaddition product reacts preferentially to form the *meso*- bis-addition product, thereby increasing the *ee* of the C_2-symmetric bis-addition product. Enantiopure 1,2-mercapto alcohols (>99% *ee*) were obtained from the *meso*-epoxide in ca. 50% overall yield by a burdensome (though effective) multistep sequence, [23].

Hou reported the use of a chiral (salen)titanium catalyst for the desymmetrization of *meso*-epoxides with thiols (Scheme 7.14). The complex, formed *in situ*

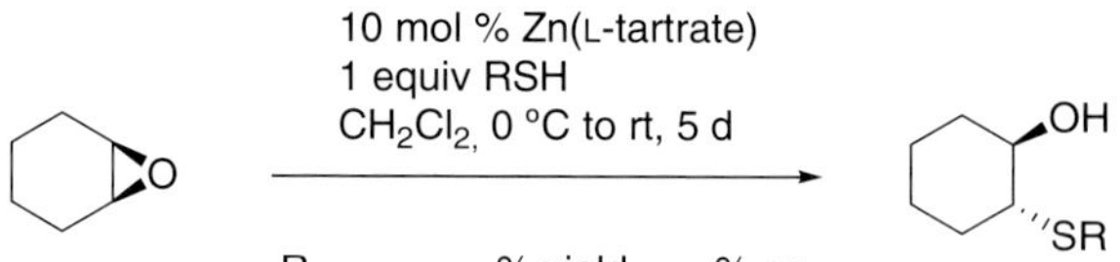

10 mol % Zn(L-tartrate)
1 equiv RSH
CH_2Cl_2, 0 °C to rt, 5 d

R	% yield	% ee
Ph	96	61
p-Tol	96	68
Bn	88	77
n-Bu	82	85
Cy	29	79

Scheme 7.10

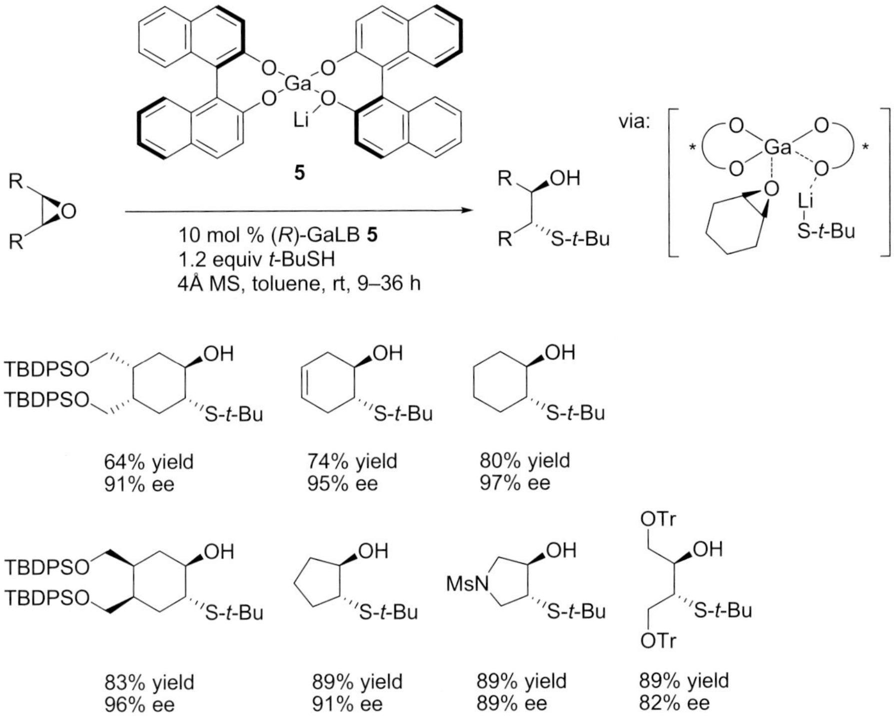

10 mol % (*R*)-GaLB **5**
1.2 equiv *t*-BuSH
4Å MS, toluene, rt, 9–36 h

Product	Yield	ee
1	64% yield	91% ee
2	74% yield	95% ee
3	80% yield	97% ee
4	83% yield	96% ee
5	89% yield	91% ee
6	89% yield	89% ee
7	89% yield	82% ee

Scheme 7.11

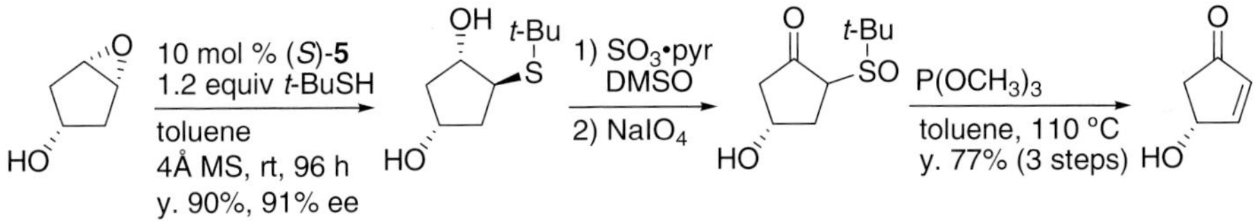

10 mol % (*S*)-**5**, 1.2 equiv *t*-BuSH, toluene, 4Å MS, rt, 96 h, y. 90%, 91% ee; 1) SO_3•pyr, DMSO; 2) $NaIO_4$; $P(OCH_3)_3$, toluene, 110 °C, y. 77% (3 steps)

Scheme 7.12

through the use of titanium tetraisopropoxide, effected the ARO of cyclohexene oxide with thiophenol, producing the mercaptoalkanol with moderate *ee* and in good yield (63% *ee*, 93% yield) [24]. Subsequently, Tang reported improved enantioselectivity with a dithiophosphorus acid as the nucleophile and the same catalyst system [25].

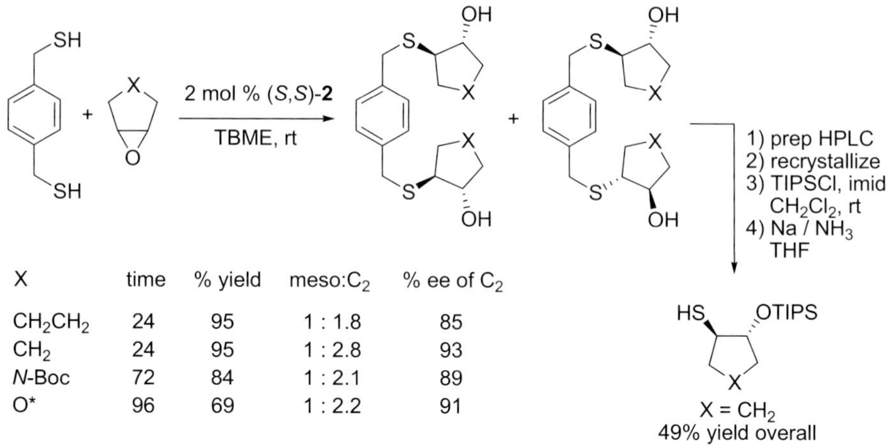

X	time	% yield	meso:C_2	% ee of C_2
CH_2CH_2	24	95	1 : 1.8	85
CH_2	24	95	1 : 2.8	93
N-Boc	72	84	1 : 2.1	89
O*	96	69	1 : 2.2	91

*A 1:1 TBME/THF solvent mixture was used with 10 mol % **2**.

Scheme 7.13

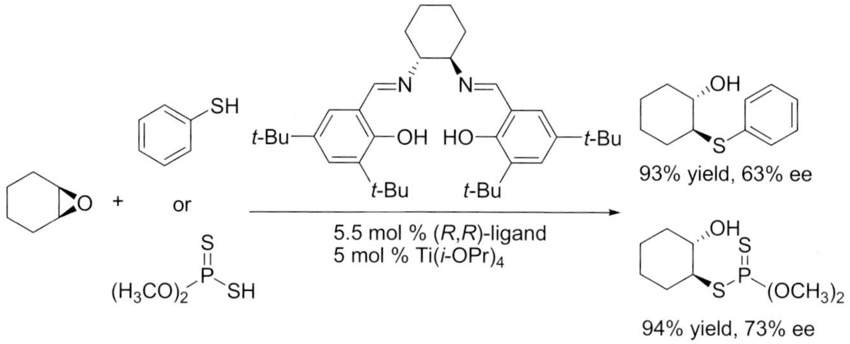

Scheme 7.14

7.2.3
Oxygen-centered Nucleophiles

Subsequent to the development of the (salen)Cr-catalyzed desymmetrization of *meso*-epoxides with azide (Scheme 7.3), Jacobsen discovered that the analogous (salen)Co(II) complex **6** promoted the enantioselective addition of benzoic acids to *meso*-epoxides to afford valuable monoprotected C_2-symmetric diols (Scheme 7.15) [26]. Under the reaction conditions, complex **6** served as a precatalyst for the (salen) Co(III)–OBz complex, which was formed *in situ* by aerobic oxidation. While the enantioselectivity was moderate for certain substrates, the high crystallinity of the products allowed access to enantiopure materials by simple recrystallization.

From the standpoints of both cost and atom economy, water is the ideal nucleophile for synthesis of enantioenriched C_2-symmetric 1,2-diols from *meso*-epoxides.

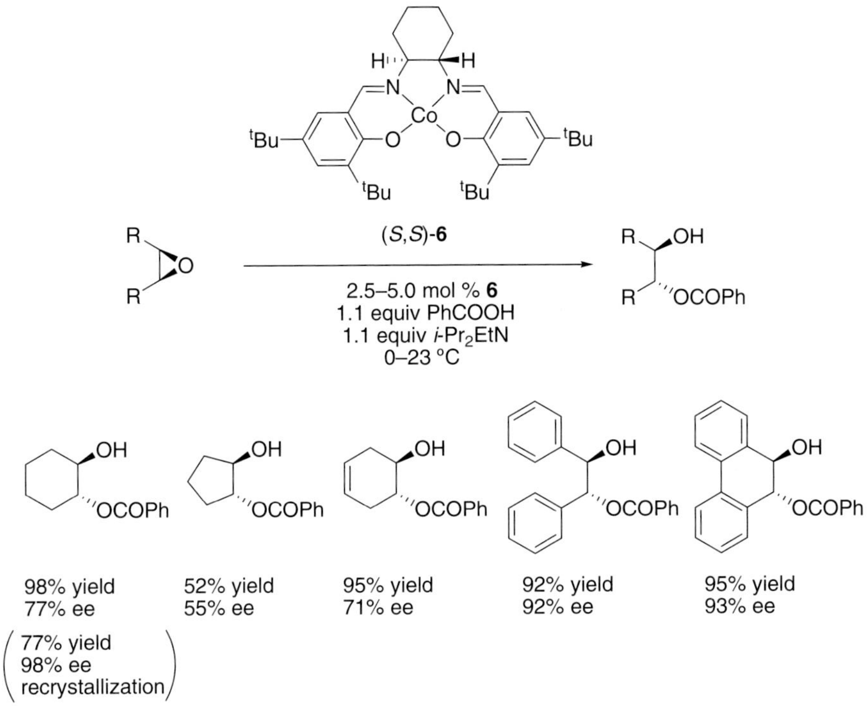

(*S*,*S*)-**6**

2.5–5.0 mol % **6**
1.1 equiv PhCOOH
1.1 equiv *i*-Pr_2EtN
0–23 °C

98% yield
77% ee
(77% yield
98% ee
recrystallization)

52% yield
55% ee

95% yield
71% ee

92% yield
92% ee

95% yield
93% ee

Scheme 7.15

Although racemic terminal epoxides undergo efficient ring-opening with kinetic resolution in the presence of (salen)Co catalysts and water (*vide infra*, Table 7.1), the analogous hydrolysis of *meso*-epoxides is promoted by the same catalysts with substantially lower reactivity and enantioselectivity. These limitations are overcome by the use of tethered catalyst systems, which dramatically increase the effective molarity of the cobalt centers and therefore help promote the cooperative bimetallic mechanism (see Scheme 7.5). Dimeric [7], polymeric [27], and dendrimeric catalysts have been developed [28], but the most efficient and most easily prepared tethered catalysts were cyclic oligomers. Several generations of these catalysts were synthesized [29], and ultimately the oligomeric (salen)Co–OTf complex **7** was found to be the most general and effective catalyst [30]. This catalyst system proved to be particularly effective for the desymmetrization of several *meso*-epoxides with water (Scheme 7.16) [31].

Although the enantioselective intermolecular addition of aliphatic alcohols to *meso*-epoxides with (salen)metal systems has not been reported, intramolecular asymmetric ring-opening of *meso*-epoxy alcohols has been demonstrated. By use of monomeric cobalt acetate catalyst **8**, several complex cyclic and bicyclic products can be accessed in highly enantioenriched form from the readily available *meso*-epoxy alcohols (Scheme 7.17) [32].

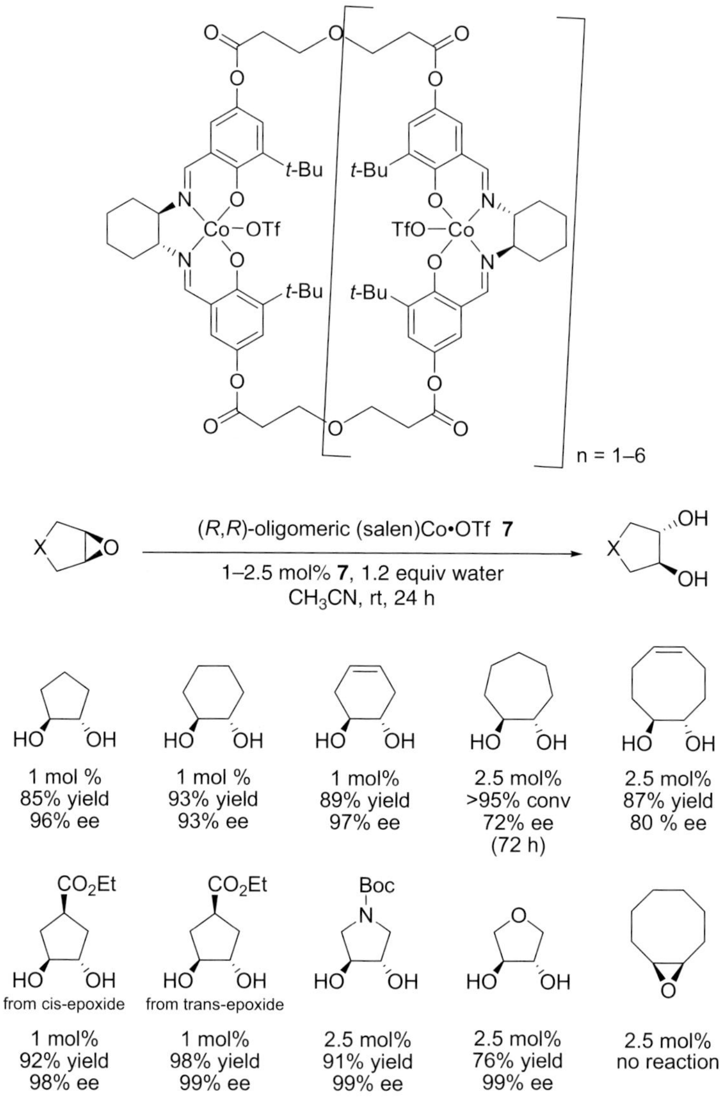

(*R*,*R*)-oligomeric (salen)Co•OTf **7**

1–2.5 mol% **7**, 1.2 equiv water
CH_3CN, rt, 24 h

n = 1–6

1 mol %	1 mol %	1 mol%	2.5 mol%	2.5 mol%
85% yield	93% yield	89% yield	>95% conv	87% yield
96% ee	93% ee	97% ee	72% ee	80 % ee
			(72 h)	

from cis-epoxide	from trans-epoxide			
1 mol%	1 mol%	2.5 mol%	2.5 mol%	2.5 mol%
92% yield	98% yield	91% yield	76% yield	no reaction
98% ee	99% ee	99% ee	99% ee	

Scheme 7.16

An impressive application of the (salen)Co-catalyzed intramolecular ARO of *meso*-epoxy alcohols in the context of total synthesis was reported recently by Danishefsky [33]. Enantioselective desymmetrization of intermediate **9** by use of the cobalt acetate catalyst **8** at low temperatures afforded compound **10**, which was obtained in 86% *ee* and >86% yield (Scheme 7.18). Straightforward manipulation of **10** eventually produced an intermediate that intersected Danishefsky's previ-

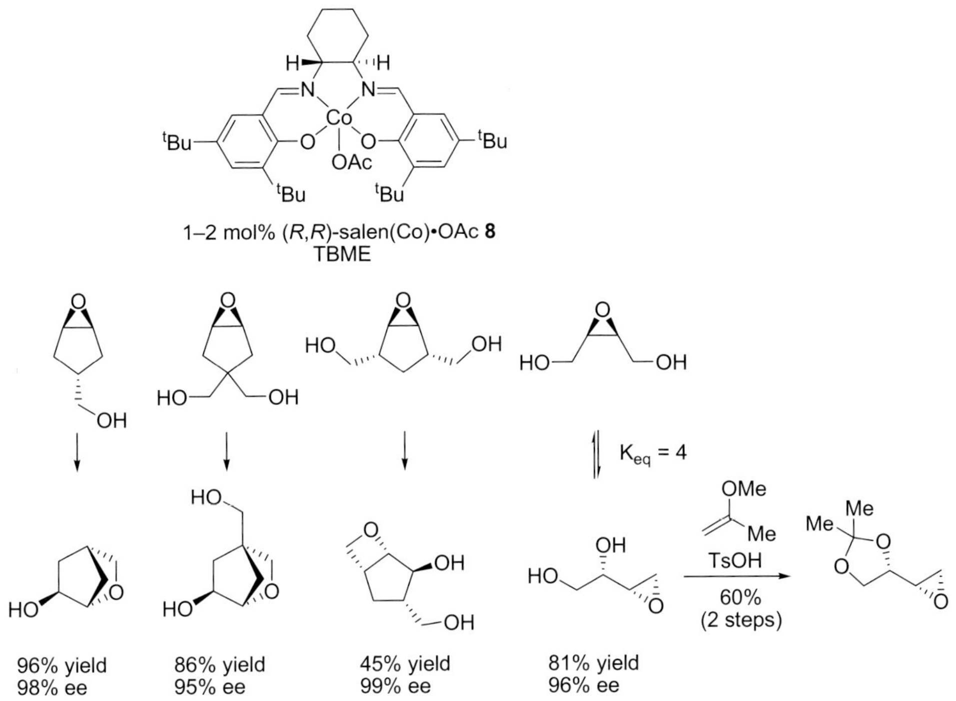

1–2 mol% (*R*,*R*)-salen(Co)•OAc **8**
TBME

96% yield
98% ee

86% yield
95% ee

45% yield
99% ee

81% yield
96% ee

K_{eq} = 4

TsOH
60%
(2 steps)

Scheme 7.17

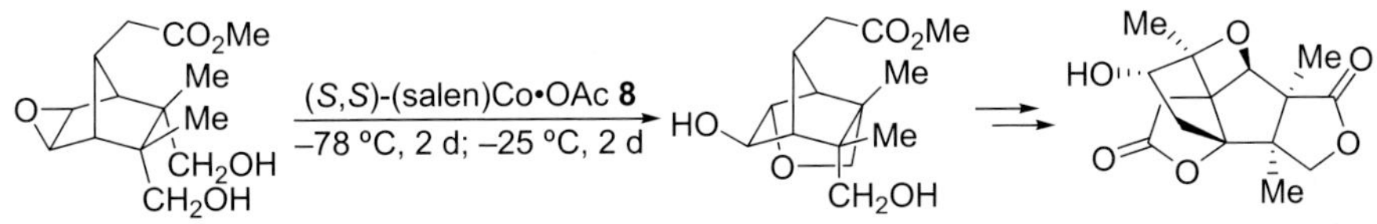

(*S*,*S*)-(salen)Co•OAc **8**
–78 °C, 2 d; –25 °C, 2 d

9

10

merrilactone A

Scheme 7.18

ously reported route [34], constituting a versatile, formal synthesis of either enantiomer of merrilactone A.

Payne rearrangement of 2,3-epoxybutane-1,4-diol (Scheme 7.17) not only provided a useful epoxydiol C4 building block, but also a rare example of an enantioselective reaction under thermodynamic control: the *meso*-starting material underwent equilibration selectively with the desired enantiomer of product. Use of the previously mentioned oligomeric catalyst **7** proceeded with 100-fold lower catalyst loading and 99% *ee* (compared to 96% *ee* with the monomeric catalyst). One-pot enantioselective Payne rearrangement and ketalization gave access to a valuable, protected building block in three steps from *cis*-butenediol [35]. The same epoxide had previously been prepared in five steps from either tartaric [36] or ascorbic acid [37].

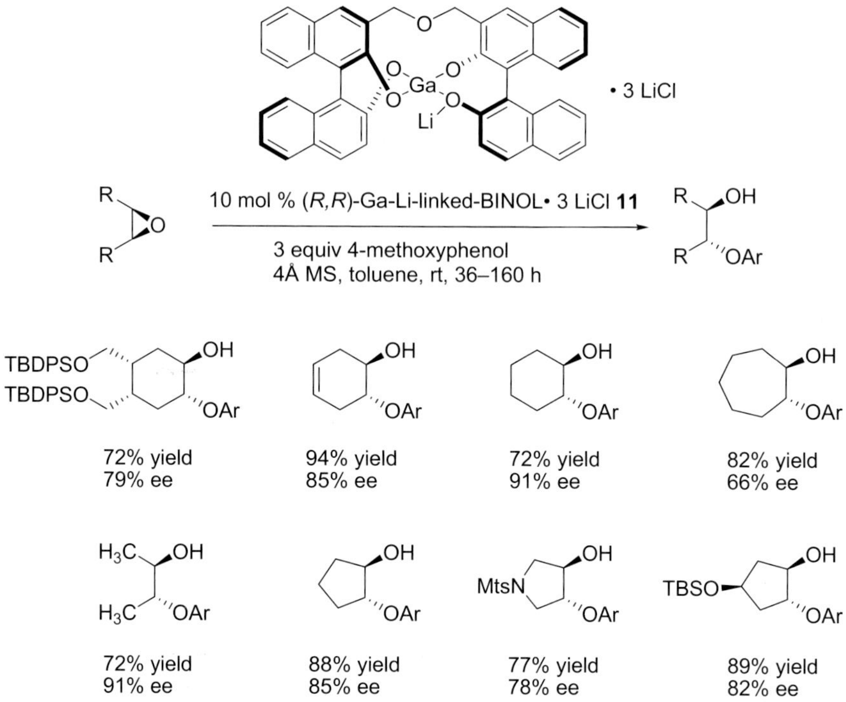

• 3 LiCl

10 mol % (R,R)-Ga-Li-linked-BINOL• 3 LiCl **11**

3 equiv 4-methoxyphenol
4Å MS, toluene, rt, 36–160 h

72% yield
79% ee

94% yield
85% ee

72% yield
91% ee

82% yield
66% ee

72% yield
91% ee

88% yield
85% ee

77% yield
78% ee

89% yield
82% ee

Scheme 7.19

Less success has been achieved to date in the addition of other oxygen-centered nucleophiles, such as phenols, to *meso*-epoxides. Shibasaki reported that the same gallium-lithium-BINOL complex as used in the enantioselective addition of thiols to *meso*-epoxides (Scheme 7.11) was also capable of catalyzing *p*-anisole addition at elevated temperatures [38]. However, high catalyst loadings (20 mol%), long reaction times, and modest yields render this process less attractive than the corresponding addition of thiols. As a result, Shibasaki investigated linked BINOL complexes, with the rationale that a more rigid ligand structure might be more stable and enantioselective. To this end, ether-linked bis(BINOL) gallium-lithium complex **11** was prepared in five linear steps from commercially available materials. The linked system allowed the epoxide ring-opening reaction to be effected with lower catalyst loading (10 mol%) than the analogous monomeric system (Scheme 7.19) [39].

An alternative method for generating enriched 1,2-diols from *meso*-epoxides consists of asymmetric copolymerization with carbon dioxide. Nozaki demonstrated that a zinc complex formed *in situ* from diethylzinc and diphenylprolinol catalyzed the copolymerization with cyclohexene oxide in high yield. Alkaline hydrolysis of the isotactic polymer then liberated the *trans* diol in 94% yield and 70% *ee* (Scheme 7.20) [40]. Coates later found that other zinc complexes such as **12** are also effective in forming isotactic polymers [41–42].

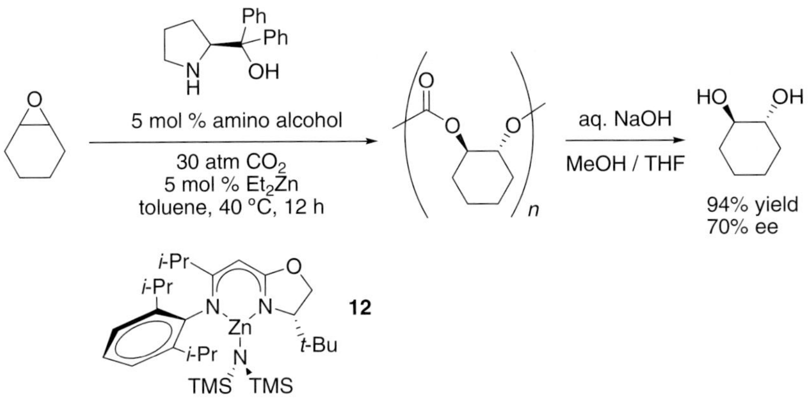

Scheme 7.20

7.2.4
Carbon-centered Nucleophiles

Although asymmetric ring-opening of epoxides coupled to C–C bond-formation is highly desirable from a synthetic standpoint, such transformations are difficult because few carbon-centered nucleophiles possess sufficient reactivity to open epoxides without effecting decomposition of the chiral metal catalysts. Nevertheless, several significant advances have been made with sp-, sp^2-, and even sp^3-hybridized carbon-centered nucleophiles. Oguni discovered that the addition of trimethylsilyl cyanide to epoxides was catalyzed by titanium tetraisopropoxide and was accelerated in the presence of achiral tridentate Schiff-base ligands [43]. Building on this result, Snapper and Hoveyda used a combinatorial approach [44] and solid-phase assembly [45] to identify an effective chiral catalyst for this transformation. After three generations of catalyst optimization, they identified a catalyst system based on ligand **13** and titanium for the desymmetrization of three representative *meso*-epoxides in the presence of TMSCN in moderate enantioselectivities and yields (Scheme 7.21).

Jacobsen developed a method employing (pybox)$YbCl_3$ for TMSCN addition to *meso*-epoxides (Scheme 7.22) [46] with enantioselectivities as high as 92%. Unfortunately, the practical utility of this method is limited because low temperatures must be maintained for very long reaction times (up to seven days). This reaction displayed a second-order dependence on catalyst concentration and a positive nonlinear effect, suggesting a cooperative bimetallic mechanism analogous to that proposed for (salen)Cr-catalyzed ARO reactions (Scheme 7.5).

The only notable success to date in the use of (salen)metal systems in catalysis of asymmetric cyanide addition to epoxides was achieved by Pietrusiewicz, who reported the aluminium-catalyzed desymmetrization of phospholene *meso*-epoxide (Scheme 7.23) in moderate *ee* [47]. Despite these significant efforts, a truly prac-

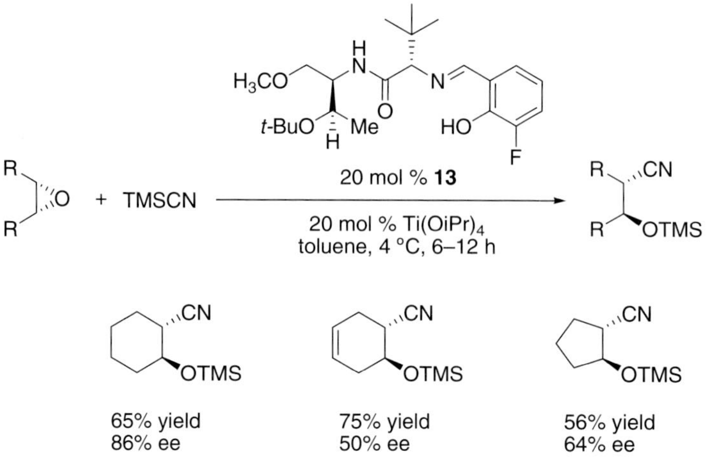

+ TMSCN

20 mol % **13**

20 mol % Ti(OiPr)$_4$
toluene, 4 °C, 6–12 h

65% yield	75% yield	56% yield
86% ee	50% ee	64% ee

Scheme 7.21

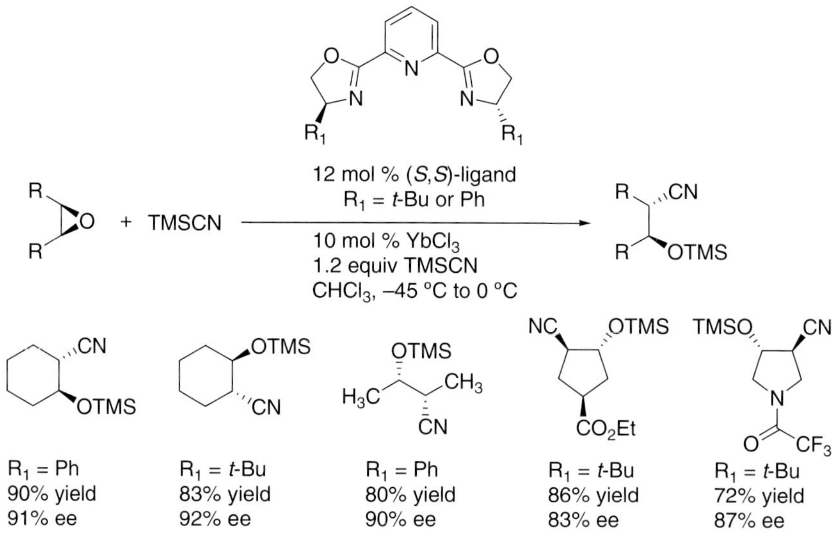

+ TMSCN

12 mol % (*S*,*S*)-ligand
R_1 = *t*-Bu or Ph

10 mol % $YbCl_3$
1.2 equiv TMSCN
$CHCl_3$, –45 °C to 0 °C

R_1 = Ph	R_1 = *t*-Bu	R_1 = Ph	R_1 = *t*-Bu	R_1 = *t*-Bu
90% yield	83% yield	80% yield	86% yield	72% yield
91% ee	92% ee	90% ee	83% ee	87% ee

Scheme 7.22

tical and general addition of cyanide to *meso*-epoxides remains an unmet challenge.

The first example of asymmetric catalytic ring-opening of epoxides with sp^2-hybridized carbon-centered nucleophiles was reported by Oguni, who demonstrated that phenyllithium and a chiral Schiff base ligand undergo reaction to form a stable system that can be used to catalyze the enantioselective addition of phenyllithium to *meso*-epoxides (Scheme 7.24) [48]. Oguni proposed that phenyllithium

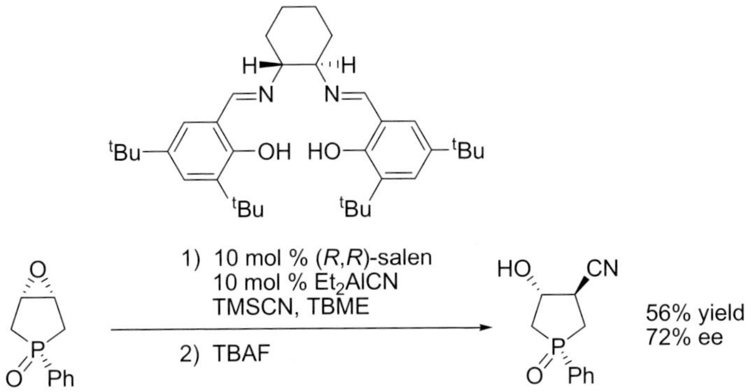

1) 10 mol % (*R*,*R*)-salen
10 mol % Et_2AlCN
TMSCN, TBME
2) TBAF

56% yield
72% ee

Scheme 7.23

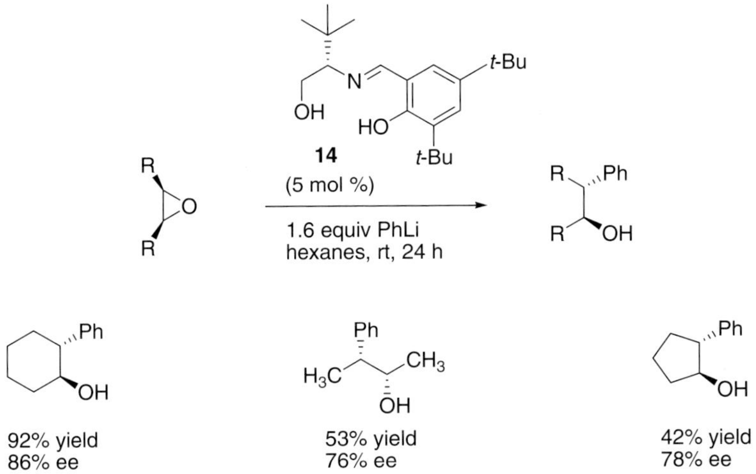

14
(5 mol %)
1.6 equiv PhLi
hexanes, rt, 24 h

92% yield
86% ee

53% yield
76% ee

42% yield
78% ee

Scheme 7.24

reacts with the ligand **14** by addition to the imine or deprotonation (or both) to form the active catalyst. Salen ligands proved ineffective.

In contrast, Cozzi and Umani-Ronchi found the (salen)Cr–Cl complex **2** to be very effective for the desymmetrization of *meso*-stilbene oxide with use of substituted indoles as nucleophiles (Scheme 7.25) [49]. The reaction is high-yielding, highly enantioselective, and takes place exclusively at sp^2-hybridized C3, independently of the indole substitution pattern at positions 1 and 2. The successful use of *N*-alkyl substrates (Scheme 7.25, entries 2 and 4) suggests that nucleophile activation does not occur in this reaction, in stark contrast with the highly enantioselective cooperative bimetallic mechanism of the (salen)Cr–Cl-catalyzed asymmetric azidolysis reaction (Scheme 7.5). However, no kinetic studies on this reaction were reported.

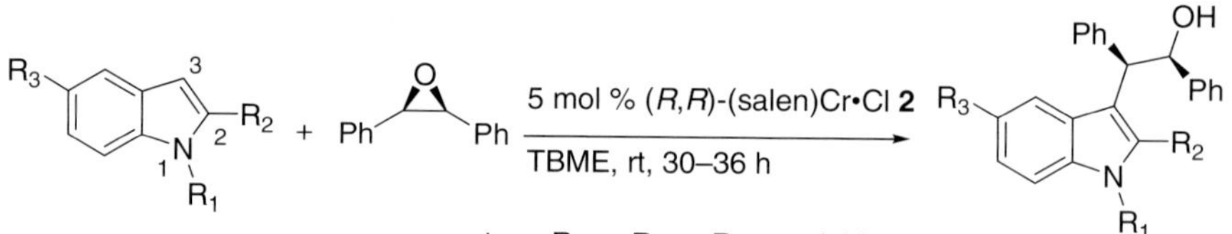

5 mol % (*R,R*)-(salen)Cr•Cl **2**
TBME, rt, 30–36 h

entry	R_1	R_2	R_3	yield	% ee
1	H	H	H	98	93
2	CH_3	H	H	96	96
3	H	CH_3	H	98	98
4	CH_3	CH_3	H	95	97
5	H	H	CH_3	95	90

Scheme 7.25

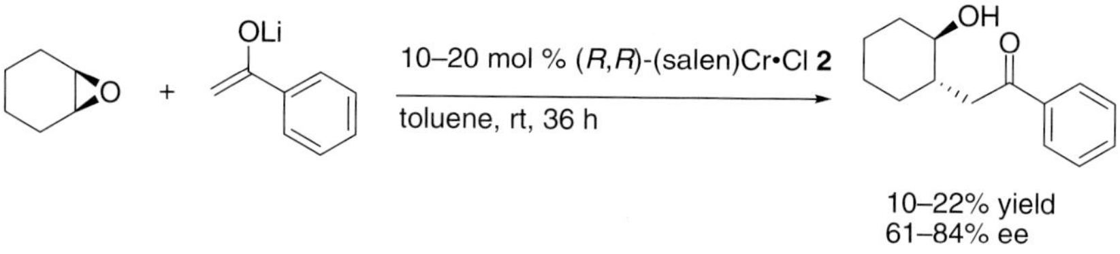

10–20 mol % (*R,R*)-(salen)Cr•Cl **2**
toluene, rt, 36 h

10–22% yield
61–84% ee

Scheme 7.26

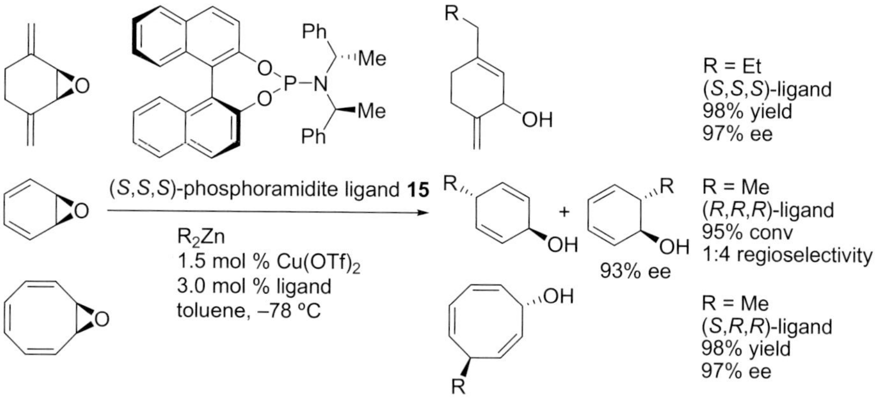

(*S,S,S*)-phosphoramidite ligand **15**
R_2Zn
1.5 mol % $Cu(OTf)_2$
3.0 mol % ligand
toluene, –78 °C

R = Et
(*S,S,S*)-ligand
98% yield
97% ee

R = Me
(*R,R,R*)-ligand
95% conv
1:4 regioselectivity
93% ee

R = Me
(*S,R,R*)-ligand
98% yield
97% ee

Scheme 7.27

Although enolates, their equivalents, and otherwise stabilized carbanions would be interesting candidates for ARO of *meso*-epoxides, no efficient catalytic method has been developed to date. Crotti reported that 20 mol% of (salen)Cr–Cl complex **2** promoted the addition of the lithium enolate of acetophenone to cyclohexene oxide with moderate *ees* (Scheme 7.26) [50]. However, the very low yields obtained suggested that catalyst turnover is very inefficient under the reaction conditions, and that the chromium complex and the lithium enolate may be incompatible.

An important advance in carbon-carbon bond-forming transformations involving asymmetric ring-opening was reported recently by Pineschi, who developed a copper-catalyzed S_N2' reaction of sp^3-hybridized carbon-centered nucleophiles to activated *meso*-epoxides [51–52]. With use of only 1.5 mol% $Cu(OTf)_2$, 3.0 mol% of phosphoramidite ligand **15**, and 1.5 equiv. of a dialkylzinc reagent, several symmetrical bis-allylic epoxides were desymmetrized with high enantioselectivity and in good to excellent yields, although in some cases with moderate regioselectivity (Scheme 7.27).

7.2.5
Halide and Hydride Nucleophiles

Halohydrins are important synthetic intermediates because they can act both as coupling partners and as surrogates for epoxides. Accordingly, the ring-opening of epoxides with halide nucleophiles stands as an important goal in asymmetric catalysis. Nugent discovered that the chiral zirconium catalyst derived from ligand **1**, previously found to promote azide addition to *meso*-epoxides (Scheme 7.2), catalyzed the highly enantioselective addition of bromide ion to several cyclic and acylic *meso*-epoxides in high yields (Scheme 7.28) [53]. This reaction required the presence of an excess of allyl bromide to effect transformation of the zirconium-azide catalyst to the putative metal-bromide complex and to suppress the formation of the azido alcohol. Haufe applied titanium-BINOL and titanium-TADDOLate complexes to catalyze chlorohydrin formation using TMSCl and Li_2CuCl_4 as chloride sources, but with 36% *ee* at best [54].

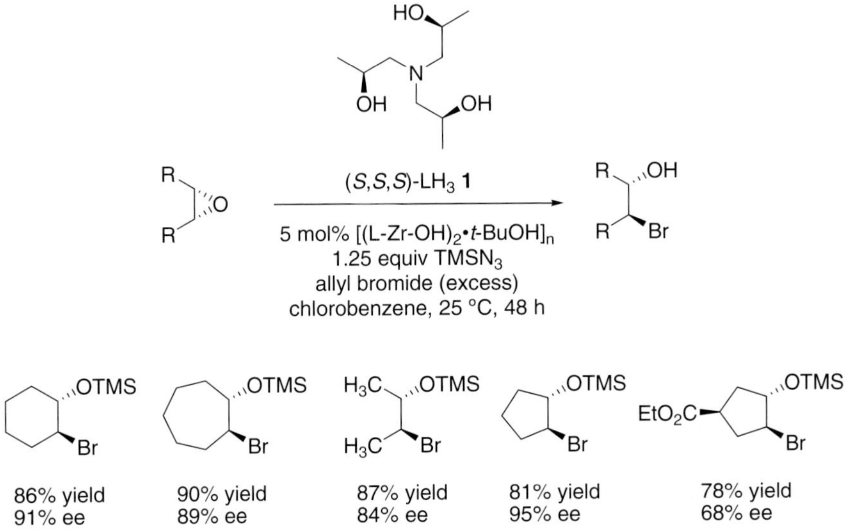

(*S,S,S*)-LH$_3$ **1**

5 mol% [(L-Zr-OH)$_2$•*t*-BuOH]$_n$
1.25 equiv TMSN$_3$
allyl bromide (excess)
chlorobenzene, 25 °C, 48 h

86% yield 91% ee	90% yield 89% ee	87% yield 84% ee	81% yield 95% ee	78% yield 68% ee

Scheme 7.28

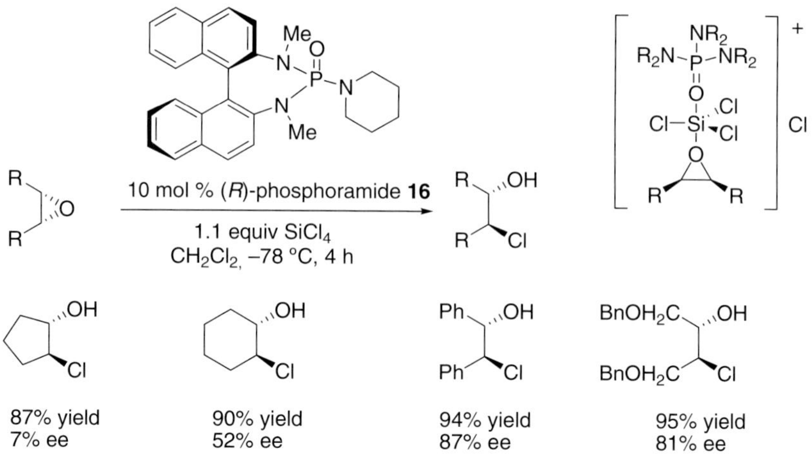

10 mol % (*R*)-phosphoramide **16**

1.1 equiv $SiCl_4$

CH_2Cl_2, –78 °C, 4 h

87% yield	90% yield	94% yield	95% yield
7% ee	52% ee	87% ee	81% ee

Scheme 7.29

Denmark and co-workers showed that the ring-opening of epoxides with $SiCl_4$ in the presence of Lewis bases (e.g., HMPA) is a general and efficient reaction [55]. Subsequently they discovered that chiral phosphoramide Lewis bases promote enantioselective halohydrin formation, although the levels of asymmetric induction are highly substrate-dependent (Scheme 7.29). The best result was achieved with *meso*-stilbene oxide and phosphoramide **16** in which the chlorohydrin was formed in 94% yield and 87% *ee*. The mechanism is thought to proceed via a cationic silicon-phosphoramide species that activates an epoxide towards chloride addition.

High enantiomeric excesses for the addition of chloride to *meso*-epoxides were also obtained with use of a planar-chiral pyridine *N*-oxide **17** developed by Fu

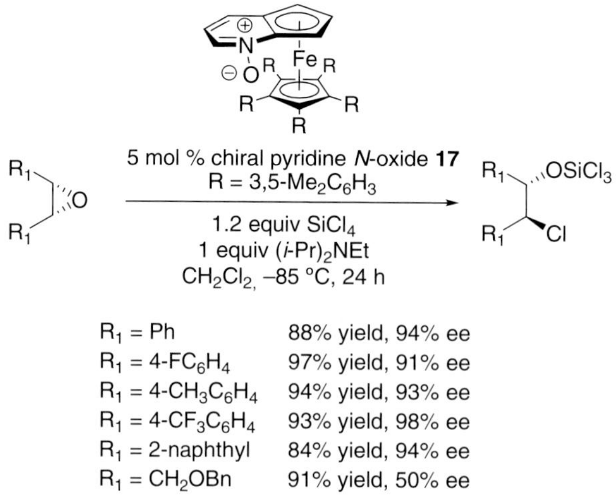

5 mol % chiral pyridine *N*-oxide **17**
R = 3,5-$Me_2C_6H_3$

1.2 equiv $SiCl_4$
1 equiv $(i\text{-}Pr)_2NEt$
CH_2Cl_2, –85 °C, 24 h

R_1 = Ph	88% yield, 94% ee
R_1 = 4-FC_6H_4	97% yield, 91% ee
R_1 = 4-$CH_3C_6H_4$	94% yield, 93% ee
R_1 = 4-$CF_3C_6H_4$	93% yield, 98% ee
R_1 = 2-naphthyl	84% yield, 94% ee
R_1 = CH_2OBn	91% yield, 50% ee

Scheme 7.30

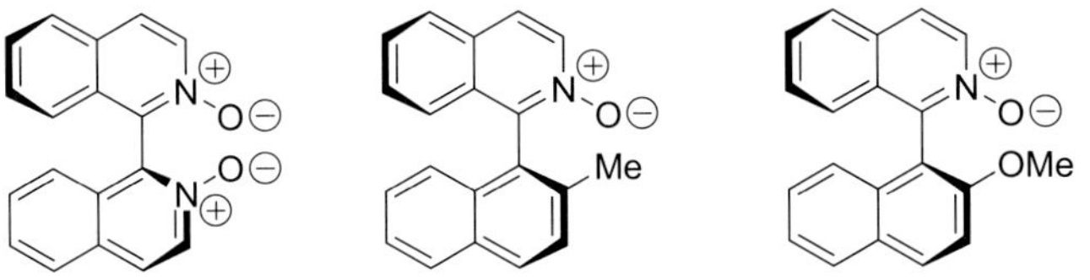

18 **19** **20**

Figure 7.3 Nakajima's chiral bipyridine *N,N'*-dioxide **18** and the less effective mono-*N*-oxide derivatives **19** and **20**.

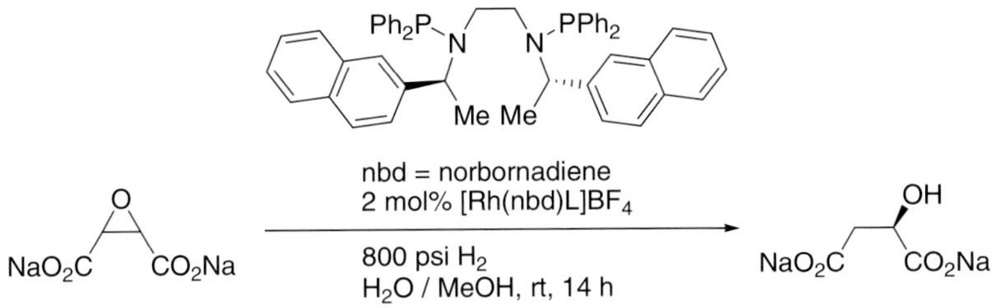
Ph2P
N
N
PPh2
Me Me
nbd = norbornadiene
2 mol% [Rh(nbd)L]BF4
800 psi H2
H2O / MeOH, rt, 14 h
O
NaO2C
CO2Na
OH
NaO2C
CO2Na

Scheme 7.31

(Scheme 7.30) [56]. In this work it was found that steric tuning of the aryl substituents of the Cp ring was key to achieving high enantioselectivity with aryl-substituted epoxides. Although *meso*-stilbene oxide derivatives underwent ring-opening with high (>90%) *ees*, the one aliphatic epoxide examined afforded much poorer results.

Nakajima reported the use of a chiral bipyridine *N,N'*-dioxide **18** in the desymmetrization of acyclic *meso* epoxides (Figure 7.3). Although the enantioselectivity was not as high as in the method developed by Fu for *meso*-stilbene oxide (90% *ee* vs. 94% *ee*), it was higher for the same aliphatic epoxide (74% *ee* vs. 50% *ee*) [57]. Nakajima showed that mono-*N*-oxide derivatives **19** and **20** were much less effective than **18** in terms of both yield and enantioselectivity, and accordingly proposed a unique mechanism for **18** involving a hexacoordinate silicon intermediate coordinated to both *N*-oxides of the catalyst.

Asymmetric hydrogenolysis of epoxides has received relatively little attention despite the utility such processes might hold for the preparation of chiral secondary alcohol products. Chan et al. showed that epoxysuccinate disodium salt was reduced by use of a rhodium norbornadiene catalyst in methanol/water at room temperature to give the corresponding secondary alcohol in 62% *ee* (Scheme 7.31) [58]. Reduction with D_2 afforded a labeled product consistent with direct epoxide C–O bond cleavage and no isomerization to the ketone or enol before reduction.

7.3 Kinetic Resolution of Racemic Epoxides

Catalytic kinetic resolution can be the method of choice for the preparation of enantioenriched materials, particularly when the racemate is inexpensive and readily available and direct asymmetric routes to the optically active compounds are lacking. However, several other criteria-including catalyst selectivity, efficiency, and cost, stoichiometric reagent cost, waste generation, volumetric throughput, ease of product isolation, scalability, and the existence of viable alternatives from the chiral pool (or classical resolution)-must be taken into consideration as well [59].

Highly enantioenriched 1-oxiranes are arguably the most useful class of epoxides for organic synthesis because of their propensity to undergo highly regioselective ring-opening reactions. The chiral pool has not proven to be a useful source of most synthetically interesting terminal epoxides. Existing methods for the synthesis of highly enantioenriched terminal epoxides are often inefficient and limited in range, and require multistep procedures. Given the ready availability of racemic terminal epoxides from inexpensive terminal olefins, and the absence of effective asymmetric epoxidation methods for these substrates, kinetic resolution strategies have proven particularly interesting for the preparation of optically active monosubstituted epoxides.

7.3.1 Nitrogen-centered Nucleophiles

Jacobsen reported the first significant advance in the kinetic resolution of epoxides by successfully extending the (salen)Cr-catalyzed azidolysis of *meso*-epoxides with $TMSN_3$ (Scheme 7.3) to monosubstituted epoxides (Scheme 7.32) [60]. The catalyst system displayed strong regioselectivity for nucleophilic attack at the sterically less hindered terminal position. The high stereoselectivity, even in the case of propylene oxide, was coupled with broad substrate generality. Some of the azido alcohol products have been converted into useful targets, including the antihypertensive agent (*S*)-2-propranolol [61] and the SIV prophylactic (*R*)-9-[2-(phosphonomethoxy)propyl]adenine [62]. Interestingly, in the resolution of epichlorohydrin with $TMSN_3$, 1,3-dichloride and 1,3-diazide byproducts were found. Furthermore, resolved epichlorohydrin (97% *ee*) underwent racemization in the presence of (salen)Cr–Cl complex **2**. These observations eventually resulted in the development of an efficient dynamic kinetic resolution of epichlorohydrin (Scheme 7.33) [63], yielding 1-azido-2-(trimethylsiloxy)-3-chloropropane in 76% yield and 97% *ee*. This intermediate was used in the asymmetric synthesis of U-100592.

Jacobsen also showed that 2,2-disubstituted epoxides underwent kinetic resolution catalyzed by (salen)Cr–N_3 complex **3** under conditions virtually identical to those employed with monosubstituted epoxides (Scheme 7.34) [64]. Several epoxides in this difficult substrate class were obtained with high *ee*s and in good yields, as were the associated ring-opened products. The kinetic resolution of TBS-

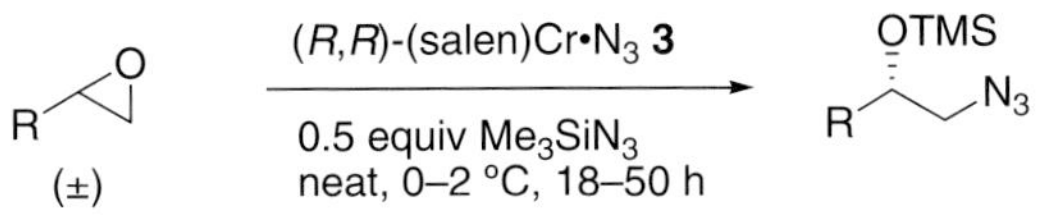

(±) → (R,R)-(salen)Cr•N_3 **3**; 0.5 equiv Me_3SiN_3; neat, 0–2 °C, 18–50 h

R	% cat	% yield	% ee	k_{rel}
CH_3	1	98	97	230
CH_2CH_3	2	83	97	140
$(CH_2)_3CH_3$	2	89	97	160
CH_2Cl	2	94	95	100
$CH_2OTBDMS$	3	96	96	150
CH_2O(1-naphthyl)	5	74	93	48
$CH_2C_6H_5$	2	94	93	71
c-C_6H_{11}	2	84	97	140
$(CH_2)_2CH{=}CH_2$	2	94	98	280
$CH(OEt)_2$	2	96	89	44
CH_2CN	2	80	92	45

yields based on the limiting reagent, $TMSN_3$

Scheme 7.32

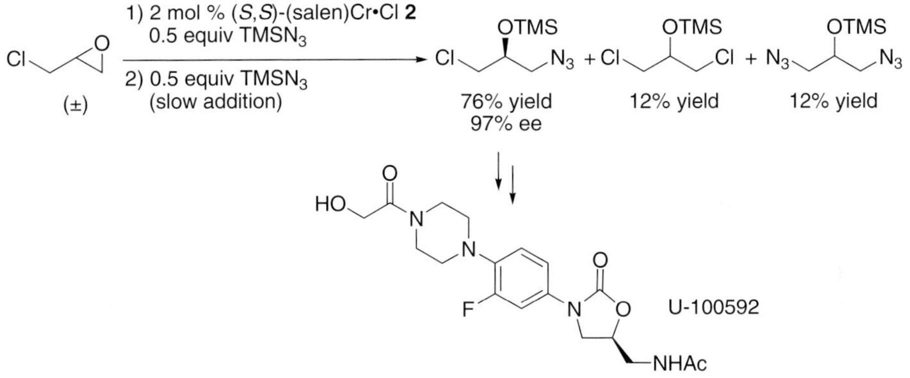

(±) 1) 2 mol % (S,S)-(salen)Cr•Cl **2**, 0.5 equiv $TMSN_3$; 2) 0.5 equiv $TMSN_3$ (slow addition)

76% yield, 97% ee; 12% yield; 12% yield

U-100592

Scheme 7.33

protected epoxide **21** was applied in an efficient total synthesis of the marine natural product taurospongin A [65]. Remarkably, this catalyst system, which discriminates effectively between a methyl group and a proton in the case of propylene oxide, also effectively differentiates between methyl and larger alkyl groups. Kinetic resolution of styrene oxide derivatives and 1,2-disubstituted epoxides by (salen)Cr-catalyzed addition of $TMSN_3$ was found to proceed with low regioselectivity, due to conflicting steric and electronic substrate biases to nucleophilic attack. However, in several cases, this could be used to advantage in the catalyst-controlled regioselective opening of enantioenriched epoxides by simple choice of catalyst enantiomer (Scheme 7.35) [66].

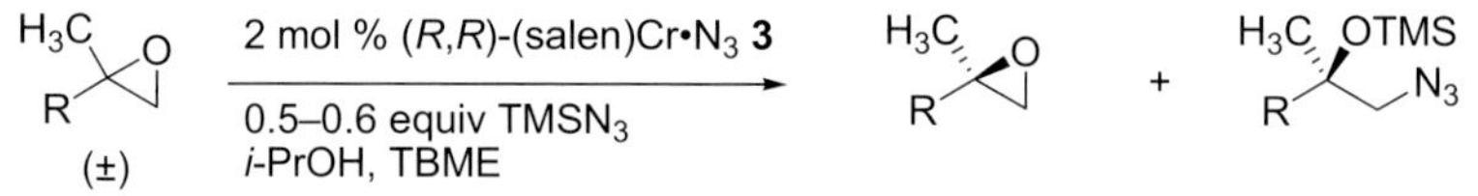

2 mol % (*R*,*R*)-(salen)Cr•N_3 **3**; 0.5–0.6 equiv $TMSN_3$; *i*-PrOH, TBME

R	% yield	% ee	% yield	% ee
$TBSOCH_2CH_2$ (**21**)	42	99	47	90
$BnOCH_2CH_2$	44	95	46	93
$CH_3(CH_2)_4$	42	99	44	95
c-C_6H_{11}	46	98	40	99
$PhCH_2CH_2$	44	97	45	92

yields based on total epoxide

Scheme 7.34

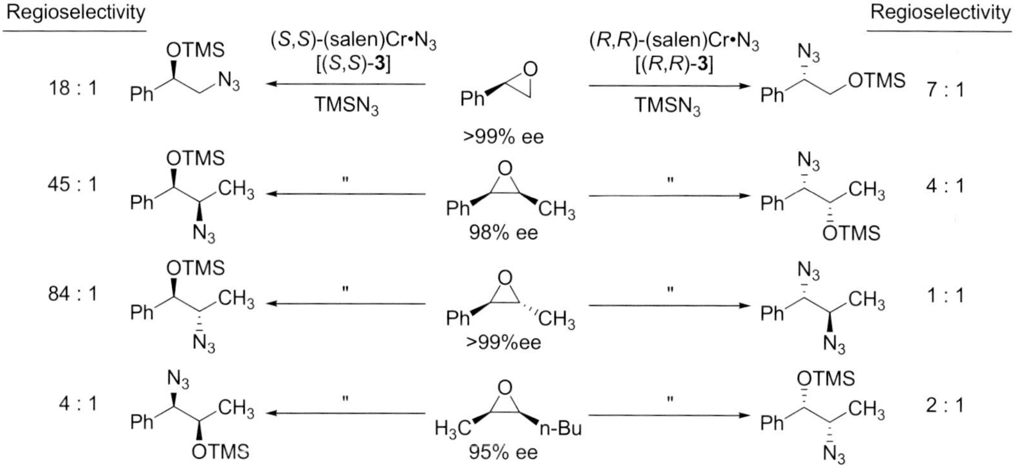

Regioselectivity	(*S*,*S*)-(salen)Cr•N_3 [(*S*,*S*)-**3**], $TMSN_3$	Epoxide	(*R*,*R*)-(salen)Cr•N_3 [(*R*,*R*)-**3**], $TMSN_3$	Regioselectivity
18 : 1		>99% ee		7 : 1
45 : 1	"	98% ee	"	4 : 1
84 : 1	"	>99%ee	"	1 : 1
4 : 1	"	95% ee	"	2 : 1

Scheme 7.35

Bartoli recently discovered that by switching from azide to *p*-anisidine as nucleophile, the ARO of racemic *trans*-β-substituted styrene oxides could be catalyzed by the (salen)Cr–Cl complex **2** with complete regioselectivity and moderate selectivity factors (Scheme 7.36) [14]. The ability to access *anti*-β-amino alcohols nicely complements the existing methods for the preparation of *syn*-aryl isoserines and related compounds [67] by asymmetric oxidation of *trans*-cinnamate derivatives [68].

The Gabriel synthesis represents another indirect but highly valuable approach to amines. Trost has demonstrated a method for the asymmetric ring-opening of butadiene monoepoxide by use of one equivalent of phthalimide, π-allylpalladium chloride dimer, and the chiral bisphosphine **22** (Scheme 7.37). The dynamic kinetic asymmetric transformation proceeded through a putative achiral intermedi-

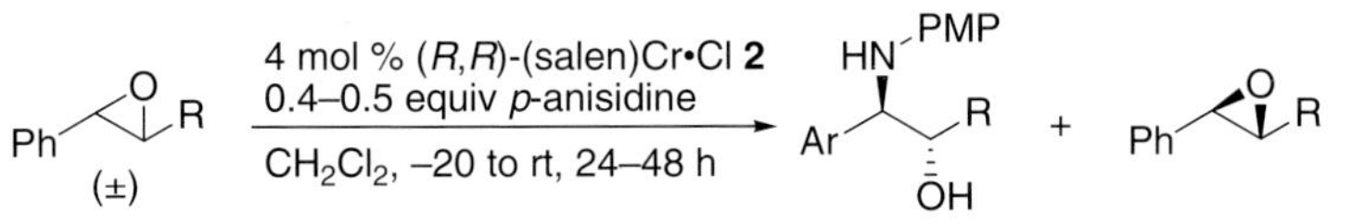

4 mol % (*R*,*R*)-(salen)Cr•Cl **2**
0.4–0.5 equiv *p*-anisidine
CH_2Cl_2, –20 to rt, 24–48 h

R	yield	% ee	% ee
Ph	60	93	–
$CH_2OTBDMS$	86	86	45
CH_2OCH_3	93	82	48
CH_2Br	95	86	53
$CH_2PO(OEt)_2$	90	77	–
CH_3	72	82	33
COOMe	93	85	51

Scheme 7.36

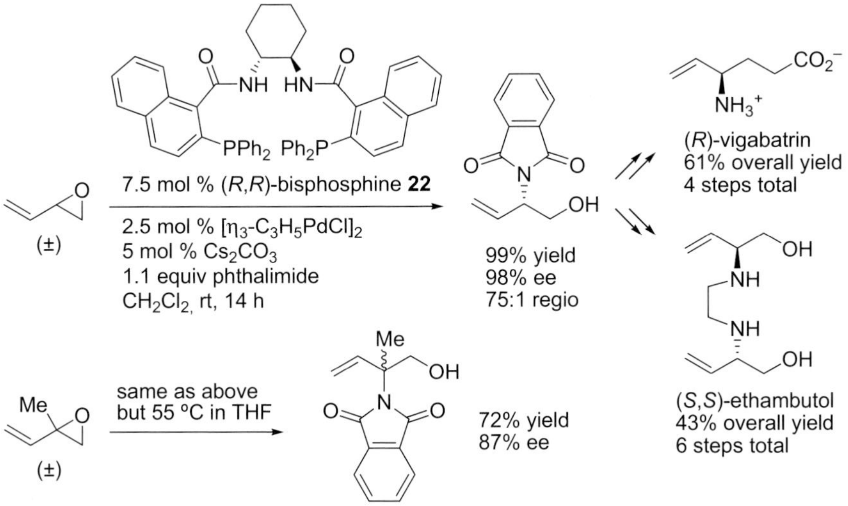

7.5 mol % (*R*,*R*)-bisphosphine **22**
2.5 mol % $[\eta_3\text{-}C_3H_5PdCl]_2$
5 mol % Cs_2CO_3
1.1 equiv phthalimide
CH_2Cl_2, rt, 14 h

99% yield
98% ee
75:1 regio

(*R*)-vigabatrin
61% overall yield
4 steps total

(*S*,*S*)-ethambutol
43% overall yield
6 steps total

same as above
but 55 °C in THF

72% yield
87% ee

Scheme 7.37

ate, to give 2-phthalimidobut-3-en-1-ol with nearly perfect regioselectivity (75:1), enantioselectivity (98% *ee*), and yield (99%) [69]. Though this method could be extended to isoprene monoepoxide (72% yield, 87% *ee*), this substrate required higher catalyst loadings and reaction temperature. The utility of this method was illustrated in the asymmetric syntheses of ethambutol, a tuberculostatic drug, as well as the antiepileptic and addiction-reducing drug vigabatrin.

One disadvantage of the ARO reactions with azide, *p*-anisidine, and phthalimide described above is that each product requires further manipulation (reduction,

β-cyclodextrin isopropylamine ← (±) → β-cyclodextrin piperidine

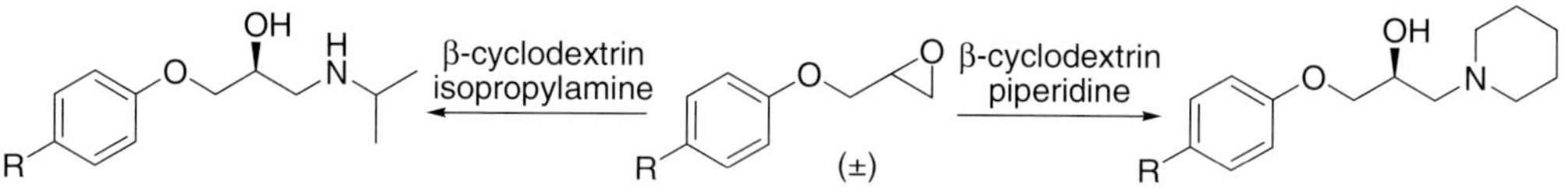

β-cyclodextrin, isopropylamine	R	β-cyclodextrin, piperidine
74% yield, 1% ee	R = H	79% yield, 74% ee
70% yield, 99% ee	R = Cl	75% yield, 99% ee
76% yield, 1% ee	R = CH_3	72% yield, 89% ee
73% yield, 4% ee	R = $CH_2CH_2OCH_3$	70% yield, 85% ee

Scheme 7.38

oxidative deprotection, or hydrazine workup) to afford the free amines, which are most often the desired products. In contrast, Rao demonstrated the direct coupling between aryl glycidyl ethers and alkyl amines promoted by stoichiometric levels of β-cyclodextrin complexes under solid-state conditions (Scheme 7.38) [70]. Surprisingly, these conditions allow for a dynamic kinetic resolution. The aminolysis has been demonstrated to be effective only with aryl glycidyl ethers capable of forming inclusion complexes with β-cyclodextrin.

Two recent reports described addition of nitrogen-centered nucleophiles in usefully protected form. Jacobsen reported that *N*-Boc-protected sulfonamides undergo poorly selective (salen)Co-catalyzed addition to racemic epoxides. However, by performing a one-pot, indirect kinetic resolution with water first (HKR, *vide infra*, Table 7.1) and then sulfonamide, it was possible to obtain highly enantiomerically enriched addition products (Scheme 7.39) [71]. These products were transformed into enantioenriched terminal aziridines in straightforward manner.

In sharp contrast, Bartoli showed that the (salen)Co catalyst system could be applied to the kinetic resolution of terminal epoxides with unprotected *tert*-butyl carbamate as nucleophile with extraordinarily high selectivity factors (Scheme 7.40) [72]. Excellent yields and selectivities are also obtained with use of ethyl, Cbz,

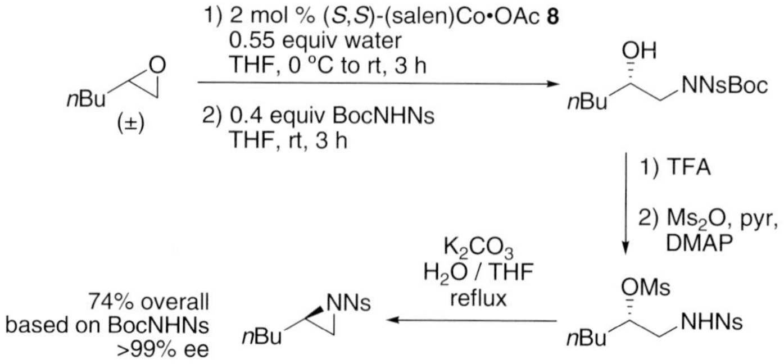

nBu (±) → 1) 2 mol % (*S*,*S*)-(salen)Co•OAc **8**, 0.55 equiv water, THF, 0 °C to rt, 3 h; 2) 0.4 equiv BocNHNs, THF, rt, 3 h → OH, nBu, NNsBoc

1) TFA; 2) Ms_2O, pyr, DMAP → OMs, nBu, NHNs

K_2CO_3, H_2O / THF, reflux → nBu, NNs

74% overall based on BocNHNs >99% ee

Scheme 7.39

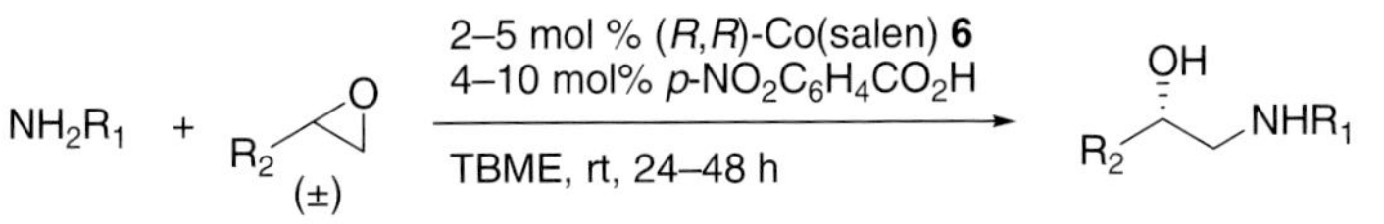

$$NH_2R_1 + R_2\text{-epoxide}\ (\pm) \xrightarrow[\text{TBME, rt, 24–48 h}]{\text{2–5 mol \% }(R,R)\text{-Co(salen) }\mathbf{6}\text{; 4–10 mol\% }p\text{-}NO_2C_6H_4CO_2H} R_2CH(OH)CH_2NHR_1$$

R_1	R_2	% yield	% ee
Boc	CH_3	99	99.3
Boc	$(CH_2)_3CH_3$	99	99.2
Boc	$(CH_2)_4CH{=}CH_2$	99	99.7
Boc	c-C_6H_{11}	84	99.9
Boc	CH_2O(1-naphthyl)	95	99.5
Boc	CH_2Cl	87	99.9
Boc	C_6H_5	90	99.9
Boc	p-BrC_6H_4	76	99.8
Boc	o-$NO_2C_6H_4$	62	99.8
Boc	CH_2OPh	99	98.9
Etoc	CH_2OPh	97	99.0
Cbz	CH_2OPh	93	99.5
Fmoc	CH_2OPh	67	99.5

yields based on carbamate

Scheme 7.40

and Fmoc carbamates, further increasing the synthetic utility of this impressive method.

7.3.2
Oxygen-centered Nucleophiles

Although several interesting nitrogen-centered nucleophiles have been developed with ARO reactions of epoxides (*vide supra*), kinetic resolutions with such reagents are unlikely to be of practical value for the recovery of enantioenriched terminal epoxides. This is due to the fact that these nucleophiles are too valuable to be discarded in a by-product of the resolution, are generally not atom-economical, and, particularly in the case of azide, may represent safety hazards.

Water may represent the ideal nucleophile for epoxide kinetic resolutions that target epoxide recovery. In addition to the nominal cost and attractive safety features of the reagent, the diol product is often valuable and readily separable from unreacted epoxide. In 1997, Jacobsen reported the hydrolytic kinetic resolution (HKR) of terminal epoxides using (salen)Co–OAc catalyst **8** [73]. This complex is now available commercially in research and large-scale quantities. With 0.55 equivalents of water, epoxides were recovered in good yield and >99% *ee*; with 0.45 equivalents of water, the diol product can be obtained in similarly good yield and high enantiomeric excess (Table 7.1). The HKR displays an extraordinarily wide substrate scope, with similarly good results having been obtained with aliphatic, α-halo, olefinic, carbonyl-containing, and aromatic terminal epoxides [74]. In the

Table 7.1

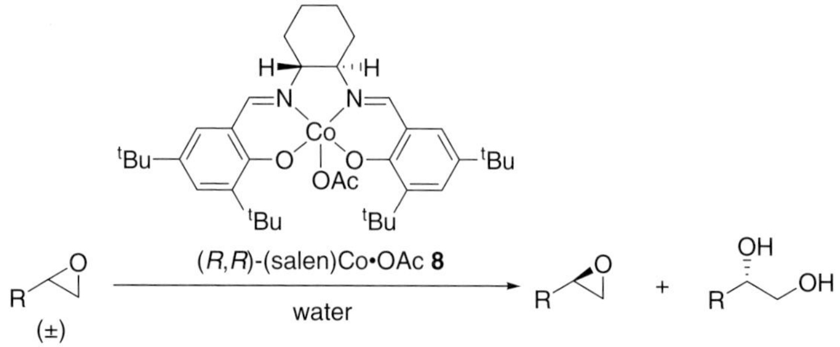

(*R*,*R*)-(salen)Co•OAc **8**

water

entry	R	cat mol %	epox % yield	epox % ee	diol % yield	diol % ee
		Aliphatic epoxides				
1	CH_3	0.2	46	>99	45	99
2	$(CH_2)_3CH_3$	0.5	43	>99	44	99
3	$(CH_2)_{11}CH_3$	0.5	42	>99	40	99
4	$(CH_3)_2CH=CH_2$	0.5	43	>99	44	99
5	CH_2Ph	0.5	46	>99	40	95
6	*c*-C6H11	0.5	44	>99	41	99
7	*t*-C4H9	2.0	41	>99	40	95
		Halogenated epoxides				
8	CH_2Cl	0.5	43	>99	40	95
9	CH_2Br (dynamic)	0.5	41	43	90	96
10	CH_2F	0.5	42	>99	38	97
11	CF_3	0.5	42	>99	42	>99
		Epoxides bearing ether and carbonyl functionality				
12	CH_2OBn	0.5	48	>99	40	95
13	CH_2OTBS	0.5	47	>99	42	98
14	CH_2OPh	0.5	47	>99	41	95
15	CH_2O(1-naphthyl)	0.5	38	>99	42	97
16	CH_2CH_2OBn	0.5	42	>99	42	95
17	oxiranyl	1.0	36	>99	36	96
18	CH_2OCO*n*-Pr	0.5	46	>99	45	43
19	CH_2CO_2Et	0.5	44	>99	41	95
20	CH_2NHBoc	2.0	36	>99	36	78
21	CO_2CH_3	2.0	43	>99	37	97
22	$COCH_3$	2.0	40	>99	40	97
23	$COCH_2CH_3$	2.0	41	>99	33	96
		Aryl, vinyl, and alkynyl epoxides				
24	C_6H_5	0.8	44	>99	42	98
25	4-ClC_6H_4	0.8	38	>99	37	94
26	3-ClC_6H_4	0.8	40	>99	44	91
27	3-$(CH_3O)C_6H_4$	0.8	41	>99	41	95
28	3-$(NO_2)C_6H_4$	0.8	38	>99	44	99
29	2-ClC_6H_4	1.5	38	>99	42	94
30	$CH=CH_2$	1.5	36	>99	38	97
31	$C\equiv C$(TBS)	0.8	41	>99	41	99

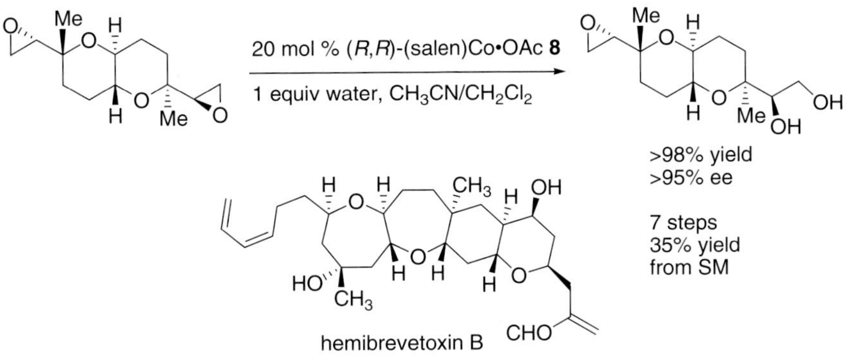

20 mol % (*R*,*R*)-(salen)Co•OAc **8**

1 equiv water, CH_3CN/CH_2Cl_2

>98% yield
>95% ee

7 steps
35% yield
from SM

hemibrevetoxin B

Scheme 7.41

case of epibromohydrin, dynamic kinetic resolution could be effected by use of 0.5 mol% catalyst **8** to yield (*R*)-1-bromopropane-2,3-diol in 96% *ee* and 90% yield [75]. The use of water as the resolving agent, low loadings of a recyclable catalyst, and ease of separation of resolved epoxide from diol have made the HKR a widely used method for preparing enantioenriched terminal epoxides. Examples of natural products synthesized with the aid of the HKR include muconin [76], amphidinolide T1 [77], cytochalasin B [78], (–)-indolizidine 223AB [79], and (+)-allosedamine [80].

The high enantioselectivity and broad substrate scope of the HKR are accompanied by an intriguing mechanistic framework involving cooperative catalysis between different catalyst species. Detailed mechanistic investigation into each of these pathways has produced new insights into cooperative catalysis and has resulted in synthetic improvements in the HKR and other ARO reactions [81].

The application of the HKR to asymmetric synthesis is not limited to the preparation of enantiopure terminal epoxides and 1,2-diols as building blocks, but may also be extended to late-stage total synthetic targets. A particularly impressive example of this was provided by Nelson in an asymmetric formal synthesis of polycyclic-ether marine natural product hemibrevetoxin B [82]. An achiral, centrosymmetric molecule containing two terminal epoxides was accessed by a bidirectional strategy. Hydrolytic desymmetrization of this compound catalyzed by (salen)Co complex **8** proceeded in high yield and with good enantioselectivity to generate an optically active compound bearing six defined stereocenters effectively from an achiral precursor (Scheme 7.41).

Jacobsen reported that phenols undergo highly efficient and enantioselective addition to terminal epoxides with (salen)Co complexes, thereby expanding the scope of the kinetic resolution with these catalysts to other oxygen-centered nucleophiles while addressing the need for a direct method for generating β-aryloxy alcohols (Table 7.2) [83]. Although kinetic resolution was satisfactory with several complexes, including the acetate catalyst, the more Lewis acidic and less nucleophilic perfluoro *tert*-butoxide complex **23** displayed superior reactivity. Like the

Table 7.2

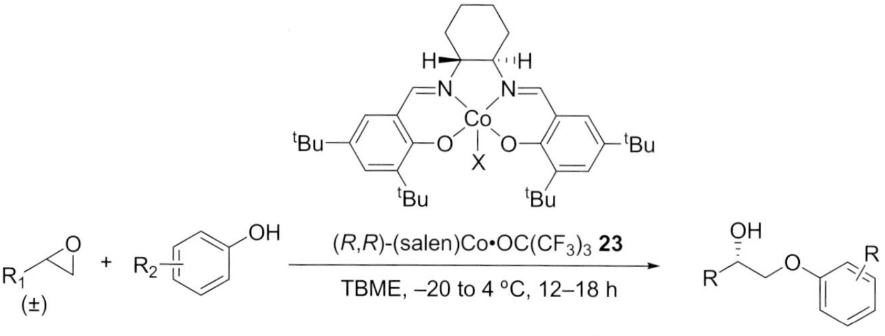

(R,R)-(salen)Co•OC(CF$_3$)$_3$ **23**, TBME, –20 to 4 °C, 12–18 h

entry	R_1	R_2	cat mol %	% yield	% ee
1	$(CH_2)_3CH_3$	*p*-CH_3	4.4	95	97
2	$(CH_2)_3CH_3$	*m*-CH_3	4.4	99	99
3	$(CH_2)_3CH_3$	*o*-CH_3	4.4	<5	–
4	$(CH_2)_3CH_3$	*p*-Br	4.4	92	99
5	$(CH_2)_3CH_3$	*o*-Br	8.8 (48 h)	98	92
6	$(CH_2)_3CH_3$	*p*-OCH_3	8.8	75	99
7	$(CH_2)_3CH_3$	*p*-NO_2	8.8	93	91
8	$(CH_2)_3CH_3$	*p*-$(CH_2)_2$NHBoc	4.4	86	99
9	$(CH_2)_3CH_3$	H	4.4	97	98
10	CH_2Cl	H	4.4	97	99
11	CH_2Br (dynamic)	H	4.0	74	99
12	CH_2O(allyl)	H	4.4	93	97
13	*c*-C_6H_{11}	H	8.8	99	97
14	$COCH_2CH_3$	H	8.8	96	96
15	CO_2CH_3	H	4.4	98	96

isolated yields based on phenol

HKR, the method is general with respect to terminal epoxides and is also quite broad in nucleophile scope, tolerating *ortho-*, *meta-*, or *para*-substituted phenols bearing electron-donating or -withdrawing functionality. Dynamic kinetic resolution using epibromohydrin is also possible with this system in the presence of catalytic LiBr (entry 11). Mechanistic investigation resulted in the discovery that a 50:50 mix of (salen)Co–Cl and (salen)Co–SbF_6 complexes is a much more reactive, and in some cases enantioselective, catalyst system for phenol addition [84].

The principle cost determinant in typical hydrolytic or phenolic resolutions is the cobalt catalyst, despite the relatively low catalyst loadings used in most cases and the demonstrated recyclability with key substrates. From this standpoint, recently developed oligomeric (salen)Co complexes, discussed earlier in this chapter in the context of the hydrolytic desymmetrization of *meso*-epoxides (Scheme 7.16), offer significant advantages for kinetic resolutions of racemic terminal epoxides (Table 7.3) [29–31]. For the hydrolytic and phenolic kinetic resolutions, the oligo-

Table 7.3

R_1-epoxide (±) + R_2OH → [(*R,R*)-oligomeric Co(salen)•OTf **7**; MS 3Å, –20 °C to rt, TBME, 12–18 h] → $R_1CH(OH)CH_2OR_2$

entry	R_1	R_2	cat mol %	% yield	% ee
		Hydrolytic Kinetic Resolution (HKR)			
1	CH_3	H	0.0003	40	>99
2	CO_2CH_3	H	0.015	44	>99
3	Ph	H	*0.04	40	>99
4	4-fluorophenyl	H	0.1	40	>99
5	vinyl	H	0.025	35	>99
6	CH_2O(allyl)	H	0.0025	43	>99
7	CH_2Cl	H	0.001	44	>99
		Phenolic Kinetic Resolution (PKR)			
8	CH_3	2-bromophenyl	0.05	96	>99
9	CO_2CH_3	2-chloro-6-methylphenyl	2.5	80	98
10	Ph	4-methylphenyl	0.075	79	98
11	CH_2O(allyl)	4-bromophenyl	0.05	87	98
12	CH_2Cl	2-bromophenyl	0.05	>99	98
13	CH_2Cl	3-acetylphenyl	0.005	90	>99
14	CH_2Cl	4-(CO_2Me)phenyl	0.0075	92	>99
15	CH_2Cl	6-(2-CO_2Et)indole	0.2	92	98
16	$(CH_2)_3CH_3$	phenyl	0.0075	92	99
17	$(CH_2)_3CH_3$	2-chlorophenyl	0.05	97	>99
18	$(CH_2)_3CH_3$	2-bromophenyl	0.075	96	99
19	$(CH_2)_3CH_3$	2-methylphenyl	0.25	89	98
20	$(CH_2)_3CH_3$	3-acetylphenyl	0.05	95	>99
21	$(CH_2)_3CH_3$	3-methylphenyl	0.0075	95	99
22	$(CH_2)_3CH_3$	4-nitrophenyl	0.15	87	97
23	$(CH_2)_3CH_3$	4-methoxyphenyl	0.05	87	99
		Alcoholic Kinetic Resolution (AKR)			
24	CH_3	4-methoxybenzyl	0.2	91	>99
25	CO_2CH_3	2-bromobenzyl	0.1	94	>99
26	Ph	allyl	0.5	80	>99
27	CH_2Cl	benzyl	0.1	98	>99
28	CH_2Cl	2-bromobenzyl	0.02	94	>99
29	CH_2Cl	2-nitrobenzyl	0.1	94	>99
30	CH_2Cl	4-methoxybenzyl	0.5	91	98
31	$(CH_2)_3CH_3$	benzyl	0.1	92	>99
32	benzyl	2-(trimethylsilyl)ethyl	0.1	>99	98
33	benzyl	methyl	0.01	92	97

HKR: 0.6 equiv H_2O relative to epoxide, neat, rt, 15–24 h, yield based on epoxide (50% max)
PKR: 2.2–2.5 equiv epoxide relative to phenol, 4 °C, CH_3CN, <24 h, yield based on phenol
AKR: 2.2–2.5 equiv epoxide relative to alcohol, 4 °C, CH_3CN, <24 h, yield based on alcohol
*(*R,R*)-oligomeric Co(salen)•OTs **21** catalyst used instead in this entry

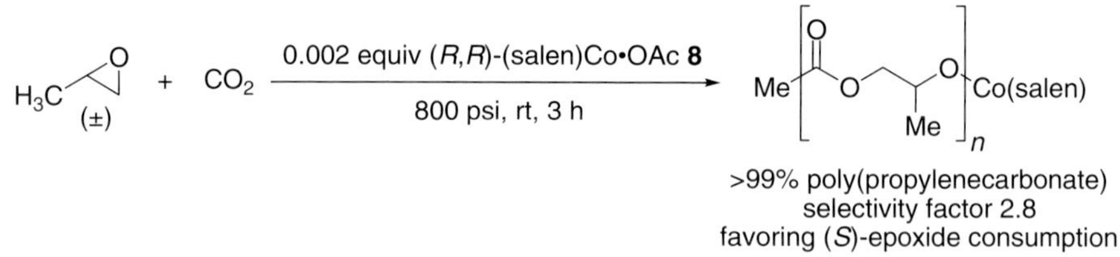

Scheme 7.42

meric catalyst **7** was found to operate effectively at catalyst loadings 20–800 times lower than required for monomeric catalysts [31]. For certain sterically hindered or electron-deficient phenols, the oligomeric catalyst also provided increased enantioselectivity. Perhaps most interestingly, the oligomeric catalyst broadened the scope of oxygen-centered nucleophiles to include primary alcohols (e. g., benzyl, allyl, trimethylsilyl, and methyl alcohols), allowing for direct synthesis of usefully monoprotected 1,2-diols.

Carbon dioxide represents another interesting reagent for kinetic resolution of epoxides because it is inexpensive, safe, and atom-economical when forming polycarbonate copolymers or optically active cyclic carbonates. Although (salen)metal complex-catalyzed copolymerization of epoxides and carbon dioxide is well established, most methods provide cyclic and acyclic carbonates as complex mixtures, and/or suffer from low regio- and stereoselectivity [85]. Recently, however, Coates has demonstrated high selectivity favoring poly(propylene carbonate) formation in the copolymerization of propylene oxide and carbon dioxide with use of (salen)Co–OAc catalyst **8** (Scheme 7.42) [86]. Copolymerization was believed to propagate through irreversible carbonate addition to epoxide, followed by carbon dioxide fixation. The selectivity factor in the kinetic resolution of racemic propylene oxide, however, was determined to be only 2.8 (compared to >500 for the HKR of this substrate with the same catalyst).

Lu has demonstrated that cyclic carbonate formation can dominate over polymer formation at much lower CO_2 pressures when the (salen)Co–OTs catalyst **24** and

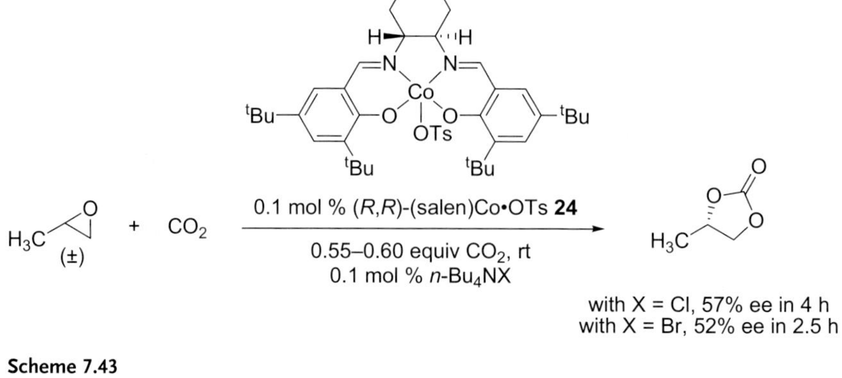

Scheme 7.43

tetrabutylammonium halide are used (Scheme 7.43) [87]. The halide was believed to act as a nucleophilic cocatalyst engaged in reversible addition to epoxide. Carboxylation of the intermediate and intramolecular elimination of the halide yields the cyclic carbonate and regenerates the catalyst/cocatalyst. With propylene oxide, the selectivity factor for this system was determined to be slightly higher than the selectivity factors obtained by Coates.

7.3.3
Carbon-centered Nucleophiles

The use of carbon-centered nucleophiles in kinetic resolutions of racemic epoxides has received much less attention than reactions using heteroatom-centered nucleophiles. Cozzi and Umani-Ronchi reported that racemic 1,2-disubstituted epoxides undergo asymmetric ring-opening with 2-substituted indoles in a valuable carbon-carbon bond-forming reaction (Scheme 7.44) [49]. The reactions displayed complete regioselectivity, favoring attack of indole C3 on the epoxide carbon bearing the aryl functionality. While most interesting for its products, this method can also be used to access resolved 1,2-disubstituted epoxides in moderate yields (~30% of the maximum 50%) and with high *ees*. Several of the epoxides investigated in this study are accessible as nonracemic mixtures by asymmetric epoxidation, however, and may be recrystallized to optical purity [88–89].

Pineschi and Feringa reported that chiral copper phosphoramidite catalysts mediate a regiodivergent kinetic resolution (RKR) of cyclic unsaturated epoxides with dialkylzinc reagents, in which epoxide enantiomers are selectively transformed into different regioisomers (allylic and homoallylic alcohols) [90]. The method was also applied to both *s-cis* and *s-trans* cyclic allylic epoxides (Schemes 7.45 and 7.46,

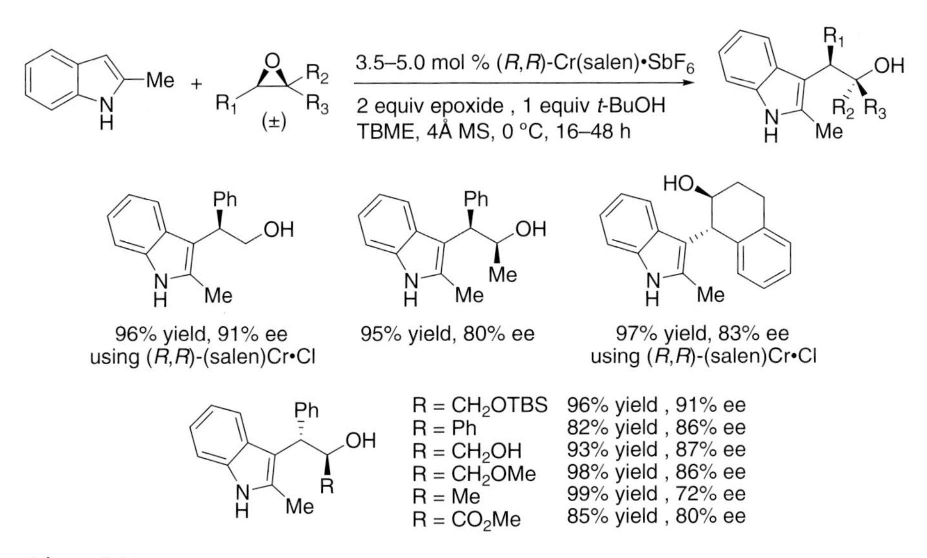

Scheme 7.44

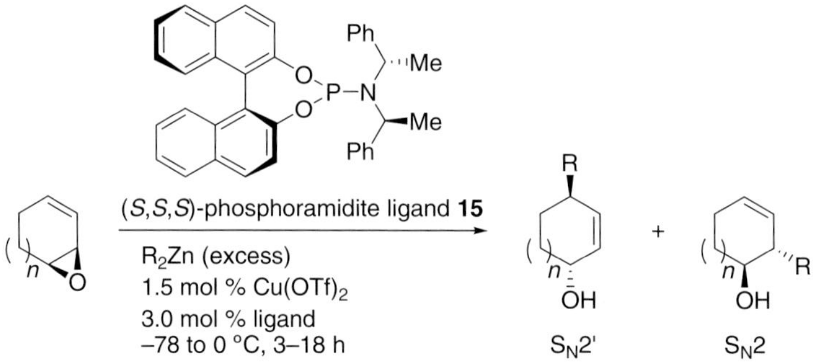

(*S*,*S*,*S*)-phosphoramidite ligand **15**

R_2Zn (excess)
1.5 mol % $Cu(OTf)_2$
3.0 mol % ligand
–78 to 0 °C, 3–18 h

n	R	S_N2' / S_N2	S_N2'	S_N2
n = 0	R = Bu	67 / 33	45% ee	91% ee
n = 1	R = Et	76 / 24	34% ee	>98% ee
n = 2	R = Me	50 / 50	>90% ee	>95% ee
n = 3	R = Me	62 / 38	76% ee	>98% ee

Scheme 7.45

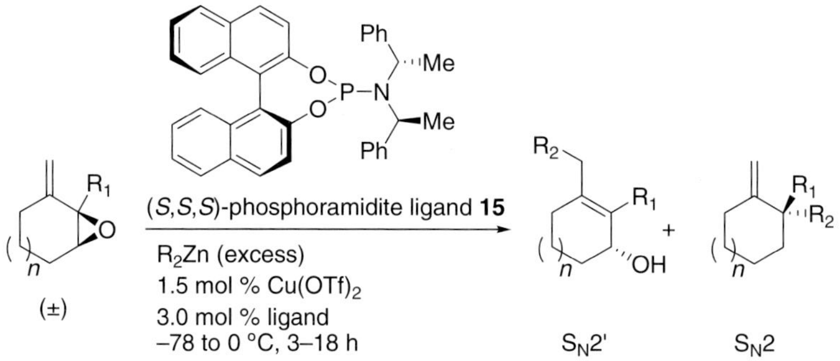

(±)

(*S*,*S*,*S*)-phosphoramidite ligand **15**

R_2Zn (excess)
1.5 mol % $Cu(OTf)_2$
3.0 mol % ligand
–78 to 0 °C, 3–18 h

n	R_1	R_2	S_N2' / S_N2	S_N2'	S_N2
n = 0	R_1 = Me	R_2 = Bu	49 / 51	96% ee	92% ee
n = 1	R_1 = Et	R_2 = Et	55 / 45	80% ee	99% ee
n = 2	R_1 = Me	R_2 = Me	23 / 77	>97% ee	32% ee
n = 3	R_1 = Bu	R_2 = Me	52 / 48	82% ee	90% ee

Scheme 7.46

respectively), and in one case was applied in the generation of an all-carbon quaternary center in high enantiomeric excess. In addition, multigram, enantioselective syntheses of *trans*-4-methylcyclohex-2-en-1-ol and 4-methylcyclohex-2-en-1-one were developed [91]. These building blocks have been used in the syntheses of several natural products, including tetronomycin [92], microcionin 2 [93], and pseudopterosin A aglycone [94] (Scheme 7.47).

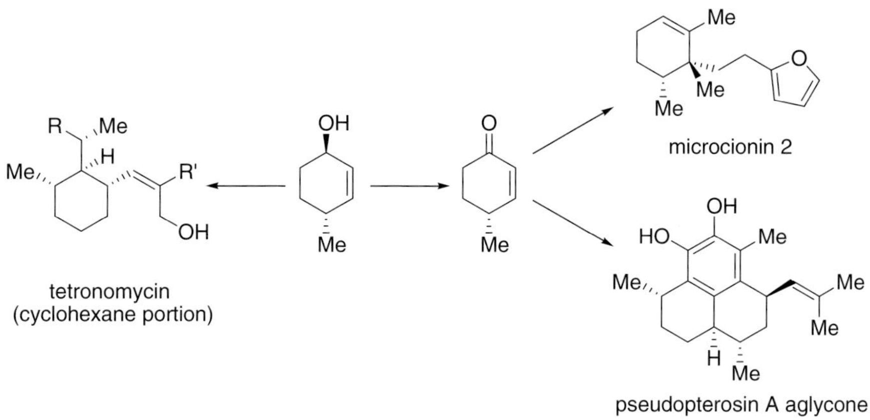

microcionin 2

tetronomycin (cyclohexane portion)

pseudopterosin A aglycone

Scheme 7.47

7.4 Enantioselective Rearrangements of Epoxides

A number of interesting and synthetically useful asymmetric ring-opening reactions of *meso*- and racemic epoxides can be performed by simple Lewis or Brønsted acid-catalyzed rearrangement. Shi found that titanium-BINOL complexes catalyze the enantioselective rearrangement of racemic enol ester epoxides into optically active α-acyloxyketones (Scheme 7.48) [95–96]. The unreacted epoxides can be recovered with excellent *ee*s and in good yields (30–40% based on epoxide). Shi showed that the titanium-catalyzed process occurred with inversion of stereochemistry while protic acids catalyzed rearrangement with retention. By combining these observations, a two-step, one-pot, parallel resolution was developed, with use of a chiral Lewis acid to effect kinetic resolution followed by an achiral Brønsted acid to convert the remaining starting material into the same product stereoisomer. This process allowed the generation of enantioenriched α-acyloxy ketones

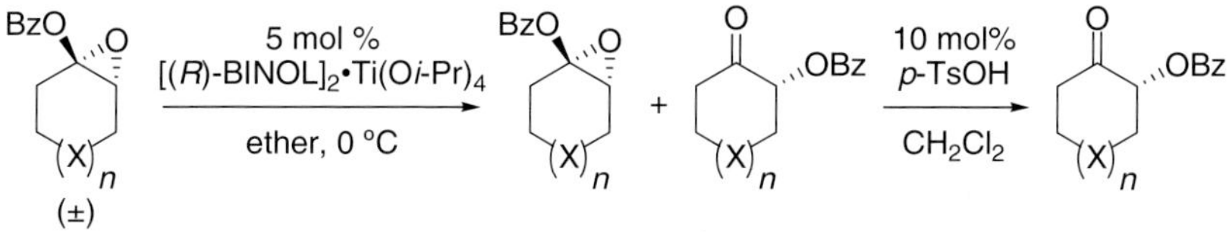

5 mol % [(*R*)-BINOL]$_2$•Ti(O*i*-Pr)$_4$, ether, 0 °C; 10 mol% *p*-TsOH, CH_2Cl_2

		Epoxide	Ketone	Product
$n = 1$	X = CH_2	97% ee	90% ee	78% yield, 93% ee
$n = 0$	X = CH_2	91% ee	92% ee	79% yield, 92% ee
$n = 2$	X = CH_2	99% ee	78% ee	83% yield, 87% ee
$n = 3$	X = CH_2	97% ee	71% ee	81% yield, 82% ee
$n = 1$	X = O	98% ee	93% ee	77% yield, 97% ee

Scheme 7.48

20 mol % Ti-(O*i*-Pr)$_4$
50 mol % (*S*)-BINOL
toluene, rt, 2–4 h

$n = 0$	R = Ph	R_1 = Ph	90% ee	61% ee
$n = 1$	R = Ph	R_1 = Ph	94% ee	60% ee
$n = 1$	R = Ph	R_1 = Bn	82% ee	48% ee
$n = 1$	R = naphthyl	R_1 = *n*-Bu	79% ee	46% ee

Scheme 7.49

in higher enantiomeric excesses and yields than could be achieved by the simple kinetic resolution.

Using the same Ti-BINOL catalyst system, Tu demonstrated that racemic α-hydroxy epoxides undergo kinetic resolution via semipinacol rearrangement to afford optically active β-hydroxy ketones [97]. The yields are also good (30–40%), but the enantioselectivity was moderate for most substrates. Furthermore, many of the starting materials are readily obtained in optically active form by asymmetric epoxidation. Nonetheless, the rearrangement provides an interesting route to enantioenriched tertiary alcohols and all-carbon quaternary centers (Scheme 7.49).

Enantioselective deprotonation/rearrangement of epoxides into allylic alcohols is an area of longstanding interest, first reported in 1980 [98]. Relatively few systems involving catalytic quantities of chiral base have been developed [99–101], although significant progress has been made recently. Asami first reported the use of a proline-derived chiral amide in the rearrangement of epoxycyclohexane, in which the allylic alcohol was obtained in 71% yield and 75% *ee* (Scheme 7.50) [102]. The limited availability of the unnatural enantiomer of the catalyst and the search for a more effective asymmetric environment prompted the investigation of various other chiral diamine catalysts and stoichiometric bases. A slightly more complicated pyrrolidinyl pyrrolidide displayed greater enantioselectivity and better reactivity (89% yield, 94% *ee*) [103]. Malhotra reported that 20 mol% of DIPAM, a

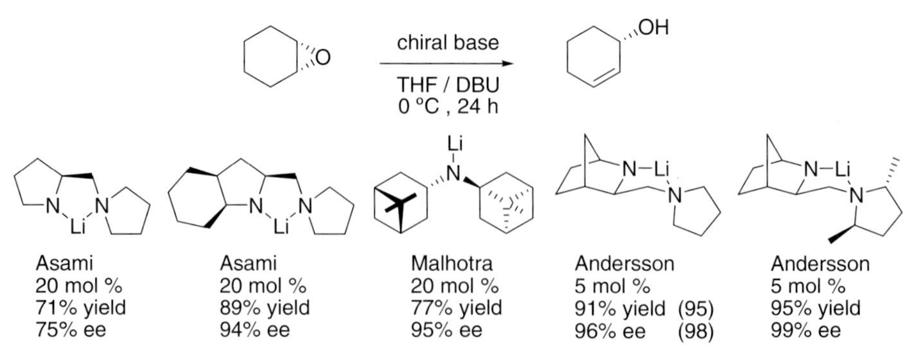

Scheme 7.50

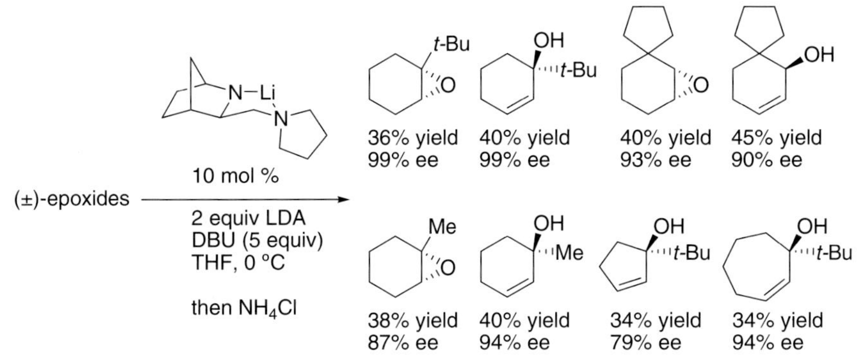

Scheme 7.51

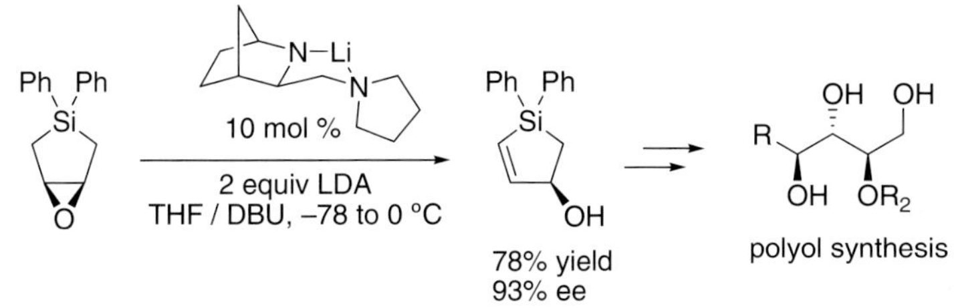

Scheme 7.52

C_2-symmetric diamine derived from α-pinene, catalyzed the reaction with similar results (77% yield, 95% *ee*) [104]. Andersson later reported that even higher selectivity could be achieved at lower catalyst loadings with use of a more rigid amide framework [105]. The corresponding amine was prepared from inexpensive starting materials in six steps [106]. Slow addition of the epoxide and the presence of various amounts of deaggregating agents, such as DBU, gave enhanced yields and enantioselectivities. Moreover, structural optimizations provided an even better catalyst for the enantioselective isomerization of several epoxides [107].

Andersson also showed that, in addition to *meso*-desymmetrization, kinetic resolution of some cyclic epoxides by use of the first-generation catalyst was also possible, giving both epoxides and allylic alcohols in good yields (Scheme 7.51) [108]. Kozmin reported the effective use of the same catalyst in the desymmetrization of diphenylsilacyclopentene oxide. The resulting products could be used in the stereocontrolled syntheses of various acyclic polyols (Scheme 7.52) [109].

7.5
Conclusion

Whereas the vast majority of interesting reactions in asymmetric catalysis involve addition to prochiral π-systems (e.g. alkenes, carbonyls, imines, π-allyl systems), epoxide ARO reactions proceed through stereospecific substitutions of σ-bonds. Nonetheless, a variety of synthetically important catalyst systems have been developed for epoxide ARO reactions over the past several years. The practical fruit of this effort is the ready availability of a rich assortment of chiral building blocks that were previously far less accessible. These include terminal epoxides, cyclic allylic alcohols, 1,2-amino alcohol derivatives, terminal aziridines, 1,2-thio alcohols, and 1,2-diols. Important challenges remain, particularly in the use of carbon-centered nucleophiles and in the ring-opening of hindered epoxides. Contrasteric nucleophilic opening of epoxides lacking strong electronic biases for internal attack represents another important and unmet challenge.

Dual activation of nucleophile and epoxide has emerged as an important mechanistic principle in asymmetric catalysis [110], and it appears to be particularly important in epoxide ARO reactions. Future work in this area is likely to build on the concept of dual substrate activation in interesting and exciting new ways.

References

1 For a review on epoxide hydrolases and related enzymes in the context of organic synthesis, see: Faber, K. *Biotransformations in Organic Chemistry*, Springer: New York; 2004.
2 Hayashi, M.; Kohmura, K.; Oguni, N. *Synlett* **1991**, 774–776.
3 Nugent, W. A. *J. Am. Chem. Soc.* **1992**, *144*, 2768–2769.
4 Martínez, L. E.; Leighton, J. L.; Carsten, D. H.; Jacobsen, E. N. *J. Am. Chem. Soc.* **1995**, *117*, 5897–5898.
5 Song, C. E.; Oh, C. R.; Roh, E. J.; Choo, D. J. *Chem. Commun.* **2000**, 1743–1744.
6 Hansen, K. B.; Leighton, J. L.; Jacobsen, E. N. *J. Am. Chem. Soc.* **1996**, *118*, 10924–10925.
7 Konsler, R. G.; Karl, J.; Jacobsen, E. N. *J. Am. Chem. Soc.* **1998**, *120*, 10780–10781.
8 Jacobsen, E. N. *Acc. Chem. Res.* **2000**, *33*, 421–431.
9 Wu, M. H.; Jacobsen, E. N. *Tetrahedron Lett.* **1997**, *38*, 1693–1696.
10 Martínez, L. E.; Nugent, W. A.; Jacobsen, E. N. *J. Org. Chem.* **1996**, *61*, 7963–7966.
11 Leighton, J. L.; Jacobsen, E. N. *J. Org. Chem.* **1996**, *61*, 389–390.
12 Kassab, D. J.; Ganem, B. *J. Org. Chem.* **1999**, *64*, 1782–1783.
13 Annis, D. A.; Helluin, O.; Jacobsen, E. N. *Angew. Chem. Int. Ed. Engl.* **1998**, *37*, 1907–1909.
14 Bartoli, G.; Bosco, M.; Carlone, A.; Locatelli, M.; Massaccesi, M.; Melchiorre, P.; Sambri, L. *Org. Lett.* **2004**, *6*, 2173–2176.
15 Hou, X.-L.; Wu, J.; Dai, L.-X.; Xia, L.-J.; Tang, M.-H. *Tetrahedron: Asymmetry* **1998**, 9, 1747–1752.
16 Sekine, A.; Ohshima, T.; Shibasaki, M. *Tetrahedron*, **2002**, *58*, 75–82.
17 Sagawa, S.; Abe, H.; Hase, Y.; Inaba, T. *J. Org. Chem.* **1999**, *64*, 4962–4965.
18 For a useful example of nonasymmetric catalytic isocyanosilylation of a *meso*-epoxide: Imi, K.; Yanagihara, N.; Utimoto, K. *J. Org. Chem.* **1987**, *52*, 1013–1016.
19 Zhu, C.; Yuan, F.; Gu, W.; Pan, Y. *Chem. Commun.* **2003**, 692–6933.
20 Yamashita, H.; Mukaiyama, T. *Chem. Lett.* **1985**, 1643–1646.
21 Iida, T.; Yamamoto, N.; Sasai, H.; Shibasaki, M. *J. Am. Chem. Soc.* **1997**, *119*, 4783–4784.

22 Shibasaki, M.; Yoshikawa, N. *Chem. Rev.* **2002**, *102*, 2187–2209.

23 Wu, M. H.; Jacobsen, E. N. *J. Org. Chem.* **1998**, *63*, 5252–5254.

24 Wu, J.; Hou, X.-L.; Dai, L.-X.; Xia, L.-J.; Tang, M.-H. *Tetrahedron: Asymmetry* **1998**, *9*, 3431–3436.

25 Li, Z.; Zhou, Z.; Li, K.; Wang, L.; Zhou, Q.; Tang, C. *Tetrahedron Lett.* **2002**, *43*, 7609–7611.

26 Jacobsen, E. N.; Kakiuchi, F.; Konsler, R. G.; Larrow, J. F.; Tokunaga, M. *Tetrahedron Lett.* **1997**, *38*, 773–776.

27 Annis, A.; Jacobsen, E. N. *J. Am. Chem. Soc.* **1999**, *121*, 4147–4154.

28 Breinbauer, R.; Jacobsen, E. N. *Angew. Chem. Int. Ed.* **2000**, *39*, 3604–3607.

29 Ready, J. M.; Jacobsen, E. N. *J. Am. Chem. Soc.* **2001**, *123*, 2687–2688. Ready, J. M.; Jacobsen, E. N. *Angew. Chem. Int. Ed.* **2002**, *41*, 1374–1377.

30 White, D. E.; Jacobsen, E. N. *Tetrahedron: Asymmetry* **2003**, *14*, 3633–3638.

31 White, D. E.; Jacobsen, E. N. Manuscript in preparation.

32 Wu, M. H.; Hansen, K. B.; Jacobsen, E. N. *Angew. Chem. Int. Ed.* **1999**, *38*, 2012–2014.

33 Meng, Z.; Danishefsky, S. J. *Angew. Chem. Int. Ed.* **2005**, *44*, 2–4.

34 Birman, V. B.; Danishefsky, S. J. *J. Am. Chem. Soc.* **2002**, *124*, 2080–2081.

35 Stevenson, C. P.; Jacobsen, E. N. Work in progress.

36 Vargeese, C.; Abushanab, E. *J. Org. Chem.* **1990**, *55*, 4400–4403.

37 Ghosh, A. K.; McKee, S. P.; Lee, H. Y.; Thompson, W. J. *J. Chem. Soc., Chem. Commun.* **1992**, 273–274.

38 Iida, T.; Yamamoto, N.; Matsunaga, S. Woo, H.-G.; Shibasaki, M. *Angew. Chem. Int. Ed.* **1998**, *37*, 2223–2226.

39 Matsunaga, S.; Das, J.; Roels, J.; Vogl, E. M.; Yamamoto, N.; Iida, T.; Yamaguchi, K.; Shibasaki, M. *J. Am. Chem. Soc.* **2000**, *122*, 2252–2260.

40 Nozaki, K.; Nakano, K.; Hiyama, T. *J. Am. Chem. Soc.* **1999**, *121*, 11008–11009.

41 Cheng, M.; Darling, N. A.; Lobkovsky, E. B.; Coates, G. W. *Chem. Commun.* **2000**, 2007–2008.

42 Coates, G. W.; Moore, D. R. *Angew. Chem. Int. Ed.* **2004**, *43*, 6618–6639.

43 Hayashi, M.; Tamura, M.; Oguni, N. *Synlett* **1992**, 663–664.

44 Cole, B. M.; Shimizu, K. D.; Krueger, C. A.; Harrity, J. P. A.; Snapper, M. L.; Hoveyda, A. H. *Angew. Chem. Int. Ed. Engl.* **1996**, *35*, 1668–1671.

45 Shimizu, K. D.; Cole, B. M.; Krueger, C. A.; Kuntz, K. W.; Snapper, M. L.; Hoveyda, A. H. *Angew. Chem. Int. Ed. Engl.* **1997**, *36*, 1704–1707.

46 Schaus, S. E.; Jacobsen, E. N. *Org. Lett.* **2000**, *2*, 1001–1004.

47 Palkulski, Z.; Pietrusiewicz, K. M. *Tetrahedron: Asymmetry* **2004**, *15*, 41–45.

48 Oguni, N.; Miyagi, Y., Itoh, K. *Tetrahedron Lett.* **1998**, *39*, 9023–9026.

49 Bandini, M.; Cozzi, P. G.; Melchiorre, P.; Umani-Ronchi, A. *Angew. Chem. Int. Ed.* **2004**, *43*, 84–87.

50 Crotti, P.; Di Bussolo, V.; Favero, L.; Macchia, F.; Pineschi, M. *Gazetta Chimica Italiana*, **1997**, *127*, 273–275.

51 Pineschi, M. *New. J. Chem.*, **2004**, *28*, 657–665.

52 Del Moro, F.; Crotti, P.; Di Bussolo, V.; Macchia, F.; Pineschi, M. *Org. Lett.* **2003**, *5*, 1971–1974.

53 Nugent, W. A. *J. Am. Chem. Soc.* **1998**, *120*, 7139–7140.

54 Bruns, S.; Haufe, G. *Tetrahedron: Asymmetry* **1999**, *10*, 1563–1569.

55 Denmark, S. E.; Barsanti, P. A.; Wong, K.-T.; Stavenger, R. A. *J. Org. Chem.* **1998**, *63*, 2428–2429.

56 Tao, B.; Lo, M. M.-C.; Fu, G. C. *J. Am. Chem. Soc.* **2001**, *123*, 353–354.

57 Nakajima, M.; Saito, M.; Uemura, M.; Hashimoto, S. *Tetrahedron Lett.* **2002**, *43*, 8827–8829.

58 Chan, A. S. C.; Coleman, J. P. *J. Chem. Soc., Chem. Commun.* **1991**, 535–536.

59 Keith, J. M.; Larrow, J. F.; Jacobsen, E. N. *Adv. Synth. Catal.* **2001**, *343*, 5–26.

60 Larrow, J. F.; Schaus, S. E.; Jacobsen, E. N. *J. Am. Chem. Soc.* **1996**, *118*, 7420–7421.

61 Wang, Z.-M.; Zhang, X.-L.; Sharpless, K. B. *Tetrahedron Lett.* **1993**, *34*, 2267–2270. Klunder, J. M.; Ko, S. Y.; Sharpless, K. B. *J. Org. Chem.* **1986**, *51*, 3710–3712. Sasai, H.; Itoh, N.; Suzuki, T.; Shibasaki, M. *Tetrahedron Lett.* **1993**, *34*, 855–858.

62 Tsai, C.-C; Follis, K. E.; Sabo, A.; Beck,

T. W.; Grant, R. F.; Bischofberger, N.; Benveniste, R. E.; Black, R. *Science* **1995**, *270*, 1197–1199.

63 Schaus, S. E.; Jacobsen, E. N. *Tetrahedron Lett.* **1996**, *37*, 7937–7940.

64 Lebel, H.; Jacobsen, E. N. *Tetrahedron Lett.* **1999**, *40*, 7303–7306.

65 Lebel, H.; Jacobsen, E. N. *J. Org. Chem.* **1998**, *63*, 9624–9625.

66 Brandes, B. D.; Jacobsen, E. N. *Synlett* **2001**, 1013–1015.

67 Reviews concerning the preparation and use of *syn*-β-amino alcohols: Bergmeier, S. C. *Tetrahedron* **2000**, *56*, 2561–2576. Ager, D. J.; Prakash, I.; Schaad, D. R. *Chem. Rev.* **1996**, *96*, 835–876.

68 Li, G.; Chang, H.-T.; Sharpless, K. B. *Angew. Chem., Int. Ed. Engl.* **1996**, *35*, 451–454.

69 Trost, B. M.; Bunt, R. C.; Lemoine, R. C.; Calkins, T. L. *J. Am. Chem. Soc.* **2000**, *122*, 5968–5976.

70 Reddy, L. R.; Bhanumathi, N.; Rao, K. R. *Chem. Commun.* **2000**, 2321–2322.

71 Kim, S. K.; Jacobsen, E. N. *Angew. Chem. Int. Ed.* **2004**, *43*, 3952–3954.

72 Bartoli, G.; Bosco, M.; Carlone, A.; Locatelli, M.; Melchiorre, P.; Sambri, L. *Org. Lett.* **2004**, *6*, 3973–3975.

73 Tokunaga, M.; Larrow, J. F.; Kakiuchi, F.; Jacobsen, E. N. *Science* **1997**, *277*, 936–938.

74 Schaus, S. E.; Brandes, B. D.; Larrow, J. F.; Tokunaga, M.; Hansen, K. B.; Gould, A. E.; Furrow, M. E.; Jacobsen, E. N. *J. Am. Chem. Soc.* **2002**, *124*, 1307–1315.

75 Furrow, M. E.; Schaus, S. E.; Jacobsen, E. N. *J. Org. Chem.* **1998**, *63*, 6776–6777 and see reference 63.

76 Schaus, S. E.; Brånalt, J.; Jacobsen, E. N. *J. Org. Chem.* **1998**, *63*, 4876–4877.

77 Colby, E. A.; O'Brien, K. C.; Jamison, T. F. *J. Am. Chem. Soc.* **2004**, *126*, 998–999.

78 Haidle, A. M.; Myers, A. G. *Proc. Natl. Acad. Sci.* **2004**, *101*, 12048–12053.

79 Smith, A. B.; Kim, D.-S. *Org. Lett.* **2004**, *6*, 1493–1495.

80 Kang, B.; Chang, S. *Tetrahedron* **2004**, *60*, 7353–7359.

81 Nielsen, L. P. C.; Stevenson, C. P.; Blackmond, D. G.; Jacobsen, E. N. *J. Am. Chem. Soc.* **2004**, *126*, 1360–1362.

82 Holland, J. M.; Lewis, M.; Nelson, A. *J. Org. Chem.* **2003**, *68*, 747–753.

83 Ready, J. M.; Jacobsen, E. N. *J. Am. Chem. Soc.* **1999**, *121*, 6086–6087.

84 Nielsen, L. P. C.; Jacobsen, E. N. work in progress.

85 Darensbourg, D. J.; Yarbrough, J. C.; Ortiz, C.; Fang, C. C. *J. Am. Chem. Soc.* **2003**, *125*, 7586–7591.

86 Qin, Z.; Thomas, C. M.; Lee, S.; Coates, G. W. *Angew. Chem. Int. Ed.* **2003**, *42*, 5484–5487.

87 Lu, X.-B.; Liang, B.; Zhang, Y.-J.; Tian, Y.-Z.; Wang, Y.-M.; Bai, C.-X.; Wang, H.; Zhang, R. *J. Am. Chem. Soc.* **2004**, *126*, 3732–3733.

88 Sharpless, K. B. *Angew. Chem. Int. Ed.* **2002**, *41*, 2024–2032.

89 Tu, Y.; Wang, Z.-X.; Shi, Y. *J. Am. Chem. Soc.* **1996**, *118*, 9806–9807.

90 Pineschi, M.; Del Moro, F.; Crotti, P.; Di Bussolo, V.; Macchia, F. *J. Org. Chem.* **2004**, *69*, 2099–2105.

91 Bertozzi, F.; Crotti, P.; Feringa, B. L.; Macchia, F.; Pineschi, M. *Synthesis* **2001** 483–486.

92 Semmelhack, M. F.; Epa, W. R.; Cheung, A. W.-H.; Gu, Y.; Kim, C.; Zhang, N.; Lew, W. *J. Am. Chem. Soc.* **1994**, *116*, 7455–7456.

93 Potvin, S.; Canonne, P. *Tetrahedron: Asymmetry* **1996**, *7*, 2821–2824.

94 Eklund, L.; Sarvary, I.; Frejd, T. *J. Chem. Soc., Perkins Trans. 1* **1996**, 303–305.

95 Feng, X.; Shu, L.; Shi, Y. *J. Am. Chem. Soc.* **1999**, *121*, 11002–11003.

96 Feng, X.; Shu, L.; Shi, Y. *J. Org. Chem.* **2002**, *67*, 2831–2836.

97 Wang, F.; Tu, Y. Q.; Fan, C. A.; Wang, S. H.; Zhang, F. M. *Tetrahedron: Asymmetry* **2002**, *13*, 395–398.

98 Whitesell, J. K.; Felman, S. W. *J. Org. Chem.* **1980**, *45*, 755–756.

99 Hodgson, D. M.; Gibbs, A. R.; Lee, G. P. *Tetrahedron* **1996**, *52*, 14361–14384.

100 Asami, M. *J. Synth. Org. Chem. Jpn.* **1996**, *54*, 188.

101 O'Brien, P. *J. Chem. Soc., Perkin Trans. 1* **1998**, 1439.

102 Asami, M.; Ishizaki, T.; Inoue, S. *Tetrahedron: Asymmetry* **1994**, *5*, 793–796.

103 Asami, M.; Suga, T.; Honda, K.; Inoue, S. *Tetrahedron Lett.* **1997**, *38*, 6425–6428.

104 Malhotra, S. V. *Tetrahedron: Asymmetry* **2003**, *14*, 645–647.

105 Magnus, A.; Bertilsson, S. K.; Andersson,

P. G. *Chem. Soc. Rev.* **2002**, *31*, 223–229 and references cited therein.

106 Södergren, M. J.; Bertilsson, S. K.; Andersson, P. G. *J. Am. Chem. Soc.* **2000**, *122*, 6610–6618.

107 Bertilsson, S. K.; Södergren, M. J.; Andersson, P. G. *J. Org. Chem.* **2002**, *67*, 1567–1573.

108 Gayet, A.; Bertillson, S.; Andersson, P. G. *Org. Lett.* **2002**, *4*, 3777–3779.

109 Liu, D.; Kozmin, S. A. *Angew. Chem. Int. Ed.* **2001**, *40*, 4757–4759.

110 Ma, J. A.; Cahard, D. *Angew. Chem. Int. Ed.* **2004**, *43*, 4566–4583.

8
Epoxides in Complex Molecule Synthesis

Paolo Crotti and Mauro Pineschi

8.1
Introduction

Epoxides (or oxiranes) constitute one of the functional groups most widely used in synthesis because, in addition to the ease with which the oxirane functionality can be introduced into a molecule, they react with a large number of nucleophiles to yield valuable bifunctional compounds in a fairly simple procedure [1]. Moreover, since very important methodologies for asymmetric epoxidations are now available [2], the opening reactions of epoxides, if carried out with rigid stereo- and regiocontrol, can be used advantageously for the construction, in a single process, of two contiguous stereochemically defined sp^3 carbon chiral centers. Epoxides can also be subjected to non-addition processes. In this case, valuable rearranged products can be obtained in a very simple manner [3a].

The aim of this contribution is to examine the most widely used methods of elaborating the oxirane functionality in the synthesis of complex molecules. In view of the extensive use of epoxides in simple transformation procedures in the early stages of total syntheses, only certain selected manipulations of epoxides representing the key steps of a total synthesis and/or their use at a late stage of the reaction sequence are considered.

8.2
Synthesis of Complex Molecules by Intramolecular Ring-opening of Epoxides with Heteronucleophiles

8.2.1
Intramolecular C–O Bond-forming Reactions

Tetrahydropyran- and tetrahydrofuran-substituted rings are common in many complex natural products, such as polyether antibiotics, displaying a wide range of biological properties [4]. An appropriate combination of OH-directed diastereoselective epoxidation of γ- and δ-alkenols [$VO(acac)_2$/TBHP or MCPBA] [5] or ster-

Aziridines and Epoxides in Organic Synthesis. Andrei K. Yudin

ISBN: 3-527-31213-7

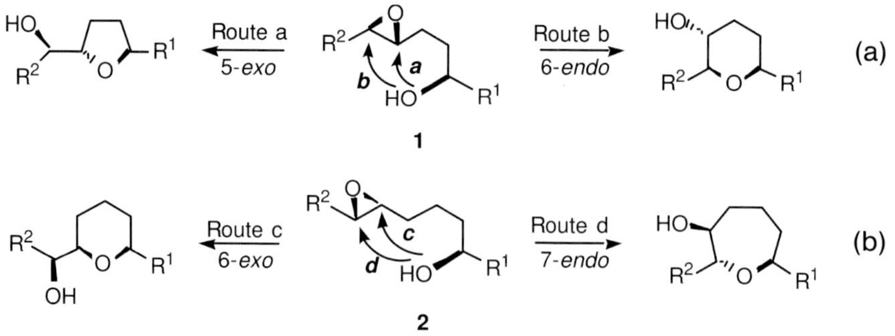

Scheme 8.1 *Endo* and *exo* cyclization modes of γ- and δ-epoxy alcohols.

eoselective epoxidation of allylic alcohols (usually Sharpless asymmetric epoxidation; SAE) [2a] followed by oxirane ring-opening by an internal *O*-nucleophile (a γ- or δ-OH group) present in the molecule is an effective process for the construction of THP and THF rings from acyclic precursors [6]. The opening process is completely stereoselective, proceeding with complete inversion of configuration at the oxirane carbon, and can be highly regioselective. As an example, if the side chain R^2 in a generic γ-epoxy alcohol **1** is an alkyl chain, a 5-*exo* cyclization process [7] is generally preferred (Route a, Eq. a, Scheme 8.1), as a consequence of well demonstrated stereoelectronic factors [8], with the formation of the smaller five-membered *O*-heterocycle (a THF ring). However, if R^2 is a group capable of stabilizing an incipient positive charge, and if the cyclization reaction is carried out under acid conditions, the intramolecular nucleophilic attack will now occur at the oxirane carbon better able to accommodate a partial carbocationic character in the transition state, thus dictating the formation of the corresponding six-membered ring (a THP ring) by a 6-*endo* cyclization pathway (Route b, Eq. a, Scheme 8.1) [7,9].

Other possible cyclizations may involve δ-epoxy alcohols such as the generic compound **2**. In this case, two pathways are theoretically possible: a more favored 6-*exo* cyclization mode (Route c) giving rise to a THP ring, or an alternative 7-*endo* cyclization mode providing an oxepane ring (Route d) (Eq. b, Scheme 8.1) [7].

8.2.1.1 Synthesis of Substituted THF Rings

For the formation of substituted THF rings (Route a, Scheme 8.1), Kishi developed a procedure based on the hydroxy-directed epoxidation of a γ-alkenol [10]. Epoxidation of bishomoallylic alcohol **3** by TBHP/VO(acac)$_2$ by this approach, followed by treatment of the intermediate epoxide **4** with acetic acid, gave the THF derivative **5** of isolasalocid A (a 5-*exo* cyclization; Scheme 8.2) [11]. Further epoxidation of **5** (a γ-alkenol) under the same conditions, followed by acetylation, afforded epoxide **6**. For the synthesis of the natural product, the configuration of epoxide **6** had to be inverted before the second cyclization reaction. Epoxide **6** was consequently hydrolyzed under acid conditions to the corresponding diol and was then selectively

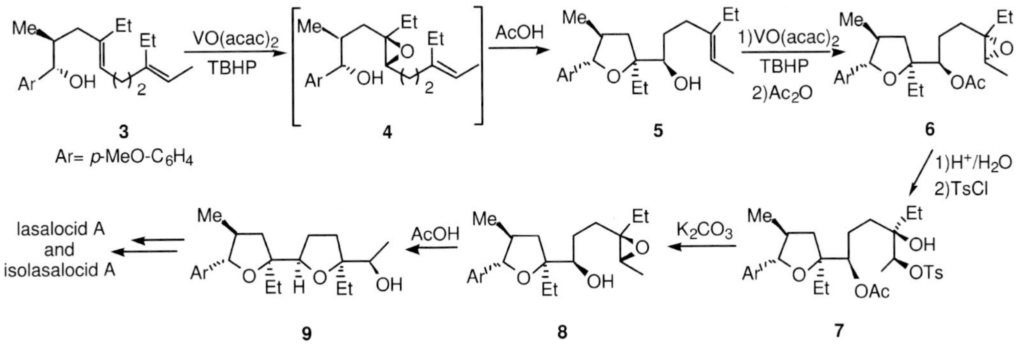

Scheme 8.2 Kishi's method for the construction of the bis(tetrahydrofuran) core of isolasalocid A.

monotosylated (7). Base-catalyzed cyclization of hydroxy tosylate **7** afforded the inverted γ-epoxy alcohol **8**. Treatment of **8** with AcOH then afforded the bis-THF derivative **9**, which was carried on into the syntheses of lasalocid A and isolasalocid A.

Similar 5-*exo* cyclization procedures have been widely utilized by Evans in his total syntheses of complex natural compounds, such as the synthesis of ionomycin [12a] and polyether antibiotic X-206 [12b]. A 5-*exo* cyclization of a γ-epoxy alcohol has also been observed under basic conditions [12c].

A dominant structural feature that appears in more than 40% of the annonaceous acetogenins [13] is, as shown in (+)-uvaricin, a ten-carbon fragment containing two adjacent THF rings flanked by two hydroxy groups, with the main variation being the relative and absolute configurations of the various oxygen-bearing stereogenic centers. Two long fatty acid chains complete the structure. In the potent antitumor agent (–)-mucocin, a THP moiety is present instead (Figure 8.1).

Two strategies for the construction of these systems are available. In the "inside-out" epoxide cascade reaction of substrates available from the "two-tartrate" approach, diene **10** was epoxidized by SAE to the diepoxide **11** (Scheme 8.3) [14]. After monotosylation, acid-catalyzed acetonide removal simultaneously opened the two oxirane rings to produce the two adjacent THF moieties (**12**) of an unnatural diastereoisomer of uvaricin. While coupling of tosylate **12** with an excess of lithium dinonylcuprate installed the necessary C_{26}–C_{34} tail, the needs for attachment of the lactone-containing head group and inversion of configuration at C_{15} were

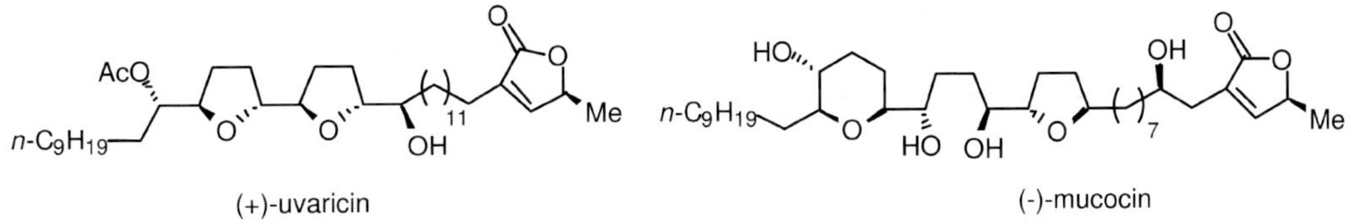

Figure 8.1 Structures of (+)-uvaricin and (–)-mucocin.

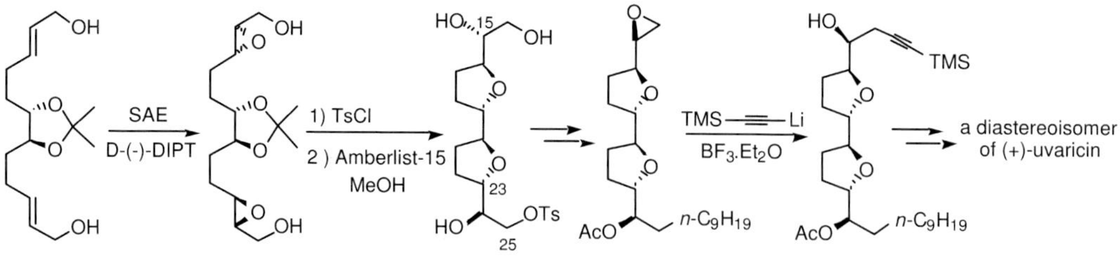

Scheme 8.3 Synthesis of an unnatural diastereisomer of (+)-uvaricin.

met by transformation of the terminal 1,2-diol moiety into the inverted terminal epoxide **13**. Ring-opening of **13** with lithium trimethylsilylacetylide by the Yamaguchi procedure [15] afforded acetylide **14**, which was subsequently transformed into the desired product (Scheme 8.3).

In Keinan's "naked carbon skeleton" strategy, all oxygen functions are selectively placed, usually by SAE and Sharpless asymmetric dihydroxylation (SAD) [2b], onto an unfunctionalized unsaturated carbon skeleton, followed by selective transformations [16]. In the total synthesis of (–)-mucocin [17], the oxygen functionalities are introduced into the tetraenediol **15** by means of SAE and SAD as shown (Scheme 8.4). Subsequently, in a later stage of the synthesis, the THP and THF rings present in the intermediate **17** simultaneously arise (by acid treatment) from the double γ-epoxy alcohol functionality present in the polyhydroxylated compound **16**. Whereas the left-hand epoxide undergoes a 6-*endo* cyclization mode by the C_{11}–OH group, due to the directing effect of the adjacent unsaturated system (see Section 8.2.1.2), the right-hand epoxide undergoes a more favored 5-*exo* cyclization by the γ-C_6–OH group (Scheme 8.4) [7, 8].

A novel epoxide cyclization approach to the formation of compound **23**, contain-

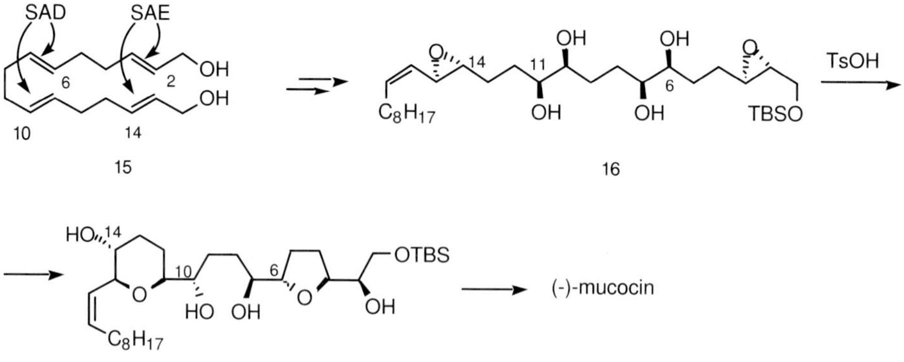

Scheme 8.4 Construction of the THP and THF rings of (–)-mucocin.

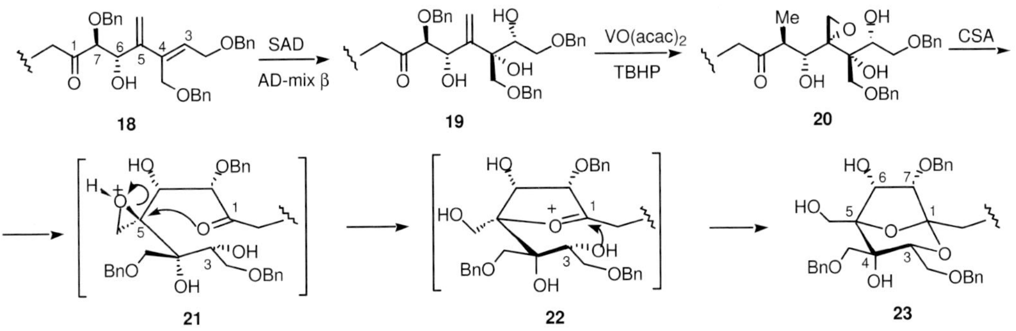

Scheme 8.5 Synthesis of the bicyclic acetal core of zaragozic acid C.

ing the bicyclic acetal skeleton of zaragozic acid C, utilizes the polyhydroxylated γ-epoxy ketone **20** (Scheme 8.5). The transformation of **20** into **23** proceeds under mild acid conditions (CSA), permitting the cyclization of the completely open-chain system present in **20** into the 2,8-dioxabicyclo[3.2.1]octane core of the natural compound. Protonation of the epoxide (shown in **21**) is followed by attack of the carbonyl oxygen, in a favored 5-*exo* cyclization process, to give the intermediate cyclic oxocarbenium ion **22**, corresponding to the THF ring of the final bicyclic core, which is subsequently attacked by the C_3–OH group to afford the bicyclic system of the natural product (Scheme 8.5) [18]. For the success of this strategy, regio- and diastereoselective introduction of the three oxygens into the starting diene **18** was obviously necessary. SAD carried out with freshly prepared enriched AD-mix β occurred only at the trisubstituted double bond, affording the corresponding triol **19**. Subsequent OH-directed epoxidation of **19** with $VO(acac)_2$/TBHP delivered epoxide **20**, the cyclization substrate, as a single isomer (Scheme 8.5) [18].

8.2.1.2 **Synthesis of Substituted THP Rings**

To circumvent the inherent tendency of γ-epoxy alcohols to cyclize in a 5-*exo* fashion under acidic conditions, Nicolaou introduced the use of an electron-rich double bond (a vinyl group or a methoxycarbonylvinyl group) adjacent to the oxirane C-O bond furthest away from the hydroxy group [9]. The π-system stabilization furnished by the unsaturated system exerts a directing effect, determining the almost exclusive formation of the six-membered ring (THP), through a 6-*endo* cyclization process (Route b, Scheme 8.1). A spectacular application of this strategy can be found in several syntheses of marine sponge toxins such as brevetoxin B and hemibrevetoxin B, which are characterized by the presence of several head-tail fused THPs (Figure 8.2).

In Nicolaou's total synthesis of brevetoxin B, for example, out of the eleven *trans*-fused six- to eight-membered cyclic ethers, THP rings B, F, G, and I are constructed by application of this methodology through cyclization of the appropriate

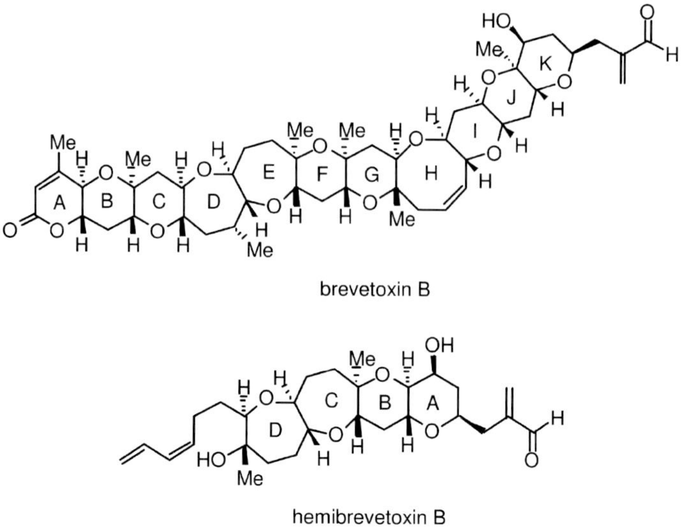

brevetoxin B

hemibrevetoxin B

Figure 8.2 Examples of natural *trans*-fused cyclic ethers: brevetoxin B and hemibrevetoxin B.

γ-epoxy alcohol [9a, 19]. In this way, ring F is constructed by exposure of the α-vinyl-γ-hydroxy epoxide **24** (R = H) to mildly acid conditions (PPTS) in CH_2Cl_2 with subsequent formation of the desired THP derivative **25** by a completely *anti*-stereoselective and 6-*endo* regioselective cyclization process. The newly formed OH group is the *O*-nucleophile for the construction of the adjacent *trans*-fused THP ring G (compound **27**), by the use of similar procedures on the intermediate α-vinyl-γ-hydroxy epoxide **26** (R = H) (Scheme 8.6).

For the construction of the I ring, the vinylic group introduced to activate the γ-hydroxy epoxide moiety of **28** towards cyclization is an acrylic ester residue, which concomitantly allows cyclization on the allylic position, with formation of the tricyclic compound **29** containing the IJK fragment of the natural product, and fur-

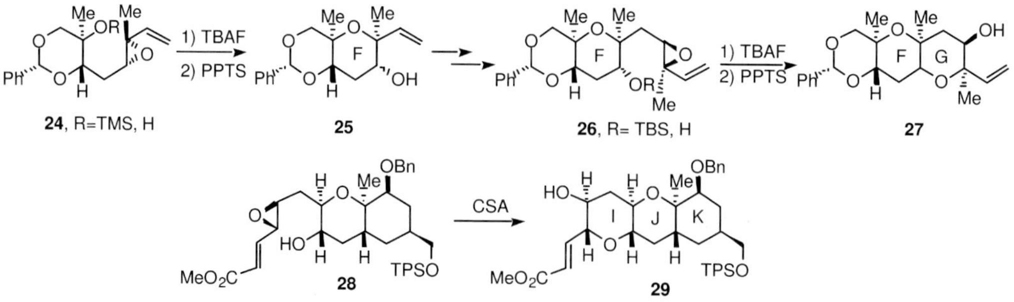

24, R=TMS, H — 1) TBAF 2) PPTS → **25** → **26**, R= TBS, H — 1) TBAF 2) PPTS → **27**

28 — CSA → **29**

Scheme 8.6 Nicolaou's approach for the construction of the F, G, and I rings of brevetoxin B.

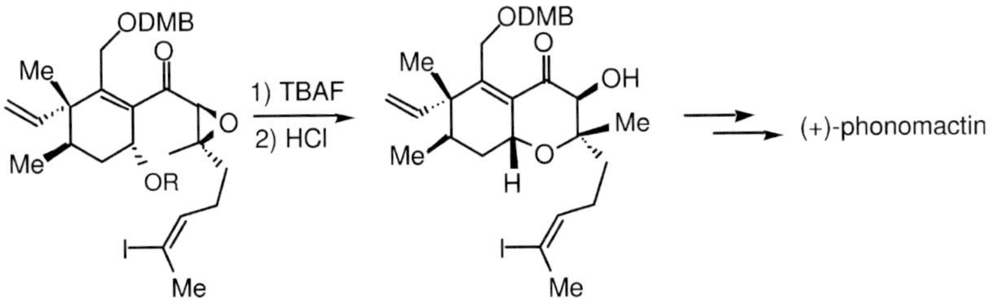

Scheme 8.7 A 6-*endo* cyclization process in a hydroxy-epoxy ketone.

ther utilization for the construction of the adjacent didehydrooxocane ring (ring H) (Scheme 8.6) [9a, 19].

A 6-*endo* cyclization by a γ-epoxy alcohol can be accomplished through the presence of an appropriately disposed electron-withdrawing group, as found in the total synthesis of (+)-phonomactin. Treatment of the hydroxy-epoxy ketone **30** (R = H) with HCl afforded the bicyclic compound **31**, containing a pyran-4-one ring, in which nucleophilic attack of the γ-OH group occurred at the oxirane carbon distal from the unfavorable electronic effect of the carbonyl group (Scheme 8.7) [20a].

An easy, silica gel-promoted 6-*endo* cyclization of γ-epoxy alcohol **32** to pyran **33**, followed by acid-catalyzed spiroketalization of the keto diol **34**, afforded the common tricyclic spiroketal fragment **35** of lituarines A, B, and C (Scheme 8.8) [20b].

In spite of the slightly electron-withdrawing effect of the $-SiMe_3$ group, α-epoxysilanes show regioselectivity in the ring-opening reaction with *O*-nucleophiles determined by exclusive attack on the oxirane carbon α to the $-SiMe_3$ group [21]. This makes the $-SiMe_3$ group an attractive candidate for achieving regioselective 6-*endo* cyclization in γ-epoxy alcohols bearing the silicon group on the oxirane carbon further away from the alcoholic functionality. This strategy was initially

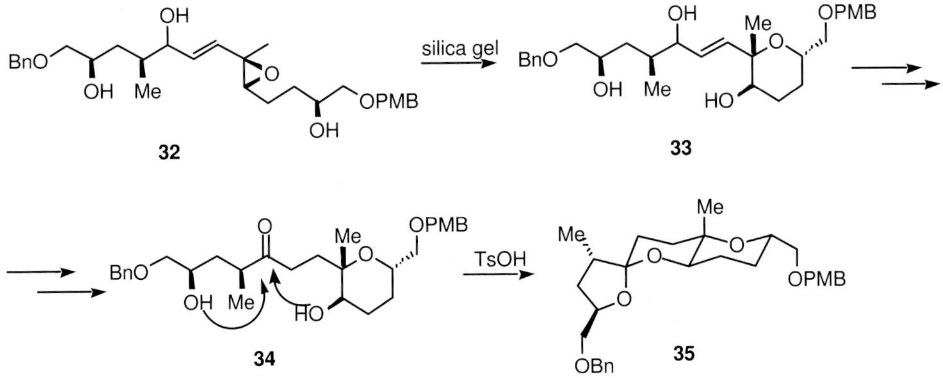

Scheme 8.8 Synthesis of the tricyclic spiroketal fragment of lituarines.

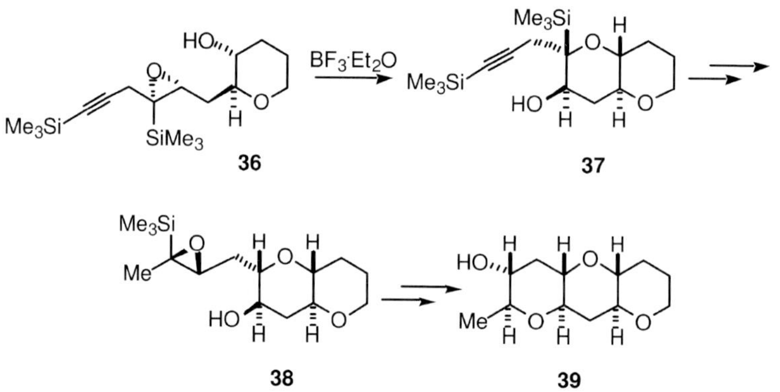

Scheme 8.9 6-*Endo* cyclization process in a Me_3Si-substituted γ-epoxy alcohol.

explored by Schaumann, who found that only *anti* (*E*)-γ-epoxy alcohols cyclized with an elevated regioselectivity to the corresponding THP derivatives through a favorable chair-like transition state [22]. Jamison's application of this procedure to the configurationally corresponding epoxy silane **36** allowed the construction of the bis-THP *trans* fused system **37** with almost complete stereo- (95% inversion) and regioselectivity (>95% *endo*-cyclization). Iterative elaboration of the trimethylsilylalkynyl group of **37** provided the new epoxysilane **38**, which cyclized with high stereo- and regioselectivity to give the tris-THP *trans* fused system **39** (Scheme 8.9) [23].

Another successful contribution to the efficient synthesis of tetrahydropyran derivatives was reported by Mukai and Hanaoka [24]. The most significant aspect of this method, based on a highly regio- and stereoselective 6-*endo* cyclization of cobalt hexacarbonyl complexes of alkynyl-substituted β- or γ-epoxy alcohols under acidic conditions, is that the reaction occurs with *retention of configuration* at the oxirane propargylic position. This procedure was used in the first total synthesis of (–)-ichthyothereol, starting from diethyl L-tartrate. By this method, *cis* epoxy alkyne **40** was converted into the corresponding dicobalt hexacarbonyl species **41**, which was then treated with a catalytic amount of $BF_3 \cdot OEt_2$ in CH_2Cl_2 at –78 °C, followed by demetalation with CAN, to give the *trans*-disubstituted tetrahydropyran **42**, which can be transformed into the natural product by a simple elaboration of the alkyne moiety (Scheme 8.10) [25].

Very recently, this procedure was used by Smith as the key step for a new construction of the novel dioxabicyclo[3.2.1]octane **46**, a subtarget in the synthesis of potent antibiotic (+)-sorangicin A [26]. Treatment of β-epoxy alcohol (+)-**43** with $Co_2(CO)_8$ (1.1 equiv.) at rt., followed at –78 °C by a catalytic amount of $BF_3 \cdot Et_2O$, yielded alkyne-cobalt complex (+)-**44**, in which the epoxide-opening at the propargylic position had occurred with retention of configuration (Scheme 8.11). Subjected to cyclization reaction conditions, both the cobalt complex and the corresponding demetalated TMS-alkyne gave very low yields of the desired 6-*exo* bicyclic

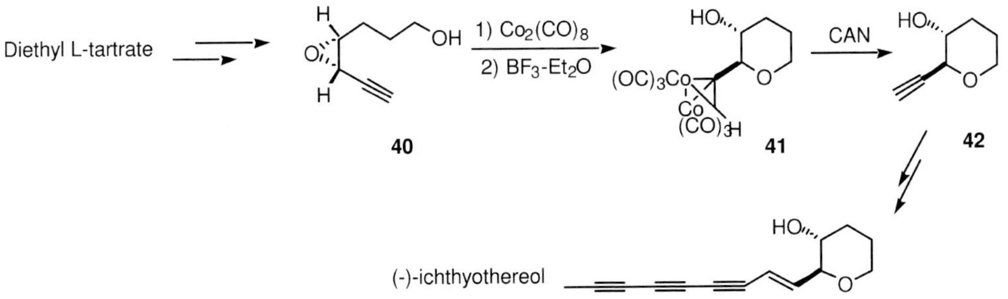

Scheme 8.10 Stereoselective total synthesis of (–)-ichthyothereol.

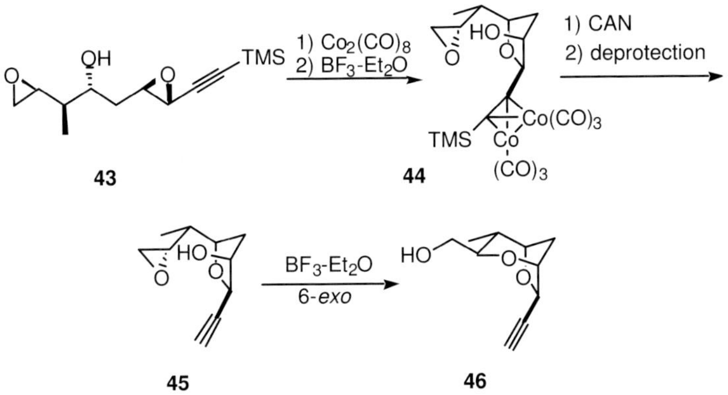

Scheme 8.11 Regio- and stereocontrolled construction of the dioxabicyclo[3.2.1]octane skeleton of (+)-sorangicin A.

product **46**, probably due to the steric bulk of the cobalt complex and TMS moieties. In fact, better results were obtained after removal of the TMS to give **45** and subsequent cyclization with a large excess of $BF_3 \cdot Et_2O$ to afford the dioxabicyclic[3.2.1] subtarget **46**, albeit not with complete regioselectivity.

Besides the numerous reactions in which an epoxide behaves as an electrophile, the transformation of a suitable epoxide into the corresponding oxiranyl anion makes it possible to use an epoxide as a nucleophile. Oxiranyl anions can be generated at –80 °/–115 °C from several precursors, with retention of configuration at the metalated carbon stereocenter, and react with reactive electrophiles such as aldehydes, ketones, Me_3SiCl, and alkyl triflates [27]. The iterative unique combination of an oxiranyl anion (A and/or B) generated by deprotonation of the corresponding *cis*-epoxide bearing a sulfonyl group, as the anion-stabilizing group, with an alkyl triflate (C and/or F), followed by 6-*endo* cyclization of the intermediate γ-epoxy alcohol (D) and stereoselective reduction of the intermediate ketone (E), results in the stereoselective synthesis of *trans*-fused THPs (G, Eq. a, Scheme 8.12). Actually, the intermediate γ-epoxy alcohol D could undergo ring-closure by

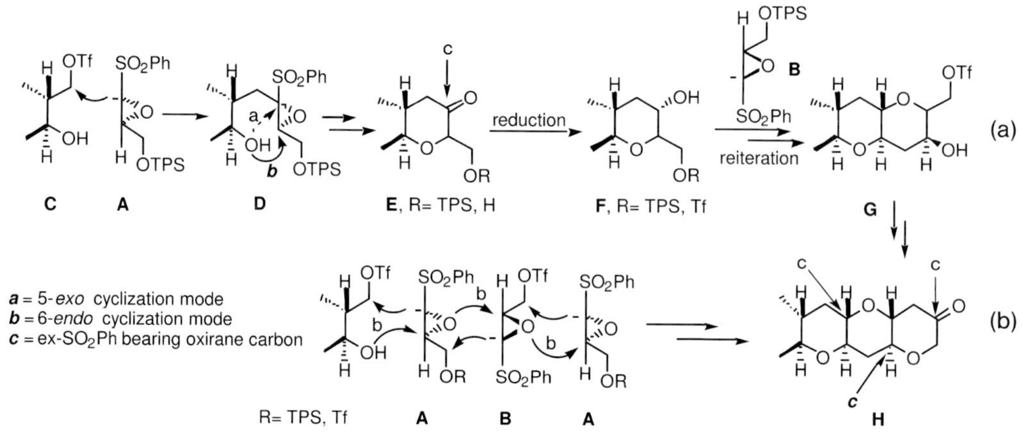

Scheme 8.12 Schematic representation of the "oxiranyl anion strategy" for the synthesis of *trans*-fused THPs.

either a 5-*exo* (Route a) or a 6-*endo* mode of cyclization (Route b) to give a THF or a THP system, respectively (Schemes 8.1 and 8.12), but the presence of the sulfonyl group would be expected to favor the 6-*endo*-mode pathway, due to its strongly electron-withdrawing nature [28]. This idea is complementary to that of the π-orbital-assisted cyclization, examined above. The freely summarized schematic representation of the reaction in Scheme 8.12 (Eq. b) shows how the oxiranyl anion strategy mimics nature, as it is thought that the obtained polycyclic structures (H) arise in nature from a cascade cyclization of a polyepoxide precursor (see Section 8.2.1.3).

The stepwise coupling of the three oxiranyl anions derived from epoxides **47**, **48**, and **49**, respectively, to triflate **50** and ring-expansion at suitable stages resulted in the construction of the *trans*-fused tetracyclic ketone **61**, containing the ABCD ring system of hemibrevetoxin B (Schemes 8.13 and 8.14). While the construction of the B ring (ketone **52**) makes use of a triflate (**50**), followed by acid-catalyzed cyclization of the intermediate γ-epoxy alcohol **51** (R = H), installation of the third ring (ketone **55**) requires the intermediate utilization of aldehyde **53**, because several attempts to couple a triflate derived from **52** with the oxiranyl anion derived from epoxide **48** were unsuccessful, due to the considerable hindrance of the angular Me group. The obtained epoxy diol **54** (R = H) is cyclized ($BF_3 \cdot Et_2O$) and then dehydroxylated (SmI_2) to give the desired ketone **55** (Scheme 8.13) [28].

For the synthesis of the complex natural product, the terminus six-membered ketone **55** had to be transformed into an oxepane ring. For this necessary transformation, the authors were attracted by the single-carbon homologation of a pyranone (a sort of ring-expansion) because, in principle, it could be used in an iterative sense at any stage of the 6-*endo* cyclization in their poly-THP-based synthetic approach for the synthesis of *trans*-fused 6,7,6 (THP-oxepane-THP) and 6,7,7 (THP-oxepane-oxepane) ring systems [28]. Treatment of ketone **55** with $TMSCHN_2$

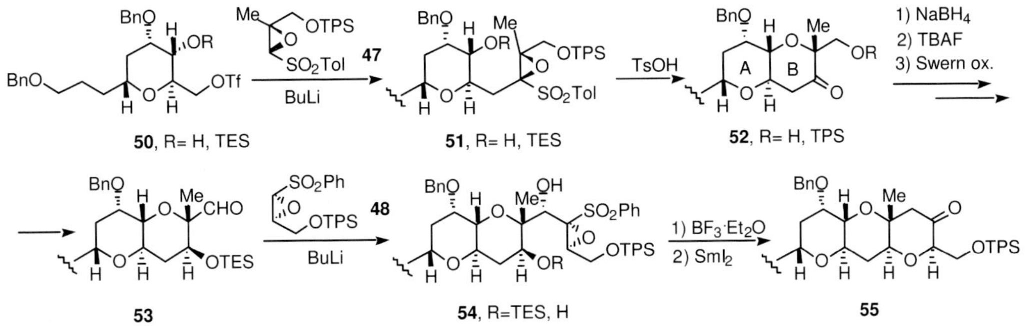

Scheme 8.13 Synthesis of the AB ring system of hemibrevetoxin B and construction of the third ring.

(trimethylsilyldiazomethane) in the presence of $BF_3 \cdot Et_2O$ gave the seven-membered ketone **56** (R = TPS) (C ring) as the main product (Scheme 8.14). Reduction of the hydroxy ketone **56** (R = H) with $Me_4NBH(OAc)_3$ afforded only diol **57**, which was subsequently transformed into triflate **58**. Treatment of **58** with the oxiranyl anion derived from epoxide **49**, containing a three-carbon side chain, under the usual conditions afforded, through epoxide **59**, the new ketone **60**. Application of the ring-expansion reaction method ($TMSCHN_2$) to **60** afforded the new tetracyclic ketone **61** (62% yield), containing the bis-oxepane ring moiety and the complete ABCD ring system of the natural compound (Scheme 8.14) [28].

In the synthesis of polyether antibiotic X-206, the terminal C_{29}–C_{37} fragment containing the THP ring (F ring, compound **64**) was efficiently established by application of SAE to unsaturated 1,7-diol **62** (an allylic alcohol and a δ-alkenol at the same time) (Scheme 8.15). During the course of this reaction, the intermediate δ-epoxy alcohol **63** underwent intramolecular attack by the C_{35}-OH group (the δ-OH group) to give the THP derivative **64** through a favored 6-*exo* cyclization (Route c, Scheme 8.1) [7, 12b]. The 1,2-diol functionality present in **64** was subsequently

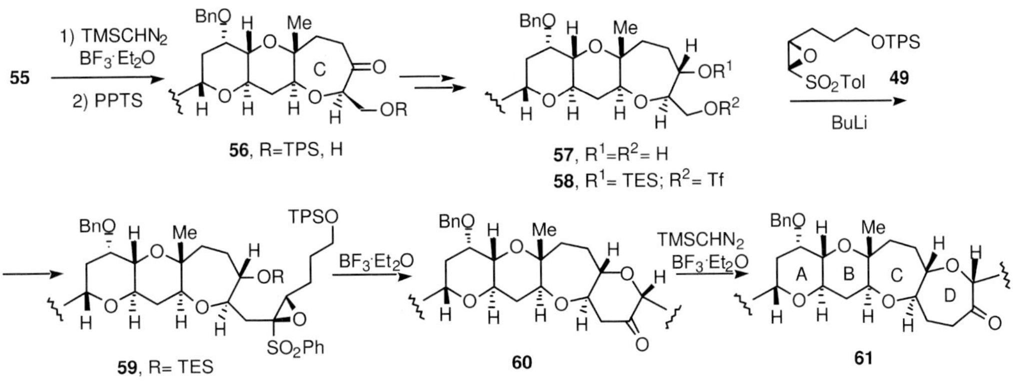

Scheme 8.14 Construction of the bis-oxepane ring moiety of hemibrevetoxin B.

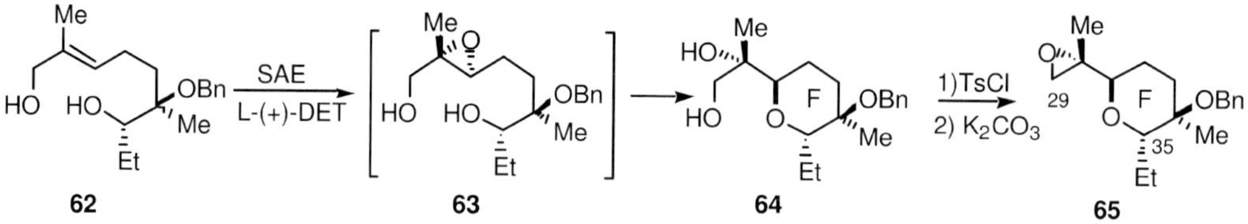

Scheme 8.15 6-*Exo* cyclization of a δ-epoxy alcohol.

utilized for the stereoselective construction (regioselective monotosylation, followed by base-catalyzed cyclization) of the corresponding terminal epoxide **65**, useful for the assembly of the two fragments C_{21}–C_{27} and C_{29}–C_{37} of the natural product (see Section 8.3.2.2) [12b]. A similar cyclization of a δ-epoxy alcohol can be also found in the synthesis of the CD ring system of hemibrevetoxin B [29a].

An elegant application of the strategy based on the appropriate use of a 6-*exo* or 6-*endo* cyclization of a δ- or γ-epoxy alcohol, respectively, to give corresponding adjacent or fused THP rings can be found in the regioselective synthesis of the ABC tris-THP core of thyrsiferol and venustatriol [29b].

The oxepane ring may be prepared by activation of the 7-*endo* δ-epoxy alcohol intramolecular opening over the 6-*exo* variant (Routes d and c, respectively, Eq. b, Scheme 8.1) through the introduction, as usual, of an electron-rich double bond at the position furthest away from the OH-bearing carbon. A synthetically useful oxepane/THP regioselectivity was obtained when vinyl and (*E*)-2-chlorovinyl substituents (R^2 group in Eq. b, Scheme 8.1) were used. However, attempts to prepare bicyclic systems containing oxepanes resulted in clean formation of the THP skeleton [30a]. A selective 7-*endo* intramolecular cyclization (92%) has been achieved in δ-epoxy alcohols bearing a methoxymethyl group through chelation of $La(OTf)_3$ between the oxygens of the epoxide and ether groups [30b].

8.2.1.3 Intramolecular 5-*exo* and 6-*endo* Cyclization of Polyepoxides

All efforts shown above for the construction of poly-THF and *trans*-fused poly-THP rings, features present in many biologically active natural products, have focused on the purely synthetic option, diverse approaches having been developed with an increasing emphasis on iterative strategies. Nature probably synthesizes these complex natural compounds in a different and rather more efficient fashion, constructing several rings and multiple stereocenters through polyene epoxidation and subsequent tandem oxacyclization of the corresponding polyepoxides [31]. Consequently, the polyepoxide cyclization cascade has been recognized as an efficient biomimetic route to complex chiral THF and THP systems [32].

As in nature, the formation of adjacent THF ring chains or fused THP ring polycyclic ethers depends on the regioselectivity (*endo* or *exo*) of the nucleophilic addition to each epoxide (Scheme 8.16).

Tandem 5-*exo*-oxacyclizations of polyepoxides to form polycyclic ethers are well known. In Evans's synthesis of lonomycin A, diepoxide **66** underwent the desired

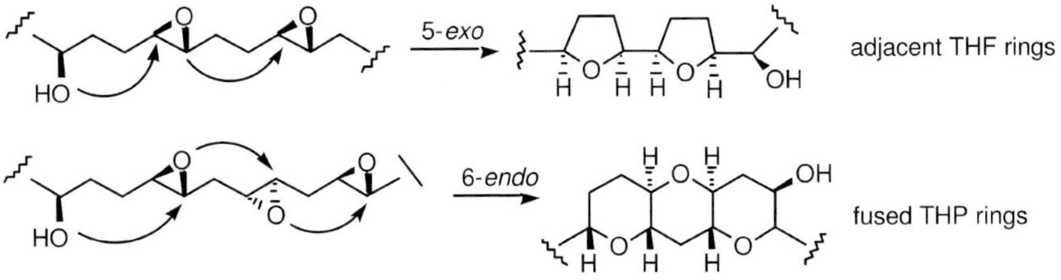

Scheme 8.16 Tandem 5-*exo*- and 6-*endo*-oxacyclization of polyepoxides.

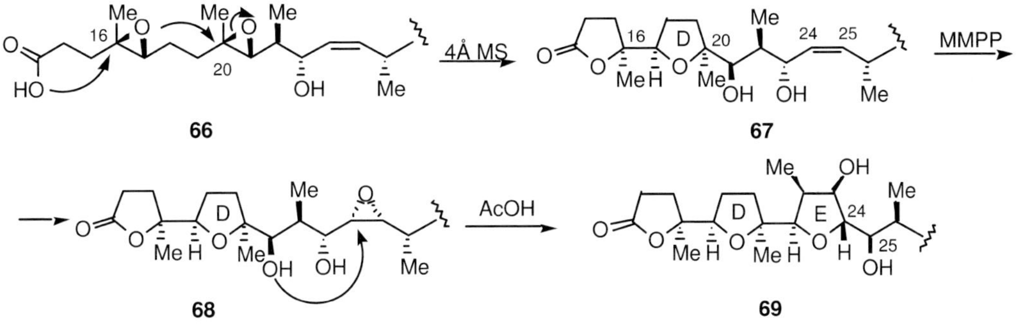

Scheme 8.17 Evans's synthesis of the DE ring system of lonomycin A.

cascade reaction, promoted by the carboxylic group within the fragment, the cyclization product **67** being obtained as the only detectable isomer containing a THF ring (the D ring of the natural product) adjacent to the γ-lactone ring. The construction of the E ring of the residual bis-tetrahydrofuran lactone **69** proceeded through the usual acid-induced hydroxy-mediated heterocyclization (5-*exo* mode) on epoxide **68**, obtained by hydroxy-directed epoxidation of **67** with magnesium monoperoxyphthalate (MMPP) (Scheme 8.17) [33].

In the synthesis of the squalenoid glabrescol (**72**; originally attributed structure), containing five adjacent (all *cis*) THF rings, the necessary precursor of the polyepoxide cascade, the pentaepoxide **71**, was achieved by epoxidation of each of the trisubstituted double bonds of the known (*R*)-2,3-dihydroxy-2,3-dihydrosqualene (**70**) by the Shi epoxidation approach (Scheme 8.18) [34]. Treatment of **71** with CSA at 0 °C and subsequent purification by column chromatography provided the pure polycyclic ether **72** by a cascade process reasonably initiated by the free secondary alcohol functionality [35a].

The structure of glabrescol was subsequently revised, and the new structure was synthesized enantioselectively through sequential hydroxy-directed *anti*-oxidative cyclization of acyclic γ-alkenols with $VO(acac)_2$/TBHP to construct the adjacent THF rings via epoxides under acid conditions [35b].

As an alternative to polyepoxide cyclization, cascade cyclization of cyclic sulfates

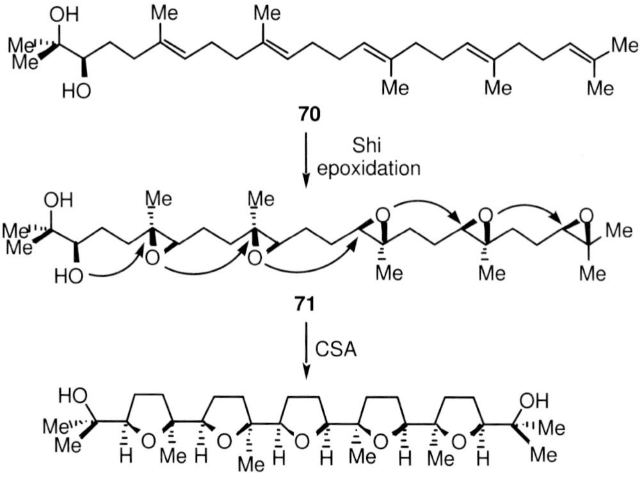

Scheme 8.18 Corey's synthesis of glabrescol (**72**; originally attributed structure) by a polyepoxide cascade.

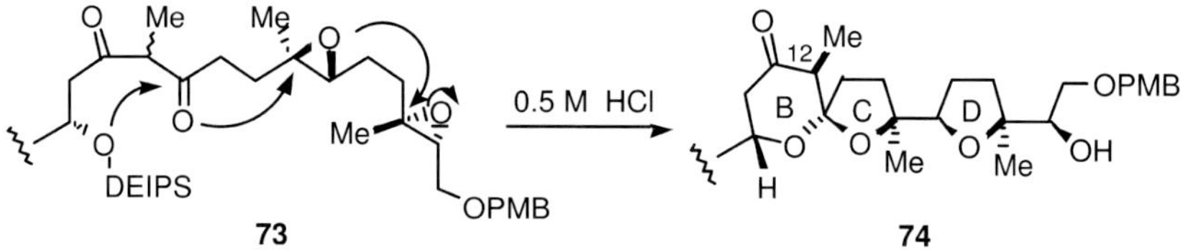

Scheme 8.19 Construction of the BCD ring system of etheromycin.

can be utilized in the synthesis of poly(tetrahydrofurans), as demonstrated in the construction of the C_{17}–C_{32} fragment of ionomycin [36].

In another interesting application of this procedure, the acid-mediated cascade cyclization of β-diketone diepoxide **73** involves the participation not only of the two oxirane rings and of the secondary alcoholic group, but also of one of the two carbonyl groups. In this way, besides the two adjacent C and D THF rings, the simultaneous construction of the spiroketal function between the B and C rings of etheromycin is obtained (compound **74**, in a 70:30 mixture with 12-*epi* compound; Scheme 8.19) [37].

Endo-regioselective oxacyclization of polyepoxides is less common; the first stereospecific tandem *endo*-regioselective biomimetic oxacyclization of polyepoxides to fused THP rings has only recently been reported [38a]. The cyclization of the hydroxy-methoxymethyl-substituted triepoxide **75** (a 1,4,7-triepoxide), promoted and directed by $La(OTf)_3$ through an intramolecular chelation and based on a procedure originally described for a monoepoxide system [38b], afforded the tri-

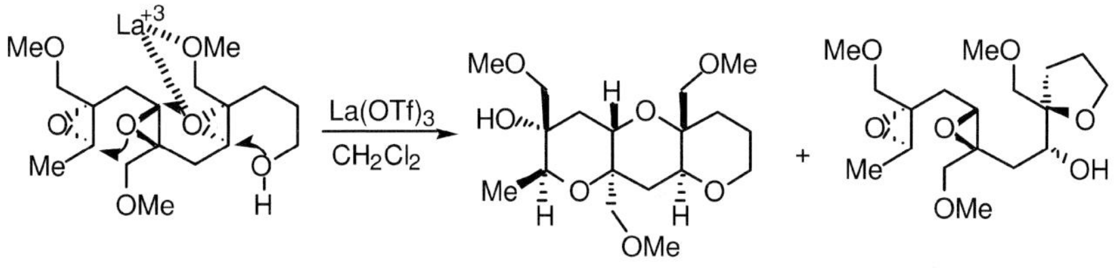

Scheme 8.20 *Endo* oxacyclization of a 1,4,7-triepoxide.

cyclic *trans*-fused ether **76** (9.3%), together with the diepoxy tetrahydrofuran **77** (15%) (Scheme 8.20). The *endo*:*exo* selectivity (77:23) in the first cyclization step turned out to be lower than in the second cyclization step (*endo*:*exo* = 92:8). The difference was attributed to inhibition of suitable chelation for the *endo*-cyclization by competitive coordination of the methoxymethyl group adjacent to the epoxide and other oxygen functional groups to the Lewis acid in the same molecule [38a].

Application of a similar procedure to the geraniol-derived 1,5-diepoxide carbonate **78** in the presence of $BF_3 \cdot Et_2O$ stereospecifically provided the fused bicyclic carbonate **82**, containing an oxepane ring, as the major cyclization product (54%). The result is consistent with a mechanism involving Lewis acid activation of the terminal epoxide (the electrophile), followed by intramolecular addition of the internal epoxide (the nucleophile; structure **79**) to give the intermediate bicyclo[4.1.0]epoxonium **80** rather than the more highly strained [3.1.0] regioisomer

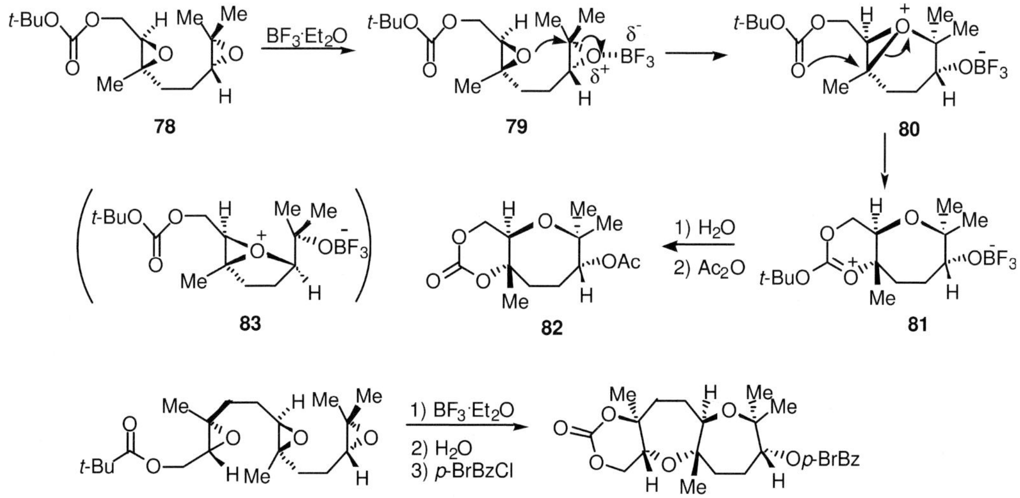

Scheme 8.21 *Endo* oxacyclization of a 1,5-diepoxide and 1,5,9-triepoxide.

83. The nucleophilic terminating carbonyl group appears to play an essential role in the effective preference for *endo*-regioselective oxacyclization by attacking the more substituted oxirane carbon of epoxonium intermediate **80** to give **81**, thus indicating the importance of the alkyl substitution at each site of nucleophilic addition. The mechanism indicates that the oxygen atoms of cyclic carbonate **82** originate from the carbonyl oxygen of the starting diepoxide (Scheme 8.21) [39].

The tandem oxacyclization of a triepoxide was also explored. On treatment with $BF_3 \cdot Et_2O$, the farnesol-derived triepoxide **84**, with a *tert*-butylcarbonate as the nucleophilic terminating functional group, afforded the tricyclic bis-oxepane **85** as the major product (52%, Scheme 8.21). The tetraepoxide derived from geranylgeraniol was similarly converted into the corresponding tetracyclic structure [39a].

8.2.2
Intramolecular C–N Bond-forming Reactions

Following Uskokovic's seminal quinine synthesis [40], Jacobsen has very recently reported the first catalytic asymmetric synthesis of quinine and quinidine. The stereospecific construction of the bicyclic framework, introducing the relative and absolute stereochemistry at the C_8– and C_9-positions, was achieved by way of the enantiomerically enriched *trans* epoxide **87**, prepared from olefin **86** by SAD (AD-mix β) and subsequent one-pot cyclization of the corresponding diol [2b]. The key intramolecular S_N2 reaction between the N_1- and the C_8-positions was accomplished by removal of the benzyl carbamate with Et_2AlCl/thioanisole and subsequent thermal cyclization to give the desired quinuclidine skeleton (Scheme 8.22) [41].

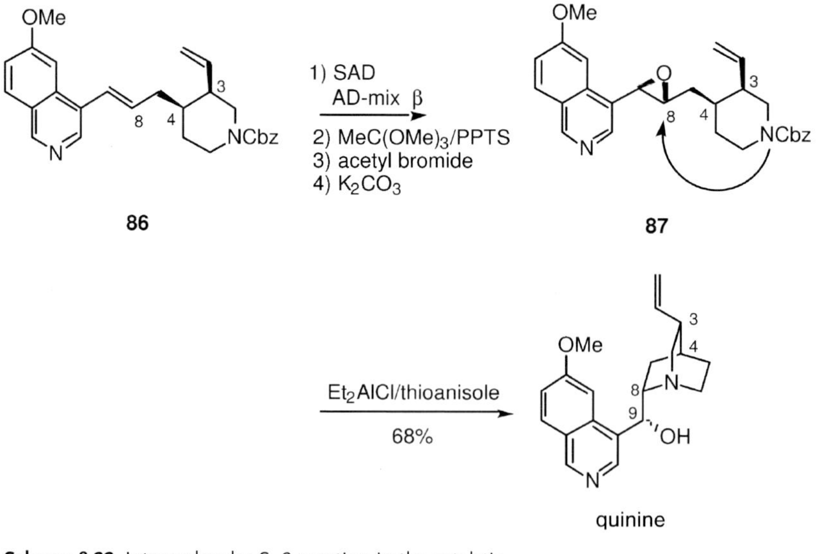

Scheme 8.22 Intramolecular S_N2 reaction in the catalytic asymmetric total synthesis of quinine.

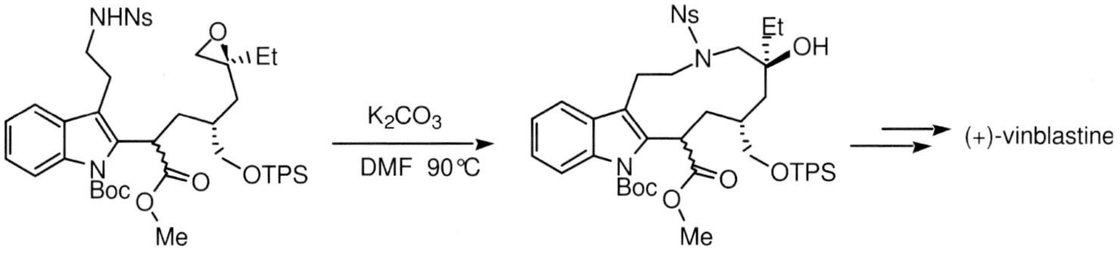

88

89

Scheme 8.23 Base-promoted macrocyclization of epoxy *N*-nosylate **88**.

Quinidine, a natural product epimeric with quinine at C_8 and C_9, was accessed through the diastereoisomeric *trans* epoxide prepared from **86** by SAD, in this case by using AD-mix α [2b, 41].

In a versatile stereocontrolled total synthesis of (+)-vinblastine, Fukuyama used a base-promoted macrocyclization of the *N*-nosylate and the terminal epoxide moiety present in **88** as one of the key steps, giving the 11-membered-ring product **89** in 82% yield (Scheme 8.23) [42].

A novel approach to azabicyclic ring systems, based on an epoxide-initiated electrophilic cyclization of an alkyl azide, has been developed by Baskaran. A new stereo- and enantioselective synthesis of the 5-hydroxymethyl azabicyclic framework **91a**, present in (+)- and (–)-indolizidines 167B and 209D, for example, was

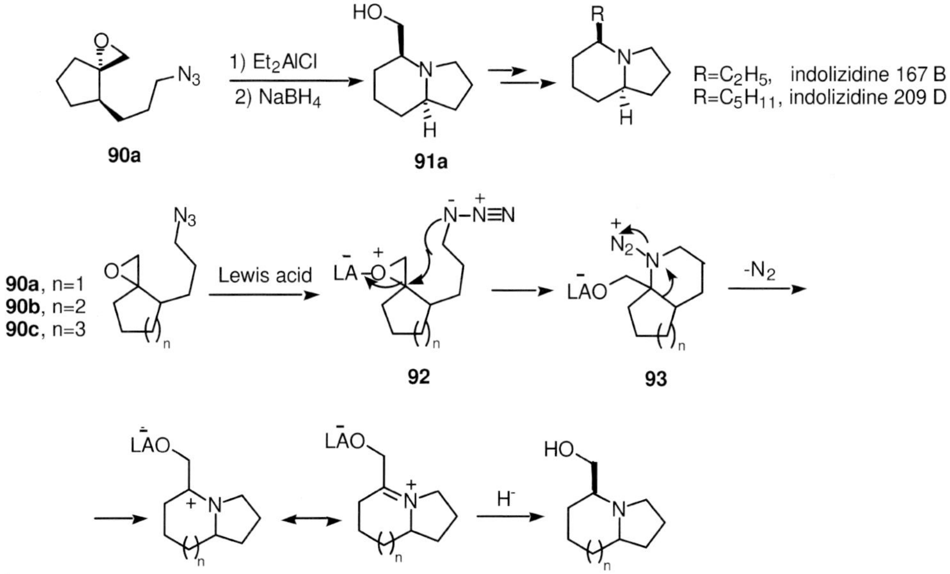

90a

91a

92

93

94

91a-c

Scheme 8.24 Epoxide-initiated electrophilic cyclization of azides.

based on the Et_2AlCl-promoted cyclization of epoxy azide **90a** and subsequent one-pot reduction with $NaBH_4$ (Scheme 8.24) [43]. However, the cyclization is limited to the *trans* isomer **90a**, which can be obtained in a pure state only by chromatography. As regards the mechanism, treatment of epoxyazides **90a-c** with a Lewis acid determines a cyclization process, as indicated in **92**, to give the aminodiazonium intermediate **93**. A subsequent intramolecular Schmidt reaction affords bicyclic iminium ion **94**, which is stereoselectively reduced by the hydride *in situ*. The generality of this methodology is further illustrated by the use of epoxy azides of different ring sizes also to give 5-hydroxymethyl azepine **91b** (42% yield) and azocine **91c** (47% yield) (Scheme 8.24) [44].

8.3 Synthesis of Complex Molecules by Ring-opening of Epoxides with C-Nucleophiles

8.3.1 Intramolecular C–C Bond-forming Reactions

Closely related to the polyepoxide cascade procedure for the synthesis of polycyclic systems is Corey's biomimetic-type, nonenzymatic, oxirane-initiated (Lewis acid-promoted) cation-olefin polyannulation. By this strategy, compound **96**, containing the tetracyclic core of scalarenedial, was constructed by exposure of the acyclic epoxy triene precursor **95** to $MeAlCl_2$-promoted cyclization reaction conditions (Scheme 8.25) [45].

Similarly, efficient tetracyclization ($MeAlCl_2$-promoted) of the bis-allylic silane/bis-epoxide **97** constitutes the key step in the synthesis of (+)-α-onocerin. In this case, because of the presence of the bis-allylic silane group, a double bis-annulation occurs, with the formation of the ethylene-bridge linked bis-decalin system present in the target compound (Scheme 8.26) [46].

Utilization of a similar [$Sc(OTf)_3$-promoted)] approach by Overman on the geranylgeraniol-derived cyclization substrate **98** provided the desired tetracyclization product **99**, in which the terminator of the cationic cyclization is an arene group. Compound **99** is then transformed into the kinesin motor protein inhibitor adociasulfate 1 (Scheme 8.27) [47].

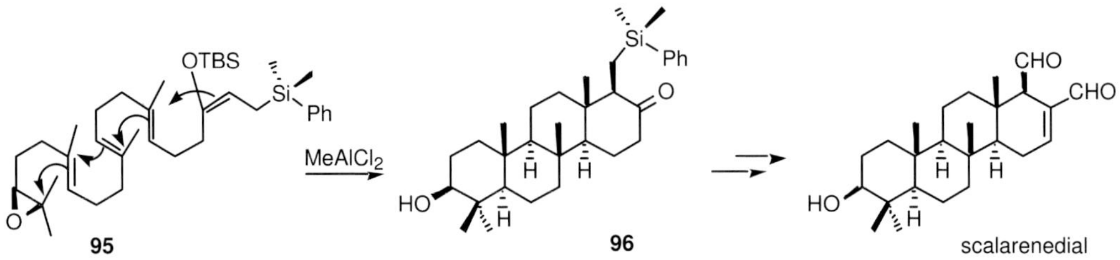

OTBS Si Ph O 95 $MeAlCl_2$ Si Ph O HO H H H 96 CHO CHO HO H H H scalarenedial

Scheme 8.25 Corey's cation-olefin polyannulation.

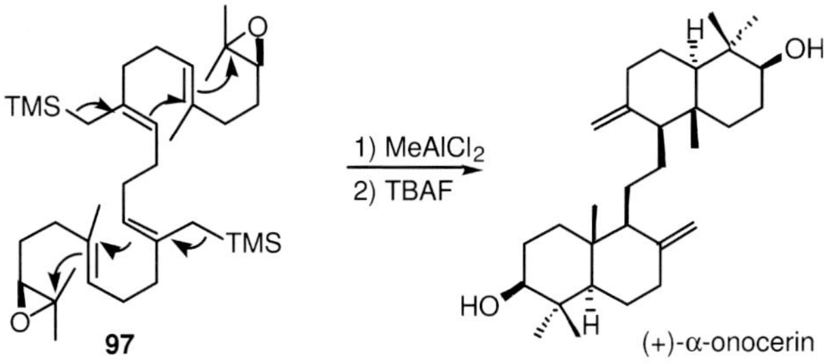

Scheme 8.26 Synthesis of (+)-α-onocerin by a double bis-annulation.

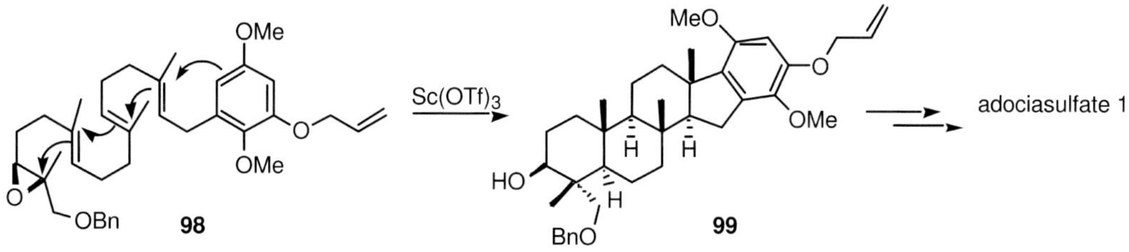

Scheme 8.27 Tetracyclization of a geranylgeraniol-derived epoxide.

Hodgson very recently reported an efficient intramolecular and completely diastereoselective cyclopropanation of bishomoallylic and trishomoallylic epoxides based on the use of α-lithiated epoxides. In a seminal paper, Crandall and Lin had reported that the reaction between *t*-BuLi and 1,2-epoxyhex-5-ene (**100**) gave, *inter alia*, small amounts of *trans*-bicyclo[3.1.0]hexan-2-ol (**102**, 9%) (Eq. a, Scheme 8.28)

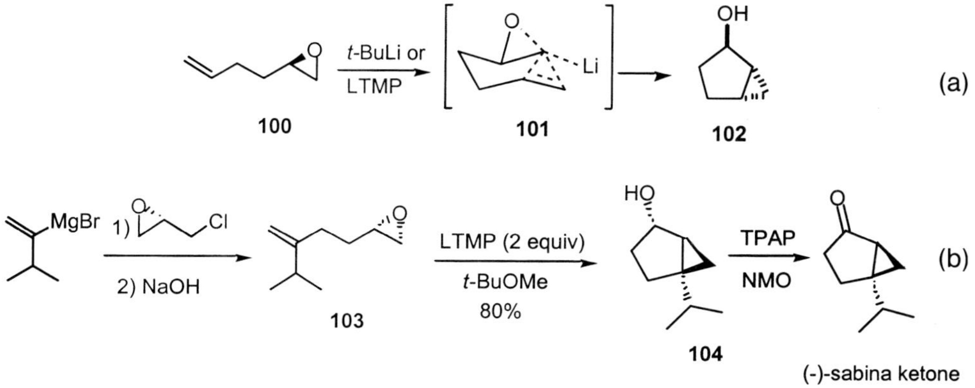

Scheme 8.28 Base-promoted intramolecular cyclopropanation of unsaturated terminal epoxides.

[48a]. The use of LTMP (2 equiv.) as a base in *t*-BuOMe allowed the intermediate α-lithiated epoxide **101** to undergo a clean intramolecular cyclopropanation with the tethered unsaturation. The utility of this chemistry was demonstrated by a novel and concise asymmetric synthesis of (–)-sabina ketone, starting from the Grignard reagent derived from 2,3-dimethylbut-1-ene and the readily available (*S*)-epichlorohydrin. The derived epoxide **103** underwent intramolecular cyclopropanation to give the fused cyclopropane **104**, which was directly oxidized to the target compound (Eq. b, Scheme 8.28) [48b].

8.3.2 Intermolecular C–C Bond-forming Reactions

8.3.2.1 Intermolecular C–C Bond-forming Reactions with Organometallic Reagents

Especially in the early steps of the synthesis of a complex molecule, there are plenty of examples in which epoxides are allowed to react with organometallic reagents. In particular, treatment of enantiomerically pure terminal epoxides with alkyl-, alkenyl-, or aryl-Grignard reagents in the presence of catalytic amounts of a copper salt, corresponding cuprates, or metal acetylides via "alanate" chemistry, provides a general route to optically active substituted alcohols useful as valuable building blocks in complex syntheses.

The nucleophilic ring-opening reactions of 2,3-epoxy alcohols (an internal epoxide) can be highly regioselective. Some of the best examples are found in ring-opening reactions with dialkyl cuprates [49]. In most cases the opening process occurs at the less hindered position, which often corresponds to the C(2) oxirane carbon (to give a 1,3-diol), suggesting that steric factors are decisive in influencing the regioselectivity of the ring-opening [50]. In some cases, however, high C-2 regioselectivity is exhibited even in the absence of a significant steric bias at either C(2) or C(3) [51].

A reiterative application of a two-carbon elongation reaction of a chiral carbonyl compound (Horner-Emmonds reaction), reduction (DIBAL) of the obtained *trans* unsaturated ester, asymmetric epoxidation (SAE or MCPBA) of the resulting allylic alcohol, and then C-2 regioselective addition of a cuprate (Me_2CuLi) to the corresponding chiral epoxy alcohol has been utilized for the construction of the polypropionate-derived chain [R-CH(Me)CH(OH)CH(Me)-R'], present as a partial structure in important natural products such as polyether, ansamycin, or macrolide antibiotics [52]. A seminal application of this procedure is offered by Kishi's synthesis of the C_{19}–C_{26} polyketide-type aliphatic segment of rifamycin S, starting from aldehyde **105** (Scheme 8.29) [53].

In some cases, if necessary, a C(3)-regioselective addition of cuprate can be accomplished by the use of a sterically demanding protecting group – such as an -OTr or -OMMTr group – on C(1), as described in the synthesis of the polypropionate segment present in (–)-amphidinolide P and (+)-amphidinolide K [54].

An unprecedented nickel-catalyzed reductive coupling between an epoxide and an alkyne to give synthetically useful homoallylic alcohols has been developed by Jamison [55a], and was recently used in a short enantioselective synthesis of am-

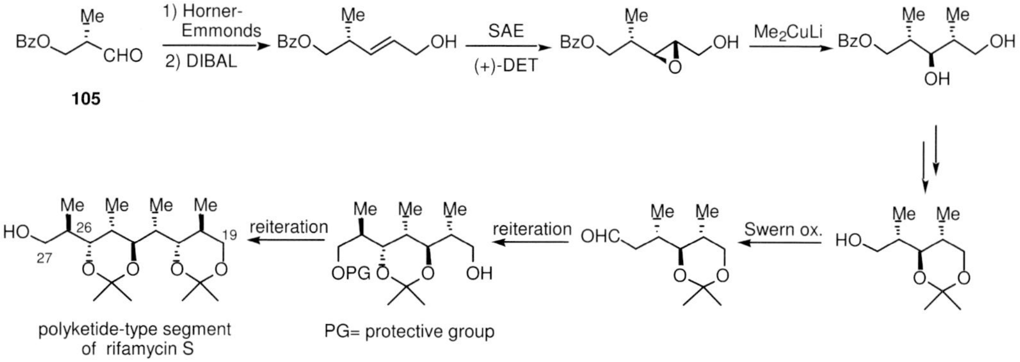

Scheme 8.29 Kishi's approach for the construction of the C_{19}–C_{26} polyketide-type segment of rifamycin S.

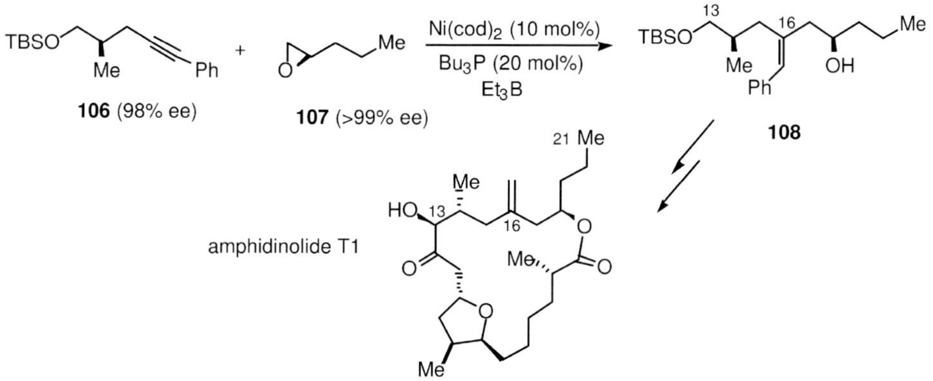

Scheme 8.30 Nickel-catalyzed reductive coupling of alkynes with epoxides.

phidinolide T1 (Scheme 8.30) [55b]. The nickel-catalyzed alkyne-epoxide reductive coupling smoothly united building blocks **106** and **107** to give the "nearly half" C_{13}–C_{21} alcoholic fragment **108** of the natural compound with a high regioselectivity.

Amos Smith III's three- and multicomponent linchpin coupling of silyl dithianes anions with terminal epoxides is a recent, very efficient route for the stereocontrolled assembly of 1,3-diol fragments [56a]. The three-component process consists of a metalation of the 2-TBS-substituted 1,3-dithiane **109** and alkylation with an epoxide, such as epoxide **110**, under solvent-controlled conditions (Et_2O or THF, the former appearing to be the solvent of choice), in which complete consumption of the initial anion is achieved prior to silyl migration (1,4-Brook rearrangement) to give the unrearranged carbinol **111**. Addition of HMPA then induces the rearrangement to afford the new silyl ether **113**, ready for a second alkylation with a different electrophile, such as epoxide **112**, to give the product of unsymmetrical bis-alkylation **114** (Scheme 8.31) [56a].

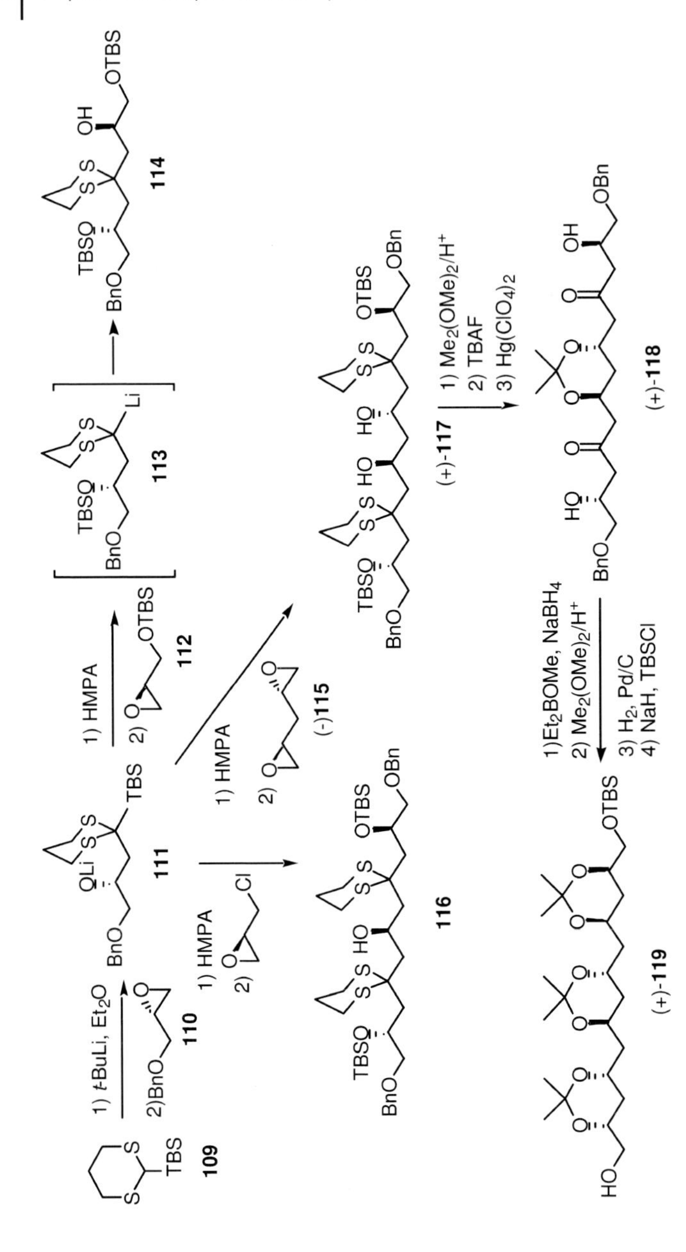

Scheme 8.31 Smith III's three- and multicomponent linchpin coupling of metalated silyl dithiane **109** with epoxides.

Terminal enantiomerically pure epoxides turned out to be excellent electrophiles for this strategy, since the completely regioselective nucleophilic attack on the primary oxirane carbon predetermines the stereogenicity of the resulting carbinol stereocenter, thus eliminating the formation of diastereoisomers. Access to all possible diastereoisomers of the 1,3,5-polyol structural feature is possible through alternation of the order of addition and the configurations of epoxides, and through stereocontrolled reduction of the ketone produced by unmasking the dithiane. By careful choice of the second electrophile (epichlorohydrin or a diepoxide), multicomponent coupling to provide exceptionally concise routes to advanced 1,3-polyol fragments present in polyene macrolides might be possible [57]. In this way, product **111** from the first alkylation (2 equiv.) is sequentially treated with HMPA and (+)-epichlorohydrin to afford the bis-dithiane carbinol **116** with construction of four carbon-carbon σ bonds in one flask (Scheme 8.31) [56a].

Similarly, in another example, alkylation of **111** with diepoxide (–)-**115** (1 equiv.) in the presence of HMPA (1.3 equiv.) furnished diol (+)-**117**. Protection of (+)-**117** to form the acetonide, removal of the silyl protecting groups (TBAF), and hydrolysis of the dithiane with $Hg(ClO_4)_2$ provided the diketone (+)-**118**. Hydroxy-directed *syn*-reduction of both carbonyl groups with $NaBH_4$ in the presence of Et_2BOMe, and triacetonide formation, followed by hydrogenolysis and monosilylation, afforded the desired Schreiber subtarget (+)-**119**, which was employed in the synthesis of (+)-mycoticins A and B (Scheme 8.31) [56b].

The multicomponent linchpin coupling of silyl dithianes with epoxides was very efficiently used to access both the AB and CD spiroketal fragments of spongistatin. In the second-generation construction of an advanced ABCD intermediate, dithiane **109** was initially alkylated with epoxide (–)-**120** in Et_2O and subsequently with epoxide (+)-**121** in the presence of HMPA to yield the coupled product (+)-**122** [58]. Upon removal of both the dithiane and silyl ethers present in (+)-**122** with $Hg(ClO_4)_2 \cdot 4H_2O$, *in situ* spiroketalization occurred, giving spiroketal (–)-**123** as the sole reaction product, which was subsequently elaborated to aldehyde (–)-**124** (Scheme 8.32). A similar procedure was utilized in the construction of the spiroketal CD fragment. In this case, the initially obtained mixture of diastereoisomeric spiroketals, favoring the undesired axial-axial stereoisomer, was equilibrated to a separable mixture of spiroketals favoring the desired axial-equatorial spiroketal. Aldol coupling of the AB and CD fragments completed the construction of the ABCD component of the natural product [58, 59].

In connection with a program directed towards the total synthesis of the 1,3-polyhydroxylated chain present in several complex natural compounds, Nicolaou reported a highly efficient method for the stereocontrolled synthesis of 1,3-diol segments, which is extendable to higher homologues by repetition [60]. The procedure is outlined in Scheme 8.33, starting from the generic allylic alcohol **125**. SAE on **125** would provide epoxide **126**, which could easily be transformed into the unsaturated epoxy ester **127** by oxidation/Horner-Emmonds olefination (two-carbon extension). This operation makes the oxirane carbon adjacent to the double bond more susceptible to nucleophilic attack by a hydride, so reductive opening (DIBAL) of **127** provides, with concomitant ester reduction, diol **128**. Pro-

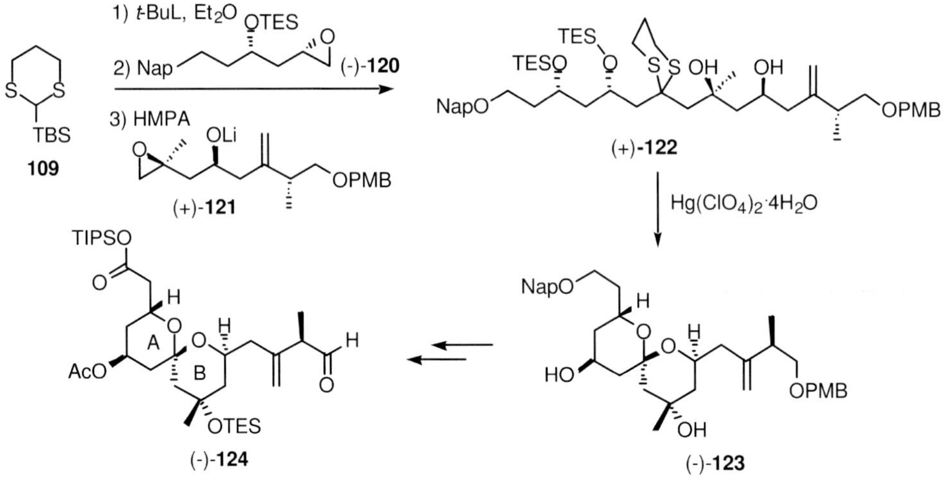

Scheme 8.32 Second-generation construction of the AB spiroketal system of spongistatin.

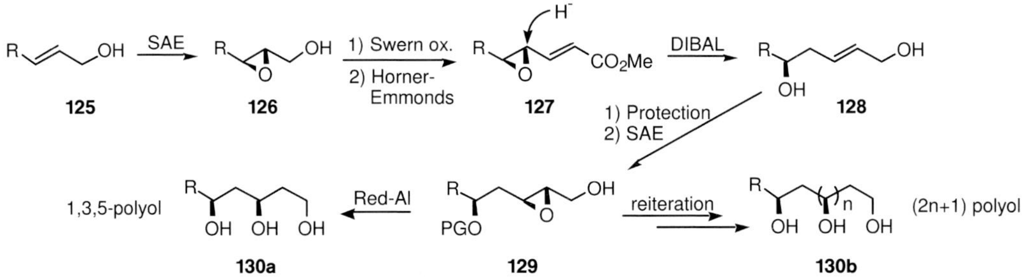

Scheme 8.33 General representation of Nicolaou's approach to (2*n*+1) polyol systems.

tection of the secondary hydroxy in **128**, followed by a second SAE reaction, would then give epoxide **129**. Regio- and stereocontrolled reductive opening of **129** with Red-Al as the reducing agent, followed by appropriate manipulation, affords the 1,3,5-polyol system **130a**, with the required stereochemistry. Reiteration of the entire process (as on **126**) on **129** would afford higher homologues (**130b**). As SAE can, in principle, provide either enantiomeric epoxide, the procedure is able to produce any stereochemical combination. It was used for the construction of two 1,3,5-polyol substructures regarded as key building blocks for the synthesis of amphoteronolide B and amphotericin B [60].

A new iterative strategy for enantio- and diastereoselective syntheses of all possible stereoisomers of 1,3-polyol arrays has been described by Shibasaki. This strategy relies on a highly catalyst-controlled epoxidation of α,β-unsaturated morpholinyl amides promoted by the Sm-BINOL-$Ph_3As{=}O$ complex, followed by the con-

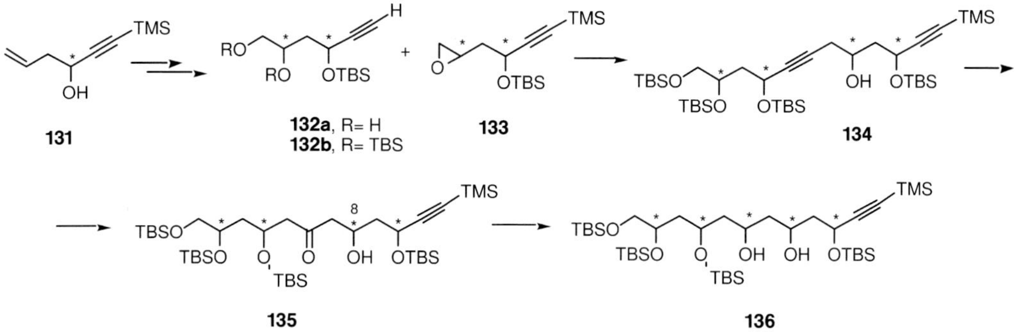

Scheme 8.34 Polyol systems by cross-coupling of six-carbon alkynol modules.

version of the amide function into a ketone and diastereoselective ketone reduction [61].

Some of the strategies examined for the construction of 1,3,5-alternating polyol chains use relatively small synthons, requiring a large number of carbon-carbon bond-forming steps. A recent new synthetic strategy for assembling polyol substructures is based on cross-couplings of six-carbon modules [62]. In this way, each of the four stereoisomers of epoxide **133** and diol **132a** is produced, with very high diastereoselectivity, from either enantiomer of unsaturated alcohol **131** (Scheme 8.34) by epoxidation followed by hydrolytic kinetic resolution (HKR) [63]. Protected **132b** (four possible stereoisomers) is coupled with epoxide **133** (four possible stereoisomers) to afford diyne **134**, with 12 carbons and four stereogenic centers (16 possible stereoisomers). Iodocyclization of the *tert*-butylcarbonate of **134**, followed by radical deiodination and basic hydrolysis of the intermediate cyclic carbonate, afforded the hydroxy ketone **135**. Reduction of **135** can be performed with either 1,3-*anti* or *syn* stereoinduction from the C_8–OH group, demonstrating the potential of this strategy for the stereoselective preparation of all 32 stereoisomers of polyol **136**. This strategy was successfully tested in the stereoselective synthesis of the alternating polyol segment corresponding to the C_{11}–C_{28} substructure of the polyene macrolide RK-397, in a process requiring only two carbon-carbon bond formation steps [62].

8.3.2.2 Addition Reactions of Metal Enolates of Non-stabilized Esters, Amides, and Ketones to Epoxides

In spite of their intrinsic synthetic potential, addition reactions of metal enolates of non-stabilized esters, amides, and ketones to epoxides are not widely used in the synthesis of complex molecules. Following the seminal work of Danishefsky [64], who introduced the use of Et_2AlCl as an efficient catalyst for the reaction, Taylor obtained valuable spiro lactones through the addition reaction of the lithium enolate of *tert*-butyl acetate to spiro-epoxides, upon treatment of the corresponding γ-

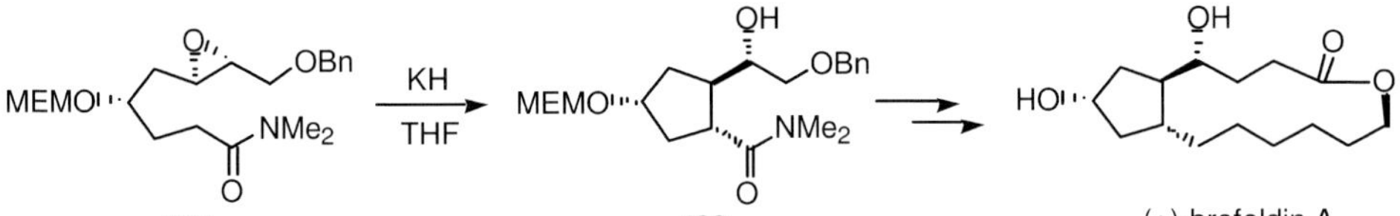

MEMO, OBn, NMe2, O, 137 — KH / THF → MEMO, OH, OBn, NMe2, O, 138 → HO, OH, O, (+)-brefeldin A

Scheme 8.35 Intramolecular cyclization of an epoxy amide.

hydroxy esters, the primary reaction product, with TsOH [65a]. Lithium enolates of esters have recently been used for the enantioselective synthesis of trisubstituted cyclopropanes [65b]. In the case of amide metal enolates, after the seminal work by Sauriol-Lord and Rathke [66], studies on the alkylation of prolinol and pseudoephedrine amides with epoxides were conducted by Askin and A. G. Meyers, providing valuable enantiomerically pure lactones as useful building blocks [67].

A direct application of the ring-opening reaction of an epoxide by a metal enolate amide for the synthesis of a complex molecule can be found in the synthesis of the trisubstituted cyclopentane core of brefeldin A (Scheme 8.35) [68a]. For this purpose, treatment of epoxy amide **137** with excess KH in THF gave a smooth cyclization to amide **138**, which was subsequently converted into the natural product. No base/solvent combination that would effect cyclization of the corresponding aldehyde or ester could be found.

Efficient coupling between a chiral 3-phenylpropionamide enolate and (*S*)-glycidyl tosylate was achieved in a practical route to the HIV-1 protease inhibitor L-735–524 [68b].

The direct addition of metal enolates of simple ketones to epoxides to give γ-hydroxy ketones is a fairly recent reaction [69]. In 1988, in the course of the total synthesis of polyether antibiotic X-206, Evans envisaged the use of the metal enolate derived from ketone **139** (X = O) with epoxide **65** for the construction of the γ-lactol corresponding to the E ring of the complex natural product (Schemes 8.36 and 8.15). However, because of the at that time established lack of reactivity of ketone enolates towards epoxides, the problem was solved by employing the more nucleophilic metalated *N,N*-dimethylhydrazone (DMH). In this context, the lithiated DMH **139** (X = $NNMe_2$) was allowed to react with epoxide **65**, containing the THP ring corresponding to the F-ring of the target molecule, to give the γ-hydroxy-DMH **140**, which intramolecularly cyclized to the bicyclic compound **141a** (Scheme 8.36). The hydrazino lactol **141a** was hydrolyzed with aqueous $NaHSO_4$ to provide lactol **141b**. Lactol **141b** was then transformed into the methyl glycoside **141c** by treatment with MeOH in the presence of PPTS at 0 °C [12b].

A similar strategy was utilized in the course of the total synthesis of ferensimycin B, for the assembly of the γ-lactol moiety corresponding to the C-ring of the natural compound. In this case, however, the construction of the new C–C bond was complicated by the low reactivity of the employed epoxy alcohol towards nucleophilic ring-opening. Conversion of the epoxy alcohol synthon into the magnesium alkoxide with EtMgBr, prior to its addition to the DMH lithium enolate,

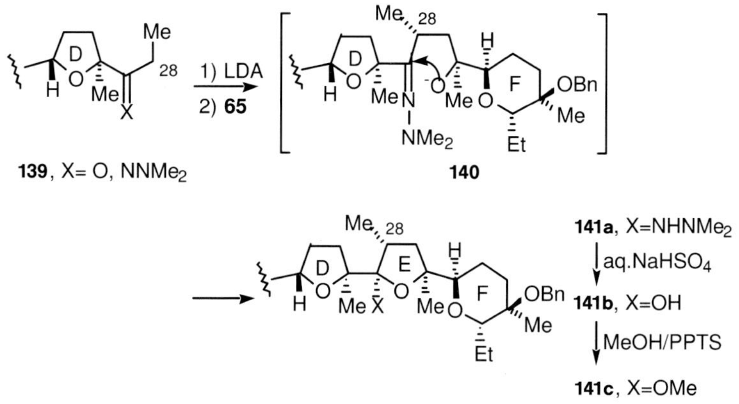

Scheme 8.36 Addition reaction of a metalated ketone-DMH to an epoxide.

activated the epoxide towards the addition reaction, all resulting in a substantial rate enhancement [70].

The lithium enolate of acetaldehyde DMH has recently been utilized in the opening reaction of the α-epoxide obtained by DMDO oxidation of enol ether **142**, to provide hemiacetal **143** after mild oxidative acid hydrolysis. The protected carbonyl functionality was subsequently used for the introduction of the *trans* enyne chain through a Wittig olefination reaction to provide alcohol **144**, which was then transformed into (+)-laurenyne (Scheme 8.37) [71].

Schreiber found that the monoalkylation of the lithium enolate of cyclononanone with propene oxide could be cleanly effected by addition of $AlMe_3$ to give the γ-hydroxy ketone **145**, a key intermediate for the synthesis of recifeiolide [69a].

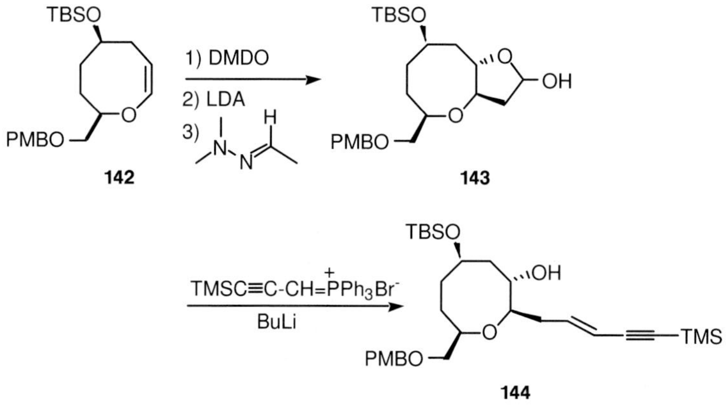

Scheme 8.37 Use of the lithium enolate of acetaldehyde DMH in an epoxide ring-opening reaction.

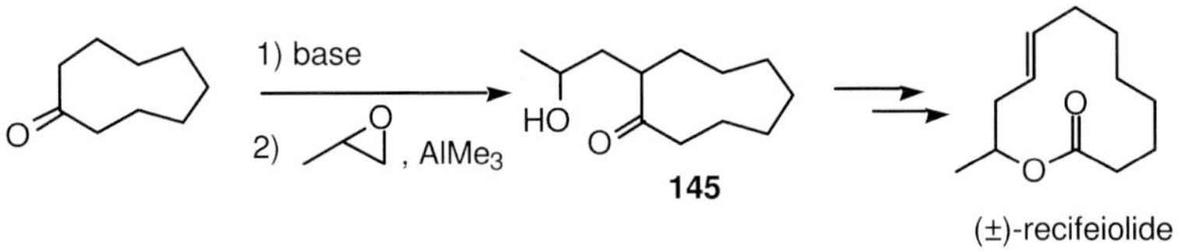

Scheme 8.38 Monoalkylation of the lithium enolate of cyclononanone with propene oxide.

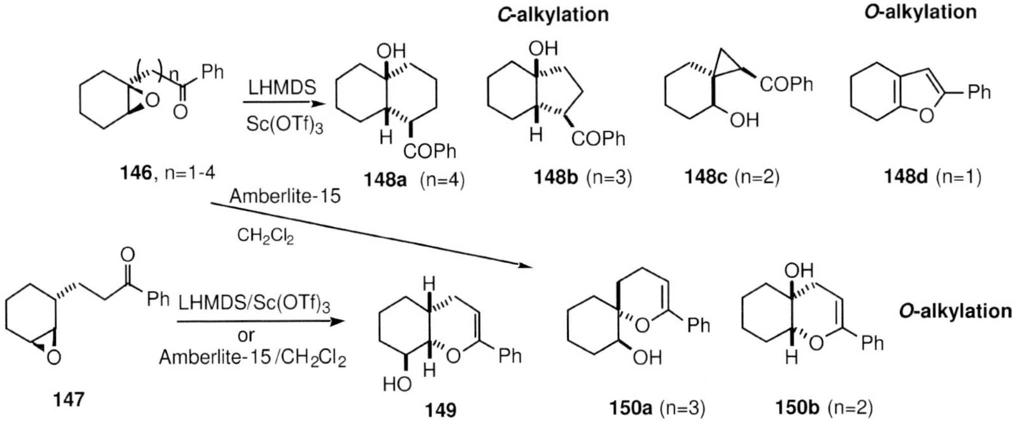

Scheme 8.39 LHMDS/$Sc(OTf)_3$ procedure for the intramolecular cyclization of epoxy ketones.

This was the first reported instance of this type of addition reaction (Scheme 8.38), though attempts to generalize this procedure were unsuccessful [12b].

The combined use of LHMDS, as the ketone enolate-generating base, and $Sc(OTf)_3$ as the necessary promoting Lewis acid were found by our group in Pisa to be a general procedure making addition reactions of lithium enolates of simple ketones to epoxides a valid tool for the construction of molecules that might otherwise be difficult to synthesize. Application of this procedure to epoxy ketones of types **146** and **147** afforded bicyclic, oxabicyclic, spiro, and dihydrobenzofuran compounds, depending on the distance between the oxirane ring and the carbonyl group, which determines the operation of intramolecular *C*- (compounds **148a-c**) or *O*-alkylation (compounds **148d** and **149**) processes. A chemoselective *O*- versus *C*-alkylation process was also achieved by the use of simple acid conditions (Amberlite-15/CH_2Cl_2) to give oxaspiro and oxabicyclic compounds (compounds **149** and **150a/150b**, Scheme 8.39) [72].

Posner recently reported a very simple and fast way to activate epoxides towards nucleophilic opening by ketone lithium enolate anions by use of $BF_3 \cdot Et_2O$ (1 equiv.) [73]. The application of this procedure to the nucleophilic opening of propene oxide with the lithium enolate of 2-cycloheptanone, obtained by the conjugate addition of trimethylstannyllithium to 2-cycloheptenone, afforded the stan-

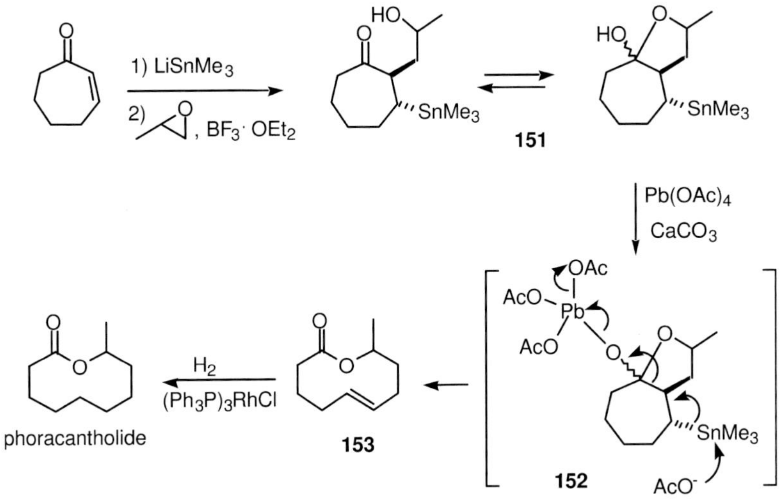

Scheme 8.40 $BF_3 \cdot Et_2O$-promoted epoxide ring-opening by ketone enolate anion.

nyl hemiketal **151**. Subsequent ring fragmentation with lead tetraacetate produced, through structure **152**, the ten-membered lactone product **153**, which was reduced to the fragrant natural product (±)-phoracantholide (Scheme 8.40) [73a]. A further application of this procedure constitutes the key step in the short synthesis of the γ-hydroxy ketone aglycon of the natural product curculigine [73b].

8.4
Epoxy Glycals

The use of glycal substrates is particularly attractive for stereoselective syntheses of carbohydrates and glycoconjugates in that both glycosidic bond formation and C(2)-functionalization of the carbohydrate donor is achieved in the process [74a]. The glycal assembly strategy developed by Danishefsky [74b] for glycosylation with simultaneous C(2)-hydroxylation involves oxidation of glycal **154** with DMDO to give the α-glycal epoxide **155** as the intermediate glycosyl donor. The subsequent ring-opening process with a nucleophilic glycosyl acceptor, carried out in the presence of a Lewis acid such as $ZnCl_2$, occurs with clean inversion of configuration and in excellent yield to give the corresponding β-glycoside. The glycosyl acceptor may be a glycal, to give a disaccharide, suitable for a reiterative process for oligosaccharide synthesis (Scheme 8.41) [74c].

The glycal epoxide method turned out to be useful for the construction of complex 2-branched β-aryl glycosides, which are salient features of the potent antibiotic vancomycin [75a]. Glycal epoxide glycosylation with sodium salts of indoles pro-

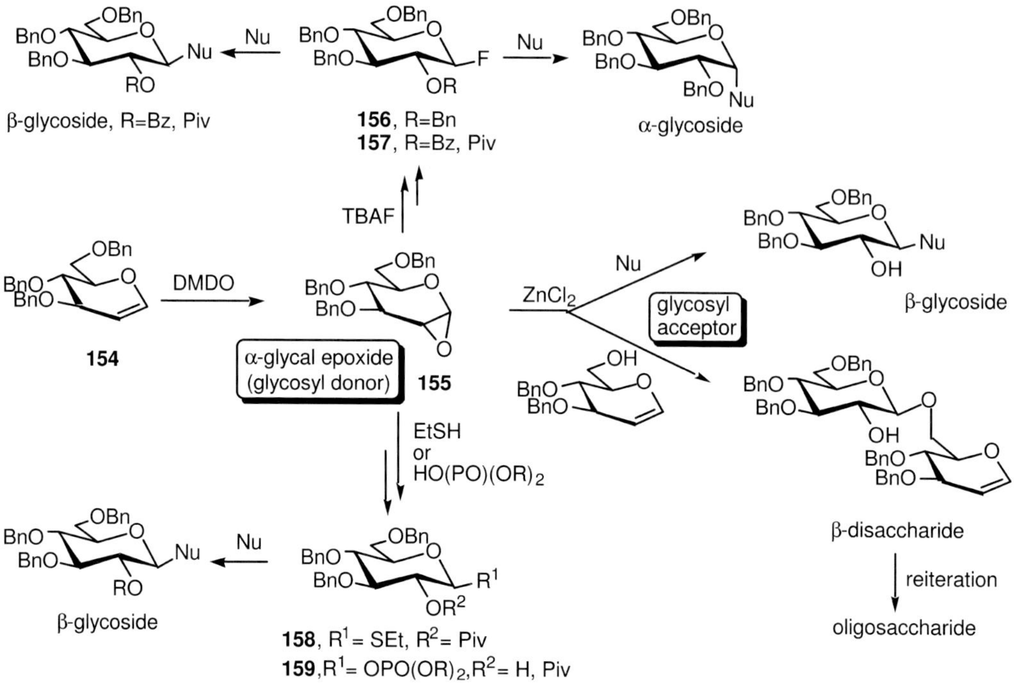

Scheme 8.41 Utilization of glycals and α-glycal epoxides in carbohydrate chemistry.

vided an improved total synthesis of the potent antitumor β-*N*-indolecarbazole glycoside rebeccamycin [75b], and a total synthesis of staurosporine [75c].

The glycal epoxides can be converted into other glycosylating agents, such as β-fluoroglycosides (**156** and **157**), useful for producing α- or β-glycosides, depending on the nature of the vicinal OR group [76], into thioethyl glycosyl donors (**158**), which are extremely powerful glycosylating agents also able to react with hindered or unreactive acceptors [77], or into glycosyl phosphate (**159**) (Scheme 8.41) [78].

The power of the reiterative strategy for oligosaccharide construction based on glycal acceptors and glycal-derived donors by epoxidation has found important application in the synthesis of Lewis- and H-type blood group determinants [79], tumor antigens [76b], and the carbohydrate domains of the enediyne antibiotics calicheamicin and esperamicin [80]. An illustrative example of this original strategy can be found in the synthesis of the saponin desgalactotigonin (Scheme 8.42) [76c]. D-Galactal derivative **160** was subjected to epoxidation (DMDO) to give epoxide **161**, which served to galactosylate tigogenin to give **162** (β-anomer), which was subsequently transformed into the tributyltin derivative **163**. Glycal **164** was oxidized with DMDO to give glycal epoxide **165** (a 4:1 mixture of α and β epoxides), which was treated with D-glucal derivative **166** to yield, after benzylation, the β-linked disaccharide **167**. Subsequent epoxidation (DMDO) afforded the corresponding α-epoxide **168**. β-Glycosylation of **163** with **168** yielded trisaccharide **169**.

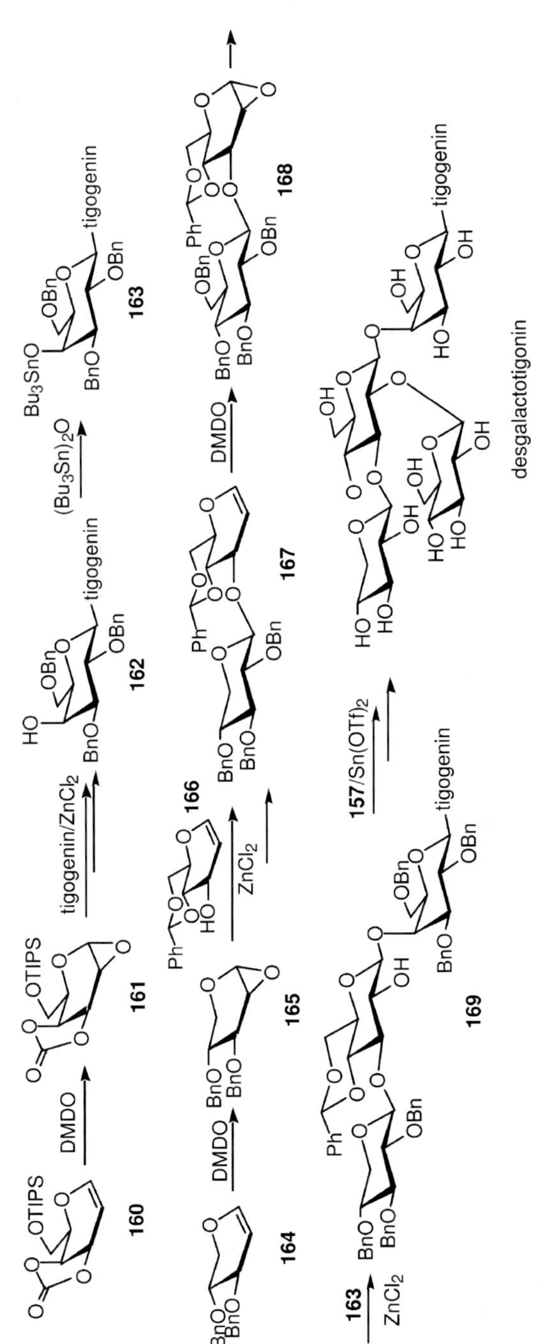

Scheme 8.42 Danishefsky's synthesis of desgalactotigonin.

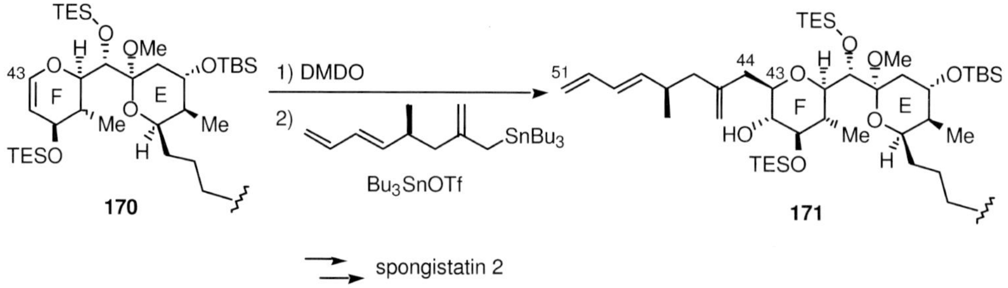

Scheme 8.43 Evanss̀ approach for the β-stereoselective introduction of the C_{44}–C_{51} side chain of spongistatin 2.

The final β-glycosylation was carried out with fluoroglycoside **157** (R = Bz) to give, after deprotection, the desired natural compound [76c].

Evans developed a new method for the synthesis of β-C-allylglycosides, based on Bu_3SnOTf-mediated ring-opening of glycal epoxides with allylstannanes as nucleophiles [81a]. This methodology has been efficiently used in the β-stereoselective introduction of the side chain (C_{44}-C_{51}) of spongistatin 2 (Scheme 8.43) [81b,c].

In a reiterative approach, enol ether epoxidation with DMDO has been coupled with C-C bond formation and ring-closing metathesis to provide *trans*-fused THP ring systems [82].

A conceptually new direct oxidative glycosylation with glycal donors, employing a reagent combination of triflic anhydride and diphenyl sulfoxide, has recently been reported by Gin [83]. This new β-glycosylation method works very well with hindered hydroxy nucleophiles, including sterically shielded carbohydrate hydroxy systems, and can be run on large scales.

8.5
Synthesis of Complex Molecules by Rearrangement Reactions of Epoxides

The rearrangement of epoxides to non-addition products is a general reaction of epoxides and can be conducted under both acid and basic conditions [3a]. Under acid conditions, treatment of an epoxide with a Lewis acid ($BF_3 \cdot Et_2O$, $MgBr_2$, $LiClO_4$, etc.) produces isomeric aldehydes and/or ketones, as found in – for example – the synthesis of analogues of azadiradione [84a], in the total synthesis of fredericamycin A [84b], or also as found under neutral conditions in the total synthesis of sordaricin [84c]. Basic conditions are able to promote the rearrangement of epoxides to allylic alcohols, as observed in the syntheses of polyketide-like macrodiolides [85a], in the formal total synthesis of roseophilin [85b], and in the total synthesis of pipermethystine [85c]. The structures and relative proportions of the rearranged products depend on the direction of the oxirane ring-opening, the relative migratory attitudes of the substituents, the type of Lewis acid, and the reaction conditions used. The reactions may be highly regio- and stereoselective. Depend-

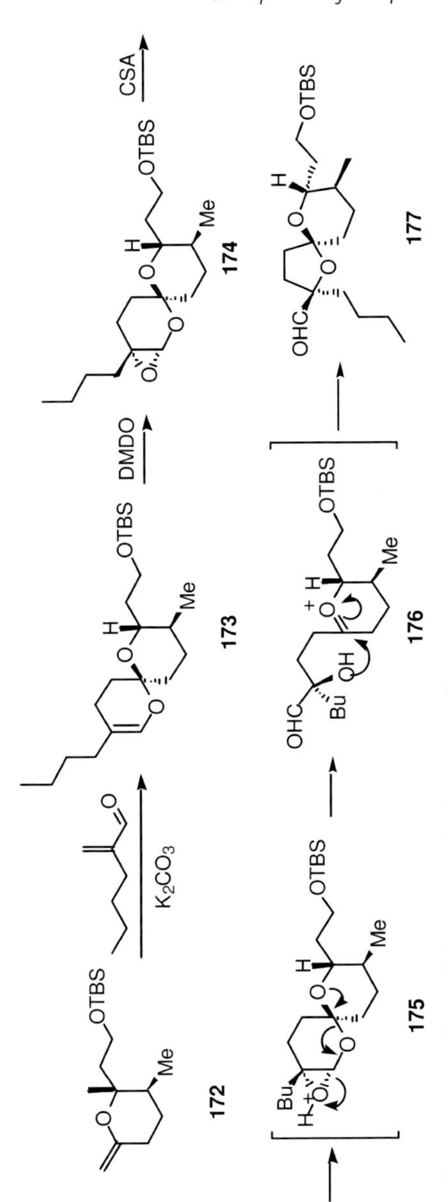

Scheme 8.44 Synthesis of the [5,6]-spiroketal segment of reveromycin B.

ing on the nature of the epoxide and of the substituents, extremely valuable cascade transformations can be achieved in a single step, even affording structures difficult to trace back to that of the starting epoxide. This is the case in the transformation of [6,6]-spiroketal **173**, derived from an inverse electron-demanding hetero-Diels-Alder reaction between butylacrolein and the *exo* enol ether **172**, into [5,6]-spiroketal **177**, in the course of the asymmetric synthesis of the spiroketal segment of reveromycin B, a process that proceeds by way of an intermediate epoxide (Scheme 8.44) [86]. Treatment of spiroketal **173** with DMDO (Danishefsky's conditions) gave the labile epoxide **174** as the only diastereoisomer, as a consequence of a stereoselective epoxidation process from the face of **173** opposite to the axial spiroketal oxygen. Acid-induced rearrangement of **174** with CSA afforded aldehyde **177**, containing the [5,6]-spiroketal moiety. The [6,6]- to [5,6]-spiroketal transformation derives from an initial cleavage of the protonated epoxide **175** to the hydroxy aldehyde-tetrahydropyranoid oxonium ion **176**, which cyclizes to the thermodynamically more stable [5,6]-spiroketal **177** [86a].

As well as the Payne rearrangement, 2,3-epoxy alcohols and some of their derivatives show some other interesting rearrangement procedures that often constitute key steps in syntheses of complex molecules.

The Suzuki-Tsuchihashi rearrangement of 2,3-epoxy alcohol derivatives is mechanistically similar to the pinacol rearrangement of 1,2-diols. Treatment of epoxy alcohol **178** or its silyl ether derivative with a Lewis acid induces an epoxide-opening, the 1,2-shift of a group (M), and carbonyl formation to give a β-hydroxy carbonyl compound (**179**) (a semipinacol reaction; Scheme 8.45) [3b]. As a result of the stereospecificity of the 1,2-shift, the epoxide stereochemistry is predictably transmitted into the aldol product. Depending on the substitution pattern of the oxirane ring, quaternary carbon centers and/or spiro compounds and ring-enlargement are effectively achieved [87].

In a formal synthesis of fasicularin, the critical spirocyclic ketone intermediate **183** was obtained by use of the rearrangement reaction of the silyloxy epoxide **182**, derived from the unsaturated alcohol **180**. Alkene **180** was epoxidized with DMDO to produce epoxy alcohol **181** as a single diastereoisomer, which was transformed into the trimethyl silyl ether derivative **182**. Treatment of **182** with $TiCl_4$ resulted in smooth ring-expansion to produce spiro compound **183**, which was subsequently elaborated to the desired natural product (Scheme 8.46) [88].

In a synthesis of (+)-asteltoxin, Cha applied the Suzuki-Tsuchihashi rearrangement to silyloxy epoxide **184** for the enantioselective construction of the unusual

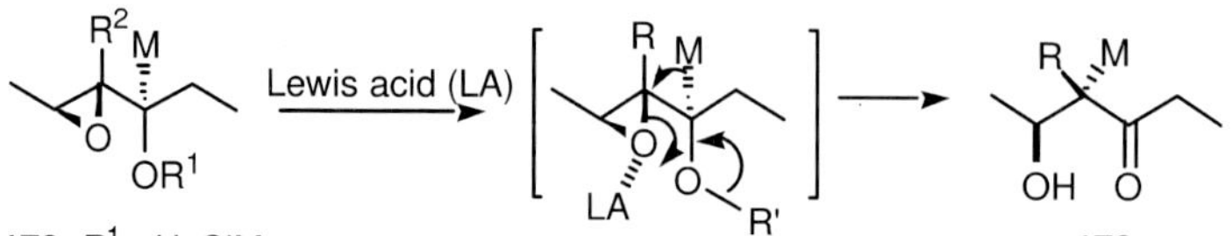

Scheme 8.45 Suzuki-Tsuchihashi rearrangement of epoxy alcohol derivatives.

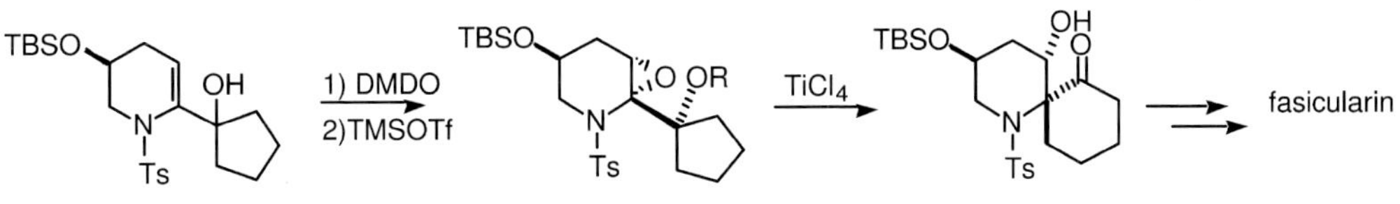

180

181, R= H
182, R= TMS

183

Scheme 8.46 Construction of the quaternary carbon center of fasicularin.

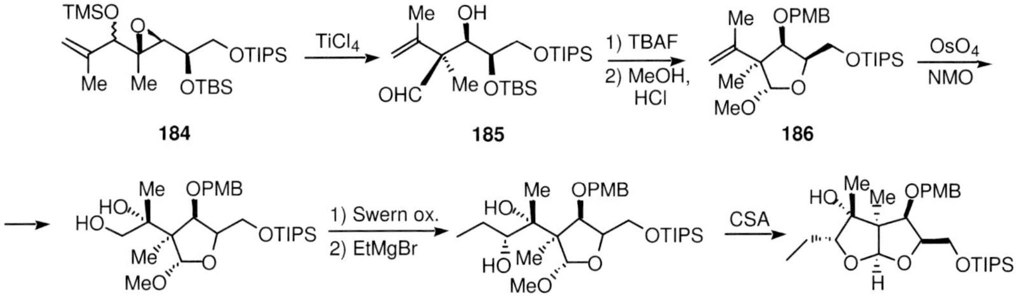

184 185 186 187 188 189

Scheme 8.47 Synthesis of the bis(tetrahydrofuran)core of (+)-asteltoxin.

bis(tetrahydrofuran) core, containing a quaternary carbon center, present in the natural product [89]. The pivotal rearrangement of the epoxy silyl ether **184** was accomplished by treatment with $TiCl_4$ to provide aldehyde **185**, which was converted into the fully protected tetrahydrofuran **186** by deprotection (TBAF) and treatment with HCl/MeOH. Subsequent diastereoselective dihydroxylation with OsO_4/NMO to produce diol **187**, followed by Swern oxidation and chelation-controlled addition of EtMgBr, allowed an efficient, stereoselective introduction of the ethyl side chain to provide alcohol **188**. Acid-induced cyclization of **188** with TsOH then afforded compound **189**, containing the bis(tetrahydrofuran) core of (+)-asteltoxin (Scheme 8.47) [89].

In the total synthesis of ingenol, the transformation of the tetracyclic carbon framework of intermediate **190** into an ingenane skeleton was achieved by stereoselective epoxidation followed by treatment of the intermediate epoxy alcohol **191** with $AlMe_3$ (Scheme 8.48). The key intermediate **192**, which has the complete ABCD ring system of ingenol and possesses two oxygen functional groups on both the A and B rings, is elegantly obtained, allowing the further stereoselective functionalization of the same rings as in the natural product [90].

Fukumoto epoxide rearrangement is worthy of note, in that a stereoselective rearrangement occurs spontaneously during the SAE of a 2-aryl-2-cyclopropylideneethanol (**194**), prepared from ketone **193**, to give the geminally substituted cyclo-

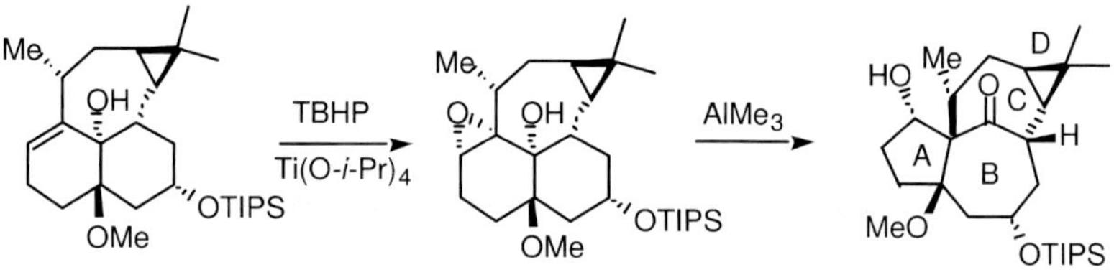

Scheme 8.48 Construction of the ingenane ABCD ring system.

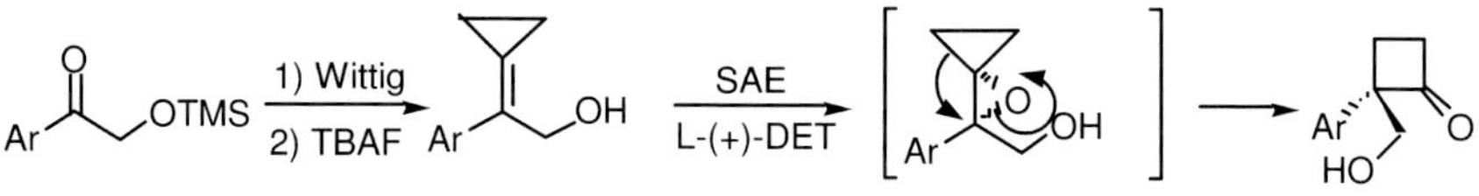

Scheme 8.49 Fukumoto rearrangement of 2-aryl-2-cyclopropylideneethanol.

butanone **196**, a valuable building block (Scheme 8.49) [91]. This tandem SAE/enantiospecific ring-expansion proceeds with complete transfer of the chirality of the epoxy alcohol generated *in situ*, and the observed stereoselectivity arises from the concerted *anti* 1,2-migration of the C-C bond of the cyclopropane ring to the epoxide moiety, as shown in **195**. The obtained cyclobutanone **196** may be subsequently subjected to ring-expansion to afford substituted cyclopentanones and γ-lactones, as found in the synthesis of (–)-aplysin [91a], (+)- and (–)-α-cuparenones [91b], and (–)-mesembrine [91c].

Highly stereoselective rearrangement reactions of epoxides can proceed either in only one step by means of a concerted mechanism, or in a stepwise manner through carbocation intermediates, as found in the biogenetic-type, acid-induced domino epoxide-opening/rearrangement/cyclization reactions of α-epoxide **197** for the synthesis of the tetracyclic core of stachyflin, a novel sesquiterpenoidal alkaloid exhibiting potent antiviral activity. It was envisioned that an acid-induced domino reaction on α-epoxide **197** or its β-diastereoisomer would proceed stereoselectively to install the required *cis*-fused AB and BC rings and the ether bond at the bridgehead of the AB ring. In this context, when α-epoxide **197** was allowed to react with $BF_3 \cdot Et_2O$ (3 equiv.), this successfully resulted in the formation of the required cyclized compound **199** in reasonable yield (41%). After epimerization of the C_3-OH carbon by oxidation, followed by diastereoselective reduction, the domino process constituted a concise route to compound **200**, containing the tetracyclic core of the target compound [92]. The authors believe that this domino transformation, represented for the sake of simplicity as a concerted mechanism in Scheme 8.50, proceeds in a stepwise manner through corresponding carbocation intermediates. Probably because of the *cis* relationship between the oxirane ring and the angular methyl group, the acid-induced domino reaction of the diaster-

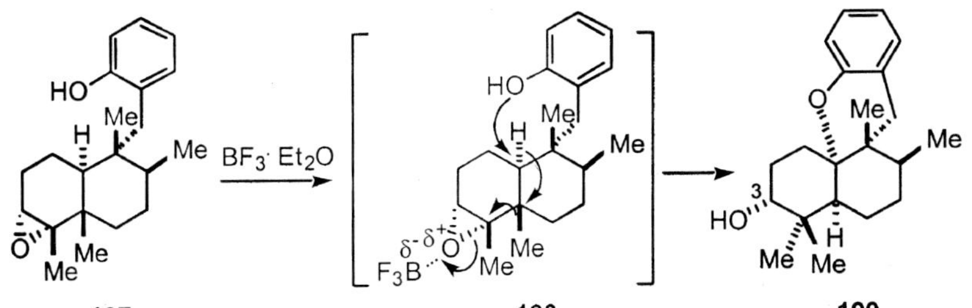

Scheme 8.50 Construction of the ABCD ring system of stachyflin by an acid-induced domino reaction.

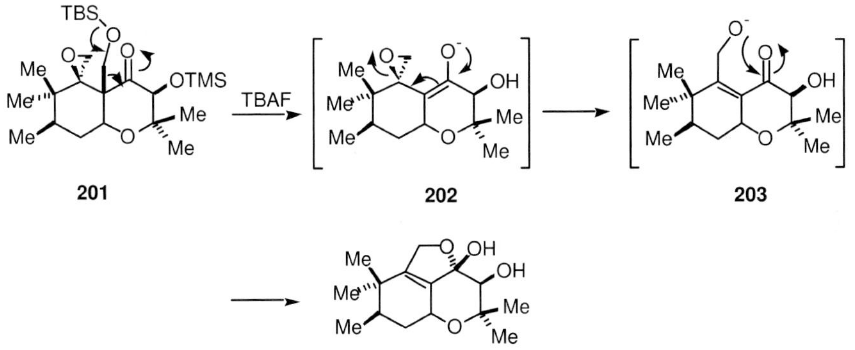

Scheme 8.51 Construction of the tricyclic core of phomactin A through a retroaldol reaction.

eoisomeric β-epoxide, which has the right stereochemistry at C_3, resulted in the direct production of the desired cyclized product **200**, but in poor yield (15%) [92].

A non-addition procedure based on a tandem retroaldol/epoxide-opening/cyclization sequence has been utilized for the elaboration of the dihydrofuran ring of the reduced furanochroman subunit of phomactin A. The strategy utilizes the *O*-TBS-protected β-epoxy-β'-hydroxy ketone **201**. Cleavage with TBAF starts the retro-

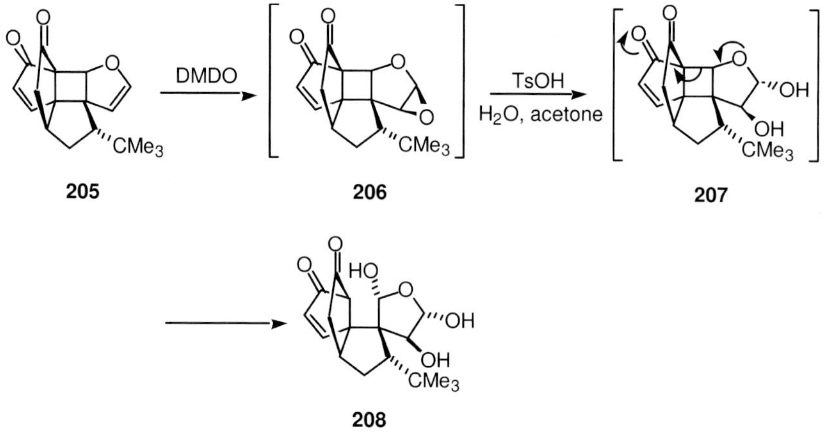

Scheme 8.52 Retro-aldol fragmentation of an epoxide-derived hemiacetal.

aldol reaction to afford enolate **202**, which isomerizes, with concomitant oxirane ring-opening, to the allylic alkoxide **203**, which in turn provides the desired hemiketal **204** upon cyclization (Scheme 8.51). Presumably, the ease of this process is enhanced by the relief of steric strain associated with loss of the hydroxymethyl substituent at the ring junction, as well as by ring-opening of the spiro epoxide [93].

A retroaldol fragmentation subsequent to the addition of *p*-TsOH and a small amount of water to epoxide **206**, obtained by oxidation of enol ether **205** with DMDO, resulted in the direct formation of dialdehyde hydrate **208**, possessing the spirostructure necessary for the construction of the fused-rings core of (±)-ginkolide B. Apparently, hydrolysis of the epoxide produces the hemiacetal **207**, which undergoes retroaldol fragmentation of the cyclobutane to afford the dialdehyde, which forms the stable hydrate **208** (Scheme 8.52) [94].

A particular case of a non-addition process of an epoxide is Eschenmoser fragmentation. This reaction is effectively a process in which the *p*-toluenesulfonylhydrazones of cyclic α-epoxy ketones are transformed, in the presence of a base such as pyridine, into terminal alkynyl ketones. In this transformation the original cyclic structure of the alkanone is completely lost, and the epoxide is isomerized into a carbonyl group (ketone). A new carbon-carbon triple bond replaces the original carbonyl group [95]. Since the alkynyl group may be seen both as the source of other functional groups (primarily a carbonyl group and a double bond) and as a site for chain-elongation, this fragmentation can occupy a central role in the synthesis of complex molecules. In the total synthesis of the alkaloid galbulimina GB 13, an Eschenmoser fragmentation was considered well suited for the transformation of the 2,3-epoxycyclohexanone moiety of the intermediate **210** into the piperidine ring present in the target compound (Scheme 8.53) [96]. As enone **209** proved to be resistant to direct epoxidation, the necessary epoxy ketone **210** was

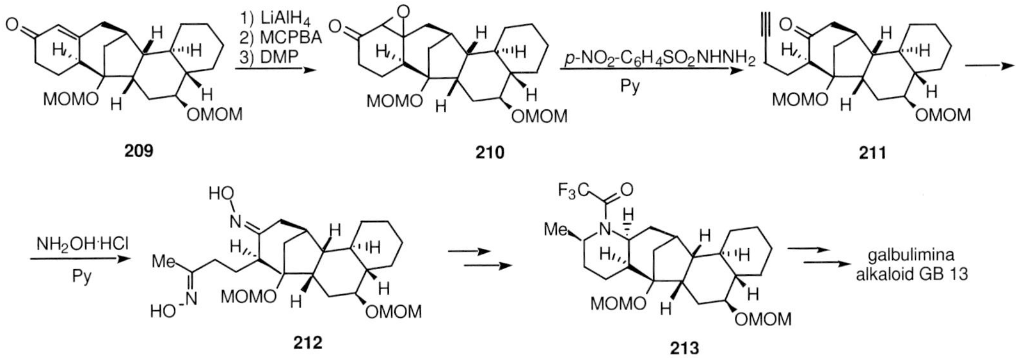

Scheme 8.53 Eschenmoser fragmentation of α-epoxy ketone 210.

prepared by a sequence involving reduction ($LiAlH_4$) to the corresponding allylic alcohol, epoxidation (MCPBA), and then oxidation (DMP).

Treatment of **210** under traditional Eschenmoser fragmentation conditions gave only low yields of the desired alkynyl ketone **211**, but this result was improved significantly by use of *p*-nitrobenzenesulfonylhydrazine in place of the commonly used *p*-toluenesulfonylhydrazine. Compound **211** was transformed into the bis-oxime **212**, reductive cyclization of which by treatment with $ZrCl_4$ and $NaBH_4$ and subsequent acylation afforded the polycyclic compound **213** with the desired all-*cis* piperidine ring stereochemistry [96].

References

1 a) J. Gorzinski Smith, *Synthesis* **1984**, 629–656. b) A. S. Rao, in *Comprehensive Organic Synthesis*; Eds.; B. M. Trost, I. Fleming, S. V. Ley; Pergamon Press: Oxford, **1991**; Vol. 7, pp. 357–387. c) G. Berti, in *Topics in Stereochemistry*; Eds; N. L. Allinger, and E. L. Eliel; Wiley Interscience: New York, **1973**; Vol. 7, p.93–251.

2 a) R. A. Johnson, K. B. Sharpless, in *Catalytic Asymmetric Synthesis;* Ed.; I. Ojima, Wiley-VCH, 2nd ed.; **2000**, pp. 231–280. b) *ibidem*, pp. 357–398. c) T. Katsuki, in *Catalytic Asymmetric Synthesis;* Ed.; I. Ojima, Wiley-VCH, 2nd ed; **2000**, pp. 287–325.

3 a) B. Rickborn, in *Comprehensive Organic Synthesis*; Eds.; B. M. Trost, I. Fleming, G. Pattenden; Pergamon Press: Oxford, **1991**; Vol. 3, pp. 733–775. b) D. J. Coveney, *ibidem*, pp. 777–801.

4 T.L.B. Boivin, *Tetrahedron* **1987**, *43*, 3309–3362.

5 A. H. Hoveyda, D. A. Evans, G. C. Fu, *Chem. Rev.* **1993**, *93*, 1307–1370.

6 P. A. Bartlett, *Tetrahedron* **1980**, *36*, 3–72.

7 J. E. Baldwin, *J. Chem. Soc. Chem. Commun.* **1976**, 734–736.

8 J. Na, K. N. Houk, C. G. Shevlin, K. D. Janda, R. A. Lerner, *J. Am. Chem. Soc.* **1993**, *115*, 8453–8454.

9 a) K. C. Nicolaou, E. J. Sorensen, *Classics in Total Synthesis*; VCH: Weinheim (FRG), **1996**; pp. 731–786. b) K. C. Nicolaou, C. V. C. Prasad, P. K. Somers, C. K. Hwang, *J. Am. Chem. Soc.* **1989**, *111*, 5330–5334.

10 T. Fukuyama, B. Vranesic, D. P. Negri, Y. Kishi, *Tetrahedron Lett.* **1978**, *19*, 2741–2744.

11 T. Nakata, G. Schmid, B. Vranesic, M. Oki-

gawa, T. Smith-Palmer, Y. Kishi, *J. Am. Chem. Soc.* **1978**, *100*, 2933–2935.

12 a) D. A. Evans, R. L. Dow, T. L. Shih, J. M. Takacs, R. Zahler, *J .Am. Chem. Soc.* **1990**, *112*, 5290–5313. b) D. A. Evans, S. L. Bender, J. Morris, *J. Am. Chem. Soc.* **1988**, *110*, 2506–2526. c) M. T. Crimmins, K. A. Emmitte, *J. Am. Chem. Soc.* **2001**, *123*, 1533–1534.

13 J. K. Rupprecht, Y.-H. Hui, J. L. McLaughlin, *J. Nat. Prod.* **1990**, *53*, 237–278.

14 T. R. Hoye, P. R. Hanson, A. C. Kovelesky, T. D. Ocain, Z. Zhuang, *J. Am. Chem. Soc.* **1991**, 113, 9369–9371. See also: T. R. Hoye, Z. Ye, *J. Am. Chem. Soc.* **1996**, *118*, 1801–1802.

15 M. Yamaguchi, I. Hirao, *Tetrahedron Lett.* **1983**, *24*, 391–394.

16 a) H. Avedissian, S. C. Sinha, A. Yazbak, A. Sinha, P. Neogi, S. C. Sinha, E. Keinan, *J. Org. Chem.* **2000**, *65*, 6035–6051. b) S. C. Sinha, E. Keinan, *J. Am. Chem. Soc.* **1993**, *115*, 4891–4892. c) A. Sinha, S. C. Sinha, S. C. Sinha, E. Keinan, *J. Org. Chem.* **1999**, *64*, 2381–2386. d) S. C. Sinha, S. C. Sinha, E. Keinan, *J. Org. Chem.* **1999**, *64*, 7067–7073.

17 P. Neogi, T. Doundoulakis, A. Yazbak, S. C. Sinha, S. C. Sinha, E. Keinan, *J. Am. Chem. Soc.* **1998**, *120*, 11279–11284. See also: P. A. Evans, V. S. Murthy, *Tetrahedron Lett.* **1999**, *40*, 1253–1256.

18 I. Paterson, K. Feβner, M. R. V. Finlay, *Tetrahedron Lett.* **1997**, *38*, 4301–4304.

19 a) K. C. Nicolaou, C.-K. Hwang, M. E. Duggan, D. A. Nugiel, Y. Abe, K. Bal Reddy, S. A. DeFrees, D. R. Reddy, R. A. Awartani, S. R. Conley, F. P. J. T. Rutjes, E. A. Theodorakis, *J. Am. Chem. Soc.* **1995**, *117*, 10227–10238. b) K. C. Nicolaou, E. A. Theodorakis, F. P. J. T. Rutjes, M. Sato, J. Tiebes, X.-Y. Xiao, C.-K. Hwang, M. E. Duggan, Z. Yang, E. A. Couladouros, F. Sato, J. Shin, H.-M. He, T. Bleckman, *J. Am. Chem. Soc.* **1995**, *117*, 10239–10251. See also: c) H. Fuwa, N. Kainuma, K. Tachibana, M. Sasaki, *J. Am. Chem. Soc.* **2002**, *124*, 14983–14992. d) I. Kadota, C. Kadowaki, C.-H. Park, H. Takamura, K. Sato, P. W. H. Chan, S. Thorand, Y. Yamamoto, *Tetrahedron* **2002**, *58*, 1799–1816.

20 a) P. J. Mohr, R. L. Halcomb, *J. Am. Chem. Soc.* **2003**, *125*, 1712–1713. b) A. B. Smith, III, M. Frohn, *Org. Lett.* **2001**, *3*, 3979–3982.

21 G. Berti, S. Canedoli, P. Crotti, F. Macchia, *J. Chem. Soc. Perkin Trans. 1* **1984**, 1183–1188. b) C. M. Robbins, G. H. Whitham, *J. Chem. Soc. Chem. Commun.* **1976**, 697–698.

22 G. Adiwidjaja, H. Flörke, A. Kirschning, E. Schaumann, *Tetrahedron Lett.* **1995**, *36*, 8771–8774.

23 T. P. Heffron, T. F. Jamison, *Org. Lett.* **2003**, *5*, 2339–2342.

24 C. Mukai, Y. Sugimoto, Y. Ikeda, M. Hanaoka, *Tetrahedron* **1998**, *54*, 823–850.

25 C. Mukai, N. Miyakoshi, M. Hanaoka, *J. Org. Chem.* **2001**, *66*, 5875–5880.

26 A. B. Smith, III, R. J. Fox, *Org. Lett.* **2004**, *6*, 1477–1480.

27 T. Satoh, *Chem. Rev.*, **1996**, *96*, 3303–3325. See also: R. Luisi, V. Capriati, C. Carlucci, L. Degennaro, S. Florio, *Tetrahedron* **2003**, *59*, 9707–9712 and references therein.

28 Y. Mori, K. Yaegashi, H. Furukawa, *J. Org. Chem.* **1998**, *63*, 6200–6209. See also: M. Inoue, S. Yamashita, A. Tatami, K. Miyazaki, M. Hirama, *J. Org. Chem.* **2004**, *69*, 2797–2804; Y. Mori, K. Nogami, H. Hayashi, R. Noyori, *J. Org. Chem.* **2003**, *68*, 9050–9060.

29 a) T. Nakata, S. Nomura, H. Matsukura, M. Morimoto, *Tetrahedron Lett.* **1996**, *37*, 217–220. b) F. E. McDonald, X. Wei, *Org. Lett.* **2002**, *4*, 593–595.

30 a) K. C. Nicolaou, C. V. C. Prasad, P. K. Somers, C. -K. Hwang, *J. Am. Chem. Soc.* **1989**, *111*, 5335–5340. b) K. Fujiwara, H. Mishima, A. Amano, T. Tokiwano, A. Murai, *Tetrahedron Lett.* **1998**, *39*, 393–396. For the synthesis of eight- and nine-membered rings from epoxy alcohols, see 30b and T. Saitoh, T. Suzuki, N. Onodera, H. Sekiguchi, H. Hagiwara, T. Hoshi, *Tetrahedron Lett.* **2003**, *44*, 2709–2712.

31 a) D. E. Cane, W. D. Celmer, J. W. Westley, *J. Am. Chem. Soc.* **1983**, *105*, 3594–3600. b) M. S. Lee, G. Qin, K. Nakanishi, M. G. Zagorski, *J. Am. Chem. Soc.* **1989**, *111*, 6234–6241.

32 R. J. Capon, R. A. Barrow, *J. Org. Chem.* **1998**, *63*, 75–83.

33 D. A. Evans, A. M. Ratz, B. E. Huff, G. S. Sheppard, *J. Am. Chem. Soc.* **1995**, *117*, 3448–3467.

34 Z.-X. Wang, Y. Tu, M. Frohn, J.-R. Zhang,

Y. Shi, *J. Am. Chem. Soc.* **1997**, *119*, 11224–11235.

35 a) Z. Xiong, E. J. Corey, *J. Am. Chem. Soc.* **2000**, *122*, 4831–4832. b) Y. Morimoto, T. Iwai, T. Kinoshita. *J. Am. Chem. Soc.* **2000**, *122*, 7124–7125.

36 T. J. Beauchamp, J. P. Powers, S. D. Rychnovsky, *J. Am. Chem. Soc.* **1995**, *117*, 12873–12874.

37 I. Paterson, R. D. Tillyer, J. B. Smaill, *Tetrahedron Lett.* **1993**, *34*, 7137–7140, and references therein.

38 a) T. Tokiwano, K. Fujiwara, A. Murai, *Synlett* **2000**, 335–338. b) K. Fujiwara, T. Tokiwano, A. Murai, *Tetrahedron Lett.* **1995**, *36*, 8063–8066.

39 a) F. E. McDonald, F. Bravo, X. Wang, X. Wei, M. Toganoh, J. R. Rodriguez, B. Do, W. A. Neiwert, K. I. Hardcastle, *J. Org. Chem.* **2002**, *67*, 2515–2523. b) F. E. McDonald, X. Wang, B. Do, K. Hardcastle, *Org. Lett.* **2000**, *2*, 2917–2919.

40 J. Gutzwiller, M. R. Uskokovic, *J. Am. Chem. Soc.* **1978**, *100*, 576–581.

41 I. T. Raheem, S. N. Goodman, E. N. Jacobsen, *J. Am. Chem. Soc.* **2004**, *126*, 706–707.

42 S. Yokoshima, T. Ueda, S. Kobayashi, A. Sato, T. Kuboyama, H. Tokuyama, T. Fukuyama, *J. Am. Chem. Soc.* **2002**, *124*, 2137–2139.

43 P. G. Reddy, B. Varghese, S. Baskaran, *Org. Lett.* **2003**, *5*, 583–585.

44 P. G. Reddy, S. Baskaran, *J. Org. Chem.* **2004**, *69*, 3093–3101.

45 E. J. Corey, G. Luo, L. S. Lin, *J. Am. Chem. Soc.* **1997**, *119*, 9927–9928. See also: a) J. Zhang, E. J. Corey, *Org. Lett.* **2001**, *3*, 3215–3216. b) E. J. Corey, G. Luo, L. S. Lin, *Angew. Chem. Int. Ed.* **1998**, *37*, 1126–1128. c) E. J. Corey, S. Lin, *J. Am. Chem. Soc.* **1996**, *118*, 8765–8766.

46 Y. Mi, J. V. Schreiber, E. J. Corey, *J. Am. Chem. Soc.* **2002**, *124*, 11290–11291.

47 M. Bogenstätter, A. Limberg, L. E. Overman, A. L. Tomasi, *J. Am. Chem. Soc.* **1999**, *121*, 12206–12207.

48 a) J. K. Crandall, L.-H. C. Lin, *J. Am. Chem. Soc.* **1967**, *89*, 4526–4527. b) D. M. Hodgson, Y. K. Chung, J.-M. Paris, *J. Am. Chem. Soc.* **2004**, *126*, 8664–8665.

49 a) C. H. Behrens, K. B. Sharpless, *Aldrichimica Acta* **1983**, *16*, 67–80. b) C. H. Behrens, K. B. Sharpless, *J. Org. Chem.* **1985**, *50*, 5696–5704.

50 a) Y. Kishi, *Aldrichimica Acta* **1980**, *13*, 23–30. b) G. H. Posner, in *Organic Reactions*; Ed.; W. G. Dauben; Wiley: New York, **1975**, Vol. 22, pp. 253–400.

51 a) A. B. Dounay, R. A. Urbanek, V. A. Frydrychowski, C. J. Forsyth, *J. Org. Chem.* **2001**, *66*, 925–938. b) M. Kobayashi, M. Kurosu, W. Wang, I. Kitagawa, *Chem. Pharm. Bull.* **1994**, *42*, 2394–2396. c) W. R. Roush, M. A. Adam, S. M. Peseckis, *Tetrahedron Lett.* **1983**, *24*, 1377–1380.

52 a) M. R. Johnson, Y. Kishi, *Tetrahedron Lett.* **1979**, *20*, 4347–4350. b) M. R. Johnson, T. Nakata, Y. Kishi, *Tetrahedron Lett.* **1979**, *20*, 4343–4346.

53 H. Nagaoka, Y. Kishi, *Tetrahedron* **1981**, *37*, 3873–3888.

54 a) D. R. Williams, K. G. Meyer, *J. Am. Chem. Soc.* **2001**, *123*, 765–766. b) D. R. Williams, B. J. Myers, L. Mi, *Org. Lett.* **2000**, *2*, 945–948.

55 a) C. Molinaro, T. F. Jamison, *J. Am. Chem. Soc.* **2003**, *125*, 8076–8077. b) E. A. Colby, K. C. O'Brien, T. F. Jamison, *J. Am. Chem. Soc.* **2004**, *126*, 998–999.

56 a) A. B. Smith, III, S. M. Pitram, A. M. Boldi, M. J. Gaunt, C. Sfouggatakis, W. H. Moser, *J. Am. Chem. Soc.* **2003**, *125*, 14435–14445. b) A. B. Smith, III, S. M. Pitram, *Org. Lett.* **1999**, *1*, 2001–2004.

57 S. D. Richnovsky, *Chem. Rev.* **1995**, *95*, 2021–2040.

58 A. B. Smith, III, V. A. Doughty, C. Sfouggatakis, C. S. Bennett, J. Koyanagi, M. Takeuchi, *Org. Lett.* **2002**, *4*, 783–786. For precedent spongistatin synthetic studies, see: A. B. Smith, III, Q. Lin, K. Nakayama, A. M. Boldi, C. S. Brook, M. D. McBriar, W. H. Moser, M. Sobukawa, L. Zhuang, *Tetrahedron Lett.* **1997**, *38*, 8675–8678 and references therein.

59 For further use of dithiane chemistry in the synthesis of spongistatin, see also: a) M. J. Gaunt, D. F. Hook, H. R. Tanner, S. V. Ley, *Org. Lett.* **2003**, *5*, 4815–4818. b) M. J. Gaunt, A. S. Jessiman, P. Orsini, H. R. Tanner, D. F. Hook, S. V. Ley, *Org. Lett.* **2003**, *5*, 4819–4822.

60 K. C. Nicolaou, R. A. Daines, J. Uenishi, W. S. Li, D. P. Papahatjis, T. K. Chakraborty, *J. Am. Chem. Soc.* **1988**, *110*, 4672–4685.

61 a) S. Tosaki, Y. Horiuchi, T. Nemoto, T. Ohshima, M. Shibasaki, *Chem. Eur. J.*

2004, *10*, 1527–1544. b) S. Tosaki, T. Nemoto, T. Ohshima, M. Shibasaki, *Org. Lett.* **2003**, *5*, 495–498.
62 a) S. A. Burova, F. E. McDonald, *J. Am. Chem. Soc.* **2004**, *126*, 2495–2500. b) S. A. Burova, F. E. McDonald, *J. Am. Chem. Soc.* **2002**, *124*, 8188–8189.
63 S. E. Schaus, B. D. Brandes, J. F. Larrow, M. Tokunaga, K. B. Hansen, A. E. Gould, M. E. Furrow, E. N. Jacobsen, *J. Am. Chem. Soc.* **2002**, *124*, 1307–1315.
64 S. Danishefsky, T. Kitahara, M. Tsai, J. Dynak , *J. Org. Chem.* **1976**, *41*, 1669–1671.
65 a) S. K. Taylor, *Tetrahedron* **2000**, *56*, 1149–1163. b) C. Agami, F. Couty, G. Evano, *Eur. J. Org. Chem.* **2002**, 29–38.
66 a) F. Sauriol-Lord, T. B. Grindley, *J. Org. Chem.* **1981**, *46*, 2833–2835. b) R. P. Woodbury, M. W. Rathke, *J. Org. Chem.* **1977**, *42*, 1688–1690.
67 a) D. Askin, R. P. Volante, K. M. Ryan, R. A. Reamer, I. Shinkai, *Tetrahedron Lett.* **1988**, *29*, 4245–4248. b) A. G. Myers, L. McKinstry, *J. Org. Chem.* **1996**, *61*, 2428–2440.
68 a) D. F. Taber, L. J. Silverberg, E. D. Robinson, *J. Am. Chem. Soc.* **1991**, *113*, 6639–6645. b) D. Askin, K. K. Eng, K. Rossen, R. M. Purick, K. M. Wells, R. P. Volante, P. J. Reider, *Tetrahedron Lett.* **1994**, *35*, 673–676.
69 a) S. L. Schreiber, *J. Am. Chem. Soc.* **1980**, *102*, 6163–6165. b) P. Crotti, V. Di Bussolo, L. Favero, M. Pineschi, M. Pasero, *J. Org. Chem.* **1996**, *61*, 9548–9552.
70 D. A. Evans, R. P. Polniaszek, K. M. DeVries, D. E. Guinn, D. J. Mathre, *J. Am. Chem. Soc.* **1991**, *113*, 7613–7630.
71 R. K. Boeckman, Jr., J. Zhang, M. R. Reeder, *Org. Lett.* **2002**, *4*, 3891–3894.
72 P. Crotti, F. Badalassi, V. Di Bussolo, L. Favero, M. Pineschi, *Tetrahedron* **2001**, *57*, 8559–8572.
73 a) G. H. Posner, Q. Wang, B. A. Halford, J. S. Elias, J. P. Maxwell, *Tetrahedron Lett.* **2000**, *41*, 9655–9659. b) G. H. Posner, J. P. Maxwell, M. Kahraman, *J. Org. Chem.* **2003**, *68*, 3049–3054.
74 a) S. J. Danishefsky, M. T. Bilodeau, *Angew. Chem. Int. Ed.* **1996**, *35*, 1380–1419. b) R. L. Halcomb, S. J. Danishefsky, *J. Am. Chem. Soc.* **1989**, *111*, 6661–6666. c) J. T. Randolph, K. F. McClure, S. J. Danishefsky, *J. Am. Chem. Soc.* **1995**, *117*, 5712–5719.
75 a) R. G. Dushin, S. J. Danishefsky, *J. Am. Chem. Soc.* **1992**, *114*, 3471–3475. b) M. Gallant, J. T. Link, S. J. Danishefsky, *J. Org. Chem.* **1993**, *58*, 343–349. c) J. T. Link, S. Raghavan, M. Gallant, S. J. Danishefsky, T. C. Chou, L. M. Ballas, *J. Am. Chem. Soc.* **1996**, *118*, 2825–2842.
76 a) D. M. Gordon, S. J. Danishefsky, *J. Am. Chem. Soc.* **1992**, *114*, 659–663. b) T. K. Park, I. J. Kim, S. Hu, M. T. Bilodeau, J. T. Randolph, O. Kwon, S. J. Danishefsky, *J. Am. Chem. Soc.* **1996**, *118*, 11488–11500. c) J. T. Randolph, S. J. Danishefsky, *J. Am. Chem. Soc.* **1995**, *117*, 5693–5700.
77 a) P. H. Seeberger, M. Eckhardt, C. E. Gutteridge, S. J. Danishefsky, *J. Am. Chem. Soc.* **1997**, *119*, 10064–10072. b) S. K. Bhattacharya, S. J. Danishefsky, *J. Org. Chem.* **2000**, *65*, 144–151.
78 O. J. Plante, R. B. Andrade, P. H. Seeberger, *Org. Lett.* **1999**, *1*, 211–214.
79 a) S. J. Danishefsky, V. Behar, J. T. Randolph, K. O. Lloyd, *J. Am. Chem. Soc.* **1995**, *117*, 5701–5711. b) J. T. Randolph, S. J. Danishefsky, *Angew. Chem. Int. Ed.* **1994**, *33*, 1470–1473.
80 a) S. A. Hitchcock, M. Y. Chu Moyer, S. H. Boyer, S. H. Olson, S. J. Danishefsky, *J. Am. Chem. Soc.* **1995**, *117*, 5750–5756. b) R. L. Halcomb, S. H. Boyer, M. D. Wittman, S. H. Olson, D. J. Denhart, K. K. C. Liu, S. J. Danishefsky, *J. Am. Chem. Soc.* **1995**, *117*, 5720–5749.
81 a) D. A. Evans, B. W. Trotter, B. Côté, *Tetrahedron Lett.* **1998**, *39*, 1709–1712. b) D. A. Evans, B. W. Trotter, B. Côté, P. J. Coleman, L. C. Dias, A. N. Tyler, *Angew. Chem. Int. Ed.* **1997**, *36*, 2744–2747. c) M. T. Crimmins, J. D. Katz, D. G. Washburn, S. P. Allwein, L. F. McAtee, *J. Am. Chem. Soc.* **2002**, *124*, 5661–5663.
82 J. D. Rainier, S. P. Allwein, *J. Org. Chem.* **1998**, *63*, 5310–5311.
83 V. Di Bussolo, Y.-J. Kim, D. Y. Gin, *J. Am. Chem. Soc.* **1998**, *120*, 13515–13516. For recent catalyzed reagent-controlled α- or β-*O*-glycosylation, see: H. Kim, H. Men, C. Lee, *J. Am. Chem. Soc.* **2004**, *126*, 1336–1337.
84 a) A. Fernández-Mateos, A. M. López Barba, E. Martin de la Nava, G. P. Coca, J. J. Pérez Alonso, A. I. R. Silvo, R. R. Gon-

zález, *Tetrahedron* **1997**, *53*, 14131–14140. b) L. N. Mander, R. J. Thomson, *Org. Lett.* **2003**, *5*, 1321–1324. c) Y. Kita, K. Higuchi, Y. Yoshida, K. Iio, S. Kitagaki, S. Akai, H. Fujioka, *Angew. Chem. Int. Ed.* **1999**, *38*, 683–686.

85 a) Q. Su, A. B. Beeler, E. Lobkovsky, J. A. Porco, Jr., J. S. Panek, *Org. Lett.* **2003**, *5*, 2149–2152. b) B. M. Trost, G. A. Doherty, *J. Am. Chem. Soc.* **2000**, *122*, 3801–3810. c) R. G. Arrayás, A. Alcudia, L. S. Liebeskind, *Org. Lett.* **2001**, *3*, 3381–3383.

86 a) A. N. Cuzzupe, C. A. Hutton, M. J. Lilly, R. K. Mann, K. J. McRae, S. C. Zammit, M. A. Rizzacasa, *J. Org. Chem.* **2001**, *66*, 2382–2393. b) A. N. Cuzzupe, C. A. Hutton, M. J. Lilly, R. K. Mann, M. A. Rizzacasa, S. C. Zammit, *Org. Lett.* **2000**, *2*, 191–194. See also: R. E. Ireland, J. D. Armstrong, III, J. Lebreton, R. S. Meissner, M. A. Rizzacasa, *J. Am. Chem. Soc.* **1993**, *115*, 7152–7165.

87 T. Saito, T. Suzuki, M. Morimoto, C. Akiyama, T. Ochiai, K. Takeuchi, T. Matsumoto, K. Suzuki, *J. Am. Chem. Soc.* **1998**, *120*, 11633–11644.

88 M. D. B. Fenster, G. R. Dake, *Org. Lett.* **2003**, *5*, 4313–4316.

89 K. D. Eom, J. V. Raman, H. Kim, J. K. Cha, *J. Am. Chem. Soc.* **2003**, *125*, 5415–5421.

90 K. Tanino, K. Onuki, K. Asano, M. Miyashita, T. Nakamura, Y. Takahashi, I. Kuwajima, *J. Am. Chem. Soc.* **2003**, *125*, 1498–1500.

91 a) H. Nemoto, M. Nagamochi, H. Ishibashi, K. Fukumoto, *J. Org. Chem.* **1994**, *59*, 74–79. b) H. Nemoto, T. Tanabe, K. Fukumoto, *J. Org. Chem.* **1995**, *60*, 6785–6790. c) H. Nemoto, H. Ishibashi, M. Nagamochi, K. Fukumoto, *J. Org. Chem.* **1992**, *57*, 1707–1712.

92 M. Nakatani, M. Nakamura, A. Suzuki, M. Inoue, T. Katoh, *Org. Lett.* **2002**, *4*, 4483–4486.

93 P. P. Seth, N. I. Totah, *Org. Lett.* **2000**, *2*, 2507–2509.

94 M. T. Crimmins, J. M. Pace, P. G. Nantermet, A. S. Kim-Meade, J. B. Thomas, S. H. Watterson, A. S. Wagman, *J. Am. Chem. Soc.* **2000**, *122*, 8453–8463.

95 A. Eschenmoser, D. Felix, G. Ohloff, *Helv. Chim. Acta* **1967**, *50*, 708–713.

96 L. N. Mander, M. M. McLachlan, *J. Am. Chem. Soc.* **2003**, *125*, 2400–2401.

9
Vinylepoxides in Organic Synthesis

Berit Olofsson and Peter Somfai

9.1
Synthesis of Vinylepoxides

Vinylepoxides can be obtained by various strategies, all with their inherent limitations. Racemic epoxidation of olefins is a straightforward route to epoxides, as pure *trans-* or *cis*-epoxides can be obtained from (*E*)- or (*Z*)-alkenes, respectively. Various oxidants – such as *m*CPBA and other peracids, H_2O_2, or $VO(acac)_2$/TBHP – can all be employed in this transformation [1].

Epoxidation of conjugated dienes can be regioselective when one double bond is more electron-rich than the other; otherwise mixtures of mono- and diepoxides will be obtained. When the alkene contains an adjacent stereocenter, the epoxidation can be diastereoselective [2]. Hydroxy groups can function as directing groups, causing the epoxidation to take place *syn* to the alcohol [2, 3].

Asymmetric epoxidations of alkenes have been intensively studied since Sharpless' initial report on asymmetric epoxidation of allylic alcohols in 1980. This reaction, discussed in Section 9.1.3, has become one of the most widely employed reactions in asymmetric synthesis, due to its reliability and high enantioselectivity [2].

Enantioselective epoxidation of unfunctionalized alkenes was until recently limited to certain *cis*-alkenes, but most types of alkenes can now be successfully epoxidized with sugar-derived dioxiranes (see Section 9.1.1.1) [2]. Selective monoepoxidation of dienes has thus become a fast route to vinylepoxides. Functionalized dienes, such as dienones, can be epoxidized with excellent enantioselectivities (see Section 9.1.2).

Alternatively, epoxides can be formed with concomitant formation of a C–C bond. Reactions between aldehydes and various carbon nucleophiles are an efficient route to epoxides, although the *cis:trans* selectivity can be problematic (see Section 9.1.4). Kinetic resolution (see Section 9.1.5.2) or dihydroxylation with sequential ring-closure to epoxides (see Section 9.1.1.3) can be employed when asymmetric epoxidation methods are unsatisfactory.

The major problem associated with the synthesis of vinylepoxides is the instability of reaction intermediates and/or products. High crude yields can be obtained by

Aziridines and Epoxides in Organic Synthesis. Andrei K. Yudin

ISBN: 3-527-31213-7

several methods, but if lengthy purification procedures are required to remove reagents (such as $Ph_3P=O$ or organic catalysts), the isolated yield often drops considerably. It is thus important to obtain high diastereoselectivity in the reaction, in order to avoid difficult isomer separations. This feature renders syntheses that yield *cis:trans* mixtures of vinylepoxides less useful. The following discussion focuses on asymmetric syntheses of vinylepoxides, although some racemic, diastereoselective epoxidations are mentioned when appropriate.

9.1.1 Vinylepoxides from Unfunctionalized Dienes

9.1.1.1 Epoxidation with Dioxiranes

Chiral ketone-catalyzed asymmetric epoxidations of olefins have been extensively studied in the last decade [4]. The reactive species in these reactions is a dioxirane, generated *in situ* from the ketone and an oxidant. The first asymmetric reaction, employing isopinocamphone as catalyst, was reported by Curci in 1984 [5]. Although the enantioselectivities were modest, the results could be improved by variation of the ketone structure. The breakthrough came in 1996, when Shi reported the fructose-derived ketone **1** as a highly efficient epoxidation catalyst (Scheme 9.1) [6]. Through small alterations of the catalyst structure, the system was subsequently shown to give good to excellent enantioselectivities and high yields in the epoxidation of a great variety of substrates, including terminal, *cis*-, *trans*-, or trisubstituted alkenes, vinylsilanes, and α,β-unsaturated esters [7].

The main drawback of the system is that the ketone catalyst slowly decomposes during the reaction, which means that 0.2–0.3 equivalents are needed for complete conversion. More robust catalysts, which can be used in 1–3 mol%, have recently been reported, but have not as yet been widely applied [8]. Ketone **1** is commercially available, or can easily be synthesized in large scale in two steps from D-fructose. *Ent*-**1** is obtained in a similar way from L-sorbose.

The catalytic system was subsequently applied to the monoepoxidation of dienes. This was potentially a difficult task, as there was a need to address the issues not only of enantioselectivity, but also of regioselectivity and monoepoxidation versus bisepoxidation. Fortunately, a wide range of dienes could be efficiently monoepoxidized by ketone **1**, which meant that a straightforward route to vinylepoxides had been developed (Table 9.1) [9].

The more electron-rich double bond was found to react more rapidly, which allowed for good regioselectivity. The regioselectivity could be further increased by

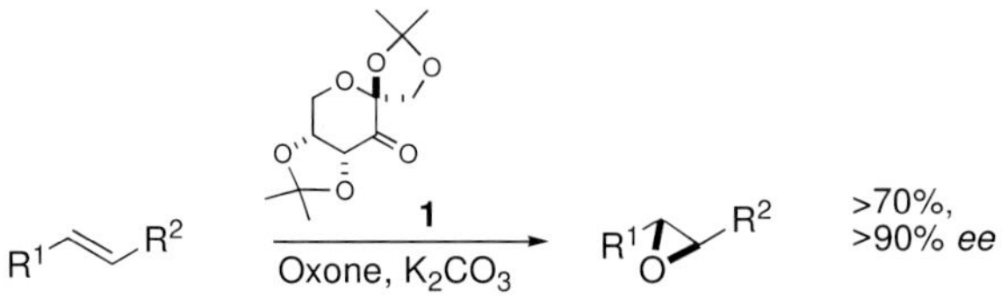

R^1, R^2 — **1**, Oxone, K_2CO_3 → epoxide; >70%, >90% *ee*

Scheme 9.1 Shi's asymmetric epoxidation with ketone **1**.

Table 9.1 A selection of dienes that could be monoepoxidized.

Entry	Diene	Epoxide	Ratio[a]	Yield (%)[b]	*Ee* (%)
1	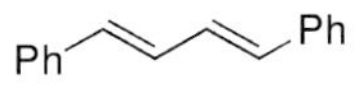 Ph Ph	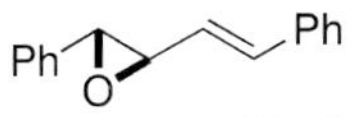 Ph O Ph	22:1[c]	77	97
2	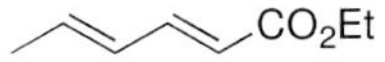 CO_2Et	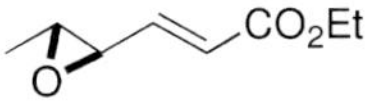 O CO_2Et	7:1	41[d]	96
3	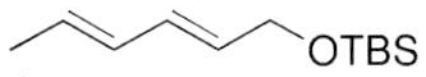 OTBS	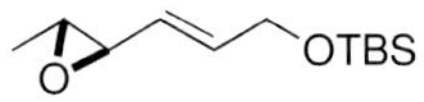 O OTBS	4.6:1	68	96
4	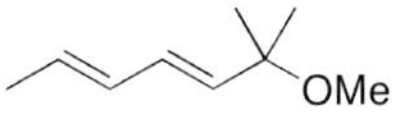 OMe	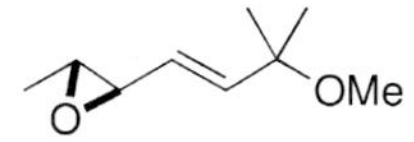 O OMe	–	65	89
5	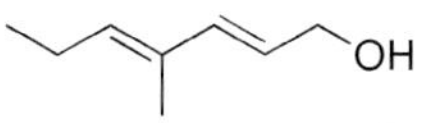 OH	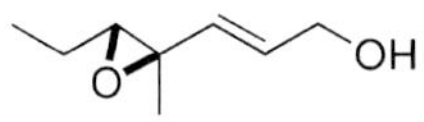 O OH	–	68[e]	90[e]
6	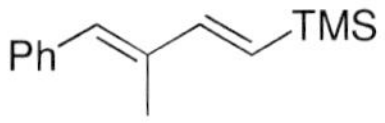 Ph TMS	 Ph O TMS	14:1	77	94

[a] Regioisomeric ratio. [b] Isolated yield of major isomer, often lowered due to instability of product. [c] Monoepoxide/bisepoxide ratio. [d] 69% conversion. [e] Increased to 81% yield and 96% *ee* with TBS-protected substrate.

introducing steric hindrance next to one olefin. Bisepoxidation could be minimized by controlling the amount of catalyst, as the first epoxide deactivated the remaining olefin by induction. The regioselectivity is complementary to the Sharpless asymmetric epoxidation (SAE) technique for dienols (see Section 9.1.2.2), as exemplified by Entries 3 and 5.

In our work on a regio- and stereodivergent route to vicinal amino alcohols, vinylepoxides were chosen as substrates [10, 11]. Initially the vinylepoxides were synthesized by the SAE + Swern/Wittig approach (see Section 9.1.3) [10]. To shorten the route for the next group of substrates we turned to the Shi epoxidation. As TBS-protected hexadienol could be monoepoxidized in good yield (Table 9.1, Entry 3), we decided to apply the reaction to benzyl-protected hexadienol **2** (Scheme 9.2a). As expected, vinylepoxide **3** was the major regioisomer and could be isolated in 66% yield and 90% *ee* [11]. Epoxidation of trisubstituted substrate **4** occurred with complete regioselectivity, reflecting the electronic influence exerted by the methyl group, and **5** was isolated in quantitative yield (Scheme 9.2b) [11].

As this synthetic strategy was clearly superior to the Swern/Wittig approach, being both shorter and having high-yielding steps, we also employed it in the synthesis of all isomers of sphingosine [12]. Substrate **6** differed from diene **2** only in the length of the carbon chain, and similar regioselectivity was to be expected. We were thus doubtful whether epoxide **7**, which was likely to be the minor isomer, could be formed in synthetically useful yields by this method. Surprisingly, epoxides **7** and **8** were formed in a 1:1 ratio, hence with far better selectivity than expected (Scheme 9.2c). The difference in regioselectivity could be due to the lipophilic alkyl chain, which might shield the 4,5-double bond. The conversion was

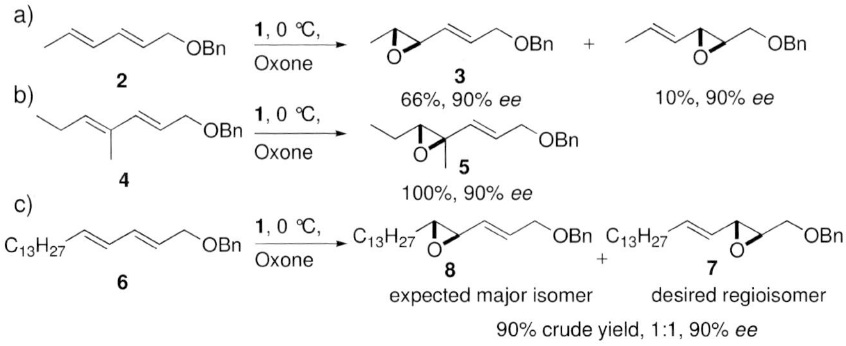

a) 2 → 1, 0 °C, Oxone → 3 66%, 90% *ee* + 10%, 90% *ee*

b) 4 → 1, 0 °C, Oxone → 5 100%, 90% *ee*

c) $C_{13}H_{27}$ 6 → 1, 0 °C, Oxone → 8 expected major isomer + 7 desired regioisomer; 90% crude yield, 1:1, 90% *ee*

Scheme 9.2 Applications of Shi's monoepoxidation.

initially poor and could not be improved by use of a longer reaction time or an increased temperature. When a stoichiometric amount of **1** was utilized, however, both enantioselectivity and yield were enhanced [12, 13].

9.1.1.2 Epoxidation with Mn-Salen Catalysts

Ten years after Sharpless's discovery of the asymmetric epoxidation of allylic alcohols, Jacobsen and Katsuki independently reported asymmetric epoxidations of unfunctionalized olefins by use of chiral Mn-salen catalysts such as **9** (Scheme 9.3) [14, 15]. The reaction works best on (*Z*)-disubstituted alkenes, although several tri- and tetrasubstituted olefins have been successfully epoxidized [16]. The reaction often requires ligand optimization for each substrate for high enantioselectivity to be achieved.

Conjugated dienes can be epoxidized to provide vinylepoxides. Cyclic substrates react with Katsuki's catalyst to give vinylepoxides with high *ees* and moderate yields [17], whereas Jacobsen's catalyst gives good yields but moderate enantioselectivities [18]. Acyclic substrates were found to isomerize upon epoxidation: (*Z,E*)-conjugated dienes reacted selectively at the (*Z*)-alkene to give *trans*-vinylepoxides (Scheme 9.4a) [19]. This feature was utilized in the formal synthesis of leukotriene A_4 methyl ester (Scheme 9.4b) [19].

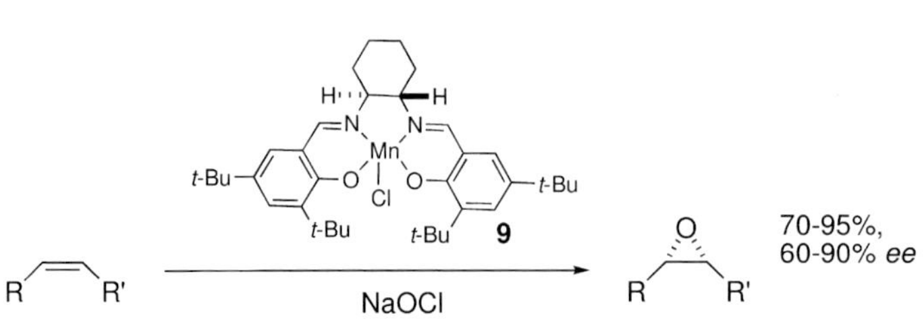

R, R' → 9, NaOCl → epoxide; 70-95%, 60-90% *ee*

Scheme 9.3 Jacobsen's asymmetric epoxidation.

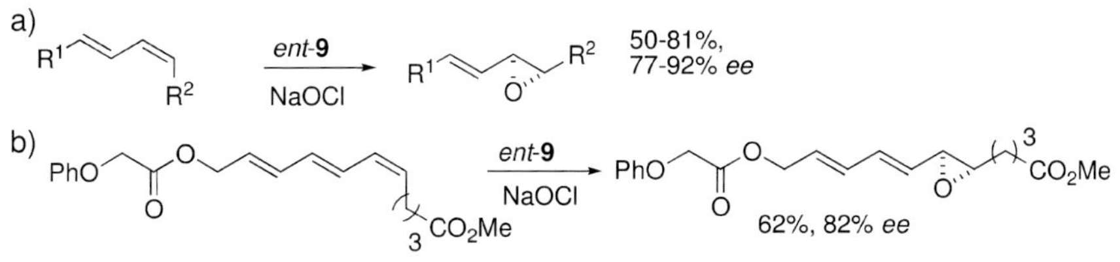

a) R^1 R^2 — *ent*-**9**, NaOCl → R^1 O R^2 50-81%, 77-92% *ee*

b) PhO O O CO_2Me 3 — *ent*-**9**, NaOCl → PhO O O O 3 CO_2Me 62%, 82% *ee*

Scheme 9.4 Monoepoxidation of (Z,E)-conjugated dienes.

9.1.1.3 Conversion of Diols into Epoxides

When asymmetric epoxidation of a diene is not feasible, an indirect route based on asymmetric dihydroxylation can be employed. The alkene is converted into the corresponding *syn*-diol with high enantioselectivity, and the diol is subsequently transformed into the corresponding *trans*-epoxide in a high-yielding one-pot procedure (Scheme 9.5) [20]. No epimerization occurs, and the procedure has successfully been applied to natural product syntheses when direct epoxidation strategies have failed [21]. Alternative methods for conversion of vicinal diols into epoxides have also been reported [22, 23].

This strategy can be applied to the synthesis of vinylepoxides, since high enantioselectivity and good regioselectivity can often be obtained in asymmetric dihydroxylation of dienes, resulting in vinylic diols [24, 25]. Transformation of the diols into epoxides thus represents an alternative route to vinylepoxides. This strategy was recently employed in the synthesis of (+)-posticlure (Scheme 9.6) [26].

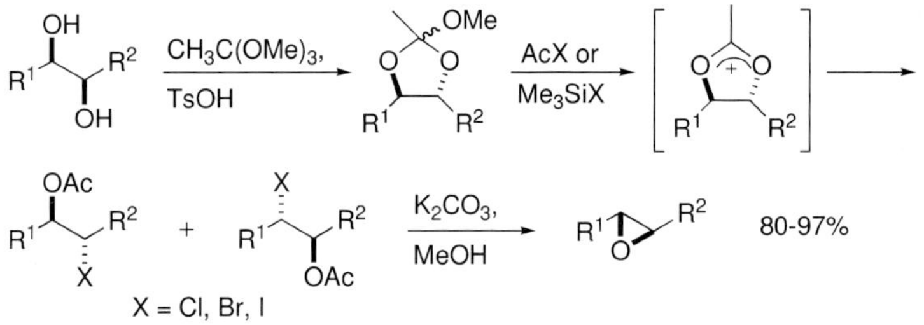

OH, OH, R^1, R^2 — $CH_3C(OMe)_3$, TsOH → OMe, O, O, R^1, R^2 — AcX or Me_3SiX → [O + O, R^1, R^2] →

OAc, X, R^1, R^2 + X, OAc, R^1, R^2 — K_2CO_3, MeOH → R^1 O R^2 80-97%

X = Cl, Br, I

Scheme 9.5 One-pot formation of epoxides from vicinal diols.

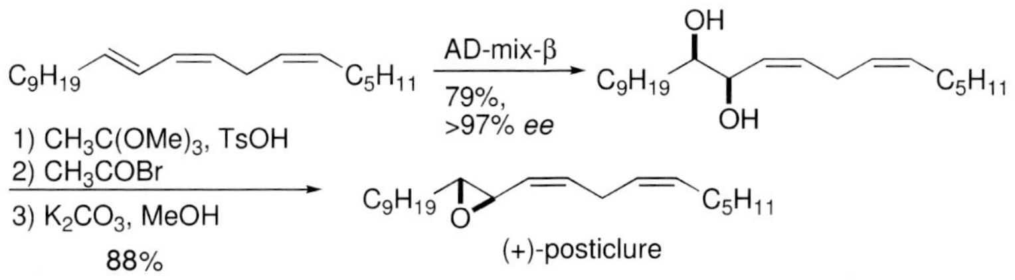

C_9H_{19} C_5H_{11} — AD-mix-β, 79%, >97% *ee* → C_9H_{19} OH OH C_5H_{11}

1) $CH_3C(OMe)_3$, TsOH
2) CH_3COBr
3) K_2CO_3, MeOH
88%
→ C_9H_{19} O C_5H_{11} (+)-posticlure

Scheme 9.6 Application to the synthesis of (+)-posticlure.

9.1.2
Vinylepoxides from Functionalized Dienes

9.1.2.1 From Dienones or Unsaturated Amides

The asymmetric epoxidation of enones with polyleucine as catalyst is called the Juliá-Colonna epoxidation [27]. Although the reaction was originally performed in a triphasic solvent system [27], phase-transfer catalysis [28] or nonaqueous conditions [29] were found to increase the reaction rates considerably. The reaction can be applied to dienones, thus affording vinylepoxides with high regio- and enantioselectivity (Scheme 9.7a) [29].

An alternative catalyst for the epoxidation of various enones, giving excellent enantioselectivities and good yields of the corresponding epoxides, was reported recently. The catalyst complex was generated from $La(O^iPr)_3$, (*R*)-BINOL, and $Ph_3As{=}O$ in a 1:1:1 ratio, but only one example with a dienone was given (Scheme 9.7b) [30]. A similar system based on gadolinium was employed for epoxidation of α,β,γ,δ-unsaturated amides, in which a range of amide-substituted vinylepoxides were synthesized (Scheme 9.7c) [31]. Again, excellent enantioselectivities and good yields were obtained, although extended reaction times were required.

A one-pot procedure from aldehydes, through Wittig olefination and a subsequent epoxidation, was also reported. Aldehydes could be converted into α,β,γ,δ-unsaturated *N*-acyl pyrroles, which were epoxidized in the same pot to give *N*-acyl pyrrole-substituted vinylepoxides [32].

It should be noted that epoxidation of a dienone with *m*CPBA or other electrophilic epoxidation reagents proceeds with complementary regioselectivity, yielding γ,δ-epoxy enones instead of the α,β-epoxy ketones discussed above. This feature has been utilized in several natural product syntheses; Scheme 9.8 demonstrates

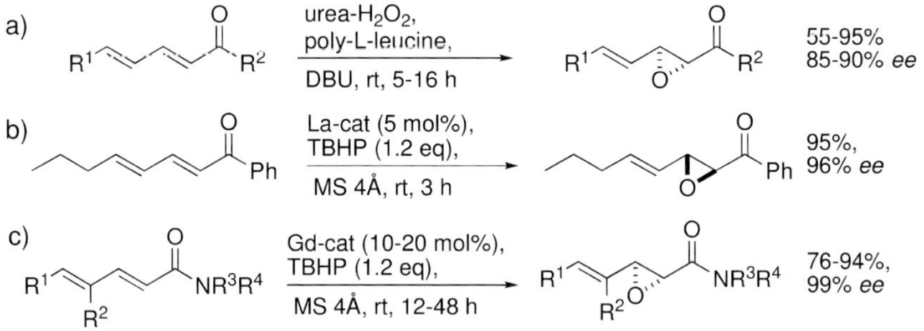

a) urea-H_2O_2, poly-L-leucine, DBU, rt, 5-16 h — 55-95%, 85-90% *ee*

b) La-cat (5 mol%), TBHP (1.2 eq), MS 4Å, rt, 3 h — 95%, 96% *ee*

c) Gd-cat (10-20 mol%), TBHP (1.2 eq), MS 4Å, rt, 12-48 h — 76-94%, 99% *ee*

Scheme 9.7 Asymmetric epoxidation of dienones.

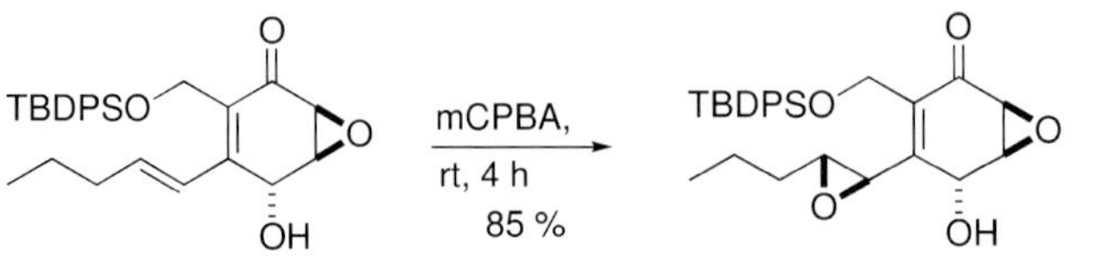

mCPBA, rt, 4 h, 85 %

Scheme 9.8 *m*CPBA epoxidation of a chiral dienone.

that *m*CPBA epoxidation can be highly regio- and diastereoselective when chiral substrates are employed [33].

9.1.2.2 From Dienols

The Sharpless asymmetric epoxidation (SAE), discussed further in Section 9.1.3, can also be applied to the epoxidation of conjugated dienols, to give a direct route to vinylepoxides (Scheme 9.9a) [34]. Although some substrates react readily, reduced reactivity is sometimes encountered, resulting in minor amounts of vinylic epoxyalcohol together with recovered dienol [11, 35]. Furthermore, the product that forms can be unstable: it can undergo a Payne rearrangement [34], for example, making it susceptible to further epoxidation (Scheme 9.9b) [35].

There are several examples of successful dienol epoxidations (Table 9.2). Catalytic SAE conditions are generally better than stoichiometric for reactive substrates (Entry 1), whilst stoichiometric conditions, on the other hand, are useful for less reactive substrates. Small variations in substrate structure can cause large differences in reactivity and product stability: pentadienol could be epoxidized in acceptable yield, whereas hexadienol gave a complex mixture of products (Entries 1, 2).

Table 9.2 Sharpless asymmetric epoxidation of dienols.

Entry	Product	Conditions	Yield (%)	*Ee* (%)	Ref.
1	 OH, O	stoichiometric	0	–	34
		catalytic	56	>91	36
2	 OH, O	catalytic	0	–	37
3	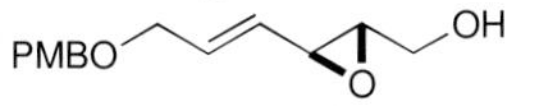 PMBO, OH, O	catalytic	67[a]	90	11
4	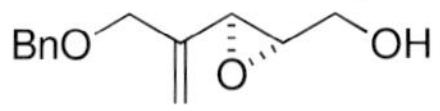 BnO, O, OH	pseudo-stoichiometric	89	97	38
5	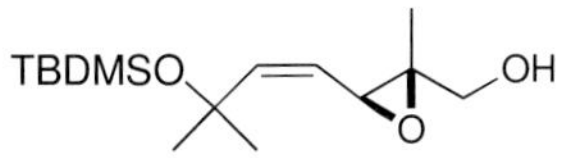 TBDMSO, OH, O	stoichiometric	75	95	39
6	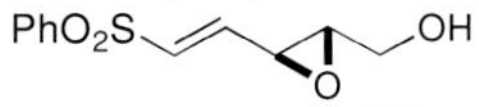 PhO_2S, OH, O	stoichiometric	85	95	40
7	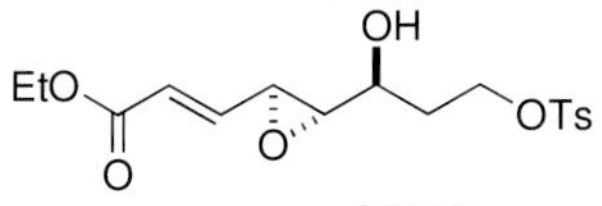 OH, EtO, O, O, OTs	stoichiometric	81–90	>95[b]	41
8	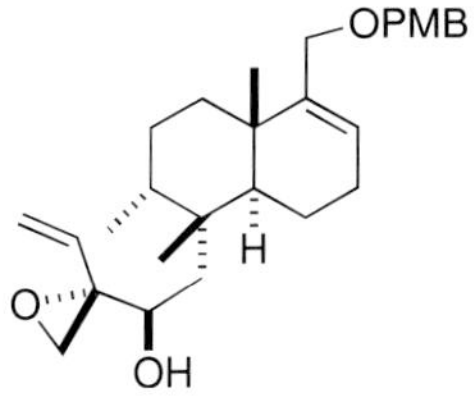 OPMB, H, O, OH	catalytic	88	86[b]	42

[a] Based on 37% recovered dienol. [b] Diastereomeric excess.

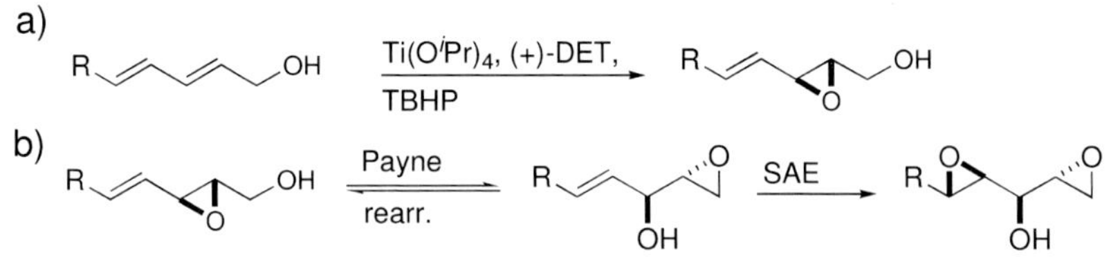

a) Ti(O^{i}Pr)$_4$, (+)-DET, TBHP

b) Payne rearr. SAE

Scheme 9.9 a) SAE on dienols. b) Payne rearrangement, resulting in byproducts.

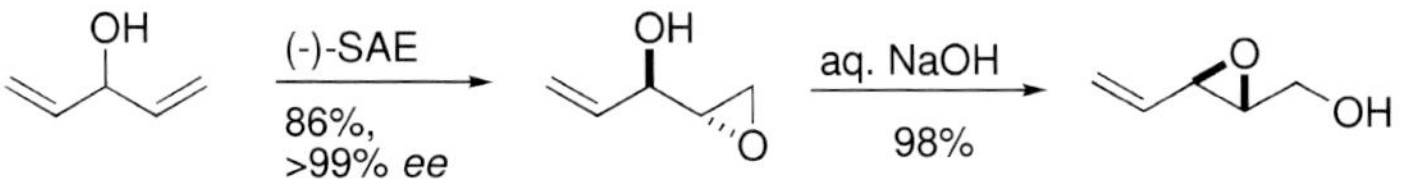

(-)-SAE 86%, >99% *ee*; aq. NaOH 98%

Scheme 9.10 Vinylepoxide synthesis from divinyl carbinol.

The epoxidation of divinyl carbinol constitutes a special case of a dienol epoxidation, as the starting diene is not conjugated (Scheme 9.10). Desymmetrization by SAE, followed by a Payne rearrangement, furnishes the vinylepoxide in high yield and with excellent enantioselectivity (compare Table 9.2, Entry 1) [43].

9.1.3 Vinylepoxides from Epoxy Alcohols

The development of Sharpless asymmetric epoxidation (SAE) of allylic alcohols in 1980 constitutes a breakthrough in asymmetric synthesis, and to date this method remains the most widely applied asymmetric epoxidation technique [34, 44]. A wide range of substrates can be used in the reaction: (*E*)-allylic alcohols generally give high enantioselectivity, whereas the reaction is more substrate-dependent with (*Z*)-allylic alcohols [34].

The classical method of making vinylepoxides thus consists of SAE followed by oxidation and Wittig olefination (Scheme 9.11) [45]. Oxidation of the epoxy alcohol was initially performed with PDC, but the Swern oxidation has become increasingly popular, thanks to its mild reaction conditions [46]. The aldehyde is subsequently subjected to Wittig or Horner-Emmons olefination to give the corresponding vinylepoxide. Drawbacks with this strategy include handling of the epoxy aldehydes, which are often unstable, volatile, or toxic. Furthermore the olefination can be low-yielding for non-stabilized ylides, as they are strong bases, and poor (*E*:*Z*)

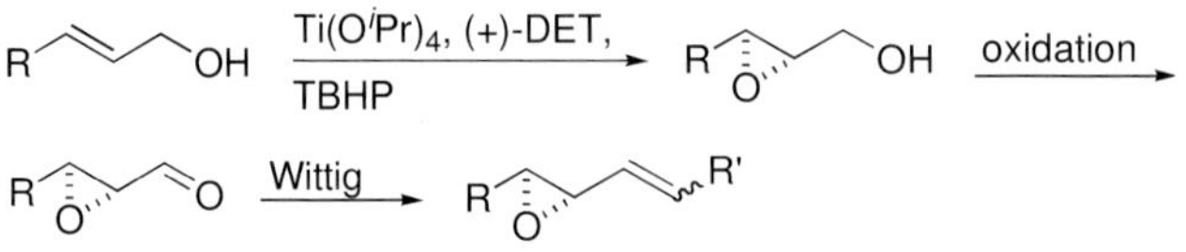

Ti(O^{i}Pr)$_4$, (+)-DET, TBHP; oxidation; Wittig

Scheme 9.11 SAE followed by oxidation and Wittig reaction.

Table 9.3 Synthesis of terminal vinylepoxides.

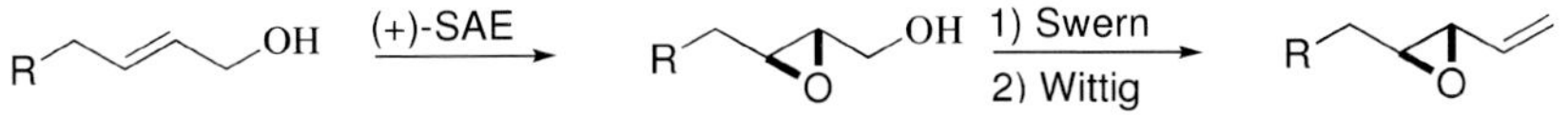

Entry	Product	SAE yield (%)	*Ee* (%)	Swern/Wittig yield (%)
1	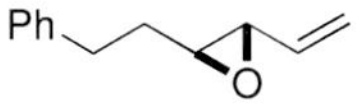	74	96	73
2		82	91	65
3	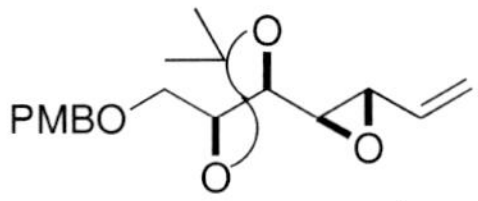	80	>90[a]	79
4	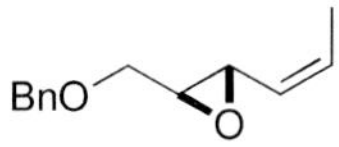	89	>95	78[b]
5	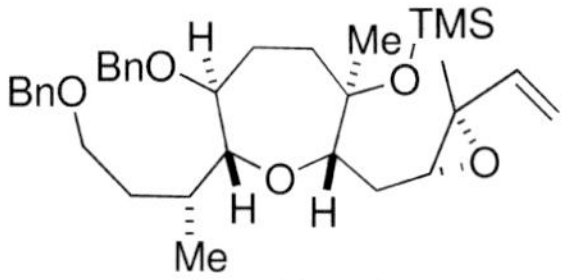	85	>95[a]	87
6	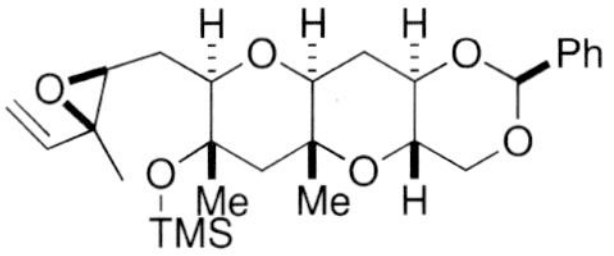	100	>95[a]	69[c]

[a] Diastereomeric excess. [b] *cis:trans* 11:1. [c] TPAP was used instead of Swern.

selectivity is often obtained [47]. Stabilized ylides usually give high yields of (*E*)-olefins [47].

In our work with aminolysis of vinylepoxides (see Section 9.2.1.1), the substrates were routinely synthesized by SAE followed by Swern/Wittig reactions (Table 9.3, Entries 1–4) [48, 49]. This procedure is well suited for terminal olefins, but disubstituted olefins can seldom be obtained with useful (*E*:*Z*) selectivities. Nakata recently synthesized some advanced intermediates towards natural products in this manner (Entries 5, 6) [50, 51].

Recently, several one-pot oxidation-Wittig procedures that circumvent the need to isolate the intermediate aldehydes have been developed. Various oxidants, including Swern [52], Dess-Martin periodinane [53], IBX [54], MnO_2 [55], and $BaMnO_4$ [56], can be used in the presence of stabilized ylides to generate α,β-unsaturated esters.

Because of the strongly basic conditions needed for Wittig reactions with *non*-stabilized ylides, one-pot reactions have not been carried out [57]. Ley's recent report of a TPAP-Wittig oxidation performed in a sequential one-pot manner is thus a promising alternative [58]. The reaction is fast and straightforward: the oxidation is performed with a stoichiometric amount of NMO and TPAP catalysis, after which the aldehyde is added to the preformed ylide (non-stabilized or stabilized). The olefins are formed in yields ranging from 60 to 94% without epimerization of any stereogenic centers. Although some of the one-pot reactions have been tested on sensitive substrates, they have not yet been applied in the synthesis of vinylepoxides.

9.1.4
Vinylepoxides from Aldehydes

The reaction between an aldehyde and a carbon nucleophile, such as a sulfur ylide, constitutes an alternative approach to the synthesis of epoxides. Since alkenes, which are the normal epoxidation substrates, are often formed from aldehydes, this approach can be highly efficient. On the other hand, the synthesis of appropriate carbon nucleophiles usually requires additional steps.

Racemic vinylepoxides can be synthesized from aldehydes in a variety of ways. Addition of a [(phenylthio)allyl]titanium reagent gives an *anti*-β-hydroxy thioether, which can be cyclized to the corresponding *trans*-vinylepoxide [59]. Chloroallyl zinc reagents can be employed to make *syn*-chlorohydrins, which give *cis*-vinylepoxides upon treatment with base (see also Section 9.1.4.1) [60, 61]. α-Halo-aldehydes or -ketones can be vinylaluminated, and the formed halohydrin ring-closed to give 2-substituted vinylepoxides [62]. Tetrasubstituted vinylepoxides can be formed from ketones in a three-step procedure involving Grignard addition, allylic bromination, and ring-closure [63]. Two recent reports on addition of diazoacetates to aldehydes have emerged (compare Scheme 9.15). Rhodium(II) acetate catalyzes this addition, giving trisubstituted vinylepoxides in good yields [64, 65].

9.1.4.1 Chloroallylboration

Asymmetric synthesis of *cis*-vinylepoxides is difficult, as most epoxidation methods give moderate enantioselectivities with (*Z*)-alkenes (see Sections 9.1.1.2 and 9.1.3). Oehlschlager's strategy towards *cis*-vinylepoxides is based on the generation of *syn*-α-vinylchlorohydrins **10**, which can be ring-closed to form epoxides in good yields (Scheme 9.12) [66]. The sequence starts with formation of chiral α-chloroallylborane **11**, which undergoes a facile [1,3]-rearrangement upon treatment with $BF_3 \cdot OEt_2$ to form the γ-chloroallylborane **12**. Addition of an aldehyde, followed by removal of the chiral borane, affords *syn*-α-chlorohydrins **10** in 90–99% *ee*s. Base-induced cyclization of **10** furnishes *cis*-vinylepoxides **13**, which can also be formed directly from **14** by oxidation under basic conditions.

Aliphatic, aromatic and vinylic aldehydes can be employed in this reaction with similar yields and enantioselectivities. When chiral aldehydes are utilized, excellent diastereoselectivity is obtained for matched cases, whereas mismatched cases yield products with moderate to good diastereoselectivity (Scheme 9.13a) [67]. The limitation of the methodology is that only terminal vinylepoxides can be obtained.

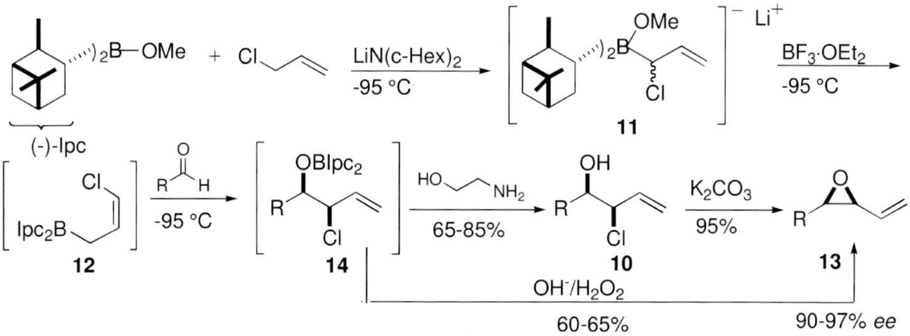

Scheme 9.12 Chloroallylboration of aldehydes.

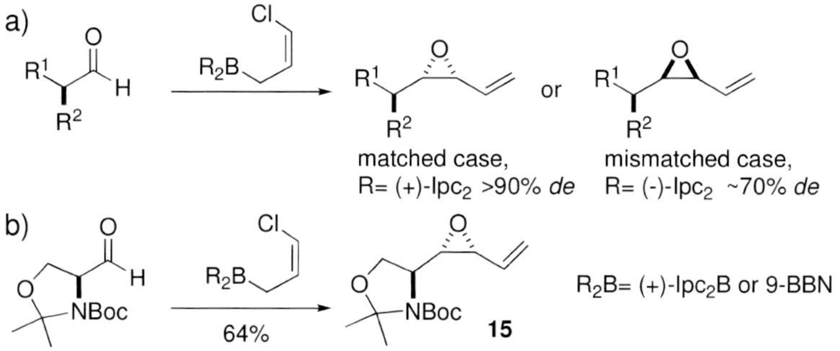

Scheme 9.13 Chloroallylboration of chiral aldehydes.

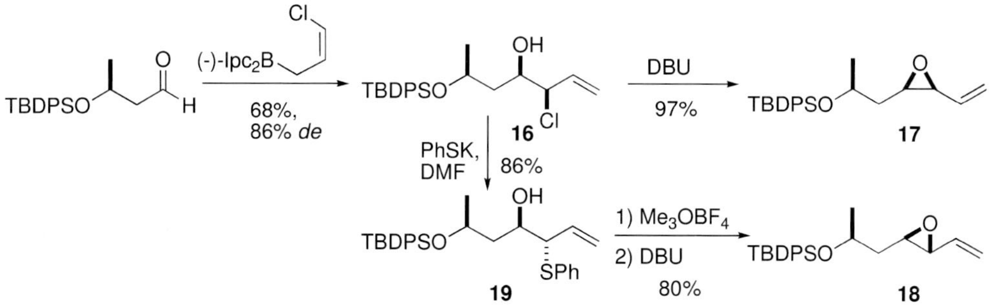

Scheme 9.14 Synthesis of *cis*- and *trans*-vinylepoxides from aldehydes.

Boland applied this methodology to Garner's aldehyde, and found the addition to be substrate-controlled rather than reagent-controlled (Scheme 9.13b) [68]. Vinylepoxide **15** could thus also be obtained with high diastereoselectivity with achiral 9-MeO-9-BBN.

Winssinger later utilized *syn*-α-vinylchlorohydrin **16** for the generation of both *cis*- and *trans*-vinylepoxides **17** and **18** (Scheme 9.14) [69]. *cis*-Epoxide **17** was formed by cyclization of **16** under basic conditions, whereas S_N2 displacement of the chloride in **16** with thiophenoxide resulted in compound **19**, which could be cyclized to *trans*-epoxide **18**.

9.1.4.2 Reaction with Sulfur Ylides

The sulfur ylide-mediated epoxidation of aldehydes has been thoroughly investigated [70, 71]. The chiral sulfur ylides reported by Aggarwal have been most broadly applicable, and a catalytic, asymmetric process yielding aromatic *trans*-epoxides has been developed [72]. In this process, the sulfur ylides are produced *in situ* from diazo compounds, generated in turn from tosylhydrazone salts (Scheme 9.15) [73].

The aldehyde structures and the tosylhydrazone salts were varied in an extensive study of scope and limitations, with use of both achiral and chiral sulfur ylides [73]. Aromatic aldehydes were excellent substrates in the reaction with benzaldehyde-derived ylides, whereas aliphatic aldehydes gave moderate yields and *trans:cis* ratios.

When α,β-unsaturated aldehydes were employed, vinylepoxides were obtained with excellent *trans:cis* ratios but in poor yields. When benzaldehyde was treated with α,β-unsaturated tosylhydrazone salts, the yields of vinylepoxides were improved but the *trans:cis* ratios dropped. When chiral sulfides were utilized, the *ees* were high with α,β-unsaturated aldehydes, whereas unsaturated, chiral sulfur ylides gave moderate *ees*, poor yields, and modest *trans:cis* ratios.

An alternative process for the synthesis of vinylepoxides was clearly needed, so reactions with stoichiometric amounts of chiral sulfide were investigated (Scheme 9.16a) [74]. Indeed, when benzyl sulfonium salt **20** was treated with unsaturated aldehydes, the *ees* and *des* were high in all cases, whereas the yields [75] were highly substrate-dependent. The same products could be formed by treatment of an unsaturated sulfonium salt with benzaldehyde, but the yields and selectivities were generally slightly lower.

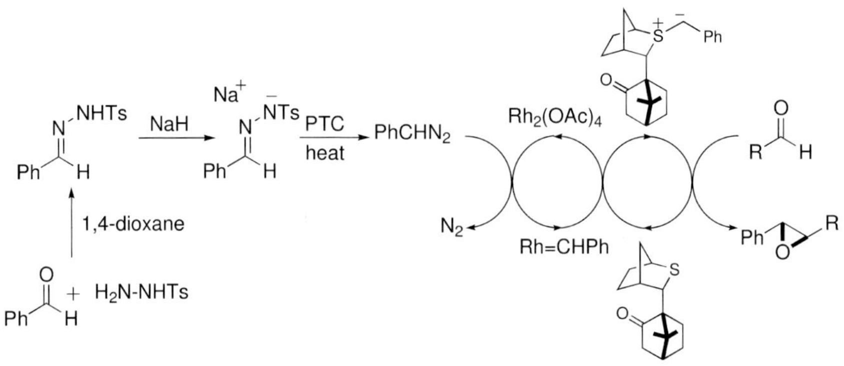

Scheme 9.15 Aggarwal's catalytic asymmetric epoxidation.

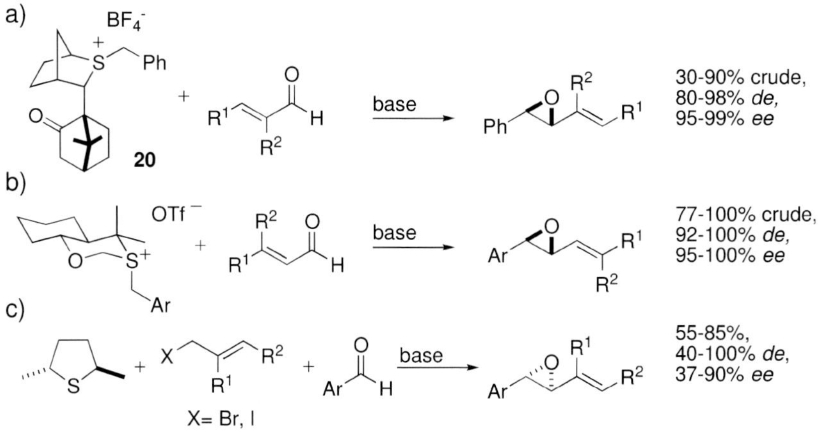

Scheme 9.16 Stoichiometric epoxidation with chiral sulfides.

Chiral sulfonium salts derived from oxathianes have been developed for stoichiometric epoxidation reactions. The sulfonium salts were deprotonated and allowed to react with α,β-unsaturated aldehydes to give *trans*-vinylepoxides with excellent *ees* and *trans*:*cis* ratios (Scheme 9.16b) [76]. The yields were generally high [75], and the best results were obtained with Ar = 4-OMePh.

Metzner and co-workers reported a one-pot epoxidation reaction in which a chiral sulfide, an allyl halide, and an aromatic aldehyde were allowed to react to give a *trans*-vinylepoxide (Scheme 9.16c) [77]. This is an efficient approach, as the sulfonium salt is formed *in situ* and deprotonated to afford the corresponding ylide, and then reacts with the aldehyde. The sulfide was still required in stoichiometric amounts, however, as the catalytic process was too slow for synthetic purposes. The yields were good and the *trans*:*cis* ratios were high when $R_1 \neq H$, but the enantioselectivities were lower than with the sulfur ylides discussed above.

The major limitation of asymmetric sulfur ylide epoxidations is that only aromatic vinylepoxides can be formed efficiently and with high selectivity. When an aliphatic aldehyde is allowed to react with a semistabilized or nonstabilized sulfur ylide, poor diastereoselectivities and yields are observed, due to problems in controlling the ylide conformation and competing ylide rearrangement reactions [71]. However, some racemic, aliphatic vinylepoxides have been successfully formed by sulfur ylide epoxidations, although varying diastereoselectivities were observed [78–80].

9.1.5
Vinylepoxides from Other Substrates

9.1.5.1 From Allenes

2,3-Allenols can be converted into terminal vinylepoxides by a Pd(0)-catalyzed insertion of aryl or vinyl halides [81]. The reactions take place with high *trans*-selectivity and in good yields (Scheme 9.17a). Chiral 2,3-allenols, which can be easily

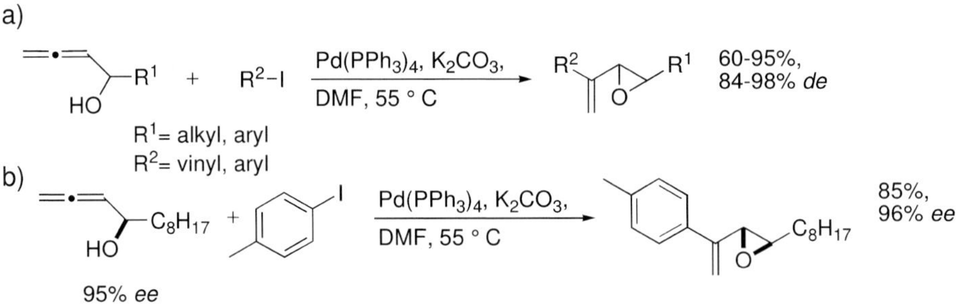
a)
R1
+
R2–I
HO
Pd(PPh3)4, K2CO3,
DMF, 55 ° C
R2
R1
O
60-95%,
84-98% de
R1= alkyl, aryl
R2= vinyl, aryl
b)
C8H17
+
HO
I
Pd(PPh3)4, K2CO3,
DMF, 55 ° C
C8H17
O
85%,
96% ee
95% ee

Scheme 9.17 Pd(0)-catalyzed epoxidation of allenes.

synthesized in one step from aldehydes [82], can be employed to obtain vinylepoxides in excellent enantiomeric excesses (Scheme 9.17b).

9.1.5.2 **Kinetic Resolution of Racemic Epoxides**

Asymmetric epoxidation of terminal olefins has remained problematic, despite the general success of the novel dioxirane-based catalysts. The enantiomeric excesses in these reactions do not usually exceed 85% (see Section 9.1.1.1). As recrystallization of epoxides can be complicated, enantiopure terminal epoxides are difficult to obtain.

One way of overcoming these problems is by kinetic resolution of racemic epoxides. Jacobsen has been very successful in applying chiral Co-salen catalysts, such as **21**, in the kinetic resolution of terminal epoxides (Scheme 9.18) [83]. One enantiomer of the epoxide is converted into the corresponding diol, whereas the other enantiomer can be recovered intact, usually with excellent *ee*. The strategy works for a variety of epoxides, including vinylepoxides. The major limitation of this strategy is that the maximum theoretical yield is 50%.

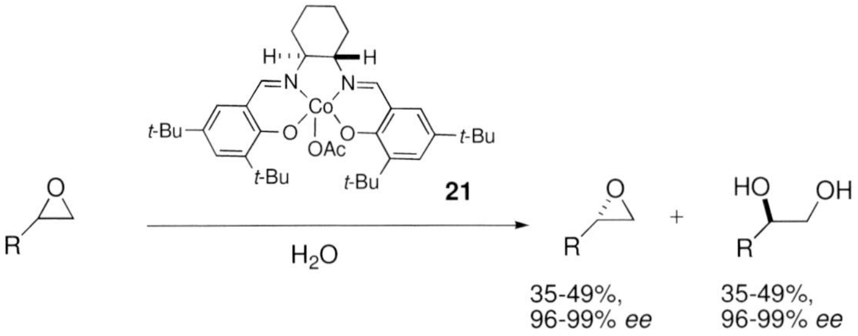
H
H
N
N
Co
O
O
OAc
t-Bu
t-Bu
t-Bu
t-Bu
21
O
R
H2O
O
R
+
HO
OH
R
35-49%,
96-99% ee
35-49%,
96-99% ee

Scheme 9.18 Kinetic resolution of terminal epoxides.

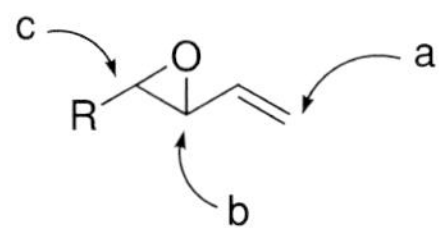

Figure 9.1 Positions for nucleophilic attack on a vinylepoxide.

9.2 Transformations of Vinylepoxides

Vinylepoxides are useful electrophiles that can react with a range of nucleophiles. Since they are a subset of allylic electrophiles, nucleophilic addition can take place through S_N2 and S_N2' attack (Figure 9.1), and the ability to control the regioselectivity is critical. Normally, soft nucleophiles prefer the S_N2' reaction variant (Route a), while hard ones participate in S_N2 attacks at the allylic position (Route b) [84], although several exceptions are known. In these nucleophilic attacks the vinyl moiety functions as a regiochemical directing element, and attack at C-4 (Route c) is not normally observed. As the epoxide moiety is a small strained-ring system, it is perhaps not surprising that vinyloxiranes participate in several useful rearrangement reactions, which are also discussed.

9.2.1 Intermolecular Opening with Oxygen and Nitrogen Nucleophiles

The regiochemistry of ring-opening of vinylepoxides with heteronucleophiles can be controlled by selecting appropriate reaction conditions. The 1,2-addition product is normally formed when the reaction is performed in the presence of Lewis or protic acids, while Pd(0)-mediated ring-openings generally favor the 1,4-product. A deviation from this selectivity pattern is found in the Pd(0)-catalyzed [2+3] additions of isocyanates, carbodiimides, and CO_2 to vinyloxiranes to give cyclic derivatives of 1,2-amino alcohols and diols.

9.2.1.1 1,2-Additions

Vinyloxiranes can be opened regio- and stereoselectively by alcohols in the presence of catalytic amounts of Lewis acids. $BF_3 \cdot Et_2O$ has proven to be the best catalyst, affording the corresponding 3-alkoxy-4-hydroxy olefins rapidly and in good yields when equivalent amounts of the epoxide and alcohol are mixed [85, 86]. Similar results have been obtained with catalytic amounts of $[Rh(CO)_2Cl]_2$, although the exact role of the catalyst is unclear [87, 88]. Both these ring-openings proceed with inversion of configuration at the allylic stereocenter, and this can also be accomplished by Lewis acid-mediated addition of acetone, directly affording the corresponding 1,3-dioxolane [89]. In contrast, Pd(0)-catalyzed opening of vinylepoxides followed by treatment with CO_2 [90] or butylboronic acid [91] affords the corresponding cyclic carbonate or boronate, respectively, with retention of configuration (Scheme 9.19). A key feature in both reactions is that the pronucleophiles

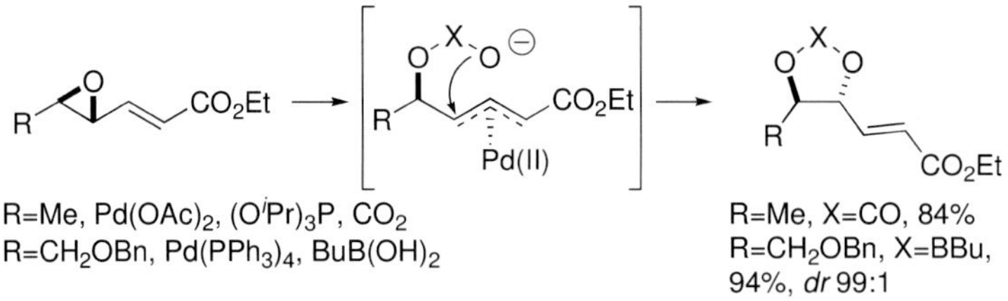

R=Me, $Pd(OAc)_2$, $(O^iPr)_3P$, CO_2
R=CH_2OBn, $Pd(PPh_3)_4$, $BuB(OH)_2$

R=Me, X=CO, 84%
R=CH_2OBn, X=BBu, 94%, *dr* 99:1

Scheme 9.19 Palladium-catalyzed opening of vinyloxiranes.

$Pd_2(dba)_3{\cdot}CHCl_3$ (1 mol%), **23** (3 mol%), Et_3B (1 mol%), PMBOH (1.1 equiv.)
69%, 98% *ee*

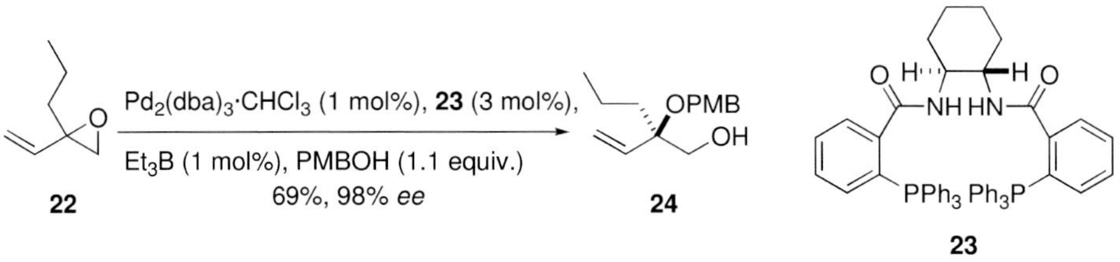

Scheme 9.20 Dynamic kinetic resolution of racemic epoxide **22**.

become anchored to the alkoxide generated by addition of Pd(0) to the oxirane, followed by intramolecular addition of the oxygen nucleophiles. This feature explains the observed regio- and stereochemical outcome.

Of particular interest is the recently described Pd(0)-catalyzed dynamic kinetic asymmetric transformation of racemic vinylepoxides (Scheme 9.20) [92]. Treatment of **22** with *p*-methoxybenzyl alcohol (PMBOH) with catalysis by $Pd_2(dba)_3 \cdot CHCl_3$, ligand **23**, and co-catalyst Et_3B, for example, gave **24** in 69% yield and with 98% *ee* (Scheme 9.20). Compound **24** was used as an intermediate in a synthesis of the nonpeptidic protease inhibitor tipranavir. Several different types of alcohols have been successfully employed in this reaction, together with butadiene monoepoxide, isoprene monoepoxide, or their derivatives [93]. For this process to be successful it is necessary that the equilibration of the diastereomeric π-allylpalladium complexes – formed by reaction between Pd(0)-**23** and **22** – should be faster than the reaction of these complexes with the nucleophile, thus allowing for a dynamic equilibrium. The role of the co-catalyst is to control the regiochemistry by acting as an intramolecular nucleophile transfer agent. Clearly, this reaction offers an efficient route from simple starting materials to attractive chiral intermediates [94].

Racemic terminal vinyloxiranes can also be kinetically resolved with water and the chiral (salen)Co^{III} complex **21**, as already shown in Scheme 9.18 [95].

The regioselective ring-opening of vinyloxiranes by nitrogen nucleophiles offers an attractive route to *vic*-amino alcohols, compounds of much recent interest. As with oxygen nucleophiles, the stereochemistry of the reaction can be controlled by choice of reaction conditions: aminolysis of **25**, for example, affords *anti*-amino alcohol **26** in excellent yield and diastereoselectivity (Scheme 9.21) [48, 96, 97], and

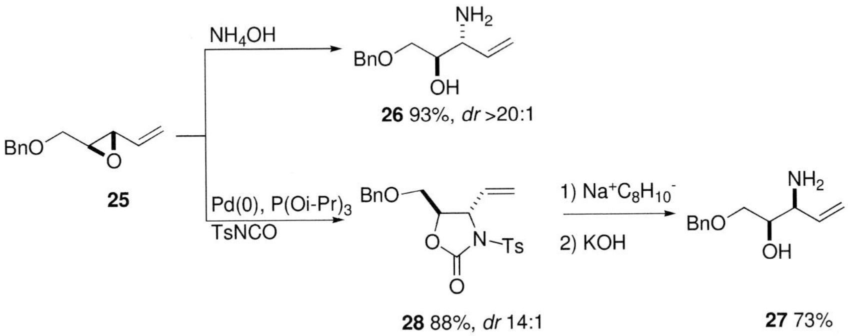

Scheme 9.21 Divergent opening of vinyloxirane **25**.

primary and aromatic amines have also been shown to be good nucleophiles in this reaction [87, 98]. The regioselectivity is usually excellent, unless the allylic carbon is quaternary, in which case a mixture of the corresponding 3,4- and 4,3-amino alcohols is obtained. Diastereomer **27** can be obtained from the same starting material by Pd(0)-mediated opening of **25** in the presence of TsNCO to give *N*-tosyl oxazolidinone **28** with retention of configuration [96, 99]. The regiochemistry is dictated by the reaction between TsNCO and the intermediate alkoxide generated by addition of Pd(0) to **25**, while the stereochemistry is the result of two consecutive S_N2 reactions [100]. Global deprotection of **28** gives *syn*-amino alcohol **27**. The approach summarized in Scheme 9.21 thus allows stereodivergent access to vinylic *vic*-amino alcohols.

Trost's group has also developed a Pd(0)-catalyzed dynamic kinetic asymmetric transformation of racemic terminal vinyloxiranes with nitrogen nucleophiles [101, 102]. For this reaction to be successful it is important that the pronucleophile should contain an acidic *N*-H bond so as to promote hydrogen bonding to the oxygen leaving group. With phthalimide as the pronucleophile and butadiene monoepoxide as the substrate, the Pd(0)-catalyzed transformation afforded the corresponding vinylglycinol derivative in high yield and with 98% *ee*.

9.2.1.2 **1,4-Additions**

1,4-Additions of oxygen and nitrogen nucleophiles to vinyloxiranes can be achieved with Pd(0) catalysis [103, 104]. Acetate, silanols, amines, sulfonamides, and azide have been used as nucleophiles, and the stereochemical outcome of these additions, where applicable, is normally the result of two consecutive S_N2 reactions. This is demonstrated by the additions of NaNHTs to vinylepoxides **29** and **30**, affording *syn*- and *anti*-amino alcohols **31** and **32**, respectively, in good yields and with high diastereoselectivities (Scheme 9.22) [105].

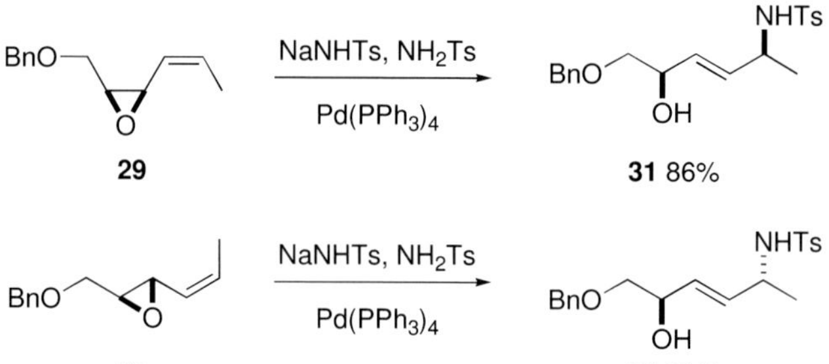

Scheme 9.22 Pd(0)-catalyzed 1,4-addition of *p*-toluenesulfonamide.

9.2.2 Intramolecular Opening with Oxygen and Nitrogen Nucleophiles

The regio- and stereoselective intramolecular opening of vinylepoxides provides an efficient route to tetrahydropyran and oxepane systems and has successfully been applied in several syntheses of natural products, a pertinent example being that of the neurotoxic marine polycyclic ether brevetoxin B [106–108]. Cyclization of hydroxy epoxide **33a** under acidic conditions results in tetrahydrofuran **34a** as the sole isomer in excellent yield (Table 9.4, Entry 1). This is the result of a favored 5-*exo* ring-closure [109, 110], in accordance with Baldwin's rules [111]. Cyclization of vinylepoxide **33b**, however, gave a mixture of **34b** and **35b**, in which the product derived from the 6-*endo* ring-closure predominated (Entry 2). This trend was also manifest in the reactions of **33c** and **33d**, which both gave the corresponding tetrahydropyrans **35c** and **35d** as the exclusive products (Entries 3 and 4). The shift from the expected 5-*exo* reaction to the 6-*endo* variant is clearly due to the in-

Table 9.4 Intramolecular opening of epoxide **33**.

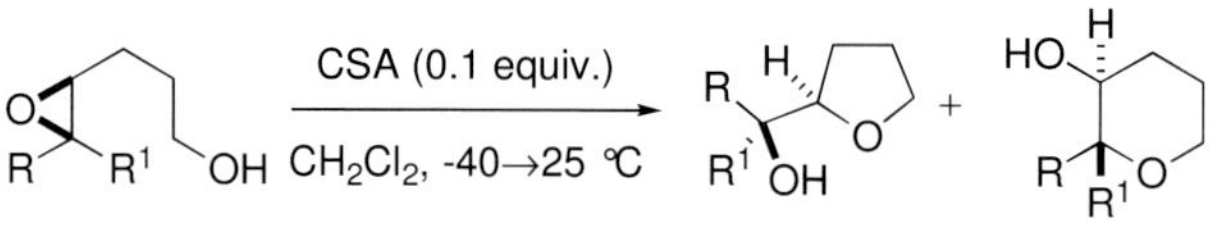

Entry	33	Ratio 34/35	Yield (%)
1	**a** R = $CH_2CH_2CO_2Me$, R^1 = H	1000:0[a]	94
2	**b** R = (*E*)-CH=$CHCO_2Me$, R^1 = H	40:60	96
3	**c** R = CH=CH_2, R^1 = H	0:100	90
4	**d** R = CH=CH_2, R^1 = Me	0:100	96

[a] Isolated as the corresponding γ-lactone.

Table 9.5 Intramolecular opening of epoxide **36**.

CSA (0.1 equiv.)
CH_2Cl_2, -40→25 °C
36 **37** **38**

Entry	36	Ratio 37/38	Yield (%)
1	**a** R = $CH_2CH_2CO_2Me$	0:100[a]	70
2	**b** R = (*E*)-CH=$CHCO_2Me$	22:78	75
3	**c** R = CH=CH_2	82:18	75
4	**d** R = (*E*)-CH=CHCl	92:8	75

[a] Isolated as the corresponding γ-lactone.

corporation of a vinyl moiety in **33b-d**. In the oxonium ions, formed by protonation of these species, the π-system can stabilize an electron-deficient orbital at the allylic position, thus promoting the formation of tetrahydropyrans. It is also clear from Table 9.4 that the more electron-rich π-systems have stronger effects than electron-deficient olefins on directing the kinetic cyclizations to the 6-*endo* reaction pathway.

In the cyclization of the corresponding *cis*-epoxides, with the aim of obtaining the corresponding *cis*-2,3-disubstituted tetrahydropyrans, a similar trend was observed. For these systems, however, the 6-*endo* pathway was less favored, which was ascribed to difficulties in attaining a TS conformation that would allow for maximum stabilization of the developing p-orbital with the adjacent π-system. Alternatively, palladium-catalyzed cyclization of the tetrabutylammonium alkoxide derived from **33b** results in the corresponding *cis*-2,3-disubstituted tetrahydropyran in excellent yield and selectivity (90%, *dr* >99:1), while the *cis*-epoxide gives stereoisomer **37b** (86%, *dr* 98:2) [112].

The cyclization of the homologous epoxide **36** under acidic conditions was also investigated (Table 9.5) [110]. As would be expected, compound **36a** reacted by a 6-*exo* cyclization to give tetrahydropyran **38a** (Entry 1). The α,β-unsaturated hydroxy epoxide **36b** gave a 1:3.5 mixture of oxepane **37b** and tetrahydropyran **38b** (Entry 2). Subjection of **36c** and **36d**, which both contain more electron-rich π-systems, to the reaction conditions resulted in preferential 7-*endo* cyclization to give **37c** and **37d**, thus confirming the powerful regiodirecting effect of the vinyl moiety (Entries 3 and 4).

In order to improve the selectivities in these reactions, styrylepoxides **39** and **40** were prepared and subjected to acid-catalyzed cyclizations (Scheme 9.23) [113]. Both compounds, however, afforded almost identical mixtures of diastereomeric oxepanes **41** and **42** (83%, *dr* 21:79 and 66%, *dr* 20:80, respectively), while no tetrahydropyran **43** was formed. Clearly, the additional stabilization provided by

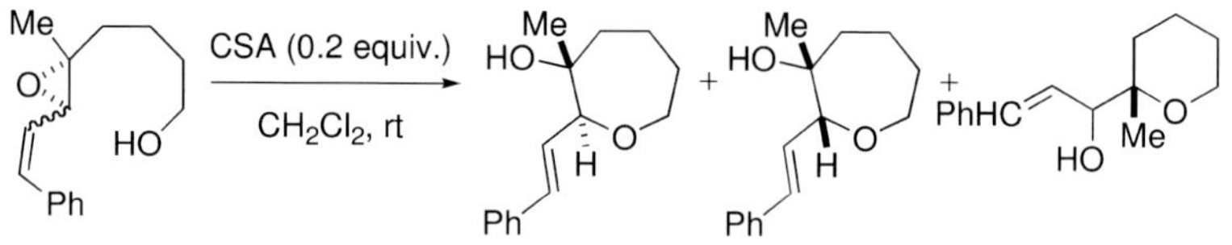

39 (*trans*)
40 (*cis*)
41
42
43

Scheme 9.23 Acid-catalyzed opening of epoxides **39** and **40**.

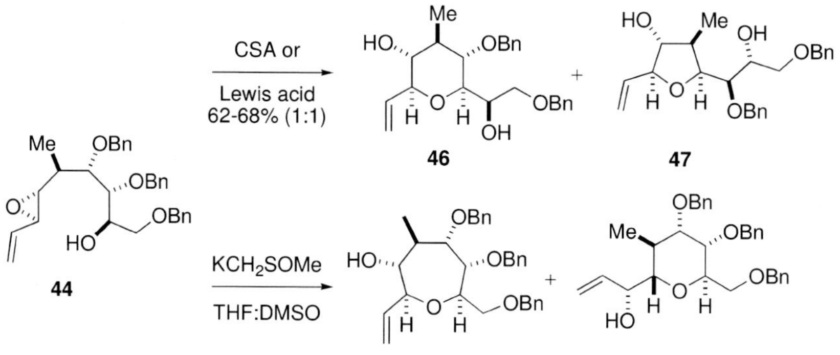

44
46
47
45 (27%)
48 (57%)

Scheme 9.24 Cyclization of epoxide **44**.

the phenyl group promotes the cyclization through an allylic cation-like intermediate in which bond rotation occurs before the desired 7-*endo* cyclization ensues.

The outcomes of intramolecular cyclizations of hydroxy vinylepoxides in more complicated systems can be difficult to predict. In a study of the synthesis of the JKLM ring fragment of ciguatoxin, epoxide **44** was prepared and subjected to acid-mediated cyclization conditions (Scheme 9.24) [114]. Somewhat surprisingly, the expected oxepane **45** was not formed, but instead a mixture of tetrahydropyran **46** and tetrahydrofuran **47** was obtained, both compounds products of attack of the C6 and C5 benzyl ether oxygens, respectively, on the allylic oxirane position (C3). Repetition of the reaction with dimsylpotassium gave a low yield of the desired **45** along with considerable amounts of tetrahydropyran **48**.

This result should be compared to the cyclization of the more advanced intermediate **49** (4:1 mixture of isomers at the acetal position; Scheme 9.25), which afforded the bicyclic oxepane **50** as the sole product in excellent yield. It was suggested that this result was due to the presence of a *trans*-fused five-membered ring, which restricts the rotation around the C7–C8 bond and thus promotes the 7-*endo* cyclization.

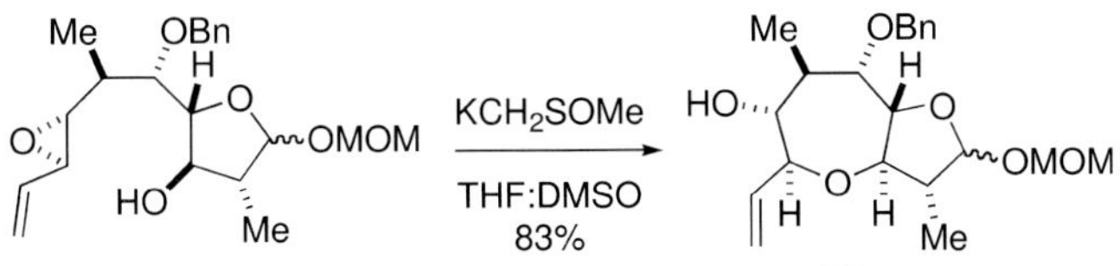

Scheme 9.25 Cyclization of epoxide **49**.

9.2.3
Opening with Carbon Nucleophiles

Addition of carbon nucleophiles to vinylepoxides is of particular importance, since a new carbon-carbon bond is formed. It is of considerable tactical value that conditions allowing for regiocontrolled opening of vinyloxiranes with this type of nucleophiles have been developed. Reactions that proceed through formation of a π-allyl metal intermediate with subsequent external delivery of the nucleophile, or that make use of a soft carbon nucleophile, generally deliver the S_N2' product. In contrast, the S_N2 variant is often the major reaction pathway when hard nucleophiles are employed. In some methods a nucleophile can be delivered selectively at either the S_N2 or S_N2' positions by changing the reaction conditions.

9.2.3.1 S_N2' Additions

1,4-Additions of carbon nucleophiles to vinylepoxides are well documented and can be accomplished by several different techniques. Palladium-catalyzed allylic alkylation of these substrates with soft carbon nucleophiles (pK_a 10–20) proceeds under neutral conditions and with excellent regioselectivities [103, 104]. The sulfone **51**, for example, was cyclized through the use of catalytic amounts of $Pd(PPh_3)_4$ and bis(diphenylphosphino)ethane (dppe) under high-dilution conditions to give macrocycle **52**, an intermediate in a total synthesis of the antitumor agent roseophilin, in excellent yield (Scheme 9.26) [115, 116].

The use of nonstabilized carbon nucleophiles in this reaction has been rare. Recently, however, it was shown that lithium ester enolates participate in Pd-catalyzed 1,4-additions to cyclic and acyclic vinyloxiranes, affording the corresponding 6-hydroxy-4-enoates in good yields and with complete regioselectivity [117, 118].

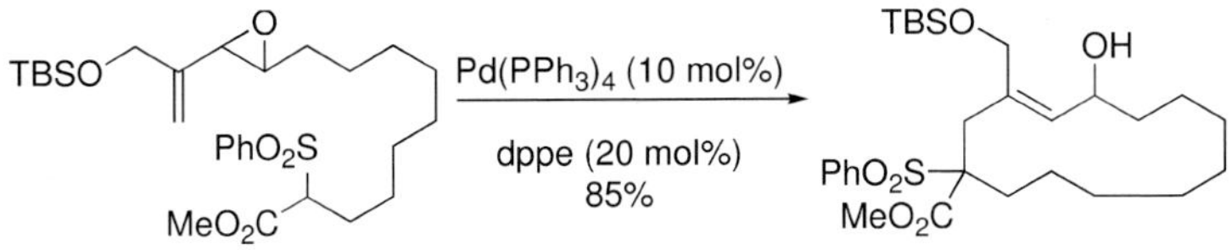

Scheme 9.26 Pd-catalyzed cyclization of compound **51**.

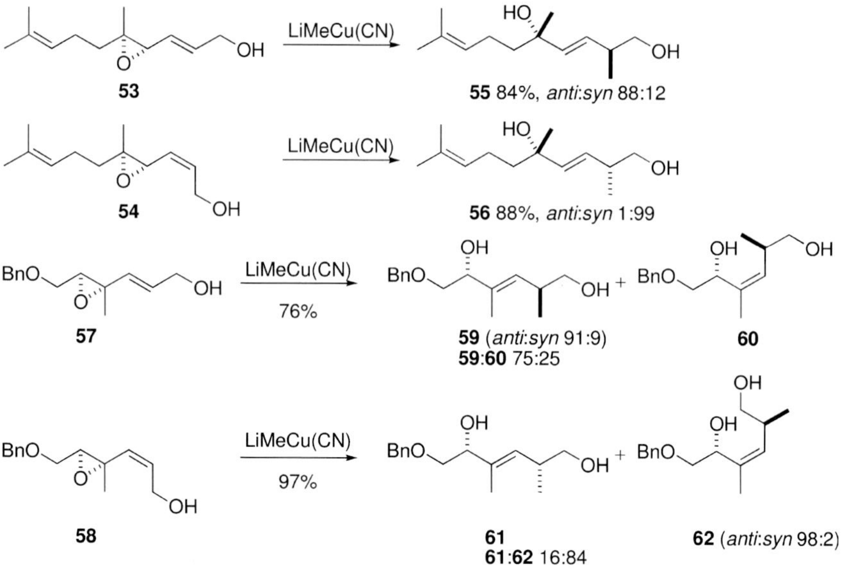

Scheme 9.27 Addition of organocuprates to vinylepoxides.

S_N2'-addition of organocopper reagents to vinyloxiranes is a complement to the Pd(0)-mediated methodology [119]. With cyclic 1,3-diene monoepoxides the reactions generally proceed in high yields and with good to excellent regioselectivities, affording the corresponding *anti*-S_N2' adducts. It should be noted, however, that the regioselectivity can be influenced both by steric hindrance in the substrate and by choice of organocuprate reagent. The situation in acyclic systems is more complicated. With substituted acyclic vinylepoxides the S_N2' pathway is usually strongly favored, although exceptions are known. Addition of LiMeCu(CN) to geranyl derivatives **53** and **54** affords compounds **55** and **56**, respectively, in good yields and with high diastereoselectivities (Scheme 9.27) [120]. In both reactions only the (*E*)-S_N2' products are detected and the stereochemical outcome is interpreted, for stereoelectronic reasons, as the result of *anti* additions to the corresponding vinylepoxide *s-trans* conformers. Addition to the trisubstituted derivatives **57** and **58** results in a more complex product distribution [121]: only the S_N2' products can be detected in both reactions and good *anti*:*syn* ratios are achieved, but the *E*:*Z* selectivities are somewhat diminished. The S_N2' additions of organocopper reagents normally proceed with *anti* stereoselection, which is believed to be due to favorable interactions between a filled d^{10} orbital of the reagent and σ^* and π^* orbitals on the substrate [122]. A qualitative rationale for these seemingly puzzling results, based on steric interactions in the reactant-like conformers, has been presented [119].

Cu(I)-catalyzed kinetic resolutions of racemic, cyclic 1,3-diene monoepoxides through the use of dialkylzinc [123] or trialkylaluminium reagents [124] have recently been described, although the efficiencies of these processes are at present modest.

Intramolecular Sm(II)-mediated coupling of ketones with distal vinylepoxides results in the corresponding carbocyclic derivatives with good to excellent diastereoselectivity [125]. When the reaction was conducted with enantioenriched substrates the products were obtained with poor *ee*s, indicating low π-facial discrimination in this reaction.

9.2.3.2 S_N2 Additions

The S_N2 reaction variant is often the major pathway when hard nucleophiles are employed. γ,δ-Epoxy acrylates can be opened regio- and stereoselectively at the γ-position with $AlMe_3/H_2O$ (Scheme 9.28) [126], and this method has been used for iterative construction of polypropionate chains and quaternary stereocenters [127]. Interestingly, water is essential for this reaction, probably due to the generation of monomeric or dimeric Al–O systems with enhanced Lewis acidity, and its exclusion results in prolonged reaction times [128]. Vinylepoxides lacking an alkoxy substituent in the C-6 or C-7 position are poor substrates in this reaction. Similar regioselectivities have been obtained with trialkylzincates [129] and with organolithium reagents in the presence of $BF_3 \cdot OEt_2$ [130].

As already seen in Section 9.2.1.1, dynamic kinetic resolution of racemic epoxides with oxygen or nitrogen nucleophiles can be highly efficient. In these cases the regiochemical outcomes were controlled by intramolecular hydrogen bonding or by use of Et_3B as a transfer agent for the nucleophile. Such strategies are not applicable with carbon nucleophiles, but it was recently shown that, when racemic epoxide **63** was treated with $Pd(dba)_3$ and ligand **23** in the presence of pronucleophile **64**, adduct **65** could be obtained in high yield and with excellent *ee* (Scheme 9.29) [131, 132]. The regiochemical outcome of this reaction was explained in terms of steric shielding of ligand **23** in the reacting π-allylpalladium complex.

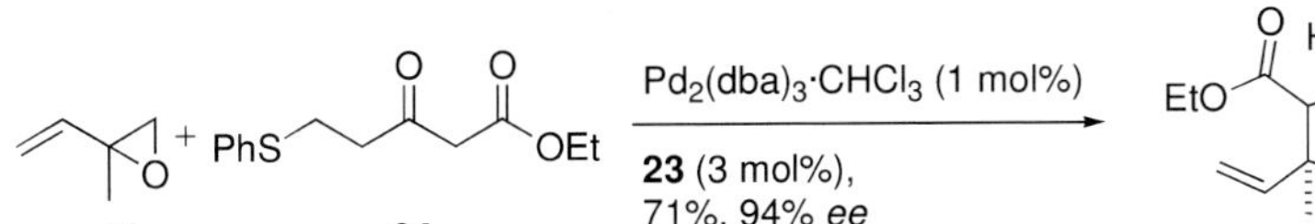

BnO ... CO_2Et → Me_3Al (10 equiv.), H_2O (6 equiv.), 96% → BnO ... OH ... CO_2Et

Scheme 9.28 Regioselective opening of a vinylepoxide with Me_3Al.

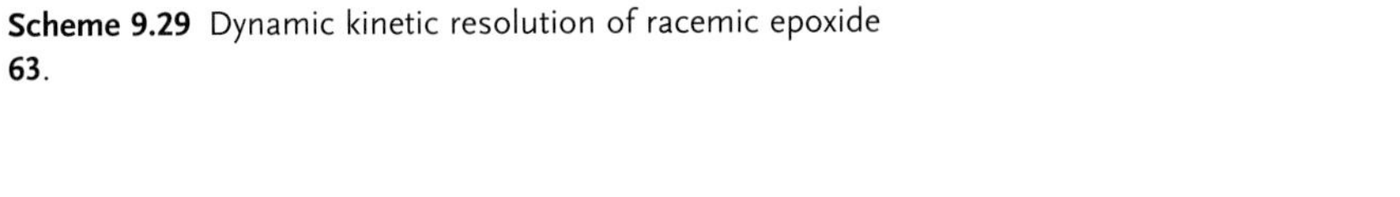

63 + PhS ... OEt **64** → $Pd_2(dba)_3 \cdot CHCl_3$ (1 mol%), **23** (3 mol%), 71%, 94% *ee* → EtO, HO, SPh, O **65**

Scheme 9.29 Dynamic kinetic resolution of racemic epoxide **63**.

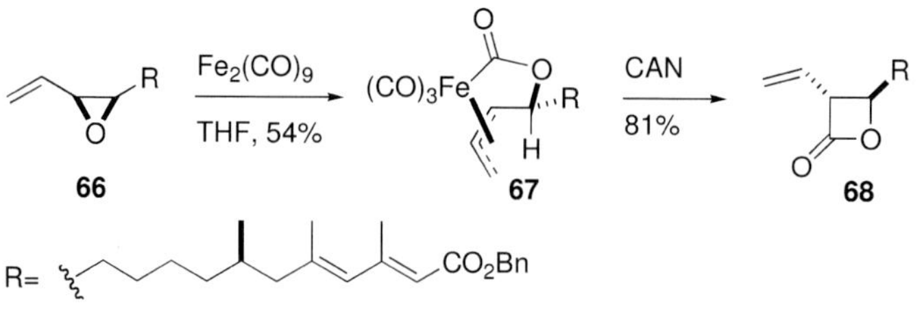

Scheme 9.30 Conversion of vinylepoxide **66** into β-lactone **68**.

Vinyloxiranes can also be converted into β-lactones (Scheme 9.30) [133, 134]. Opening of **66** with $Fe_2(CO)_9$ resulted in the (π-allyl)tricarbonyliron derivative **67** in good yield, together with a minor diastereomer (not shown). Oxidative cleavage of **67** then gave β-lactone **68**, which was used as a key intermediate in the preparation of the cholesterol biosynthesis inhibitor 1233A.

Other methods for the regioselective S_N2-opening of vinyloxiranes include intramolecular enolate addition for formation of cyclohexane systems [135, 136] and Friedel-Crafts alkylations [49, 137, 138].

9.2.3.3 **Regiodivergent Additions**

A valuable addition to the available methods for addition of carbon nucleophiles to vinyloxiranes are techniques that allow for selective additions to S_N2 and S_N2' positions. While $AlMe_3$ preferentially reacts by the S_N2 pathway, for example, LiMeCuCN affords the 1,4-addition product [126, 139]. It has also been shown that the regiochemical outcomes in the addition of dithianes are controlled by steric factors [140]: sterically unencumbered dithiane anions add by an S_N2 mechanism while those with a large steric bulk react by the S_N2' pathway, the regioselectivity being excellent for both routes. Finally, it is possible to control the regiochemistry in the addition of ethoxyacetylene to vinylepoxides by varying the counterion of the nucleophile. When lithium ethoxyacetylide together with $BF_3 \cdot OEt_2$ is employed, complete S_N2 regioselectivity is observed, while treatment of the same substrate with the corresponding alkynylalane gives the S_N2' product [141].

9.2.4
Rearrangement Reactions

Vinylepoxides participate in acid-catalyzed anionotropic rearrangements to afford aldehydes or ketones in good yields, the ratios depending on the migratory aptitudes of the involved groups and the rates of migration (Scheme 9.31) [142, 143]. When R is an allylic, benzylic, or silylethyl moiety, this group migrates and an aldehyde is normally obtained. With groups of lower migratory aptitude, such as phenethyl or cyclohexylmethyl, hydride migration to form the corresponding ketone can compete and mixtures are obtained. The reaction proceeds with inversion

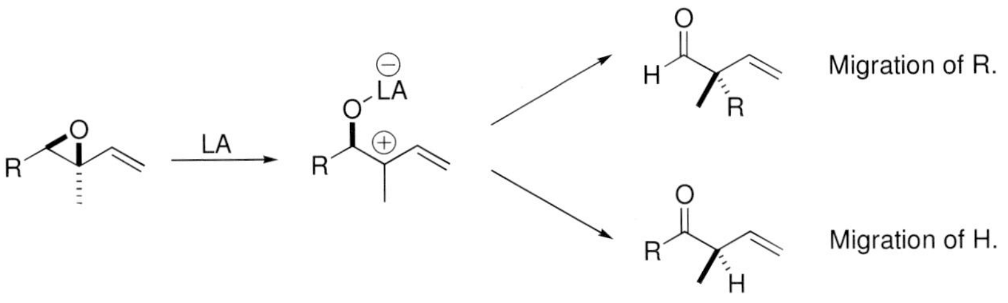

Migration of R.

Migration of H.

Scheme 9.31 Lewis acid-mediated rearrangement of vinylepoxides.

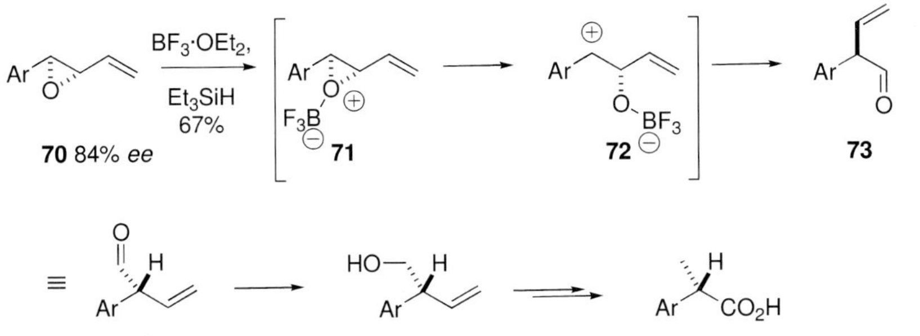

70 84% *ee*, $BF_3{\cdot}OEt_2$, Et_3SiH, 67%, **71**, **72**, **73**, **74** 84% *ee*, (*S*)-ibuprofen (**69**)

Scheme 9.32 Synthesis of (*S*)-ibuprofen (**69**) from vinylepoxide **70**; Ar = 4-^{i}BuC$_6$H$_4$.

at the quaternary center and the *ee* of the starting material is normally retained in the product, although exceptions are known [144].

In a synthesis of (*S*)-ibuprofen (**69**) by this methodology, epoxide **70** was treated with $BF_3 \cdot OEt_2$ in the presence of a reducing agent (Scheme 9.32) [145]. The initially formed complex **71** is assumed to ring-open selectively to give carbenium ion **72**, after which the allyl group migrates with inversion of the original stereochemistry at the benzylic position to form aldehyde **73**. *In situ* reduction of **73** then affords alcohol **74**, which is converted into **69** by standard techniques. It should be noted that, had **71** ring-opened to form the allylic carbonium ion, this species would have rearranged to give *ent*-**73** and hence *ent*-**74**. This result implies a greater stability of the benzylic cation than of the allylic one, if it is assumed that the two groups have similar migratory aptitudes.

The aldehydes formed in this rearrangement can be trapped either with nucleophiles [146, 147] or, by allowing them to isomerize into the corresponding dienolate, with electrophiles (Scheme 9.33) [148]. $BF_3 \cdot OEt$-mediated rearrangement of epoxide **75** affords aldehyde **76**, which can be trapped *in situ* by the potassium salt of (*Z*)-crotyl trifluoroborate to give **77** in good yield and with high *dr* (Route a). It was noted that in the absence of an external nucleophile, **76** dimerized to give the corresponding δ-hydroxy-α,β-unsaturated aldehyde. This outcome was interpreted

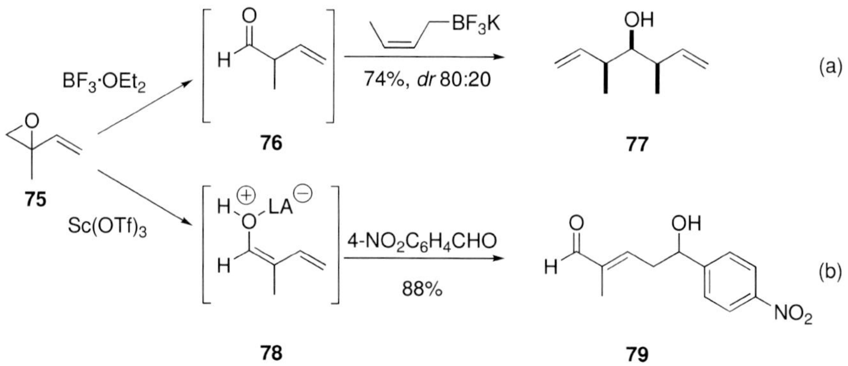

Scheme 9.33 Lewis acid-mediated transformations of epoxide 75.

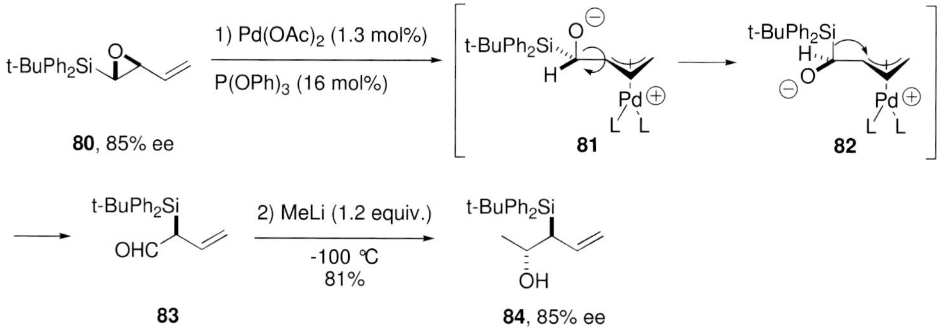

Scheme 9.34 Palladium-mediated rearrangement of epoxide 80.

by assuming that the initially formed aldehyde **76** isomerizes into dienolate **78**, with a subsequent reaction between **78** and **76** giving the observed product. Indeed, after optimization of the reaction conditions, in which $Sc(OTf)_3$ was identified as the Lewis acid of choice, it was found that efficient cross-couplings with several nonenolizable aldehydes could be achieved (Route b) [148]. By employing imines as electrophiles, this method was developed into an efficient vinylogous Mannich reaction, affording δ-amino-α,β-unsaturated aldehydes [149, 150].

In the presence of catalytic amounts of Pd(0), silicon-substituted vinyloxiranes can rearrange into the corresponding α-silyl-β,γ-unsaturated aldehydes (Scheme 9.34) [151]. Treatment of **80** with $Pd(OAc)_2$ and $P(OPh)_3$ results in the formation of π-allylpalladium complex **81**. Bond rotation to give **82**, followed by migration of the silyl moiety, affords aldehyde **83**, which is trapped *in situ* to provide the Felkin-Anh product **84**. The reaction proceeds with retention of configuration and the *ee* of the starting material is retained in the product. The size of the silicon substituents is critical for the outcome of the reaction, as is the choice of ligands on palladium.

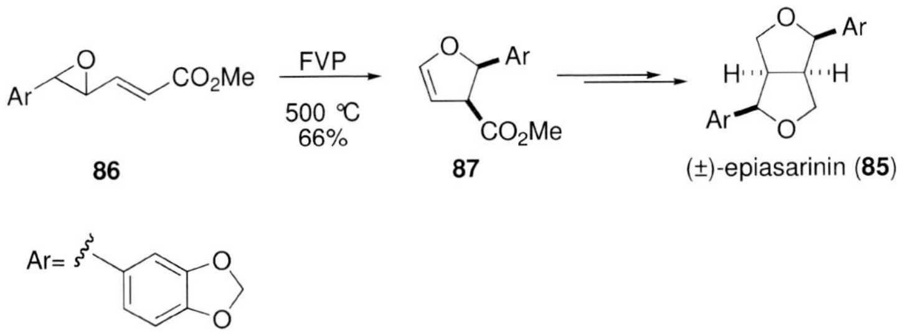

Scheme 9.35 Flash vacuum pyrolysis of compound **86**.

Small groups on silicon favor the formation of the corresponding silyl dienol ether, formed by a Brook rearrangement of **81**, and this reaction pathway is minimized by employment of $P(OPh)_3$.

Vinylepoxides can also be ring-expanded into the corresponding dihydropyrans [152], a reaction that generally proceeds with good regio- and diastereoselectivity [153]. In an efficient synthesis of the furofuran (±)-epiasarinin (**85**), flash vacuum pyrolysis (FVP) of **86**, to give dihydropyran **87** in good yield and with high selectivity (*cis:trans* 8:1), was used as a key step (Scheme 9.35) [154]. The reaction is believed to proceed through conrotatory ring-opening followed by a disrotatory cyclization to give the observed product.

9.2.5 Hydrogenolysis

Pd(0)-catalyzed hydrogenolysis of vinylepoxides offers an attractive regio- and diastereoselective route to homoallylic alcohols (Scheme 9.36) [104, 155, 156]. Thus, hydrogenolysis of (*E*) olefin **88** affords *syn* isomer **89** with inversion of configuration at the allylic carbon, while subjection of (*Z*) isomer **90** to identical reaction conditions results in the *anti* isomer **91**. The outcomes of these reactions are ex-

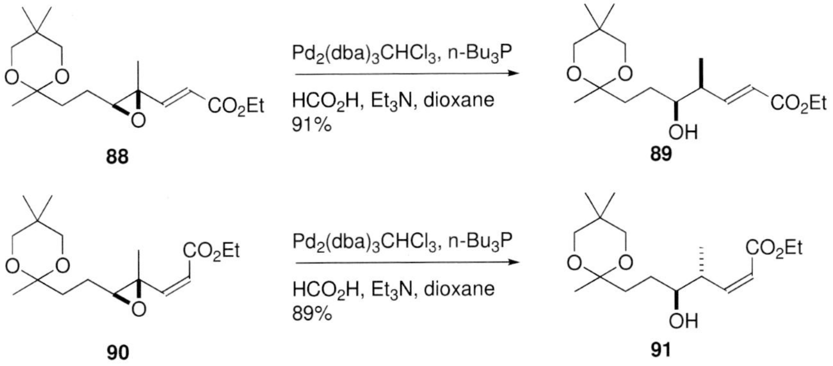

Scheme 9.36 Hydrogenolysis of vinylepoxides **88** and **90**.

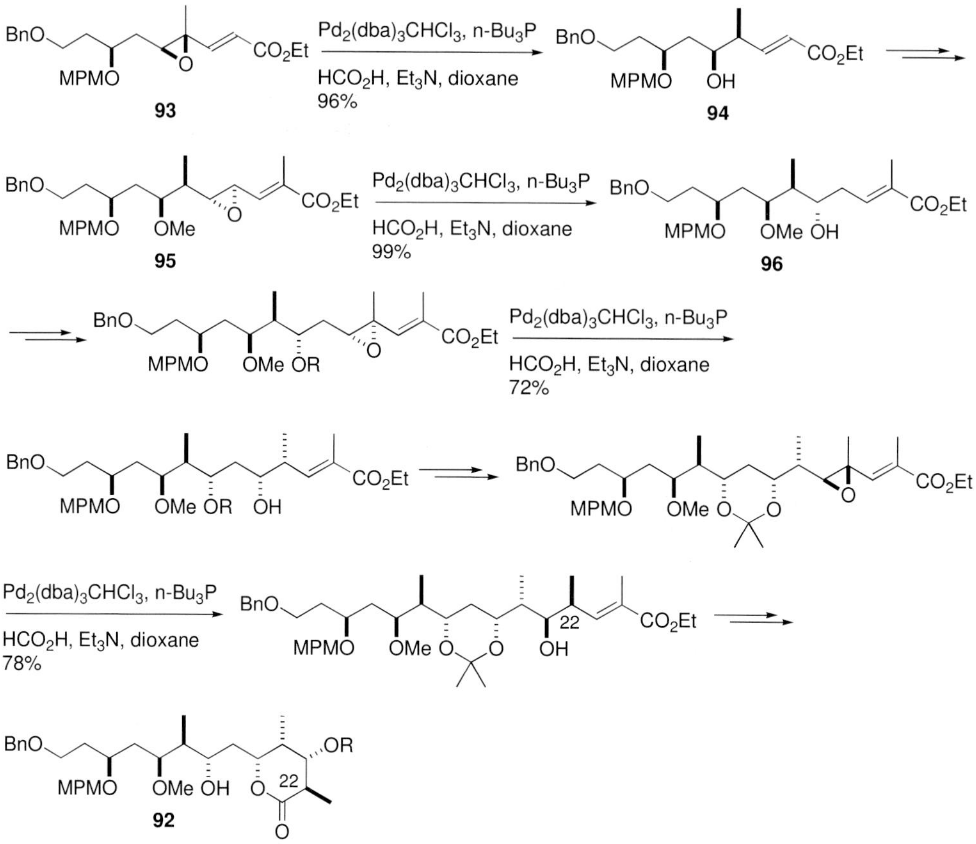

Scheme 9.37 Iterative hydrogenolysis of vinylepoxides; R = TBDMS.

plained by assuming attack of Pd(0) on each vinylepoxide to form the corresponding π-allylpalladium intermediate. Displacement of a ligand from this complex with formate anion gives a new π-allylpalladium formate complex, which undergoes a decarboxylation/hydride transfer. The hydride is then intramolecularly delivered to the sterically most hindered position of the π-allylpalladium moiety, giving the product with inversion of the stereochemistry at the allylic position. In the case of the (*Z*) olefins, the initially formed *anti* π-allylpalladium complex isomerizes through a π-σ-π mechanism to the thermodynamically more favored *syn* complex, and decarboxylation/hydride transfer from this intermediate gives the product with retention of stereochemistry [157].

Both the solvent and ligand were found to play pivotal roles in the reaction. In dioxane the byproduct formation was minimized while acceptable rates and yields were retained. In these studies it was also shown that both *n*-Bu_3P and Ph_3P are efficient ligands for the hydrogenolysis, but it is important that the ligand:Pd(0) ratio be kept below 1 [158].

When hydrogenolysis of vinylepoxides is used sequentially, it allows for the controlled formation of 1,3-polyols. In the synthesis of the C11–C23 fragment **92** of preswinholide A, hydrogenolysis of (*E*) olefin **93** gave the *syn* isomer **94** (Scheme 9.37) [159]. Methylation, reduction, epoxidation, oxidation, and olefination of this material then gave vinylepoxide **95**, which was subjected to hydrogenolysis to afford **96** in excellent yield. Repetition of this sequence ultimately afforded the desired derivative **94**.

As an interesting complement to the above technique, vinyloxiranes can be reduced with SmI_2 to give the corresponding (*E*)-allylic alcohol [160].

9.3 Conclusions

Excellent techniques for the enantioselective synthesis of vinylepoxides are now available. Depending on the substrate structure, a range of synthetic strategies can be employed. In view of the impressive progress in this area, it can be foreseen that the scope of the methods, preferentially efficient one-step processes, should be expandable to encompass broader ranges of substrates. Vinylepoxides have been used as substrates in a number of interesting transformations. As they are ambident electrophiles, much of their chemistry is concerned with controlling the regiochemistry in nucleophilic addition reactions, and conditions for achieving this are available for both hetero- and carbon nucleophiles. An interesting area of development is the Pd(0)-mediated dynamic kinetic resolution of racemic vinylepoxides in the presence of suitable pronucleophiles, allowing for highly efficient routes, with high enantioselectivities, from simple starting materials to complex structures.

The developments highlighted in this chapter show that vinylepoxides are readily available starting materials. Several techniques for their regio- and stereoselective synthesis and derivatization exist, and it is consequently to be expected that the use of vinylepoxides as substrates in organic synthesis will increase.

References

1 Larock, R. C., in *Comprehensive Organic Transformations*; 2nd Ed.; Wiley-VCH: New York, **1999**, pp. 915–927.
2 Gawley, R. E., Aubé, J., *Principles of Asymmetric Synthesis*; Pergamon: Oxford, **1996**.
3 Huang, J.-M., Xu, K.-C., Loh, T.-P. *Synthesis* **2003**, 755–764.
4 Frohn, M., Shi, Y. *Synthesis* **2000**, 1979–2000.
5 Curci, R., Fiorentino, M., Serio, M. R. *Chem. Commun.* **1984**, 155–156.
6 Tu, Y., Wang, Z.-X., Shi, Y. *J. Am. Chem. Soc.* **1996**, *118*, 9806–9807.
7 Wu, X.-Y., She, X., Shi, Y. *J. Am. Chem. Soc.* **2002**, *124*, 8792–8793 and references therein.
8 Tian, H., She, X., Shi, Y. *Org. Lett.* **2001**, *3*, 715–718.
9 Frohn, M., Dalkiewicz, M., Tu, Y., Wang, Z.-X., Shi, Y. *J. Org. Chem.* **1998**, *63*, 2948–2953.
10 Olofsson, B., Khamrai, U., Somfai, P. *Org. Lett.* **2000**, *2*, 4087–4089.

11 Olofsson, B., Somfai, P. *J. Org. Chem.* **2002**, *67*, 8574–8583 and unpublished results.

12 Olofsson, B., Somfai, P. *J. Org. Chem.* **2003**, *68*, 2514–2517.

13 Vinylepoxides are known to be unstable on silica, and separation of epoxides **7** and **8** was possible only at the expense of decreased yield. To avoid this loss, we separated the isomers at a later stage in the synthesis.

14 Zhang, W., Loebach, J. L., Wilson, S. R., Jacobsen, E. N. *J. Am. Chem. Soc.* **1990**, *112*, 2801–2803.

15 Irie, R., Noda, K., Ito, Y., Matsumoto, N., Katsuki, T. *Tetrahedron Lett.* **1990**, *31*, 7345–7348.

16 Brandes, B. D., Jacobsen, E. N. *J. Org. Chem.* **1994**, *59*, 4378–4380.

17 Mikame, D., Hamada, T., Irie, R., Katsuki, T. *Synlett* **1995**, 827–828.

18 Chang, S. K., Heid, R. M., Jacobsen, E. N. *Tetrahedron Lett.* **1994**, *35*, 669–672.

19 Chang, S. K., Lee, N. H., Jacobsen, E. N. *J. Org. Chem.* **1993**, *58*, 6939–6941.

20 Kolb, H. C., Sharpless, K. B. *Tetrahedron* **1992**, *48*, 10515–10530.

21 Raheem, I. T., Goodman, S. N., Jacobsen, E. N. *J. Am. Chem. Soc.* **2004**, *126*, 706–707.

22 Liang, J., Moher, E. D., Moore, R. E., Hoard, D. W. *J. Org. Chem.* **2000**, *65*, 3143–3147.

23 Chang, H.-T., Sharpless, K. B. *J. Org. Chem.* **1996**, *61*, 6456–6457.

24 Xu, D., Crispino, G. A., Sharpless, K. B. *J. Am. Chem. Soc.* **1992**, *114*, 7570–7571.

25 Becker, H., Soler, M. A., Sharpless, K. B. *Tetrahedron* **1995**, *51*, 1345–1376.

26 Fernandes, R. A., Kumar, P. *Tetrahedron* **2002**, *58*, 6685–6690.

27 Colonna, S., Molonari, H., Banfi, S., Juliá, S., Masana, J., Alvarez, A. *Tetrahedron* **1983**, *39*, 1635–1641.

28 Geller, T., Gerlach, A., Krüger, C. M., Militzer, H.-C. *Tetrahedron Lett.* **2004**, *45*, 5065–5067.

29 Allen, J. V., Bergeron, S., Griffiths, M. J., Mukherjee, S., Roberts, S. M., Williamson, N. M., Wu, L. E. *J. Chem. Soc., Perkin Trans. 1* **1998**, 3171–3179.

30 Nemoto, T., Ohshima, T., Yamaguchi, K., Shibasaki, M. *J. Am. Chem. Soc.* **2001**, *123*, 2725–2732.

31 Kakei, H., Nemoto, T., Ohshima, T., Shibasaki, M. *Angew. Chem. Int. Ed.* **2004**, *43*, 317–320.

32 Kinoshita, T., Okada, S., Park, S.-R., Matsunaga, S., Shibasaki, M. *Angew. Chem. Int. Ed.* **2003**, *42*, 4680–4684.

33 Li, C., Pace, E. A., Liang, M.-C., Lobkovsky, E., Gilmore, T. D., Porco, J. A., Jr. *J. Am. Chem. Soc.* **2001**, *123*, 11 308–11309.

34 Katsuki, T., Martin, V. S., in *Organic Reactions*; Paquette, L. A., Ed.; John Wiley & Sons, Inc.: **1996**; Vol. 48, pp. 1–285.

35 Bernet, B., Vasella, A. *Tetrahedron Lett.* **1983**, *24*, 5491–5494.

36 Werschofen, S., Scharf, H.-D. *Synthesis* **1988**, 854–858.

37 Olofsson, B.; Somfai, P. Unpublished results.

38 Weigand, S., Bruckner, R. *Synlett* **1997**, 225–228.

39 Díez Martin, D., Marcos, I. S., Basabe, P., Romero, R. E., Moro, R. F., Lumeras, W., Rodríguez, L., Urones, J. G. *Synthesis* **2001**, 1013–1022.

40 Díez, D., Beneitez, T., Marcos, I. S., Garrido, N. M., Basabe, P., Urones, J. G. *Tetrahedron Asymmetry* **2002**, *13*, 639–646.

41 Kim, N.-S., Choi, J.-R., Cha, J. K. *J. Org. Chem.* **1993**, *58*, 7096–7099.

42 Xiang, A. X., Watson, D. A., Ling, T., Theodorakis, E. A. *J. Org. Chem.* **1998**, *63*, 6774–6775.

43 Romero, A., Wong, C.-H. *J. Org. Chem.* **2000**, *65*, 8264–8268 and references therein.

44 Gao, Y., Hanson, R. M., Klunder, J. M., Ko, S. Y., Masamune, H., Sharpless, K. B. *J. Am. Chem. Soc.* **1987**, *109*, 5765–5780.

45 Corey, E. J., Mehrotra, M. M. *Tetrahedron Lett.* **1986**, *27*, 5173–5176.

46 Díez-Martin, D., Kotecha, N. R., Ley, S. L., Mantegani, S., Menéndez, J. C., Organ, H. M., White, A. D., Banks, J. B. *Tetrahedron* **1992**, *48*, 7899–7938.

47 Maryanoff, B. E., Reitz, A. B. Chem. *Rev.* **1989**, *89*, 863–927.

48 Lindström, U. M., Olofsson, B., Somfai, P. *Tetrahedron Lett.* **1999**, *40*, 9273–9276.

49 Lindström, U. M., Somfai, P. *Synthesis* **1998**, 109–117.

50 Sakamoto, Y., Matsukura, H., Nakata, T. *Org. Lett.* **2001**, *3*, 2749–2752.

51 Matsuo, G., Hori, N., Matsukura, H.,

Nakata, T. *Tetrahedron Lett.* **2000**, *41*, 7677–7680.

52 Ireland, R. E., Norbeck, D. W. J. *Org. Chem.* **1985**, *50*, 2198–2200.

53 Barrett, A. G. M., Hamprecht, D., Ohkubo, M. J. *Org. Chem.* **1997**, *62*, 9376–9378.

54 Maiti, A., Yadav, J. S. *Synth. Commun.* **2001**, *31*, 1499–1506.

55 Wiei, X., Taylor, R. J. K. *Tetrahedron Lett.* **1998**, *39*, 3815–3818.

56 Shuto, S., Niizuma, S., Matsuda, A. J. *Org. Chem.* **1998**, *63*, 4489–4493.

57 Activated alcohols can be converted into olefins with MnO2 and nonstabilized ylides, see Blackburn, L.; Pei, C., Taylor, R. J. K. *Synlett* **2002**, 215–218.

58 Maccoss, R. N., Balskus, E. P., Ley, S. V. *Tetrahedron Lett.* **2003**, *44*, 7779–7781.

59 Hertweck, C., Boland, W. Eur. *J. Org. Chem.* **1998**, 2143–2148.

60 Mallaiah, K., Satayanarayana, J., Ila, H.,Junjappa, H. *Tetrahedron Lett.* **1993**, *34*, 3145–3148.

61 For a similar approach, see Lautens, M.; Maddess, M. L.; Sauer, E. L. O., Quellet, S. G. *Org. Lett.* **2002**, *4*, 83-86 and references therein.

62 Ramachandran, P. V., Rudd, M. T., Burghardt, T. H., Reddy, M. V. R. J. *Org. Chem.* **2003**, *68*, 9310–9316.

63 Cookson, R. C., Crumbie, R. L. *Tetrahedron Lett.* **1985**, *26*, 3377–3380.

64 Doyle, M. P., Hu, W., Timmons, D. J. *Org. Lett. 2001, 3*, 933–935.

65 Davies, H. M. L., Demeese, J. *Tetrahedron Lett.* **2001**, *42*, 6803–6805.

66 Hu, S., Jayaraman, S., Oehlschlager, A. C. *J. Org. Chem.* **1996**, *61*, 7513–7520.

67 Hu, S., Jayaraman, S., Oehlschlager, A. C. *J. Org. Chem.* **1998**, *63*, 8843–8849.

68 Hertweck, C., Goerls, H., Boland, W. *Chem. Commun.* **1998**, 1955–1956.

69 Barluenga, S., Lopez, P., Moulin, E., Winssinger, N. *Angew. Chem. Int. Ed.* **2004**, *43*, 3467–3470.

70 Li, A.-H., Dai, L.-X., Aggarwal, V. K. *Chem. Rev.* **1997**, *97*, 2341–2372.

71 Aggarwal, V. K., Winn, C. L. *Acc. Chem. Res.* **2004**, *37*, 611–620.

72 Aggarwal, V. K., Ford, J. G., Fonquerna, S., Adams, H., Jones, R. V. H., Fieldhouse, R. *J. Am. Chem. Soc.* **1998**, *120*, 8328–8339.

73 Aggarwal, V. K., Alonso, E., Bae, I., Hynd, G., Lydon, K. M., Palmer, M. J., Patel, M., Porcelloni, M., Richardson, J., Stenson, R. A., Studley, J. R., Vasse, J.-L., Winn, C. L. *J. Am. Chem. Soc.* **2003**, *125*, 10926–10940.

74 Aggarwal, V. K., Bae, I., Lee, H.-Y., Richardson, J., Williams, D. T. *Angew. Chem. Int. Ed.* **2003**, *42*, 3274–3278.

75 Crude yields are given, as aryl vinylepoxides are difficult to purify without degradation.

76 Sollaidié-Cavallo, A., Bouérat, L., Roje, M. *Tetrahedron Lett.* **2000**, *41*, 7309–7312.

77 Zanardi, J., Lamazure, D., Minière, S., Reboul, V., Metzner, P. *J. Org. Chem.* **2002**, *67*, 9083–9086.

78 Zhou, Y.-G., Li, A.-H., Hou, X.-L., Dai, L.-X. *Chem. Commun.* **1996**, 1353–1354.

79 Li, K., Deng, X.-M., Tang, Y. *Chem. Commun.* **2003**, 2074–2075.

80 Fürstner, A., Gastner, T. *Org. Lett.* **2000**, *2*, 2467–2470 and references therein.

81 Ma, S., Zhao, S. *J. Am. Chem. Soc.* **1999**, *121*, 7943–7944.

82 Corey, E. J., Yu, C. M., Lee, D. H. *J. Am. Chem. Soc.* **1990**, *112*, 878–879.

83 Schaus, S. E., Brandes, B. D., Larrow, J. F., Tokunaga, M., Hansen, K. B., Gould, A. E., Furrow, M. E., Jacobsen, E. N. *J. Am. Chem. Soc.* **2002**, *124*, 1307–1315.

84 Jaime, C., Ortuno, R., M., Font, J. *J. Org. Chem.* **1988**, *53*, 139–141.

85 Prestat, G., Baylon, C., Heck, M.-P., Mioskowski, C. *Tetrahedron Lett.* **2000**, *41*, 3829–3831.

86 Heck, M.-P., Baylon, C., Nolan, S. P., Mioskowski, C. *Org. Lett.* **2001**, *3*, 1989–1991.

87 Fagnou, K., Lautens, M. *Org. Lett.* **2000**, *2*, 2319–2321.

88 For one example of Pd(0)-catalyzed 1,2-addition of p-methoxyphenol to 1-methyl-1-oxirane, see: Goujon, J.-Y., Duval, A., Kirscleger. B. *J. Chem. Soc., Perkin Trans. 1* **2002**, 496–499.

89 Wershofen, S., Scharf, H. D. *Synthesis* **1988**, 854–858.

90 Trost, B. M., Angle, S. R. *J. Am. Chem. Soc.* **1985**, *107*, 6123–6124.

91 Hirai, A., Yu, X.-Q., Tonooka, T., Miyashita, M. *Chem. Commun* **2003**, 2482–2483.

92 Trost, B. M., Tang, W. *Org. Lett.* **2001**, *3*, 3409–3411.

93 Trost, B. M., Mceachern, E. J., Toste, F. D. *J. Am. Chem. Soc.* **1998**, *120*, 12702–12703.

94 For other synthetic applications, see: Trost, B. M., Tang, W., Schulte, J. L. *Org. Lett.* **2000**, *2*, 4013–4015; Trost, B. M., Brown, B. S., McEachern, E. J., Kuhn, O. *Chem. Eur. J.* **2003**, *9*, 4442–4451; Trost, B. M. *J. Org. Chem.* **2004**, *69*, 5813–5837.

95 Schaus, S. E., Brandes, B. D., Larrow, J. F., Tokunaga, M., Hansen, K. B., Gould, A. E., Furrow, M. E., Jacobsen, E. N. *J. Am. Chem. Soc.* **2002**, *124*, 1307–1315.

96 Olofsson, B., Somfai, P. *J. Org. Chem.* **2002**, *67*, 8574–8583.

97 Romero, A., Wong, C.-H. *J. Org. Chem.* **2000**, *65*, 8264–8268.

98 Tang, M., Pyne, S. G. *J. Org. Chem.* **2003**, *68*, 7818–7824.

99 Trost, B. M., Sudhakar, A. R. *J. Am. Chem. Soc.* **1987**, *109*, 3792–3794.

100 For Pd(0)-catalyzed addition of carbodiimides to vinyloxiranes, see: Larksarp, C., Alper, H. *J. Org. Chem.* **1998**, *63*, 6229–6233.

101 Trost, B. M., Bunt, R. C. *Angew. Chem. Int. Ed. Engl.* **1996**, *35*, 99-102.

102 Trost, B. M., Bunt, R. C., Lemoine, R. C., Calkins, T. L. *J. Am. Chem. Soc.* **2000**, *122*, 5968–5976.

103 Godleski, S. A, in *Comprehensive Organic Synthesis*; Trost, B. M. and Fleming, I., Eds.; Pergamon: Oxford, **1991**; Vol. 4, pp. 585–661.

104 Tsuji, J, in *Palladium Reagents and Catalysts*; Wiley: Chichester, **1995**, pp. 376–377.

105 Pettersson-Fasth, H., Riesinger, S. W., Bäckvall, J. E. *J. Org. Chem.* **1995**, *60*, 6091–6096.

106 Nicolaou, K. C., Rutjes, F. P. J. T., Theodorakis, E. A., Tiebes, J., Sato, M., Untersteller, E. *J. Am. Chem. Soc.* **1995**, *117*, 1173–1174.

107 Nicolaou, K. C., Shi, G.-Q., Gunzner, J. L., Gartner, P., Wallace, P. A., Ouellette, M. A., Shi, S., Bunnage, M. E., Agrios, K. A., Veale, C. A., Hwang, C.-K., Hutchinson, J., Prasad, C. V. C., Ogilvie, W. W., Yang, Z. *Chem. Eur. J.* **1999**, *5*, 628–645.

108 Nicolaou, K. C., Sorensen, E. J. *Classics in Total Synthesis*; VCH: Weinheim, **1996**.

109 Nicolaou, K. C., Duggan, M. E., Hwang, C.-K., Somers, P. K. *J. Chem. Soc., Chem. Commun.* **1985**, 1359–1362.

110 Nicolaou, K. C., Prasad, C. V. C., Somers, P. K., Hwang, C.-K. *J. Am. Chem. Soc.* **1989**, *111*, 5330–5334.

111 Baldwin, J. E. *J. Chem. Soc., Chem. Commun.* **1976**, 734–736.

112 Suzuki, T., Sato, O., Hirama, M. *Tetrahedron Lett.* **1990**, *31*, 4747–4750.

113 Matsukura, H., Morimoto, M., Koshino, H., Nakata, T. *Tetrahedron Lett.* **1997**, *38*, 5545–5548.

114 Sasaki, M., Inoue, M., Takamatsu, K., Tachibana, K. *J. Org. Chem.* **1999**, *64*, 9399–9415.

115 Fürstner, A., Weintritt, H. *J. Am. Chem. Soc.* **1998**, *120*, 2817–2825.

116 Trost, B. M. *Angew. Chem. Int. Ed. Engl.* **1989**, *28*, 1173–1192.

117 Elliott, M. R., Dhimane, A.-L., Malacria, M. *J. Am. Chem. Soc.* **1997**, *119*, 3427–3428.

118 Elliott, M. R., Dhimane, A.-L., Malacria, M. *Tetrahedron Lett.* **1998**, *39*, 8849–8852.

119 Marshall, J. A. *Chem. Rev.* **1989**, *89*, 1503–1511.

120 Marshall, J. A., Trometer, J. D., Cleary, D. G. *Tetrahedron* **1989**, *45*, 391–402.

121 Marshall, J. A., Trometer, J. D., Blough, B. E., Crute, T. D. *J. Org. Chem.* **1988**, *53*, 4274–4282.

122 Corey, E. J., Boaz, N. W. *Tetrahedron Lett.* **1984**, *25*, 3063–3066.

123 Badalassi, F., Crotti, P., Macchia, F., Pineschi, M., Arnold, A., Feringa, B. L. *Tetrahedron Lett.* **1998**, *39*, 7795–7798.

124 Equey, O., Alexakis, A. *Tetrahedron: Asymmetry* **2004**, *15*, 1531–1536.

125 Molander, G. A., Shakya, S. R. *J. Org. Chem.* **1996**, *61*, 5885–5894.

126 Miyashita, M., Hoshino, M., Yoshikoshi, A. *J. Org. Chem.* **1991**, *56*, 6483–6485.

127 Ishibashi, N., Miyazawa, M., Miyashita, M. *Tetrahedron Lett.* **1998**, *39*, 3775–3778.

128 Abe, N., Hanawa, H., Maruoka, K., Sasaki, M., Miyashita, M. *Tetrahedron Lett.* **1999**, *40*, 5369–5372.

129 Equey, O., Vrancken, E., Alexakis, A. Eur. *J. Org. Chem.* **2004**, 2151–2159.

130 Alexakis, A., Vrancken, E., Mangeney, P., Chemla, F. *J. Chem. Soc., Perkin Trans. 1* **2000**, 3352–3353.

131 Trost, B. M., Jiang, C. J. *Am. Chem. Soc.* **2001**, *123*, 12907–12908.

132 Trost, B. M., Jiang, C. *Org. Lett.* **2003**, *5*, 1563–1565.

133 Ley, S. V., Cox, L. R., Meek, G. *Chem. Rev.* **1996**, *96*, 423–442.

134 Bates, R. W., Fernandez-Megia, E., Ley, S. V., Ruck-Braun, K., Tilbrook, D. M. G. *J. Chem. Soc. Perkin Trans. 1* **1999**, 1917–1925.

135 Stork, G., Kobayashi, Y., Suzuki, T., Zhao, K. *J. Am. Chem. Soc.* **1990**, *112*, 1661–1663.

136 Stork, G., Zhao, K. *J. Am. Chem. Soc.* **1990**, *112*, 5875–5876.

137 Ono, M., Suzuki, K., Akita, H. *Tetrahedron Lett.* **1999**, *40*, 8223–8226.

138 Nagumo, S., Miyoshi, I., Akita, H., Kawahara, N. *Tetrahedron Lett.* **2002**, *43*, 2223–2226.

139 Hirai, A., Matsui, A., Komatsu, K., Tanino, K., Miyashita, M. *Chem. Commun.* **2002**, 1970–1971.

140 Smith, A. B., Iii, Pitram, S. M., Gaunt, M. J., Kozmin, S. A. *J. Am. Chem. Soc.* **2002**, *124*, 14516–14517.

141 Restorp, P., Somfai, P. *Chem. Commun.* **2004**, 2086–2087.

142 Rickborn, B, in *Comprehensive Organic Synthesis*; Trost, B. M. and Fleming, I., Eds.; Pergamon: Oxford, 1991; Vol. 3, pp. 733–775.

143 Jung, M. E., D'amico, D. C. *J. Am. Chem. Soc.* **1995**, *117*, 7379–7388.

144 Tonder, J. E., Tanner, D. *Tetrahedron* **2003**, *59*, 6937–6945.

145 Jung, M. E., Anderson, K. L. *Tetrahedron Lett.* **1997**, *38*, 2605–2608.

146 Wipf, P., Xu, W. *J. Org. Chem.* **1993**, *58*, 825–826.

147 Oh, B. K., Cha, J. H., Cho, Y. S., Choi, K. I., Koh, H. Y., Chang, M. H., Pae, A. N. *Tetrahedron Lett.* **2003**, *44*, 2911–2913.

148 Lautens, M., Ouellet, S. G., Raeppel, S. *Angew. Chem. Int. Ed.* **2000**, *39*, 4079–4082.

149 Lautens, M., Tayama, E., Nguyen, D. *Org. Lett.* **2004**, *6*, 345–347.

150 Lautens, M., Tayama, E., Nguyen, D. *Tetrahedron Lett.* **2004**, *45*, 5131–5133.

151 Le Bideau, F., Gilloir, F., Nilsson, Y., Aubert, C., Malacria, M. *Tetrahedron* **1996**, *52*, 7487–7510.

152 For a recent review, see: Hudlicky, T., Reed, J. W, in *Comprehensive Organic Synthesis*; Trost, B. M., Fleming, I., Eds.; Pergamon: Oxford, 1991; Vol. 5, pp. 899–970.

153 Eberbach, W., Burchard, B. *Chem. Ber.* **1978**, *111*, 3665–3698.

154 Aldous, D. J., Dalencon, A. J., Steel, P. G. *Org. Lett.* **2002**, *4*, 1159–1162.

155 Oshima, M., Yamazaki, H., Shimizu, I., Nisar, M., Tsuji, J. *J. Am. Chem. Soc.* **1989**, *111*, 6280–6287.

156 Tsuji, J., Mandai, T. *Synthesis* **1996**, 1-24.

157 For a discussion of the stereochemical outcome in the hydrogenolysis of (Z)-olefins, see: Noguchi, Y., Yamada, T., Uchiro, H., Kobayashi, S. *Tetrahedron Lett.* **2000**, *41*, 7493–7497.

158 For the use of dimethylamine-borane as an alternative hydride source, see: David, H., Dupuis, L., Guillerez, M.-G., Guibe, F. *Tetrahedron Lett.* **2000**, *41*, 3335–3338.

159 Nagasawa, K., Shimizu, I., Nakata, T. *Tetrahedron Lett.* **1996**, *37*, 6881–6884.

160 Molander, G. A., La Belle, B. E., Hahn, G. *J. Org. Chem.* **1986**, *51*, 5259–5264.

10
The Biosynthesis of Epoxides

Sabine Grüschow and David H. Sherman

10.1
Introduction

Epoxides are found in thousands of biological molecules and constitute vital functional entities. They can impart localized structural rigidity, confer cytotoxicity through their role as alkylating agents, or act as reactive intermediates in complex synthetic sequences. The widespread occurrence of epoxides is contrasted by only a handful of aziridines that are known to date. In this chapter we would like to introduce the different mechanisms by which enzymes produce epoxides.

The majority of epoxides are derived from reactions with molecular oxygen. However, dioxygen is kinetically inert towards organic molecules in the absence of a suitable catalyst. This stems from the electron distribution within O_2. In the ground state, dioxygen has two unpaired electrons in the $2\pi^*$ orbitals (triplet oxygen). Organic molecules, on the other hand, usually have paired electrons and hence exist in singlet ground states. Reactions between singlet-state molecules and triplet-state molecules to give products with paired electrons are spin-forbidden, so enzymes need to provide mechanisms through which to adjust the spin states of O_2 or of its substrates. Transfer of a single electron to $^3O_2$ to give superoxide is a spin-allowed reaction, for example. The reductant can be a metal or an organic molecule with an accessible radical form and the resulting superoxide can then participate in one- or two-electron reactions [1, 2].

It may be helpful to clarify a few points regarding enzyme terminology before we start. In this chapter we refer to enzymes that catalyze the formation of epoxides as epoxidases. The same term is also used occasionally in the literature to describe epoxide-hydrolyzing enzymes. Oxidases are enzymes that perform reactions resulting in a net oxidation of the substrate, whereas oxygenases are a subset of this family that insert oxygen into their substrate molecules. Mono- and dioxygenases are defined by transferring one or two atoms, respectively, from molecular oxygen into their substrates.

Aziridines and Epoxides in Organic Synthesis. Andrei K. Yudin

ISBN: 3-527-31213-7

10.2
Cytochrome P450 Monooxygenases

10.2.1
Mechanism of Cytochrome P450 Monooxygenases

Cytochrome P450 monooxygenases are characterized through the presence of the heme (protoporphyrin IX) prosthetic group (Scheme 10.1) that is coordinated to the enzyme through a conserved cysteine ligand. They have obtained their name from the signature absorption band with a maximum near 450 nm in the difference spectrum when incubated with CO. The absorption arises from the Soret π–π^* transition of the ferrous protoporphyrin IX-CO complex.

Cytochrome P450s are ubiquitous in both eukaryotic and prokaryotic cells and catalyze a large number of different reactions. The reactions cytochrome P450 enzymes are probably best known for are hydroxylations. Liver microsomal P450s, for example, play a crucial role in the metabolism of hydrophobic xenobiotics. Through the addition of hydroxy groups these compounds are rendered water-soluble and can subsequently be excreted. A large proportion of P450 enzymes, particularly of eukaryotic origin, are membrane-bound and are thus very challenging to isolate and purify. Most of the mechanistic investigations have therefore been carried out with the soluble P450cam (Scheme 10.2), a bacterial P450 monooxygenase from *Pseudomonas putida*. P450cam is readily purified in sufficient quantities for biochemical investigations and it is also the first enzyme of its class for which the crystal structure has been solved [3].

The overall hydroxylation or epoxidation reaction catalyzed by cytochrome P450s involves the insertion of one oxygen atom, derived from molecular oxygen, into a C–H bond or into the π-system of an olefin, with the concomitant reduction of the

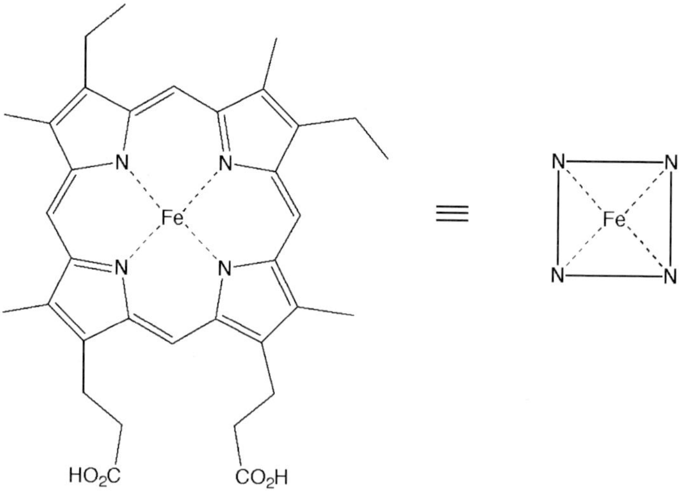

N N Fe N N ≡ N N Fe N N

HO_2C CO_2H

Scheme 10.1 Structure of the heme prosthetic group.

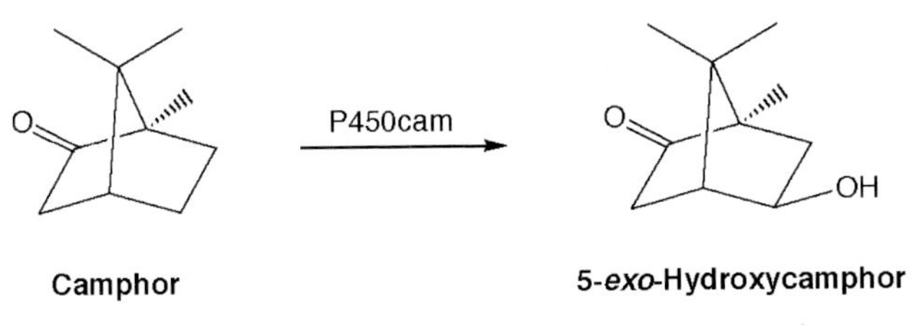

Scheme 10.2 Cytochrome P450cam-catalyzed hydroxylation of camphor.

Mitochondrial and bacterial system:

NAD(P)H → [FAD] (Ferredoxin reductase) → [Fe_2-S_2] (Ferredoxin) → P450

Microsomal system:

NAD(P)H → [FAD → FMN] (NADPH-cytochrome P450 reductase) → P450

Scheme 10.3 Electron-transport systems associated with cytochrome P450 monooxygenases. Arrows indicate electron transfer.

other oxygen atom to water. The electrons needed to reduce the second oxygen atom have to be supplied through an electron-transport system, with NAD(P)H as the ultimate reducing agent. Most bacterial and mitochondrial P450s utilize a two-component system that consists of a flavoprotein whose FAD cofactor will accept the two-electron reducing equivalent from the nicotinamide substrate, and an iron-sulfur protein that will relay one-electron equivalents from reduced flavin to the P450 enzyme (Scheme 10.3). In contrast, microsomal P450s utilize a single flavoprotein containing both FAD and FMN as cofactors to transfer the electrons from NAD(P)H to the heme. In both systems, the individual components usually reside on different polypeptide chains. However, there are some exceptions to this rule, mostly of bacterial origin, in which the reductase is fused to the P450 to form a single, self-sufficient enzyme. The first such example to be discovered is P450 BM3 from the bacterium *Bacillus megaterium*. P450 BM3 is responsible for long–chain fatty acid oxidation and, in fact, best resembles a ligated microsomal P450 system [4].

The P450 reaction cycle (Scheme 10.4) starts with four "stable" intermediates that have been characterized by spectroscopic methods. The resting state of the enzyme is a six-coordinate, low-spin ferric state (complex **I**) with water (or hydroxide) coordinated *trans* to the cysteinate ligand. The spin state of the iron changes to high-spin upon substrate binding and results in a five-coordinate ferric ion (com-

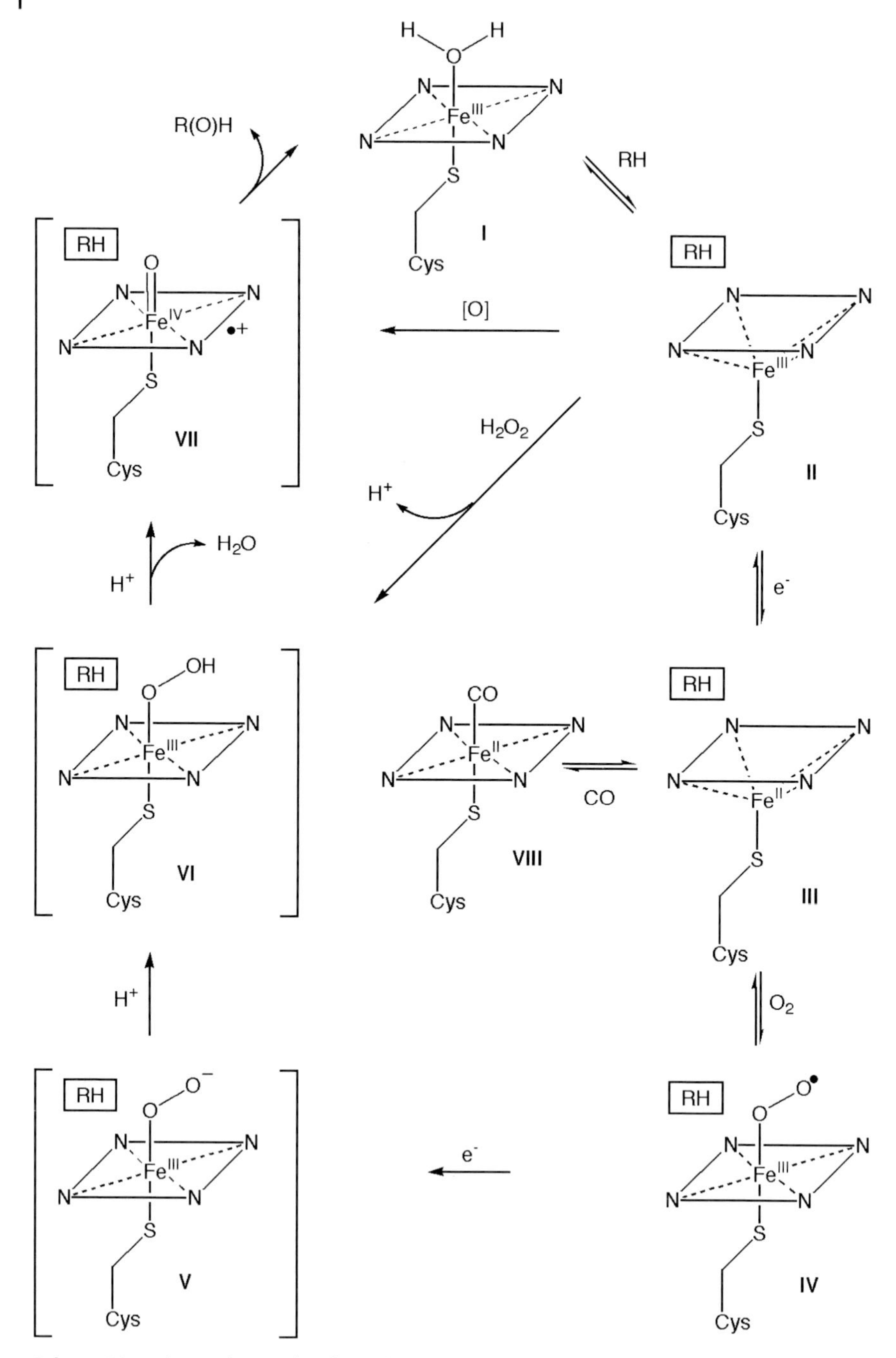

Scheme 10.4 The catalytic cycle of cytochrome P450. Only one possible valence structure of the oxoferrous species **IV** has been depicted for clarity. See text for details.

plex **II**) through loss of the water ligand. The increase in redox potential of the heme upon transition to the high-spin state facilitates electron transfer from the electron-transport system to form complex **III**. Molecular oxygen binds to the ferrous iron complex **III** to give the oxo-ferric species **IV**. One more electron is then transferred from the electron-transport system, followed by addition of two protons with concomitant release of water and generation of a high-valent oxo-iron species (complex **VII**). This species is most commonly depicted with a porphyrin radical cation as shown in Scheme 10.4. The reduction step that yields complex **V** represents the rate-limiting step of the reaction cycle.

When working with new P450 monooxygenases, particularly from less studied organisms, the associated electron-transport system is frequently unknown. In the absence of this electron-transport system, the cytochrome P450 enzyme is only capable of catalyzing a single turnover *in vitro* which is inadequate for many biochemical studies. However, there are ways to circumvent the issue of finding the endogenous reductase. Firstly, the endogenous electron-transport system can sometimes be substituted by one from a different organism (the commercially available spinach ferredoxin/ferredoxin reductase pair is frequently used). Secondly, numerous oxygen atom donors can replace the two electrons from NAD(P)H as well as O_2 by directly reacting with the high-spin ferric heme (complex **II**) to give the reactive iron-oxo species **VII** (Scheme 10.4). These oxygen donors include oxidizing agents such as alkyl peroxides, peracids, sodium chlorite, and sodium periodate. Lastly, some P450 enzymes are able to accomplish turnover in the presence of peroxide by the peroxide shunt shown in Scheme 10.4.

Most studies on the mechanism of oxygen insertion from the iron-oxo intermediate into the substrate have dealt with the hydroxylation of alkanes. There is still ongoing debate over whether the reaction proceeds through a substrate cation or radical intermediate, and there is evidence for each pathway. The same is true for the cytochrome P450-catalyzed epoxidation of double bonds. It has to be noted that the same cytochrome P450 enzyme is often capable of performing both hydroxylation and epoxidation reactions depending on the nature of the substrate – the oxygen may be inserted into the C–H σ bond of an alkane to give a hydroxylated product or it may be inserted into the π bond of an olefin to yield the epoxide. Also, there is some indication that maybe both radical and cationic mechanisms can apply for the same P450 enzyme and that the major pathway chosen may primarily depend on the nature of the substrate under investigation [1]. As for the mechanism of epoxidation specifically, the initial step is thought to be the formation of a charge-transfer complex between the oxo-iron species **VII** and the olefin (Scheme 10.5) [5, 6]. Oxygen insertion occurs with retention of configuration; a *cis*-olefin, for example, will give a *cis*-epoxide. Even though this is consistent with a concerted mechanism, a stepwise mechanism is also possible [1, 6]. Electron transfer within the charge-transfer complex would initially give a substrate-centered radical cation, which could then proceed to give a radical or cationic σ-complex [6]. The stepwise process is supported by the observation that epoxidation of some olefins yields side products; terminal olefins, for example, can act as suicide substrates through formation of an irreversibly *N*–alkylated porphyrin with concomi-

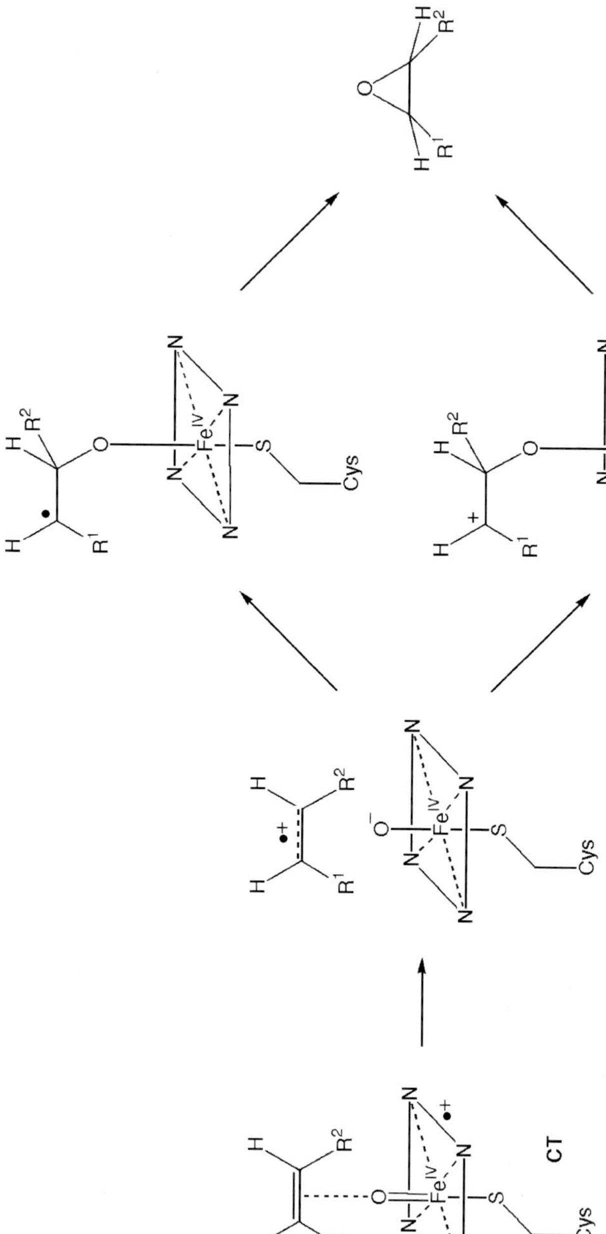

Scheme 10.5 Tentative mechanism for cytochrome P450-catalyzed epoxidation of a double bond. The reactive iron-oxo species **VII** (see Scheme 10.4) reacts with the olefin to give a charge transfer (CT) complex. This complex then resolves into the epoxide either through a radical or through a cationic intermediate.

tant demetalation [7, 8]. The porphyrin adduct is not produced exclusively, but alongside the epoxide (Scheme 10.6). Trichloroethylene, on the other hand, mainly yields trichloroacetaldehyde whereas the epoxide is formed in small amounts. More importantly, trichloroethylene oxide is not a precursor for aldehyde formation [9]. The formation of the porphyrin adduct can be explained equally well by radical or cationic intermediates, but the presence of a cationic intermediate better accounts for group migration (Scheme 10.6).

Cytochrome P450 enzymes have been the subject of a number of recent reviews in which their mechanism and scope of action are covered in much detail [1, 6, 10, 11]. The reader is referred to these articles for a more thorough account of the mechanism and reactivity of cytochrome P450 enzymes, while we present a few representative examples of cytochrome P450-catalyzed epoxidation below. The enzymes we chose are all involved in the biosynthesis of polyketide natural products. Polyketides are a large, structurally diverse family of compounds and have provided a wealth of therapeutically useful drugs and drug leads.

10.2.2 **Epothilones**

Epothilones – such as paclitaxel (Taxol®), cryptophycin, and laulimalide (Scheme 10.7) – exhibit strong cytotoxicity by stabilizing microtubule arrays resulting in G2-M arrest, apoptosis, and cell death [12, 13]. The major advantage of epothilone over paclitaxel is its potency against P–glycoprotein-expressing multiple-drug-resistant cell lines [14]. This activity has made epothilone and its derivatives very attractive for further investigation as anticancer agents [13, 15]. Biosynthetically, epothilones originate from a hybrid polyketide synthase (PKS)/nonribosomal peptide synthetase (NRPS) system in which the 16-membered macrolactone core is solely PKS-derived whilst the pendant thiazole moiety stems from the NRPS portion.

The epothilone gene cluster has been identified and isolated from the myxobacterium *Sorangium cellulosum* [16, 17]. As PKS- and NRPS-derived natural products constitute a large class of bioactive compounds, we will use the example of epothilone to demonstrate the biosynthetic logic by which the core structures are assembled. In essence, PKS-derived compounds are formed through the polymerization of acetate units, whereas NRPS-derived compounds are oligopeptides. Both PKSs and NRPSs act as multifunctional enzyme complexes in which all of the intermediates are covalently attached to the protein. The final product is thus assembled from a starter molecule in repetitive elongation steps, during which the intermediates are passed down the enzymatic assembly line. In this process, modules are those entities responsible for catalyzing one cycle of elongation, whilst domains are the individual catalytic units, each with its specific function. The linear product obtained in this fashion is typically cleaved from the enzyme through cyclization to form the macrolactone. Other PKS or NRPS systems release their products through hydrolysis or reductively to yield linear products. This core structure can then be further modified through tailoring enzymes such as methyl-

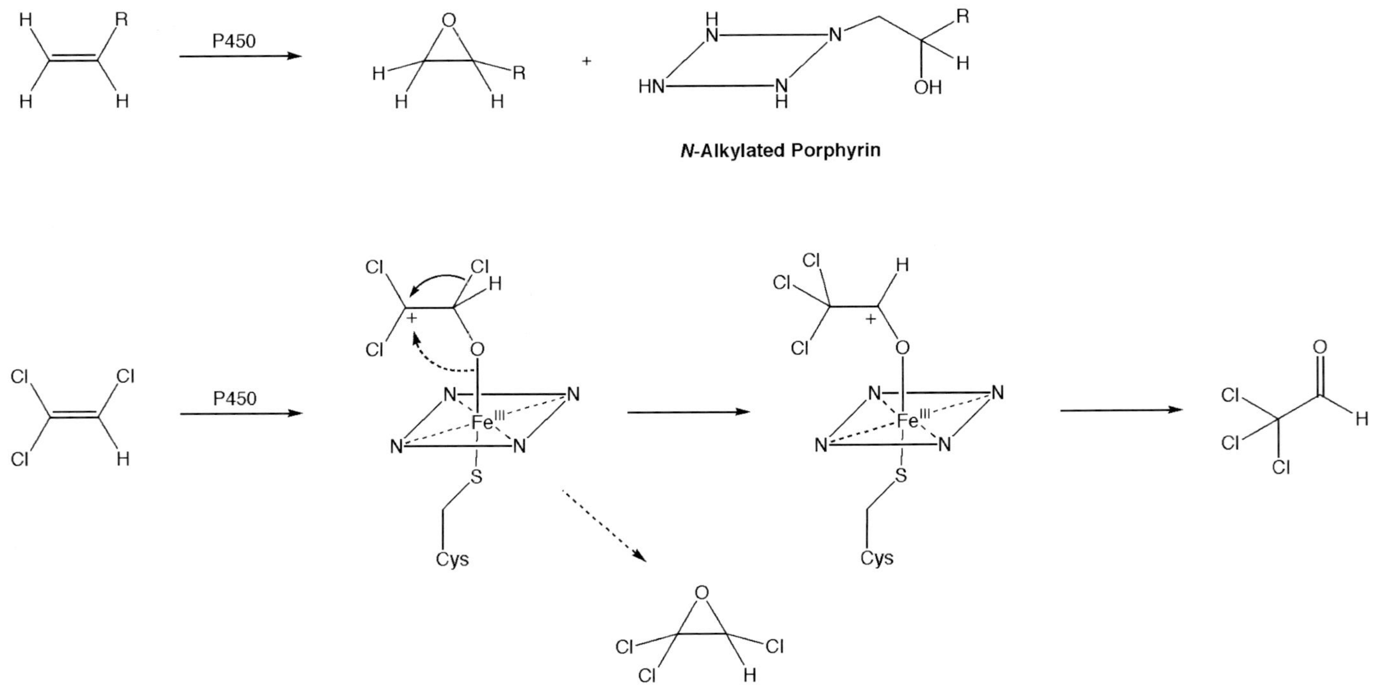

P450

N-Alkylated Porphyrin

P450

Scheme 10.6 Byproducts of cytochrome P450 oxygenase catalyzed epoxidation. Top: *N*-alkylation of the porphyrin ring. Bottom: group migration to give aldehydes.

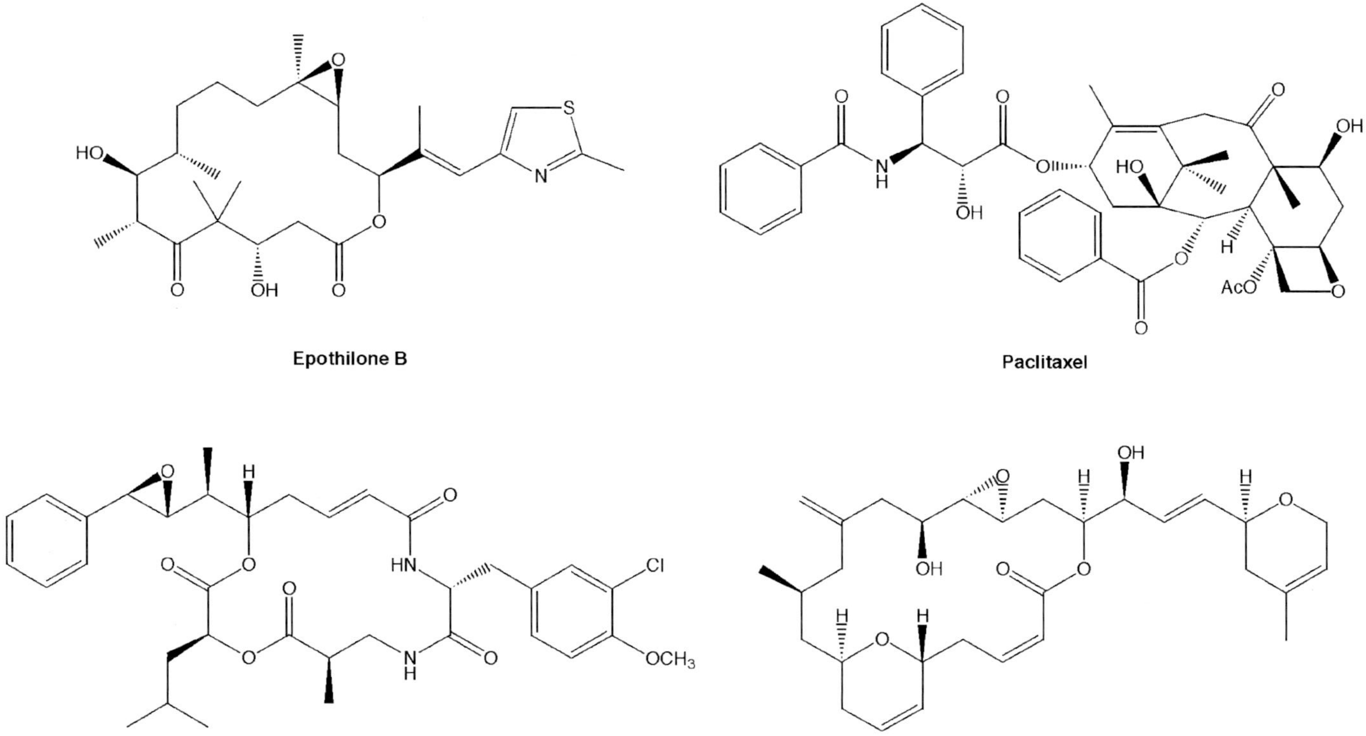

Paclitaxel

Laulimalide

Epothilone B

Cryptophycin-1

Scheme 10.7 Microtubule-binding natural products.

transferases or glycosyltransferases. For a more detailed discussion on PKS, NRPS, or hybrid PKS/NRPS systems the reader is referred to references [18–20].

The biosynthesis of epothilone is outlined in Scheme 10.8. Briefly, the aglycone is synthesized by loading of an acetyl starter unit on the first PKS module. The downstream NRPS module then attaches cysteine to the starter unit, with concomitant dehydration and oxidation of the formed thiazoline ring to produce a thiazole moiety. The remaining steps leading to the aglycone are carried out by several multimodular PKS enzymes to give epothilone D. A gene encoding a cytochrome P450 enzyme (P450epoK) is also found in the cluster. P450epoK was shown to convert epothilone D to epothilone B *in vitro* by epoxidation of the C12=C13 double bond.[17] Spinach ferredoxin and spinach ferredoxin:NADP$^+$ oxidoreductase were capable of replacing the unidentified endogenous electron-transport system necessary for turnover. P450epoK has been crystallized in the absence and in the presence of substrate or product and represents the only instance of a cytochrome P450 enzyme with native epoxidase activity for which the structure has been solved.[21] The closest homologue of P450epoK with a known structure is the cytochrome P450 hydroxylase from the erythromycin pathway, P450eryF [22], that is responsible for the introduction of the C6 hydroxy group into the macrolide ring (Scheme 10.9) [23, 24]. The overall fold of P450epoK is the same as for other P450 enzymes. Major differences, however, are observed in the substrate-binding pocket: the volume of the active site cavity in P450epoK is significantly greater than that in P450cam [25, 26] (240 Å^3 in P450cam and 1060 Å^3 in P450epoK) to accommodate the much larger substrate.

In both P450eryF and P450epoK the substrates are oriented roughly perpendicular to the heme plane. However, the macrolide ring system of epothilone is rotated by approximately 90° relative to erythromycin in order to accommodate the thiazole substituent. Most of the hydrogen bonds between the protein and its ligand are mediated through water molecules. Amino acids lining the recognition pocket are mainly small, nonpolar residues such as alanine or glycine. One notable exception is the Arg71 residue of P450epoK. This amino acid provides the only sidechain involved in the hydrogen bonding network, whereas the other interactions

Scheme 10.8 Biosynthesis of epothilone. Individual PKS domains are represented as circles and individual NRPS domains as hexagons. Acyl carrier proteins (ACPs) and thiolation domains (T) are posttranslationally modified by a phosphopantetheinyl group to which the biosynthetic intermediates are covalently bound throughout the chain assembly. The thioesterase domain (TE) cyclizes the fully assembled carbon chain to give the 16-membered lactone. Following dehydration of C12–C13 to give epothilones C and D, the final step in epothilone biosynthesis is the epoxidation of the C12=C13 double bond by the cytochrome P450 enzyme P450epoK. KS: ketosynthase; KS(Y): active-site tyrosine mutant of KS; AT: acyltransferase; C: condensation domain; A: adenylation domain; ER: enoyl reductase; DH: dehydratase; KR: ketoreductase; MT: methyltransferase; Ox: oxidase. ▶

EpoA EpoB EpoC EpoD EpoE EpoF

Loading NRPS Module 1 Module 2 Module 3 Module 4 Module 5 Module 6 Module 7 Module 8

P450epoK

R = H: Epothilone C
R = CH_3: Epothilone D

R = H: Epothilone A
R = CH_3: Epothilone B

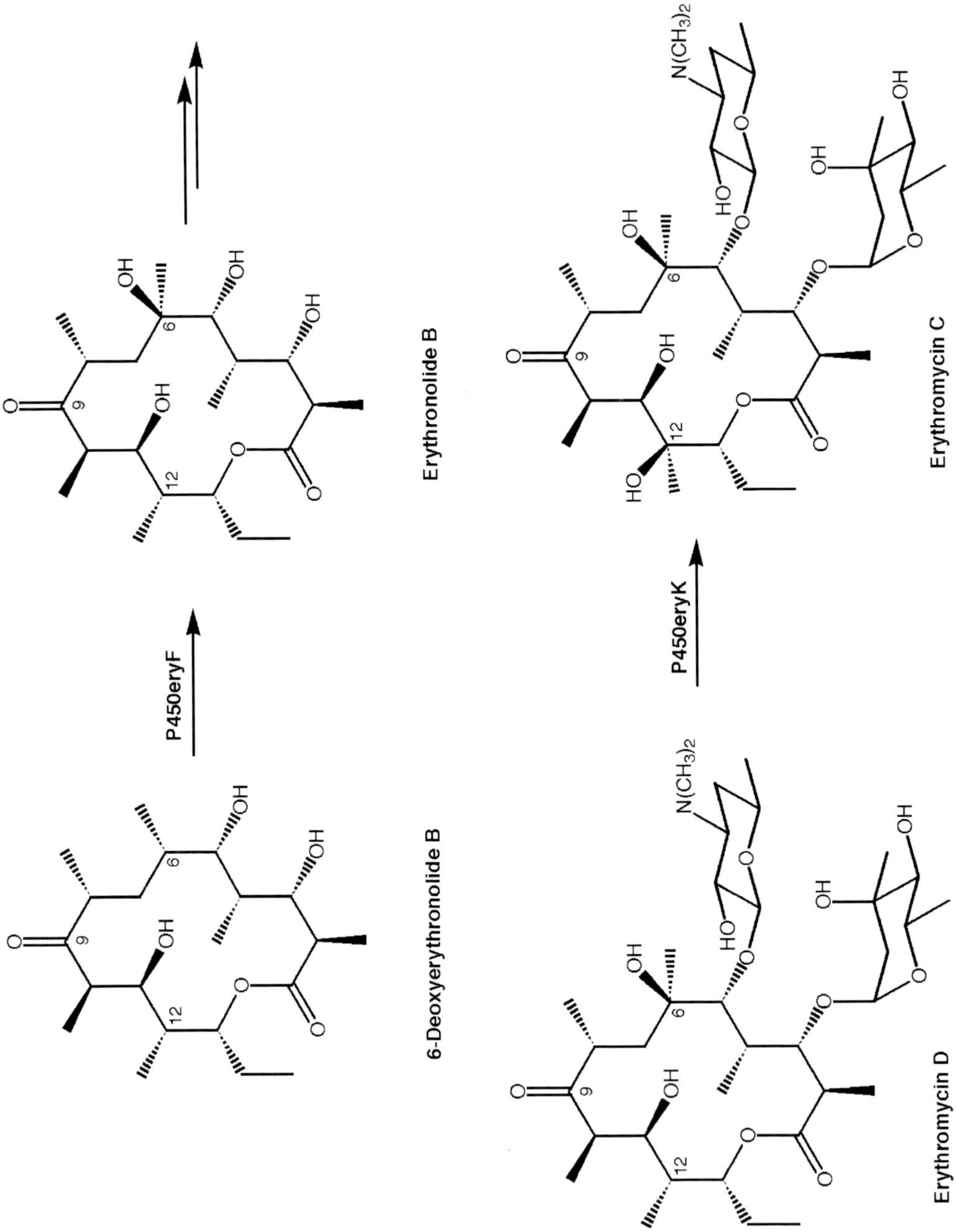

6-Deoxyerythronolide B → P450eryF → Erythronolide B →

Erythromycin D → P450eryK → Erythromycin C

Scheme 10.9 Final steps in the biosynthesis of erythromycin.

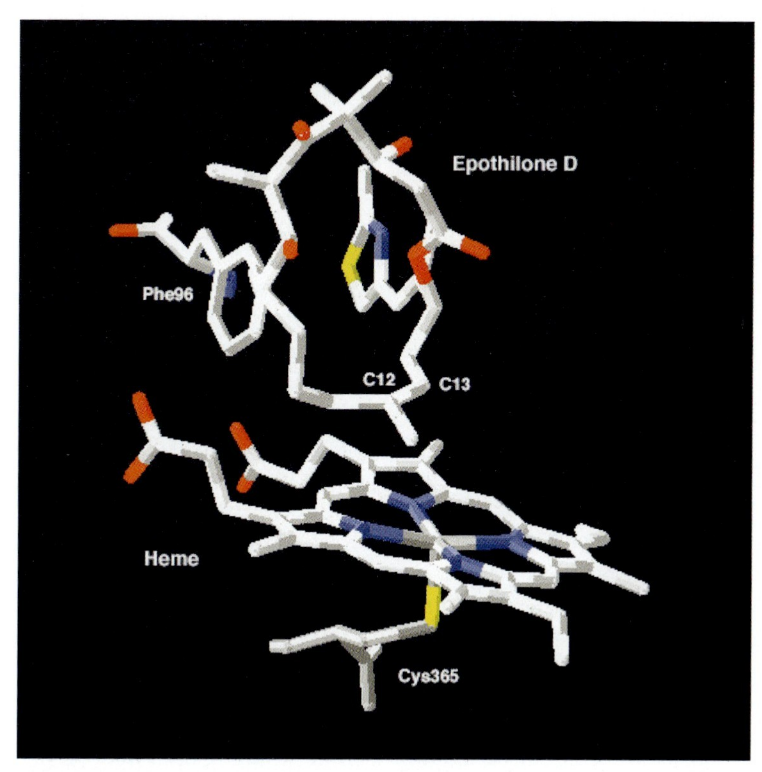

Figure 10.1 Conformation of epothilone D bound in the P450epoK active site. The figure was prepared from PDB entry 1PKF using Swiss-PdbViewer [27]. The heme prosthetic group is shown with its cysteinate ligand (Cys365) bound. For clarity, amino acids lining the active site have been omitted apart from Phe96.

all involve the protein backbone. The aromatic ring of Phe96 and the thiazole ring are suitably positioned for π-π stacking. This interaction, together with the Arg71 hydrogen bond to the thiazole ring (mediated by a water molecule), is likely to be critical for substrate specificity. As can be seen in Figure 10.1, the C12=C13 double bond of epothilone D is close to the iron site with the π-system perpendicular to the heme plane. This orientation is perfectly suited for oxygen insertion by the oxo-iron species (complex **VII** in Scheme 10.4) and also accounts for the stereochemical outcome of the reaction.

From the active site topology it seems that there is room for substrate flexibility. Indeed, experiments with the closely related P450eryF have demonstrated that some substitutions within the macrolactone ring of the substrate are possible [28]; for example, reduction of the C9 oxo to the hydroxy group is well tolerated. However, any changes with impact on the overall conformation of the substrate, thus changing the trajectory between the reactive C–H bond and the iron-bound oxygen, decrease the reactivity significantly [28]. In a different study, a single amino acid substitution, A245T, conferred the ability to oxidize sterols on P450eryF [29]. This example serves to demonstrate that simple changes can significantly alter the substrate specificity of P450 enzymes. It should be possible to apply the same strategy to other cytochrome P450 monooxygenases such as P450epoK in order to obtain catalysts for the epoxidation of complex molecules.

10.2.3 Mycinamicin

Mycinamicin II (Scheme 10.10) is a 16-membered macrolide antibiotic with strong antibacterial activity against Gram-positive bacteria. The gene cluster responsible for the biosynthesis of mycinamicin has been isolated from the actinomycete *Micromonospora griseorubida* [30]. It incorporates the expected type I PKS genes as well as genes responsible for the biosynthesis of the sugar substituents and those responsible for other tailoring functions including the cytochrome P450 homologue *mycG*. Mycinamicin I (Scheme 10.10) is isolated from *Micromonospora griseorubida* alongside mycinamicin II at a ratio of 1:5. The two compounds differ from each other in that the latter contains a hydroxy group at C14 that is missing in mycinamicin I.

Bioconversion studies in *Micromonospora griseorubida* have led to a biosynthetic scheme in which the mycinamicin aglycon is first glycosylated with desosamine at the C5 hydroxy group, C21 is then hydroxylated, and the didesmethylmycinose is subsequently attached to the newly formed hydroxy group. Methylation of this sugar gives mycinamicin IV [31]. Hydroxylation at C14 and epoxidation of the C12=C13 double bond were studied with a mutant strain of the producing organism that accumulated mycinamicin IV [32]. Mycinamicin IV differs from the final product mycinamicin II in that it lacks both the C14 hydroxy group and the C12–C13 epoxide. Introduction of the cytochrome P450 gene *mycG* alone was sufficient to restore the production of mycinamicins I and II, at the same relative ratio as observed in the wild-type organism [32]. It can therefore be concluded that the *mycG* product, P450mycG, harbors both hydroxylation and epoxidation functions.

To account for the presence of mycinamicin I and mycinamicin II, two separate pathways starting with mycinamicin IV ("upper" and "lower" pathway in Scheme 10.10) can be envisaged. Binding of mycinamicin IV to P450mycG brings either the C14–H bond or the π-system of the C12=C13 double bond into close proximity to the heme iron. The former substrate orientation results in the hydroxylated product mycinamicin V, whereas the latter substrate orientation results in the epoxidized product mycinamicin I. Mycinamicin V is also a substrate for P450-mycG and is epoxidized to the main product mycinamicin II. The epoxide mycinamicin I, however, is a poor substrate for the cytochrome monooxygenase and hence also accumulates in the producing organism. The bifunctionality of P450-mycG therefore seems to arise from a certain plasticity of its active site towards the nature of the substrate and towards orientation of the substrate.

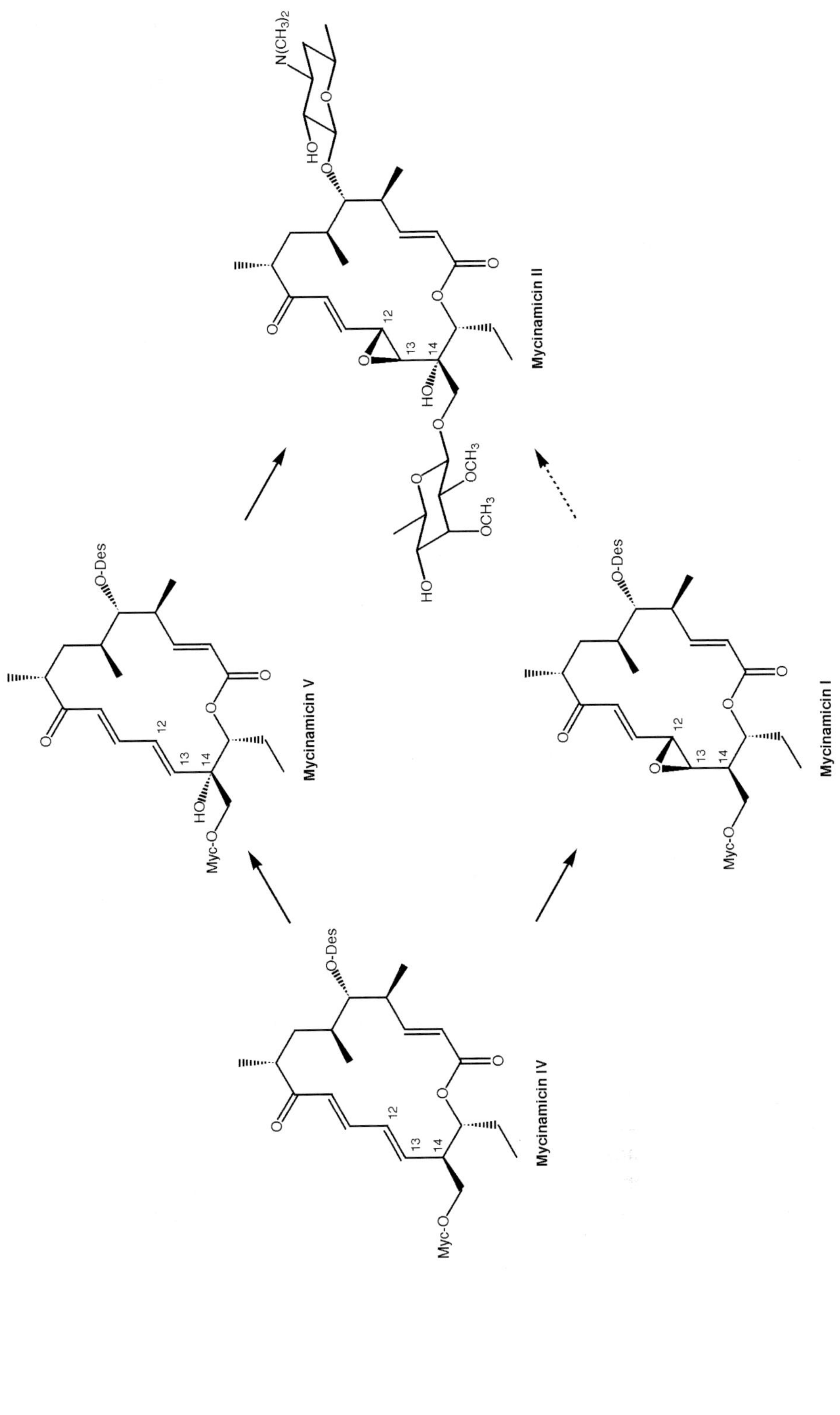

Mycinamicin IV

Mycinamicin V

Mycinamicin I

Mycinamicin II

Scheme 10.10 Final steps in mycinamicin biosynthesis. Des: desosamine; Myc: mycinose.

Apart from its bifunctionality, P450mycG also differs from P450epoK discussed above in the size of the substrate it utilizes. Whereas P450epoK epoxidizes a non-glycosylated macrolactone, P450mycG acts on a glycosylated macrolactone. This requires the active site of P450mycG to be significantly larger than that of P450-epoK in order to accommodate the bulky sugar substituents. In this respect, P450mycG is similar to P450eryK from erythromycin biosynthesis (Scheme 10.9) as the latter also acts on the glycosylated macrolide ring to introduce the C12 hydroxy group. P450eryK should not be confused with P450eryF, also mentioned above, for which the crystal structure has been solved.

10.2.4
Griseorhodin A

Griseorhodin A is a member of the rubromycin family of compounds (Scheme 10.11), most of which are strong inhibitors of human telomerase and of retroviral reverse transcriptase [33, 34]. The presence of the rarely observed spiroketal moiety gives the structure an axial chirality and has been shown to be very important for the biological activity of these compounds [33]. The biosynthetic gene cluster of griseorhodin A has been isolated from the bacterium *Streptomyces* sp. JP95, found associated with the marine invertebrate *Aplidium lenticulum* [35]. Unlike the type I PKS-derived macrolide antibiotics discussed above, the core structure of the rubromycin family of natural products is biosynthesized by iteratively acting type II PKS systems. Type II PKSs catalyze Claisen condensations by a single discrete ketosynthase that is used repetitively on the growing polyketide chain. The chain extensions occur without reductive steps and the final polyketide product is aromatized and released from the ACP with the assistance of accessory enzymes (see [36] for a review on type II PKSs).

A possible biosynthetic pathway to griseorhodin A is outlined in Scheme 10.12. The most striking feature in the biogenesis of all rubromycin antibiotics is undoubtedly the formation of the spiroketal moiety. Formation of the spiroketal is accompanied by loss of two carbon atoms from a polyketide intermediate such as collinone or a closely related compound [37, 38]. The α-hydroxyketone formed after loss of the two-carbon unit is a likely intermediate in the biosynthetic pathway of the rubromycin family of natural products. Ketal formation from this precursor would provide a suitable substrate for epoxidation of the olefin to give griseorhodin A. The gene cluster contains a gene pair encoding for a cytochrome P450 and for a ferredoxin. The associated gene products are the best candidates for the epoxide formation and a dedicated electron-transport system, respectively [35]. The epoxidation in griseorhodin A biosynthesis probably occurs at a late stage. In this way, the same assembly logic can be applied to all rubromycins with the structural variations between the individual members being evoked through the action of a small set of tailoring enzymes.

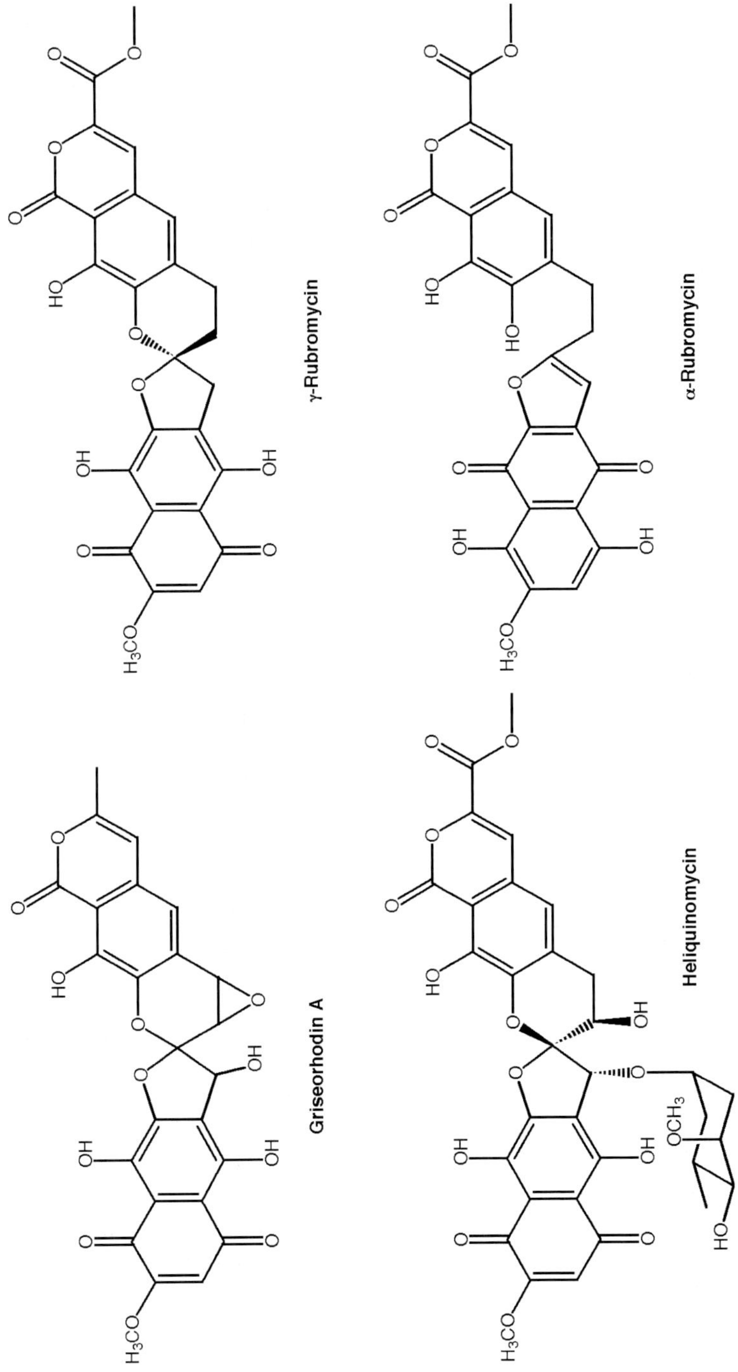

γ-Rubromycin

α-Rubromycin

Griseorhodin A

Heliquinomycin

Scheme 10.11 The rubromycin family of natural products.

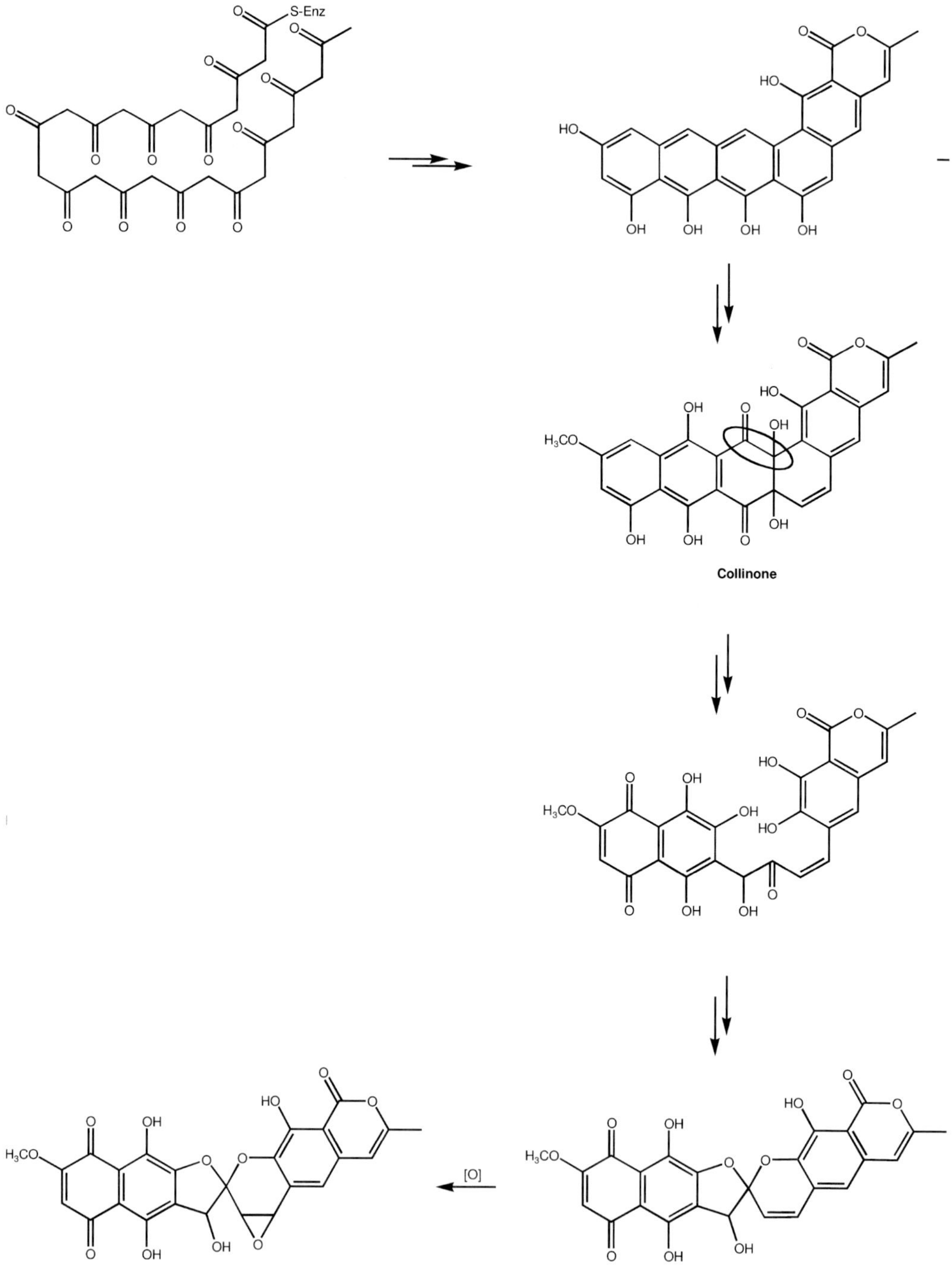

Collinone

[O]

Scheme 10.12 Proposed biosynthetic pathway leading to griseorhodin A.

10.2.5 Hedamycin

The rubromycins constitute an unusual structural class, due to the presence of the spiroketal moiety. A more typical example of a type II PKS-derived natural product is the antitumor agent hedamycin (Scheme 10.13) [39, 40]. Hedamycin belongs to a family of closely related compounds, the pluramycins. These antibiotics have the

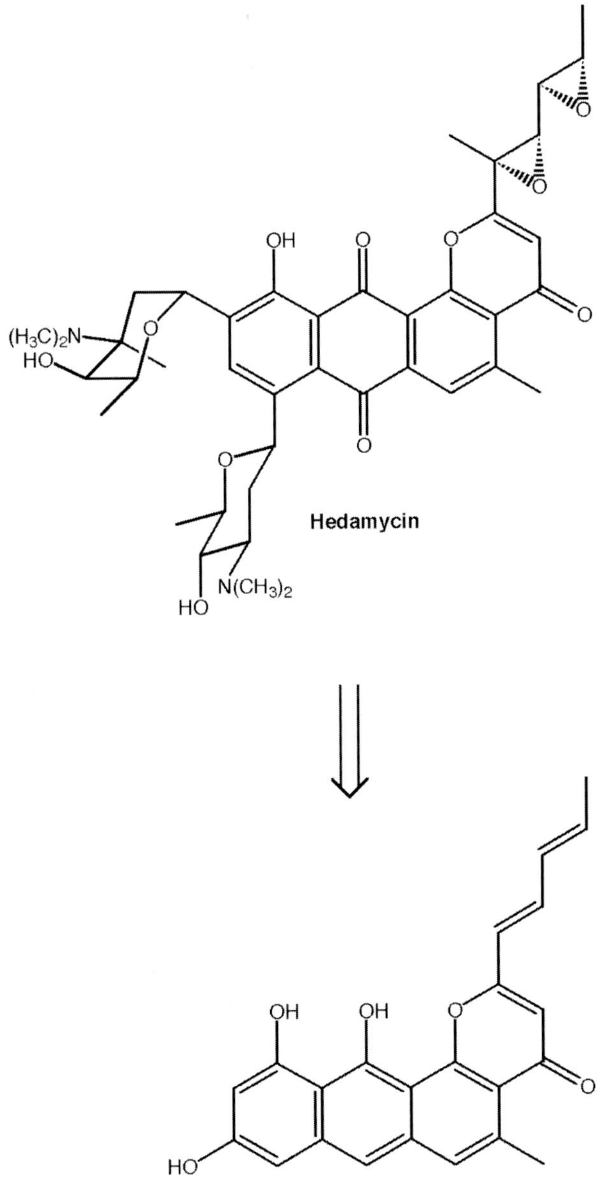

Hedamycin

Scheme 10.13 Structures of hedamycin and its type II PKS-derived precursor.

tetracyclic core structure in common but differ in the natures of their sugar substituents and in the identities of the alkyl sidechains.

The most unusual feature of the hedamycin biosynthetic pathway is the cooperation of type I and type II PKS systems to provide the aglycone. An iteratively acting type I PKS is implicated in synthesizing the hexenoate moiety utilized as the starter unit for the type II PKS system [41]. The hexenyl group also provides the double bonds for installment of the two epoxide rings. The timing of the epoxidation or the identity of the enzyme(s) involved have not yet been established experimentally, but two candidate genes responsible for the epoxidations can be assigned through analysis of the hedamycin gene cluster isolated from *Streptomyces griseoruber* [41]. The first gene encodes a cytochrome P450 homologue and the second a flavin-dependent monooxygenase; the latter enzyme class is discussed in Section 10.3. The point that remains to be clarified is whether two different enzymes are indeed involved in carrying out the individual oxygenations. With both sites of epoxidation so close to each other it seems at least intuitively possible that just one enzyme might suffice. This would be a situation similar to that seen in mycinamicin biosynthesis where P450mycG catalyzes two oxidations at neighboring positions.

10.3 Flavin-dependent Epoxidases

10.3.1 Squalene Epoxidase

Sterols, produced by most eukaryotes and a handful of bacteria, are vital components of cell membranes (cholesterol in mammals and ergosterol in fungi, for example) and are also important signaling molecules or hormones (such as brassinosteroids in higher plants) [42–44]. The biosynthetic pathway leading to different sterols is a branch of the isoprenoid pathway and is initially shared by all eukaryotes but diverges after the synthesis of 2,3-oxidosqualene. This compound is produced through epoxidation of the triterpenoid squalene by the enzyme squalene epoxidase (also referred to as 2,3–squalene monooxygenase) as shown in Scheme 10.14, and is the direct precursor for cyclization to give the sterol framework [43, 45]. For cyclization, the triterpenoid precursor is molded into prechair and preboat conformations. Opening of the epoxide gives a tertiary cation ideally placed for electrophilic addition to the C6=C7 double bond initiating a series of ring-forming reactions [46]. Interestingly, bacterial squalene cyclization is an anaerobic process; instead of using an epoxide to generate the initial carbocation, bacterial squalene cyclases protonate the C2=C3 double bond of squalene [46]. Squalene epoxidases from different organisms can be selectively inhibited [47, 48] which has resulted in the development of antifungal agents such as the allylamine terbinafine [49]. Furthermore, human squalene epoxidase is being investigated as a target for the design of hypercholesterolemic drugs [50].

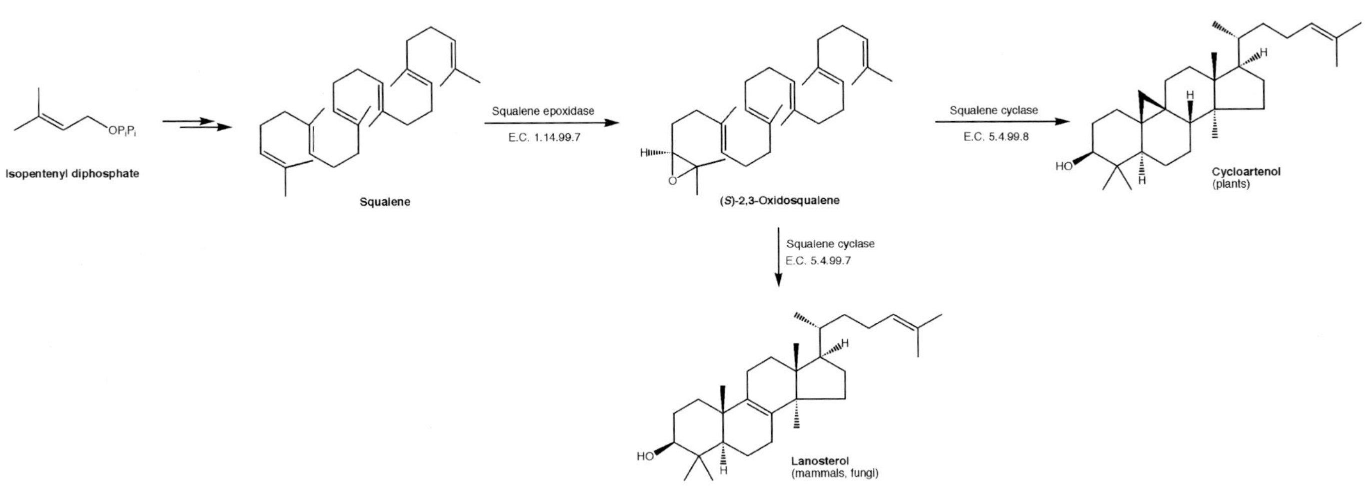

Scheme 10.14 Partial biosynthetic pathway leading to sterol natural products.

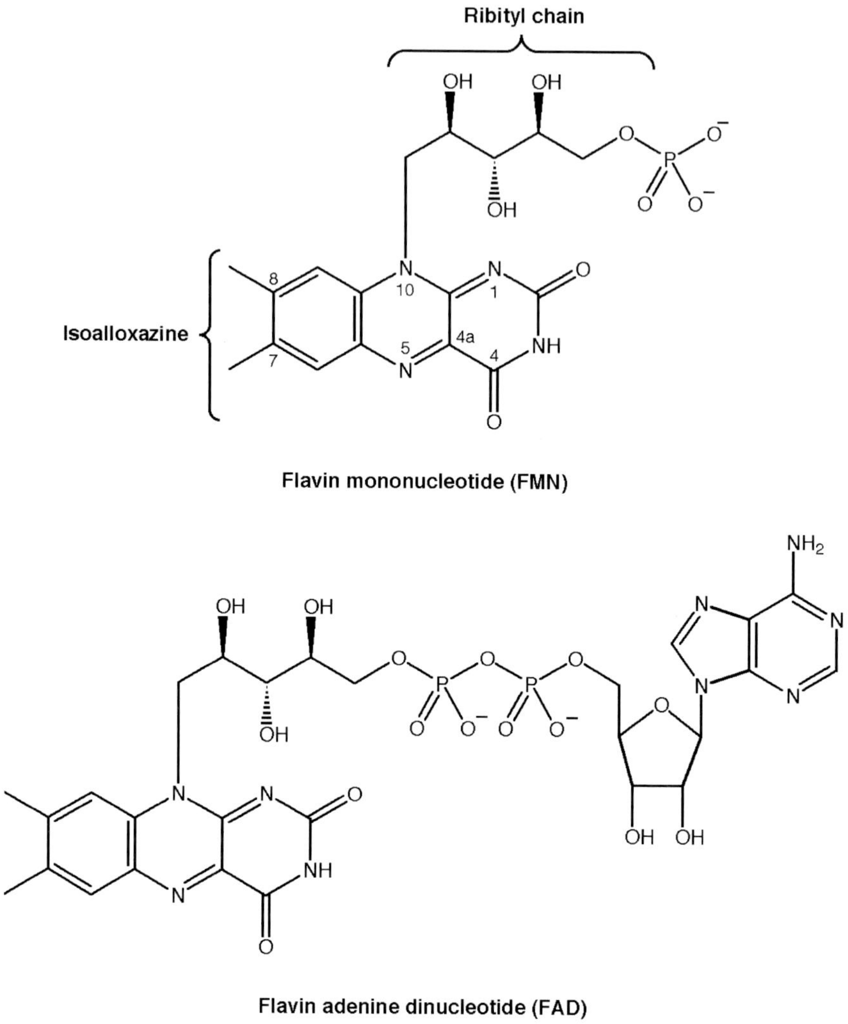

Ribityl chain

Isoalloxazine

Flavin mononucleotide (FMN)

Flavin adenine dinucleotide (FAD)

Scheme 10.15 Chemical structures of FMN and FAD.

Squalene epoxidase, like most enzymes responsible for the later steps of sterol biosynthesis [43, 51], is membrane-bound which makes its purification in native form challenging. The purification is additionally complicated by the presence of a large number of cytochrome P450 and other enzymes that have similar hydrophobicity and size as squalene epoxidase and are hence difficult to remove [52]. Most studies have been carried out with rat liver microsome squalene epoxidase either partially purified or as a homogenate of the cell membrane fraction. *In vitro* reconstitution of squalene epoxidase activity is absolutely dependent on molecular oxygen, NADPH, FAD, and NADPH-cytochrome c reductase [52, 53]. In this respect, squalene epoxidase resembles the cytochrome P450 enzymes described above. Instead of the heme-iron prosthetic group, however, squalene epoxidase utilizes flavin for activation of molecular oxygen [54, 55].

Flavin-dependent enzymes, or flavoproteins, constitute a large family of enzymes with a multitude of activities reflecting the chemical versatility of the flavin cofactor. Two chemically distinct forms of flavin – flavin mononucleotide (FMN) and flavin adenine dinucleotide (FAD, Scheme 10.15) – exist, and most enzymes will discriminate between them. Apart from electron-transport reactions, flavoproteins are involved in dehydrogenations, hydroxylations, and disulfide reductions to name but a few [56]. In contrast to the nicotinamide cofactors, which are obligate two-electron donors/acceptors, or cytochrome P450s, which are obligate one-electron donors/acceptors, flavins can act as both one- and two-electron donors/acceptors (Scheme 10.16). The cytochrome P450 electron-transport systems (Section 10.2.1) make use of this property to enable communication between the NAD(P)H donor and the cytochrome P450 acceptor. Flavoprotein monooxygenases are characterized by the use of NAD(P)H to reduce enzyme-bound FAD and by the subsequent reaction of reduced flavin with molecular oxygen to form the C4a peroxide via a caged radical pair [56]. Flavin peroxide in its deprotonated form is a good nucleophile and is used, for example, by bacterial luciferase. Bacterial luci-

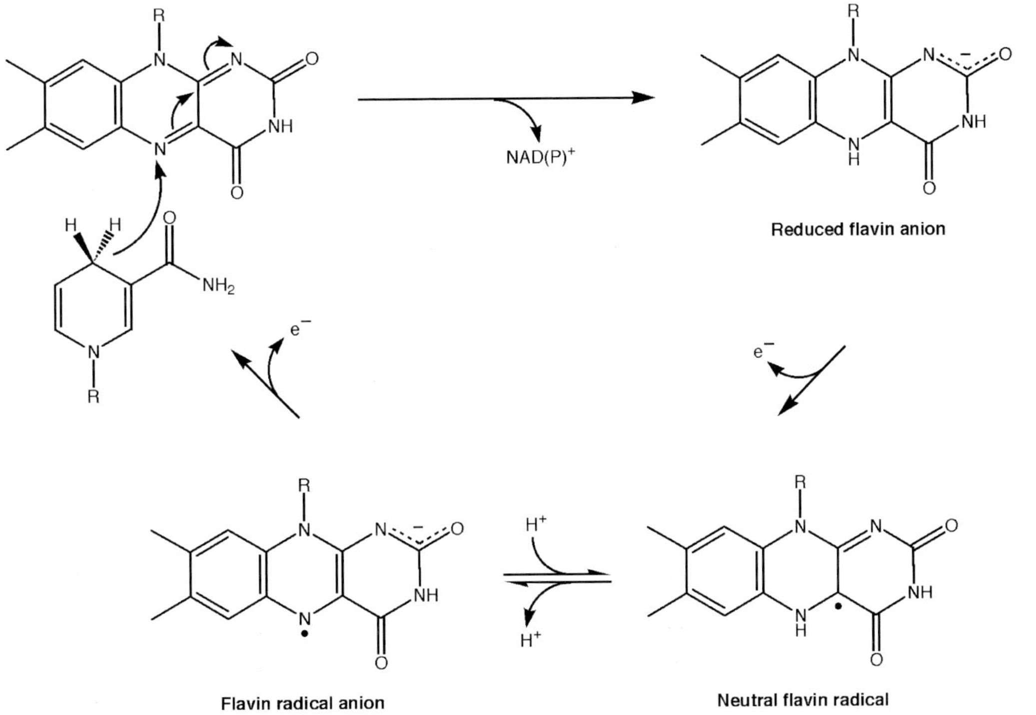

NAD(P)$^+$

Reduced flavin anion

Flavin radical anion

Neutral flavin radical

Scheme 10.16 Redox cycle of flavins. The cycle is depicted with a two-electron reduction of flavin by NAD(P)H and two one-electron oxidations.

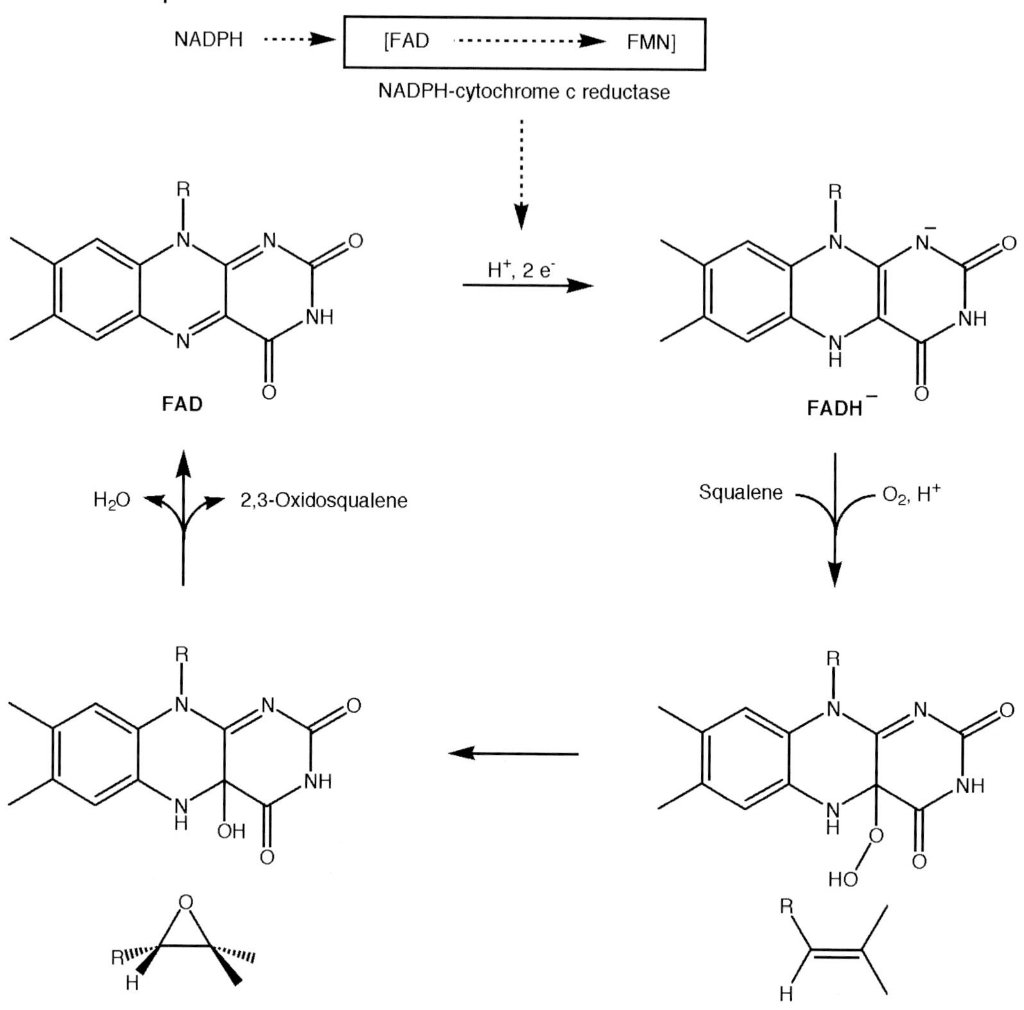

Scheme 10.17 Reaction cycle of the flavin-dependent squalene monooxygenase. Dashed arrows indicate electron transport.

ferases catalyze the oxidation of long-chain aldehydes to the corresponding acids with concomitant emission of light [57, 58]. Flavin peroxide in its protonated form, on the other hand, is a good electrophile and is employed in, for example, aromatic hydroxylases. Aromatic hydroxylases introduce additional hydroxy groups into phenols and play an important role in soil detoxification [59]. In bifunctional flavoprotein monooxygenases both FAD and NAD(P)H are bound on the same polypeptide chain [56]. The term "bifunctional" refers to the ability of these enzymes to generate the reduced flavin and to perform the oxidation of the substrate at the same time. Monofunctional flavoprotein monooxygenases, on the other hand, require an external factor to produce the reduced flavin [56]. This factor is another flavoprotein, NAD(P)H:flavin reductase.

Let us return to the specific case of squalene monooxygenase. It was found that FAD binds to the enzyme with a dissociation constant in the low micromolar range [54, 55]. The reducing equivalents are mediated through NADPH-cytochrome c reductase and, as already indicated above, this component is absolutely required to reconstitute epoxidase activity *in vitro* [52, 53]. As such, squalene epoxidase is related to monofunctional flavoprotein monooxygenases but makes use of an unusual redox partner. A tentative reaction mechanism is shown in Scheme 10.17. After electron transfer from the associated reductase oxygen binds to the reduced flavin to form the peroxide as is observed for other flavin-dependent monooxygenases. This represents the reactive oxygen species. The subsequent steps in the reaction cycle are analogous to those of aromatic hydroxylases. The distal oxygen of the peroxide is inserted into the π-system of the substrate olefin and oxidized flavin is regenerated through loss of water. No detailed mechanistic studies have yet been carried out, hence it is not known whether oxygen insertion occurs in a concerted fashion or if it is a stepwise process. Knowledge about the three-dimensional structure of squalene epoxidase could shed light on these questions; no crystal structure has been solved yet, but progress is being made in identifying residues involved in FAD and in substrate binding by site-directed mutagenesis and photoaffinity labeling with substrate analogues [54, 60]. These studies should help in elucidating the active site geometry.

To summarize, squalene epoxidase is a flavoprotein capable of catalyzing the insertion of oxygen into the 2,3-double bond of squalene to give 2,3-oxidosqualene, with the second oxygen atom from O_2 being reduced to water. The reducing equivalents necessary for this transformation are relayed from NADPH through NADPH-cytochrome c reductase to the flavin cofactor of the epoxidase.

10.3.2 Styrene Epoxidase

A number of microorganisms (*Pseudomonas, Nocardia, Xanthobacter* species and others) are capable of utilizing styrene as the sole source of carbon. Two catabolic pathways for styrene have been described: the first involves epoxidation as the first step (Scheme 10.18), while the second pathway starts with the dihydroxylation of the aromatic ring [61, 62]. Epoxidation of styrene in the Gram-negative bacterium *Pseudomanas* is achieved through a two-component system consisting of a flavin-dependent monooxygenase (StyA) and a dedicated NADH:flavin reductase (StyB) [63–65]. Epoxide formation is likely to proceed as shown for squalene epoxidase in Scheme 10.17. The StyA-catalyzed epoxidation of styrene yields (*S*)-epoxystyrene in over 99% enantiomeric excess [66]. The main differences from the squalene epoxidase system are the class of flavin reductase employed and the fact that StyA and StyB are soluble proteins and not membrane-bound. The NADH:flavin reductase StyB is unusual in that it seems to bind FAD after NADH has bound to the enzyme [64]. Typical flavoproteins bind their flavin cofactor very tightly, so that it

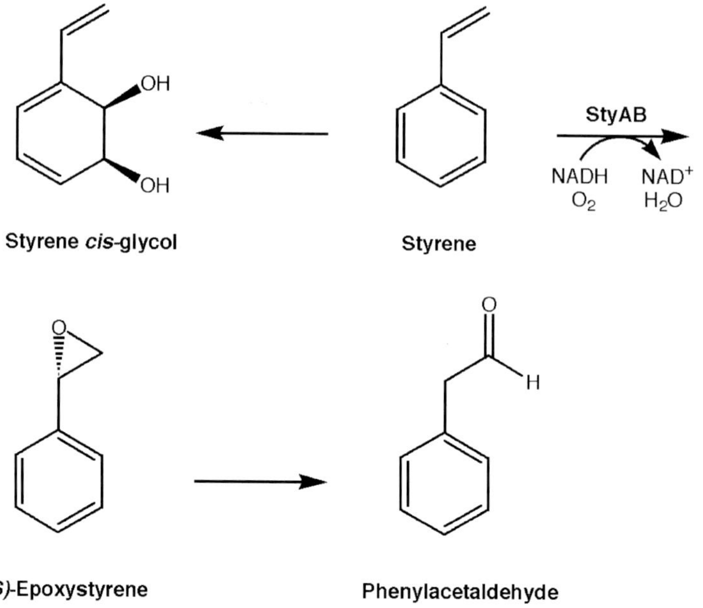

Styrene *cis*-glycol

Styrene

StyAB

NADH O_2 NAD$^+$ H_2O

(*S*)-Epoxystyrene

Phenylacetaldehyde

Scheme 10.18 First steps in the degradation of styrene.

actually copurifies with the enzyme [67]. This is not the case for squalene epoxidase [68], styrene epoxidase, and StyB [64]. Investigation into the kinetics of StyAB-catalyzed epoxidation has resulted in a model in which StyB reduces FAD to $FADH_2$ which diffuses to the monooxygenase StyB where it serves to activate O_2 for epoxide formation [64]. This model is controversial, however, and will need to be further substantiated as free $FADH_2$ is very reactive towards molecular oxygen [56, 67].

Styrene is also metabolized by eukaryotes [69–71]. In mammals, styrene metabolism serves primarily to counteract the toxic effects of styrene [71], whereas a few fungal strains can utilize the compound as carbon source for growth [70]. The toxicity of styrene and its metabolite epoxystyrene in humans, combined with its widespread use as a precursor in the manufacturing of polymers, has spurred efforts to utilize bacterial and fungal cells for bioremediation. Even though the first step of styrene metabolism in eukaryotes is also the formation of the epoxide, fungal and mammalian cells employ cytochrome P450 monooxygenases (Section 10.2.1) to catalyze this reaction [69, 72]. It is quite remarkable but not uncommon to find that the same chemical transformation can be achieved by different mechanisms.

Both compounds discussed in this section, oxidosqualene and epoxystyrene, are intermediates in metabolic pathways. That is to say that the epoxide is not found in the final product but rather serves to impart a specific reactivity to the molecule that is vital for the subsequent reaction step. The epoxide in oxidosqualene can be viewed as a masked cation that is required in order to initiate a series of C–C bond-

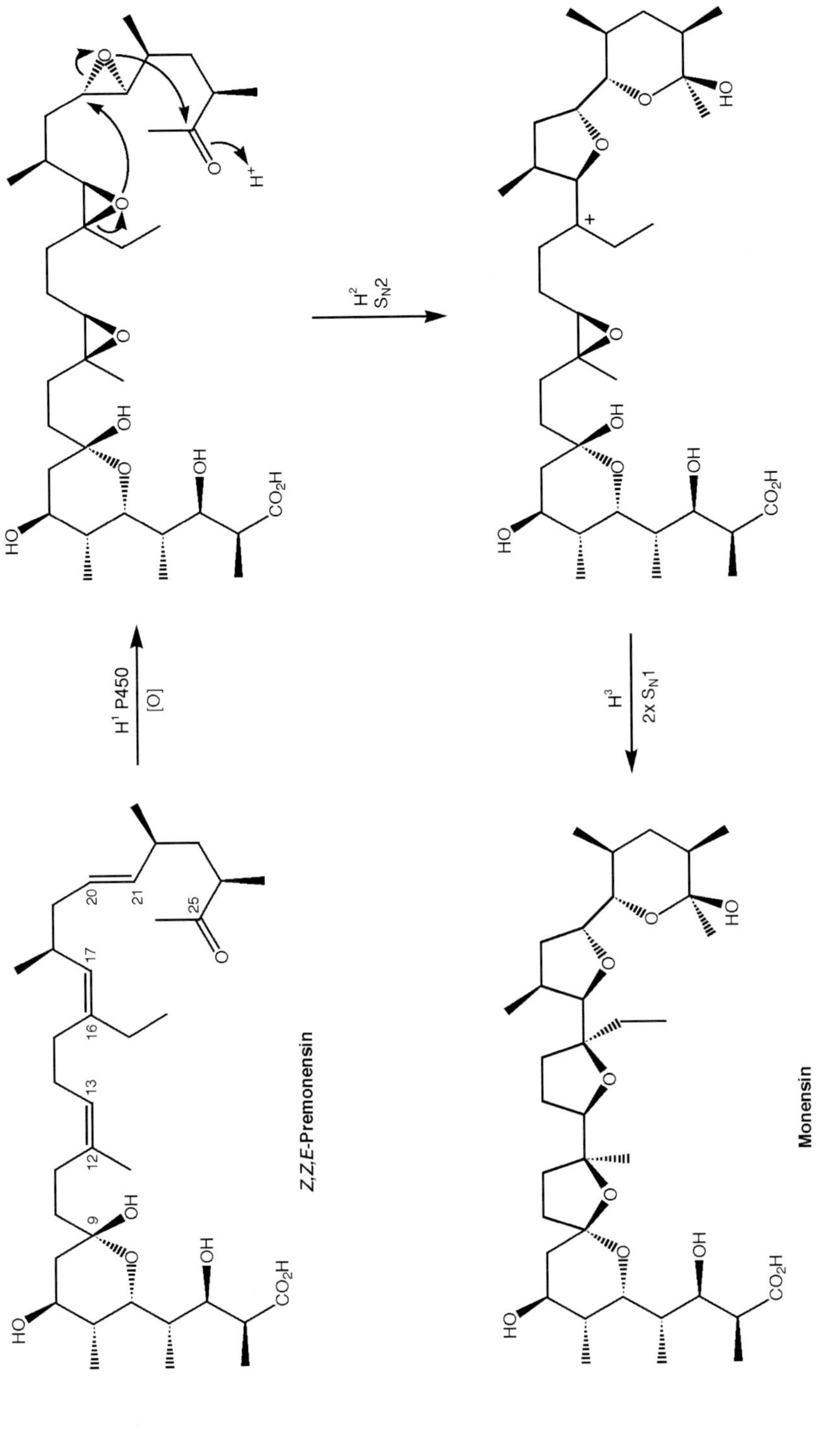

Z,Z,E-Premonensin

H^1 P450
[O]

H^2
S_N2

H^3
2x S_N1

Monensin

Scheme 10.19 Ether formation in monensin biosynthesis.

forming reactions eventually affording the steroid carbon framework. In styrene metabolism, on the other hand, the epoxide is easily converted into the aldehyde by isomerization. The formation of pyran and furan rings in the antibiotic monensin (Scheme 10.19) is also thought to involve epoxide intermediates [73]. In this case, the double bonds of the linear type I PKS product are epoxidized by a cytochrome P450 enzyme. During subsequent ring-formation, the epoxide groups not only provide cations or cation equivalents for ring-closure but they also furnish the ether oxygen atoms [73, 74]. In other words, both the electrophile and the nucleophile are supplied by individual epoxides. Thus, both the steroid pathway and the polyether pathway employ the same synthetic strategy to achieve the synchronized installment of multiple C–C or C–O bonds, respectively.

10.4 Dioxygenases

Like the examples above, dihydroxyacetanilide epoxidase (DHAE) uses an olefin as the substrate for epoxidation. Its mechanism, however, is fundamentally different from those of cytochrome P450 or flavin-dependent enzymes. Dihydroxyacetanilide is an intermediate in the biosynthesis of the epoxyquinones LL-C10037α, an antitumor agent produced by the actinomycete *Streptomyces* LL-C10037 [75, 76], and MM14201, an antibiotic produced by *Streptomyces* MPP 3051 (Scheme 10.20) [77]. The main structural difference between the two antibiotics lies in the opposite stereochemistry of the oxirane ring.

The enzymes responsible for epoxide formation in the LL-C10037α pathway (DHAE I) as well as in the MM14201 pathway (DHAE II) have been purified from their producing organisms [78]. Both enzymes use dihydroxyacetanilide as substrate. While DHAE I produces (5*R*,6*S*)-2-acetamido-5,6-epoxy-1,4-benzoquinone as the sole product, DHAE II produces the enantiomer in agreement with the final stereochemistry of their respective products [79]. Unlike any of the epoxidases discussed so far, DHAE does not require the addition of any cofactors other than molecular oxygen for its activity to be reconstituted *in vitro* [78]. In contrast to cytochrome P450 enzymes and squalene epoxidase, DHAE is a dioxygenase, as it transfers both atoms of molecular oxygen to the substrate [80]. No reducing system is required for enzyme turnover as all electrons needed for the reduction of O_2 are directly derived from the substrate. In order to demonstrate the incorporation of both oxygen atoms into the product, the reaction has to be coupled to the next step in the biosynthetic pathway – the reduction of the carbonyl to the hydroxy group [80]. This procedure is necessary because of the fast exchange of the C4 oxygen with solvent water that would wash out any incorporated ^{18}O label [81]. A mechanism involving a dioxetane intermediate has been proposed for the DHAE-catalyzed epoxidation of dihydroxyacetanilide that is based on the incorporation of ^{18}O into the epoxide as well as into the hydroxy group (Scheme 10.21) [80].

The formation of the epoxide moiety found in a number of compounds from the manumycin group of antibiotics (Scheme 10.22) might proceed through a similar

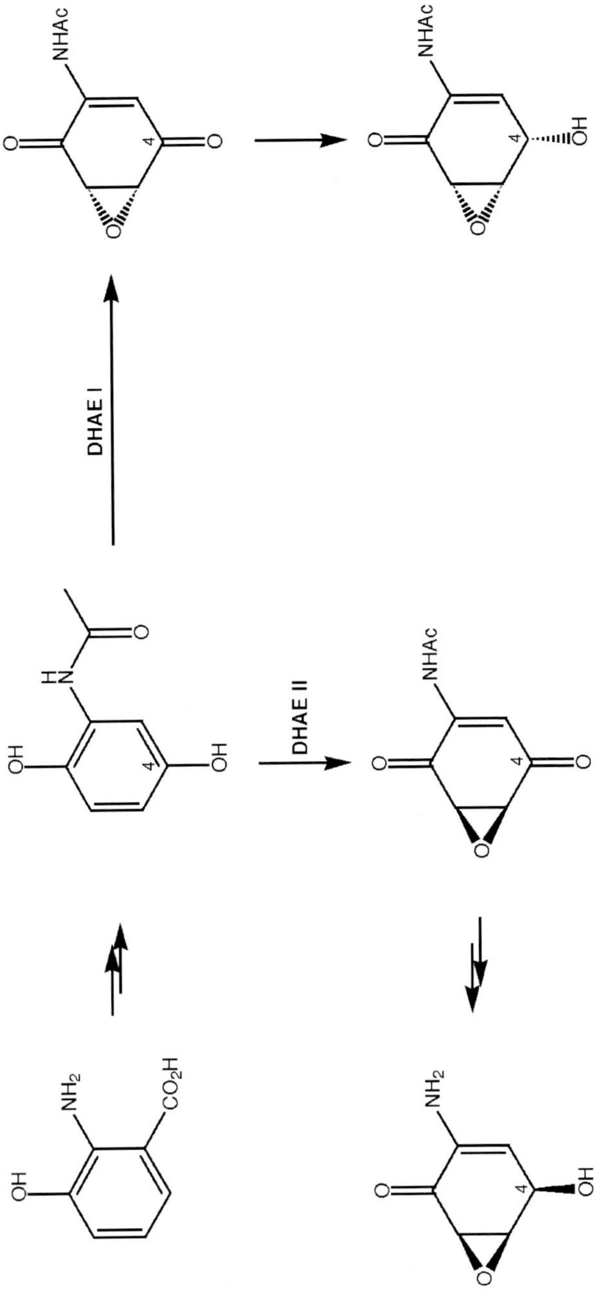

DHAE I

DHAE II

MM14201

LL-C10037α

Scheme 10.20 Biosynthesis of LL-C10037α and MM14201.

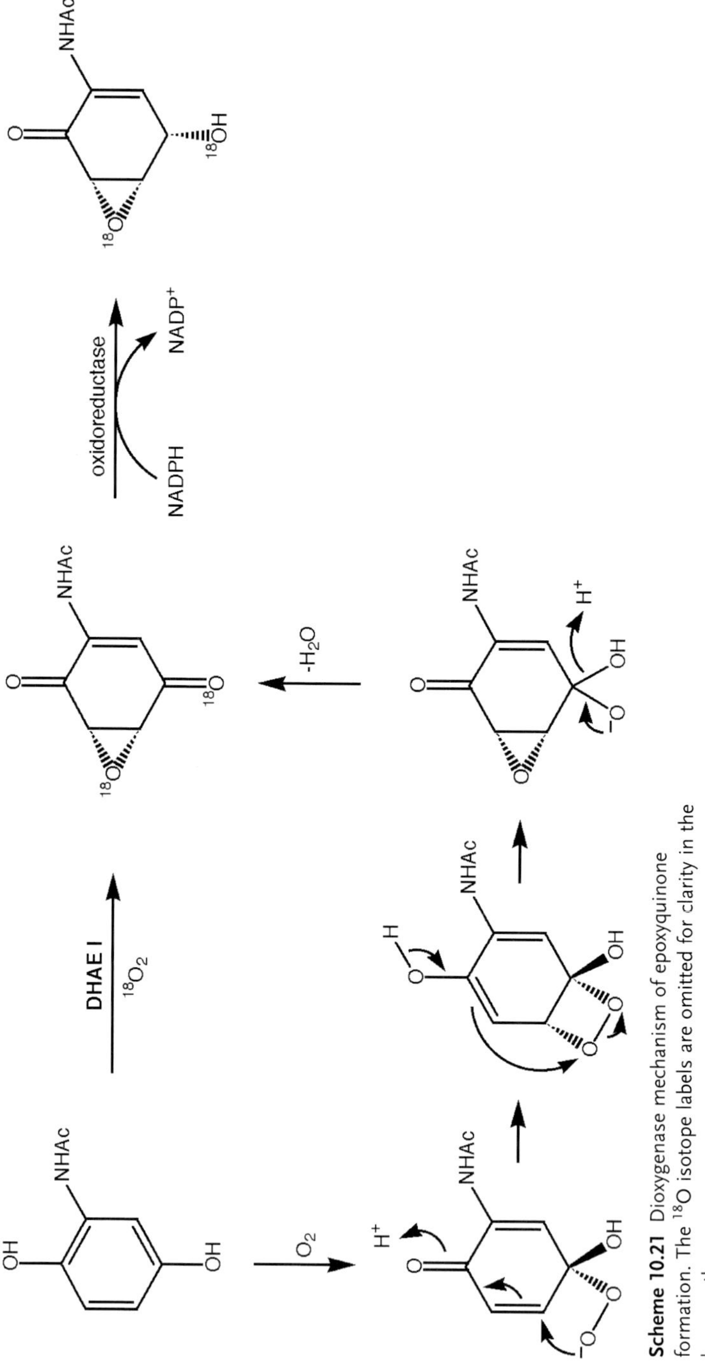

Scheme 10.21 Dioxygenase mechanism of epoxyquinone formation. The ^{18}O isotope labels are omitted for clarity in the lower pathway.

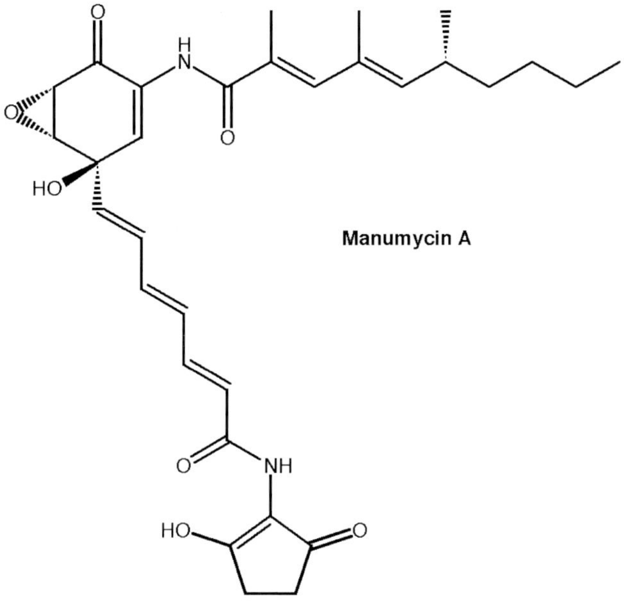

Manumycin A

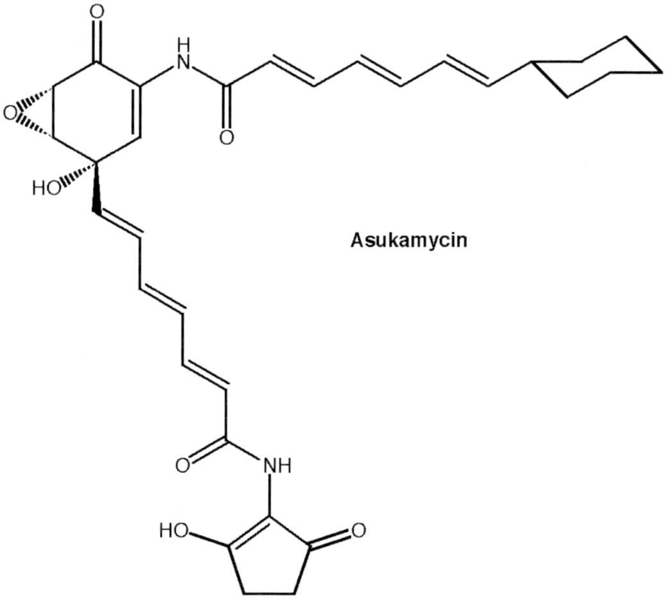

Asukamycin

Scheme 10.22 Chemical structures of manumycin and asukamycin.

dioxygenase mechanism [82, 83]. The manumycin group is a small and discrete class of natural products that to date comprises 23 secondary metabolites all of which are produced by actinomycetes. Manumycin A, for example, acts as an inhibitor of Ras farnesyltransferase [84]. The structural element that defines the manumycin group is the *m*-C_7N unit linking the unsaturated upper and lower carbon chains. The central *m*-C_7N varies in its stereochemistry and in the presence or absence of the epoxide ring [82].

When the producing organism is cultured under $^{18}O_2$ atmosphere, ^{18}O is found both in the epoxide and in the *m*-C_7N hydroxy group.[85] Furthermore, protoasukamycin is efficiently epoxidized to asukamycin under fermentation conditions (Scheme 10.23) [86]. However, the cofactor requirement has not yet been established, as all experiments have been carried out with whole cells and not with purified enzymes. In addition, the above findings could be explained by two independent monooxygenase reactions. The main concern with a dioxygen mechanism, though, is that the epoxide (or any epoxide-derived group) and the neighboring hydroxy group should have *syn* configuration. Even though this holds true for asukamycin and a large number of other manumycins, manumycins A and B, which are the major metabolites in their producing organism, have *anti* stereochemistry. A dioxygen mechanism could still apply in the latter case if epimerization is part of the biosynthetic pathway [83].

It has been speculated that DHAE I might utilize the same mechanism as other enzymes that epoxidize hydroquinones, such as vitamin K-dependent glutamate carboxylase (Scheme 10.24) [87]. Vitamin K-dependent γ-carboxylation of specific glutamate residues is an important posttranslational modification in proteins, and the carboxylation of glutamate residues is coupled to epoxidation of vitamin K [88]. It has been established that vitamin K-dependent glutamate carboxylase has dioxygenase rather than monooxygenase activity [89, 90]. Dioxygenation of reduced vitamin K is thought to proceed through a dioxetane intermediate, cleavage of the O–O bond furnishing the epoxide and an alkoxide anion. This strong base is thought to be responsible for deprotonation of the glutamate residue at the γ-position, thus driving the carboxylation reaction [91].

Vitamin K-dependent glutamate carboxylase and DHAE are referred to in the literature as being cofactor-free. This also implies the absence of metal ions from the active site, which would have important mechanistic implications. It is true that no metals have to be added to the purified or partially purified enzymes in order for catalytic activity to be observed. However, the activity of both enzymes is stimulated by the addition of divalent cations. Carboxylase activity is affected most by the presence of Mn^{2+} in relatively high concentrations [92]. DHAE I-catalyzed epoxidation is enhanced by the addition of Mn^{2+}, Ni^{2+}, and Co^{2+} [78]. Interestingly, no such effect is observed for DHAE II. Both DHAE enzymes lose their activity when treated with the chelating agent 1,10-phenanthroline, but only the DHAE I apoenzyme can be reconstituted by the addition of metal ions [78]. It is not clear whether metal ions are directly involved in dioxygen activation or if they might have a structural function. The role of the metal is particularly difficult to determine for the vitamin K-dependent glutamate carboxylase. On one hand, two differ-

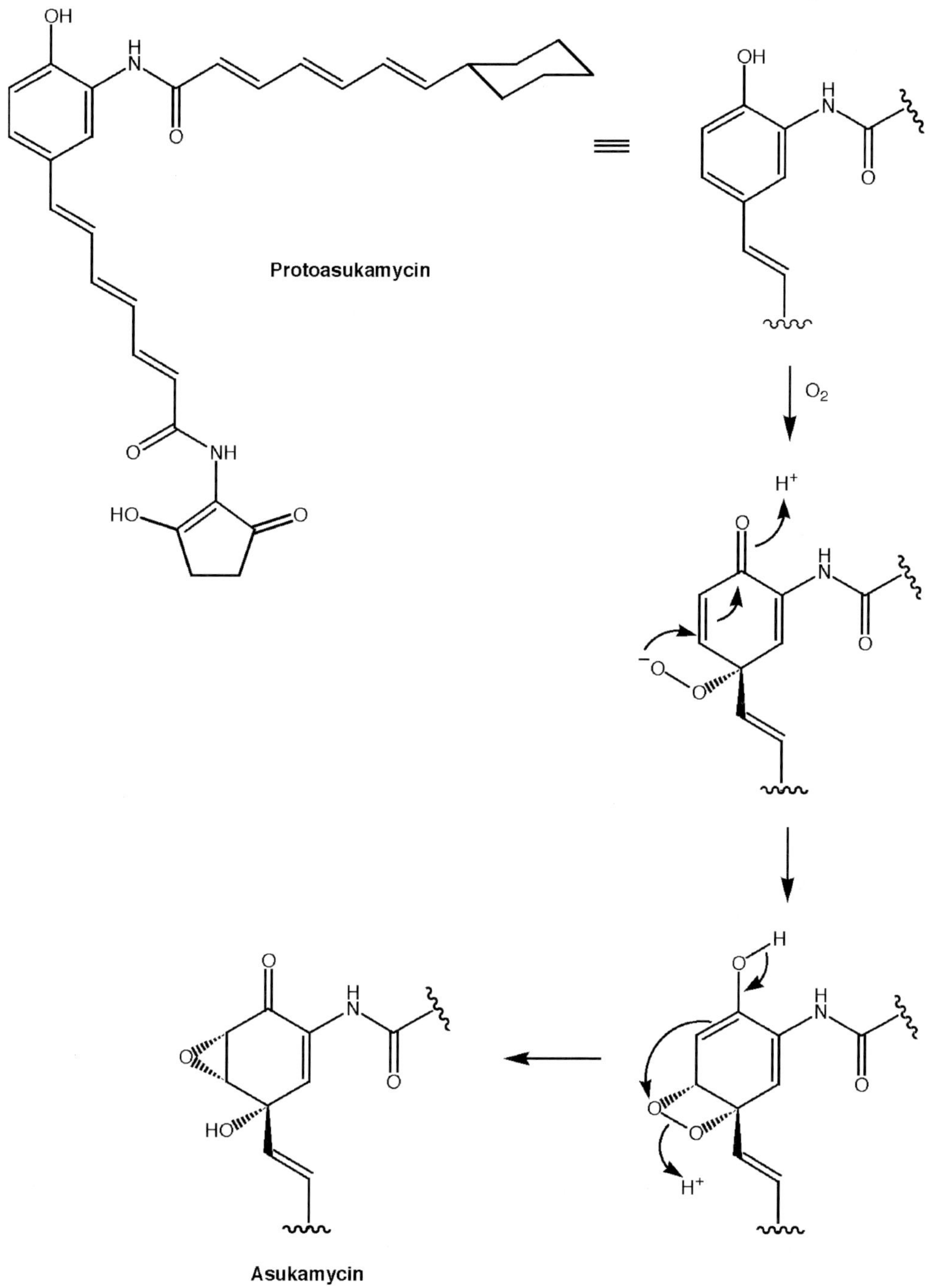

Protoasukamycin

O_2

Asukamycin

Scheme 10.23 Possible dioxygenase mechanism for the transformation of protoasukamycin to asukamycin.

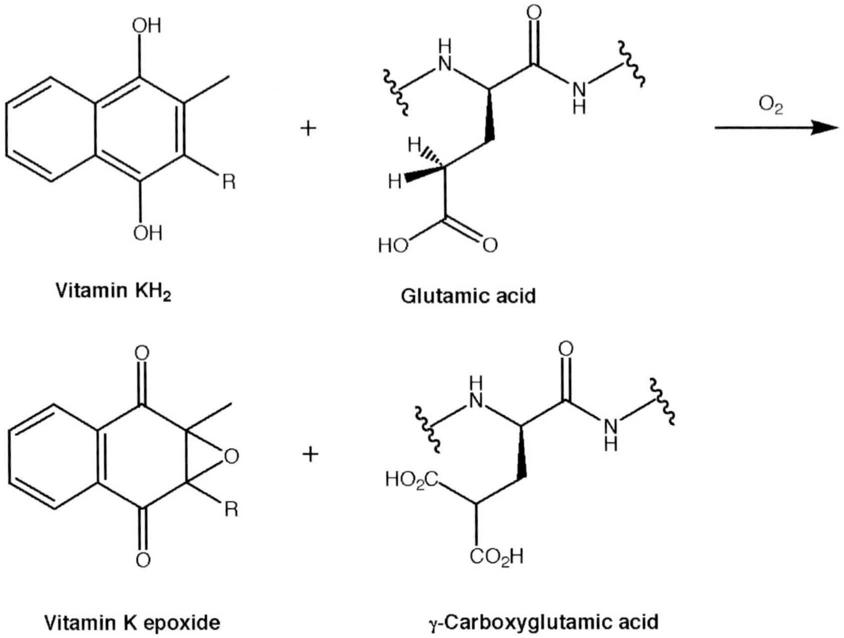

Vitamin KH_2 + Glutamic acid $\xrightarrow{O_2}$ Vitamin K epoxide + γ-Carboxyglutamic acid

Scheme 10.24 Reaction catalyzed by vitamin K-dependent glutamate carboxylase.

ent reactions are catalyzed by this enzyme: the O_2-dependent epoxidation of reduced vitamin K hydroquinone and the carboxylation of glutamate. On the other hand, the enzyme has to be studied in the presence of micelles in order for it to be functional, hence there are many possible points of involvement for metal ions in the glutamate carboxylase system. The question is how triplet oxygen can react with singlet organic compounds in the absence of any cofactors. To date, only very few oxygenases are known that are capable to act in the absence of cofactors [93, 94], and all of them act on phenolic compounds that can access stable radical forms. We can therefore propose a mechanism in which direct electron transfer from the substrate anion to O_2 yields a caged radical pair (Scheme 10.25). This mechanism is analogous to the reaction between reduced flavin anion and dioxygen (Scheme 10.17). The subsequently formed peroxide anion is a suitable nucleophile for intramolecular addition to give the dioxetane intermediate. Both reduced vitamin K hydroquinone and dihydroxyacetanilide could follow the same reaction pathway, making the presence of a metal ion redundant. Other mechanisms have also been proposed, but will not be discussed here [93]. A recent review on dioxygenases gives an excellent account on the mechanism and occurrence of this class of enzymes [2].

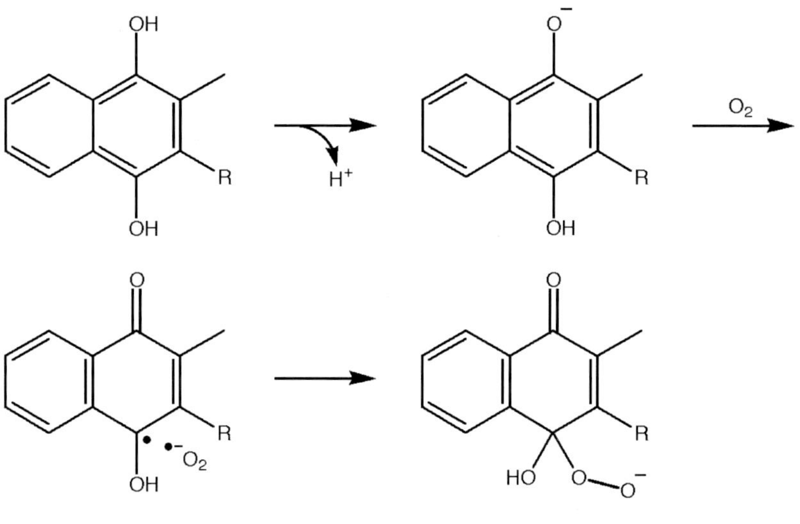

caged radical pair

Scheme 10.25 Dioxygen activation in cofactor-free enzymes.

10.5
Epoxidation through Dehydrogenation

10.5.1
Fosfomycin

Fosfomycin is an antibiotic produced by several *Streptomyces* species [95, 96] as well as by the Gram-negative *Pseudomonas syringiae* and *Pseudomonas viridiflava*.[97, 98] As an analogue of phosphoenolpyruvate, it irreversibly inhibits UDP-*N*-acetylglucosamine-3-*O*-enolpyruvyltransferase (MurA), the enzyme that catalyzes the first step in peptidoglycan biosynthesis [99].

Despite its seemingly simple structure, fosfomycin has attracted considerable interest from a biosynthetic point of view. The epoxide functionality, and particularly the extremely rare C–P bond, makes fosfomycin an attractive target for biosynthetic investigations. Early steps in the biosynthesis of fosfomycin are shared with the bialaphos pathway (Scheme 10.26) [100], whilst the formation of the epoxide ring is the final step in fosfomycin biosynthesis. The isolation of the fosfomycin gene clusters from *Streptomyces wedmorensis* and *Pseudomonas syringiae* PB-5123 has facilitated the discovery of the corresponding enzyme (*S*)-HPP epoxidase (HppE) [101, 102]. Labeling experiments have established that the epoxide oxygen is derived not from molecular oxygen but rather from the hydroxy group of (*S*)-2-hydroxypropylphosphonic acid ((*S*)-HPP) [103]. Furthermore, *cis*-propenylphosphonic acid (PPA) is not a precursor of fosfomycin (Scheme 10.26). [104]

Recombinant (*S*)-HPP epoxidase (HppE), purified as the *apo*-protein, can be

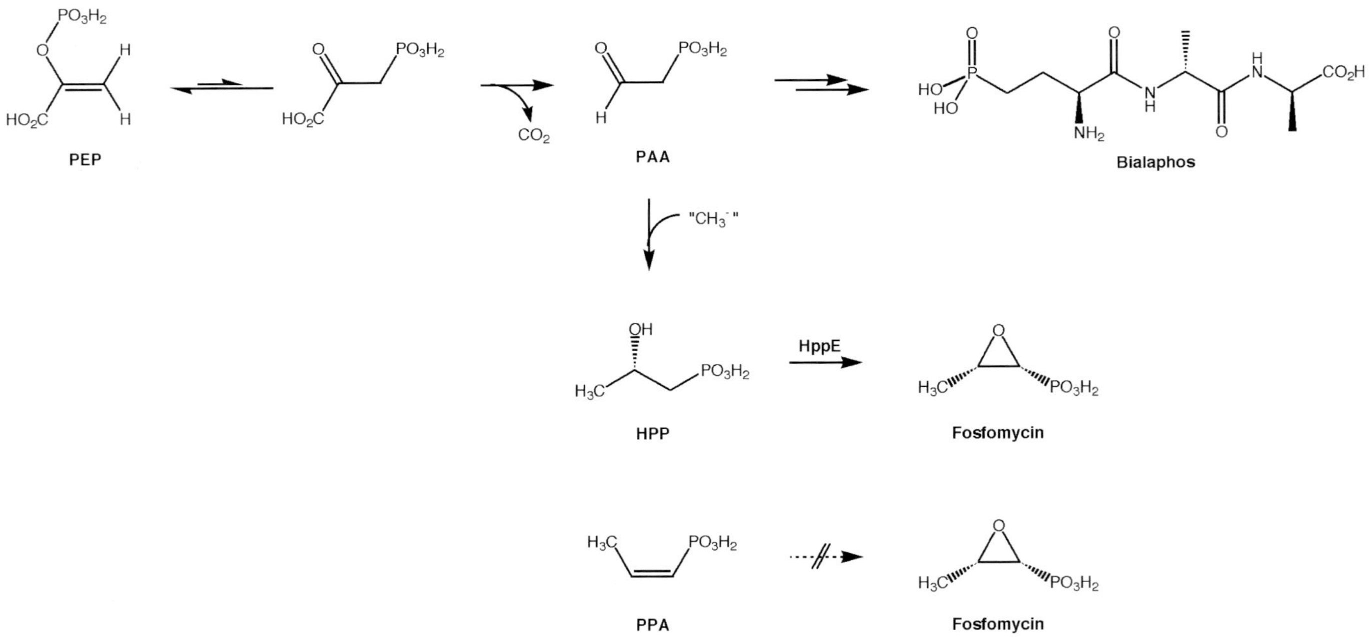

PEP

PAA

CO_2

"CH_3^-"

HppE

Bialaphos

HPP

Fosfomycin

PPA

Fosfomycin

Scheme 10.26 Partial biosynthetic pathway of fosfomycin and bialaphos. Both pathways use a homologous set of enzymes for the synthetic steps leading from phosphoenolpyruvate (PEP) to phosphonoacetaldehyde (PAA). The conversion of hydroxypropylphosphonic acid (HPP) to fosfomycin is catalyzed by the epoxidase HppE. Propenylphosphonic acid (PPA), however, is not converted to fosfomycin.

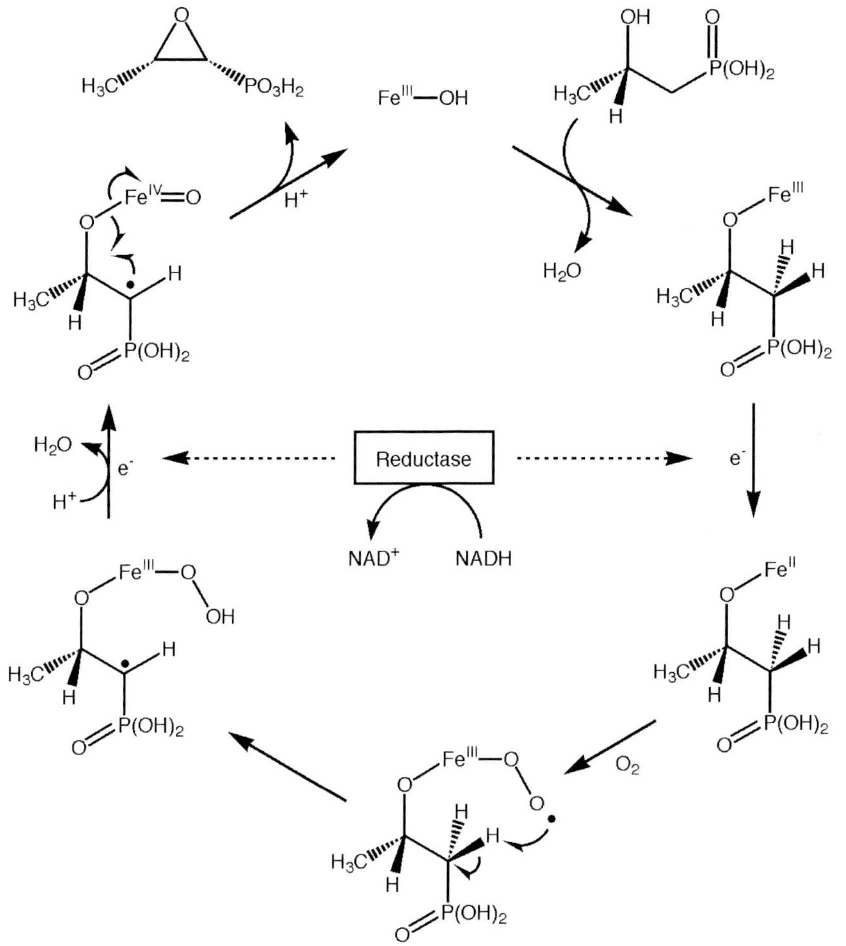

Scheme 10.27 Catalytic cycle of HppE. Dashed arrows indicate electron transport. In this scheme HPP binds to iron^{III}. After a one-electron reduction, dioxygen binds and reoxidizes the iron center. The peroxide radical is capable of stereospecifically abstracting the (*pro-R*) hydrogen. Another one-electron reduction is required to reduce one peroxide oxygen to water. Epoxide formation is mediated by the resulting iron^{IV}-oxo species.

reconstituted with $Fe(NH_4)_2(SO_4)_2$ to give one iron center per enzyme monomer [105]. Even though O_2 is not incorporated into the product, no conversion of (*S*)-2-HPP is observed in the absence of molecular oxygen [105]. Enzyme activity is independent of organic co-substrates (such as ascorbate, α-ketoglutarate) normally associated with non-heme iron oxygenases. [105] The additional electrons needed to complete the four-electron reduction of O_2 to H_2O are provided by NADH. *In vitro*, the electron flow from NADH to HppE is greatly enhanced by the presence of an exogenous protein reductase [105]. The use of E_3, an NADH-dependent

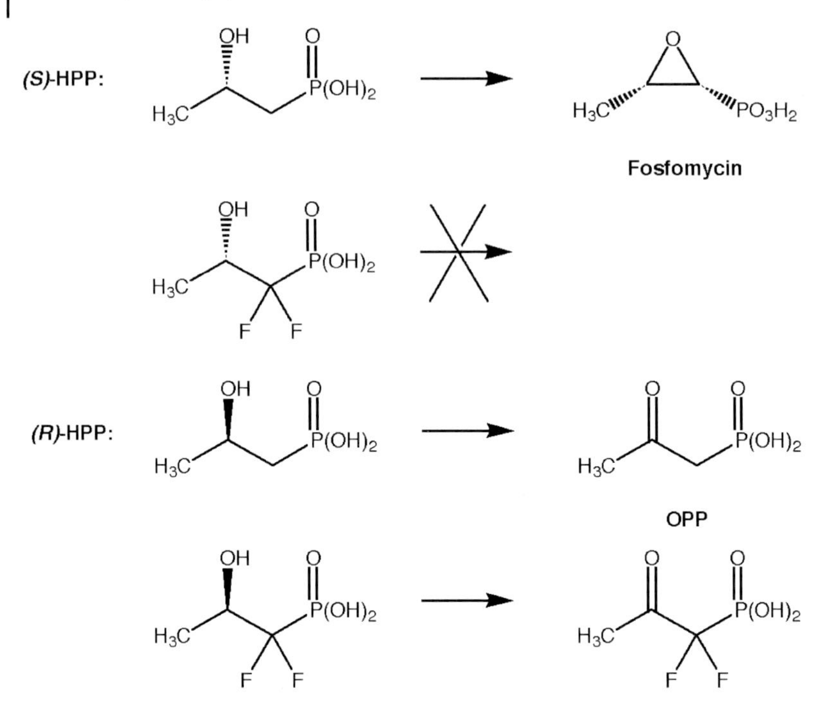

Scheme 10.28 Reaction of HppE with substrate analogues.

[2Fe-2S]-containing flavoprotein from the 3,6-dideoxyhexose biosynthetic pathway of *Yersinia pseudotuberculosis* [106], has been successful as a substitute for the as yet unidentified endogenous reductase [105].

HppE formally catalyzes the dehydrogenation of (*S*)-2-HPP to fosfomycin. The stereochemical course of this reaction has been investigated through the use of stereospecifically deuterated HPP. Findings from these studies show that only the pro-(*R*) hydrogen at C1 is abstracted [107]. The same researchers also report the isolation of small amounts of the *trans* isomer of fosfomycin from the culture broth [108]. This compound could be formed by rotation around the C1–C2 bond at the C1-radical stage [107]; a tentative reaction cycle is shown in Scheme 10.27. Interestingly, HppE is capable of oxidizing both the native substrate (*S*)-HPP and also its enantiomer (*R*)-HPP. Whereas the (*S*)-enantiomer yields the epoxide fosfomycin, however, the (*R*)-enantiomer is converted to 2-oxopropylphosphonic acid (OPP) instead (Scheme 10.28) [109]. A mechanism in which the carbonyl group originates from hydroxylation to form the C2-*gem*-diol and subsequent dehydration can be excluded, as no labeled oxygen from $^{18}O_2$ is incorporated [109]. Clues about the mechanism of OPP formation have been provided in experiments with 1,1-difluorinated HPP (Fl-HPP) as substrate. Only (*R*)-Fl-HPP gives rise to product, whereas the (*S*) enantiomer acts as an inhibitor of the enzyme (Scheme 10.28) [109]. The inhibitory action of (*S*)-1,1-difluoroHPP agrees well with homolytic cleavage of the C1–H bond. The most likely mechanism for the reaction with (*R*)-

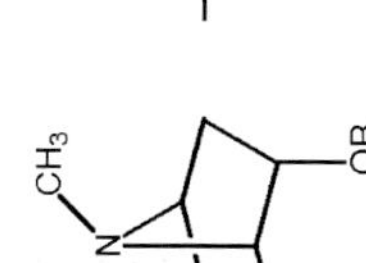

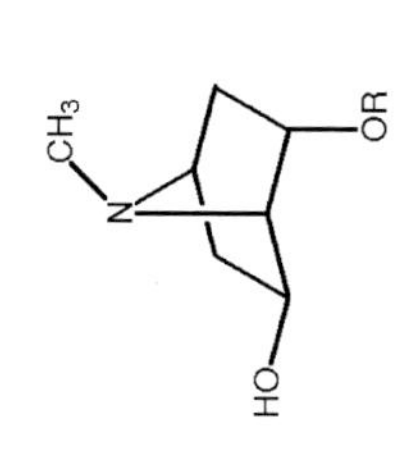

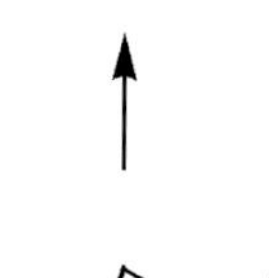

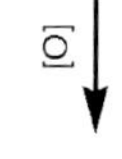

Scheme 10.29 Hydroxylation and epoxidation affording scopolamine.

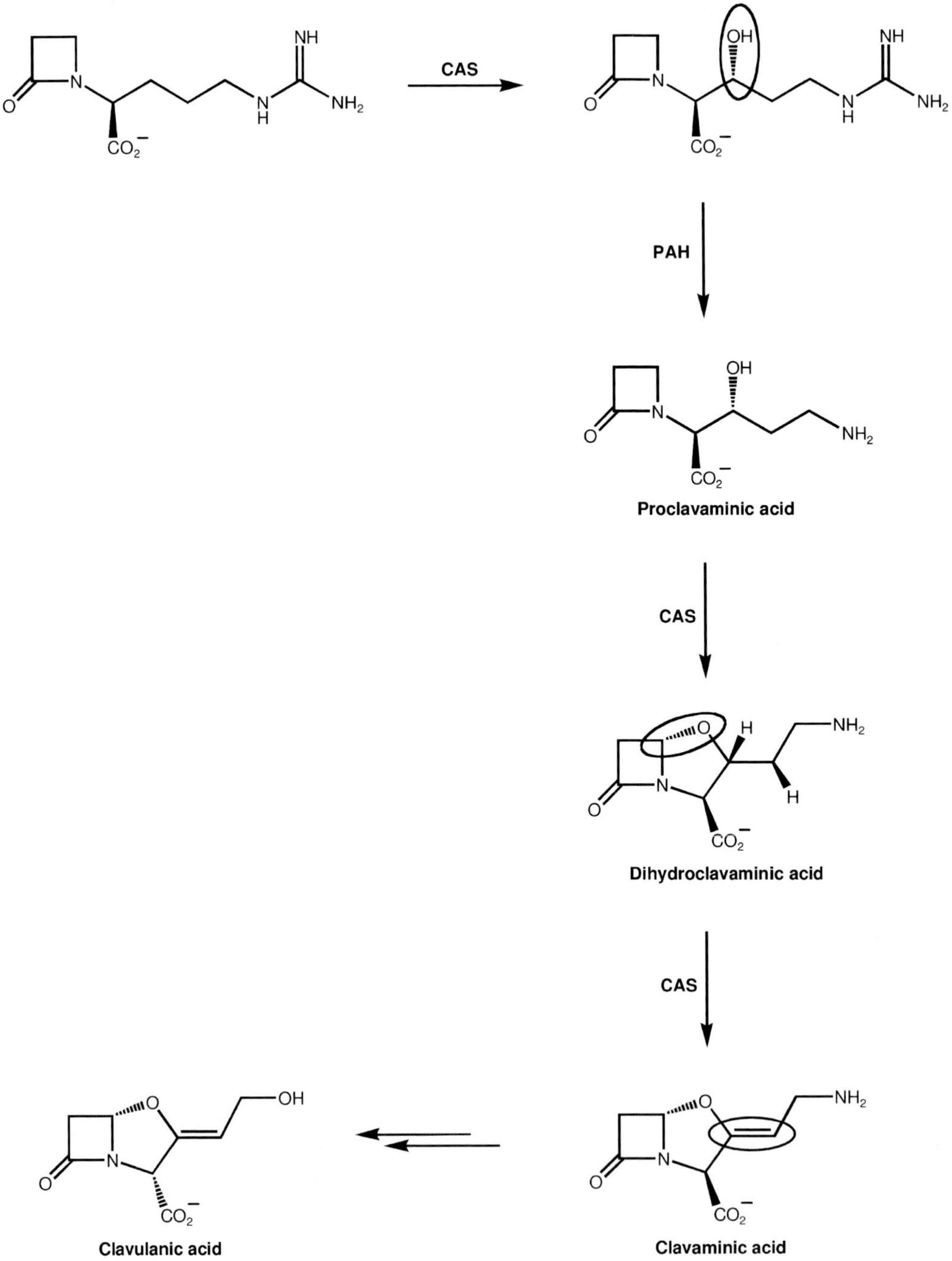

CAS

PAH

Proclavaminic acid

CAS

Dihydroclavaminic acid

CAS

Clavulanic acid

Clavaminic acid

Scheme 10.30 Part of clavulanic acid biosynthesis. Bonds installed by clavaminic acid synthase (CAS) are circled. CAS: clavaminic acid synthase. PAH: proclavaminic acid amidino hydrolase.

HPP or (*R*)-1,1-difluoroHPP is one involving abstraction of the C2 hydrogen instead [109]. Consequently, the above findings not only support a radical- based mechanism but they also indicate that different reaction pathways are used for the individual enantiomers.

Nothing is known about the identity of the iron species responsible for dehydrogenation of the substrate. Iron-oxo species such as Fe^{IV}=O or Fe^{III}-OOH are postulated as the oxidants in most heme or non-heme iron oxygenases. It has to be considered that any mechanistic model proposed must account not only for the observed stereochemistry but also for the lack of hydroxylation activity and its inability to convert the olefinic substrate. Furthermore, no HppE sequence homologue is to be found in protein databases. Further studies should shed more light on the mechanism with which this unique enzyme operates.

10.5.2 Scopolamine

We conclude this section by describing an enzyme that belongs to the important and versatile class of α-keto acid-dependent enzymes. Hyoscyamine 6β-hydroxylase is involved in the biosynthesis of the tropane alkaloid scopolamine in plants [110]. This bifunctional enzyme first hydroxylates hyoscyamine and subsequently forms the epoxide to give scopolamine (Scheme 10.29) [110]. The second part of the reaction is formally similar to the epoxide formation in fosfomycin biosynthesis, as the epoxide oxygen atoms in both reactions are derived from the hydroxy groups. Unlike HppE, however, this oxygenase also efficiently converts the olefinic substrate analogue 6,7-dehydrohyoscyamine into scopolamine *in vitro* (Scheme 10.29) [111]. Furthermore, hyoscyamine 6β-hydroxylase requires the keto acid α-ketoglutarate as co-substrate [112].

Other examples of α-keto acid-dependent enzymes are mammalian proline hydroxylase and bacterial clavaminate synthase [113]. The latter enzyme is of particular interest as it is responsible for the catalysis of three individual steps in the biosynthesis of the β-lactamase inhibitor clavulanic acid (Scheme 10.30).

Enzymes of the α-keto acid-dependent oxygenase family are non-heme iron proteins. They generate the reactive iron-oxygen species, believed to be Fe^{IV}=O, with the aid of the α-ketoglutarate co-substrate, which is converted into succinate and CO_2 in the process [113–115]. Kinetic studies point towards ordered and sequential binding of the reaction partners, with α-ketoglutarate binding first, followed by the reaction substrate, and finally by molecular oxygen [116]. The first product to be released is CO_2 (or carbonate). Succinate and the reaction product are released subsequently, with the order of release varying between different enzymes (Scheme 10.31) [117]. The mechanistic steps following formation of the ternary enzyme-substrate-cosubstrate complex are not yet well understood [115].

A complete cycle of iron-oxo generation and substrate oxidation with the consumption of one equivalent O_2 and one equivalent α-ketoglutarate is necessary for each of the reactions catalyzed by clavaminate synthase. The different substrates utilized by clavaminate synthase adopt slightly different positions relative to the

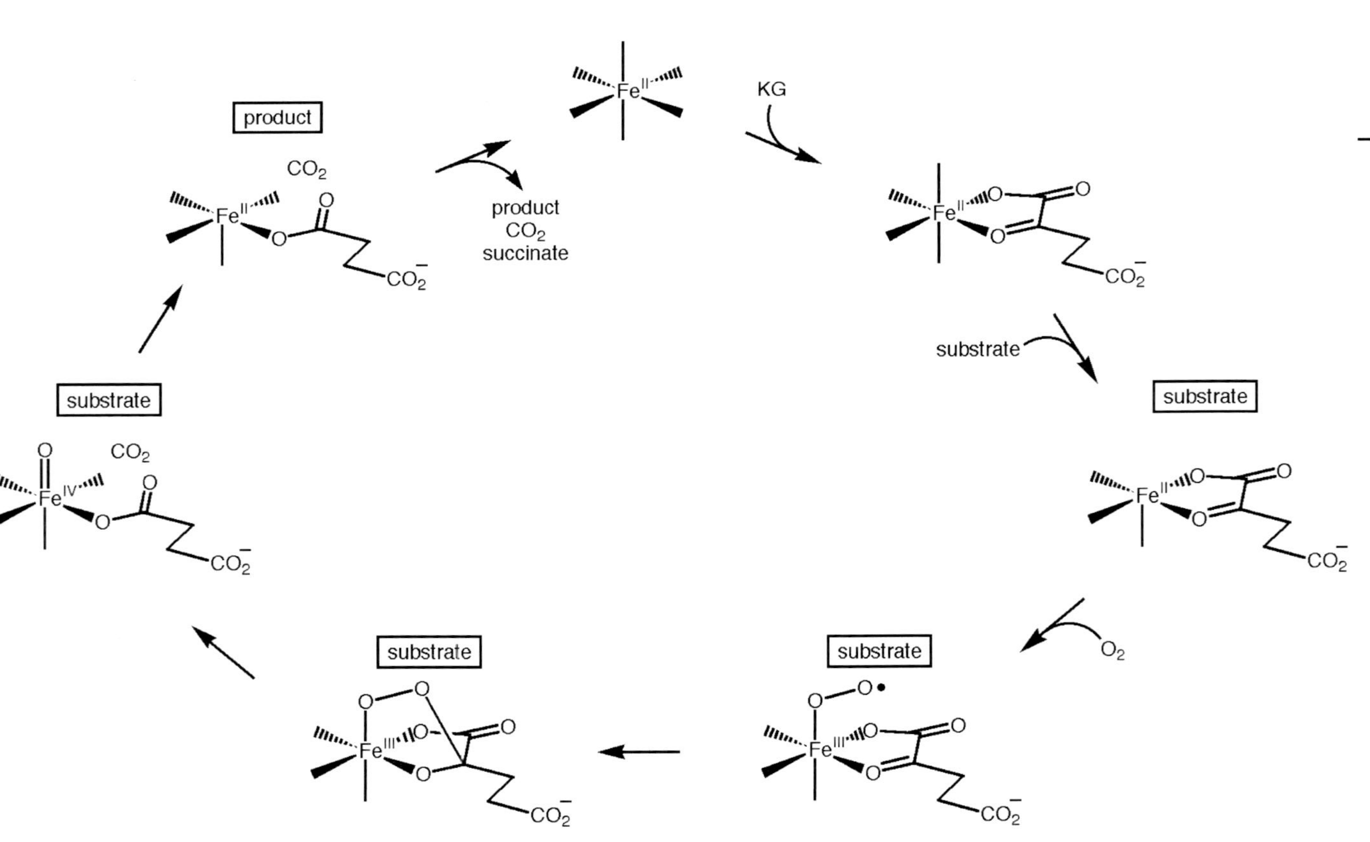

Scheme 10.31 Reaction cycle of KG-dependent (KG = α-ketoglutarate) enzymes. Metal ligands from protein side chains and water are omitted for clarity. One of the oxygens of O_2 is incorporated into succinate. The other oxygen is either incorporated into the product or reduced to water depending on the nature of the reaction.

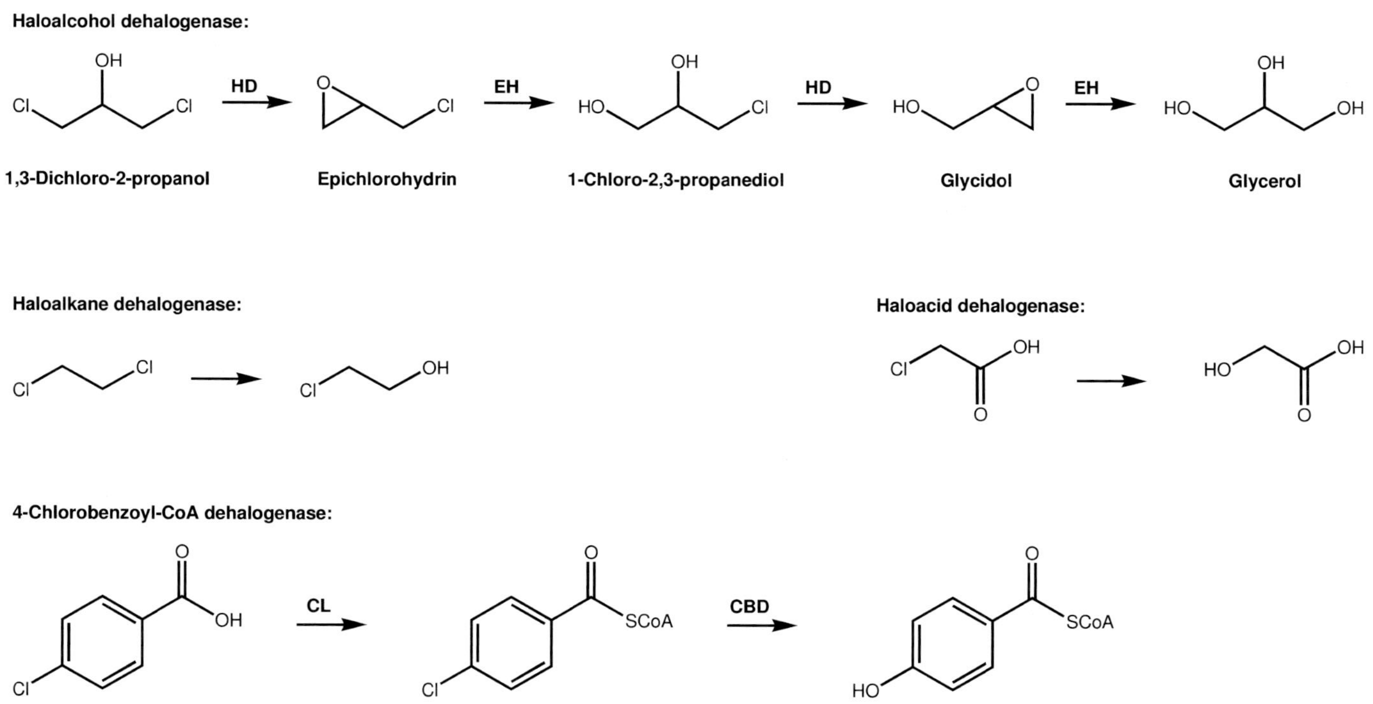

Haloalcohol dehalogenase:

1,3-Dichloro-2-propanol → (HD) Epichlorohydrin → (EH) 1-Chloro-2,3-propanediol → (HD) Glycidol → (EH) Glycerol

Haloalkane dehalogenase:

Haloacid dehalogenase:

4-Chlorobenzoyl-CoA dehalogenase:

CL

CBD

Scheme 10.32 Examples of reactions catalyzed by different classes of dehalogenases. HD: haloalcohol dehalogenase; EH: epoxide hydrolase; CL: *p*-chlorobenzoyl-CoA ligase; CBD: *p*-chlorobenzoyl-CoA dehalogenase.

metal center, determining which site in the substrate is poised for reaction [118, 119]. Thus, the substrate conformation in the enzyme active site is the main factor that influences the further fate of the reaction. As the first two reactions catalyzed by clavaminate synthase are analogous to the hydroxylation-epoxidation sequence catalyzed by hyoscyamine 6β-hydroxylase, it is plausible that both enzymes employ similar strategies to carry out multiple reactions.

Keto acid-dependent enzymes are involved in numerous reactions and are found in prokaryotes as well as eukaryotes. Apart from scopolamine and clavulanic acid biosynthesis, α-keto acid-dependent enzymes are also found in the biosynthetic

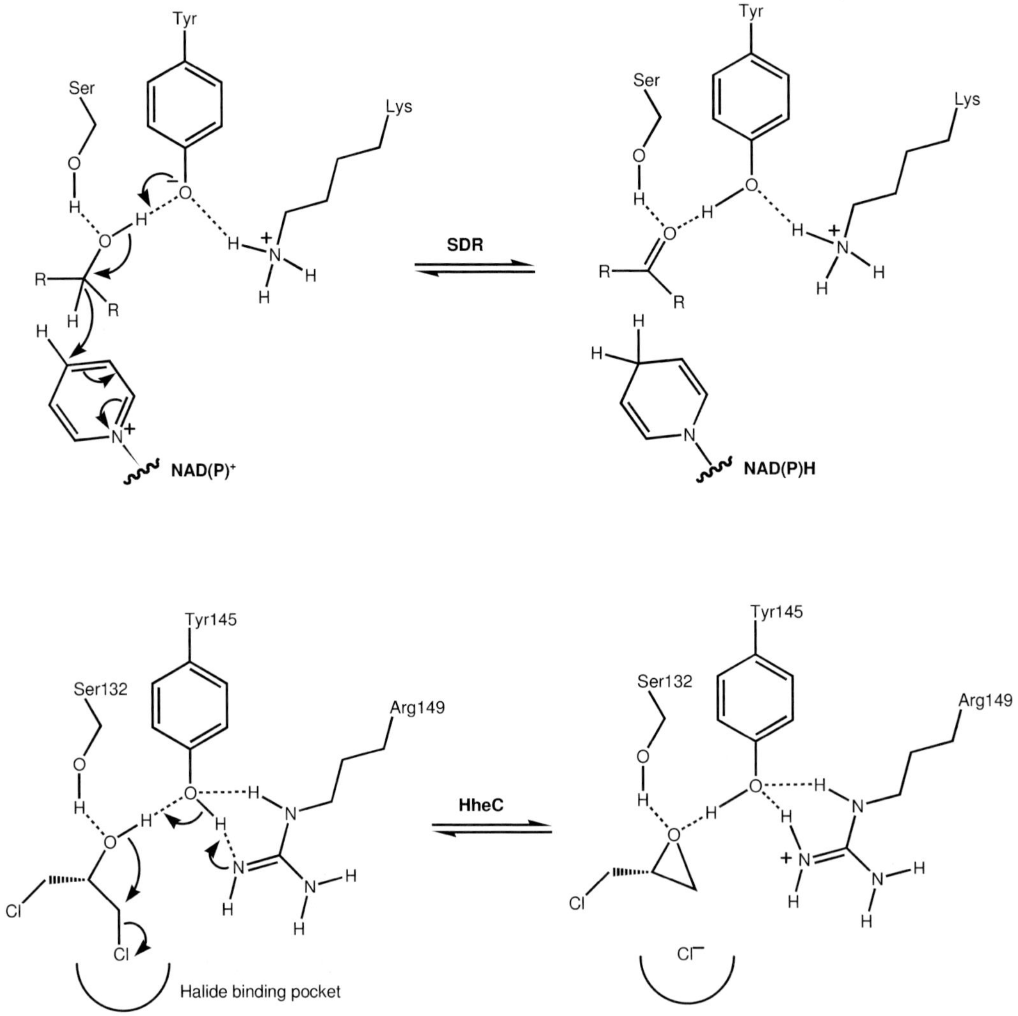

Scheme 10.33 Schematic representation of the mechanism of SDRs (top) and the halohydrin dehalogenase HheC (bottom). Scheme adapted from Reference 129.

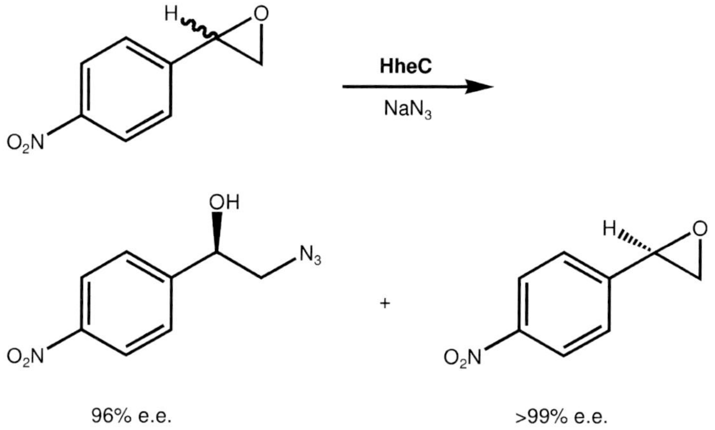

HheC

NaN_3

96% e.e. >99% e.e.

Scheme 10.34 HheC-catalyzed azidolysis of an epoxystyrene derivative.

pathways of the plant hormone gibberellin and the β-lactam cephalosporin. Some posttranslational modifications of proteins, such as the hydroxylation of prolyl and lysyl residues, are also catalyzed by α-keto acid-dependent enzymes. These are but a few examples. Several reviews covering this family of proteins and related enzymes have been published [113–115, 117, 120, 121].

10.6 Dehalogenases

In the previous sections, epoxidation was accompanied by a net oxidation of the substrate. We would like to conclude our discussion of various mechanisms of epoxide formation by presenting one example in which the substrate does not undergo a change in redox state.

Certain microorganisms possess the ability to utilize organic halides as their sole carbon source and have developed a dedicated set of catabolic enzymes for this purpose. Dehalogenation frequently represents the first step in the degradation of halogenated compounds [122, 123]. Different classes of dehalogenases that differ in their substrate specificities, their mechanisms, and in their structures have been identified (Scheme 10.32) [124]. One particular class of dehalogenases, the haloalcohol or halohydrin dehalogenases, convert 1,2-haloalcohols into the corresponding epoxides through displacement of the halogen by the vicinal hydroxy group in an S_N2-like substitution [122]. This reaction is reversible and can proceed in a stereoselective manner. The degree of stereoselectivity is dependent on the substrate and on the specific enzyme under investigation [125–128].

The crystal structure of the halohydrin dehalogenase from the soil bacterium *Agrobacterium radiobacter* AD1, HheC, has been solved [129]. HheC is structurally related to the family of NAD(P)H-dependent short-chain dehydrogenases/reductases (SDRs) [130, 131]. Members of the SDR family use the Ser-Tyr-Lys catalytic triad to catalyze the dehydrogenation of alcohols or the reduction of ketones. The active site serine and tryrosine residues hydrogen bond to the substrate carbonyl or hydroxy group as shown schematically in Scheme 10.33. The nicotinamide cofactor is suitably positioned to accept or donate a hydride to the substrate. Haloalcohol dehalogenases have lost the ability to bind the nicotinamide cofactor, and in its place a halide binding pocket has been created. The catalytic triad of the SDR family is also present in haloalcohol dehalogenases and performs analogous functions in both enzyme classes (Scheme 10.33) [132]. Activation of the alcohol functionality for nucleophilic displacement of the halide is achieved through a hydrogen bonding network with the residues of the catalytic triad. The expelled halide makes extensive contacts with residues lining the binding pocket; there are several interactions with backbone nitrogen atoms and to aromatic side chains, and a water molecule that bridges the carbonyl and nitrogen of a *cis*-peptide bond in the back of the cavity is also in contact with the halide ion [129].

The reversibility of halohydrin dehalogenase-catalyzed reactions has been used for the regioselective epoxide-opening with "nonnatural" nucleophiles (an example is given in Scheme 10.34) [133]. The stereoselectivity of the enzyme results in the resolution of the racemic substrate. At the same time, the regioselectivity imposed by the active site geometry yields the anti-Markovnikov product. [128]

10.7 Summary and Outlook

Enzymes catalyze the formation of epoxides mainly through oxidative processes by two fundamentally different routes. In the first, oxygen is inserted from O_2 into a double bond. The molecular oxygen required for this reaction is activated either by a metal ion or flavin. In rare cases, the substrate itself can substitute for the cofactor requirement. In the second route, a substrate hydroxy group is inserted into the vicinal C–H bond. This reaction is formally a dehydrogenation and requires metal catalysis. There are fewer examples of enzymatic formation of epoxides through nonoxidative reactions. One particular class of dehalogenases, for instance, uses 1,2-haloalcohols as substrates and displaces chloride by the vicinal alcohol group in an S_N2-like mechanism.

Numerous oxygenases have been successfully employed as catalysts in organic syntheses to provide enantiomerically pure epoxides for use as chiral starting materials [134]. Interestingly, the enzymes that have been used are mostly hydroxylases and not epoxidases [135], but, as pointed out earlier, monooxygenases can show alternative reactivity when challenged with nonphysiological substrates. The ω-hydroxylation system from *Pseudomonas oleovorans* catalyzes the terminal hydroxylation of fatty acids and alkanes [136], but has also found use in the conversion of the corresponding olefins to epoxides [135]. Cytochrome P450cam from *Pseudomonas putida* (Scheme 10.2) has been utilized in a similar fashion [137]. The cytochrome P450 monooxygenase from *Bacillus megaterium* (Section 10.2.1) has become a very attractive platform for protein engineering and protein evolution as its electron-transport system is covalently attached to the monooxygenase [138]. It remains to be seen if enzymes from biosynthetic pathways, such as the epothilone (Section 10.2.2) or the mycinamicin epoxidase (Section 10.2.3), will provide alternative platforms from which to develop catalysts for the stereoselective oxygenation of structurally complex molecules.

References

1 M. Sono, M. P. Roach, E. D. Coulter, J. H. Dawson, *Chem. Rev.* 1996, *96*, 2841–2887.
2 T. D. H. Bugg, *Tetrahedron* **2003**, *59*, 7075–7101.
3 T. L. Poulos, B. C. Finzel, A. J. Howard, *J. Mol. Biol.* **1987**, *195*, 687–700.
4 L. O. Narhi, A. J. Fulco, *J. Biol. Chem.* **1986**, *261*, 7160–7169.
5 D. Ostovic, T. C. Bruice, *Acc. Chem. Res.* **1992**, *25*, 314–320.
6 F. P. Guengerich, *Arch. Biochem. Biophys.* **2003**, *409*, 59–71.
7 P. R. Ortiz de Montellano, *Cytochrome P450: Structure, Mechanism, and Biochemistry*, 2nd ed., Plenum, New York, **1995**.
8 P. R. Ortiz de Montellano, K. L. Kunze, B. A. Mico, *Mol. Pharmacol.* **1980**, *18*, 602–605.
9 R. E. Miller, F. P. Guengerich, *Biochemistry* **1982**, *21*, 1090–1097.
10 F. P. Guengerich, *Chem. Res. Toxicol.* **2001**, *14*, 611–650.
11 B. Meunier, S. P. de Visser, S. Shaik, *Chem. Rev.* **2004**, *104*, 3947–3980.
12 D. S. Su, A. Balog, D. F. Meng, P. Bertinato, S. J. Danishefsky, Y. H. Zheng, T. C. Chou, L. F. He, S. B. Horwitz, *Angew. Chem. Int. Ed. Engl.* **1997**, *36*, 2093–2096.
13 R. J. Kowalski, P. Giannakakou, E. Hamel, *J. Biol. Chem.* **1997**, *272*, 2534–2541.
14 D. M. Bollag, P. A. McQueney, J. Zhu, O. Hensens, L. Koupal, J. Liesch, M. Goetz, E. Lazarides, C. M. Woods, *Cancer Res.* **1995**, *55*, 2325–2333.
15 K. C. Nicolaou, F. Roschangar, D. Vourloumis, *Angew. Chem. Int. Ed. Engl.* **1998**, *37*, 2015–2045.
16 I. Molnár, T. Schupp, M. Ono et al., *Chem. Biol.* **2000**, *7*, 97–109.
17 L. Tang, S. Shah, L. Chung, J. Carney, L. Katz, C. Khosla, B. Julien, *Science* **2000**, *287*, 640–642.
18 J. Staunton, K. J. Weissman, *Nat. Prod. Rep.* **2001**, *18*, 380–416.
19 H. D. Mootz, D. Schwarzer, M. A. Marahiel, *ChemBioChem* **2002**, *3*, 491–504.
20 L. H. Du, C. Sánchez, B. Shen, *Metab. Eng.* **2001**, *3*, 78–95.
21 S. Nagano, H. Y. Li, H. Shimizu, C. Nishida, H. Ogura, P. R. O. de Montellano, T. L. Poulos, *J. Biol. Chem.* **2003**, *278*, 44886–44893.
22 J. F. Andersen, C. R. Hutchinson, *J. Bacteriol.* **1992**, *174*, 725–735.
23 J. Cupp-Vickery, R. Anderson, Z. Hatziris, *Proc. Natl. Acad. Sci. U.S.A.* **2000**, *97*, 3050–3055.
24 J. R. Cupp-Vickery, T. L. Poulos, *Nat. Struct. Biol.* **1995**, *2*, 144–153.
25 T. L. Poulos, B. C. Finzel, I. C. Gunsalus, G. C. Wagner, J. Kraut, *J. Biol. Chem.* **1985**, *260*, 6122–6130.
26 T. L. Poulos, B. C. Finzel, A. J. Howard, *Biochemistry* **1986**, *25*, 5314–5322.
27 N. Guex, M. C. Peitsch, *Electrophoresis* **1997**, *18*, 2714–2723.
28 J. F. Andersen, K. Tatsuta, H. Gunji, T. Ishiyama, C. R. Hutchinson, *Biochemistry* **1993**, *32*, 1905–1913.
29 H. Xiang, R. A. Tschirret-Guth, P. R. O. de Montellano, *J. Biol. Chem.* **2000**, *275*, 35999–36006.
30 Y. Anzai, N. Salto, M. Tanaka, K. Kinoshita, Y. Koyama, F. Kato, *FEMS Microbiol. Lett.* **2003**, *218*, 135–141.
31 H. Suzuki, S. Takenaka, K. Kinoshita, T. Morohoshi, *J. Antibiot.* **1990**, *43*, 1508–1511.
32 M. Inouye, Y. Takada, N. Muto, T. Beppu, S. Horinouchi, *Mol. Gen. Genet.* **1994**, *245*, 456–464.

33 T. Ueno, H. Takahashi, M. Oda, M. Mizunuma, A. Yokoyama, Y. Goto, Y. Mizushina, K. Sakaguchi, H. Hayashi, *Biochemistry* **2000**, *39*, 5995–6002.

34 M. E. Goldman, G. S. Salituro, J. A. Bowen, J. M. Williamson, D. L. Zink, W. A. Schleif, E. A. Emini, *Mol. Pharmacol.* **1990**, *38*, 20–25.

35 A. Y. Li, J. Piel, *Chem. Biol.* **2002**, *9*, 1017–1026.

36 B. Shen, *Topics Curr. Chem.* **2000**, *209*, 1–51.

37 C. Puder, S. Loya, A. Hizi, A. Zeeck, *Eur. J. Org. Chem.* **2000**, 729–735.

38 R. Martin, O. Sterner, M. A. Alvarez, E. de Clercq, J. E. Bailey, W. Minas, *J. Antibiot.* **2001**, *54*, 239–249.

39 W. T. Bradner, B. Heinemann, A. Gourevitch, *Antimicrob. Agents Chemother.* **1966**, 6, 613–8.

40 H. Schmitz, K. E. Crook, Jr., J. A. Bush, *Antimicrob. Agents Chemother.* **1966**, *6*, 606–12.

41 T. Bililign, C. G. Hyun, J. S. Williams, A. M. Czisny, J. S. Thorson, *Chem. Biol.* **2004**, *11*, 959–969.

42 G. J. Schroepfer, *Ann. Rev. Biochem.* **1981**, *50*, 585–621.

43 N. D. Lees, M. Bard, D. R. Kirsch, *Crit. Rev. Biochem. Mol. Biol.* **1999**, *34*, 33–47.

44 H. Schaller, *Plant Physiol. Bioch.* **2004**, *42*, 465–476.

45 T. J. Bach, P. Benveniste, *Prog. Lipid Res.* **1997**, *36*, 197–226.

46 I. Abe, M. Rohmer, G. D. Prestwich, *Chem. Rev.* **1993**, *93*, 2189–2206.

47 R. S. Burden, D. T. Cooke, G. A. Carter, *Phytochemistry* **1989**, *28*, 1791–1804.

48 E. I. Mercer, *Prog. Lipid Res.* **1993**, *32*, 357–416.

49 S. Jain, V. N. Sehgal, *Int. J. Dermatol.* **2000**, *39*, 412–423.

50 A. Chugh, A. Ray, J. B. Gupta, *Prog. Lipid Res.* **2003**, *42*, 37–50.

51 K. Bloch, *Science* **1965**, *150*, 19–28.

52 T. Ono, K. Nakazono, H. Kosaka, *Biochim. Biophys. Acta* **1982**, *709*, 84–90.

53 H. H. Tai, K. Bloch, *J. Biol. Chem.* **1972**, *247*, 3767–3773.

54 H.-K. Lee, P. Denner-Ancona, J. Sakakibara, T. Ono, G. D. Prestwich, *Arch. Biochem. Biophys.* **2000**, *381*, 43–52.

55 B. P. Laden, Y. Z. Tang, T. D. Porter, *Arch. Biochem. Biophys.* **2000**, *374*, 381–388.

56 V. Massey, *Biochem. Soc. Trans.* **2000**, *28*, 283–296.

57 M. J. Cormier, B. L. Strehler, *J. Am. Chem. Soc.* **1953**, *75*, 4864–4865.

58 W. D. McElroy, J. W. Hastings, V. Sonnenfeld, J. Coulombre, *Science* **1953**, *118*, 385–386.

59 B. Entsch, D. P. Ballou, V. Massey, *J. Biol. Chem.* **1976**, *251*, 2550–2563.

60 H. K. Lee, Y. F. Zheng, X. Y. Xiao, M. Bai, J. Sakakibara, T. Ono, G. D. Prestwich, *Biochem. Bioph. Res. Comm.* **2004**, *315*, 1–9.

61 M. H. Fu, M. Alexander, *Environ. Sci. Technol.* **1992**, *26*, 1540–1544.

62 A. M. Warhurst, K. F. Clarke, R. A. Hill, R. A. Holt, C. A. Fewson, *Appl. Environ. Microb.* **1994**, *60*, 1137–1145.

63 F. Beltrametti, A. M. Marconi, G. Bestetti, C. Colombo, E. Galli, M. Ruzzi, E. Zennaro, *Appl. Environ. Microb.* **1997**, *63*, 2232–2239.

64 K. Otto, K. Hofstetter, M. Röthlisberger, B. Witholt, A. Schmid, *J. Bacteriol.* **2004**, *186*, 5292–5302.

65 S. Hartmans, M. J. van der Werf, J. A. M. de Bont, *Appl. Environ. Microb.* **1990**, *56*, 1347–1351.

66 S. Panke, B. Witholt, A. Schmid, M. G. Wubbolts, *Appl. Environ. Microb.* **1998**, *64*, 2032–2043.

67 F. Müller (Ed.), *Chemistry and Biochemistry of Flavoenzymes*, CRC Press, Boca Raton, **1991**.

68 T. Ono, Y. Imai, *Methods Enzymol.* **1985**, *110*, 375–380.

69 M. Salmona, J. Pachecka, L. Cantoni, G. Belvedere, E. Mussini, S. Garattini, *Xenobiotica* **1976**, *6*, 585–591.

70 H. H. J. Cox, J. H. M. Houtman, H. J. Doddema, W. Harder, *Appl. Microbiol. Biotechnol.* **1993**, *39*, 372–376.

71 T. Watabe, M. Isobe, T. Sawahata, K. Yoshikawa, S. Yamada, E. Takabatake, *Scan. J. Work Environ. Health* **1978**, *4 Suppl 2*, 142–155.

72 H. H. J. Cox, B. W. Faber, W. N. M. van Heiningen, H. Radhoe, H. J. Doddema, W. Harder, *Appl. Environ. Microb.* **1996**, *62*, 1471–1474.

73 M. Oliynyk, C. B. W. Stark, A. Bhatt et al., *Mol. Microbiol.* **2003**, *49*, 1179–1190.

74 A. A. Ajaz, J. A. Robinson, *J. Chem. Soc. Chem. Comm.* **1983**, 679–680.

75 M. D. Lee, A. A. Fantini, G. O. Morton, J. C. James, D. B. Borders, R. T. Testa, *J. Antibiot.* **1984**, *37*, 1149–1152.
76 B. Shen, Y. G. Whittle, S. J. Gould, D. A. Keszler, *J. Org. Chem.* **1990**, *55*, 4422–4426.
77 S. J. Box, M. L. Gilpin, M. Gwynn, G. Hanscomb, S. R. Spear, A. G. Brown, *J. Antibiot.* **1983**, *36*, 1631–1637.
78 B. Shen, S. J. Gould, *Biochemistry* **1991**, *30*, 8936–8944.
79 S. J. Gould, B. Shen, *J. Am. Chem. Soc.* **1991**, *113*, 684–686.
80 S. J. Gould, M. J. Kirchmeier, R. E. LaFever, *J. Am. Chem. Soc.* **1996**, *118*, 7663–7666.
81 Y. G. Whittle, S. J. Gould, *J. Am. Chem. Soc.* **1987**, *109*, 5043–5044.
82 I. Sattler, R. Thiericke, A. Zeeck, *Nat. Prod. Rep.* **1998**, *15*, 221–40.
83 Y. Hu, C. R. Melville, S. J. Gould, H. G. Floss, *J. Am. Chem. Soc.* **1997**, *119*, 4301–4302.
84 M. Hara, K. Akasaka, S. Akinaga, M. Okabe, H. Nakano, R. Gomez, D. Wood, M. Uh, F. Tamanoi, *Proc. Natl. Acad. Sci. U.S.A.* **1993**, *90*, 2281–2285.
85 R. Thiericke, A. Zeeck, A. Nakagawa, S. Omura, R. E. Herrold, S. T. S. Wu, J. M. Beale, H. G. Floss, *J. Am. Chem. Soc.* **1990**, *112*, 3979–3987.
86 Y. Hu, H. G. Floss, *J. Am. Chem. Soc.* **2004**, *126*, 3837–44.
87 B. Furie, B. A. Bouchard, B. C. Furie, *Blood* **1999**, *93*, 1798–1808.
88 J. W. Suttie, *Ann. Rev. Biochem.* **1985**, *54*, 459–477.
89 P. Dowd, S. W. Ham, R. Hershline, *J. Am. Chem. Soc.* **1992**, *114*, 7613–7617.
90 A. Kuliopulos, B. R. Hubbard, Z. Lam, I. J. Koski, B. Furie, B. C. Furie, C. T. Walsh, *Biochemistry* **1992**, *31*, 7722–7728.
91 P. Dowd, R. Hershline, S. W. Ham, S. Naganathan, *Science* **1995**, *269*, 1684–1691.
92 A. E. Larson, J. W. Suttie, *FEBS Lett.* **1980**, *118*, 95–98.
93 S. Fetzner, *Appl. Microbiol. Biotechnol.* **2002**, *60*, 243–257.
94 C. C. Tseng, F. H. Vaillancourt, S. D. Bruner, C. T. Walsh, *Chem. Biol.* **2004**, *11*, 1195–1203.
95 T. O. Rogers, J. Birnbaum, *Antimicrob. Agents Chemother.* **1974**, *5*, 121–132.
96 D. Hendlin, E. O. Stapley, M. Jackson et al., *Science* **1969**, *166*, 122–123.
97 J. Shoji, T. Kato, H. Hinoo, T. Hattori, K. Hirooka, K. Matsumoto, T. Tanimoto, E. Kondo, *J. Antibiot.* **1986**, *39*, 1011–1012.
98 N. Katayama, S. Tsubotani, Y. Nozaki, S. Harada, H. Ono, *J. Antibiot.* **1990**, *43*, 238–246.
99 F. M. Kahan, J. S. Kahan, P. J. Cassidy, H. Kropp, *Ann. N. Y. Acad. Sci.* **1974**, *235*, 364–386.
100 H. Seto, T. Kuzuyama, *Nat. Prod. Rep.* **1999**, *16*, 589–596.
101 T. Hidaka, M. Goda, T. Kuzuyama, N. Takei, M. Hidaka, H. Seto, *Mol. Gen. Genet.* **1995**, *249*, 274–280.
102 T. Kuzuyama, T. Seki, S. Kobayashi, T. Hidaka, H. Seto, *Biosci. Biotech. Bioch.* **1999**, *63*, 2222–2224.
103 F. Hammerschmidt, G. Bovermann, K. Bayer, *Liebigs Ann. Chem.* **1990**, 1055–1061.
104 H. Seto, T. Hidaka, T. Kuzuyama, S. Shibahara, T. Usui, O. Sakanaka, S. Imai, *J. Antibiot.* **1991**, *44*, 1286–1288.
105 P. H. Liu, A. M. Liu, F. Yan, M. D. Wolfe, J. D. Lipscomb, H. W. Liu, *Biochemistry* **2003**, *42*, 11577–11586.
106 V. P. Miller, J. S. Thorson, O. Ploux, S. F. Lo, H. W. Liu, *Biochemistry* **1993**, *32*, 11934–11942.
107 A. Woschek, F. Wuggenig, W. Peti, F. Hammerschmidt, *ChemBioChem* **2002**, *3*, 829–835.
108 B. P. Simov, F. Wuggenig, M. Lammerhofer, W. Lindner, E. Zarbl, F. Hammerschmidt, *Eur. J. Org. Chem.* **2002**, 1139–1142.
109 Z. B. Zhao, P. H. Liu, K. Murakami, T. Kuzuyama, H. Seto, H. W. Liu, *Angew. Chem. Int. Ed. Engl.* **2002**, *41*, 4529–4532.
110 T. Hashimoto, J. Matsuda, Y. Yamada, *FEBS Lett.* **1993**, *329*, 35–39.
111 T. Hashimoto, Y. Yamada, *Eur. J. Biochem.* **1987**, *164*, 277–285.
112 T. Hashimoto, J. Kohno, Y. Yamada, *Phytochemistry* **1989**, *28*, 1077–1082.
113 A. G. Prescott, M. D. Lloyd, *Nat. Prod. Rep.* **2000**, *17*, 367–383.
114 L. Que, R. Y. N. Ho, *Chem. Rev.* **1996**, *96*, 2607–2624.
115 C. J. Schofield, Z. H. Zhang, *Curr. Opin. Struct. Biol.* **1999**, *9*, 722–731.

116 K. I. Kivirikko, T. Pihlajaniemi, *Adv. Enzymol. Rel. Areas Mol. Biol.* **1998**, *72*, 325–98.

117 C. Schofield, L. C. Hsueh, Z. H. Zhang, J. K. Robinson, I. Clifton, K. Harlos, *J. Inorg. Biochem.* **1999**, *74*, 49–49.

118 Z. Zhang, J. Ren, K. Harlos, C. H. McKinnon, I. J. Clifton, C. J. Schofield, *FEBS Lett.* **2002**, *517*, 7–12.

119 Z. H. Zhang, J. S. Ren, D. K. Stammers, J. E. Baldwin, K. Harlos, C. J. Schofield, *Nat. Struct. Biol.* **2000**, *7*, 127–133.

120 M. J. Ryle, R. P. Hausinger, *Curr. Opin. Chem. Biol.* **2002**, *6*, 193–201.

121 L. Que, R. P. Hausinger, E. L. Hegg, R. Y. N. Ho, A. Liu, M. P. Mehn, M. J. Ryle, *J. Inorg. Biochem.* **2001**, *86*, 388–388.

122 S. Fetzner, *Appl. Microbiol. Biotechnol.* **1998**, *50*, 633–657.

123 D. B. Janssen, F. Pries, J. R. van der Ploeg, *Ann. Rev. Microbiol.* **1994**, *48*, 163–191.

124 R. M. de Jong, B. W. Dijkstra, *Curr. Opin. Struct. Biol.* **2003**, *13*, 722–730.

125 H. M. S. Assis, A. T. Bull, D. J. Hardman, *Enzyme Microb. Technol.* **1998**, *22*, 545–551.

126 T. Nakamura, T. Nagasawa, F. Yu, I. Watanabe, H. Yamada, *Tetrahedron* **1994**, *50*, 11821–11826.

127 J. H. Lutje Spelberg, J. E. T. van Hylckama Vlieg, T. Bosma, R. M. Kellogg, D. B. Janssen, *Tetrahedron: Asymmetry* **1999**, *10*, 2863–2870.

128 L. Tang, J. H. Lutje Spelberg, M. W. Fraaije, D. B. Janssen, *Biochemistry* **2003**, *42*, 5378–5386.

129 R. M. de Jong, J. J. W. Tiesinga, H. J. Rozeboom, K. H. Kalk, L. Tang, D. B. Janssen, B. W. Dijkstra, *EMBO J.* **2003**, *22*, 4933–4944.

130 N. Tanaka, T. Nonaka, T. Tanabe, T. Yoshimoto, D. Tsuru, Y. Mitsui, *Biochemistry* **1996**, *35*, 7715–7730.

131 A. Yamashita, H. Kato, S. Wakatsuki, T. Tomizaki, T. Nakatsu, K. Nakajima, T. Hashimoto, Y. Yamada, J. Oda, *Biochemistry* **1999**, *38*, 7630–7637.

132 J. E. T. van Hylckama Vlieg, L. Tang, J. H. Lutje Spelberg, T. Smilda, G. J. Poelarends, T. Bosma, A. E. J. van Merode, M. W. Fraaije, D. B. Janssen, *J. Bacteriol.* **2001**, *183*, 5058–5066.

133 J. H. Lutje Spelberg, J. E. T. van Hylckama Vlieg, L. Tang, D. B. Janssen, R. M. Kellogg, *Org. Lett.* **2001**, *3*, 41–43.

134 K. Faber, R. V. A. Orru in *Enzyme Catalysis in Organic Chemistry*, 2nd ed., K. Drauz, H. Waldmann (Eds.), Wiley-VCH, Weinheim, **2002**, Vol. II, 579–608.

135 S. Flitsch, G. Grogan, D. Ashcroft in *Enzyme Catalysis in Organic Synthesis*, 2nd ed.; K. Drauz, H. Waldmann (Eds), Wiley-VCH, Weinheim, **2002**, Vol. III, 1065–1108.

136 S. W. May, B. J. Abbott, *J. Biol. Chem.* **1973**, *248*, 1725–1730.

137 S. X. Jin, T. M. Makris, T. A. Bryson, S. G. Sligar, J. H. Dawson, *J. Am. Chem. Soc.* **2003**, *125*, 3406–3407.

138 C. S. Miles, T. W. B. Ost, M. A. Noble, A. W. Munro, S. K. Chapman, *Biochim. Biophys. Acta* **2000**, *1543*, 383–407.

11
Aziridine Natural Products – Discovery, Biological Activity and Biosynthesis

Philip A. S. Lowden

11.1
Introduction and Overview

In contrast with the vast number of epoxide natural products and metabolites, there are only a very small number of known naturally occurring compounds that contain an aziridine ring. The best known and first to be discovered are the mitomycin antitumor antibiotics, known to possess an aziridine ring since the 1960s, and several others have been isolated since. There are even a few examples of naturally occurring azirines. Where the mode of action of these compounds has been studied in detail, the electrophilic nature of the aziridine ring has been shown to play a key role in the molecular mechanism. The biochemical reactions used for formation of the aziridine ring remain a mystery, although recent progress in the genetics of mitomycin biosynthesis has brought us a large step closer to understanding how this occurs. While most enzymatic epoxide formation occurs by insertion of an oxygen atom from O_2 into an alkene, it seems unlikely that aziridines are formed in a similar manner from N_2, nitrogen being extremely unreactive towards electrophiles and towards one- and two-electron reduction. Although nitrogen fixation can be achieved by bacteria, the enzyme system involved, nitrogenase, is rather inefficient, requires ATP, and is composed of several metal clusters that are rapidly deactivated by O_2 [1].

Although few in number, these molecules are worth studying to define the role of the aziridine functional group in their bioactivity, and because new and interesting biochemical reactions are bound to be revealed in their biosynthetic pathways. The aziridine ring has the potential to play a useful and versatile role in the design and construction of bioactive agents, and a deep understanding of the biosynthetic enzymes should allow for biosynthetic engineering of designed aziridine natural products. This chapter summarizes the current state of knowledge of the occurrence, mode of action, and biosynthesis of aziridine-containing natural products. It does not cover in detail the large amount of effort devoted to synthesis of the natural products and related structures, as this is reviewed elsewhere.

Aziridines and Epoxides in Organic Synthesis. Andrei K. Yudin

ISBN: 3-527-31213-7

11.2
Mitomycins and Related Natural Products

11.2.1
Discovery and Anticancer Properties

Mitomycins A (**1**; Scheme 11.1) and B (**2**) were isolated from *Streptomyces caespitosus* by Hata et al. at the Kitasato Institute in Japan in 1956 [2, 3] and were found to have potent antibiotic and antitumor activity. In 1958 the Kyowa Hakko Kogyo pharmaceutical company isolated mitomycin C (**3**) from the same organism [4]. Porfiromycin (**4**) was isolated in 1960 from *Streptoverticillatium ardus* [5], and in 1962 all four mitomycins, as well as the inactive analogue mitiromycin (**5**), were isolated from *Streptomyces verticillatus* [6]. The structures of the mitomycins were established soon after their discovery [7–11] and the stereochemical relationships between them were later corrected by high-resolution X-ray crystallography [12] and by further chemical studies [13].

Mitomycin C was found to have broad activity against a range of tumors and has been used clinically since the early 1960s [14, 15]. It causes many specific cellular effects, including inhibition of DNA synthesis, recombination, chromosome breakage, sister chromatid exchange, induction of DNA repair, and induction of

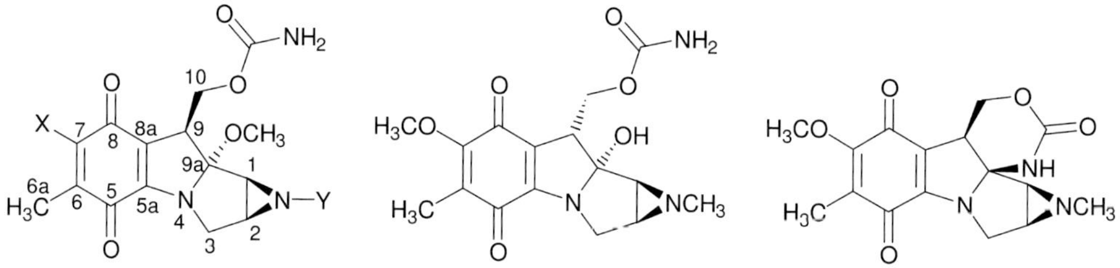

1, X = OCH_3, Y = H
3, X = NH_2, Y = H
4, X = NH_2, Y = CH_3

2

5

Scheme 11.1

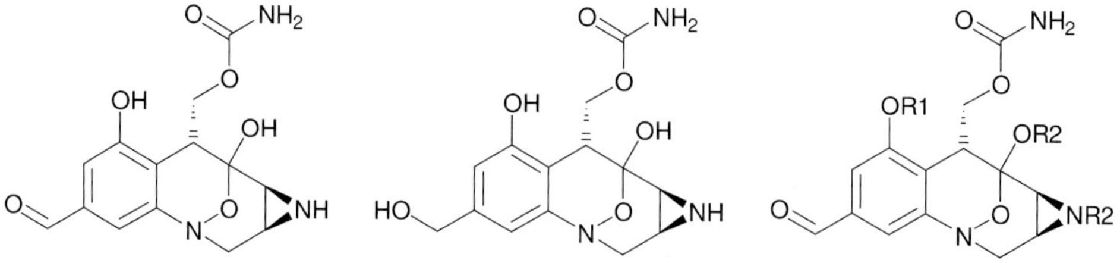

6

7

8, R1 = R2 = Ac
9, R1 = CH_3, R2 = Ac

Scheme 11.2

apoptosis [16]. Hundreds of analogues have been synthesized and evaluated for anticancer activity [17].

The antitumor antibiotics FR900482 (**6**; Scheme 11.2) and FR66979 (**7**) were isolated from *Streptomyces sandaensis* 6897 at the Fujisawa Pharmaceutical Company in Japan, and their structures are clearly related to those of the mitomycins [18–22]. The semisynthetic derivatives FK973 (**8**) [23] and FK317 (**9**) [24] have been prepared; they have improved pharmacological properties and have been investigated in clinical trials for application as anticancer drugs. They are promising candidates, since they are more potent and less toxic than mitomycin C.

11.2.2
Mode of Action

The molecular basis of the cytotoxicity of the mitomycins has been the subject of thorough investigation for many years, and has been reviewed in great detail [16, 25–27]. The key aspects are summarized here. Early investigations revealed that the mitomycins formed interstrand crosslinks in DNA [28], as well as monoalkylated adducts [29–32], and that these alkylation events were responsible for cytotoxicity, with one crosslink per genome sufficient to kill bacteria [29]. Alkylation and crosslinking are sequence-specific. Monoalkylation shows selectivity for guanine in the sequence 5'-CpG-3' (CpG) [33–35], while crosslinking is completely specific for CpG · GpC duplex sequences [36–38]. It has also been demonstrated that the efficiency and guanine selectivity of alkylation and crosslinking by mitomycin C is enhanced when the target guanines are base-paired with 5-methylcytosine [39–42]. This has been ascribed to the electron-donating effect of the methyl group being transmitted through the hydrogen bond to N-2 of guanine, thus increasing its basicity [43, 44].

The mitomycins do not react directly with DNA, but require prior activation by reduction of the quinone. This property of bioreductive activation has inspired the design and development of synthetic anticancer drugs that are also activated by reduction, as this is expected to confer a degree of tumor selectivity [45, 46]. Many solid tumors are short of oxygen relative to normal tissue, so reductive activation of the mitomycins and other bioreductive drugs can proceed in tumors, while it is inhibited by the oxidizing environments in normal tissues.

The requirement for reduction prior to DNA alkylation and crosslinking was first demonstrated by Iyer and Szybalski in 1964 [29], and can be induced both by chemical reducing agents such as sodium dithionite and thiols *in vitro* and by various reductive enzymes such as DT-diaphorase (NAD(P)H-quinone oxidoreductase) *in vitro* and *in vivo* [47]. Much work to characterize the mechanism of reductive activation and alkylation has been carried out, principally by the Tomasz and Kohn groups, and Figure 11.1 illustrates a generally accepted pathway for mitomycin C [16, 48–50] based on these experiments, which is very similar to the mechanism originally proposed by Iyer and Szybalski [29].

Two-electron reduction of the quinone provides the hydroquinone **10**, which may also be formed by two sequential one-electron reductions. It has also been

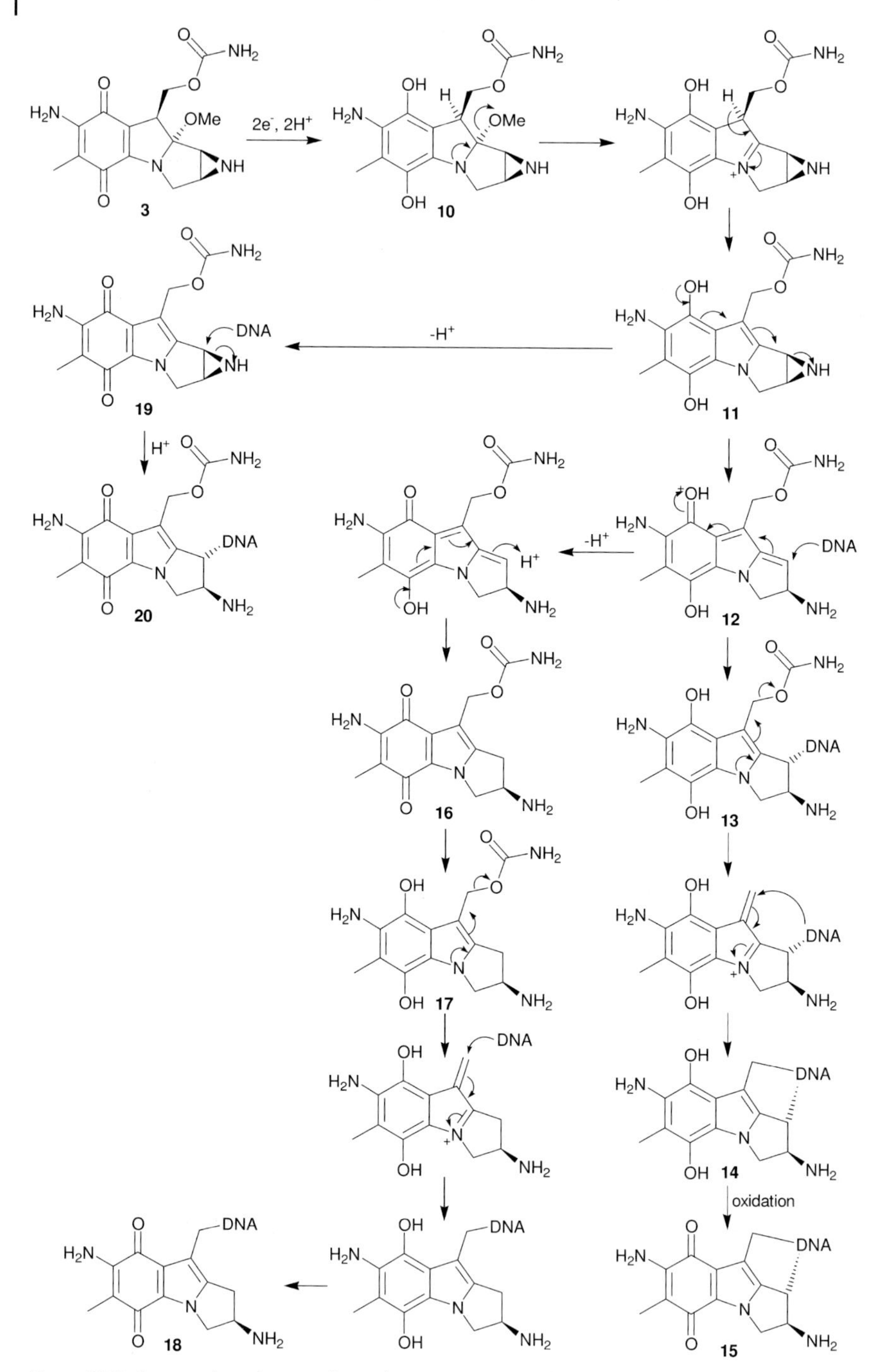

Figure 11.1 Proposed mechanism for reductive activation of and DNA alkylation/crosslinking by mitomycin C.

proposed that one-electron reduction to the semiquinone radical is sufficient for activation [51]. With the N-4 lone pair now removed from conjugation with the quinone C=O, elimination of methanol can occur to form *leuco*-2,7-diaminomitosene (**11**), followed by aziridine-opening to form reactive intermediate **12**. Species **12** can now alkylate DNA at the 2-NH_2 group of guanine to form monoadduct **13**, which can then alkylate a second guanine N-2 by loss of the carbamate to form crosslinked adduct **14**, which is then trapped by oxidation to **15**. Intermediate **12** can alternatively be inactivated by deprotonation and tautomerization to form the mitomycin C metabolite 2,7-diaminomitosene (**16**), which must be reactivated by quinone reduction to **17**, which can then form DNA monoadduct **18** (bound at either N-2 or N-7 of guanine) in a fashion similar to that seen in the formation of the crosslinked adduct. Monoalkylation generally occurs by reoxidation of **11** to form aziridinomitosene **19**, which can alkylate N-2 of guanine to form adduct **20**. While it is quite certain that quinone reduction is necessary before loss of methanol can occur, it has been noted that the reduced state may not be required for subsequent alkylation steps [52].

The different possible adducts formed between mitomycin C and DNA have been isolated by degradation of DNA after *in vitro* alkylation/crosslinking reactions and structurally characterized. Monoadduct **21** (Scheme 11.3), derived from alkylation at C-1 only [53], and monoadducts **22** [54] and **23** [55, 56] (derived from C-10 alkylation by **16** at N-7 or N-2 of guanine, respectively) have been isolated, together with bisadducts **24** [57] and **25** [58], derived from interstrand and intrastrand crosslinks, respectively, and adduct **26** [59], formed by addition of a molecule of water to C-10 instead of the second guanine. All of these adducts have also been isolated from DNA after *in vivo* crosslinking [60, 61].

Three-dimensional structures of adducts **21** [62], **22** [63], and **24** [64] in the context of short oligonucleotide duplexes have been determined by two-dimensional NMR and molecular dynamics. The conformations found by NMR are largely similar to those previously derived from molecular modeling studies [36, 53, 57, 65–67] and provide great insight into the efficiency and sequence-specificity of alkylation and crosslinking. The crosslinked structure revealed that mitomycin C was bound in the minor groove of DNA and that very little structural perturbation of DNA was required for crosslinking to occur. The structure containing monoadduct **21** revealed the origin of the specificity for crosslinking at the CpG · GpC sequence. A hydrogen bond was observed between the carbamoyl oxygen atom directly attached to C-10 of mitomycin C and the NH_2 group of the guanine residue on the opposite strand. This aligns the mitomycin C molecule in an optimal orientation for crosslinking.

Like the mitomycins, FR900482 (**6**), FR66979 (**7**), FK973 (**8**), and FK317 (**9**) have also been shown to crosslink DNA both *in vivo* [68–70], and *in vitro* after reductive activation [71–76] with selectivity for the 5'-CpG-3' sequence [77]. The mechanism outlined in Figure 11.2 was originally proposed by Goto and Fukuyama [78] and has been verified by the experimental work of Williams and Hopkins [71–77, 79]. Reduction of the N–O bond produces intermediate **27**, which can lose a molecule of water to form **28**, which reacts with DNA by a mechanism similar to that found

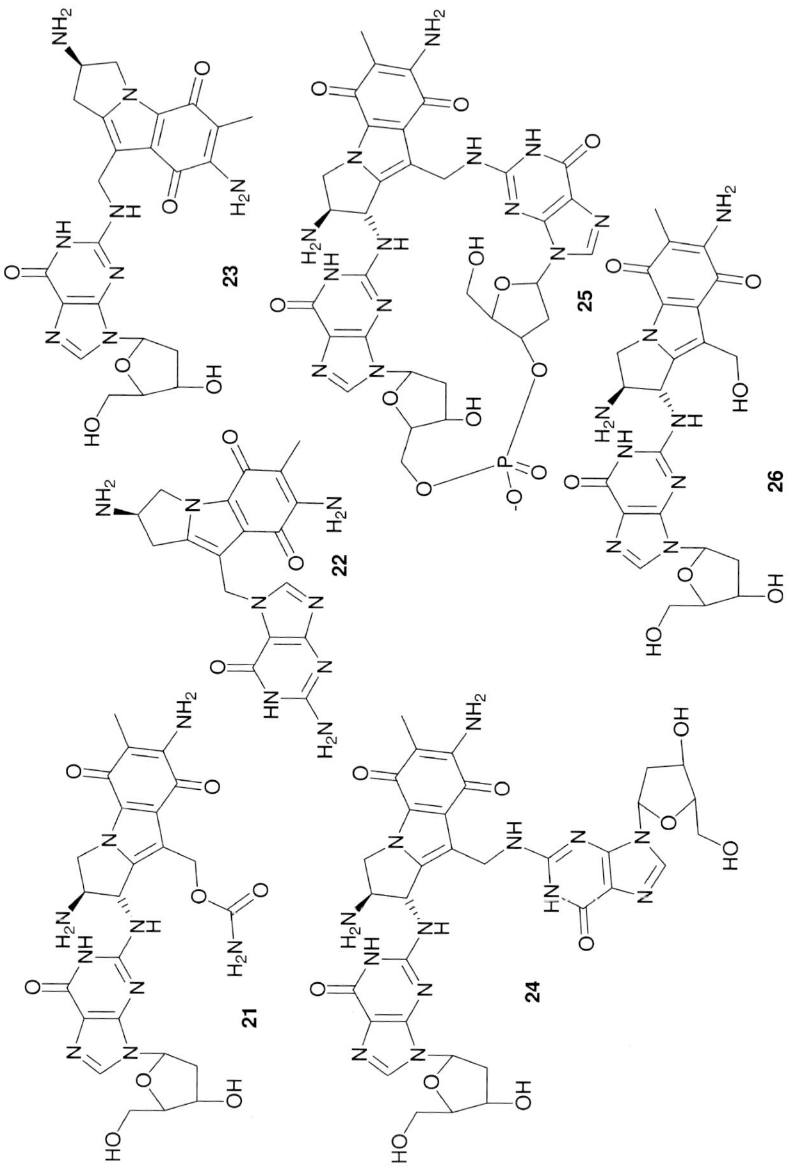

21

22

23

24

25

26

Scheme 11.3

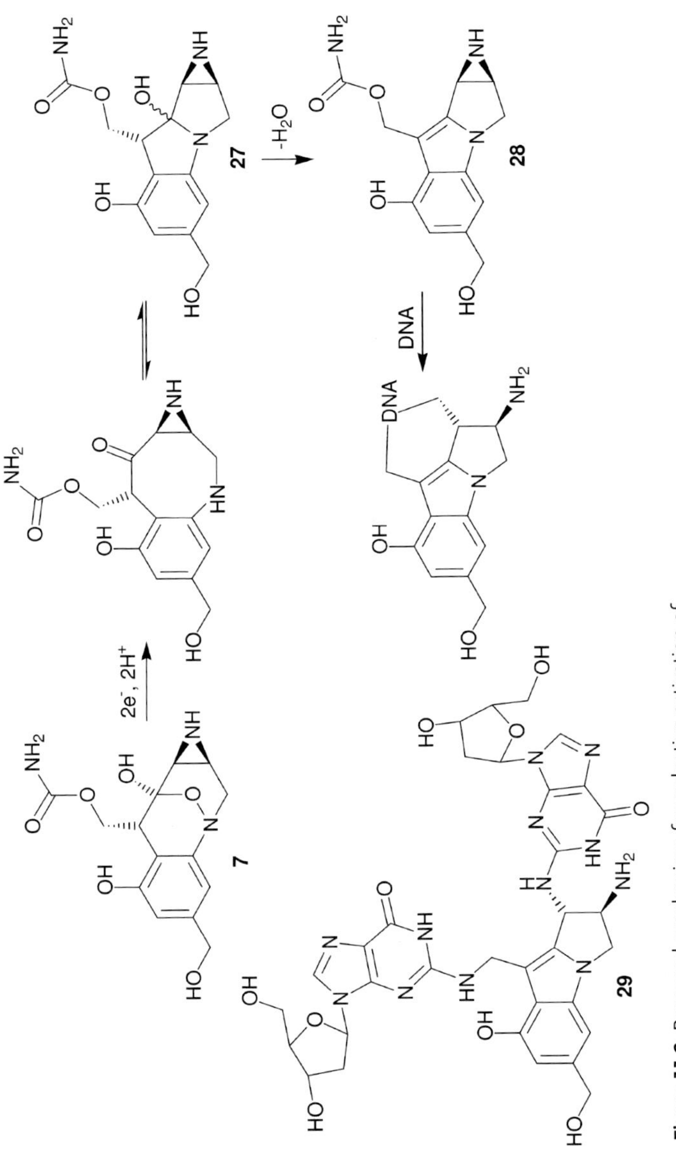

Figure 11.2 Proposed mechanism for reductive activation of and DNA alkylation/crosslinking by FR66979.

for the mitomycins. The crosslinked adduct **29** has been isolated from *in vitro* reactions of reductively activated **7** and DNA [79]. A DNA monoadduct of **7** has also been identified and shown to arise from alkylation of N-2 of guanine, with lower selectivity for 5'-CpG-3' than observed with mitomycin C [75, 80]. The lower toxicity of **6** and **7** is believed to be due to the fact that their activation does not require reduction of a quinone, which can result in superoxide production and consequent DNA strand cleavage and lipid oxidation [81].

Early on, **6** and **7** were shown to form DNA-protein crosslinks *in vivo* [69, 70], and this has been confirmed *in vitro* through **7**-induced crosslinking of synthetic oligonucleotides to a peptide corresponding to the DNA minor groove binding domain of the oncogenic protein HMG A1 [82, 83]. Crosslinking of this protein to DNA was subsequently confirmed *in vivo* for **6** and **9** [84, 85]. FK317 (**9**) is the most promising clinical candidate, since it does not induce Vascular Leak Syndrome, a serious side effect that caused FK973 (**8**) to be withdrawn from clinical trials. It has recently been reported that FR900482 (**6**) induces necrotic cell death, whereas **9** induces apoptotic cell death at higher concentrations, and that this may explain the lower toxicity of **9** [85].

11.2.3 Biosynthesis

Early studies on the biosynthesis of the mitomycins by Kirsch and Korshalla and the groups of Vining and Hornemann established that the *O*- and *N*-methyl groups, but not the *C*-methyl groups, of mitomycins A [86, 87], B [87], and C [87, 88] and of porfiromycin [87] are derived from methionine (**30**) via *S*-adenosylmethionine, shown by feeding of [*methyl*-^{14}C]-L-methionine to *Streptomyces verticillatus*. The carbamoyl groups of all four mitomycins were labeled with [*guanidino*-^{14}C]- L-arginine (**31**) [87] and [*ureido*-^{14}C]-L-citrulline (**32**) [89], and subsequent feeding of [NH_2CO-^{13}C,^{15}N]-L-citrulline [90] demonstrated intact incorporation of the ureido group of citrulline into mitomycins A and B. This suggests that the arginine dihydrolase pathway is utilized and that arginine is first hydrolyzed to citrulline, which is then converted by ornithine transcarbamoylase into carbamoyl phosphate, which acts as the immediate precursor to the mitomycins. These results are summarized for mitomycin C in Figure 11.3.

Mitomycins were efficiently labeled by [1-^{14}C]-D-glucosamine (**33**), with degradative studies indicating incorporation into the non-quinone C_6 chain [87]. Subsequent work demonstrated that the amino group of glucosamine is incorporated into the aziridine ring [91]. Labels from [1-^{14}C,^{15}N]-D-glucosamine [91] or from [1-^{13}C,^{15}N]-D-glucosamine [89] were both incorporated into mitomycin B at comparable levels, indicating intact incorporation, and the locations of the ^{13}C and ^{15}N labels were confirmed from the fragmentation pattern in the mass spectrum. Label from [6-^{14}C]-D-glucosamine was incorporated into mitomycins, and degradation revealed labeling predominantly at C-10. These results are summarized for mitomycin C in Figure 11.4.

^{14}C-labeled D-glucose (**34**) [87, 88] and D-ribose [87] were efficiently incorporated

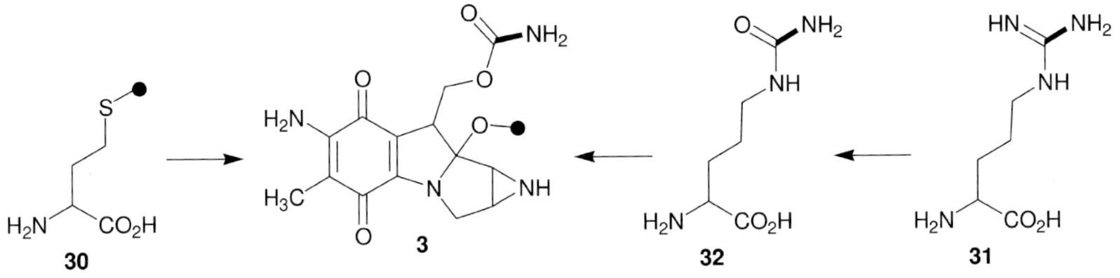

Figure 11.3 Biosynthetic origin of the O-methyl and carbamoyl groups of mitomycin C.

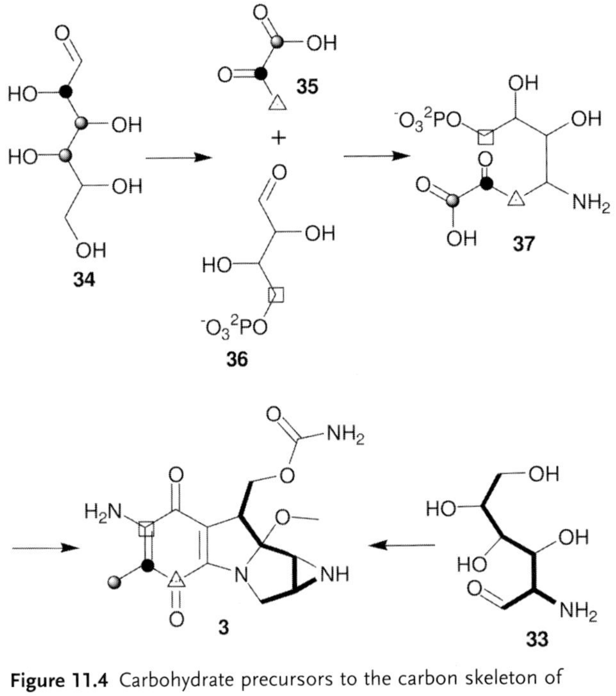

Figure 11.4 Carbohydrate precursors to the carbon skeleton of mitomycin C.

into mitomycins, with ribose labeling the quinone ring. Neither aromatic amino acids nor shikimic acid were incorporated to any significant extent, precluding a late branch from the shikimate pathway. Hornemann et al. demonstrated specific labeling of the C-6 methyl group of mitomycin C by [1-^{14}C]-pyruvate (**35**) and [3,4-^{14}C]-D-glucose and of C-6 by [2-^{14}C]-pyruvate and [2-^{14}C]-D-glucose (Figure 11.4) [92]. They concluded that this would be consistent with an early branch from the shikimate pathway. They later observed labeling at C-7 of mitomycins by [4-^{14}C]-D-erythrose (**36**), but not by sodium [3-^{14}C]-pyruvate, which labeled mitomycins elsewhere in the quinone, presumably at C-5 (Figure 11.4) [93]. [7-^{14}C]-

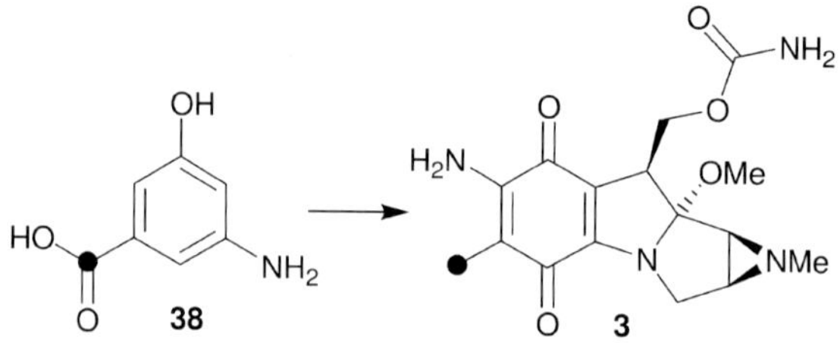

Figure 11.5 3-Amino-5-hydroxybenzoic acid as a precursor to the porfiromycin C quinone.

3-Dehydroquinic acid and [1,6-^{14}C]-D-shikimic acid methyl ester were not incorporated, indicating a very early branch from the shikimate pathway. The intermediacy of 4-amino-3,4-dideoxy-D-*arabino*-heptulosonic acid 7-phosphate (**37**) was proposed, consistent with later findings on the role of the variant aminoshikimate pathway [94].

Rickards et al. established that the quinone ring of the mitomycins is derived from the unusual amino acid 3-amino-5-hydroxybenzoic acid (AHBA, **38**) [95], which also acts as an intermediate in the biosynthesis of a range of natural products, such as the ansamycin antibiotics [96], that contain a 'C_7N' structural fragment [94]. They fed [*carboxy*-^{13}C]-**38** to *S. verticillatus* and observed specific incorporation of ^{13}C into the C-6 methyl group of porfiromycin (Figure 11.5) [95]. They also succeeded in isolating **38** from *S. verticillatus*, confirming its presence as a free intermediate [97].

These findings support Hornemann's earlier proposal that the quinone moiety of the mitomycins was derived from an early branch of the shikimate pathway. The biosynthesis of **38** has been extensively characterized, especially in the rifamycin producer *Amycolatopsis mediterranei* [96]. It has been shown to arise from a parallel aminoshikimate pathway (Figure 11.6) [98, 99], in which ammonia is incorporated from the amide nitrogen of glutamine by way of the imine **39** during the formation of 3,4-dideoxy-4-amino-D-*arabino*-heptulosonic acid 7-phosphate (aminoDAHP, **37**), which is then processed through aminodehydroquinic acid (**40**) and aminodehydroshikimic acid (**41**), which acts as the substrate for the pyridoxal phosphate-dependent enzyme AHBA synthase. This enzyme has been purified from *A. mediterranei*, cloned, and sequenced [100], and its crystal structure has been determined [101]. The sequence of AHBA synthase was used to locate and sequence the entire biosynthetic gene cluster for rifamycin [102], which included a dedicated set of genes for biosynthesis of AHBA by the aminoshikimate pathway. These have been studied in more detail through mutation and heterologous expression [103].

A number of mitomycin analogues have been prepared by precursor-directed biosynthesis [104]. A range of amines were fed to *S. caespitosus*, and novel derivatives of mitomycin C (type I analogues) and mitomycin B (type II analogues) were identified and in some cases (**42–46** and **52–56**; Scheme 11.4) isolated and characterized. Antibiotic and antitumor activities were comparable to those of mitomycin C, with the type I analogues more active than the type II analogues.

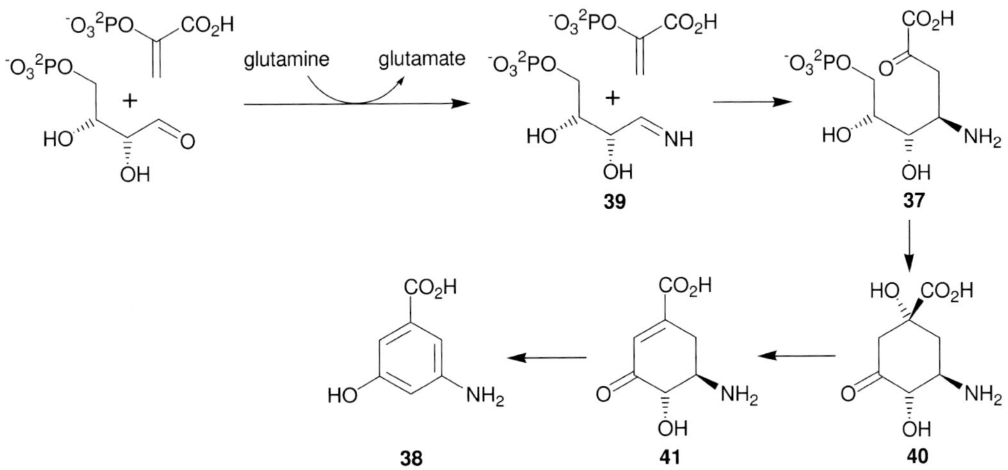

Figure 11.6 The aminoshikimate pathway to 3-amino-5-hydroxybenzoic acid.

Type I analogues
42, R = Me
43, R = Et
44, R = Pr
45, R = propargyl
46, R = 2-methylallyl
47, R = allyl
48, R = 2-chloroethyl
49, R = 3-chloropropyl
50, R = Bn
51, R = Ph

Type II analogues
52, R = Me
53, R = Et
54, R = Pr
55, R = propargyl
56, R = 2-methylallyl
57, R = allyl
58, R = 2-chloroethyl
59, R = 3-chloropropyl
60, R = Bn
61, R = Ph

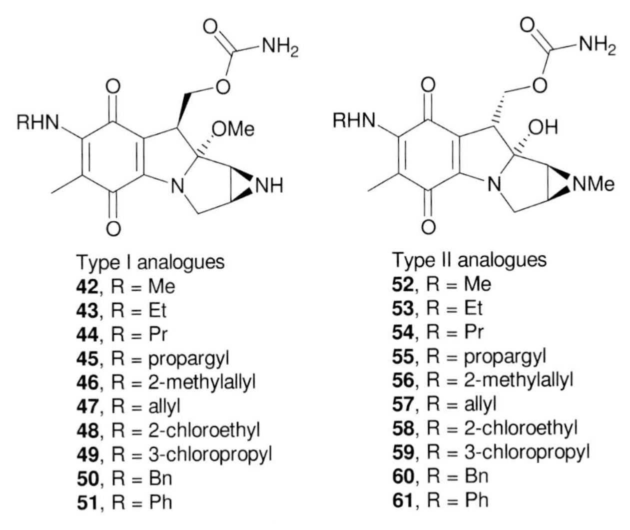

Scheme 11.4

Sherman's group have made great advances in understanding of the genetics and enzymology of mitomycin production. They have cloned and sequenced the gene cluster for biosynthesis of mitomycin C in *Streptomyces lavendulae* [105] and have made progress in probing the functions of these genes in the biosynthetic pathway.

Their strategy was firstly to identify genes conferring resistance to mitomycin upon the producing organism, since antibiotic resistance genes are often clustered with the biosynthetic genes. They successfully cloned and sequenced a mitomycin C resistance gene (*mcr*) with sequence similarity to FAD-dependent oxidoreductases [106], and subsequently demonstrated that the MCR protein possesses a covalently bound FAD cofactor and confers resistance to mitomycin C by oxidizing the reductively activated form of the antibiotic [107].

They identified a second mitomycin resistance determinant (*mrd*) and found that the MRD protein bound unreduced mitomycin C noncovalently, thus preventing it from becoming activated [108]. The crystal structure of MRD has since been determined [109]. Surprisingly, MRD has been found to have weak quinone reductase activity and actually reduces mitomycin C to its activated form, although reduction is slow enough for the drug-binding to confer resistance [110]. Sequencing of the biosynthetic gene cluster revealed a third resistance gene (*mct*), which was characterized as a putative mitomycin C transport protein [111]. It is membrane-bound and confers mitomycin C resistance when coexpressed with *mrd*. These two resistance genes are believed to act together as a mitomycin-specific drug export system.

Parallel advances in the genetic analysis of ansamycin biosynthesis provided an additional route to the mitomycin biosynthetic gene cluster through the availability of the sequence of the AHBA synthase (*rif*K) from the rifamycin producer *Amycolatopsis mediterranei* [100]. This was used to probe for homologous genes in the *S. lavendulae* genome, and a gene (*mit*A) with 71% identity to *rif*K was cloned and sequenced [112]. Disruption abolished mitomycin C production, confirming that *mit*A is required for mitomycin C biosynthesis.

The *mrd* (but not *mcr*) and *mit*A genes were found to be linked in one clone of a cosmid library of *S. lavendulae*. Sequencing around this region provided the gene cluster for mitomycin C biosynthesis, which contains 47 genes, many of which could be assigned functions on the basis of homology with known genes. These functional assignments are summarized in Table 11.1. The role of the cluster in mitomycin C biosynthesis was confirmed through gene disruptions, resulting in reduction or abolition of mitomycin C production.

Seven genes, including AHBA synthase, homologous to the genes (*rif*G-*rif*N) required for AHBA biosynthesis in *Amycolatopsis mediterranei* were found. None of the genes in the cluster could be assigned to aminoDAHP synthase, but a gene homologous to the aminoDAHP synthase in *A. mediterranei* and unlinked to the mitomycin C cluster was found by hybridization.

A chitinase is found at one end of the gene cluster, as would be expected for supply of glucosamine for mitomycin biosynthesis. No genes could confidently be unambiguously assigned a role in formation of the mitosane core structure. A glycosyltransferase that may be involved in joining the amino group of AHBA to C-1 of glucosamine was found, however, together with a gene homologous to the berberine bridge C-C bond-forming enzyme, which may be required to form the unusual linkage between C-3 of AHBA and C-5 of glucosamine. A number of subsequent tailoring steps, summarized in Figure 11.7, are required to form the fully functionalized mitomycin structure.

A number of genes can be assigned roles in tailoring of the mitosane core

Table 11.1 Predicted sizes and proposed functions of each protein product of the mitomycin C biosynthetic gene cluster.

Gene	Amino acids	Proposed function
mitT	270	aminoquinate dehydrogenase (RifI homologue)
mitS	315	kinase (RifN homologue)
mitR	514	McrA homologue
mitQ	164	putative regulator
mitP	343	aminoDHQ synthase (RifG homologue)
mitO	163	unknown
mitN	275	methyltransferase
mitM	283	methyltransferase
mitL	520	esterase
mitK	346	F420-dependent H_4MPT reductase
mitJ	235	phosphatase (RifM homologue)
mitI	290	unknown
mitH	382	F420-dependent H_4MPT reductase
mitG	404	oxidoreductase (RifL homologue)
mitF	257	reductase
mite	707	CoA ligase
mitD	383	unknown
mitC	260	unknown
mitB	272	glycosyltransferase
mitA	388	AHBA synthase (RifK homologue)
mmcA	514	unknown
mmcB	93	acyl carrier protein
mmcC	470	unknown
mmcD	611	methyltransferase
mmcE	359	H_4MPT:CoM methyltransferase
mmcF	145	aminoDHQ dehydratase (RifJ homologue)
mmcG	177	unknown
mmcH	254	unknown
mmcI	264	F420-dependent H_4MPT reductase
mmcJ	274	F420-dependent H_4MPT reductase
mmcK	460	unknown
mmcL	511	aldehyde dehydrogenase
mmcM	472	McrA homologue
mmcN	395	cytochrome P450 hydroxylase
mmcO	474	unknown
mrd	130	mitomycin resistance determinant
mmcP	443	unknown
mmcQ	123	unknown
mmcR	351	*O*-methyltransferase
mmcS	546	carbamoyltransferase
mmcT	568	hydroxylase
mmcU	160	sulfate adenylate transferase unit I
mmcV	319	sulfate adenylate transferase unit II
Mct	484	mitomycin C translocase
mmcW	163	repressor
mmcX	413	unknown
mmcY	271	chitinase

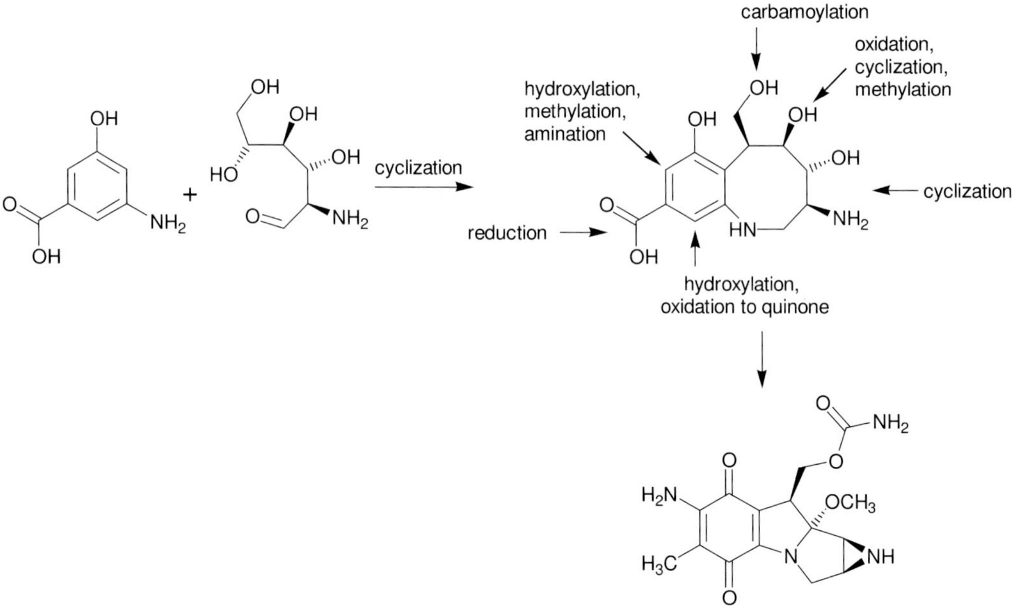

Figure 11.7 Summary of the biosynthetic pathway to mitomycin C.

structure. Four coenzyme F420-dependent tetrahydromethanopterin reductase genes and one tetrahydromethanopterin-CoM methyltransferase gene similar to genes involved in reduction of CO_2 to CH_4 by methanogenic bacteria were found. These may be involved in reduction of the carboxyl group of AHBA to the C-6 methyl group. Three SAM-dependent methyltransferases that would be predicted to be involved in addition of *O*- and *N*-methyl groups were identified. Two cytochrome P450 genes were found, consistent with the requirement for hydroxylation at C-5 and C-7, and a putative carbamoyltransferase was also identified.

Although none of the biosynthetic genes could confidently be assigned a role in forming the aziridine ring, a reasonable hypothesis is that cyclization occurs through an S_N2 reaction in which the C-2 amine from glucosamine acts as the nucleophile and C-3 from glucosamine the electrophile, perhaps after conversion of the glucosamine C-3 OH into a better leaving group X (Figure 11.8). This has precedent in the biotransformation of vicinal hydroxyamines to aziridines via intermediate sulfate esters (see Section 11.4.7), and the enzymatic formation of epoxides from 1,2-haloalcohols (see Chapter 10 by Grüschow and Sherman).

Sherman et al. have analyzed the function of the MitM methyltransferase in more detail [113]. They knocked out the MitM gene and found that mitomycin C production was abolished, and were able to isolate a small quantity of 9a-demethyl mitomycin A (**62**) from this mutant (Figure 11.9). MitM was expected to act as an *O*-methyltransferase, but surprisingly recombinant MitM converted **62** into 9-*epi*-mitomycin B (**63**), hence acting as an aziridine *N*-methyltransferase. Mitomy-

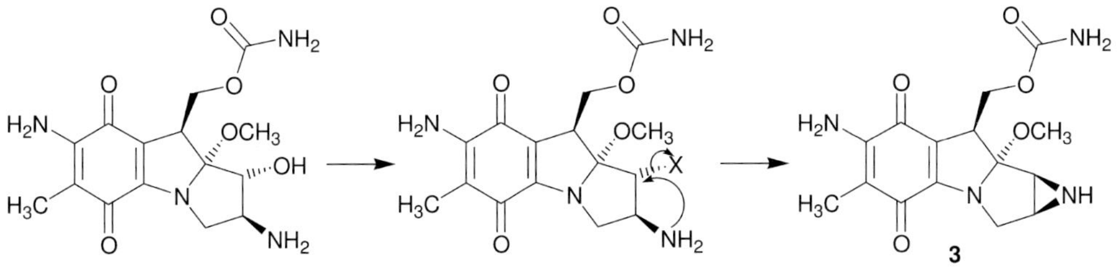

Figure 11.8 Hypothetical mechanism for aziridine formation in mitomycin C; X = leaving group.

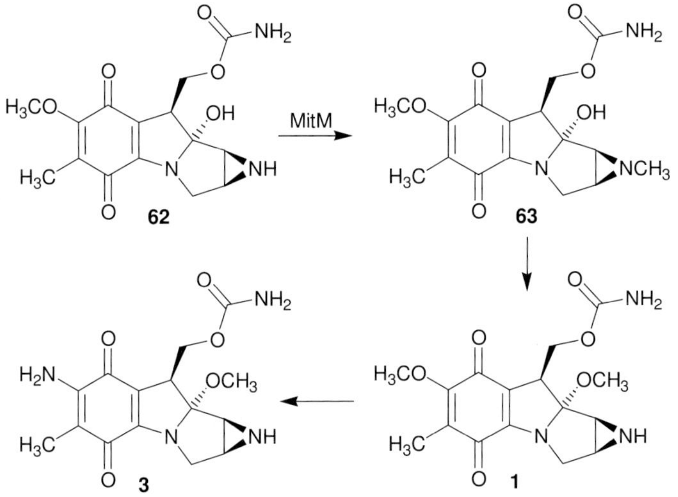

Figure 11.9 Role of the MitM methyltransferase in mitomycin C biosynthesis.

cins A (**1**), B (**2**), and C (**3**) were tested as substrates, with only **1** being methylated. It seems that **62** and **63** may be intermediates in the pathway to mitomycin C, involving an intramolecular methyl transfer from the aziridine to the C-9a hydroxy group to form mitomycin A, followed by amination. Further work is required to elucidate the precise order of biotransformations fully.

There has been one report on the biosynthesis of FR900482 [114]. Radiolabeled D-glucosamine **33** and AHBA **38** were efficiently incorporated into **6**, and D-[1-^{13}C]-glucosamine was incorporated in the expected orientation (Figure 11.10). The biosynthetic gene cluster for FR900482 has been isolated from *Streptomyces sandaensis* and indicates a strong biosynthetic relationship with mitomycin C (Y. Mao, D.H. Sherman, unpublished results).

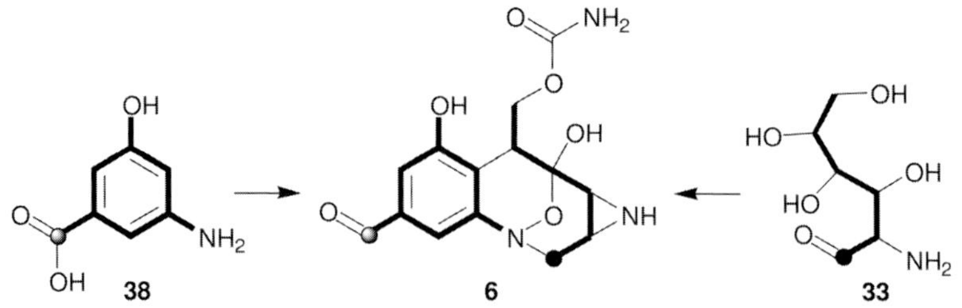

Figure 11.10 Biosynthetic precursors to the carbon skeleton of FR900482.

11.3 The Azinomycins

11.3.1 Discovery and Anticancer Properties

Azinomycins A and B (**64** and **65**; Scheme 11.5) [115, 116] were isolated in 1986 from *Streptomyces griseofuscus* at the SS pharmaceutical company in Japan and were found to have potent *in vitro* cytotoxicities [117] and to display significant *in vivo* antitumor activity [118]. Since their discovery [117] and structure determination [119], their unusual molecular framework has attracted the attention of many chemists interested in their mode of action and in developing synthetic approaches to them [115, 116]. The first total synthesis of azinomycin A was achieved in 2000 by Coleman et al. [120].

However, the story of these unusual molecules goes back to 1954, with the discovery of carzinophilin [121], an antitumor agent extracted from *Streptomyces sahachiroi*, by Hata et al., the same group that discovered the first of the mitomycins. Despite industrial interest in carzinophilin [122] and much effort over several years [123–129] its structure long remained unknown until structures **66** (Scheme 11.6) [127] and **67** [128, 129] were proposed in the early 1980s. Several years later, it was revealed by Armstrong that these structures were incorrect and that carzinophilin is in fact identical to azinomycin B [130]. Since then, most researchers have referred to **65** as azinomycin B, whether it is isolated from *S. griseofuscus* or from *S. sahachiroi*.

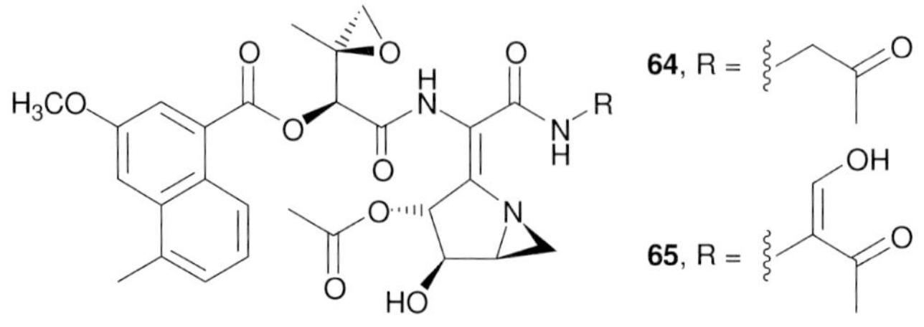

Scheme 11.5

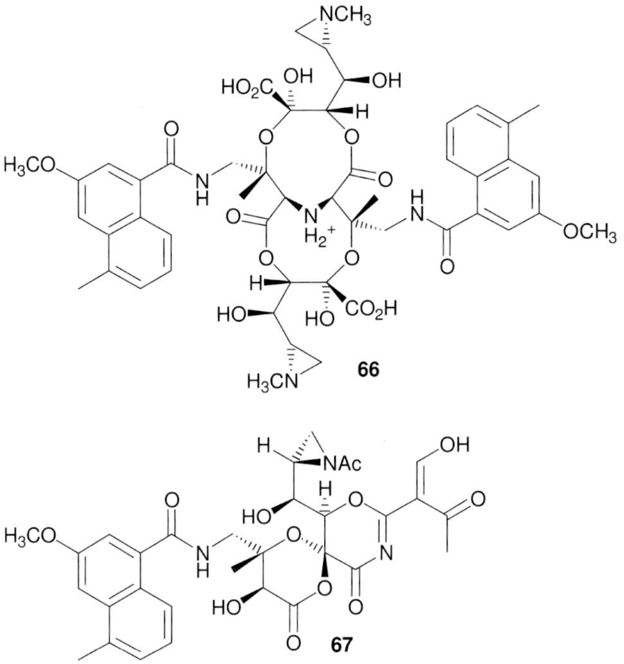

Scheme 11.6

The azinomycins display *in vitro* cytotoxic activity against the L5178Y tumor cell line, azinomycin A being more potent, as well as antibiotic activity against both Gram positive and Gram negative bacteria, although they were inactive against yeast or fungi [117]. Subsequent *in vivo* studies showed that the azinomycins could prolong survival of mice after intraperitoneal inoculation with the tumors P388 leukemia, B-16 melanoma. and Ehrlich carcinoma [119]. Previously, carzinophilin had been found to inhibit tumor growth and to prolong host survival when administered to rats inoculated with Yoshida sarcoma, Ehrlich carcinoma, or ascitic hepatoma [121].

There were some early clinical studies of carzinophilin, both alone [131, 132] and in combination with mitomycin C, in humans [133]. Despite showing promising reduction in the number of cancer cells, there were significant toxic side effects and clinical application has not been pursued any further.

11.3.2 Mode of Action

Early studies by Terawaki and Greenberg on the antibiotic activity of carzinophilin established that it inhibited DNA synthesis but not RNA or protein synthesis in *E. coli* strain B_0 and in *Bacillus subtilis* [134]. They also found that exposure to carzinophilin removed the transforming capacity of *B. subtilis* DNA [135]. They established that carzinophilin formed interstrand crosslinks in DNA, and that crosslinking disappeared on post-treatment incubation, implying that the crosslinking reaction is reversible.

Lown and Majumdar conducted further *in vitro* studies of the reaction between carzinophilin and DNA [136]. They confirmed that it induces interstrand crosslinking of DNA and showed this to be accompanied by alkylation of DNA, indicated by reduced ethidium fluorescence. Interestingly, they found that, although crosslinking was reversible, alkylation was not. They also found that prolonged incubation of a solution of carzinophilin resulted in destruction of its crosslinking activity, but not its alkylation activity. It was found that alkylation and crosslinking did not result in loss of either purine or pyrimidine bases, and that alkylation and crosslinking increased with increasing GC content, implying a preference for alkylation of guanine. Alkylation and crosslinking were shown to be strongly pH-dependent and favored at lower pH. It was proposed that this could explain the tumor selectivity of carzinophilin, since tumor cells have a lower pH than normal cells.

Armstrong and coworkers conducted detailed studies to establish the precise nature of the interstrand crosslinks formed by azinomycin B [137], treating azinomycin B with ^{32}P-end-labeled, short, synthetic DNA duplexes consisting of strands of different lengths and monitoring crosslinking through observation of higher molecular weight species in denaturing polyacrylamide gel electrophoresis. Piperidine-induced backbone cleavage at alkylated sites allowed them to locate crosslinking sites on each strand. They found that crosslinking occurred between a guanine and either a guanine or an adenine residue two base pairs away on the opposite strand. The crosslinking yield was influenced by the nature of the intervening base pair, with sequences possessing a GC base pair between the crosslink sites showing more efficient crosslinking than with AT. Unlike in the case of the mitomycins, no crosslinking was observed between adjacent base pairs at GC sequences. Monoalkylation was observed at all guanine residues, but not at adenine residues. They also substituted deoxyguanosine with 7-deaza-deoxyguanosine (**Z**; Scheme 11.7) at the crosslink site and observed no alkylation or crosslinking at this position. Substitution with deoxyinosine (**I**) resulted in alkylation and crosslinking, providing strong evidence that the azinomycins react with DNA at the N-7 position of guanine, and hence in the major groove.

A crosslinking mechanism based on these data was proposed. Initial monoalkylation at guanine is followed by a template-directed alkylation at either guanine or adenine two base pairs away on the opposite strand. In support of this, Armstrong

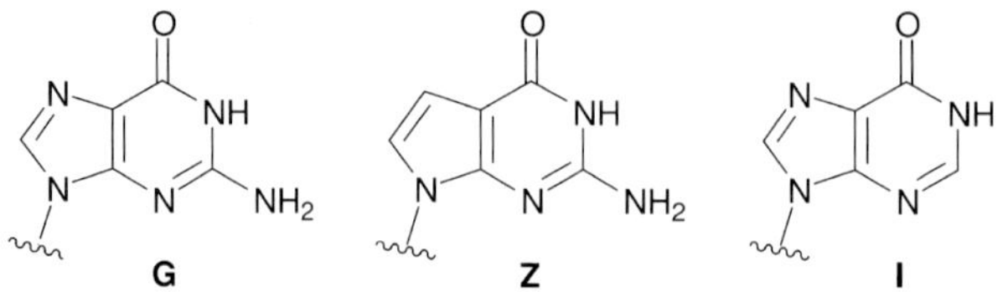

O NH N N NH2 **G**

O NH N N NH2 **Z**

O NH N N N **I**

Scheme 11.7

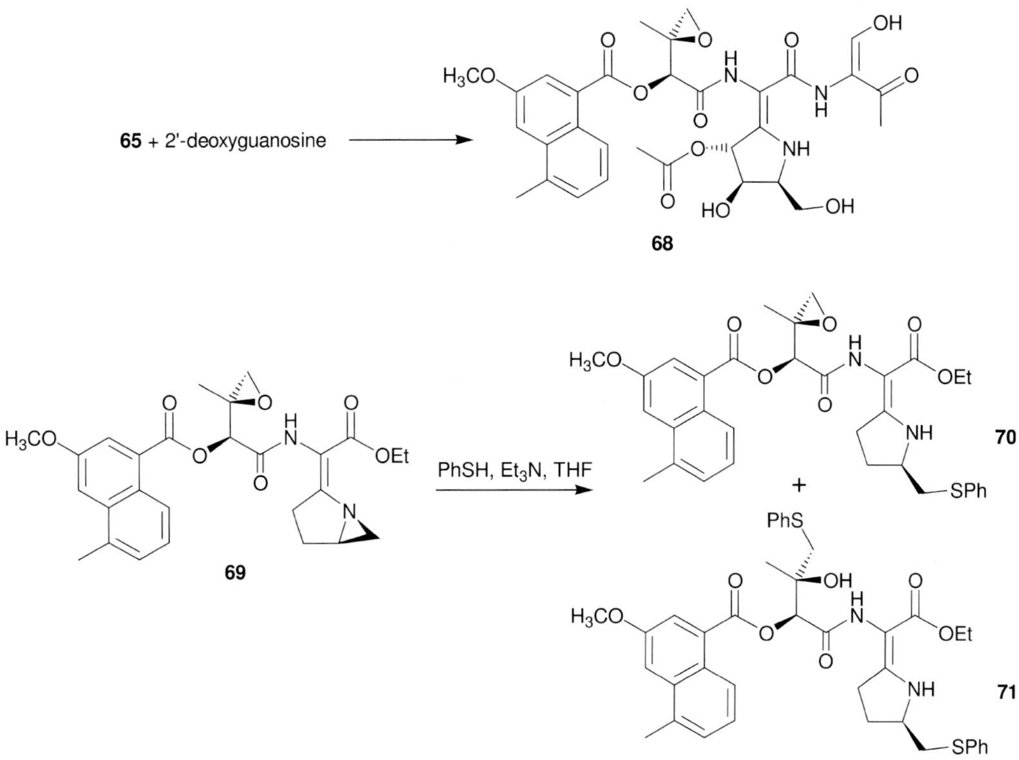

Figure 11.11 Reactivity of the aziridine and epoxide rings of azinomycin B.

separately reported the isolation and identification – by HPLC and FAB MS – of a guanine monoadduct of azinomycin B after treatment with calf thymus DNA [138]. Attempts to confirm the alkylation position on azinomycin B unambiguously by treatment with deoxyguanosine failed, due to rapid hydrolysis of the azinomycin B, but the hydrolysis product was identified as **68** (Figure 11.11), implying that the aziridine ring is more electrophilic. Additional studies by Terashima et al. provided evidence that the initial monoalkylation involves the aziridine ring of azinomycin B. Treatment of synthetic azinomycin analogue **69** with thiophenol resulted in initial opening of the aziridine ring to form **70**, followed by epoxide-opening to form **71** [139].

Saito et al. achieved the first direct confirmation of double alkylation of purine bases by azinomycin B [140]. They incubated azinomycin B with the self-complementary DNA duplex d(TAGCTA)$_2$ and monitored the reaction by HPLC and ion spray MS. They observed initial formation of a monoadduct that was then converted into a crosslinked bisadduct. The crosslink position was identified as between the guanine of one strand and the 5'-adenine on the other strand by thermolytic depurination. Further decomposition prevented structural analysis of the azi-

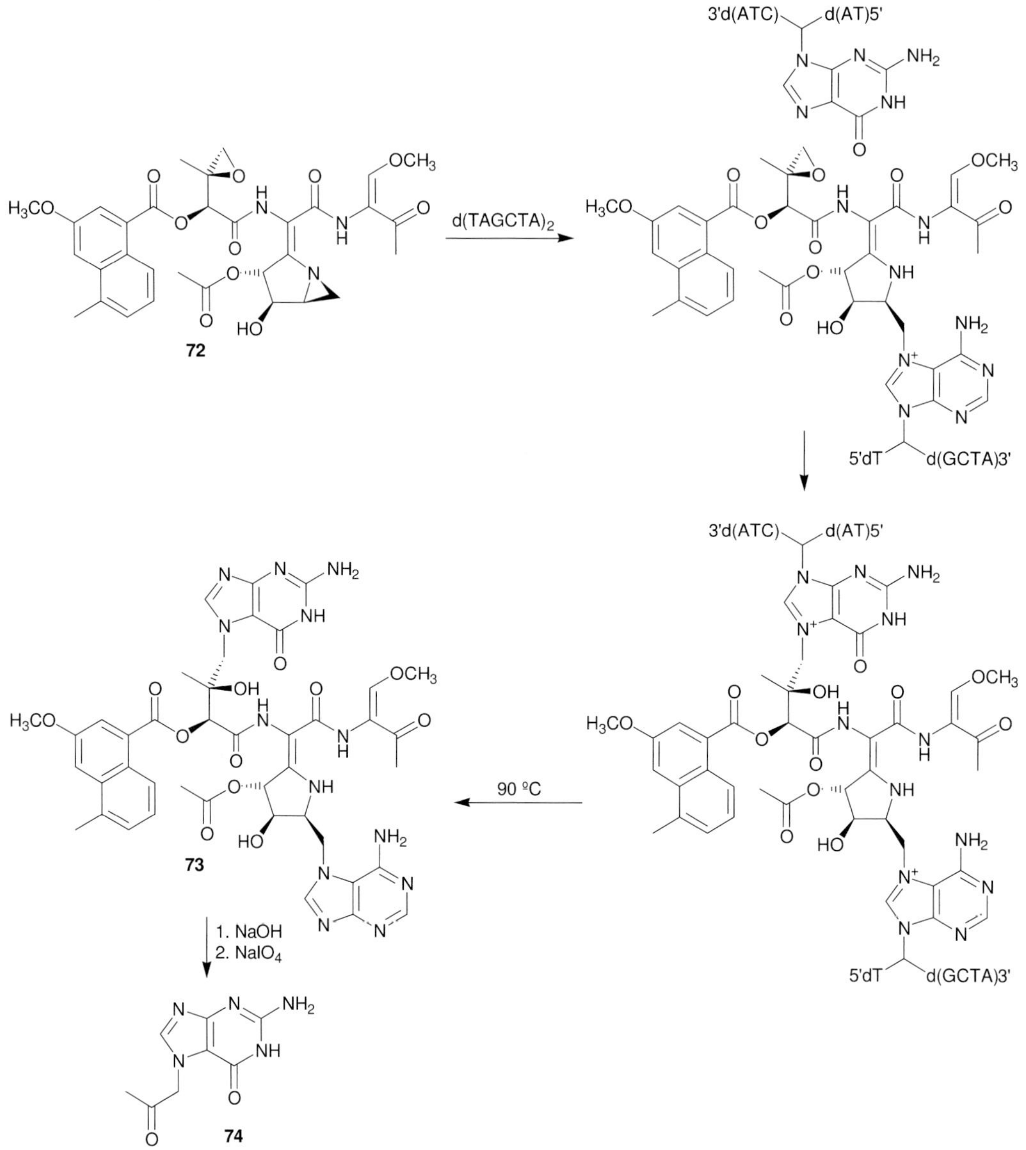

Figure 11.12 Generation and isolation of a DNA crosslinked adduct with 4-*O*-methylazinomycin B.

nomycin B-base adducts, but treatment of the methylated analogue **72** with DNA and subsequent thermolysis provided a new product **73**, which could be isolated and characterized (Figure 11.12). MS and ^{1}H NMR suggested that **73** was formed by nucleophilic addition of adenine and guanine to the aziridine and epoxide groups of **72**. The regiochemistry of addition was determined by chemical degradation of **73** to **74**.

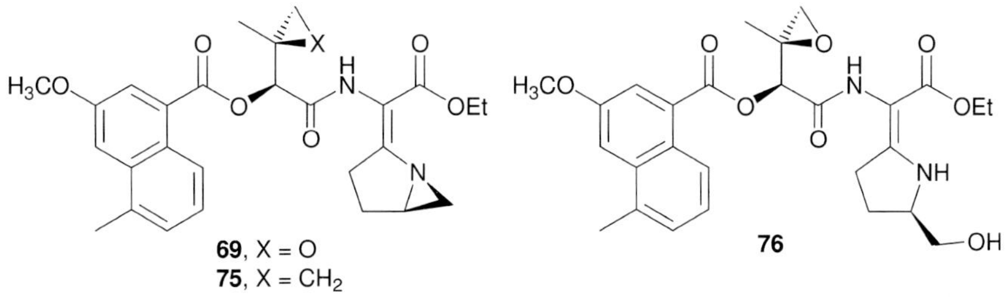

Scheme 11.8

It was concluded that the crosslinking reaction proceeded through initial alkylation of N-7 of adenine by the aziridine ring, followed by guanine N-7 alkylation by the epoxide (Figure 11.12). This seems inconsistent with Armstrong's earlier conclusions, but the very short oligonucleotides used by Saito et al. may influence the DNA alkylation chemistry.

Shipman et al. have provided further evidence for the role of the aziridine and epoxide groups in DNA crosslinking through synthesis and analysis of designed analogues [141]. Compound **69** (Scheme 11.8) caused efficient DNA crosslinking, while **75** and **76**, lacking the epoxide and aziridine rings, respectively, did not.

Coleman and Alcaro have conducted molecular modeling studies of the interaction between azinomycin B and DNA, constituting an important complement to the experimental studies. They initially developed an improved set of suitable force field parameters and applied these to energy minimization of azinomycin B, using positional constraints based on DNA crosslink geometry to filter the results [142]. They found that the global minimum conformation was compatible with crosslink formation, implying a degree of preorganisation for crosslinking. They then modeled the crosslinking reaction of azinomycin B with the six base pair duplex studied experimentally by Saito et al. [143]. The DNA conformation was modeled as either native B-DNA or as B-DNA with a preformed intercalation site. They performed Monte Carlo simulations on all four possible monoalkylated intermediates and found the monoadduct formed by reaction between the aziridine and the internal adenine to be most favored for crosslink formation, consistent with the experimental results of Saito et al., for both intercalative and nonintercalative binding modes. They have more recently extended their modeling studies through further refinement of the force field parameters to simulate the sequence selectivity of crosslinking [144]. Their results are consistent with the qualitative findings of Armstrong et al. that crosslinking is favored between two guanine residues and is significantly favored with a GC base pair between the crosslink positions. The modeling study revealed structural features likely to be responsible for this sequence selectivity. Monoadducts that are predicted to form crosslinks more easily possess favorable hydrogen bonding networks, and the presence of an intervening AT base pair leads to steric hindrance between the azinomycin B molecule and the C-5 methyl group of thymine. Again, the results were consistent with either an intercalative or a nonintercalative mode of binding.

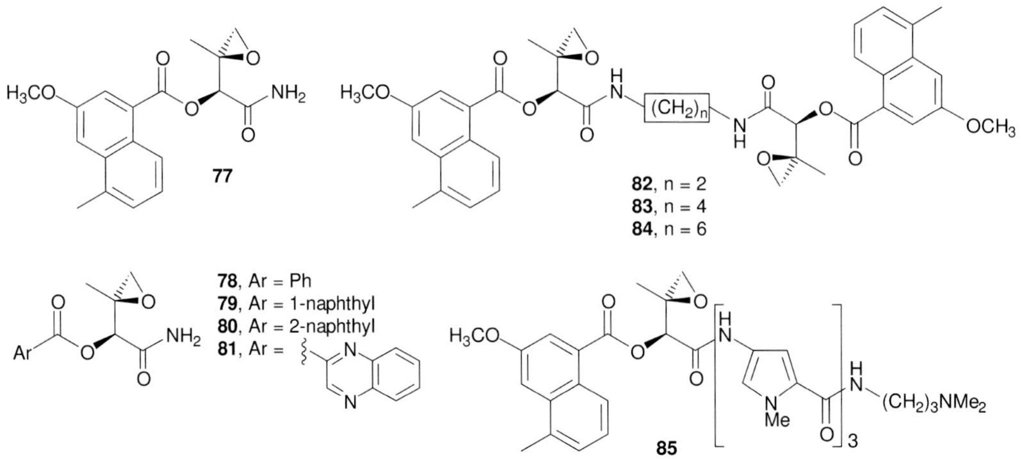

77

82, n = 2
83, n = 4
84, n = 6

78, Ar = Ph
79, Ar = 1-naphthyl
80, Ar = 2-naphthyl
81, Ar =

85

Scheme 11.9

Although most work to elucidate the mode of action of the azinomycin natural products has focussed on their DNA crosslinking activity, it seems that this is not strictly necessary for cytotoxicity and that monoalkylation may suffice. Truncated azinomycin **77** (Scheme 11.9) was isolated along with azinomycins A and B from *Streptomyces griseofuscus* [118, 119]. Although it was not originally identified as having biological activity, it was later found to display potent cytotoxicity against P388 murine leukemia [139].

A number of analogues of **77** have been synthesized and tested for cytotoxic activity. Shipman et al. prepared **78–81** to probe the nature of the aromatic group [145]. All were less active than **77**, with the naphthyl analogues **79** and **80** displaying the best cytotoxicity. Surprisingly, **81**, containing the known intercalating quinoxaloyl group, was completely inactive. Synthesis of a cyclopropane analogue of **77** confirmed the importance of the epoxide for biological activity. Preliminary DNA alkylation studies indicated that cytotoxicity correlated with DNA binding. The same group prepared dimeric analogues **82–84** and studied their DNA crosslinking and cytotoxicity [146]. All three crosslinked DNA, with **83** being the most effective. All three also displayed potent cytotoxicity against a range of tumor cell lines, broadly comparable with that for **77**. Shibuya et al. [147] have prepared compounds such as **85**, hybrids between **77** and DNA-targeting lexitropsins, and shown that they possess DNA-cleaving activity.

Zang and Gates have studied **77** in detail and confirmed that it acts as an efficient DNA alkylating agent with selectivity for guanine residues [148]. They also performed a number of assays to test whether **77** and the nonalkylating analogues **86** and **87** (Scheme 11.10) bound to DNA by intercalation. They observed a red shift in the UV/Vis spectrum upon titration of **77** or **86** with DNA, an increase in fluorescence of **87** in the presence of DNA, induction of DNA unwinding by **86**, and a small increase in viscosity of a DNA solution after addition of **77** or **87**.

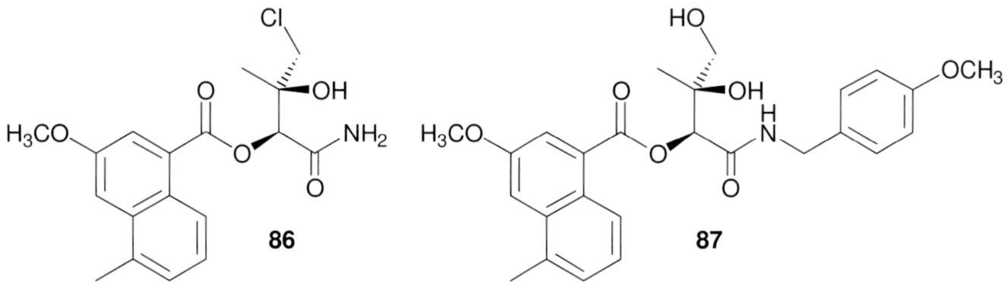

Scheme 11.10

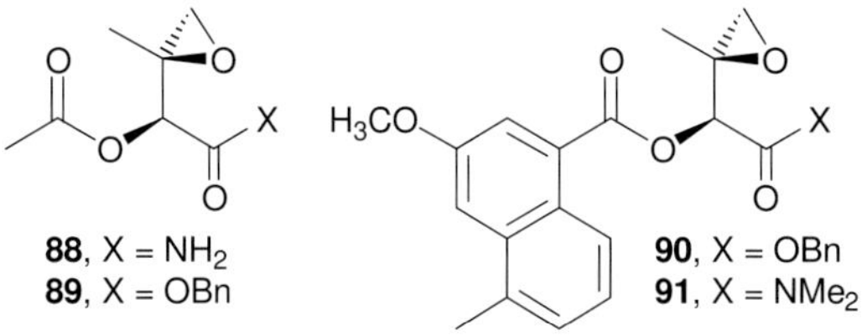

Scheme 11.11

Equilibrium dialysis was used to measure a binding constant of ~10^3 M^{-1} between **87** and DNA. They interpreted these results as being consistent with intercalation of DNA by **77**, indicating that azinomycin B itself may also intercalate.

Coleman et al. have performed a preliminary structure/activity study for a series of analogues of **77** [149]. They found that removal of the naphthoate moiety (**88**; Scheme 11.11) dramatically reduced the yield of DNA alkylation, while replacement of the NH_2 group with *O*-benzyl (**89** and **90**) abolished DNA alkylation completely. Compound **91** alkylated DNA with reduced efficiency, so this effect is not simply due to a requirement for a hydrogen bond donor at this position. Perhaps the amide is required at this position to increase the ability of the C=O to act as a hydrogen bond acceptor. Importantly, they found a strong correlation between the extent of *in vitro* DNA alkylation and cell culture cytotoxicity.

Coleman et al. have also reported a detailed investigation of DNA alkylation by azinomycin B, using synthetic oligonucleotides containing a 5'-Pu-N-Py-3' target sequence flanked on each side by six unreactive AT base pairs [150]. They conducted a thorough quantitative analysis of sequence selectivity and regiochemistry for monoalkylation and crosslinking and determined that crosslink formation was more rapid between two guanines than between guanine and adenine, as predicted from the greater nucleophilicity of guanine at N-7. They also found that crosslinking yield was significantly lower for an intervening AT base pair than for GC, as predicted from their modeling studies, most probably due to steric hindrance between azinomycin B and the C-5 methyl group of thymine and favorable hydrogen bonding in the more efficiently crosslinked products. Kinetic analysis of both steps of the reaction allowed determination of the orientation of crosslinking. For the most reactive sequence, 5'-d(GGC · CCG)-3', it was found that monoalk-

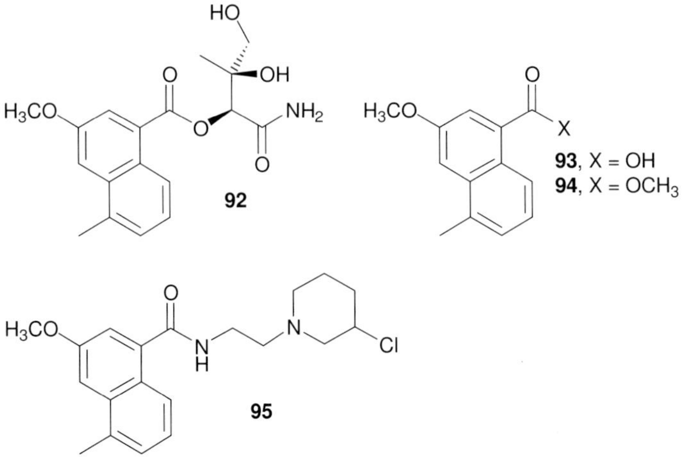

Scheme 11.12

ylation occurred almost exclusively on the 5'-d(GGC)-3' strand, presumably through the aziridine ring, consistently with the greater nucleophilicity of guanine when on the 5' side of an adjacent guanine.

Further experiments were conducted in order to analyze the potential role of intercalation in noncovalent binding of azinomycin B to DNA. No evidence for intercalation of azinomycin B or non-alkylating partial structure **92** (Scheme 11.12) was found by viscometry or fluorescence contact energy transfer. There was also no evidence for intercalation by the alkylating partial structure **77** found by fluorescence contact energy transfer. In addition, there was no sign of intercalation when naphthoic acid **93** or ester **94** were studied in a DNA unwinding assay. These results are at odds with those of Zang and Gates, though the experimental details and compounds tested in the two studies are not identical.

Searcey et al. have just published additional work that supports an intercalative mode of binding, at least for partial azinomycin structures. They performed DNA unwinding assays with **77** and designed prodrug analogue **95** and conclude that both compounds bind weakly to DNA through intercalation [151].

These contrasting results for partial azinomycin structures are confusing, but may be due to subtle differences in experimental design. However, the results of Coleman et al. on azinomycin B itself provide considerable evidence that its binding to DNA does not involve intercalation, and that the naphthoate moiety is involved in more general hydrophobic interactions.

Very recently, the Shipman group have made a further step towards a comprehensive structure/activity profile for noncovalent interactions between azinomycin B and DNA [152]. They synthesized simplified azinomycin analogues **69** and **96–98** (Scheme 11.13), retaining both the epoxide and aziridine alkylating functionalities, with systematically altered substitution on the naphthoate fragment, and analyzed their DNA crosslinking by gel electrophoresis. They found that cross-

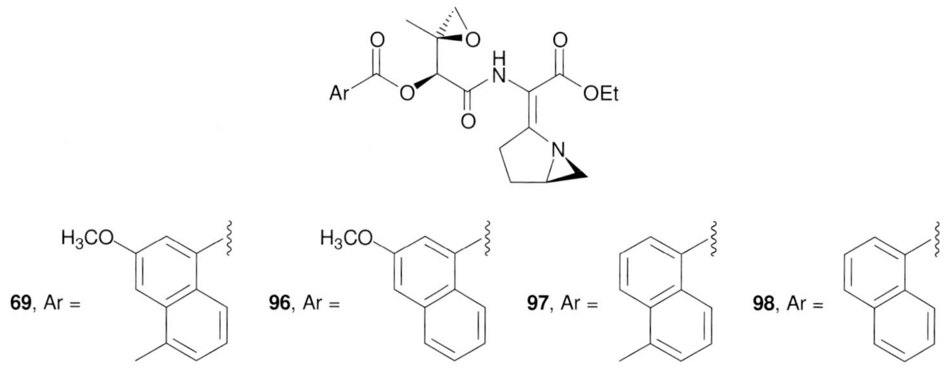

69, Ar = **96**, Ar = **97**, Ar = **98**, Ar =

Scheme 11.13

linking efficiency followed the trend **69** ~ **97** > **96** > **98**, demonstrating that the C5' methyl group probably plays an important role in the binding between the azinomycins and DNA. The precise nature of this binding interaction remains unknown, however.

Although great progress has been made in study of the mode of action of the azinomycins, a full understanding of their interaction with DNA will require future structural characterization of azinomycin-DNA crosslinks by NMR or X-ray crystallography.

11.3.3 Biosynthesis

The first biosynthetic studies on azinomycin B have recently been reported, with the initial goal being to confirm a proposed polyketide origin for the naphthoate fragment. C1-, C2-, and doubly labeled sodium acetate were fed to the azinomycin producer *S. sahachiroi* and incorporation was analyzed by ^{13}C NMR [153]. Clear labeling was observed in the naphthoate fragment (Figure 11.13), consistent with a polyketide pathway in which one molecule of acetyl-CoA is condensed with five molecules of malonyl-CoA to form the naphthoate carbon skeleton. Labeling was also clearly observed in the aziridine and enol fragments, consistent with the involvement of Krebs cycle intermediates involved in amino acid biosynthesis.

The enol fragment was labeled in a tail-to-tail manner consistent with an origin from oxaloacetate **99** via aspartic acid and threonine **100** (Figure 11.14). Five of the seven carbons of the aziridine fragment were labeled by acetate, with a tail-to-tail arrangement as would be expected were they to be derived from α-ketoglutarate **101** (Figure 11.14). Much remains to be discovered about the steps required to elaborate α-ketoglutarate into the aziridine fragment of azinomycin B, not least the origin of the aziridine ring itself.

No significant labeling was observed in the epoxide fragment, although a low degree of labeling that could not be interpreted was observed after feeding of high

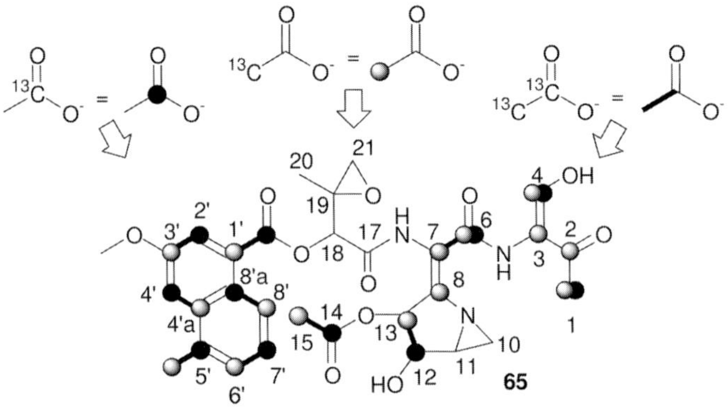

Figure 11.13 Incorporation of ^{13}C-labeled acetate into azinomycin B.

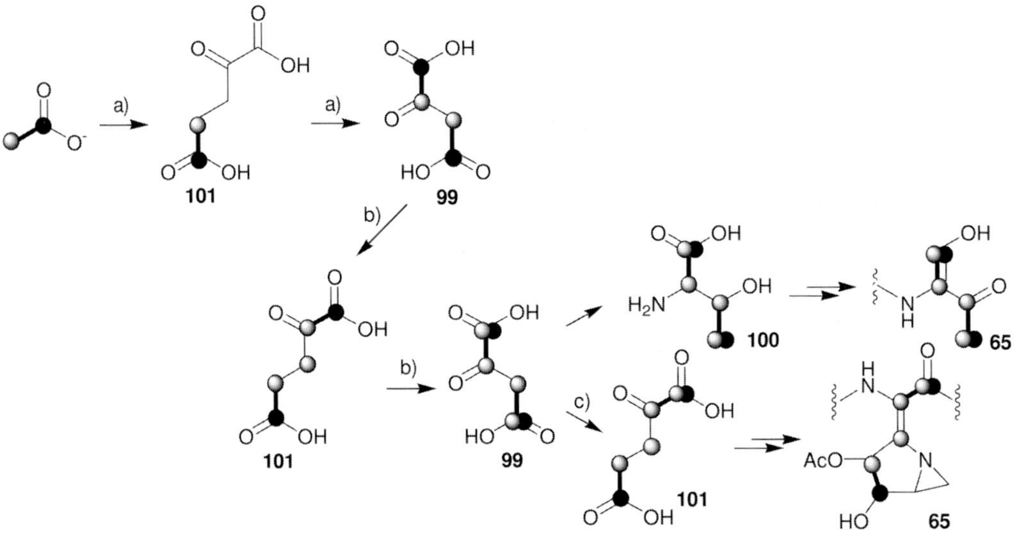

Figure 11.14 Predicted labeling of the enol and aziridine fragments, via oxaloacetate (**99**) and threonine (**100**) for the enol fragment, and via α-ketoglutarate (**101**) for the aziridine fragment. a) First pass through Krebs cycle. b) Second pass through Krebs cycle. c) Third and subsequent passes through Krebs cycle.

doses of labeled sodium acetate. Further experiments with other primary metabolic precursors are required to discover the origin of this structural unit.

The acetate labeling results clearly demonstrated a polyketide origin for the naphthoate fragment. This resulted in the hypothesis that the first enzyme-free intermediate in azinomycin biosynthesis would be naphthoate **102**, with condensation to form a polyketone chain, reduction, cyclization, and dehydration/aromati-

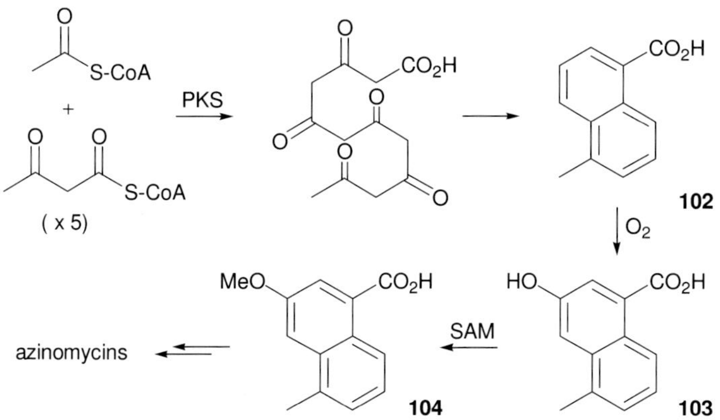

Figure 11.15 Proposed pathway to the naphthoate fragment of azinomycin B.

zation occurring on a polyketide synthase (Figure 11.15). The methoxy group would be introduced by oxygenation and methylation at some point later in the biosynthetic pathway. To test whether **102** was an intermediate, and to investigate at which stage the methoxy group was introduced, Corre et al. synthesized potential naphthoate intermediates **102–104** in deuterium-labeled form, fed them to *S. sahachiroi*, and analyzed incorporation by mass spectrometry and ^{2}H NMR [154]. Deuterium was incorporated from each of **102–104** very efficiently, implying that all are true biosynthetic intermediates and that the methoxy group is added to the naphthalene before the rest of the molecule is attached. The early stages of azinomycin biosynthesis are therefore believed to proceed as outlined in Figure 11.15. Intermediate **102** is produced by a polyketide synthase (PKS), then oxygenated (most probably with O_2, catalyzed by a cytochrome P450), and methylated with *S*-adenosyl methionine (SAM). The expected origin of the methoxy group was confirmed by the efficient and specific incorporation of ^{13}C from the methyl group of methionine [153].

Further support for this pathway was provided by competition feeding studies. If **104** were *not* a true biosynthetic intermediate and were incorporated due to some flexibility in the biosynthetic enzymes, then its incorporation would be expected to be reduced by an equivalent concentration of the true substrate **102** or **103**. If it *were* a real intermediate, then it would be expected that incorporation of earlier intermediates would be reduced by an equivalent concentration of **104**. Precursor **104**, labeled with deuterium at the *O*-methyl group, was co-fed with an approximately equal concentration of either **102** or **103** labeled with deuterium at the C-5 methyl group, and relative incorporation levels were compared by measurement of the CD_3 peak intensities in the ^{2}H NMR spectrum of the labeled azinomycin B. In each case, there was approximately twice as much deuterium labeling at the *O*-methyl group as at the other methyl group, consistently with **104** being a true biosynthetic intermediate.

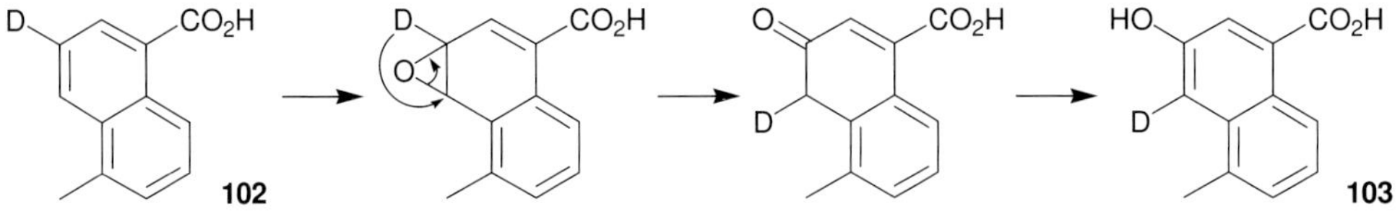

Figure 11.16 Proposed 'NIH shift' mechanism for hydroxylation of the azinomycin naphthoate.

The mechanism of hydroxylation of **102** to **103** was probed by synthesis and feeding of intermediate **102** labeled with deuterium at C-3 [154]. Significant incorporation of the deuterium was observed by mass spectrometry, even though it is located at the position in the precursor where the hydroxy group is introduced. Analysis by ^{1}H and ^{2}H NMR revealed that azinomycin B was labeled specifically at the C-4' position, meaning that a regioselective 1,2 hydrogen shift had occurred. This result is consistent with the involvement of an 'NIH shift' mechanism for hydroxylation of aromatic rings (Figure 11.16) [155]. This was first observed in the oxidation of aromatic amino acids, although at least one other example has been found for its involvement in polyketide biosynthesis, during the construction of aflatoxin B1 [156].

It is tempting to speculate about some features of the genetics and enzymology of azinomycin biosynthesis. Similar substituted naphthoate fragments are found in the enediyne DNA-cleaving agents neocarzinostatin chromophore (**105**; Scheme 11.14) [157, 158] and N1999A2 (**106**) [159]. The neocarzinostatin biosynthetic gene cluster in *Streptomyces carzinostaticus* has recently been sequenced [160], revealing the existence of an iterative Type I polyketide synthase believed to be required for synthesis of the naphthoate fragment. The gene for this synthase

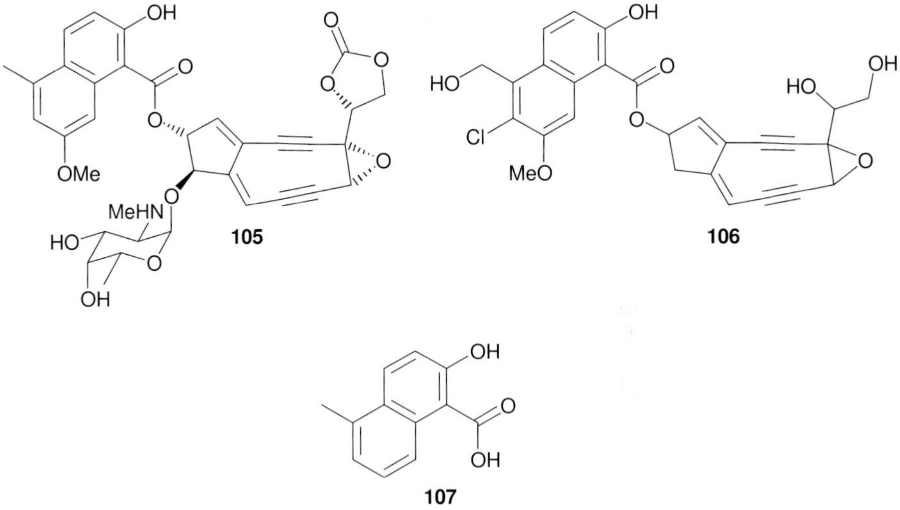

Scheme 11.14

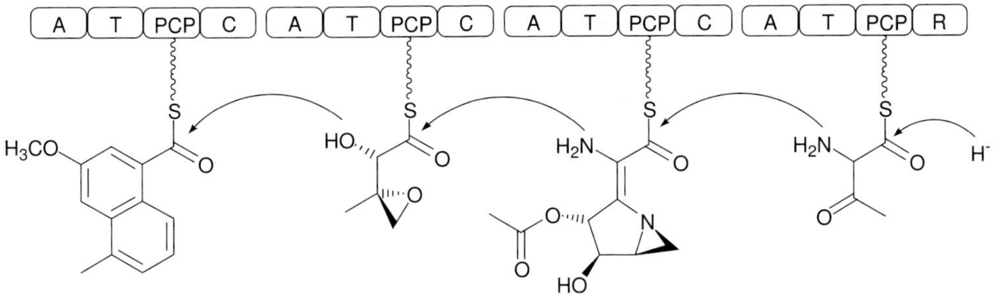

Figure 11.17 Hypothetical organization of an azinomycin B peptide synthetase.

has recently been cloned and expressed in *Streptomyces lividans* [161]. Although a naphthoate product **107** was produced upon introduction of this gene, confirming its role as a polyketide synthase, it lacked one of the expected hydroxy groups. As for azinomycin B, this must arise in a separate oxygenation reaction.

It is very likely that a similar Type I polyketide synthase constructs the naphthoate fragment of azinomycin B. This will be a very interesting enzyme to study, since it will need to perform an unprecedented three regioselective reduction reactions, as well as controlling the polyketide chain length and directing its cyclization.

A very attractive prospect is that the four separate fragments of azinomycin B are combined by a nonribosomal peptide synthetase (NRPS). This class of multifunctional enzymes is responsible for the assembly of a vast range of bioactive natural products consisting of small peptides with varying degrees of further modification by cyclization, oxidation, glycosidation, etc. [162]. The enzymes consist of a series of modules, each of which is responsible for the recognition and incorporation of a specific amino acid, hydroxy acid, or carboxylic acid. Each module contains an adenylation (A) domain that selects its substrate and activates it as an acyl adenylate by reaction with ATP. The substrate is then transferred to the phosphopantetheinyl cofactor of a peptidyl carrier protein (PCP) domain catalyzed by a thiolation (T) domain to form an enzyme-bound thioester. A condensation (C) domain then catalyzes amide or ester formation between two substrates covalently bound to adjacent modules. Figure 11.17 shows the organization of a hypothetical synthetase for azinomycin B production; it may be that the formyl group of the enol fragment is generated by direct reduction of the enzyme-bound thioester by a reductase (R) domain, as observed for myxochelin [163]. The stage at which the crucial epoxide and aziridine functionalities are introduced – whether they are present before the four different fragments are combined, or whether they are elaborated at a late stage in the biosynthesis – cannot yet be predicted.

11.4 Other Aziridine Natural Products

11.4.1 Ficellomycin

Ficellomycin (**108**; Scheme 11.15) was isolated from *Streptomyces ficellus* during a screen for antibiotics at Upjohn in 1976 [164]. It inhibits growth of Gram positive bacteria *in vitro* and showed *in vivo* effectiveness against *Staphylococcus aureus* infections in mice. The structure of **108** was determined by 2D NMR in 1989 [165]; it shares the 1-azabicyclo[3.1.0]hexane ring system with the azinomycins, although it is more reduced. There are still some ambiguities with respect to the stereochemistry of ficellomycin. Although the L stereochemistry of the valine residue was unequivocally determined by degradation, and the relative stereochemistry of the ring substituents could be deduced from NMR coupling constants, the stereochemistry of the α-carbon of the aziridine-containing amino acid relative to the ring system and to the valine, and also the absolute stereochemistry of the ring system, have not yet been determined.

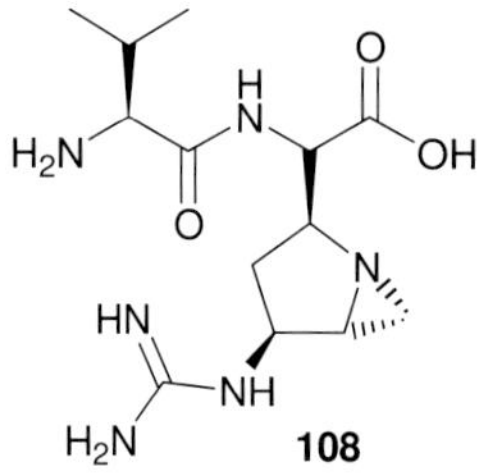

Scheme 11.15

Ficellomycin was found to inhibit semiconservative DNA replication in *Eschirichia coli*, and this was found not to be due to direct inhibition of DNA polymerase [166]. It has been suggested that ficellomycin may exert its biological activity by alkylation of DNA [165], in common with the azinomyins. The biosynthesis of ficellomycin has not been studied, but it seems highly probable that its 1-azabicyclo[3.1.0]hexane ring system will arise from a pathway related to that for the azinomycins.

11.4.2 593A/NSC-135758

Compound 593A (**109**; Scheme 11.16) was isolated from *Streptomyces griseoluteus* by a group at Merck, Sharp, and Dohme in 1970 [167]. Its structure was proposed in 1973 [168] and confirmed by X-ray crystallography in 1976 [169, 170]. The trivial name streptolutine was proposed for the amino acid **110**, so **109** would conse-

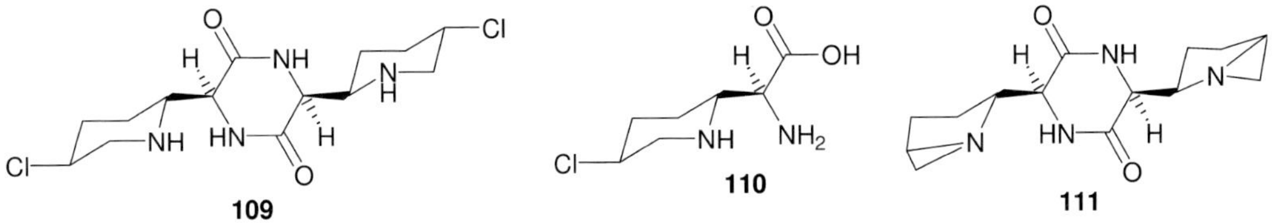

Scheme 11.16

quently be *cyclo*-(streptolutyl-streptolutyl) [169]. Compound **109** inhibits growth of a range of different cancer cell lines, and was taken forward to clinical trials for leukemia and ovarian, cervical, and uterine cancers during the 1970s and 1980s [171, 172], although it has not been further developed for clinical use.

During mode of action studies *in vitro*, **109** was found to inhibit DNA synthesis, but not RNA or protein synthesis, and to act as an alkylating agent [173–175]. A bis-aziridine derivative **111** was prepared and its *in vitro* biological activity was found to be ten times weaker than that of **109** [174]. This may be due to inactivation of **111** by hydrolysis, leaving open the possibility that **109** is actually a prodrug, and that the bis-aziridine **11** is formed *in situ* and is the active alkylating species. Indeed, it may be that **109** is merely a byproduct of the isolation procedure, which involves treatment with hydrochloric acid to form the dihydrochloride salt, and that **111** is the true natural product and is generated by a biosynthetic route similar to that for ficellomycin.

11.4.3 Dicarboxyaziridine and Miraziridine A

(2*S*,3*S*)-Dicarboxyaziridine (**112**) was isolated from a *Streptomyces* strain in 1975 and found to have moderate antibacterial activity against *Aeromonas salmonecida* [176]. Subsequent studies showed that **112** acts as a competitive inhibitor of fumarase, through mimicry of a carbanionic transition state [177]. No biosynthetic studies have been reported for **112**, but it is conceivable that it may arise from cyclization of (3*R*)-hydroxyaspartic acid (Figure 11.18).

More recently, miraziridine A (**113**) was isolated from a marine sponge related to *Theonella mirabilis* and shown to inhibit the cysteine protease cathepsin B. It has been shown that the aziridine ring plays a key role in this biological activity and gives rise to irreversible inhibition of cathepsins B and L, presumably through

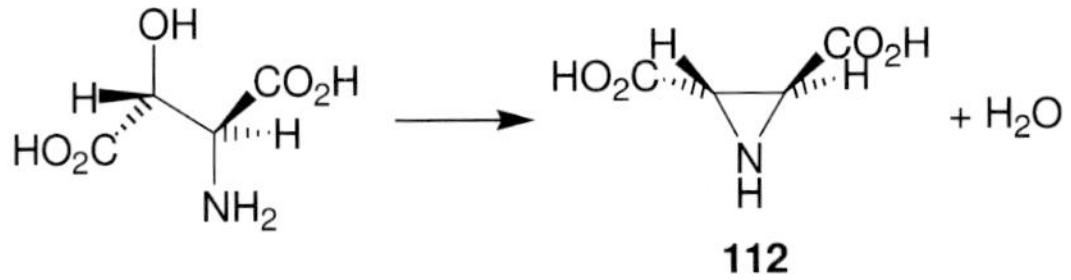

Figure 11.18 Hypothetical biosynthesis of (2*S*,3*S*)-dicarboxyaziridine.

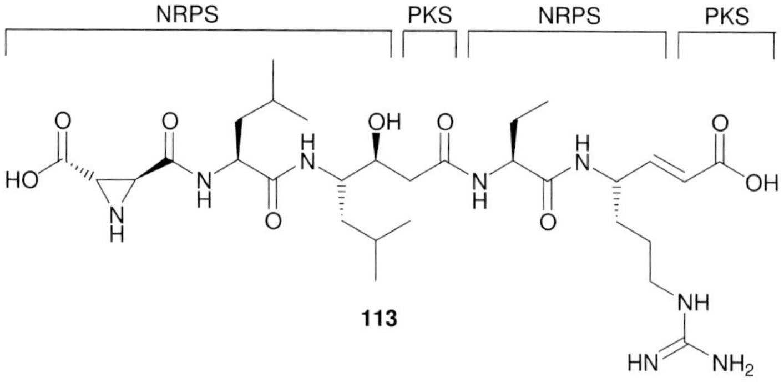

NRPS PKS NRPS PKS

113

Figure 11.19 Hypothetical biosynthetic organization for miraziridine A.

alkylation of the active site cysteine by the aziridine [178]. Compound **113** prepared by total synthesis was also shown to inhibit the serine protease trypsin and the aspartyl protease pepsin, probably due to the vinylogous arginine and the central statin moieties, respectively [179]. The dicarboxyaziridine residue has also been incorporated as an alkylating functional group into synthetic cysteine protease inhibitors [180, 181].

It is likely that the biosynthesis of **113** is directed by a hybrid polyketide synthase/nonribosomal peptide synthetase enzyme system, as indicated in Figure 11.19.

11.4.4 Azicemicins

Azicemicins A (**114**) and B (**115**) were isolated in 1993 from a species of *Amycolatopsis* related to *Amycolatopsis sulphurea* [182–184], and shown to display moderate growth inhibition of Gram positive bacteria and mycobacteria, though the mode of action was not investigated. No biosynthetic studies have been reported, but an obvious possibility is the use of serine as a starter unit for an aromatic polyketide synthase, with cyclization through an S_N2 reaction occurring at some stage (Figure 11.20).

11.4.5 Maduropeptin

Maduropeptin is a complex of macromolecular antitumor agents isolated from *Actinomadura madurae* by scientists at Bristol-Myers-Squibb [185]. Maduropeptin exhibited antibiotic activity against Gram positive bacteria and strong cytotoxicity against murine melanoma B16-F10 and human colon carcinoma HCT-116 tumor cells *in vitro*, and also prolonged the survival time of mice implanted with P388 leukemia and B16 melanoma. The active components were found to consist of an

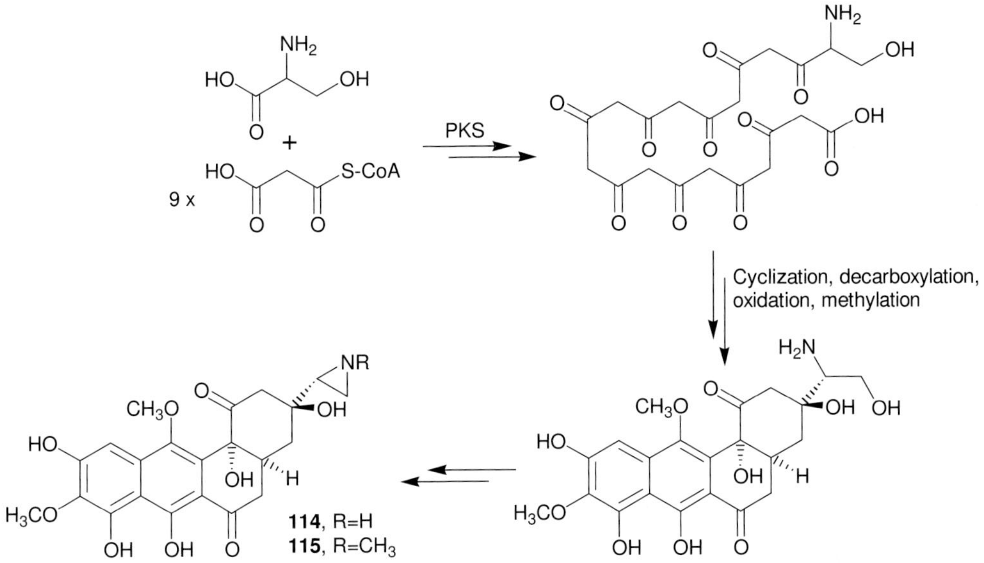

Figure 11.20 Hypothetical biosynthesis of the azicemicins.

unstable chromophore bound to a 32 kDa acidic protein. The most stable and most active component, maduropeptin A_1, was studied further to determine the structure of the chromophore [186] and to elucidate the mode of action [187].

The tightly bound chromophore could be extracted from the protein with methanol [186], and the major component of the extract was determined to have the enediyne structure **116** (Figure 11.21), related to chromophores of other chromoprotein antitumor agents such as neocarzinostatin. Additional minor components were extracted, variously containing an OH group instead of OMe attached to the enediyne core, with Cl instead of OMe when chloride was present in the buffer salt, or with OEt instead of OMe when ethanol was used for the extraction. Another byproduct was isolated in the form of structure **117**, consistent with a facile cycloaromatization reaction as observed for all other enediyne antibiotics. Surprisingly, **117** also displayed antibiotic and antitumor activity, perhaps due to alkylation of DNA or protein by the aziridine. The interpretation of these results was that **116** and the other enediyne byproducts were merely artifacts of the extraction procedure and that the true structure of the maduropeptin chromophore is the aziridine **118**.

The mode of action of maduropeptin has been studied [187]. Like the other enediyne antibiotics, maduropeptin cleaves double-stranded DNA. A mixture of single and double strand breaks is observed, with double strand cleavage staggered by two base pairs, with a degree of sequence selectivity. There is evidence that the maduropeptin chromophore acts in the minor groove of DNA and damages DNA though cycloaromatization to a benzenoid diradical and abstraction of the 4'-H atom of deoxyribose, with subsequent radical-induced cleavage of the sugar phos-

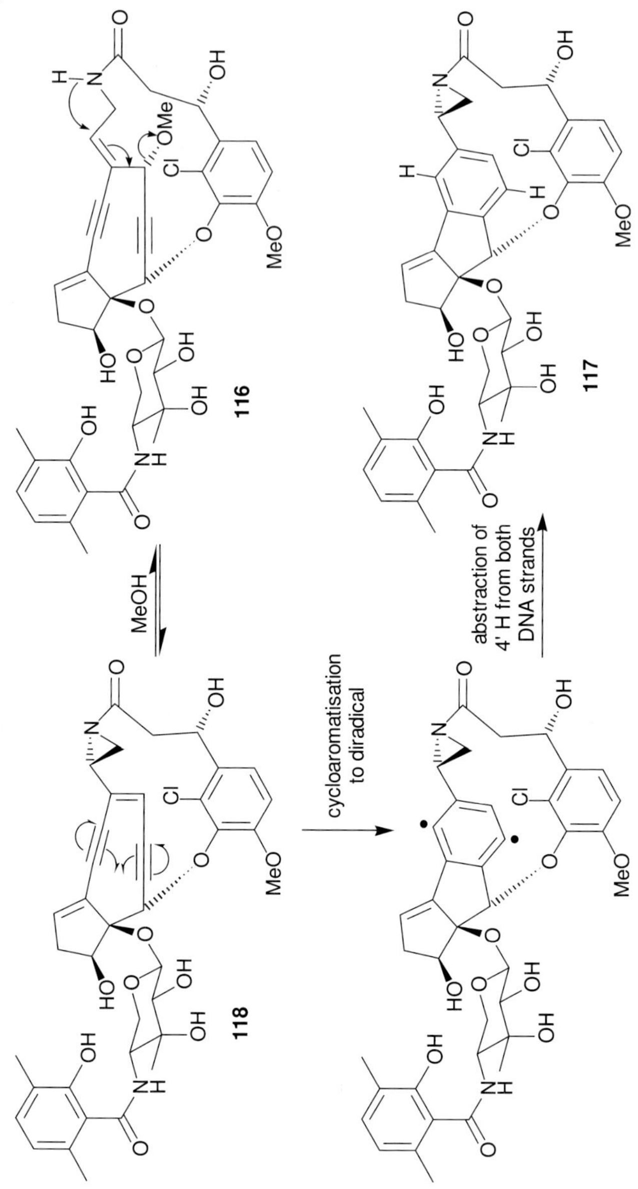

Figure 11.21 Mode of action of maduropeptin.

phate backbone (Figure 11.21). This mechanism is similar to those of the other enediyne antibiotics, although maduropeptin, unlike the majority of the enediynes, does not require reductive activation.

Compound **116** cleaves DNA in the same way as maduropeptin, though with 100-fold less potency. It is suggested that **116** and the other isolation artifacts act as prodrugs of the active form **118**, through nucleophilic attack of the amide nitrogen on the enediyne ring system to form the aziridine and to release methanol (Figure 11.20). Interestingly, the protein component of maduropeptin exhibits protease activity against histones, which may facilitate access of the chromophore "warhead" to DNA [187].

The biosynthesis of maduropeptin has not been studied in detail, but an iterative type I polyketide synthase gene predicted to be responsible for forming the enediyne core structure has been identified [188].

11.4.6
The Madurastatins

Madurastatins A1 (**119**; Scheme 11.17), A2 (**120**), A3 (**121**), B1 (**122**), and B2 (**123**) were isolated in 2004 from a clinical isolate of a pathogenic actinomycete, a strain of *Actinomadura madurae* [189]. The stereochemistry of the *N*-methyl-*N*-hydroxyornithine residue remains unknown, but it was determined that the serine and aziridine residues are D, and the rest are L.

Only the major components **119** and **120** exhibited activity against *Micrococcus luteus*, implying that the aziridine ring is required for antibiotic activity. Madur-

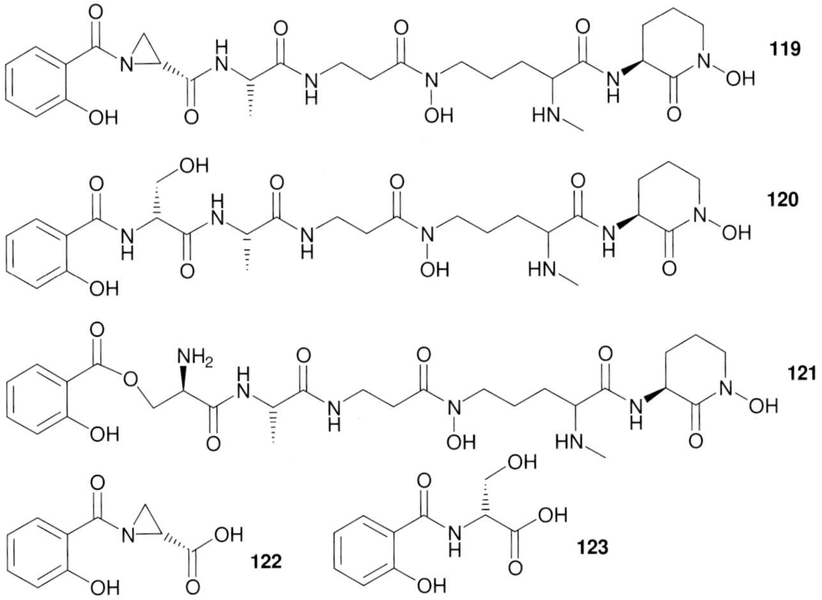

Scheme 11.17

astatin A1 was found readily to form an Fe(III) complex, inactive as an antibiotic, upon treatment with ferric chloride. This implies that the principal role of the madurastatins is as siderophores, acting to sequester iron from the environment into the bacterial cell. Since pathogenic bacteria have to compete for iron with their host and have to colonize iron-poor environments, siderophores act as virulence factors, and inhibition of siderophore biosynthesis may be a fruitful strategy for antibiotic development.

It is likely that the madurastatins are biosynthesized on a nonribosomal peptide synthetase, from salicylic acid as the starter acid. L-Serine is probably the precursor to the aziridine moiety, with epimerization occurring on the enzyme-bound amino acid as found for other nonribosomal peptides, with aziridine formation occurring at a late stage. Compounds **120** and **123** could therefore be biosynthetic precursors to **119** and **122**, respectively.

11.4.7
Aziridine Metabolites from Amino Alcohols

It has been proposed that aziridines may be more widespread in biological systems than is generally realized [190]. Many drugs such as ephedrine (**124**; Scheme 11.18) and pronethalol (**125**) and endogenous metabolites such as adrenaline (**126**) contain a β-aminoalcohol moiety, which may act as a precursor to an aziridine metabolite that may explain the known carcinogenicity of some of these compounds such as pronethalol.

Bicker and Fischer incubated ephedrine isomer **127** with a rat liver homogenate to which ATP and sulfate were added [191]. Formation of aziridine **128**, requiring

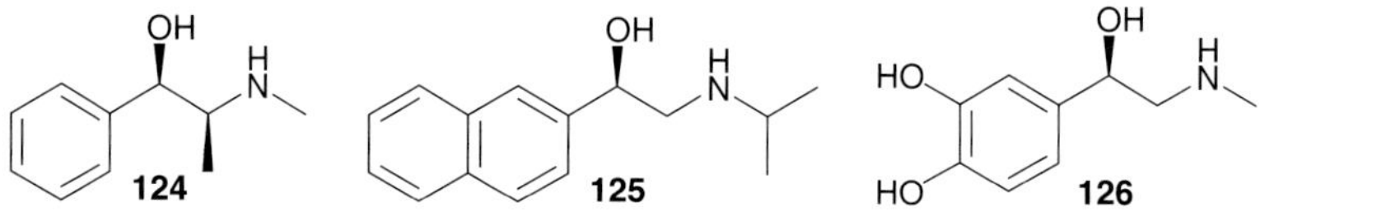

Scheme 11.18

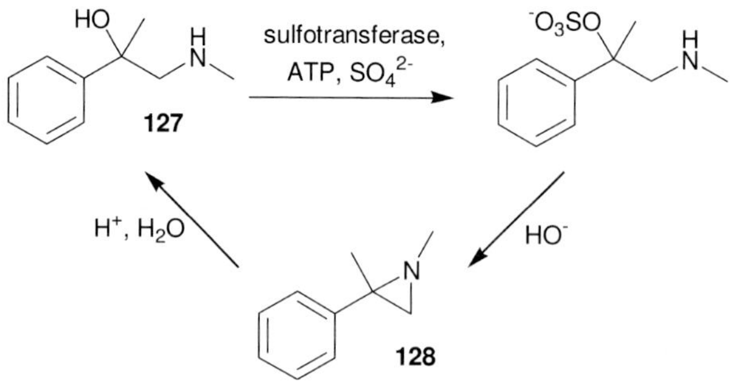

Figure 11.22 Biochemical formation of aziridines in rat liver homogenate.

the presence of the ATP and the sulfate, was observed by thin layer chromatography. It was reasoned that aziridine formation occurred (either spontaneously or enzyme-catalyzed) after enzymatic sulfation of the hydroxy group by phosphoadenosinephosphosulfate (PAPS) formed *in situ* (Figure 11.21). Although no further studies on enzymatic aziridine formation have been reported, it remains an intriguing possibility that this underlies the toxicity of many drugs or that transformation of endogenous β-amino alcohols may be a contributory factor in cancer. Another exciting possibility is that this mechanism may be general and may be involved in the biosynthesis of the aziridine rings of all the natural products discussed in this chapter.

11.4.8 Azirine and Diazirine Natural Products

Azirinomycin (**129**; Scheme 11.19) was isolated as a broad-spectrum antibiotic from *Streptomyces aureus* by workers at Merck, Sharp, and Dohme in 1971 [192, 193]. Although it has good *in vitro* activity against both Gram positive and Gram negative bacteria, it is highly toxic and has no *in vivo* activity, probably due to its chemical instability.

Dysidazirine **130** was isolated by Ireland in 1988 from a Fijian sample of the marine sponge *Dysidea fragilis* [194] as the major lipophilic component of the sponge and was shown to possess cytotoxic activity against L1210 cells, as well as

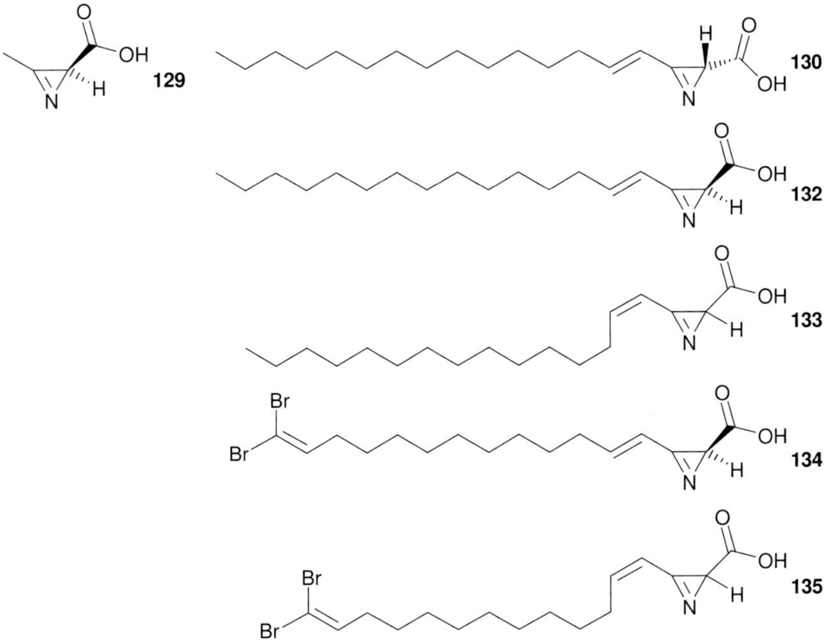

Scheme 11.19

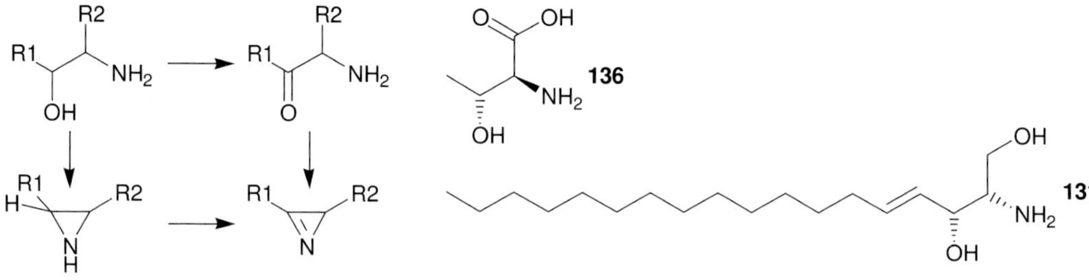

Figure 11.23 Hypothetical biosynthesis of azirine natural products.

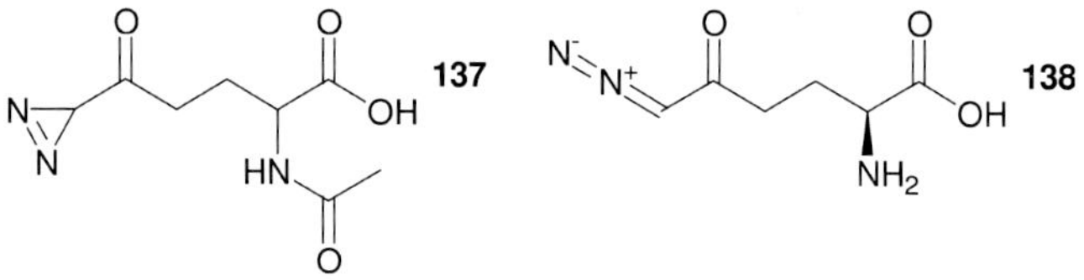

Scheme 11.20

inhibiting growth of Gram negative bacteria and yeast. It was noted that **130** is structurally similar to sphingosine **131**, a well known component of glycolipids, which may act as a biosynthetic precursor to **130** (Figure 11.23). In 1995, Faulkner et al. reported the isolation of further azirine metabolites from *Dysidea fragilis*, collected at Pohnpei [195]. As well as the enantiomer and (*Z*) isomer of dysidazirine (**132** and **133**), bromine-containing (4*E*)-(*S*)-antazirine (**134**) and (4*Z*)-antazirine (**135**) were isolated, although none of the new compounds displayed any antibiotic activity.

No biosynthetic experiments have been reported for these compounds, but they probably all share the same biosynthetic mechanism. One possibility is that they are generated by cyclization of an α-amino-β-keto carboxyl intermediate that would arise from threonine (**136**) and sphingosine (**131**) for **139** and **130**, respectively (Figure 11.23). Alternatively, cyclization may precede oxidation, with an aziridine intermediate being formed.

One example of a naturally occurring diazirine, duazomycin A (**137**; Scheme 11.20), has been reported, isolated in 1985 from a *Streptomyces* species during a screen for herbicidal compounds [196]. It was found to inhibit *de novo* starch synthesis and it was suggested that this is due to direct inhibition of protein synthesis. Duazomycin A is structurally related to 6-diazo-5-oxo-L-norleucine (**138**), also reported as a natural product from *Streptomyces* [197], which acts as a glutamine antagonist and inhibits purine biosynthesis [198].

References

1 D.C. Rees, J.B. Howard, *Curr. Opin. Chem. Biol.* **2000**, *4*, 559–566.
2 T. Hata, Y. Sano, R. Sugawara, A. Matsumae, K. Kanomori, T. Shima, T. Hoshi, *J. Antibiot.* **1956**, *9*, 141–146.
3 T. Hata, R. Sugawara, *J. Antibiot.* **1956**, *9*, 147–151.
4 S. Wakaki, H. Marumo, K. Tomoka, G. Shimizu, E. Kata, H. Kamada, S. Kudo, Y. Fugimoto, *Antibiot. Chemother.* **1958**, *8*, 228–240.
5 C. DeBoer, A. Dietz, N.E. Lummis, G.M. Savage, *Antimicrob. Agents. Ann. 1960.* **1961**, *8*, 17–22.
6 D.V. Lefemine, M. Dann, F. Barbatschi, W.K. Hausmann, V. Zbinovsky, P. Monnikendam, J. Adam, N. Bohonos, *J. Am. Chem. Soc.* **1962**, *84*, 3184–3185.
7 J.S. Webb, D.B. Cosulich, J.H. Mowat, J.B. Patrick, R.W. Broschard, W.E. Meyer, R.P. Williams, C.F. Wolf, W. Fulmor, C. Pidacks, J.E. Lancaster, *J. Am. Chem. Soc.* **1962**, *84*, 3185–3187.
8 J.S. Webb, D.B. Cosulich, J.H. Mowat, J.B. Patrick, J.E. Lancaster, R.W. Broschard, W.E. Meyer, R.P. Williams, C.F. Wolf, W. Fulmor, C. Pidackas, *J. Am. Chem. Soc.* **1962**, *84*, 3187–3188.
9 C.L. Stevens, K.G. Taylor, M.E. Munk, W.S. Marshall, K. Noll, G.D. Shah, L.G. Shah, K. Uzu, *J. Med. Chem.* **1965**, *8*, 1–10.
10 A. Tulinsky, *J. Am. Chem. Soc.* **1962**, *84*, 3188–3190.
11 A. Tulinsky, J.H. Van Den Hende, *J. Am. Chem. Soc.* **1967**, *89*, 2905–2911.
12 K. Shirahata, N. Hirayama, *J. Am. Chem. Soc.* **1983**, *105*, 7199–7200.
13 U. Hornemann, M.J. Heins, *J. Org. Chem.* **1985**, *50*, 1301–1302.
14 S.K. Carter, S.T. Crooke (eds.), *Mitomycin C: Current Status and New Developments*, Academic Press, New York (**1979**).
15 I.C. Henderson, *Oncology* **1991**, *1*, 1–83.
16 M. Tomasz, Y. Palom, *Pharmacol. Ther.* **1997**, *76*, 73–87.
17 M. Kasai, H. Arai, *Exp. Opin. Ther. Patents* **1995**, *5*, 757–770.
18 M. Iwami, S. Kiyoto, H. Terano, M. Kohsaka, H. Aoki, H. Imanaka, *J. Antibiot.* **1987**, *40*, 589–593.
19 S. Kiyoto, T. Shibata, M. Yamashita, T. Komori, M. Okuhara, H. Terano, M. Kohsaka, H. Aoki, H. Imanaka, *J. Antibiot.* **1987**, *40*, 594–599.
20 I. Uchida, S. Takase, H. Kayakiri, S. Kiyoto, M. Hashimoto, T. Tada, S. Koda, Y. Morimoto, *J. Am. Chem. Soc.* **1987**, *109*, 4l08–4109.
21 K. Shimomura, O. Hirai, T. Mizota, S. Matsumoto, J. Mori, F. Shibayama, H. Kikuchi, *J. Antibiot.* **1987**, *40*, 600–606.
22 O. Hirai, K. Shimomura, T. Mizota, S. Matsumoto, J. Mori, H. Kikuchi, *J. Antibiot.* **1987**, *40*, 607–611.
23 K. Shimomura, T. Manda, S. Mukumoto, K. Masuda, T. Nakamura, T. Mizota, S. Matsumoto, F. Nishigaki, T. Oku, J. More, F. Shibayama, *Cancer Res.* **1988**, *48*, 1116–1172.
24 Y. Naoe, M. Inami, S. Matsumoto, F. Nishigaki, S. Tsujimoto, I. Kawamura, K. Miyayasu, T. Manda, K. Shimomura, *Cancer Chemother. Pharmacol.* **1998**, *42*, 31–36.
25 M. Tomasz, *Chem. & Biol.* **1995**, *2*, 575–579.
26 S.R. Rajski, R.M. Williams, *Chem. Rev.* **1998**, *98*, 2723–2795.
27 S.E. Wolkenberg, D.L. Boger, *Chem. Rev.* **2002**, *102*, 2477–2495.
28 V.N. Iyer, W. Szybalski, *Proc. Natl. Acad. Sci. U.S.A.* **1963**, *50*, 355–362
29 V.N. Iyer, W. Szybalski, *Science* **1964**, *145*, 55–58.
30 A. Weissbach, A. Lisio, *Biochemistry* **1965**, *4*, 1063–1071.
31 M. Tomasz, C.M. Mercado, J. Olson, N. Chatterjee, *Biochemistry* **1974**, *13*, 4878–4887.
32 J.W. Lown, A. Begleiter, D. Johnson, A.R. Morgan, *Can. J. Biochem.* **1976**, *54*, 110–119.
33 V. Li, H. Kohn, *J. Am. Chem. Soc.* **1991**, *113*, 275–283.
34 S. Kumar, R. Lipman, M. Tomasz, *Biochemistry* **1992**, *31*, 1399–1407.
35 H. Kohn, V. Li, P. Schiltz, M. Tang, *J. Am. Chem. Soc.* **1992**, *114*, 9218–9220.
36 S.P. Teng, S.A. Woodson, D.M. Crothers, *Biochemistry*, **1989**, *28*, 3901–3907.
37 J.T. Millard, M.F. Weidner, S. Raucher,

P. B. Hopkins, *J. Am. Chem. Soc.* **1990**, *112*, 3637–3641.
38 H. Borowy-Borowski, R. Lipman, M. Tomasz, *Biochemistry* **1990**, *29*, 2999–3004.
39 J. T. Millard, T. M. Beachy, *Biochemistry* **1993**, *32*, 12850–12856.
40 W. S. Johnson, Q.-Y. He, M. Tomasz, *Bioorg. Med. Chem.* **1995**, *3*, 851–860.
41 V.-S. Li, M. Reed, Y. Zheng, H. Kohn, M. Tang, *Biochemistry* **2000**, *39*, 2612–2618.
42 V.-S. Li, M. Tang, M. Tomasz, *Bioorg. Med. Chem.* **2001**, *9*, 863–873.
43 A. Das, K. S. Tang, S. Gopalakrishnan, M. J. Waring, M. Tomasz, *Chem. Biol.* **1999**, *6*, 461–471.
44 J. J. Dannenberg, M. Tomasz, *J. Am. Chem. Soc.* **2000**, *122*, 2062–2068.
45 H. W. Moore, *Science* **1977**, *297*, 527–532.
46 A. C. Sartorelli, W. F. Hodnick, M. F. Belcourt, M. Tomasz, B. Haffty, J. J. Fischer, S. Rockwell, *Oncol Res* **1994**, *6*, 501–508.
47 J. Cummings, V. J. Spanswick, M. Tomasz, J. F. Smyth, *Biochem. Pharmacol.* **1998**, *56*, 405–414.
48 G. Suresh Kumar, R. Lipman, J. Cummings, M. Tomasz, *Biochemistry* **1997**, *36*, 14128–14136.
49 P. Schiltz, H. Kohn, *J. Am. Chem. Soc.* **1993**, *115*, 10510–10518.
50 A. S. Prakash, H. Beall, D. Ross, N. W. Gibson, *Biochemistry* **1993**, *32*, 5518–5525.
51 C. Cera, M. Egbertson, S. P. Teng, D. M. Crothers, S. J. Danishefsky, *Biochemistry* **1989**, *28*, 5665–5669.
52 V.-S. Li, D. Choi, M. Tang, H. Kohn, *J. Am. Chem. Soc.* **1996**, *118*, 3765–3766.
53 M. Tomasz, D. Chowdary, R. Lipman, S. Shimotakahara, D. Veiro, V. Walker, G. L. Verdine, *Proc. Natl. Acad. Sci. U.S.A.* **1986**, *83*, 6702–6706.
54 G. Suresh Kumar, S. M. Musser, J. Cummings, M. Tomasz, *J. Am. Chem. Soc.* **1996**, *118*, 9209–9217.
55 B. S. Iyengar, R. T. Dorr, N. G. Shipp, W. A. Remers, *J. Med. Chem.* **1990**, *33*, 253–257
56 Y. Palom, M. F. Belcourt, S. M. Musser, A. C. Sartorelli, S. Rockwell, M. Tomasz, *Chem. Res. Toxicol.* **2000**, *13*, 479–488.
57 M. Tomasz, R. Lipman, D. Chowdary, J. Pawlak, G. L. Verdine, K. Nakanishi, *Science* **1987**, *235*, 1204–1208.
58 R. Bizanek, B. F. McGuiness, K. Nakanishi, M. Tomasz, *Biochemistry* **1992**, *31*, 3084–3091.
59 M. Tomasz, R. Lipman, B. F. McGuiness, K. Nakanishi, *J. Am. Chem. Soc.* **1988**, *110*, 5892–5896.
60 R. Bizanek, D. Chowdary, H. Arai, M. Kasai, C. S. Hughes, A. C. Sartorelli, S. Rockwell, M. Tomasz, *Cancer Res.* **1993**, *53*, 5127–5134.
61 Y. Palom, M. F. Belcourt, G. Suresh Kumar, H. Arai, M. Kasai, A. C. Sartorelli, S. Rockwell, M. Tomasz, *Oncol. Res.* **1998**, *10*, 509–521.
62 M. Sastry, R. Fiala, R. Lipman, M. Tomasz, D. J. Patel, *J. Mol. Biol.* **1995**, *247*, 338–359.
63 D. Norman, D. Live, M. Sastry, R. Lipman, B. E. Hingerty, M. Tomasz, S. Broyde, D. J. Patel, *Biochemistry* **1990**, *29*, 2861–2876.
64 G. Subramaniam, M. M. Paz, G. Suresh Kumar, A. Das, Y. Palom, C. C. Clement, D. J. Patel, M. Tomasz, *Biochemistry* **2001**, *40*, 10473–10484.
65 W. A. Remers, S. N. Rao, T. P. Wunz, P. Kollman, *J. Med. Chem.* **1988**, *31*, 1612–1620.
66 S. K. Arora, M. B. Cox, P. Arjunan, *J. Med. Chem.* **1990**, *33*, 3000–3008.
67 P. B. Hopkins, J. T. Millard, J. Woo, M. F. Weidner, J. J. Kircher, S. T. Sigurdsson, S. Raucher, *Tetrahedron* **1991**, *47*, 2475–2489.
68 K. Masuda, T. Makamura, K. Shimomura, H. Terano, M. Kohsaka, *J. Antibiot.* **1988**, *41*, 1497–1499.
69 K. Masuda, T. Nakamura, T. Mizota, J. Mori, K. Shimomura, *Cancer Res.* **1988**, *48*, 5172–5177.
70 T. Nakamura, K. Masuda, S. Matsumoto, T. Oku, T. Manda, J. Mori, K. Shimomura, *Jpn. J. Pharmacol.* **1989**, *49*, 317–324.
71 J. Woo, S. T. Sigurdsson, P. B. Hopkins, *J. Am. Chem. Soc.* **1993**, *115* 1199–1200.
72 H. Huang, S. R. Rajski, R. M. Williams, P. B. Hopkins, *Tetrahedron Lett.* **1994**, *35*, 9669–9672.
73 M. M. Paz, P. B. Hopkins, *Tetrahedron Lett.* **1997**, *38*, 343–346.
74 M. M. Paz, P. B. Hopkins, *J. Am. Chem. Soc.* **1997**, *119*, 5999–6005.
75 R. M. Williams, S. R. Rajski, S. B. Rollins, *Chem. & Biol.* **1997**, *4*, 127–137.

76 R. M. Williams, P. Ducept, *Biochemistry* **2003**, *42*, 14696–14701.
77 R. M. Williams, S. R. Rajski, *Tetrahedron Lett.* **1993**, *34*, 7023–7026.
78 T. Fukuyama, S. Goto, *Tetrahedron Lett.* **1989**, *30*, 6491–6494.
79 H. Huang, T. K. Pratum, P. B. Hopkins, *J. Am. Chem. Soc.* **1994**, *116*, 2703–2709.
80 M. M. Paz, S. T. Sigurdsson, P. B. Hopkins, *Bioorg. Med. Chem.* **2000**, *8*, 173–179.
81 A. C. Sartorelli, C. A. Pristos, *Cancer Res.* **1986**, *46*, 3528–3532.
82 S. R. Rajski, S. B. Rollins, R. M. Williams, *J. Am. Chem. Soc.* **1998**, *120*, 2192–2193.
83 S. R. Rajski, R. M. Williams, *Bioorg. Med. Chem.* **2000**, *8*, 1331–1342.
84 L. Beckerbauer, J. J. Tepe, J. Cullison, R. Reeves, R. M. Williams, *Chem. Biol.* **2000**, *7*, 805–812.
85 L. Beckerbauer, J. J. Tepe, R. A. Eastman, P. Mixter, R. M. Williams, R. Reeves, *Chem. Biol.* **2002**, *9*, 427–441.
86 E. J. Kirsch, J. D. Korshalla, *J. Bacteriol.* **1964**, *87*, 247–255.
87 U. Hornemann, J. C. Cloyd, *J. Chem. Soc. Chem. Commun.* **1971**, 301–302.
88 G. S. Bezanson, L. C. Vining, *Can. J. Biochem.* 1971, *49*, 911–918.
89 U. Hornemann, J. P. Kehrer, C. S. Nunez, R. L. Ranieri, *J. Am. Chem. Soc.* **1974**, *96*, 320–322.
90 U. Hornemann, J. H. Eggert, *J. Antibiot.* **1975**, *28*, 841–843.
91 U. Hornemann, M. J. Aikman, *J. Chem. Soc. Chem. Commun.* **1973**, 88–89.
92 U. Hornemann, J. P. Kehrer, J. H. Eggert, *J. Chem. Soc. Chem. Commun.* **1974**, 1045–1046.
93 U. Hornemann, J. H. Eggert, D. P. Honor, *J. Chem. Soc. Chem. Commun.* **1980**, 11–13.
94 H. G. Floss, *Nat. Prod. Rep.* **1997**, *14*, 433–452.
95 M. G. Anderson, J. J. Kibby, , R. W. Rickards, J. M. Rothschild, *J. Chem. Soc. Chem. Commun.* **1980**, 1277–1278.
96 J. J. Kibby, I. A. McDonald, R. W. Rickards, *J. Chem. Soc. Chem. Commun.* **1980**, 768–769.
97 J. J. Kibby, R. W. Rickards, *J. Antibiot.* **1981**, *34*, 605–607.
98 C.-G. Kim, A. Kirschning, P. Bergon, Y. Ahn, J. J. Wang, M. Shibuya, H. G. Floss, *J. Am. Chem. Soc.* **1992**, *114*, 4941–4943.
99 C.-G. Kim, A. Kirschning, P. Bergon, P. Zhou, E. Su, B. Sauerbrei, S. Ning, Y. Ahn, M. Breuer, E. Leistner, H. G. Floss, *J. Am. Chem. Soc.* **1996**, *118*, 7486–7491.
100 C.-G. Kim, T.-W. Yu, C. B. Fryhle, S. Handa, H. G. Floss, *J. Biol. Chem.* **1998**, *273*, 6030–6040.
101 J. C. Eads, M. Beeby, G. Scapin, T. W. Yu, H. G. Floss, *Biochemistry* **1999**, *38*, 9840–9849.
102 P. R. August, L. Tang, Y. J. Yoon, S. Ning, R. Müller, T.-W. Yu, M. Taylor, D. Hoffmann, C.-G. Kim, X. Zhang, C. R. Hutchinson, H. G. Floss, *Chem. & Biol.* **1998**, *5*, 69–79.
103 T.-W. Yu, R. Müller, M. Müller, X. Zhang, G. Draeger, C.-G. Kim, E. Leistner, H. G. Floss, *J. Biol. Chem.* **2001**, *276*, 12546–12555.
104 C. A. Claridge, J. A. Bush, T. W. Doyle, D. E. Nettleton, J. E. Mosely, D. Kimbali, M. F. Kammer, J. Veitch, *J. Antibiot.* **1986**, *39*, 437–446.
105 Y. Mao, M. Varoglu, D. H. Sherman, *Chem. & Biol.* **1999**, *6*, 251–263.
106 P. R. August, M. C. Flickinger, D. H. Sherman, *J. Bacteriol.* **1994**, *176*, 4448–4454.
107 D. A. Johnson, P. R. August, C. Shackleton, H.-w. Liu, D. H. Sherman, *J. Am. Chem. Soc.* **1997**, *119*, 2576–2577.
108 P. J. Sheldon, D. A. Johnson, P. R. August, H.-w. Liu, D. H. Sherman, *J. Bacteriol.* **1997**, *179*, 1796–1804.
109 T. W. Martin, Z. Dauter, Y. Devedjiev, P. Sheffield, F. Jelen, M. He, D. H. Sherman, J. Otlewski, Z. S. Derewenda, U. Derewenda, *Structure*, **2002**, *10*, 933–942.
110 M. He, P. J. Sheldon, D. H. Sherman, *Proc. Natl. Acad. Sci. U.S.A.* **2001**, *98*, 926–931.
111 P. J. Sheldon, Y. Mao, M. He, D. H. Sherman, *J. Bacteriol.* **1999**, *181*, 2507–2512.
112 Y. Mao, M. Varoglu, D. H. Sherman, *J. Bacteriol.* **1999**, *181*, 2199–2208.
113 M. Varoglu, Y. Mao, D. H. Sherman, *J. Am. Chem. Soc.* **2001**, *123*, 6712–6713.
114 T. Fujita, S. Takase, T. Otsuka, H. Terano, M. Kohsaka, *J. Antibiot.* **1988**, *41*, 392–394.
115 For a review, see T. J. Hodgkinson, M. Shipman, *Tetrahedron* **2001**, *57*, 4467–4488.

116 For a review, see M. Casely-Hayford, M. Searcey, 'The azinomycins. Discovery, synthesis and DNA-binding studies.' in *DNA and RNA binders. From small molecules to drugs*, Wiley-VCH, Weinheim, **2002**. pp. 676–696.
117 K. Nagaoka, M. Matsumoto, J. Onoo, K. Yokoi, S. Ishizeki, T. Nakashima, *J. Antibiot.* **1986**, *39*, 1527–1532.
118 S. Ishizeki, M. Ohtsuka, K. Irinoda, K.-I. Kukita, K. Nagaoka and T. Nakashima, *J. Antibiot.* **1987**, *40*, 60–65.
119 K. Yokoi, K. Nagaoka, T. Nakashima, *Chem. Pharm. Bull.* **1986**, *34*, 4554–4561.
120 R. S. Coleman, J. Li, A. Navarro, *Angew. Chem. Int. Ed.* **2001**, *40*, 1736.
121 T. Hata, F. Koga, Y. Sano, K. Kanamori, A. Matsumae, R. Sunagawa, T. Hoshi, T. Shima, S. Ito, S. Tomozawa, *J. Antibiot. Ser. A*, **1954**, *7*, 107–112.
122 H. Kamada, S. Wakaki, Y. Fujimoto, K. Tomioka, S. Ueyama, H. Marumo, K. Uzo, *J. Antibiot.* **1955**, *8*, 187–188.
123 M. Tanaka, T. Kishi, Y. Maruta, *J. Antibiot. Ser. B* **1959**, *12*, 361–364.
124 M. Tanaka, T. Kishi, Y. Maruta, *J. Antibiot. Ser. B* **1960**, *13*, 177–181.
125 M. Onda, Y. Konda, S. Omura, T. Hata, *J. Antibiot.* **1969**, *22*, 42–44.
126 M. Onda, Y. Konda, S. Omura, T. Hata, *Chem. Pharm. Bull.* **1971**, *19*, 2013–2019.
127 J. W. Lown, C. C. Hanstock, *J. Am. Chem. Soc.* **1982**, *104*, 3213–3224.
128 M. Onda, Y. Konda, A. Hatano, T. Hata, S. Omura, *J. Am. Chem. Soc.* **1983**, *105*, 6311–6312.
129 M. Onda, Y. Konda, A. Hatano, T. Hata, S. Omura, *Chem. Pharm. Bull.* **1984**, *32*, 2995–3002.
130 E. J. Moran, R. W. Armstrong, *Tetrahedron Lett.* **1991**, *32*, 3807–3810.
131 N. Shimada, M. Uekusa, T. Denda, Y. Ishii, T. Iizuka, Y. Sato, T. Hatori, M. Fukui, M. Sudo, *J. Antibiot. Ser. A* **1955**, *8*, 67–76.
132 B. A. Stoll, *Cancer* **1960**, *13*, 439–441.
133 C. Hossenlopp, *J. Antibiot.* **1961**, *14*, 289–297.
134 A. Terawaki, J. Greenberg, *Nature* **1966**, *209*, 481–484.
135 A. Terawaki, J. Greenberg, *Biochim. Biophys. Acta* **1966**, *119*, 59–64.
136 J. W. Lown, K. C. Majumdar, *Can. J. Biochem.* **1977**, *55*, 630
137 R. W. Armstrong, M. E. Salvati, M. Nguyen, *J. Am. Chem. Soc.* **1992**, *114*, 3144–3145.
138 M. E. Salvati, E. J. Moran, R. W. Armstrong, *Tetrahedron Lett.* **1992**, *33*, 3711–3714.
139 M. Hashimoto, M. Matsumoto, K. Yamada, S. Terashima, *Tetrahedron Lett.* **1994**, *35*, 2207–2210.
140 T. Fujiwara, I. Saito, *Tetrahedron Lett.* **1999**, *40*, 315–318.
141 J. A. Hartley, A. Hazrati, L. R. Kelland, R. Khanim, M. Shipman, F. Suzenet, L. F. Walker, *Angew. Chem. Int. Ed.* **2000**, *39*, 3467–3470.
142 S. Alcaro, R. S. Coleman, *J. Org. Chem.* **1998**, *63*, 4620–4625.
143 S. Alcaro, R. S. Coleman, *J. Med. Chem.* **2000**, *43*, 2783–2788.
144 S. Alcaro, F. Ortuso, R. S. Coleman, *J. Med. Chem.* **2002**, *45*, 861–870.
145 T. J. Hodgkinson, L. R. Kelland, M. Shipman, F. Suzenet, *Bioorg. Med. Chem. Lett.* **2000**, *10*, 239–241.
146 J. A. Hartley, A. Hazrati, T. J. Hodgkinson, L. R. Kelland, R. Khanim, M. Shipman, F. Suzenet, *Chem. Commun.* **2000**, 2325–2326.
147 K. Shishido, S. Haruna, H. Iitsuka, M. Shibuya, *Heterocycles* **1998**, *49*, 109–112.
148 H. Zang, K. S. Gates, *Biochemistry* **2000**, *39*, 14968–14975.
149 R. S. Coleman, C. H. Burk, A. Navarro, R. W. Brueggemeier, E. S. Diaz-Cruz, *Org. Lett.* **2002**, *4*, 3545–3548.
150 R. S. Coleman, R. J. Perez, C. H. Burk, A. Navarro, *J. Am. Chem. Soc.* **2002**, *124*, 13008–13017.
151 M. A. Casely-Hayford, K. Pors, L. H. Patterson, C. Gerner, S. Neidle, M. Searcey, *Bioorg. Med. Chem. Lett.* **2005**, *15*, 653–656.
152 C.A.S. Landreau, R. C. LePla, M. Shipman, A.M.Z. Slawin, J. A. Hartley, *Org. Lett.* **2004**, *6*, 3505–3507.
153 C. Corre, P.A.S. Lowden, *Chem. Commun.* **2004**, 990–991.
154 C. Corre, C.A.S. Landreau, M. Shipman, P.A.S. Lowden, *Chem. Commun.* **2004**, 2600–2601.
155 G. Guroff, J. W. Daly, D. M. Jerina, J. Renson, B. Witkop, S. Udenfriend, *Science* **1967**, *157*, 1524–1530.
156 T. J. Simpson, A. E. de Jesus, P. S. Steyn,

R. Vleggaar, *J. Chem. Soc. Chem. Commun.* **1983**, 338–340.
157 N. Ishida, K. Miyuzaki, K. Kumagai, M. Rikimaru, *J. Antibiot.* **1965**, *18*, 68–76.
158 K. Edo, M. Mizugaki, Y. Koide, H. Seto, K. Furihata, N. Otaken, N. Ishida, *Tetrahedron Lett.* **1985**, *26*, 331–334.
159 T. Ando, M. Ishii, T. Kajiura, T. Kameyama, K. Miwa, Y. Sugiura, *Tetrahedron Lett.* **1998**, *39*, 6495–6498.
160 W. Liu, K. Nonaka, L.P. Nie, J. Zhang, S.D. Christenson, J. Bae, S.G. Van Lanen, E. Zazopoulos, C.M. Farnet, C.F. Yang, B. Shen, *Chem. Biol.* **2005**, *12*, 293–302.
161 B. Sthapit, T.-J. Oh, R. Lamichhane, K. Liou, H.C. Lee, C.-G. Kim, J.K. Sohng, *FEBS Lett.* **2004**, *566*, 201–206.
162 D. Schwarzer, R. Finking, M.A. Marahiel, *Nat. Prod. Rep.* **2003**,*20*, 275–287.
163 N. Gaitatzis, B. Kunze, R. Müller, *Proc. Natl. Acad. Sci. U.S.A.* **2001**, *98*, 11136–11141.
164 A.D. Argoudelis, F. Reusser, H.A. Whaley, L. Baczynskyj, S.A. Mizsak, R.J. Wnuk, *J. Antibiot.* **1976**, *29*, 1001–1006.
165 M.-S. Kuo, D.A. Yurek, S.A. Miszak, *J. Antibiot.* **1989**, *42*, 357–360.
166 F. Reusser, *Biochemistry* **1977**, *16*, 3406–3412.
167 C.O. Gitterman, E.L. Rickes, D.E. Wolf, J. Madas, S.B. Zimmerman, T.H. Stoudt, T.C. Demny, *J. Antibiot.* **1970**, *23*, 305–310.
168 B.H. Arison, J.L. Beck, *Tetrahedron* **1973**, *29*, 2743–2746.
169 G.R. Pettit, R.B. Von Dreele, D.L. Herald, M.T. Edgar, H.B. Wood Jr., *J. Am. Chem. Soc.* **1976**, *98*, 6742–6743.
170 R.B. Von Dreele, *Acta Cryst. B* **1981**, *37*, 93–98.
171 G.S. Tarnowski, F.A. Schmid, D.J. Hutchison, C.C. Stock, *Cancer Chemother. Rep.* **1973**, *57*, 21–27.
172 M.A. Dimopoulos, J.C. Yau, S.D. Huan, S. Jagannath, G. Spitzer, J.A. Spinolo, G.K. Zagars, C.F. Lemaistre, K.A. Dicke, A.R. Zander, *Am. J. Hematol.* **1994**, *46*, 82–86.
173 G.P. Wheeler, V.H. Bono, B.J. Bowdon, D.J. Adamson, R.W. Brockman, *Cancer Treat. Rep.* **1976**, *60*, 1307–1316.
174 R.W. Brockman, S.C. Shaddix, M. Williams, R.F. Struck, *Cancer Treat. Rep.* **1976**, *60*, 1317–1324.
175 F.M. Schabel Jr., M.W. Trader, W.R. Laster Jr., S.C. Shaddix, R.W. Brockman, *Cancer Treat. Rep.* **1976**, *60*, 1325–1333.
176 H. Naganawa, N. Usui, T. Takita, M. Hamada, H. Umezawa, *J. Antibiot.* **1975**, *28*, 828–829.
177 J. Greenhut, H. Umezawa, F.B. Rudolph, *J. Biol. Chem.* **1985**, *260*, 6684–6686.
178 Y. Nakao, M. Fujita, K. Warabi, S. Matsunaga, N. Fusetani, *J. Am. Chem. Soc.* **2000**, *122*, 10462–10463.
179 N. Schashke, *Bioorg. Med. Chem. Lett.* **2004**, *14*, 855–857.
180 T. Schirmeister, M. Peric, *Bioorg. Med. Chem.* **2000**, *8*, 1281–1291.
181 C. Gelhaus, R. Vicik, R. Hilgenfeld, C.L. Schmidt, M. Leippe, T. Schirmeister, *Biol. Chem.* **2004**, *385*, 435–438.
182 T. Tsuchida, H. Iinuma, *J. Antibiot.* **1993**, *46*, 1772–1773.
183 T. Tsuchida, H. Iinuma, N. Kinoshita, T. Ikeda, T. Sawa, M. Hamada, T. Takeuchi, *J. Antibiot.* **1995**, *48*, 217–221.
184 T. Tsuchida, R. Sawa, Y. Takahashi, H. Iinuma, T. Sawa, H. Naganawa, T. Takeuchi, *J. Antibiot.* **1995**, *48*, 1148–1152.
185 M. Hanada, H. Ohkuma, T. Yonemoto, K. Tomita, M. Ohbayashi, H. Kamei, T. Miyaki, M. Konishi, H. Kawaguchi, S. Forenza, *J. Antibiot.* **1991**, *44*, 403–414.
186 D.R. Schroeder, K.L. Colson, S.E. Klohr, N. Zein, D.R. Langley, M.S. Lee, J.A. Matson, T.W. Doyle, *J. Am. Chem. Soc.* **1994**, *116*, 9351–9352.
187 N. Zein, W. Solomon, K.L. Colson, D.R. Schroeder, *Biochemistry* **1995**, *34*, 11591–11597.
188 W. Liu, J. Ahlert, Q. Gao, E. Wendt-Pienkowski, B. Shen, J.S. Thorson, *Proc. Natl. Acad. Sci. U.S.A.* **2003**, *100*, 11959–11963.
189 K.-I. Harada, K. Tomita, K. Fujii, K. Masuda, Y. Mikami, K. Yazawa, H. Komaki, *J. Antibiot.* **2004**, *57*, 125–135.
190 R. Howe, *Nature* **1965**, *207*, 594.
191 U. Bicker, W. Fischer, *Nature* **1974**, *249*, 344–345.
192 E.O. Stapley, D. Hendlin, M. Jackson, A.K. Miller, S. Hernandez, J.M. Mata, *J. Antibiot.* **1971**, *24*, 42–47.

193 T.W. Miller, E.W. Tristram, F.J. Wolf, *J. Antibiot.* **1971**, *24*, 48–50.

194 T.F. Molinski, C.M. Ireland, *J.Org. Chem.* **1988**, *53*, 2103–2105.

195 C.E. Salomon, D.H. Williams, D.J. Faulkner, *J. Nat. Prod.* **1995**, *58*, 1463–1466.

196 T. Kida, H. Shibai, *Agric. Biol. Chem.* **1985**, *49*, 3231–3237.

197 H.W. Dion, S.A. Fusari, Z.L. Jakubowski, J.G. Zora, Q.R. Bartz, *J. Am. Chem. Soc.* **1956**, *78*, 3075–3077.

198 A.J. Tomisek, M.R. Reid, *J. Biol. Chem.* **1962**, *237*, 807–811.

12
Epoxides and Aziridines in Click Chemistry

Valery V. Fokin and Peng Wu

12.1
Introduction

The success of a search for novel compounds with desired properties, whether in the fields of pharmaceuticals, agrochemicals, or new materials, is often dependent on the degree of diversity of the building blocks that are available: the greater the variety of structures and functional groups that can be employed in the construction of candidate compounds, the more likely it is that useful function will be discovered. However, while the throughput of assays and automated compound-handling techniques have advanced with the lightning speed during the last decade, methods that allow practical synthesis of truly diverse collections of compounds still leave much to be desired. The problems with transformations employed in the diversity setting often arise from at least one, and usually a combination, of the following: limited scope, hard-to-obtain starting materials, requirements for protecting groups, and difficult purifications. The lesson is clear: with an enormous diversity space to explore (it has been estimated that the number of small drug-like compounds may be as high as 10^{64}) [1], the size of a given collection becomes much less important than its reach, which is in turn dictated by the synthetic methods available for its assembly.

The concept of click chemistry began to take shape more than a decade ago in the laboratories of K. B. Sharpless. The first written account was published in 2001 [2]. Although the concept has been refined since then, the guiding principle has remained: click chemistry relies on the best connecting reactions that fulfil the most stringent criteria of utility and convenience. Thus, it is simultaneously enabled and constrained by a handful of near-perfect click reactions (Scheme 12.1). Click reactions are modular, use only readily available reagents, and exhibit wide scope, giving consistently high yields with a variety of starting materials. They are easy to perform (i. e., insensitive to oxygen or water, so protecting groups and offensive organic solvents are not necessary). Workup and product isolation is usually simple, without the need for chromatographic purification. A defining attribute of a click reaction is a high thermodynamic driving force, usually greater than 20 kcal · mol^{-1}, which results in the spontaneous formation of a desired

Aziridines and Epoxides in Organic Synthesis. Andrei K. Yudin

ISBN: 3-527-31213-7

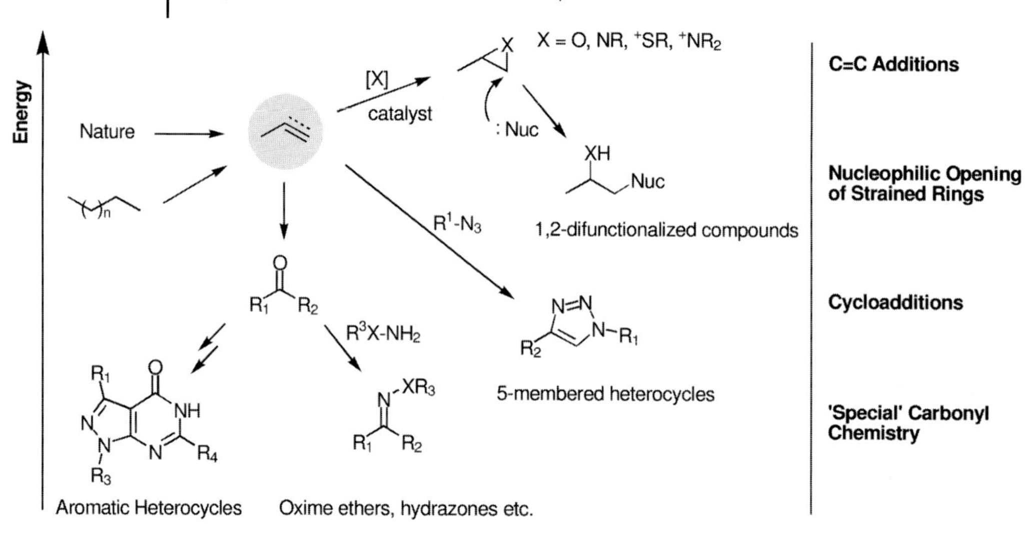

Scheme 12.1 Click chemistry: energetically driven linking reactions.

product. Such "spring-loaded" processes proceed rapidly to completion and also tend to be highly selective for a single product. The majority of click reactions form carbon-heteroatom bonds and are tolerant of – and frequently accelerated by – water or aqueous solvents [3]. The following carbon-heteroatom bond-forming reactions comprise the most common examples:

- additions to carbon-carbon multiple bonds, especially oxidation of olefins, such as epoxidation, dihydroxylation, aziridination, and aminohydroxylation;
- nucleophilic substitution chemistry, particularly ring-opening reactions of strained heterocyclic electrophiles such as epoxides, aziridines, aziridinium ions, and episulfonium ions;
- cycloadditions of unsaturated species, especially 1,3-dipolar cycloaddition reactions.

It seems fair to say that the history of synthetic organic chemistry has been dominated by the goal of imitating nature's biosynthetic pathways. Today, the art of organic synthesis has evolved to a point where even the most complex molecules created by living systems have been, or can be, made in the laboratory. However, the bioactive natural products that have most intrigued synthetic organic chemists have frameworks of extended C–C bonds that are often exceedingly difficult to construct, even though nature tackles this problem seemingly effortlessly. Why do organic chemists end up with reaction sequences so starkly different from nature's for the same synthetic targets? The answer is rooted in appreciation of the fact that organic chemists and nature begin from different *starting materials* and use different *methods for controlling reaction selectivity*. Nature ultimately starts her syntheses from carbon dioxide, water, ammonia, and phosphate, and has therefore evolved

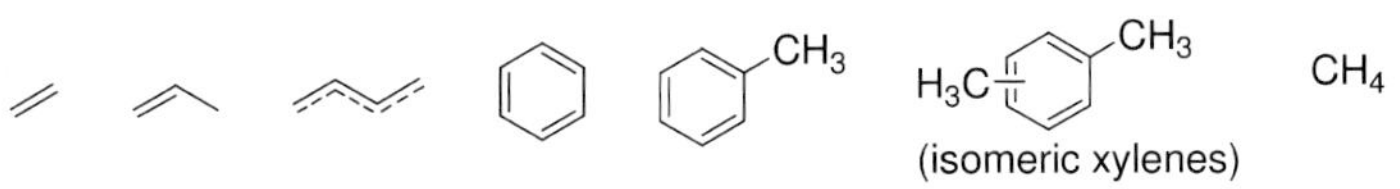

Scheme 12.2 Seven hydrocarbons are the progenitors of most organic compounds produced by the chemical industry.

exquisite enzymatic pathways to construct C–C bonds based on the mostly reversible, aldol-type transformations of the carbonyl group and to control each step through small incremental energy bursts (ca. 3 kcal · mol^{-1}) derived from the hydrolysis of ATP. Lacking her synthetic stratagems for channeling reactions in the desired pathway, organic chemists had no choice but to resort to aggressive reagents, such as organolithiums and enolates, which require the use of aprotic solvents and protecting groups to achieve the same goals.

Fortunately, we have a different set of carbon starting materials available to us, derived from nature but inaccessible to her: hydrocarbons, which are more advanced than carbon dioxide. Most come from the oil left for us by prehistoric lifeforms. The petrochemical industry cracks it into smaller unsaturated blocks containing at most eight carbon atoms, from which almost 90% (by weight) of all useful synthetic organics are made (Scheme 12.2) [4].

Endowed with about 20–25 kcal · mol^{-1} of free energy above that for the corresponding saturated hydrocarbon, olefins are ideal starting materials for organic synthesis. Their potential is unlocked by their oxidative functionalization. The broad scope and high yields of many olefin oxidation processes make these transformations the most fundamental enablers of click chemistry. The oxidation step generates highly reactive, yet stable, building blocks, such as epoxides and aziridines, or even more energetic intermediates, such as aziridinium and episulfonium ions – all nearly perfect for click chemistry transformations (Scheme 12.3). Both the oxidations to the intermediates *and* their subsequent reactions with nucleophiles are stereospecific, and are therefore very predictable. To sum up, the combination of oxidative olefin functionalization and subsequent manipulation of the resulting products allows one to proceed from a system of high energy to a system of lower energy in a *stepwise* fashion, thereby enabling the introduction of new elements of diversity in every step, each with a strong driving force, making for mild reaction conditions and reliably high yields. This sequence of oxidative creation/nucleophilic quenching of reactive electrophiles is at the heart of click chemistry and is the subject of the discussion that follows.

Although beyond the scope of the present discussion, another key realization that has shaped the definition of click chemistry in recent years was that while olefins, through their selective oxidative functionalization, provide convenient access to reactive modules, the assembly of these energetic blocks into the final structures is best achieved through cycloaddition reactions involving carbon-heteroatom bond formation, such as [1,3]-dipolar cycloadditions and hetero-Diels-Alder reactions. The copper(I)-catalyzed cycloaddition of azides and terminal alkynes [5] is arguably the most powerful and reliable way to date to "stitch" a broad variety

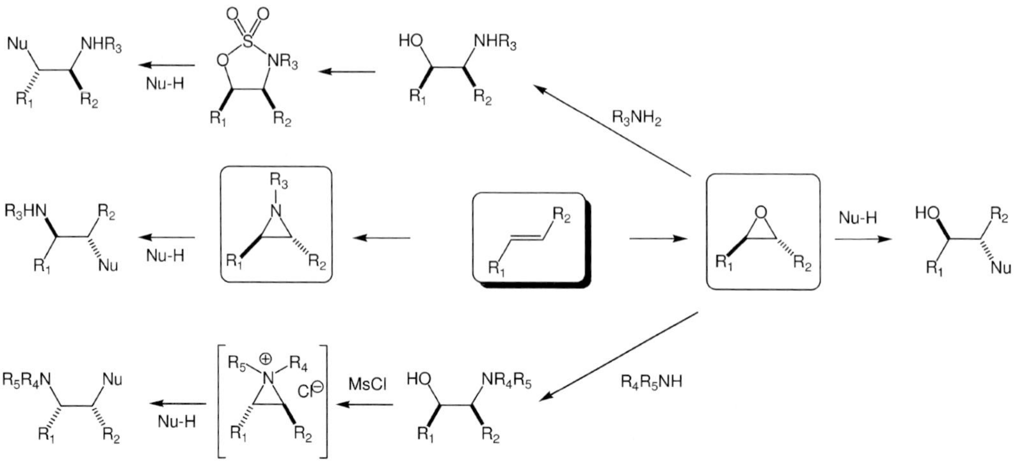
Nu
NHR3
R1
R2
Nu-H
NR3
HO
NHR3
R3NH2
R3HN
R2
Nu
R3
N
R1
R2
Nu-H
HO
R2
R5R4N
Nu
R5
R4
Cl
MsCl
NR4R5
R4R5NH

Scheme 12.3

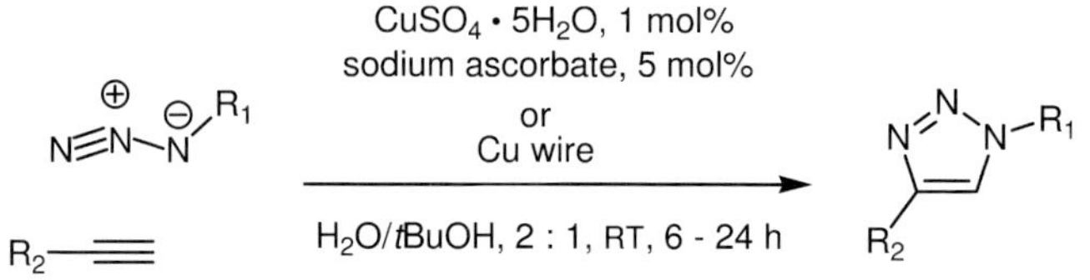
CuSO4 • 5H2O, 1 mol%
sodium ascorbate, 5 mol%
or
Cu wire
H2O/tBuOH, 2 : 1, RT, 6 - 24 h
R1
R2

Scheme 12.4 Copper(I)-catalyzed synthesis of 1,2,3-triazoles.

of blocks irreversibly by means of the 1,2,3-triazole connection, a link that is notably stable and inert to severe hydrolytic, as well as redox, conditions (Scheme 12.4). Although both alkynes and azides are highly energetic species, their chemoselectivity profiles are quite narrow; that is, "orthogonal" to an unusually broad range of reagents, solvents, and other functional groups. This allows reliable and clean sequential transformations of broad scope without the need for any protecting groups. This catalytic variant of the Huisgen synthesis of 1,2,3-triazoles has allowed a number of applications, ranging from medicinal chemistry and molecular biology to materials science and electrochemistry. The impact of this transformation on drug discovery has recently been reviewed [6]. In this light, the paramount importance of transformations that allow efficient introduction of azide and alkyne moieties into various structures is easily appreciated, and nucleophilic openings of epoxides, aziridines, and aziridinium ions are among the most convenient processes to achieve this goal.

12.2 Epoxides in Click Chemistry

Epoxides are often encountered in nature, both as intermediates in key biosynthetic pathways and as secondary metabolites. The selective epoxidation of squalene, resulting in 2,3-squalene oxide, for example, is the prelude to the remarkable olefin oligomerization cascade that creates the steroid nucleus [7]. Tetrahydrodiols, the ultimate products of metabolism of polycyclic aromatic hydrocarbons, bind to the nucleic acids of mammalian cells and are implicated in carcinogenesis [8]. In organic synthesis, epoxides are invaluable building blocks for introduction of diverse functionality into the hydrocarbon backbone in a 1,2-fashion. It is therefore not surprising that chemistry of epoxides has received much attention [9].

The past thirty years have witnessed great advances in the selective synthesis of epoxides, and numerous regio-, chemo-, enantio-, and diastereoselective methods have been developed. Discovered in 1980, the Katsuki-Sharpless catalytic asymmetric epoxidation of allylic alcohols, in which a catalyst for the first time demonstrated both high selectivity and substrate promiscuity, was the first practical entry into the world of chiral 2,3-epoxy alcohols [10, 11]. Asymmetric catalysis of the epoxidation of unfunctionalized olefins through the use of Jacobsen's chiral [(salen)Mn^{III}] [12] or Shi's chiral ketones [13] as oxidants is also well established. Catalytic asymmetric epoxidations have been comprehensively reviewed [14, 15].

12.2.1 Synthesis of Epoxides

Owing to the aforementioned importance of epoxides, methods for their preparation have been the focus of intense studies during the last five decades. While this vast area is addressed in more details elsewhere in this volume, brief comments about procedures that have been found particularly convenient in click chemistry context are offered below.

Epoxidation of olefins with meta-chloroperbenzoic acid, (MCPBA) remains to this day among the most widely used methods for research-scale applications [16]. Discovered by Nikolai Prilezahev in 1909 [17], it became popular only decades later, mostly through the works of Daniel Swern in the 1940s [18]. Despite its simplicity, and not unlike most epoxidation methods in use today, it suffers from undesired epoxide opening caused by the slight acidity of the reaction milieu. Although acid-catalyzed side reactions can sometimes be minimized by use of buffered systems [19], the necessity for stringent pH control for the preparation of acid-sensitive epoxides significantly limits substrate scope and overall efficiency of this process [20].

Dimethyldioxirane DMDO discovered by Murray and coworkers, is a superior choice for the epoxidation of most olefins, giving comparable or higher yields than *m*-CPBA-based epoxidation [21]. Proceeding rapidly under neutral and mild conditions, it is especially well suited for the synthesis of sensitive epoxides of enol esters, enol lactones [22], and enol ethers [23]. The reaction is stereospecific, gen-

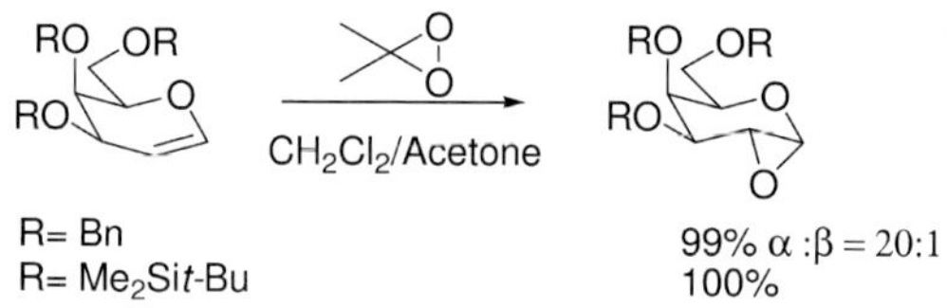

RO OR / RO / O — (dimethyldioxirane) → CH_2Cl_2/Acetone → RO OR / RO / O / O

R= Bn — 99% α :β = 20:1
R= Me_2Sit-Bu — 100%

Scheme 12.5 Direct epoxidation of glycals.

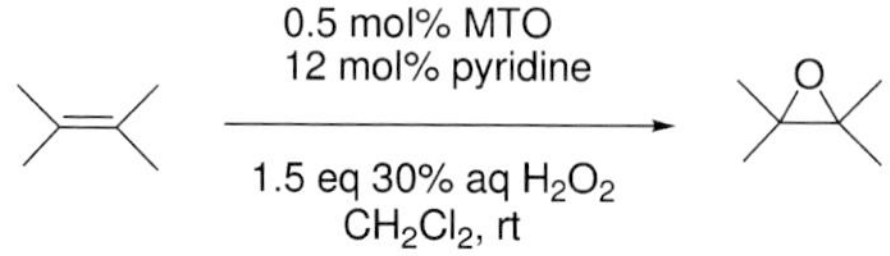

0.5 mol% MTO
12 mol% pyridine
1.5 eq 30% aq H_2O_2
CH_2Cl_2, rt

Scheme 12.6 MTO/H_2O_2/pyridine epoxidation.

erally furnishing the desired epoxides in almost quantitative yields. The procedure is experimentally convenient, and in most cases removal of the solvent generates pure products. An elegant synthesis of β-linked oligosaccharides by this method was accomplished by Danishefsky (Scheme 12.5) [24].

After Herrmann's groups, discovery of its remarkable catalytic properties [25], methyltrioxorhenium (MTO) has emerged as a useful catalyst for a number of organic transformations, epoxidations in particular. MTO and hydrogen peroxide as a terminal oxidant provide efficient conversion of olefins into their corresponding epoxides. Because of the Lewis acidic nature of the Re center, however, only those epoxides uncommonly resistant to ring-opening could be obtained with the original procedure. The utility of this transformation was greatly expanded by the subsequent discovery that pyridine ligands not only exhibit a remarkable acceleration effect on the reaction rate but also prevent acid-catalyzed epoxide-opening [26, 27]. In addition, the catalyst lifetime is significantly increased when sufficient pyridine ligand (3–12 mol%) is provided in aprotic, noncoordinating solvents (Scheme 12.6). Taken together, the reactivity of MTO/H_2O_2 closely resembles that of dimethyldioxirane, but the former is far more convenient and inexpensive to generate.

The MTO/H_2O_2/pyridine system enjoys a broad substrate scope and has become the method of choice for the epoxidation of di-, tri-, and tetrasubstituted olefins. As an added benefit, it gives high diastereoselectivities for a number of cyclic dienes (Table 12.1).

Further improvement of this process was accomplished by Yudin and Sharpless by use of the stoichiometric oxidant bis(trimethylsilyl) peroxide (BTSP, Scheme 12.7) [28].

This highly active epoxidation system, based on the controlled hydrolysis of BTSP with a catalytic amount of water, maximizes the formation of the Re monoperoxide complex at the expense of the more thermodynamically stable bis(peroxide) (Scheme 12.8). BTSP is very stable and can be prepared in molar amounts

Table 12.1 MTO/pyridine-catalyzed epoxidation of dienes.[a]

Entry	Substrate	Product	Time (h)	Ratio monoepoxide: diepoxide	Selectivity (%)[c]	Diastereomeric ratio for diepoxide
1[b]		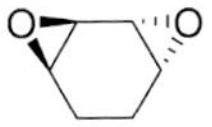	3	1.2:1	95	99% *anti*
2		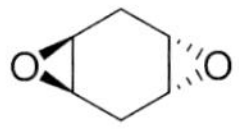	6	1:100	99	96:4 (*anti*:*syn*)
3[b]	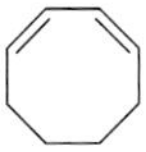	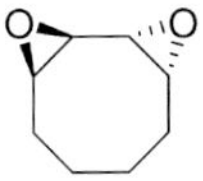	7	1:1.3	99	99% *anti*
4	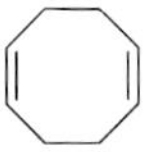	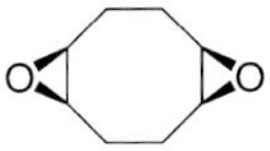	5	1:100	97	99% *syn*

[a] Reaction conditions: $c_{substrate} = 2$ mol · L^{-1}, 0.5 mol% MTO, 12 mol% pyridine, 2.5 equiv. 50% aq. H_2O_2, CH_2Cl_2,, r. t. water bath, 2 mmol scale, selectivities and rates determined by ^1H NMR; [b] Reaction performed at 10 °C. [c] Chemoselectivity with respect to the epoxidation reaction.

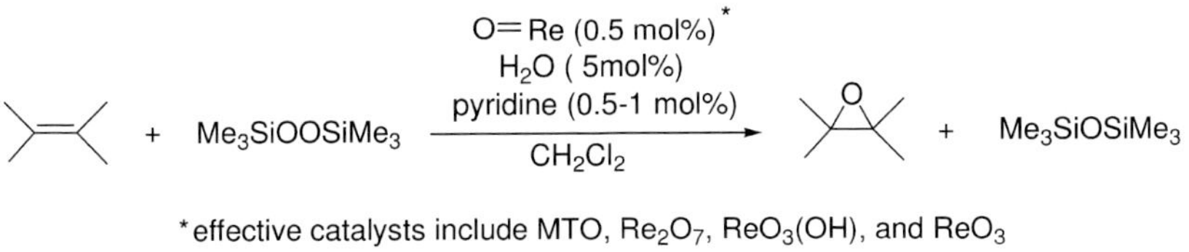

O=Re (0.5 mol%)*
H_2O (5mol%)
pyridine (0.5-1 mol%)
+ $Me_3SiOOSiMe_3$ → (CH_2Cl_2) + $Me_3SiOSiMe_3$

*effective catalysts include MTO, Re_2O_7, $ReO_3(OH)$, and ReO_3

Scheme 12.7 Re/BTSP epoxidation.

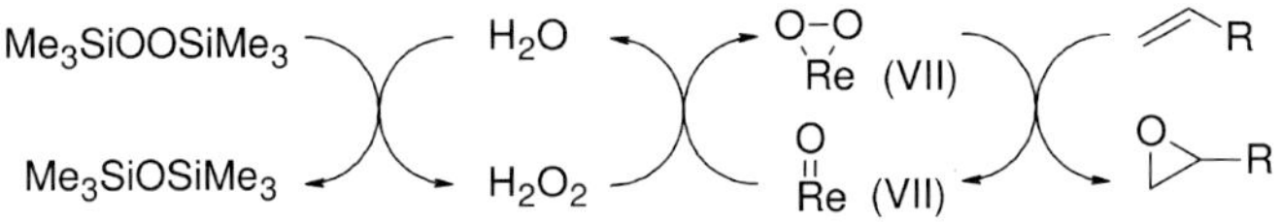

$Me_3SiOOSiMe_3$ H_2O O–O Re (VII) R
$Me_3SiOSiMe_3$ H_2O_2 O Re (VII) O R

Scheme 12.8 Catalytic cycle of Re/BTSP epoxidation.

by treatment of the urea/hydrogen peroxide adduct with bis(trimethylsilyl)urea. Another advantage of this method is that the expensive and relatively sensitive organorhenium complex MTO can be replaced by the stable and easily handled rhenium oxides (e.g., Re_2O_7, $ReO_3(OH)$, and ReO_3).

Table 12.2 Epoxidation of olefins with bis(trimethylsilyl) peroxide (BTSP) catalyzed by high-valent oxorhenium derivatives.[a, b]

Entry	Alkene	Catalyst precursor/[c] solvent	Yield (%)	Time (h)
1	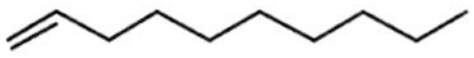	A/CH_2Cl_2	94	14
2	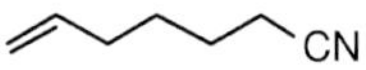 CN	A/CH_2Cl_2	92	18
3	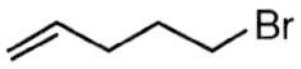 Br	B/CH_2Cl_2	82	9
4	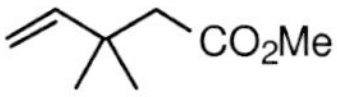 CO_2Me	B/CH_2Cl_2	77	15
5	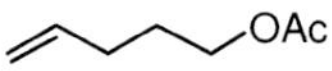 OAc	B/CH_2Cl_2	82	3
6	 Ph	B/CH_2Cl_2	88	9
7	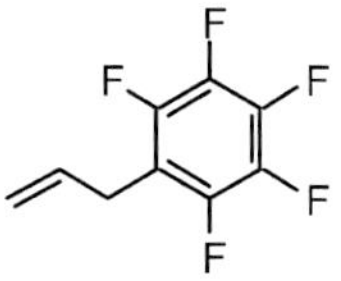 F, F, F, F, F	C/CH_2Cl_2	90	15
8	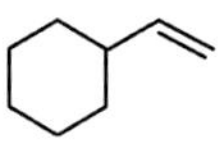	A/CH_2Cl_2	95	7
9	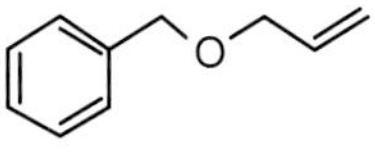 O	B/CH_2Cl_2	95	8
10		D/CH_2Cl_2 E/CH_2Cl_2	88 80	12 18
11	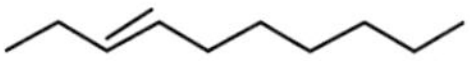	B/CH_2Cl_2 F/CH_2Cl_2	90 85	9 9
12	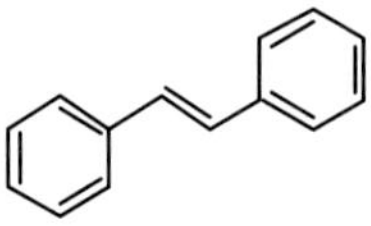	A/THF	96	10
13	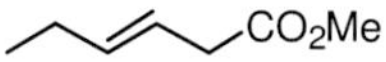 CO_2Me	B/CH_2Cl_2	90	3

Table 12.2 (continued).

Entry	Alkene	Catalyst precursor/[c] solvent	Yield (%)	Time (h)
14	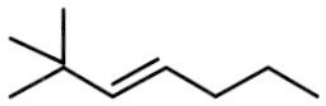	B/CH_2Cl_2	70	4
15	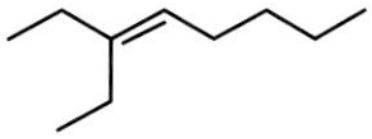	B/CH_2Cl_2	96	4
16	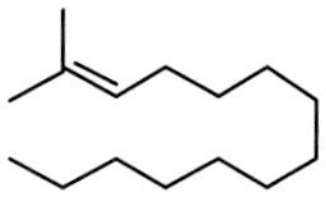	B/CH_2Cl_2	81	14
17	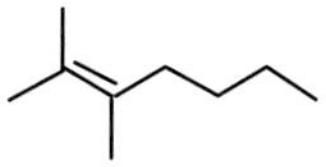	B/CH_2Cl_2	82	14
18	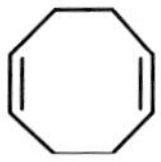	A/CH_2Cl_2	82	10

[a] 10 mmol scale. [b] 1.5 equiv. BTSP per double bond was used.
[c] A: Re_2O_7 (0.5 mol%), H_2O (5 mol%). B: ReO_3 (0.5 mol%), H_2O (5 mol%). C: ReO_3 (0.5 mol%), H_2O (1 mol%). D: Re_2O_7 (0.5 mol%), pyridine (1 mol%), H_2O (5 mol%). E: 2py/$HReO_4$ (0.5 mol%), H_2O (5 mol%). F: MTO(0.5 mol%), H_2O (5 mol%).

The procedure is experimentally simple, and the workup involves only the destruction of the traces of hydrogen peroxide with manganese dioxide and evaporation of the hexamethyldisiloxane. Pyridine additives serve to buffer the highly acidic rhenium species and to shut down the detrimental acid-catalyzed epoxide-opening pathways. The scope of this transformation is best appreciated through the examples presented in Table 12.2 [28].

12.2.2
Nucleophilic Opening of Epoxides

The chief advantage in nucleophilic openings of epoxides and aziridines is that competing elimination processes are stereoelectronically disfavored [29], resulting in higher yields and easier product isolation. These "fusion" reactions can be performed in the absence of solvents, causing the mixture of starting materials to become a pure product, which only needs to be collected. Naturally, care must be exercised in large-scale reactions with many epoxides and *N*-sulfonyl-substituted

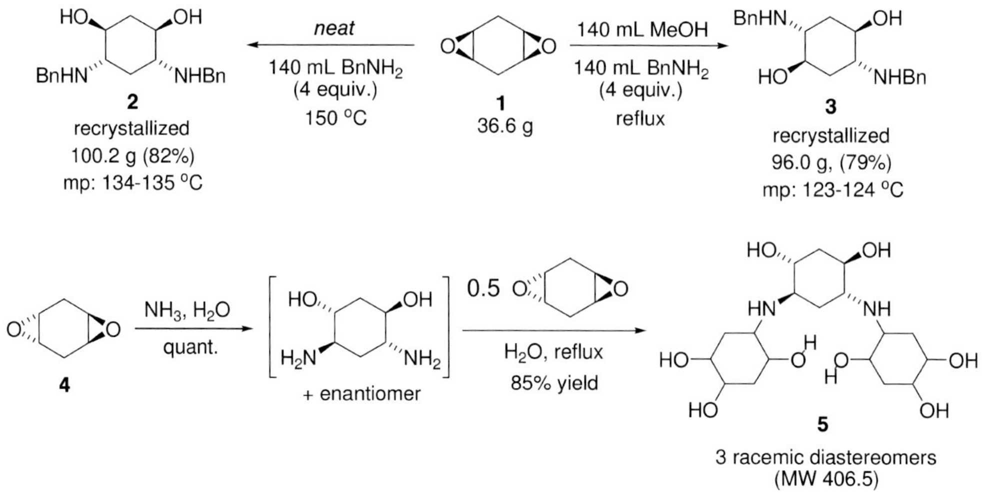

Scheme 12.9 Reactions of cyclohexadiene diepoxides with amines [2].

aziridines, since the opening reactions are exothermic and, when performed neat, are often autocatalytic. In a number of cases the regioselectivity can be controlled by the choice of solvent, a perfect example being the reaction between *cis*-cyclohexadiene diepoxide **1** [30] and amines [31] (Scheme 12.9, top). In the absence of solvents, the diepoxide **1** reacted with amines in 1,3-fashion to give the amino alcohol **2**, while dilution with protic solvents caused complete reversal of regioselectivity, providing the 1,4-product **3**. In both cases the products were isolated and purified by crystallization, thereby allowing preparation of the diamino diols on a large scale. Regioselectivity is similarly well controlled for the ring opening of the *trans* diepoxide **4**; the example in Scheme 12.9, bottom, shows the one-pot "stitching" of three diepoxide units with two equivalents of ammonia to give **5** as a mixture of only three diastereomers.

The 1,3-selectivity in the neat reaction probably stems from the interplay of the following effects: the strong preference for *trans*-diaxial epoxide opening, plus the possibility of the intramolecular epoxide activation by the hydroxy group released in the first step, on a playing field governed by the Curtin-Hammett principle. The presence of a protic solvent has such a dramatic effect because it increases the energy difference between the two chair conformers by better solvating the polar groups while simultaneously inhibiting intramolecular direction from the resident hydroxy groups.

The other two cyclohexadiene diepoxides (there are four, not counting enantiomers) are also readily available, along with the two diepoxides of cyclohexatriene (formally, benzene diepoxides), obtained by way of benzoquinone [32]. Each of these diepoxides gave an excellent yield of the expected diazido diol upon ring-opening with azide; remarkably, every case provided greater than 95% regioselectivity. These six diepoxides likewise undergo ring-opening with excess benzyla-

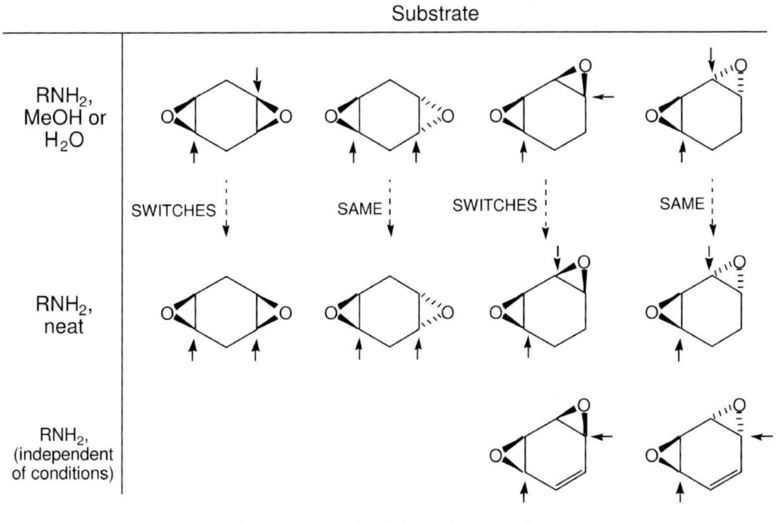

Substrate

RNH$_2$, MeOH or H$_2$O

SWITCHES SAME SWITCHES SAME

RNH$_2$, neat

RNH$_2$, (independent of conditions)

Scheme 12.10 Selectivity of reactions of cyclohexadiene and benzene diepoxides with amines [2].

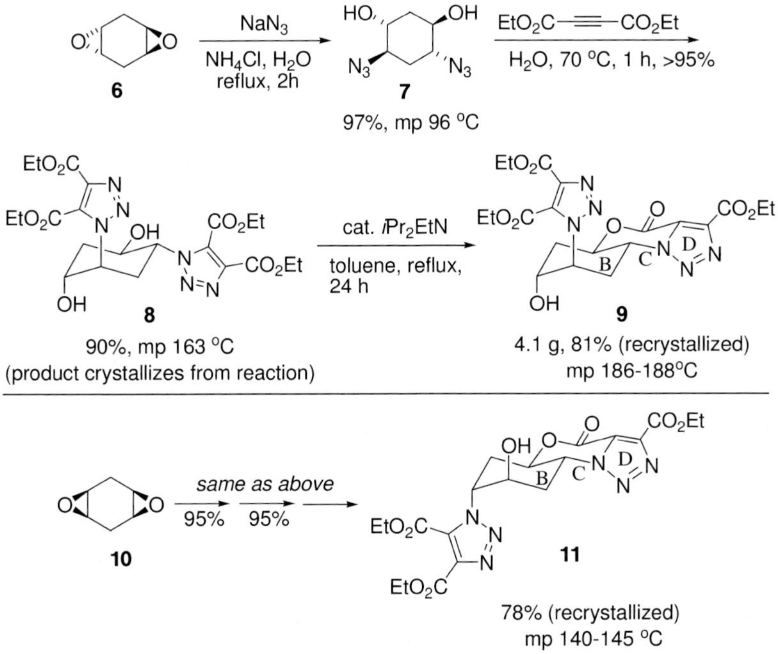

6 — NaN_3 / NH_4Cl, H_2O reflux, 2h → 7 (97%, mp 96 °C) — $EtO_2C{-}{\equiv}{-}CO_2Et$ / H_2O, 70 °C, 1 h, >95% →

8 — 90%, mp 163 °C (product crystallizes from reaction) — cat. *i*Pr_2EtN / toluene, reflux, 24 h → 9 — 4.1 g, 81% (recrystallized) mp 186-188°C

10 — *same as above* 95% 95% → 11 — 78% (recrystallized) mp 140-145 °C

Scheme 12.11 [2]

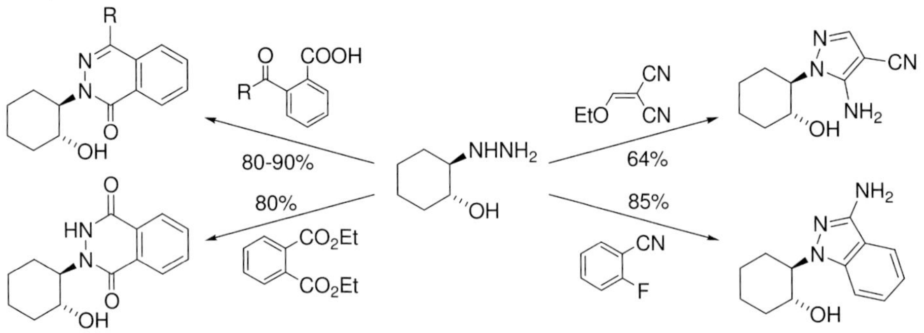

Scheme 12.12 Synthesis of heterocycles from cyclohexene oxide.

mine in the presence and in the absence of added water with interesting regiochemical control (Scheme 12.10). Thus, 1,3-cyclohexadiene *syn*-diepoxide undergoes 1,3-opening with neat benzylamine and the 1,4-attack in the presence of protic cosolvent. The *anti*-diepoxide isomer is attacked in 1,3-fashion under either set of conditions. Both cyclohexatriene diepoxides open only at the allylic positions regardless of the conditions.

Combined with 1,3-dipolar cycloadditions, the nucleophilic opening of epoxides has been used for rapid assembly of complex structures. One example is the formation of the tricyclic molecule **9** (Scheme 12.11) in just three steps, conducted sequentially in one-pot fashion, starting from diepoxide **6**. The nucleophilic opening of **6** with buffered azide was highly regioselective, resulting in the formation of the crystalline azido alcohol **7** in excellent yield. The bis-1,2,3-triazole **8**, formed by the 1,3-dipolar addition between the diazide and diethyl acetylenedicarboxylate, was collected from the reaction mixture by filtration. Thermal lactonization then broke the C_2 symmetry of the system to furnish lactone **9**, the three rings of which resemble the B, C, and D rings found in steroids.

Through the same three-step sequence the analogous *cis*-diepoxide **10** [33] gave the related lactone **11**, identical to **9** except that the two B-ring substituents have switched positions. All steps for both sequences proceeded in excellent yield.

The potential of such reaction sequences for the generation of molecular diversity was also demonstrated by the synthesis of a library of heterocycles. Epoxide ring-opening with hydrazine and subsequent condensation with β-diketones or other bifunctional electrophiles gave rise to a variety of functionalized heterocyclic structures in high purity [34]. A selection based on the substrate derived from cyclohexene oxide is shown in Scheme 12.12.

12.3 Aziridines in Click Chemistry

Aziridines, the aza analogues of epoxides, are spring-loaded electrophiles ideal for the critical intermolecular reactions that anneal blocks together [35]. Through variation of the substituent on the ring nitrogen, aziridines allow much greater product diversity than epoxides. Their ring-opening reactivity can be modulated over a broad range [36] and, most importantly, the nitrogen substituent and the nature of the solvent can be used to control the regioselectivity of ring opening in unsymmetrical cases [37].

12.3.1 Synthesis of Aziridines

Although aziridines can be prepared by direct oxidation of olefins [38], general and reliable methods available to a synthetic chemist are conspicuous by their rarity. This is in contrast to the situation for three-membered rings containing nitrogen's immediate first row neighbors, carbon and oxygen, since a number of good synthetic methods for the direct cyclopropanation and epoxidation of olefins are available. Aziridines have attracted much attention, especially in the last decade, so there is little doubt that new practical methods for their preparation will be developed and the existing ones will be improved. Meanwhile, aziridines can be very efficiently obtained by manipulation of epoxides [39] or amino alcohols [40] and through haloazidation of olefins followed by reductive cyclization [41].

Two methods that are particularly convenient for large-scale synthesis of aziridines are discussed below. Both utilize readily available chloramine salts, such as chloramine-T, as sources of nitrogen. The first method involves direct olefin aziridination catalyzed by phenyltrimethylammonium tribromide ($PhNMe_3^+Br_3^-$; PTAB) [42]. In the second method, 1,2-hydroxysulfonamides, conveniently obtained by osmium-catalyzed aminohydroxylation of olefins, are converted into aziridines by one-pot cyclodehydration.

12.3.1.1 Bromine-catalyzed Aziridination of Olefins with Chloramines

In the presence of a catalytic amount (10%) of PTAB and anhydrous Chloramine-T (1.1 equiv.), a variety of olefins have been readily converted into the corresponding aziridines in acetonitrile at room temperature (Scheme 12.13). This method exhibits broad substrate scope, and the yields are usually high (Table 12.3) [42].

In many cases products were isolated as off-white solids after simple recrystalli-

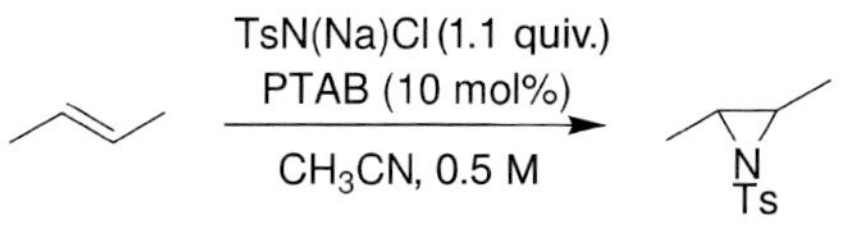

TsN(Na)Cl (1.1 quiv.)
PTAB (10 mol%)
CH_3CN, 0.5 M
N
Ts

Scheme 12.13 Bromine-catalyzed aziridination of olefins.

Table 12.3 Aziridination of olefins with TsNClNa/PTAB.[a]

Entry	Olefin	product	Yield (%)[b]	m.p. (°C)
1	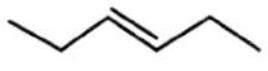	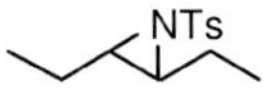 NTs	93 (90)	69–70
2	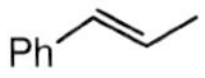 Ph	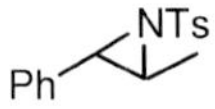 Ph NTs	76 (62)	85–87
3	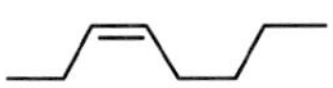	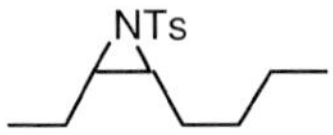 NTs	95 (88)	oil
4	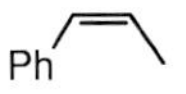 Ph	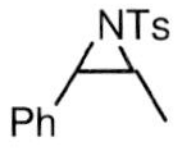 Ph NTs	89 (72)	82–83
5		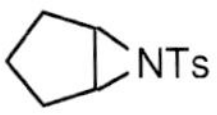 NTs	86 (80)	71–72
6	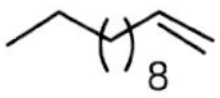 8	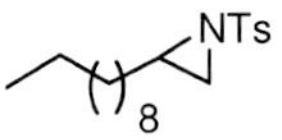 8 NTs	54	oil
7	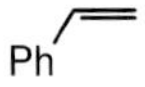 Ph	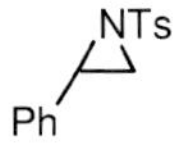 Ph NTs	68 (65)	88–89
8	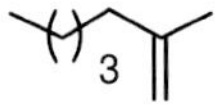 3	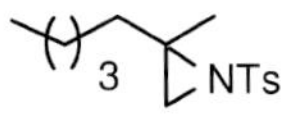 3 NTs	76 (60)	oil
9	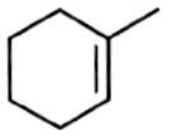	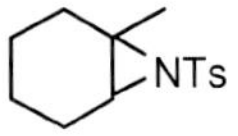 NTs	51	90–91

[a] All reactions were performed on a 3 mmol scale. [b] Isolated yields from silica gel column chromatography after 12 h. Yields in parentheses were obtained with TsNClNa · $3H_2O$.

zation. 1,2-Disubstituted olefins and 2- or 3-substituted allylic alcohols (Table 12.4) were especially good substrates for this catalytic process, providing the corresponding aziridines stereospecifically and in excellent yields. As an example, *cis*- and *trans*-β-methylstyrene exclusively gave the *cis*- and the *trans*-aziridines, respectively.

The commercially available chloramine-T trihydrate (TsNNaCl · $3H_2O$) could also be used directly as the oxidant, although slightly more dilute concentrations (0.2 M vs. 0.5 M) had to be employed to ensure comparable yields. The applicability of this "trihydrate" version to large-scale syntheses was demonstrated by the aziridination of cyclopentene on a 0,5 mol scale reaction, providing 6-tosylazabicyclo-[3.1.0]hexane in 80% isolated yield (Scheme 12.14).

Despite the effectiveness of chloramine-T in this new method, removal of the toluenesulfonyl group from the newly introduced nitrogen substituent requires harsh conditions. The finding that the *N*-chloramine salt of *tert*-butylsulfonamide is also an efficient nitrogen source and the terminal oxidant for aziridination of

Table 12.4 Bromine-catalyzed aziridination of allylic alcohols with anhydrous TsNClNa.[a]

Entry	Olefin	Product	Yield (%)[b]
1	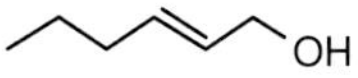 OH	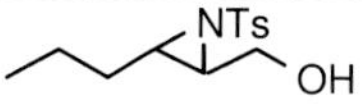 NTs, OH	97
2	 OH	 NTs, OH	95
3	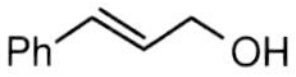 Ph, OH	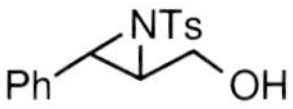 Ph, NTs, OH	70
4	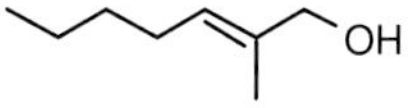 OH	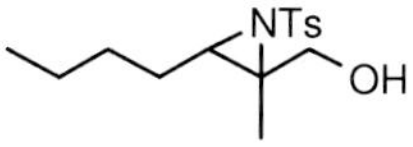 NTs, OH	73
5	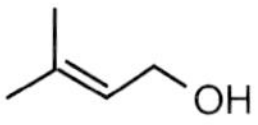 OH	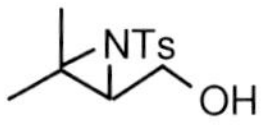 NTs, OH	94
6	 OH	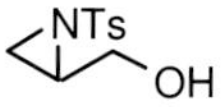 NTs, OH	30
7	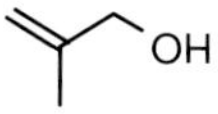 OH	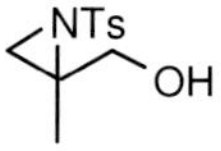 NTs, OH	70
8	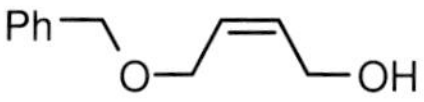 Ph, O, OH	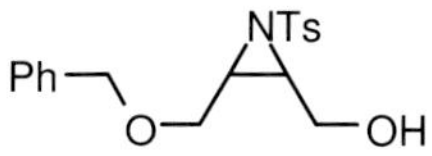 Ph, O, NTs, OH	80
9	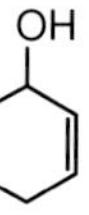 OH	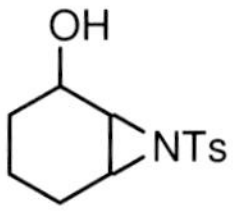 OH, NTs	87[c]

[a] All reactions were performed on a 2 mmol scale. [b] Isolated yields from silica gel column chromatography after 12 h. [c] *Syn/anti* = 2.5/1.0.

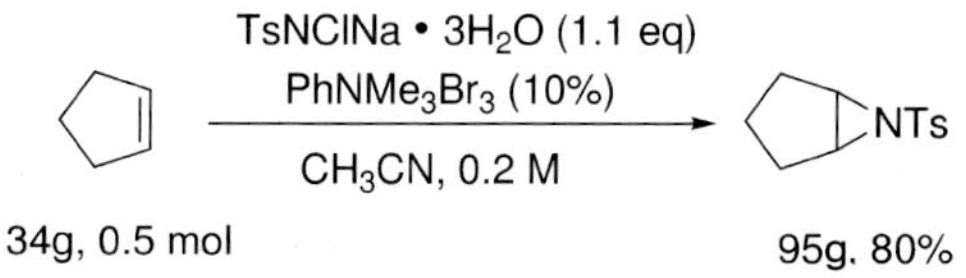

TsNClNa • $3H_2O$ (1.1 eq)
$PhNMe_3Br_3$ (10%)
CH_3CN, 0.2 M
NTs

34g, 0.5 mol

95g, 80%

Scheme 12.14 Aziridination of cyclohexene.

olefins, closely resembling chloramine-T in these catalytic reactions, was therefore a noteworthy improvement [43]. The *tert*-butylsulfonyl group in the products or their derivatives can be easily removed under mild acidic conditions, allowing facile liberation of the amino group [44].

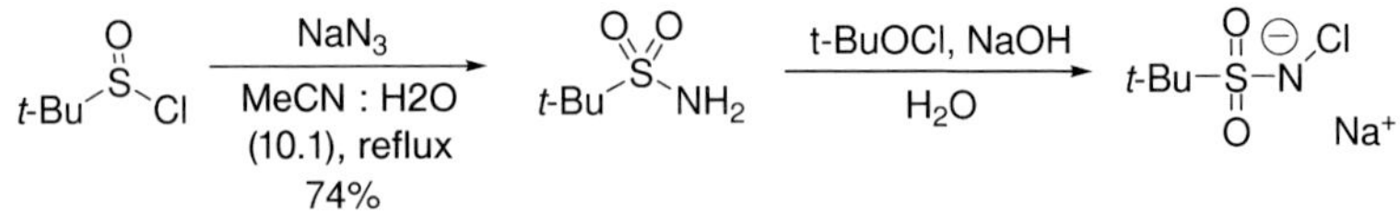

t-Bu–S(O)–Cl → (NaN$_3$; MeCN : H2O (10.1), reflux; 74%) → t-Bu–S(O)$_2$–NH$_2$ → (t-BuOCl, NaOH; H_2O) → t-Bu–S(O)$_2$–N$^-$(Cl) Na$^+$

Scheme 12.15 Synthesis of *N*-chloro-*N*-sodio-*tert*-butylsulfonamide.

Table 12.5 PTAB-catalyzed aziridinations with BusN(Cl)Na.

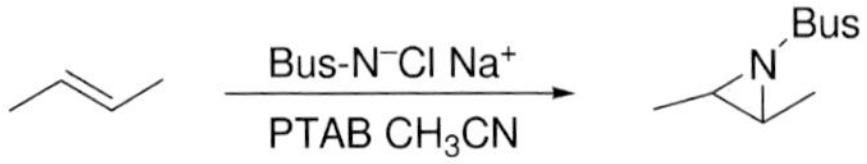

Alkene → (Bus-N$^-$Cl Na$^+$; PTAB CH_3CN) → N-Bus aziridine

Entry	Alkene	Product	Yield, %
1		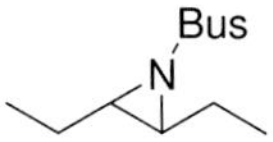 Bus, N	93
2	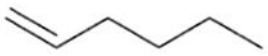	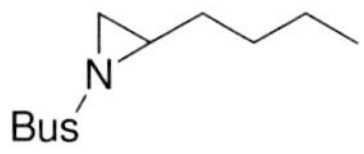 N, Bus	95
3		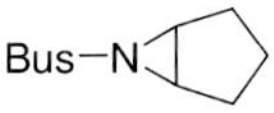 Bus–N	82
4	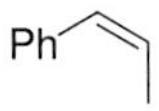 Ph	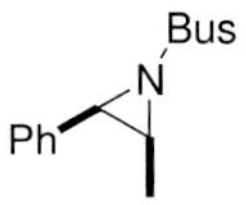 Bus, N, Ph	95
5	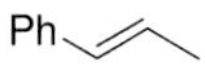 Ph	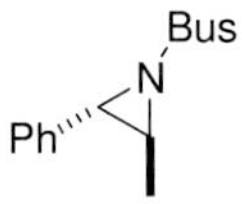 Bus, N, Ph	92
6	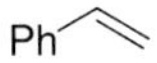 Ph	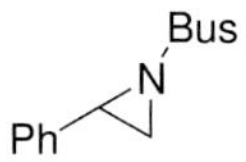 Bus, N, Ph	87
7		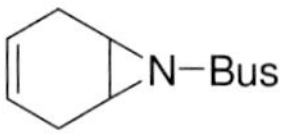 N–Bus	24
8	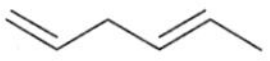	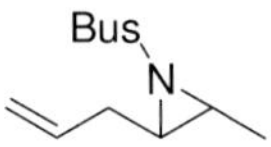 Bus, N	65
9	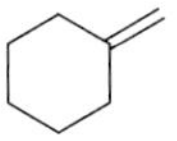	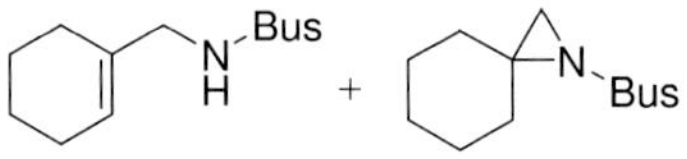 N(H)Bus + N-Bus	24:49

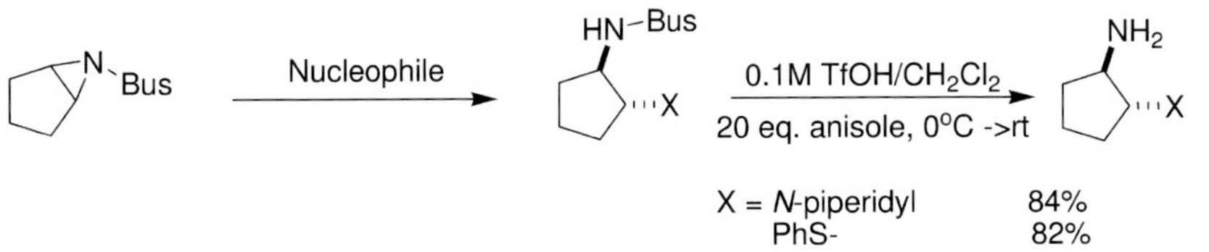

Bus | Nucleophile | HN–Bus | X | 0.1M TfOH/CH_2Cl_2 | 20 eq. anisole, 0°C ->rt | NH_2 | X

X = *N*-piperidyl 84%
PhS- 82%

Scheme 12.16 Nucleophilic opening and deprotection of Bus-aziridines.

tert-Butylsulfonamide, the precursor to *N*-chloro-*N*-sodio-*tert*-butylsulfonamide (BusNClNa), can be prepared by dropwise addition of neat sulfinyl chloride into acetonitrile containing suspended sodium azide and a small amount of water at reflux. Subsequent treatment of the *tert*-butylsulfonamide product with *tert*-butyl hypochlorite and sodium hydroxide generates the corresponding chloramine salt (Scheme 12.15), which can be then used directly as the nitrogen source in the PTAB-catalyzed aziridinations.

As Table 12.5 demonstrates, simple unfunctionalized olefins gave high yields of aziridines, which, in general, were highly crystalline. Recrystallization gave analytically pure products, eliminating the necessity for chromatography. The Bus protecting group in the products of nucleophilic ring-opening was easily removed by treatment with a methylene chloride solution of triflic acid in the presence of anisole (Scheme 12.16) [44]. The resulting free amines were separated from the side products by acid/base extractions.

12.3.2.2 Aminohydroxylation followed by Cyclodehydration

The most useful indirect method for the preparation of aziridines is, arguably, by activation of the hydroxy group of a vicinal amino alcohol, followed by base-promoted ring-closure [45]. Vicinal hydroxysulfonamides, versatile synthetic equivalents of 1,2-aminoalcohols, can be obtained in one step by osmium-catalyzed aminohydroxylation of α,β-unsaturated amides and in two steps from the salts of α,β-unsaturated carboxylic acids [46, 47]. These ligand-independent aminohydroxylation processes are superior to the asymmetric aminohydroxylations with alkaloid ligands (except for enantioselectivity, of course) in the following ways: *i*) they require much lower loading of the osmium catalyst (0.1–1.0 mol% as opposed to 4–5 mol%), *ii*) they need only a stoichiometric amount of the nitrogen source (cf. the >3 moles needed in the AA), *iii*) diol formation, an inevitable side reaction under standard A and AA conditions, and which in the worst cases accounts for as much as 40% of the product, has not been observed, *iv*) a range of solvents (water/*tert*-butanol and water/acetonitrile, for example) can be employed for the reaction, but very importantly, it often proceeds just as well in water without any organic cosolvent, *v*) scale-up is facilitated by the ability to run the reaction at high concentrations (0.3–0.8 M in olefin being typical), whereas the AA requires concentrations <0.1 M for optimal results, and finally, *vi*) since the only co-product is sodium chloride, product isolation is very easy, for most of the α,β-hydroxyaminoacid de-

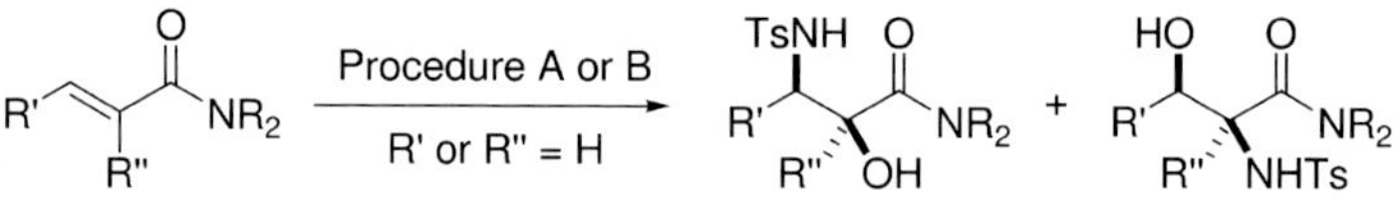

Procedure A (products insoluble in the reaction mixture):
(1) 0.1-1.0 mol% OsO_4 or $K_2OsO_2(OH)_4$, 1.1-1.25 equiv TsNClNa · $3H_2O$, 1:1 t-BuOH-H_2O, 0.5-1 M, RT;
(2) Filtration

Procedure B (products soluble in the reaction mixture):
(1) 0.1-1.0 mol% OsO_4 or $K_2OsO_2(OH)_4$, 1.0 equiv TsNClNa · $3H_2O$, 1:1 MeCN-H_2O, 0.5 M, RT;
(2) Na_2SO_3

Scheme 12.17 Aminohydroxylation of α,β-unsaturated amides.

rivatives precipitate upon acidification and are collected by simple filtration as pure materials.

Two procedures have been developed for the aminohydroxylation of α,β-unsaturated amides: Procedure A for products that are insoluble in the reaction mixture and Procedure B for soluble products (Scheme 12.17) [48]. These differ only in that the former requires a 10–25% excess of chloramine-T and *t*-BuOH as the cosolvent, while the latter uses only one equivalent of the chloramine salt and MeCN as the cosolvent. The excess of chloramine-T in Procedure A allows better turnover near the end of the reaction, and the trace amount of *p*-toluenesulfonamide by-product can be removed by recrystallization. However, elimination of the necessity to remove *p*-toluenesulfonamide far outweighed the inconvenience of slightly longer reaction times needed in procedure **B** without the use of excess chloramine salt.

As can be seen from the examples in Tables 12.6–12.8, excellent yields of the hydroxysulfonamide products were obtained from each of the olefins examined. It is noteworthy that these high yields were achieved at high substrate concentrations and room temperature in the presence of as little as 0.10 mol% catalyst.

Both regioisomers were observed in aminohydroxylation of almost all the substrates that were examined. By taking advantage of their high combined yields, as well as the racemic nature of the aminohydroxylation products, a one-pot, two-step synthesis of sulfonyl aziridines through the cyclodehydration of hydroxysulfonamides was developed (Scheme 12.18).

The strategy involved the activation of the hydroxy group as the mesylate in the first step and subsequent ring-closure with DBU as the base. Small-scale reactions

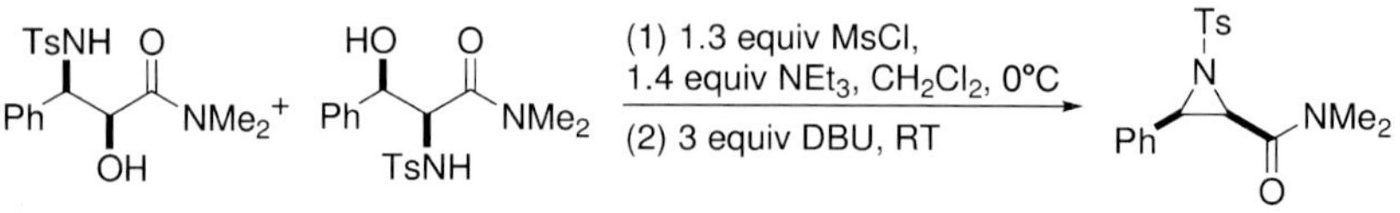

(1) 1.3 equiv MsCl, 1.4 equiv NEt_3, CH_2Cl_2, 0°C
(2) 3 equiv DBU, RT

Scheme 12.18 Synthesis of *N*-sulfonyl aziridines from 1,2-hydroxysulfonamides with DBU as a base.

Table 12.6 Aminohydroxylation of unsubstituted cinnamamides by Procedure A [46].

Entry	Substrate	Scale (mmol)	Catalyst (mol%)	Combined yield (regioselectivity)[a]
1	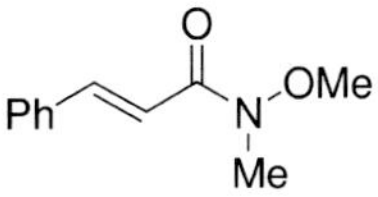 Ph, O, N, OMe, Me	50	OsO_4 (0.1 %)	93 % (87:13)
2	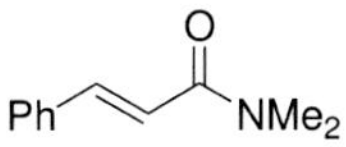 Ph, O, NMe_2	571	$K_2OsO_2(OH)_4$ (0.25 %)	95 % (72:28)
3	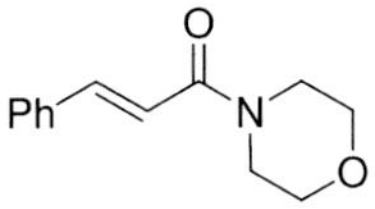 Ph, O, N, O	143	$K_2OsO_2(OH)_4$ (0.25 %)	97 % (68:32)
4	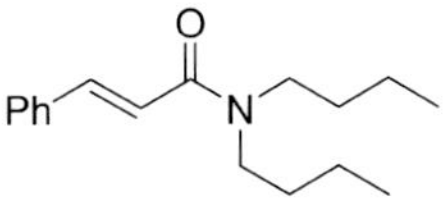 Ph, O, N	100	$K_2OsO_2(OH)_4$ (0.25 %)	96 % (71:29)
5	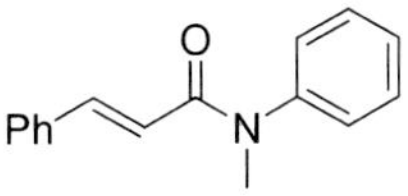 Ph, O, N	100	$K_2OsO_2(OH)_4$ (0.25 %)	97 % (89:11)
6	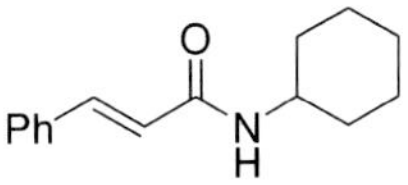 Ph, O, N, H	184	$K_2OsO_2(OH)_4$ (0.25 %)	96 % (58:42)
7	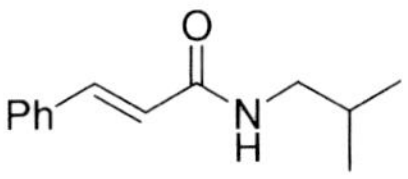 Ph, O, N, H	200	$K_2OsO_2(OH)_4$ (0.25 %)	95 % (50:50)
8	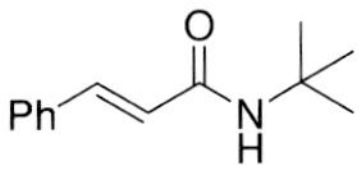 Ph, O, N, H	100	$K_2OsO_2(OH)_4$ (0.25 %)	94 % (62:38)

[a] Yield is of the mixture of regioisomers isolated by filtration. Regioselectivity refers to the ratio of **12**:**13**, determined by ^1H NMR spectroscopy.

Table 12.7 Aminohydroxylation of substituted cinnamamides by Procedure A [46].

Entry	Substrate	Scale (mmol)	Catalyst (mol%)	Combined yield (regioselectivity)[a]
1	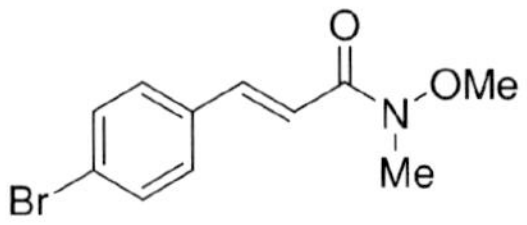 Br, O, N, OMe, Me	20	OsO_4 (0.25%)	95% (79:21)
2	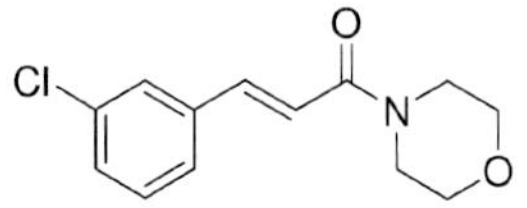 Cl, O, N, O	60	$K_2OsO_2(OH)_4$ (0.25%)	97% (64:36)
3	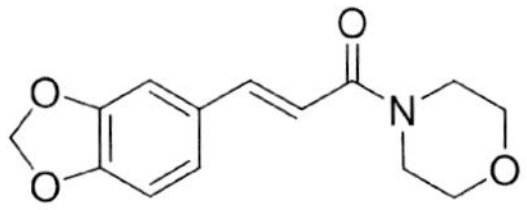 O, O, O, N, O	58	$K_2OsO_2(OH)_4$ (0.25%)	80% (67:33)[b]
4	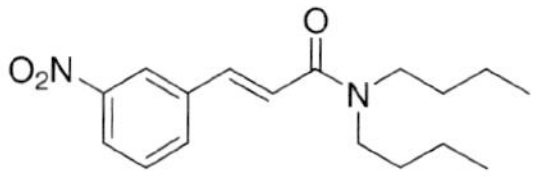 O_2N, O, N	73	$K_2OsO_2(OH)_4$ (0.25%)	92% (80:20)
5	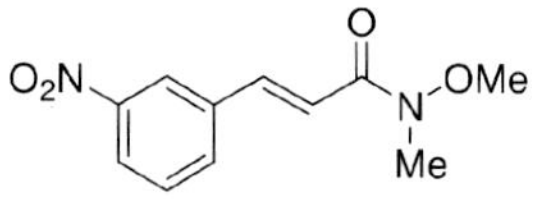 O_2N, O, N, OMe, Me	20	OsO_4 (0.5%)	95% (85:15)

[a] Yield is of the mixture of regioisomers isolated by filtration. Regioselectivity refers to the ratio of **12**:**13**, determined by ^{1}H NMR spectroscopy. [b] Yield is based on RCO_2H for the three-step sequence: $RCO_2H \rightarrow RCOCl \rightarrow RCONR_2 \rightarrow$ regioisomeric hydroxy-sulfonamides.

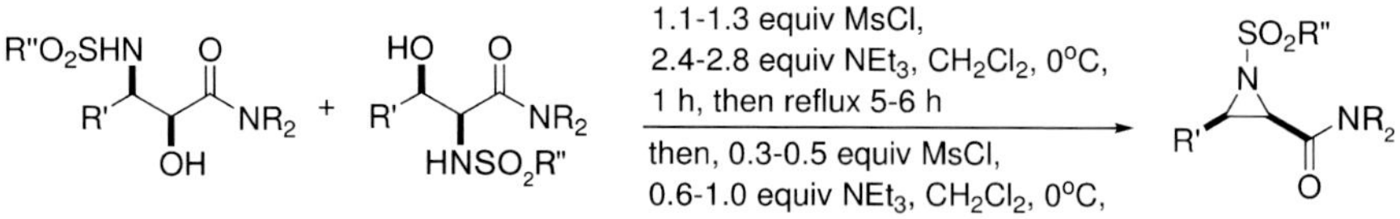

R"O_2SHN, O, R', NR_2, OH (**14**) + HO, O, R', NR_2, HNSO$_2$R" (**15**) → 1.1-1.3 equiv MsCl, 2.4-2.8 equiv NEt_3, CH_2Cl_2, 0°C, 1 h, then reflux 5-6 h; then, 0.3-0.5 equiv MsCl, 0.6-1.0 equiv NEt_3, CH_2Cl_2, 0°C, then, reflux 5-6 h → SO_2R", N, R', NR_2, O (**16**)

Scheme 12.19 Optimized synthesis of *N*-sulfonyl aziridines from 1,2-hydroxysulfonamides.

(1–5 mmol) gave sulfonyl aziridines in excellent yields (80–95%). Process optimization to avoid the use of the large excess of DBU and to adapt the method for large-scale preparations was carried out, resulting in the replacement of DBU with triethylamine (Scheme 12.19). The hydroxysulfonamide isomers were treated with triethylamine and mesyl chloride in two portions. After the reaction was complete, a small amount of silica gel (2–3 g for 40–50 g of aziridine) was added to the

Table 12.8 Aminohydroxylation of unsaturated amides by Procedure B [46].

Entry	Substrate	Scale (mmol)	Catalyst (mol%)	Combined yield (regioselectivity)[a]
1	O NMe$_2$	49	$K_2OsO_2(OH)_4$ (0.5%)	99% (91:9)
2	O Me N O	10	$K_2OsO_2(OH)_4$ (0.5%)	97% (42:58)
3	O NMe$_2$	10	$K_2OsO_2(OH)_4$ (1%)	90% (>20:1)
4	O N OMe Me	10[b]	OsO_4 (0.5%)	99% (75:25)

[a] Yield is of the mixture of **12** and **13** isolated by filtration. Regioselectivity refers to the ratio of **12**:**13**, determined by ^{1}H NMR spectroscopy. [b] *t*-BuOH was used as the cosolvent.

concentrated mixture. The impurities retained in the silica gel were removed by filtration. The isolated product, as a solid, could be further purified by trituration with diethyl ether or recrystallization from ethanol, if necessary. When *p*-nitrobenzenesulfonyl aziridines were prepared, the final recrystallization was either omitted or performed from ethyl acetate, since these aziridines are known to solvolyze in boiling alcohols [49].

The generality of this method for large-scale preparations was demonstrated for the syntheses of toluenesulfonyl aziridines from aminohydroxylation products of *N*-substituted cinnamamides (Table 12.9). The results of synthesis of other sulfonyl aziridines are summarized in Tables 12.10 and 12.11.

Table 12.9 Large-scale synthesis of toluenesulfonyl.[a]

Entry	Substrate ratio, (method)[b]	Aziridine	Scale (mmol)	Yield[c] (mp)
1	72:28 (Method B)	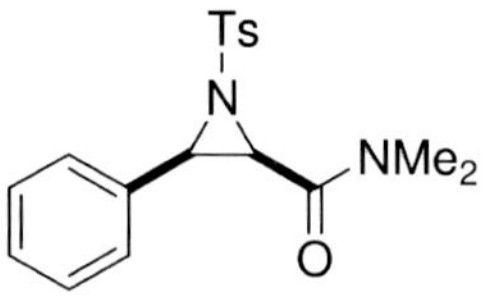	138	87% (130–131 °C)
2	68:32 (Method B)	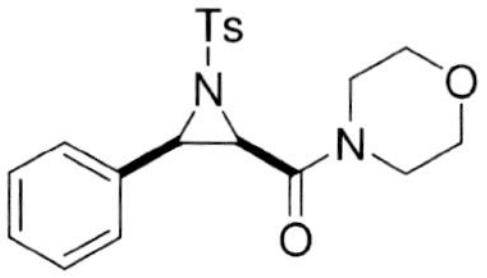	124	84% (132–133 °C)
3	89:11 (Method B)	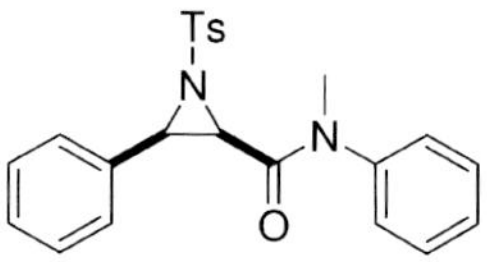	95	82% (141–142 °C)
4	69:31 (Method B)	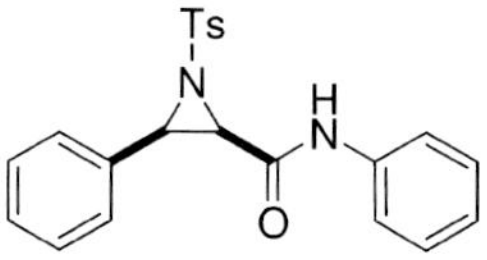	36	80% (199–202 °C)

[a] Reaction conditions according to Schemes 12.18 and 12.19. [b] Method A: DBU, Method B: excess MsCl and Et_3N. [c] Yield of the pure aziridine after recrystallization from EtOH. [d] Yield of the pure aziridine after flash chromatography.

Table 12.10 Toluenesulfonyl aziridines.[a]

Entry	Substrate ratio, (method)[b]	Aziridine	Yield[c] (mp)
1	79:21 (Method A)	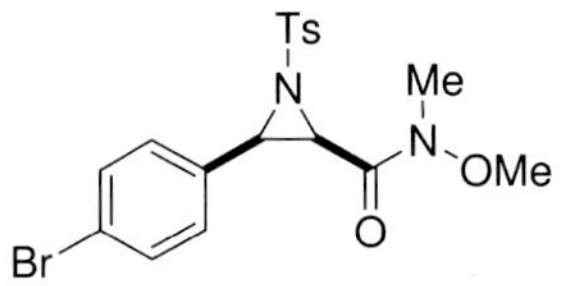	97 %[d] (50–53 °C)
2	64:36 (Method B)	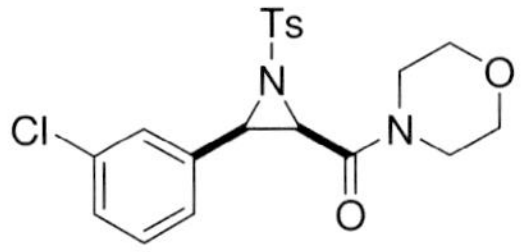	77 %[c] (124–126 °C)
3	85:15 (Method A)	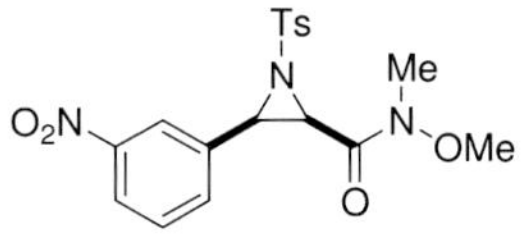	94 %[d] (127–128 °C)
4	80:20 (Method B)	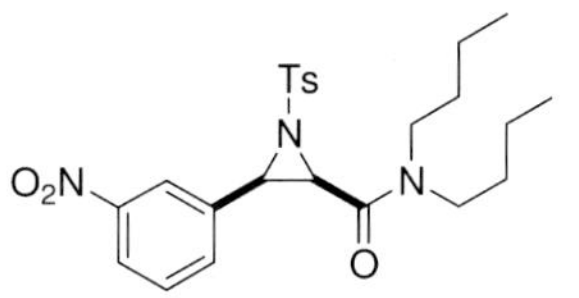	63 %[c] (114–116 °C)
5	91:9 (Method A)	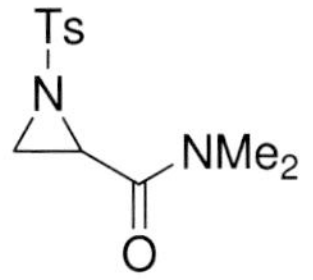	81 %[d] (oil)
6	42:58 (Method A)	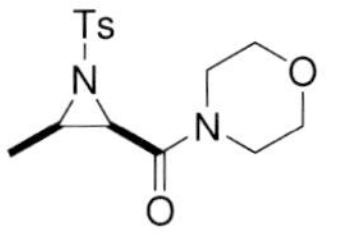	92 %[d] (151–152 °C)
7	75:25 (Method A)	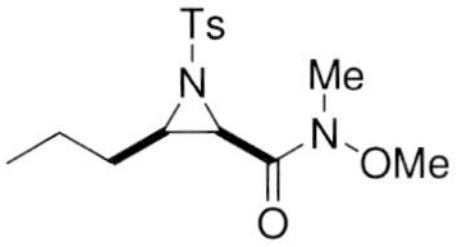	87 %[d] (oil)

[a] Reaction conditions according to Schemes 12.18 and 12.19. [b] Method A: DBU, Method B: excess MsCl and Et_3N. [c] Yield of the pure aziridine after recrystallization from EtOH. [d] Yield of the pure aziridine after flash chromatography.

Table 12.11 Arylsulfonyl aziridines[a]

Entry	Substrate ratio	Aziridine	Yield (mp)
1	96:4	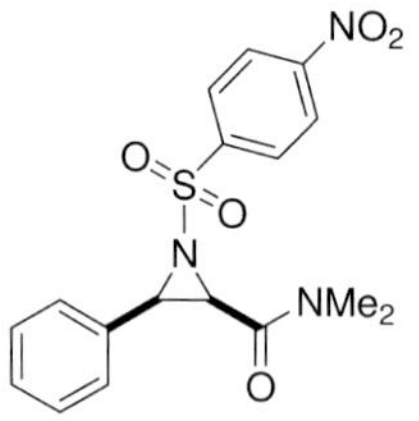	90 %[d] (138–140 °C)
2	73:27	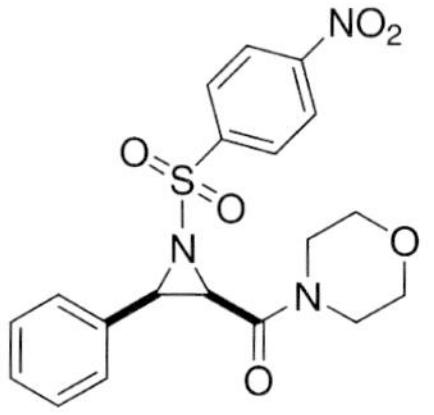	95 %[d] (166–168 °C)
3	56:44	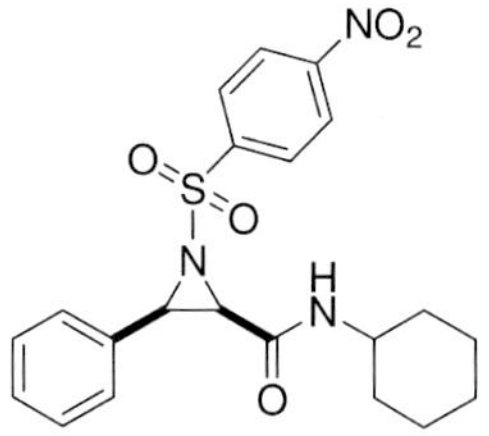	71 %[c] (185–187 °C)
4	55:45	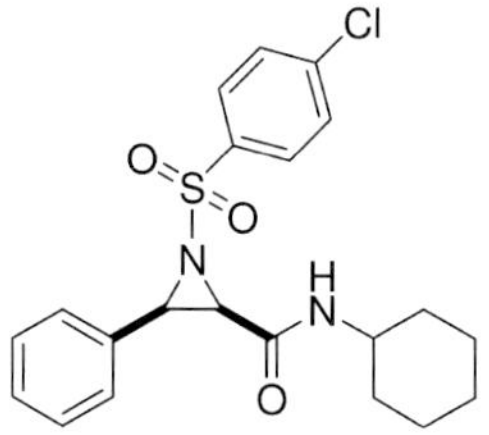	65 %[b] (187–188 °C)
5	55:45	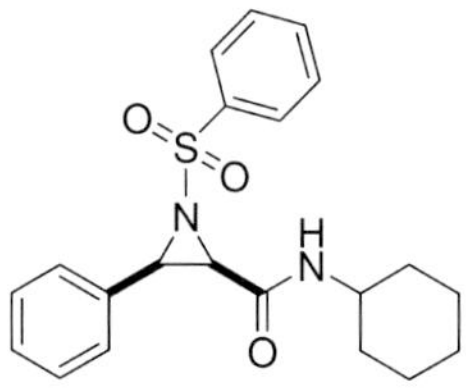	56 %[b] (146–147 °C)

[a] Reaction conditions according to Scheme 12.19. [b] Yield of the pure aziridine after recrystallization from EtOH. [c] Yield of the pure aziridine after recrystallization from EtOAc. [d] Yield of the pure aziridine after trituration with Et_2O.

12.3.2 Nucleophilic Opening of Aziridines

As mentioned above, the regioselectivity of ring-opening of unsymmetrical aziridines can be controlled by nitrogen substituent and the nature of the solvent. As an example, Stamm [37] observed that when a tosyl-protected aziridine was treated with thiophenol, attack at the less hindered carbon occurred, whereas when the benzoyl analogue was used, addition at the more substituted carbon was favored (Scheme 12.20). Effects deriving from the tendency of sulfonamides to be pyramidal at nitrogen and amides to be planar cause *N*-sulfonyl aziridines to furnish the regioisomer derived from attack of the nucleophile at the sterically less hindered center, while *N*-acyl aziridines favor the opposite regioisomer.

Similarly, when aziridines **16** were opened with thiolates, exclusive addition at the carbon proximal to the carbonyl was observed (Scheme 12.21). The best results were obtained when DMF was used as the solvent.

The ring-opening reactions of aziridines **16** with amines deserve special attention. Unlike ring-openings with the other nucleophiles examined, they do not require added base, and they do not produce stoichiometric byproducts (as with azide). In certain cases, they do not even require a solvent, as exemplified by the ring-opening of aziridine **18** (Table 12.12, entries 1–3). In addition, the reactions

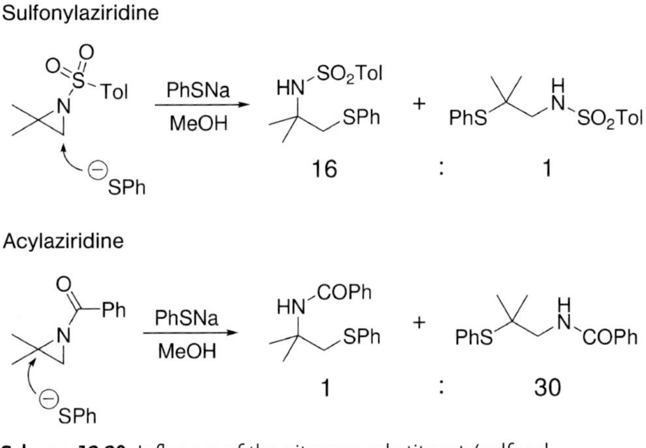

Scheme 12.20 Influence of the nitrogen substituent (sulfonyl, acyl) on the regioselectivity of aziridine opening.

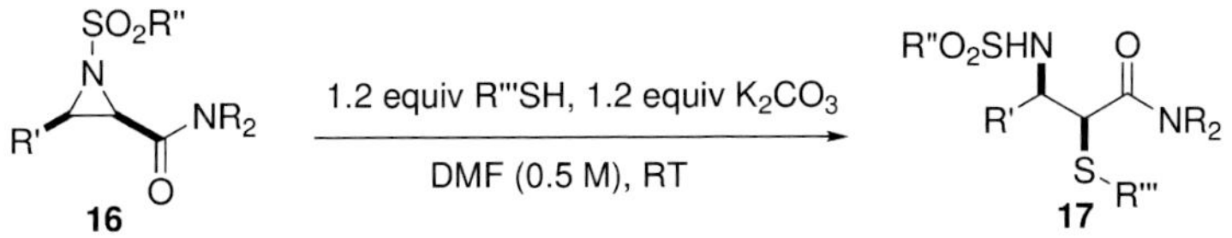

Scheme 12.21 Opening of sulfonyl aziridines with thiolates.

Table 12.12 Nucleophilic ring-opening of aziridine **18** with amines.

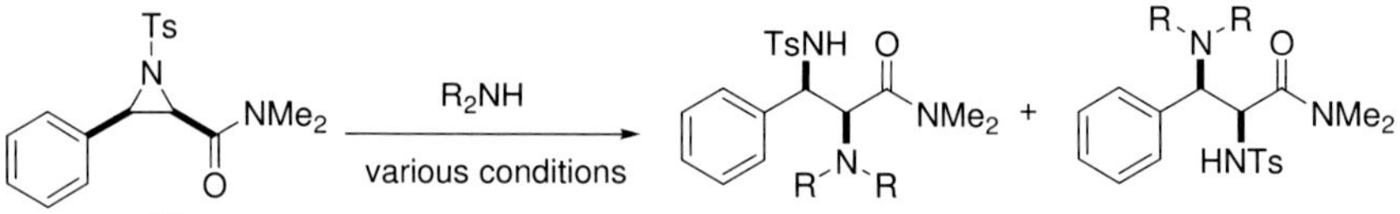

$$\mathbf{18} \xrightarrow[\text{various conditions}]{R_2NH} \mathbf{19} + \mathbf{20}$$

Entry	Amine	Regioselectivity (19:20) reaction method[a]	Product, yield (mp)[b]
1	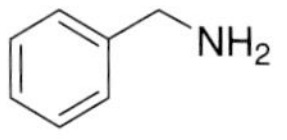 benzylamine ($PhCH_2NH_2$)	9:91 Method A	**20.1**, 67% (153–154 °C)
2	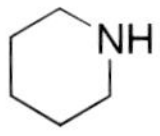 piperidine	9:91 Method A	**20.2**, 71% (175–176 °C)
3	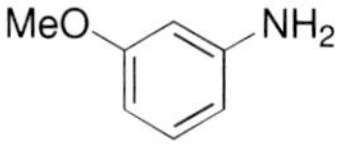 3-methoxyaniline	7:93 Method A	**20.3**, 71% (149–152 °C)
4	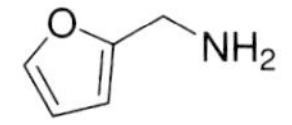 furfurylamine	13:87 Method B	**20.4**, 69% (160–161 °C)
5	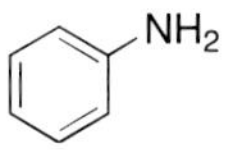 aniline	10:90 Method C	**20.5**, 65% (155–156 °C)

[a] Reaction conditions: **A**: *ca.* 5 equiv. amine, no solvent, 100 °C. **B**: 1.2 equiv. amine, *n*-PrOH (1 M), 80 °C; **C**: 1.2 equiv. amine, DMF (1 M), 80 °C. [b] Yield and melting point of the pure product **20** after recrystallization from *i*-PrOH (Entries 3–5) or EtOAc/hexanes (Entries 1,2).

are insensitive to air and water, which makes them ideal candidates for parallel synthesis applications.

Further investigations of the ring-opening of aziridine **18** with amines revealed that regioisomer **20** was formed preferentially in most polar protic solvents. In a subsequent study, the regioselective ring-opening of aziridines **21–23** with amines was exploited in the preparation of a chemical library of α,β-diaminoamides (Scheme 12.22), which was accomplished by automated parallel solution-phase techniques [50]. The insoluble nature of the ring-opened products in *i*-PrOH makes filtration an ideal choice for product isolation. The amines were chosen from a basic 45-member palette of 15 piperazines and morpholines, 18 alkyl primary amines, and 11 anilines [48].

Of the 135 possible ring-opened products, 102 crystallized and were collected by

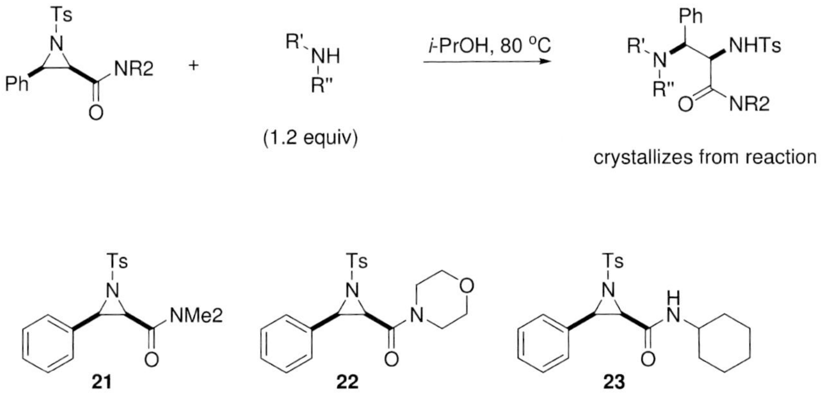

Scheme 12.22 Synthesis of a library of diamines from sulfonyl aziridines.

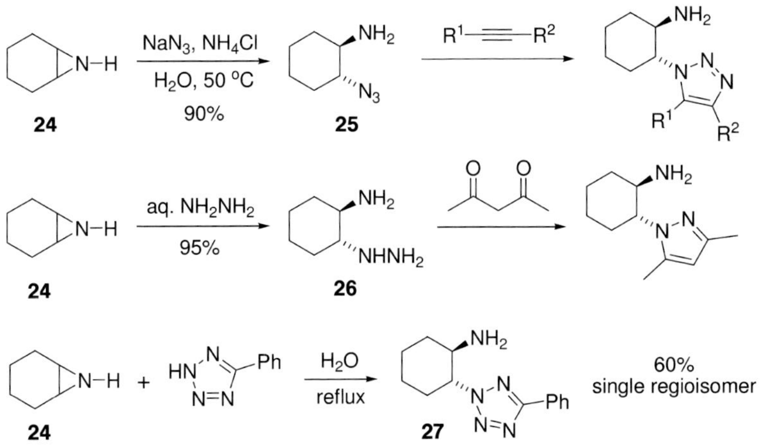

Scheme 12.23 Nucleophilic ring-opening of unactivated aziridines.

filtration. Most of them were obtained with purities greater than 95%. The isolated yields for this subset of the library were mostly above 60%.

Unactivated aziridines, such as **24**, are not as reactive as their *N*-sulfonyl analogues. Nevertheless, in aqueous conditions they react with different nucleophiles, as Scheme 12.23 illustrates. Treatment with buffered azide at 50 °C gave **25** in 90% yield. Hydrazine proved potent even at room temperature and **26** was formed in 95% yield, while phenyltetrazole required heating at reflux in water. The resulting amines participated in dipolar cycloadditions with alkynes and condensations with β-diketones.

12.4 Aziridinium Ions in Click Chemistry

Aziridinium ions have received much attention in biological studies [51]. Their intermediacy was postulated in the DNA cross-linking induced by nitrogen mustard agents [52], but aziridinium ions did not find much use in organic synthesis until recent reports described methods for their generation under mild conditions followed by capture with nucleophiles [53]. These facile openings are in direct contrast to the ring openings of their parent aziridines, the activation of which usually requires protonation or Lewis acid complexation.

12.4.1 Generation of Aziridinium Ions

Perhaps the most straightforward method for the generation of aziridinium ions is to transform the hydroxy group of a vicinal tertiary amino alcohol into a good leaving group. By this strategy, epoxy ester **28** was transformed into the chloroamines **31** via the intermediate amino alcohols **29** and **30** as outlined in Scheme 12.24. Treatment of epoxy ester **28** with secondary amines (morpholine, 1-phenylpiperazine, diallylamine, and dibenzylamine) at reflux in EtOH generated a mixture of amino alcohols **29** and **30** (Table 12.13). Mesylation of the crude mixtures of **29** and **30** converged both regioisomers to the rearranged chloroamines **31** in near quantitative yields. The structure of chloroamine **31a** was confirmed by X-ray analysis. The stable, crystalline β-chloroamine **31** then serves as the source for regeneration, on demand, of aziridinium ion intermediate **32**.

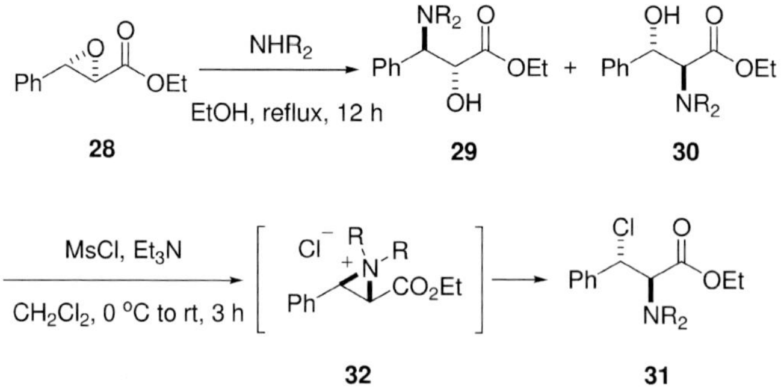

Ph OEt O O **28** — NHR_2, EtOH, reflux, 12 h → Ph NR_2 O OEt OH **29** + Ph OH O OEt NR_2 **30**

— $MsCl$, Et_3N, CH_2Cl_2, 0 °C to rt, 3 h → [Cl^- R ^+N R Ph CO_2Et **32**] → Ph Cl O OEt NR_2 **31**

Scheme 12.24 Generation of aziridinium ions from epoxides.

Table 12.13 Synthesis of chloroamines **31**.

Entry	—NR$_2$	Amino alcohol 29 & 30 Yield [%][a]; 29:30[b]	Chloroamine 31; Yield [%][a]
1	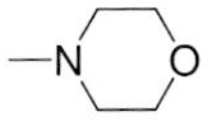 —N O	96; (87:13)	**31a**; 96
2	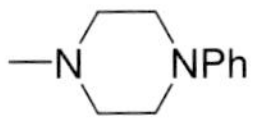 —N NPh	99; (87:13)	**31b**; 99
3	—N(allyl)$_2$	94; (88:12)	**31c**; 99
4	—NBn$_2$	97; (88:12)	**31d**; 94

[a] Isolated yield. [b] Analysis by HPLC (Zorbax SB-C18 reverse analytical column, 150 × 4.6 mm; Gradient eluent 80/20 to 0/100 H_2O/MeCN containing 0.1% TFA, 0.5 mL/min for 15 min) by intergration of absorption at 254 nm.

12.4.2
Nucleophilic Opening of Aziridinium Ions

As illustrated above, highly activated aziridinium systems can be easily generated *in situ* from β-halo amines in the course of neighboring group-assisted nucleophilic substitutions. All these substitutions are stereospecific and proceed with double inversion, and hence net retention of configuration when the nucleophilic attack occurs at the center bearing the leaving group. Inversion of stereochemistry at both carbons was observed in cases in which the amino group underwent a 1,2-shift. Furthermore, with chloride as the counterion, the ring-opening of aziridinium exhibits high yield and selectivity in the aqueous medium. In fact, these

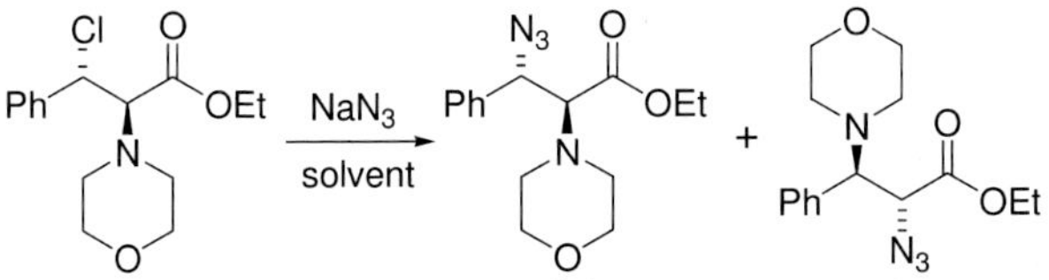

NaN$_3$, solvent → **C-3** + **C-2**

solvent, temperature, time	yield (%)	(**C-3**:**C-2**)
CH_3CN, 60 °C, 12 h	85	(60:40)
EtOH/H_2O (4/1), 60 °C, 12 h	97	(88:12)
EtOH/H_2O (1/4), 60 °C, 6 h	92	(93:7)
H_2O, 60 °C, 6 h	>99	(95:5)

Scheme 12.25 Nucleophilic opening of aziridinium ions with azide.

reactions benefit from being performed "on water", without any organic co-solvent. The beneficial effects of this approach are demonstrated in Scheme 12.25, with azide acting as a nucleophile. As the water content of the reaction mixture increases, both yield and selectivity are improved.

Taken together, the broad scope of these reactions and the easy access to the starting materials from the corresponding epoxides makes aziridinium ions ideal click intermediates.

12.4.2.1 Synthesis of Diamino Esters and β-Lactams

α,β-Diamino esters are key intermediates in the synthesis of β-lactams [54], but the common route to α,β-diamino esters – direct condensation of ester enolates with imines [55, 56] – is complicated by the poor stereochemical outcome, and the reaction often proceeds directly on to the β-lactam. With amines as nucleophiles, ring-opening of aziridium ion **32** provides an alternative stereo- and regioselective route to α,β-diamino esters. Excellent results were obtained with a wide variety of amines, and the ring-opening products at the benzylic position were formed preferentially, with regioselectivities greater than 91%. In addition, the reactions were stereospecific in all cases, giving only the *anti* (*erythro*) relationship (Table 12.14).

Table 12.14 α,β-Diamino esters via aziridinium ion **32**.

entry	NuH	yield of 33[a] [%]
1	n-butylamine ($CH_3CH_2CH_2CH_2NH_2$)	85
2	isopropylamine ($(CH_3)_2CHNH_2$)	88
3	tert-butylamine ($(CH_3)_3CNH_2$)	80
4	piperidine	84
5	1-phenylpiperazine	93
6	aniline ($C_6H_5NH_2$)	79
7	NH_3	81

[a] Isolated yield for both regioisomers after filtration and drying to constant weight.

Table 12.15 Synthesis of β-lactams.

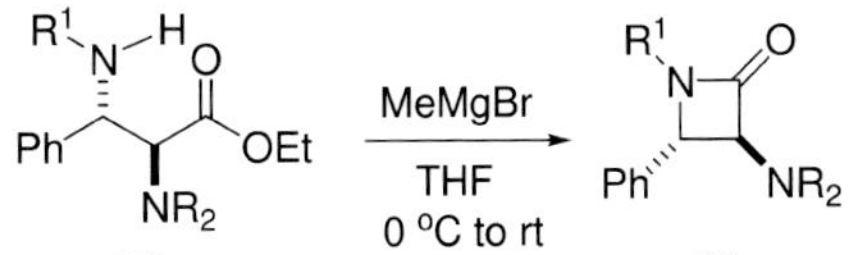

33 → MeMgBr, THF, 0 °C to rt → **34**

Entry	R^1	$—NR_2$	Product Yield [%][a]
1	*t*-Bu	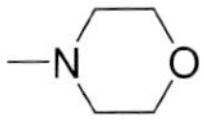 —N O	**34a**
2		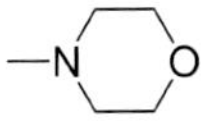 —N O	**34b**
3	*n*-Bu	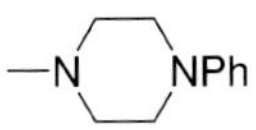 —N NPh	**34c**

[a]Isolated yield.

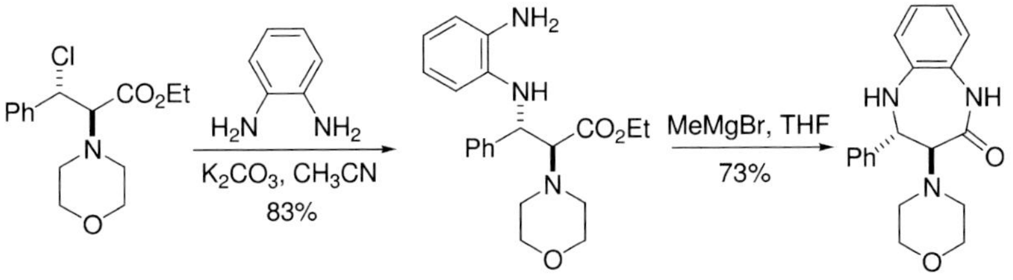

31a → H_2N, NH_2; K_2CO_3, CH_3CN, 83% → **35** → MeMgBr, THF, 73% → **36**

Scheme 12.26 Synthesis of benzodiazepine derivatives.

Ring-opening products derived from primary amines are attractive precursors for the preparation of β-lactams [57]. With methylmagnesium bromide as the base, diamino esters **33** cyclized readily in THF and stereospecifically generated *anti*-3-amino-β-lactams **34** (Table 12.15).

Another example of the utility of this transformation is the facile preparation of 1,5-benzodiazepine derivatives (Scheme 12.26).

12.4.2.2 **Synthesis of Pyrazolo[1,2-α]pyrazoles**

Aziridinium ion-based click chemistry provides convenient access to pyrazolo[1,2-α]pyrazoles, active inhibitors of penicillin-binding proteins [58, 59]. Ring-opening of aziridinium ions **32** at the benzylic position with hydrazine, followed by intramolecular cyclization, gave pyrazolidin-3-ones **37** in excellent yields (Scheme 12.27). Heating of the hydrazides **37** with aromatic aldehydes at reflux in absolute

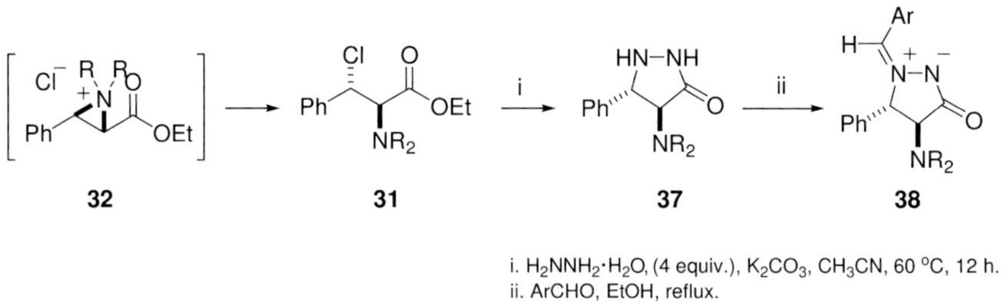

32
31
37
38
i. $H_2NNH_2 \cdot H_2O$, (4 equiv.), K_2CO_3, CH_3CN, 60 °C, 12 h.
ii. ArCHO, EtOH, reflux.

Scheme 12.27 Synthesis of azomethine ylides **38**.

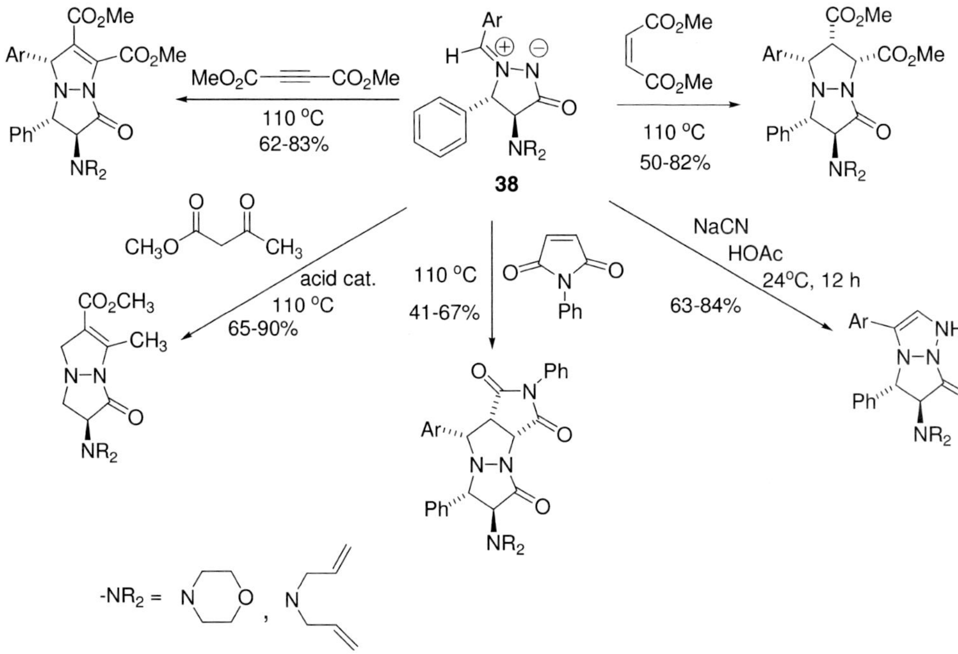

38
110 °C
62-83%
110 °C
50-82%
acid cat.
110 °C
65-90%
110 °C
41-67%
NaCN
HOAc
24°C, 12 h
63-84%

Scheme 12.28 Synthesis of pyrazolo[1,2-α]pyrazoles from azomethine ylides.

ethanol in the presence of catalytic CF_3COOH furnished pyrazolidin-1-ium-2-ides **38**, the 1,3-dipoles for the creation of the pyrazolo[1,2-α]pyrazole skeleton [60].

Azomethine ylides **38** reacted with a variety of dipolarophiles to generate cycloadducts in a highly diastereo- and regioselective manner. The novel shapes, along with the rich array of functionalities displayed by these products, provide opportunities for the creation of chemical diversity (Scheme 12.28).

References

1 R. S. M. Bohacek, W. C. Guida, *Med. Res. Rev.* **1996**, 16 (1), 3–50.
2 H. C. Kolb, M. G. Finn, K. B. Sharpless, *Angew. Chem. Int. Ed.* **2001**, 40 (11), 2004–2021.
3 S. Narayan, J. Muldoon, H. C. Kolb, M. G. Finn, V. V. Fokin, K. B. Sharpless *Angew. Chem. Int. Ed.*, **2005**, 44 (21), 3275.
4 H. A. Wittcoff, B. G. Reuben, in *Industrial Organic Chemicals*, Wiley, New York, **1996**.
5 V. V. Rostovtsev, L. G. Green, V. V. Fokin, K. B. Sharpless, *Angew. Chem. Int. Ed.* **2002**, *41*, 2596.
6 H. C. Kolb, K. B. Sharpless, *Drug Discov. Today*, **2003**, 8, 1128–1137.
7 B. Kadis, *J. Steroid Biochem.* , **1978**, **9**, 75
8 R. G. Harvey, *Synthesis*, **1986**, 605.
9 E. G. Lewars, in *Comprehensive Heterocyclic Chemistry* (Ed. A. R. Katritzky), Pergamon Press, Oxford, **1984**, vol. 7. p. 95; *Modern Oxidation Methods*, (Ed. J. E. Bäckvall), Wiley, New York, 2004.
10 T. Katsuki, K. B. Sharpless *J. Am. Chem. Soc.* **1980**, *102*, 5974.
11 For reviews see: (a) M. G. Finn, K. B. Sharpless, *Asymmetric Synth.* **1985**, *5*, 247–308; (b) K. B. Sharpless, *Chemistry in Britain* **1986**, *22*, 38–40, 43–44; (c) R. A. Johnson, K. B. Sharpless, in *Catalytic Asymmetric Synthesis* (Ed. I. Ojima), VCH, New York **1993**, chapter 4. 1
12 E. N. Jacobsen, in *Catalytic Asymmetric Synthesis* (Ed. I. Ojima), VCH, New York **1993**, chapter 4. 2
13 M. Frohn, Y. Shi, *Synthesis* **2000**, 1979.
14 (a) I. Erden, in "*Comprehensive Heterocyclic Chemistry II*" (Ed. A. Padwa), Pergamon, New York, **1996**, Vol. 1A, pp. 97–144. (b) K. B. Sharpless, T. R. Verhoeven, *Aldrichimica Acta*, **1979**, *12*, 63
15 (a) T. Katsuki, *Catalytic Asymmetric Synthesis (2nd Edition)* **2000**, 287–325; (b) L. Pu. *Tetrahedron Asymm*, **1998**, *9*, 1457
16 A. S. Rao, in *Comprehensive Organic Synthesis* **1992** (Ed. B. M. Trost), Pergamon Press, Oxford, chapter 3. 1
17 N. Prilezhaev, *Ber.* **1909**, 42, 4811.
18 D. Swern, *Chem. Rev.* **1949**, *45*, 16; *Org. React.* **1953**, *7*, 378.
19 (a) F. Camps; J. Coll; A. Messeguer; F. Pujol *J. Org. Chem.* **1982**, *47*, 5402; (b) M. Imuta; H. Ziffer *J. Org. Chem.* **1979**, *44*, 1351; (c) W. K. Anderson; T. Veysoglu *J. Org. Chem.* **1973**, *38*, 2267.
20 The Payne epoxidation with benzonitrile/hydrogen peroxide is also an efficient epoxidation process. It is often the method of choice for industrial batch-type applications, but on a small scale the need for continuous pH control is inconvenient. G. B. Payne, *Tetrahedron* **1962**, *18*, 763
21 R. W. Murray, S. Singh, *Org. Syntheses* **1996**, *74*, 91.
22 W. Adam, L. Hadjiarapoglou, V. Jagar, B. Seidel, *Tetrahedron Lett.* **1989**, *30*, 4223
23 W. Adam, L. Hadjiarapoglou, X. Wang, *Tetrahedron Lett.* **1989**, *30*, 6497
24 R. L. Halcomb, S. J. Danishefsky, J. Am. Chem. Soc. 1989, 111, 6661.
25 (a) W. A. Herrmann, R. W. Fischer, D. W. Marz. *Angew. Chem.* , *Int. Ed. Engl.* **1991**, *30*, 1638. (b) W. A. Herrmann; R. W. Fischer; M. U. Rauch; W. Scherer *J. Mol. Catal.* **1994**, *86*, 243. (c) W. A. Herrmann *J. Organomet. Chem.* **1995**, *500*, 149.
26 J. Rudolph, K. L. Reddy, J. P. Chiang, K. B. Sharpless, *J. Am. Chem. Soc.* **1997**, *119*, 6189.
27 C. Copéret, H. Adolfsson, K. B. Sharpless, *Chem. Comm.* , **1997**, *16*, 1565.
28 (a) A. K. Yudin, K. B. Sharpless, *J. Am. Chem. Soc.* **1997**, *119*, 11536; (b) A. K. Yudin; J. P. Chiang, H. Adolfsson, C. Copéret, *J. Org. Chem.* **2001**, *66*, 4713.
29 D. Seebach, J. D. Aebi, M. Gander-Coquoz, R. Naef, *Helv. Chim. Acta* **1987**, *70*, 1194.
30 Synthesis of *cis*-cyclohexadiene dioxide: (a) G. Kavadias, S. Velkof, B. Belleau, *Can. J. Chem.* **1978**, *56*, 404–409; (b) T. Suami, S. Ogawa, H. Uchino, Y. Funaki, *J. Org. Chem.* **1975**, *40*, 456–461; (c) T. W. Craig, G. R. Harvey, G. A. Berchtold, *J. Org. Chem.* **1967**, *32*, 3743–3749.
31 (a) J. P. Chiang, Ph. D. Dissertation, The Scripps Research Institute, **2001**; (b) G. Kavadias, R. Droghini, *Can. J. Chem.* **1979**, *57*, 1870–1876; F. Haviv, B. Belleau, *Can. J. Chem.* **1978**, *56*, 2677–2680; T. Suami, S. Ogawa, H. Uchino, Y. Funaki, *J. Org. Chem.* **1975**, *40*, 456–461.
32 a) H. -J. Altenbach, H. Stegelmeier, E. Vo-

gel, *Tetrahedron Lett.* **1978**, *36*, 3333–3336. (b) E. Vogel, H. -J. Altenbach, E. Schmidbauer, *Angew. Chem. Int. Ed.* **1973**, *12*, 838–839. (c) C. -H. Foster, G. -A. Berchtold, *J. Org. Chem.* **1975**, *40*, 3743–3746. (d) H. -J. Altenbach, E. Vogel, *Angew. Chem. Int. Ed.* **1972**, *11*, 937–939

33 Synthesis of *cis*-1,4-cyclohexadiene epoxides: (a) T. W. Craig, G. R. Harvey, G. A. Berchtold, *J. Org. Chem.* **1967**, *32*, 3743–3749; (b) G. Kavadias, S. Velkof, B. Belleau, *Can. J. Chem.* **1978**, *56*, 404–409; (c) T. Suami, S. Ogawa, H. Uchino, Y. Funaki, *J. Org. Chem.* **1975**, *40*, 456–461. For ring-opening reactions of six-membered ring di- and triepoxides, see: (d) R. Kuehlmeyer, R. Keller, R. Schwesinger, T. Netscher, H. Fritz, H. Prinzbach, *Chem. Ber.* **1984**, *117*, 1765–1800; (e) J. Schubert, R. Keller, R. Schwesinger, H. Prinzbach, *Chem. Ber.* **1983**, *116*, 2524–2545; (f) C. Rucker, H. Muller-Botticher, W. D. Braschwitz, H. Prinzbach, U. Reifenstahl, H. Irngartinger, *Liebigs Ann.* **1997**, 967–989; (g) S. Kagabu, C. Kaiser, R. Keller, P. G. Becker, K. H. Mueller, L. Knothe, G. Rihs, H. Prinzbach, *Chem. Ber.* **1988**, *121*, 741–756.

34 E. Stevens, A. Gontcharov, K. B. Sharpless, unpublished results.

35 (a) W. H. Pearson, B. W. Lian, S. C. Bergmeier, in "*Comprehensive Heterocyclic Chemistry II*" (Ed. A. Padwa), Pergamon, New York, **1996**, Vol. 1A, pp. 1–61.

36 For a study on the influence of Lewis acids on the nucleophilic opening of *N*-acyl aziridines vs. their rearrangement to oxazolines, see: D. Ferraris, W. J. Drury III, C. Cox, T. Lectka, *J. Org. Chem.* **1998**, *63*, 4568–4569.

37 a) H. Stamm, *J. Prakt. Chem.* **1999**, *341*, 319–331; b) K. Bellos, H. Stamm, *J. Org. Chem.* **1995**, *60*, 5661–5666.

38 Synthesis by transition metal-catalyzed aziridination of olefins, *cf.* H. Kwart, A. A. Kahn, *J. Am. Chem. Soc.* **1967**, *89*, 1951–1953; D. Mansury, J. -P. Mahy, D. Annie, B. Gustave, P. Battioni, *J. Chem. Soc., Chem. Comm.* **1984**, 1161–1163; D. A. Evans, M. M. Faul, M. T. Bilodeau, *J. Org. Chem.* **1991**, *56*, 6744–6746; E. Vedejs, H. Sano, *Tetrahedron Lett.* **1992**, *33*, 3261–3264; M. J. Sodergren, D. A. Alonso, A. V. Bedekar, P. G. Andersson, *Tetrahedron Lett.* **1997**, *38*, 6897–6900; asymmetric transition metal-catalyzed aziridination of olefins: D. A. Evans, M. M. Faul, M. T. Bilodeau, B. A. Anderson, D. M. Barnes, *J. Am. Chem. Soc.* **1993**, *115*, 5328–5329; R. E. Rowenthal, S. Masamune, *Tetrahedron Lett.* **1991**, *32*, 7373–7376; Z. Li, K. B. Conser, E. N. Jacobson, *J. Am. Chem. Soc.* **1993**, *115*, 5326–5327; H. Nishikori, T. Katsuki, *Tetrahedron Lett.* **1996**, *37*, 9245–9248; aziridination of olefins with PhI=NTs: Y. Yamada, T. Yamamoto, *Chem. Lett.* **1975**, 361.

39 a) D. Tanner, C. Birgersson, *Tetrahedron Lett.* **1991**, *32*, 2533–2536; b) D. Tanner, A. Almario, T. Högberg, *Tetrahedron* **1995**, *51*, 6061–6070; c) Z. Wang, L. S. Jimenez, *J. Org. Chem.* **1996**, *61*, 816–818; d) P. Besse, H. Veschambre, R. Chenevert, M. Dickman, *Tetrahedron: Asymmetry* **1994**, *5*, 1727–1744; e) P. Wipf, P. C. Fritch, *J. Org. Chem.* **1994**, *59*, 4875–4886; f) V. C. O. Njar, R. W. Hartmann, C. H. Robinson, *J. Chem. Soc. , Perkin Trans. 1*, **1995**, 985–991; g) P. Crotti, V. Di Bussolo, L. Favero, F. Macchia, M. Pineschi, *Tetrahedron: Asymmetry* **1996**, *7*, 779–786; h) N. A. J. M. Sommerdijk, P. J. J. A. Buynsters, H. Akdemir, D. G. Geurts, R. J. M. Nolte, B. Zwanenburg, *J. Org. Chem.* **1997**, *62*, 4955–4960.

40 a) H. Wenker, *J. Am. Chem. Soc.* **1935**, *57*, 2328; b) P. A. Leighton, W. Perkins, M. L. Renquist, *J. Am. Chem. Soc.* **1947**, *69*, 1540; c) C. F. H. Allen, F. W. Spangler, E. R. Webster, *Org. Synth. Coll. Vol. IV*, **1963**, 433–435; d) A. Galindo, L. Orea, D. Gnecco, R. G. Enríquez, R. A. Toscano, W. F. Reynolds, *Tetrahedron: Asymmetry* **1997**, *8*, 2877–2879.

41 a) I. -C. Chiu, H. Kohn, *J. Org. Chem.* **1983**, *48*, 2857–2866; b) D. van Ende, A. Krief, *Angew. Chem. Int. Ed. Engl.* **1974**, *13*, 279–280.

42 J. U. Jeong, B. Tao, I. Sagasser, H. Henniges, K. B. Sharpless, *J. Am. Chem. Soc.* **1998**, *120*, 6844–6845

43 A. V. Gontcharov, H. Liu, K. B. Sharpless, *Org. Lett.* **1999**, *1*, 783–786.

44 P. Sun, S. M. Weinreb, *J. Org. Chem.* **1997**, *62*, 8604

45 (a) W. H. Pearson, B. W. Lian, S. C. Bergmeier, in "*Comprehensive Heterocyclic Chemistry II*" (Ed. A. Padwa), Pergamon,

New York, **1996**, Vol. IA, pp. 1–60. (b) D. Tanner, *Angew. Chem. Int. Ed.* **1994**, *33*, 599. (c) Osborn, H. M. I. Sweeney, J. *Tetrahedron: Asymmetry*, **1997**, *8*, 1693.

46 A. E. Rubin, K. B. Sharpless, *Angew. Chem. Int. Ed.* **1997**, *36*, 2637–2640.

47 (a) W. Pringle, K. B. Sharpless, *Tetrahedron Lett.* **1999**, *40*, 5150–5154; (b) V. V. Fokin, K. B. Sharpless; *Angew. Chem. Int. Ed.* **2001**, 40, 3455–3457

48 A. E. Rubin, Ph. D. Thesis, The Scripps Research Institute (USA), **1999**.

49 P. E. Maligres, M. M. See, D. Askin, P. J. Reider, *Tetrahedron Lett.* , **1997**, *38*, 5253.

50 (a) Special issue on solution phase combinatorial chemistry: *Tetrahedron*, **1998**, 54. (b) F. Balkenhohl, C. von dem Bussche-Hünnefeld, A. Lansky, C. Zechel, *Angew. Chem. Int. Ed.* **1996**, *35*, 2288.

51 S. M. Rink, M. S. Solomon, M. J. Taylor, S. B. Rajur, L. W. McGlaughlin, P. B. Hopkins, *J. Am. Chem. Soc.* **1993**, *115*, 2551 and references cited therein.

52 G. B. Jones, J. E. Mathews, *Bioorg. Med. Chem. Lett.* **1995**, *5*, 93

53 a) K. Weber, S. Kuklinski, P. Gmeiner, *Org. Lett.* **2000**, *2*, 647; b) S. E. de Sousa, P. O'Brien, P. Poumellec, *J. Chem. Soc., Perkin Trans. 1.* **1998**, 1483; c) Q. Liu, A. P. Marchington, C. M. Rayner, *Tetrahedron* **1997**, *53*, 15729; d) Q. Liu, A. P. Marchington, N. Boden, C. M. Rayner, *J. Chem. Soc. , Perkin Trans. 1.* **1997**, 511; e) C. M. Rayner, *Synlett* **1997**, 11; f) P. F. Richardson, L. T. J. Nelson, K. B. Sharpless, *Tetrahedron Lett.* **1995**, *36*, 9241; g) M. Okuda, K. Tomioka, *Tetrahedron Lett.* **1994**, *35*, 4585; h) P. Gmeiner, D. Junge, A. Kaertner, *J. Org. Chem.* **1994**, *59*, 6766; i) Q. Liu, M. J. Simms, N. Boden, C. M. Rayner, *J. Chem. Soc. , Perkin Trans 1.* **1994**, 1363; j) J. Freedman, M. J. Vaal, E. W. Huber, *J. Org. Chem.* **1991**, *56*, 670; k) D. R. Williams, D. L. Brown, J. W. Benbow, *J. Am. Chem. Soc.* **1989**, *111*, 1923; l) T. Rosen, S. W. Fesik, D. T. W. Chu, A. G. Parnet, *Synthesis* **1988**, 40; m) D. Tanner, P. Somfai, *Tetrahedron* **1986**, *42*, 5657.

54 (a) *Chemistry and Biology of β-Lactam Antibiotics* (Eds. R. B. Morin, M. Gorman), Academic Press: New York, **1982**; Vols. 1–3. (b) G. A. Koppel, in *Small Ring Heterocycles-Azetidines, Lactams, Diazetidines and Diaziridines* (Ed. A. Hassner), Wiley: New York, **1982**. (c) W. Dürckheimer, J. Blumbach, R. Lattrel, K. H. Scheunemann, *Angew. Chem. , Int. Ed. Engl.* **1985**, *24*, 180.

55 (a) V. Dryanska, I. Pashkuleva, D. Tasheva, *Synth. Commun.* **1997**, *27*, 1849. (b) F. H. van der Steen, H. Kleijn, T. B. H. Jastrzebski, G. van Koten, *J. Org. Chem.* **1991**, *56*, 5147.

56 For reviews of the ester enolate-imine condensation, see: (a) M. J. Brown, *Heterocycles* **1989**, *29*, 2225. (b) D. J. Hart, D. -C. Ha, *Chem. Rev.* **1989**, *89*, 1447. (c) F. H. van der Steen, G. van Koten, *Tetrahedron* **1991**, *47*, 7503.

57 T. -H. Chuang, K. B. Sharpless, *Org. Letters*, **1999**, *1*, 1435–1437.

58 a) J. Svete, A. Preseren, B. Stanovnik, L. Golic, S. Golic-Grdadolnik, *J. Heterocyclic Chem.* **1997**, *34*, 1323; b) L. N. Jungheim, S. K. Sigmund, *J. Org. Chem.* **1987**, *52*, 4007; c) L. N. Jungheim, *Tetrahedron Lett.* **1989**, *30*, 1889.

59 N. E. Allen, J. N. Hobbs, Jr. , E. Wu in *"Antibiotic Inhibition of Bacterial Cell Surface Assembly and Function"*, (Eds. P. Actor, L. Daneo-Moore, M. L. Higgins, M. R. J. Satton, G. D. Shockman), American Society for Microbiology, Washington D. C., **1988**, p. 569.

60 T. -H. Chuang, K. B. Sharpless, *Helv. Chim. Acta*, **2000**, 83, 1734.

Index

Aziridines and Epoxides in Organic Synthesis. Andrei K. Yudin

ISBN: 3-527-31213-7